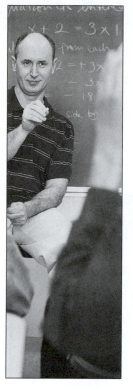

Success in Mathematics is just a click away

With *WileyPLUS*, students and instructors will experience success in the classroom.

When students succeed in your course—when they stay on-task and make the breakthrough that turns confusion into confidence—they are empowered to realize the possibilities for greatness that lie within each of them. We know your goal is to create an environment where students reach their full potential and experience the exhilaration of academic success that will last them a lifetime. *WileyPLUS* can help you reach that goal.

Wiley**PLUS** is an online suite of resources—including the complete text—that will help your students:

- come to class better prepared for your lectures
- get immediate feedback and context-sensitive help on assignments and quizzes
- track their progress throughout the course

"I just wanted to say how much this program helped me in studying… I was able to actually see my mistakes and correct them. … I really think that other students should have the chance to use *WileyPLUS*."

Ashlee Krisko, *Oakland University*

www.wileyplus.com

88% of students surveyed said it improved their understanding of the material. *

FOR INSTRUCTORS

WileyPLUS is built around the activities you perform in your class each day. With WileyPLUS you can:

Prepare & Present
Create outstanding class presentations using a wealth of resources such as PowerPoint™ slides, image galleries, interactive simulations, and more. You can even add materials you have created yourself.

Create Assignments
Automate the assigning and grading of homework or quizzes by using the provided question banks, or by writing your own.

Track Student Progress
Keep track of your students' progress and analyze individual and overall class results.

Now Available with WebCT and eCollege!

"It has been a great help, and I believe it has helped me to achieve a better grade."

Michael Morris,
Columbia Basin College

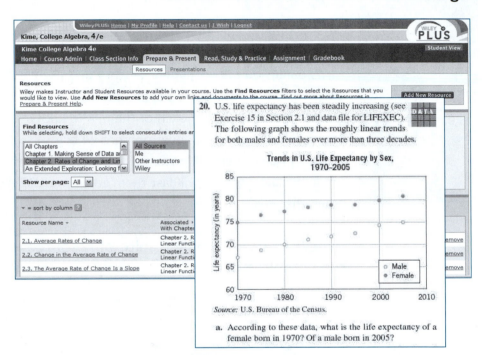

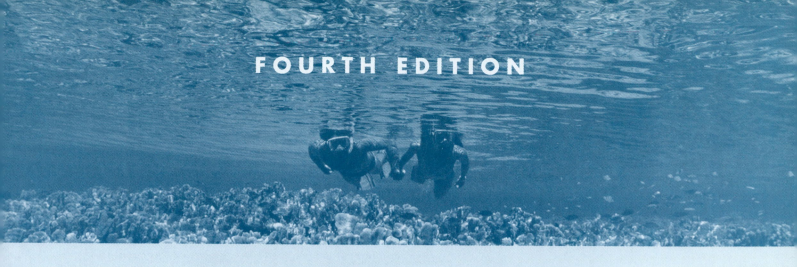

FOURTH EDITION

EXPLORATIONS IN COLLEGE ALGEBRA

LINDA ALMGREN KIME
JUDITH CLARK

University of Massachusetts, Boston, Retired

BEVERLY K. MICHAEL

University of Pittsburgh

in collaboration with

Norma M. Agras *Miami Dade College*

Robert F. Almgren *Courant Institute, New York University*

Linda Falstein *University of Massachusetts, Boston, Retired*

Meg Hickey *Massachusetts College of Art*

John A. Lutts *University of Massachusetts, Boston*

Peg Kem McPartland *Golden Gate University, Retired*

Jeremiah V. Russell *University of Massachusetts, Boston; Boston Public Schools*

software developed by

Hubert Hohn *Massachusetts College of Art*

Funded by a National Science Foundation Grant

WILEY

JOHN WILEY & SONS, INC.

Publisher	Laurie Rosatone
Acquisitions Editor	Jessica Jacobs
Assistant Editor	Michael Shroff
Editorial Assistant	Jeffrey Benson
Marketing Manager	Jaclyn Elkins
Production Manager	Dorothy Sinclair
Senior Production Editor	Sandra Dumas
Design Director	Harry Nolan
Senior Designer	Madelyn Lesure
Senior Media Editor	Stefanie Liebman
Production Management	Publication Services
Bicentennial Logo Design	Richard J. Pacifico

This book was set in 10/12 Times Roman by Publication Services, and printed and bound by Courier (Westford). The cover was printed by Courier (Westford).

This book is printed on acid-free paper. ∞

To order books or for customer service call 1-800-CALL-WILEY(225-5945).

ISBN 978-0471-91688-8

Printed in the United States of America.

10 9 8 7 6 5 4 3 2 1

To our students, who inspired us.

A Letter from a Student

My name is Lexi Fournier and I am a freshman here at Pitt. This semester I am enrolled in the Applied Algebra course using "Explorations in College Algebra." Before coming to Pitt, I had taken numerous math courses varying from algebra to calculus, all of which produced frustration, stress, and a detestation for math as a subject. When I was told that I was required to take a math course here, I was livid. I am a pre-law and creative writing major; why do I need math? My adviser calmed me by informing me of this new math class aimed at teaching non-math/science majors the basic skills they will need in everyday life.

At first I was skeptical, but I'm writing to you now to emphatically recommend this course. What I have learned thus far in this course have been realistic math skills presented in a "left brain" method that fosters confidence and motivation. For once in my career as a student, math is relatable. The concepts are clear and realistic (as opposed to the abstract, amorphous topics addressed in my earlier math classes). I look forward to this class. I enjoy doing my homework and projects because I feel that the lessons are applicable to my life and my future and because I feel empowered by my understanding.

This course is a vital addition to the math department. It has altered my view on the subject and stimulated an appreciation for what I like to call "everyday math."

It is my belief that many students will find the class as encouraging and helpful as I have. Thank you for your attention.

Sincerely,

Lexi Fournier
Student, University of Pittsburgh

PREFACE

This text was born from a desire to reshape the college algebra course, to make it relevant and accessible to all of our students. Our goal is to shift the focus from learning a set of discrete mechanical rules to exploring how algebra is used in the social and physical sciences. Through connecting mathematics to real-life situations, we hope students come to appreciate its power and beauty.

Guiding Principles

The following principles guided our work.

- Develop mathematical concepts using real-world data and questions.
- Pose a wide variety of problems designed to promote mathematical reasoning in different contexts.
- Make connections among the multiple representations of functions.
- Emphasize communication skills, both written and oral.
- Facilitate the use of technology.
- Provide sufficient practice in skill building to enhance problem solving.

Evolution of Explorations in College Algebra

The fourth edition of *Explorations* is the result of a 15-year long process. Funding by the National Science Foundation enabled us to develop and publish the first edition, and to work collaboratively with a nationwide consortium of schools. Faculty from selected schools in the consortium continued to work with us on the second, third, and now the fourth editions. During each stage of revision we solicited extensive feedback from our colleagues, reviewers and students.

Throughout the text, families of functions are used to model real-world phenomena. After an introductory chapter on data and functions, we first focus on linear and exponential functions, since these are the two most commonly used mathematical models. We then discuss logarithmic, power, quadratic, and polynomial functions and finally turn to ways to extend and combine all the types of functions we've studied to create new functions.

The text adopts a problem-solving approach, where examples and exercises lie on a continuum from open-ended, nonroutine questions to problems on algebraic skills. The materials are designed for flexibility of use and offer multiple options for a wide range of skill levels and departmental needs. The text is currently used in small classes, laboratory settings, and large lectures, and in both two- and four-year institutions.

Special Features and Supplements

An instructor is free to choose among a number of special features. The Instructor's Teaching Manual provides support for using these features and includes sample test questions. The Instructor's Solutions Manual contains answers to the even exercises and even problems in the Chapter Reviews. Both manuals are available free for adopters either online at *www.wiley.com/college/kimeclark* or in hardcopy by contacting your local Wiley representative.

Exploring Mathematical Ideas

 Explorations   These are open-ended investigations designed to be used in parallel with the text. They appear at the end of each chapter and in two chapter-length Extended Explorations.

New! Chapter Review: Putting It All Together   Each review contains problems that apply all of the basic concepts in the chapter. The answers to the odd-numbered problems are in the back of the text.

Check Your Understanding   A set of mostly true/false questions at the end of each chapter (with answers in the back of the text) offer students a chance to assess their understanding of that chapter's mathematical ideas.

 *Something to Think About*   Provocative questions, posed throughout the text, can be used to generate class discussion or for independent inquiry.

60-Second Summaries   Short writing assignments in the exercises and Explorations ask students to succinctly summarize their findings.

 Readings   A variety of articles related to topics covered in the text are available on the course website at *www.wiley.com/college/kimeclark*.

Using Technology

COURSE WEBSITE   Technology is not required to teach this course. However, we provide numerous resources, described below, for teaching with technology at the course website *www.wiley.com /college/kimeclark*.

Graphing Calculator Manual   The manual offers step-by-step instructions for using the TI83/TI84 family of calculators that are coordinated with the chapters in the text. It is free on the course website or at a discount when packaged in hardcopy with the text.

 Interactive Software for Mac and PC   Programs for visualizing mathematical concepts, simulations, and practice in skill building are available on the course website. They may be used in classroom demonstrations or a computer lab, or downloaded for student use at home.

 Excel and TI83/TI84 Graph Link Files   Data files containing all the major data sets used in the text are available on the course website.

Practice in Skill Building

Algebra Aerobics   These collections of skill-building practice problems are integrated throughout each chapter. Answers for all Algebra Aerobics problems are in the back of the text.

WileyPLUS   This is a powerful online tool that provides a completely integrated suite of teaching and learning resources in one easy-to-use website. It offers an online assessment system with full gradebook capabilities, which contains algorithmically generated skill-building questions from the Algebra Aerobics problems and the exercises in each chapter. Faculty can view the online demo at *www.wiley.com/college/wileyplus*.

The Fourth Edition

Overall Changes

Extensive faculty reviews guided our work on the fourth edition. The sequence of the chapters remains the same as in the third edition, but we have included

- New chapter reviews, called "Putting It All Together," with problems that bring together the major concepts of the chapter.
- A relocation of exercises from the end of the chapter to the end of each section.
- Expanded coverage of several topics, including function notation, range and domain, piecewise linear functions (including absolute value and step functions), rational functions, composition, and inverse functions.
- Extensive updates of the data sets.
- Revisions to many chapters for greater clarity.
- Many new problems and exercises, ranging from basic algebraic manipulations to real-world applications.

Detailed Changes

CHAPTER REVIEWS: "PUTTING IT ALL TOGETHER" appear at the end of each chapter.

CHAPTER 1: Making Sense of Data and Functions has a new section on the language of functions, with expanded coverage of function notation, domain, and range. Boxes have been added to highlight important concepts.

CHAPTER 2: Rates of Change and Linear Functions has a new subsection on piecewise linear functions, including the absolute value function and step functions.

EXTENDED EXPLORATION: Looking for Links between Education and Earnings uses an updated data set from the U.S. Census about 1000 individuals.

CHAPTER 5: Growth and Decay: An Introduction to Exponential Functions has an expanded discussion on constructing an exponential function given its doubling time or half-life.

CHAPTER 7: Power Functions has an added discussion of asymptotes for negative integer power functions.

CHAPTER 8: Quadratics, Polynomials, and Beyond has changed the most. The old Section 8.6 has been expanded and broken up into three sections. Section 8.6, "New Functions from Old," discusses the effect of stretching, compressing, shifting, reflecting, or rotating a function. Section 8.7, "Combining Two Functions," includes the algebra of functions and an expanded subsection on rational functions. Section 8.8, "Composition and Inverse Functions," extends the coverage of these topics.

Acknowledgments

We wish to express our appreciation to all those who helped and supported us during this extensive collaborative endeavor. We are grateful for the support of the National Science Foundation, whose funding made this project possible, and for the generous help of our program officers then, Elizabeth Teles and Marjorie Enneking. Our original Advisory Board, especially Deborah Hughes Hallett and Philip Morrison, and our original editor, Ruth Baruth provided invaluable advice and encouragement.

Over the last 15 years, through five printings (including a rough draft and preliminary edition), we worked with more faculty, students, TAs, staff, and administrators than we can possibly list here. We are deeply grateful for supportive colleagues at our own University. The generous and ongoing support we received from Theresa Mortimer, Patricia Davidson, Mark Pawlak, Maura Mast, Dick Cluster, Anthony Beckwith, Bob Seeley, Randy Albelda, Art MacEwan, Rachel Skvirsky, Brian Butler, among many others, helped to make this a successful project.

We are deeply indebted to Ann Ostberg and Rebecca Hubiak for their dedicated search for mathematical errors in the text and solutions, and finding (we hope) all of them. A text designed around the application of real-world data would have been impossible without the time-consuming and exacting research done by Patrick Jarrett, Justin Gross, and Jie Chen. Edmond Tomastik and Karl Schaffer were gracious enough to let us adapt some of their real-world examples in the text.

One of the joys of this project has been working with so many dedicated faculty who are searching for new ways to reach out to students. These faculty, and their teaching assistants and students all offered incredible support, encouragement, and a wealth of helpful suggestions. In particular, our heartfelt thanks to members of our original consortium: Sandi Athanassiou and all the wonderful TAs at University of Missouri, Columbia; Natalie Leone, University of Pittsburgh; Peggy Tibbs and John Watson, Arkansas Technical University; Josie Hamer, Robert Hoburg, and Bruce King, past and present faculty at Western Connecticut State University; Judy Stubblefield, Garden City Community College; Lida McDowell, Jan Davis, and Jeff Stuart, University of Southern Mississippi; Ann Steen, Santa Fe Community College; Leah Griffith, Rio Hondo College; Mark Mills, Central College; Tina Bond, Pensacola Junior College; and Curtis Card, Black Hills State University.

The following reviewers' thoughtful comments helped shape the fourth edition: Mark Gïnn, Appalachian State; Ernie Solheid, California State University, Fullerton; Pavlov Rameau, Florida International University; Karen Becker, Fort Lewis College; David Phillips, Georgia State University; Richard M. Aron and Beverly Reed, Kent State University; Nancy R. Johnson, Manatee Community College; Lauren Fern, University of Montana; Warren Bernard, Linda Green, and Laura Younts, Santa Fe Community College; Sarah Clifton, Southeastern Louisiana University; and Jonathan Prewett, University of Wyoming.

We are especially indebted to Laurie Rosatone at Wiley, whose gracious oversight helped to keep this project on track. Particular thanks goes to our new editors Jessica Jacobs, Acquisitions Editor; John-Paul Ramin, Developmental Editor; Michael Shroff, Assistant Editor; and their invaluable assistant Jeffrey Benson. It has been a great pleasure, both professionally and personally, to work with Maddy Lesure on her creative cover design and layout of the text. "*Explorations*" and the accompanying media would never have been produced without the experienced help from Sandra Dumas, Dorothy Sinclair, and Stefanie Liebman. Kudos to Jan Fisher at Publication Services. Throughout the production of this text, her cheerful attitude and professional skills made her a joy to work with. Over the years many others at Wiley have been extraordinarily helpful in dealing with the myriad of endless details in producing a mathematics textbook. Our thanks to all of them.

Our families couldn't help but become caught up in this time-consuming endeavor. Linda's husband, Milford, and her son Kristian were invaluable scientific and, more importantly, emotional resources. They offered unending encouragement and sympathetic shoulders. Judy's husband, Gerry, become our Consortium lawyer, and her daughters, Rachel, Caroline, and Kristin provided support, understanding, laughter, editorial help and whatever was needed. Beverly's husband, Dan, was patient and understanding about the amount of time this edition took. Her daughters Bridget and Megan would call from college to cheer her on and make sure she was not getting too stressed! All our family members ran errands, cooked meals, listened to our concerns, and gave us the time and space to work on the text. Our love and thanks.

Finally, we wish to thank all of our students. It is for them that this book was written.

Linda, Judy, and Bev

P.S. We've tried hard to write an error-free text, but we know that's impossible. You can alert us to any errors by sending an email to ***math@wiley.com***. Be sure to reference *Explorations in College Algebra*. We would very much appreciate your input.

COVERING THE CONTENTS

The following flow chart suggests some alternative paths through the chapters that have worked successfully for others.

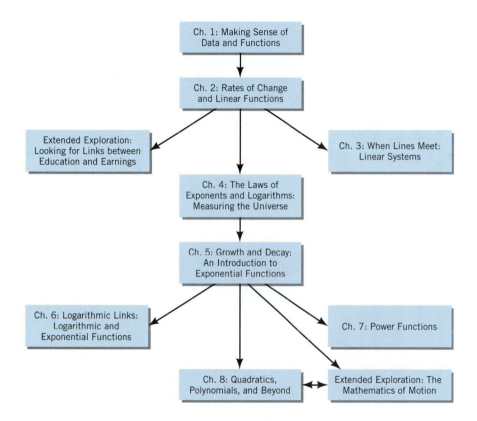

The straight vertical path through Chapters 1, 2, 4, 5, and 8, covering linear, exponential, quadratic and other polynomial functions, indicates the core content of the text. You may choose to cover these chapters in depth, spending time on the explorations, readings and student discussions, writing, and presentations. Or you may pick up the pace and include as many of the other chapters and Extended Explorations as is appropriate for your department's needs.

TABLE OF CONTENTS

CHAPTER 3

CHAPTER 4

CHAPTER 5

GROWTH AND DECAY: AN INTRODUCTION TO EXPONENTIAL FUNCTIONS

See *www.wiley.com/college/kimeclark* for Course Software, Anthology of Readings, Excel and Graph Link data files, and the Graphing Calculator Manual. The Instructor's Teaching Manual and Instructor's Solutions Manual are also available on the site, but password protected to restrict access to Instructors.

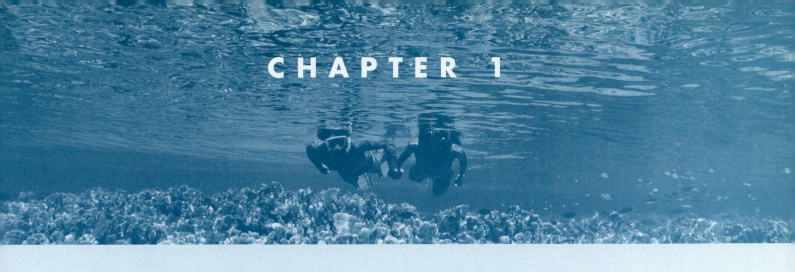

MAKING SENSE OF DATA AND FUNCTIONS

OVERVIEW

How can you describe patterns in data? In this chapter we explore how to use graphs to visualize the shape of single-variable data and to show changes in two-variable data. Functions, a fundamental concept in mathematics, are introduced and used to model change.

After reading this chapter, you should be able to

- describe patterns in single- and two-variable data

- construct a "60-second summary"

- define a function and represent it in multiple ways

- identify properties of functions

- use the language of functions to describe and create graphs

1.1   *Describing Single-Variable Data*

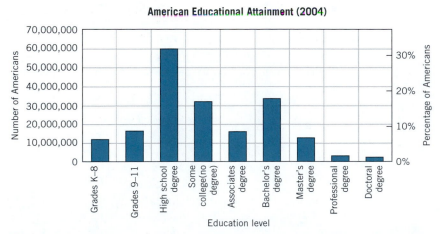

Exploration 1.1 provides an opportunity to collect your own data and to think about issues related to classifying and interpreting data.

This course starts with you. How would you describe yourself to others? Are you a 5-foot 6-inch, black, 26-year-old female studying biology? Or perhaps you are a 5-foot 10-inch, Chinese, 18-year-old male English major. In statistical terms, characteristics such as height, race, age, and major that vary from person to person are called *variables*. Information collected about a variable is called *data*.[1]

Some variables, such as age, height, or number of people in your household, can be represented by a number and a unit of measure (such as 18 years, 6 feet, or 3 people). These are called *quantitative variables*. For other variables, such as gender or college major, we use categories (such as male and female or biology and English) to classify information. These are called *categorical* (or *qualitative*) data. The dividing line between classifying a variable as categorical or quantitative is not always clear-cut. For example, you could ask individuals to list their years of education (making education a quantitative variable) or ask for their highest educational category, such as college or graduate school (making education a categorical variable).

Many of the controversies in the social sciences have centered on how particular variables are defined and measured. For nearly two centuries, the categories used by the U.S. Census Bureau to classify race and ethnicity have been the subject of debate. For example, Hispanic used to be considered a racial classification. It is now considered an ethnic classification, since Hispanics can be black, or white, or any other race.

Visualizing Single-Variable Data

Humans are visual creatures. Converting data to an image can make it much easier to recognize patterns.

Bar charts: How well educated are Americans?

Categorical data are usually displayed with a bar chart. Typically the categories are listed on the horizontal axis. The height of the bar above a single category tells you either the *frequency count* (the number of observations that fall into that category) or the *relative frequency* (the percentage of total observations). Since the relative size of the bars is the same using either frequency or relative frequency counts, we often put the two scales on different vertical axes of the same chart. For example, look at the vertical scales on the left- and right-hand sides of Figure 1.1, a bar chart of the educational attainment of Americans age 25 or older in 2004.

Figure 1.1 Bar chart showing the education levels for Americans age 25 or older.
Source: U.S. Bureau of the Census, *www.census.gov*.

---

[1]*Data* is the plural of the Latin word *datum* (meaning "something given")—hence one datum, two data.

The vertical scale on the left tells us the number (the frequency count) of Americans who fell into each educational category. For example, in 2004 approximately 60 million Americans age 25 or older had a high school degree but never went on to college.

It's often more useful to know the percentage (the relative frequency) of all Americans who have only a high school degree. Given that in 2004 the number of people 25 years or older was approximately 186,877,000 and the number who had only a high school degree was approximately 59,810,000, then the percentage of those with only a high school degree was

$$\frac{\text{Number with only a high school degree}}{\text{Total number of people age 25 or older}} = \frac{59,810,000}{186,877,000}$$

$$\approx 0.32 \text{ (in decimal form) or } 32\%$$

The vertical scale on the right tells us the percentage (relative frequency). Using this scale, the percentage of Americans with only a high school degree was about 32%, which is consistent with our calculation.

EXAMPLE 1 **What does the bar chart tell us?**

a. Using Figure 1.1, estimate the number and percentage of people age 25 or older who have bachelor's degrees, but no further advanced education.

b. Estimate the total number of people and the percentage of the total population age 25 or older who have at least a high school education.

c. What doesn't the bar chart tell us?

d. Write a brief summary of educational attainment in the United States.

SOLUTION a. Those with bachelor's degrees but no further education number about 34 million, or 18%.

b. Those who have completed a high school education include everyone with a high school degree up to a Ph.D. We could add up all the numbers (or percentages) for each of those seven categories. But it's easier to subtract from the whole those who do not meet the conditions, that is, subtract those with either a grade school or only some high school education from the total population (people age 25 or older) of about 187 million.

	Grade School + Some High School	= Total without High School Degree
Number (approx.)	12 million + 16 million	= 28 million
Percentage (approx.)	6% + 9%	= 15%

The number of Americans (age 25 or over) with a high school degree is about 187 million − 28 million = 159 million. The corresponding percentage is about 100% − 15% = 85%. So more than four out of five Americans 25 years or older have completed high school.

c. The bar chart does not tell us the total size of the population or the total number (or percentage) of Americans who have a high school degree. For example, if we include younger Americans between age 18 and age 25, we would expect the percentage with a high school degree to be higher.

d. About 85% of adult Americans (age 25 or older) have at least a high school education. The breakdown for the 85% includes 32% who completed high school but did not go on, 43% who have some college (up to a bachelor's degree), and about 10% who have graduate degrees. This is not surprising, since the United States population ranks among the mostly highly educated in the world.

An important aside: What a good graph should contain

When you encounter a graph in an article or you produce one for a class, there are three elements that should always be present:

1. An informative title that succinctly describes the graph
2. Clearly labeled axes (or a legend) including the units of measurement (e.g., indicating whether age is measured in months or years)
3. The source of the data cited in the data table, in the text, or on the graph

See the program "F1: Histograms."

Histograms: What is the distribution of ages in the U.S. population?

A histogram is a specialized form of a bar chart that is used to visualize single-variable quantitative data. Typically, the horizontal axis on a histogram is a subset of the real numbers with the unit (representing, for example, number of years) and the size of each interval marked. The intervals are usually evenly spaced to facilitate comparisons (e.g., placed every 10 years). The size of the interval can reveal or obscure patterns in the data. As with a bar chart, the vertical axis can be labeled with a frequency or a relative frequency count. For example, the histogram in Figure 1.2 shows the distribution of ages in the United States in 2005.

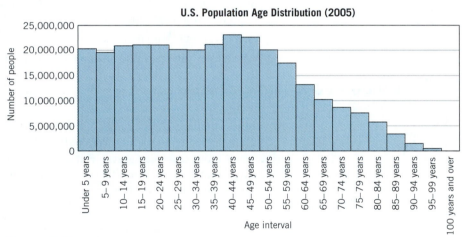

Figure 1.2 Age distribution of the U.S. population in 5-year intervals.
Source: U.S. Bureau of the Census, *www.census.gov.*

EXAMPLE 2

What does the histogram tell us?

a. What 5-year age interval contains the most Americans? Roughly how many are in that interval? (Refer to Figure 1.2.)

b. Estimate the number of people under age 20.

c. Construct a topic sentence for a report about the U.S. population.

SOLUTION

a. The interval from 40 to 44 years contains the largest number of Americans, about 23 million.

b. The sum of the frequency counts for the four intervals below age 20 is about 80 million.

c. According to the U.S. Census Bureau 2005 data, the number of Americans in each 5-year age interval remained fairly flat up to age 40, peaked between ages 40 to 50, then fell in a gradual decline.

EXAMPLE 3   Describe the age distribution for Tanzania, one of the poorest countries in the world (Figure 1.3).

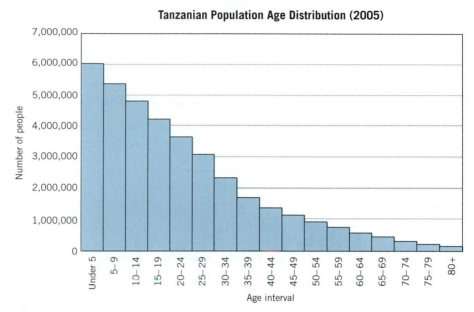

Tanzanian Population Age Distribution (2005)

Figure 1.3 The age distribution in 2005 of the Tanzanian population in 5-year intervals.
Source: U.S. Bureau of the Census, International Data Base, April 2005.

SOLUTION   The age distributions in Tanzania and the United States are quite different. Tanzania is a much smaller country and has a profile typical of a developing country; that is, each subsequent 5-year interval has fewer people. For example, there are about 6 million children 0 to 4 years old, but only about 5.3 million children age 5–9 years, a drop of over 10%. For ages 35 to 39 years, there are only about 1.7 million people, less than a third of the number of children between 0 and 4 years. Although the histogram gives a static picture of the Tanzanian population, the shape suggests that mortality rates are much higher than in the United States.

SOMETHING TO THINK ABOUT

What are some trade-offs in using pie charts versus histograms?

Pie charts: Who gets the biggest piece?

Both histograms and bar charts can be transformed into pie charts. For example, Figure 1.4 shows two pie charts of the U.S. and Tanzanian age distributions (both now divided into 20-year intervals). One advantage of using a pie chart is that it clearly shows the size of each piece relative to the whole. Hence, they are usually labeled with percentages rather than frequency counts.

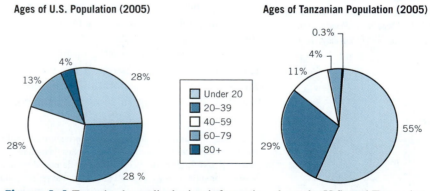

Ages of U.S. Population (2005)   **Ages of Tanzanian Population (2005)**

Under 20
20–39
40–59
60–79
80+

Figure 1.4 Two pie charts displaying information about the U.S. and Tanzanian age distributions.
Source: U.S. Bureau of the Census, *www.census.gov.*

In the United States the first three 20-year age intervals (under 20, 20–39, and 40–59 years) are all approximately equal in size and together make up about 84% of the population. Those 60 and older represent 17% of the population. Note that the percentages add up to more than 100% due to rounding.

In Tanzania, the proportions are entirely different. Over half of the population are under 20 years and more than 80% are under 40 years old. Those 60 and older make up less than 5% of Tanzania's population.

Mean and Median: What Is "Average" Anyway?

In 2005 the U.S. Bureau of the Census reported that the mean age for Americans was 37.2 and the median age was 36.7 years.

> **The Mean and Median**
>
> The *mean* is the sum of a list of numbers divided by the number of terms in the list.
>
> The *median* is the middle value of an ordered numerical list; half the numbers lie at or below the median and half at or above it.

The mean age of 37.2 represents the sum of the ages of every American divided by the total number of Americans. The median age of 36.7 means that if you placed all the ages in order, 36.7 would lie right in the middle; that is, half of Americans are younger than or equal to 36.7 and half are 36.7 or older.

In the press you will most likely encounter the word "average" rather than the term "mean" or "median."[2] The term "average" is used very loosely. It usually represents the mean, but it could also represent the median or something much more vague, such as the "average" American household. For example, the media reported that:

- The *average* American home now has more television sets than people. . . . There are 2.73 TV sets in the typical home and 2.55 people.[3]

- The *average* American family now owes more than $9,000 in credit debt. . . . and is averaging about seven cards.[4]

The significance of the mean and median

The median divides the number of entries in a data set into two equal halves. If the median age in a large urban housing project is 17, then half the population is 17 or under. Hence, issues such as day care, recreation, and education should be high priorities with the management. If the median age is 55, then issues such as health care and wheelchair accessibility might dominate the management's concerns.

The median is unchanged by changes in values above and below it. For example, as long as the median income is larger than the poverty level, it will remain the same even if all poor people suddenly increase their incomes up to that level and everyone else's income remains the same.

The mean is the most commonly cited statistic in the news media. One advantage of the mean is that it can be used for calculations relating to the whole data set. Suppose a corporation wants to open a new factory similar to its other factories. If the managers know the mean cost of wages and benefits for employees,

---

SOMETHING TO THINK ABOUT

If someone tells you that in his town "all of the children are above average," you might be skeptical. (This is called the "Lake Wobegon effect.") But could most (more than half) of the children be above average? Explain.

See the reading "The Median Isn't the Message" to find out how an understanding of the median gave renewed hope to the renowned scientist Stephen J. Gould when he was diagnosed with cancer.

---

[2]The word "average" has an interesting derivation according to Klein's etymological dictionary. It comes from the Arabic word *awariyan,* which means "merchandise damaged by seawater." The idea being debated was that if your ships arrived with water-damaged merchandise, should you have to bear all the losses yourself or should they be spread around, or "averaged," among all the other merchants? The words *averia* in Spanish, *avaria* in Italian, and *avarie* in French still mean "damage."

[3]*Source:* USA Today, *www.usatoday.com/life/television/news/2006.*

[4]*Source:* Newsweek, *www.msnbc.msn.com/id/14366431/site/newsweek/.*

they can make an estimate of what it will cost to employ the number of workers needed to run the new factory:

total employee cost $=$ (mean cost for employees) $\cdot$ (number of employees)

The mean, unlike the median, can be affected by a few extreme values called *outliers*. For example, suppose Bill Gates, founder of Microsoft and the richest man in the world, were to move into a town of 10,000 people, all of whom earned nothing. The median income would be \$0, but the mean income would be in the millions. That's why income studies usually use the median.

EXAMPLE 4   **"Million-dollar Manhattan apartment? Just about average"**
According to a report cited on *money.cnn.com*, in 2006 the median price of purchasing an apartment in Manhattan was \$880,000 and the mean price was \$1.4 million. How could there be such a difference in price? Which value do you think better represents apartment prices in Manhattan?

SOLUTION   Apartments that sold for exorbitant prices (in the millions) could raise the mean above the median. If you want to buy an apartment in Manhattan, the median price is probably more important because it tells you that half of the apartments cost \$880,000 or less.

An Introduction to Algebra Aerobics

In each section of the text there are "Algebra Aerobics" with answers in the back of the book. They are intended to give you practice in the algebraic skills introduced in the section and to review skills we assume you have learned in other courses. These skills should provide a good foundation for doing the exercises at the end of each chapter. The exercises include more complex and challenging problems and have answers for only the odd-numbered ones. We recommend you work out these Algebra Aerobics practice problems and then check your solutions in the back of this book. The Algebra Aerobics are numbered according to the section of the book in which they occur.

Algebra Aerobics 1.1

1. Fill in Table 1.1. Round decimals to the nearest thousandth.

Fraction	Decimal	Percent
$\frac{7}{12}$		
	0.025	
		2%
$\frac{1}{200}$		
	0.35	
		0.8%

Table 1.1

2. Calculate the following:
   a. A survey reported that 80 people, or 16% of the group, were smokers. How many people were surveyed?
   b. Of the 236 students who took a test, 16.5% received a B grade. How many students received a B grade?
   c. Six of the 16 people present were from foreign countries. What percent were foreigners?

3. When looking through the classified ads, you found that 16 jobs had a starting salary of \$20,000, 8 had a starting salary of \$32,000, and 1 had a starting salary of \$50,000. Find the mean and median starting salary for these jobs.[5]

[5]Recall that given the list of numbers 9, 2, −2, 6, 5, the *mean* = the sum 9 + 2 + (−2) + 6 + 5 divided by 5 (the number of items in the list) = 20/5 = 4; the *median* = the middle number of the list in ascending order −2, 2, 5, 6, 9, which is 5. If the list had an even number of elements—for example, −2, 2, 5, 6—the median would be the mean of the middle two numbers on the ordered list, in this case (2 + 5)/2 = 7/2 = 3.5.

4. Find the mean and median grade point average (GPA) from the data given in Table 1.2.

GPA	Frequency Count
1.0	56
2.0	102
3.0	46
4.0	12

Table 1.2

5. Figure 1.5 presents information about the Hispanic population in the United States from 2000 to 2005.

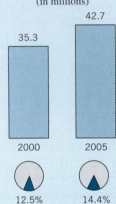

Hispanic Population in the United States (2000–2005)
(in millions)

Figure 1.5 Change in the Hispanic population in the United States.
Source: U.S. Bureau of the Census, *www.census.gov.*

a. What does the bar chart tell you that the pie chart does not?

b. Using the bar and pie charts, what was the U.S. population in 2000? In 2005?

6. a. Fill in Table 1.3. Round your answers to the nearest whole number.

Age	Frequency Count	Relative Frequency (%)
1–20		38
21–40	35	
41–60	28	
61–80		
Total	137	

Table 1.3

b. Calculate the percentage of the population who are over 40 years old.

7. Use Table 1.3 to create a histogram and pie chart.

8. From the histogram in Figure 1.6, create a frequency distribution table. Assume that the total number of people represented by the histogram is 1352. (*Hint:* Estimate the relative frequencies from the graph and then calculate the frequency count in each interval.)

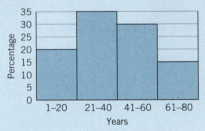

Figure 1.6 Distribution of ages (in years).

9. Calculate the mean and median for the following data:
a. $475, $250, $300, $450, $275, $300, $6000, $400, $300
b. 0.4, 0.3, 0.3, 0.7, 1.2, 0.5, 0.9, 0.4

10. Explain why the mean may be a misleading numerical summary of the data in Problem 9(a).

Exercises for Section 1.1

1. Internet use as reported by teenagers in 2006 in the United States is shown in the accompanying graph.
  a. What percentage of 13- to 17-year-old females spend at least 3 hours per day on the internet outside of school?

  b. What percentage of 13- to 17-year-old males spend at least 3 hours per day on the internet outside of school?

  c. What additional information would you need in order to find out the percentage of 13 to 17 year olds who spend at least 3 hours per day on the internet?

**Time Spent Per Day on the Internet Outside of School
13–17 Year Olds in U.S.**

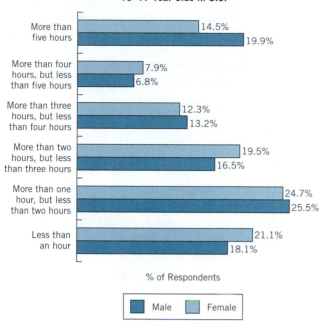

More than five hours	14.5% / 19.9%
More than four hours, but less than five hours	7.9% / 6.8%
More than three hours, but less than four hours	12.3% / 13.2%
More than two hours, but less than three hours	19.5% / 16.5%
More than one hour, but less than two hours	24.7% / 25.5%
Less than an hour	21.1% / 18.1%

% of Respondents

■ Male   ■ Female

Chart 1 – Time Spent Online Outside of School
Source: BURST Research, May 2006

2. The accompanying bar chart shows the five countries with the largest populations in 2006.

The Five Most Populous Countries (2006)

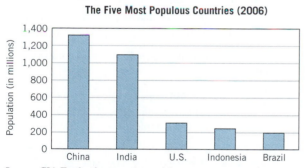

Source: CIA Factbook, *www.cia.gov/cia.*

a. What country has the largest population, and approximately what is its population, size?

b. The population of India is projected in the near future to exceed the population of China. Given the current data, what is the minimum number of additional persons needed to make India's population larger than China's?

c. The world population in 2006 was estimated to be about 6.5 billion. Approximately what percentage of the world's population live in China? In India? In the United States?

3. In 2003 some taxpayers received $300–600 tax rebates. Congress approved this spending as a means to stimulate the economy. According to a May 2003 ABC News/*Washington Post* poll, the accompanying pie chart shows how people would use the money.

Where the Money Will Go in 2003

Source: ABCNEWS/*Washington Post* poll. May, 2003.

a. What is the largest category on which people say they will spend their rebates? Why does the category look so much larger than its actual relative size?

b. What might make you suspicious about the numbers in this pie chart?

4. The point spread in a football game is the difference between the winning team's score and the losing team's score. For example, in the 2004 Super Bowl game, the Patriots won with 32 points versus the Carolina Panthers' 29 points. So the point spread was 3 points.

a. In the accompanying bar chart, what is the interval with the most likely point spread in a Super Bowl? The least likely?

Point Spreads in 39 Super Bowl Games

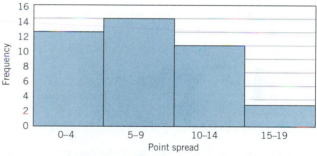

Source: www.docsports.com/point-spreads-for-every-super-bowl.html.

b. What percentage of these Super Bowl games had a point spread of 9 or less? Of 14 or less?

5. Given here is a table of salaries taken from a survey of recent graduates (with bachelor degrees) from a well-known university in Pittsburgh.

Salary (in thousands)	Number of Graduates Receiving Salary
21–25	2
26–30	3
31–35	10
36–40	20
41–45	9
46–50	1

a. How many graduates were surveyed?

b. Is this quantitative or qualitative data? Explain.

5. (continued)

  c. What is the relative frequency of people having a salary between \$26,000 and \$30,000?

  d. Create a histogram of the data.

6. The accompanying bar chart shows the predictions of the U.S. Census Bureau about the future racial composition of American society. Hispanic origin may be of any race, so the other categories may include people of Hispanic origin.

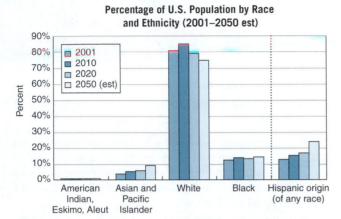

Percentage of U.S. Population by Race and Ethnicity (2001–2050 est)

Sources: U.S. Bureau of the Census, *Statistical Abstract of the United States: 2002.*

  a. Estimate the following percentages:

   i. Asian and Pacific Islanders in the year 2050

   ii. Combined white and black population in the year 2020

   iii. Non-Hispanic population in the year 2001

  b. The U.S. Bureau of the Census has projected that there will be approximately 392,031,000 people in the United States in the year 2050. Approximately how many people will be of Hispanic origin in the year 2050?

  c. Write a topic sentence describing the overall trend.

7. Shown is a pie chart of America's spending patterns at the end of 2006.

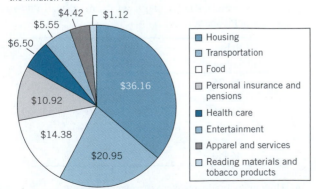

How Americans Spend Their Money

How the average American spends \$100, as measured in late 2006. The Labor Department uses this survey—along with price samples—to calculate the inflation rate.

\$4.42   \$1.12
\$5.55
\$6.50

$36.16, $20.95, $14.38, $10.92

Housing, Transportation, Food, Personal insurance and pensions, Health care, Entertainment, Apparel and services, Reading materials and tobacco products

Sources: U.S. Bureau of the Census, *Statistical Abstract of the United States: 2006.*

  a. In what single category did Americans spend the largest percentage of their income? Estimate this percentage.

  b. According to this chart, if an American family has an income of \$35,000, how much of it would be spent on food?

  c. If you were to write a newspaper article to accompany this pie chart, what would your opening topic sentence be?

8. Attendance at a stadium for the last 30 games of a college baseball team is listed as follows:

5072	3582	2504	4834	2456	3956
2341	2478	3602	5435	3903	4535
1980	1784	1493	3674	4593	5108
1376	978	2035	1239	2456	5189
3654	3845	673	2745	3768	5227

Create a histogram to display these data. Decide how large the intervals should be to illustrate the data well without being overly detailed.

9. a. Compute the mean and median for the list: 5, 18, 22, 46, 80, 105, 110.

  b. Change one of the entries in the list in part (a) so that the median stays the same but the mean increases.

10. Suppose that a church congregation has 100 members, each of whom donates 10% of his or her income to the church. The church collected \$250,000 last year from its members.

  a. What was the mean contribution of its members?

  b. What was the mean income of its members?

  c. Can you predict the median income of its members? Explain your answer.

11. Suppose that annual salaries in a certain corporation are as follows:

Level I (30 employees)	\$18,000
Level II (8 employees)	\$36,000
Level III (2 employees)	\$80,000

Find the mean and median annual salary. Suppose that an advertisement is placed in the newspaper giving the average annual salary of employees in this corporation as a way to attract applicants. Why would this be a misleading indicator of salary expectations?

12. Suppose the grades on your first four exams were 78%, 92%, 60%, and 85%. What would be the lowest possible average that your last two exams could have so that your grade in the class, based on the average of the six exams, is at least 82%?

13. Read Stephen Jay Gould's article "The Median Isn't the Message" and explain how an understanding of statistics brought hope to a cancer victim.

14. a. On the first quiz (worth 25 points) given in a section of college algebra, one person received a score of 16, two people got 18, one got 21, three got 22, one got 23, and one got 25. What were the mean and median of the quiz scores for this group of students?

14. (continued)

    b. On the second quiz (again worth 25 points), the scores for eight students were 16, 17, 18, 20, 22, 23, 25, and 25.

        i. If the mean of the scores for the nine students was 21, then what was the missing score?

        ii. If the median of the scores was 22, then what are possible scores for the missing ninth student?

15. Why is the mean age larger than the median age in the United States? What prediction would you make for your State? What predictions would you make for other countries? Check your predictions with data from the U.S. Census Bureau at *www.census.gov.*

16. Up to and including George W. Bush, the ages of the last 15 presidents when they first took office[6] were 56, 55, 51, 54, 51, 60, 62, 43, 55, 56, 52, 69, 64, 46, 54.

    a. Find the mean and median ages of the past 15 presidents when they took office.

    b. If the mean age of the past 16 presidents is 54.94, at what age did the missing president take office?

    c. Beginning with age 40 and using 5-year intervals, find the frequency count for each age interval.

    d. Create a frequency histogram using your results from part (c).

17. Herb Caen, a Pulitzer Prize–winning columnist for the *San Francisco Chronicle*, remarked that a person moving from state A to state B could raise the average IQ in both states. Is he right? Explain.

18. Why do you think most researchers use median rather than mean income when studying "typical" households?

19. According to the 2000 U.S. Census, the median net worth of American families was $55,000 and the mean net worth was $282,500. How could there be such a wide discrepancy?

20. Read the *CHANCE News* article and explain why the author was concerned.

21. The Greek letter Σ (called sigma) is used to represent the sum of all of the terms of a certain group. Thus, $a_1 + a_2 + a_3 + \cdots + a_n$ can be written as

$$\sum_{i=1}^{n} a_i$$

which means to add together all of the values of a_i from a_1 to a_n.

    a. Using Σ notation, write an algebraic expression for the mean of the five numbers x_1, x_2, x_3, x_4, x_5.

    b. Using Σ notation, write an algebraic expression for the mean of the n numbers $t_1, t_2, t_3, \ldots, t_n$.

    c. Evaluate the following sum:

$$\sum_{k=1}^{5} 2k$$

[6]*http://www.campvishus.org/PresAgeDadLeft.htm#AgeOffice.*

22. The accompanying table gives the ages of students in a mathematics class.

Ages of Students

Age Interval	Frequency Count
15–19	2
20–24	8
25–29	4
30–34	3
35–39	2
40–44	1
45–49	1
Total	21

    a. Use this information to estimate the mean age of the students in the class. Show your work. (*Hint:* Use the mean age of each interval.)

    b. What is the largest value the actual mean could have? The smallest? Why?

23. (Use of calculator or other technology recommended.) Use the following table to generate an estimate of the mean age of the U.S. population. Show your work. (*Hint:* Replace each age interval with an age approximately in the middle of the interval.)

Ages of U.S. Population in 2004

Age (years)	Population (thousands)
Under 10	39,677
10–19	41,875
20–29	40,532
30–39	41,532
40–49	45,179
50–59	35,986
60–74	31,052
75–84	12,971
85 and over	4,860
Total	293,655

Source: U.S. Bureau of the Census, *Statistical Abstract of the United States: 2006.*

24. An article titled "Venerable Elders" (*The Economist,* July 24, 1999) reported that "both Democratic and Republican images are selective snapshots of a reality in which the median net worth of households headed by Americans aged 65 or over is around double the national average—but in which a tenth of such households are also living in poverty." What additional statistics would be useful in forming an opinion on whether elderly Americans are wealthy or poor compared with Americans as a whole?

25. Estimate the mean and median from the given histogram. (See hint in Exercise 23.) The program "F4: Measures of Central Tendency" in *FAM1000 Census Graphs* can help you understand the mean and median and their relationship to histograms.

Monthly Allowance of Junior High School Students

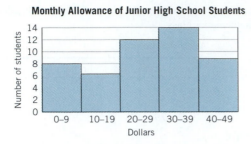

26. Choose a paragraph of text from any source and construct a histogram of word lengths (the number of letters in the word). If the same word appears more than once, count it as many times as it appears. You will have to make some reasonable decisions about what to do with numbers, abbreviations, and contractions. Compute the mean and median word lengths from your graph. Indicate how you would expect the graph to be different if you used:

   a. A children's book   **c.** A medical textbook

   b. A work of literature

27. (Computer and course software required.) Open up the program "F1: Histograms" in *FAM1000 Census Graphs* in the course software. The 2006 U.S. Census data on 1000 randomly selected U.S. individuals and their families are imbedded in this program. You can use it to create histograms for education, age, and different measures of income. Try using different interval sizes to see what patterns emerge. Decide on one variable (say education) and compare the histograms of this variable for different groups of people. For example, you could compare education histograms for men and women or for people living in two different regions of the country. Pick a comparison that you think is interesting. Create a possible headline for these data. Describe three key features that support your headline.

28. Population pyramids are a type of chart used to depict the overall age structure of a society. Use the accompanying population pyramids for the United States to answer the following questions.

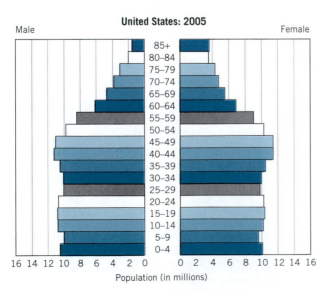

Source: U.S. Bureau of Census, International Data Base, *www.census.gov.*

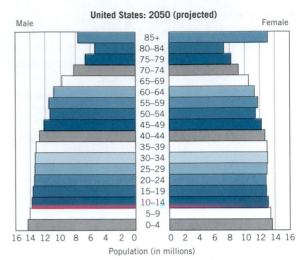

Source: U.S. Bureau of the Census, International Data Base, *www.census.gov.*

   a. Estimate the number of:

   i. Males who were between the ages 35 and 39 years in 2005.

   ii. Females who were between the ages 55 and 59 years in 2005.

   iii. Males 85 years and older in the year 2050; females 85 years and older in the year 2050.

   iv. All males and females between the ages of 0 and 9 years in the year 2050.

   b. Describe two changes in the distribution of ages from the year 2005 to the predictions for 2050.

29. The accompanying population pyramid shows the age structure in Ghana, a developing country in Africa, for 2005. The previous exercise contains a population pyramid for the United States, an industrialized nation, for 2005. Describe three major differences in the distribution of ages in these two countries in 2005.

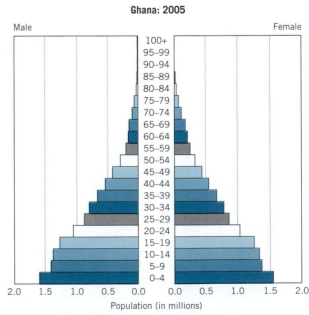

Source: U.S. Census Bureau, International Data Base, *www.census.gov.*

1.2 *Describing Relationships between Two Variables*

By looking at two-variable data, we can learn how change in one variable affects change in another. How does the weight of a child determine the amount of medication prescribed by a pediatrician? How does median age or income change over time? In this section we examine how to describe these changes with graphs, data tables, written descriptions, and equations.

Visualizing Two-Variable Data

E X A M P L E   1

Scatter Plots

Table 1.4 shows data for two variables, the year and the median age of the U.S. population. Plot the data in Table 1.4 and then use your graph to describe the changes in the U.S. median age over time.

Median Age of the U.S. Population, 1850–2050*

Year	Median Age	Year	Median Age
1850	18.9	1950	30.2
1860	19.4	1960	29.5
1870	20.2	1970	28.0
1880	20.9	1980	30.0
1890	22.0	1990	32.8
1900	22.9	2000	35.3
1910	24.1	2005	36.7
1920	25.3	2010	36.0
1930	26.4	2025	37.5
1940	29.0	2050	38.1

DATA

MEDAGE
Excel and graph link files for the median age data are called MEDAGE.

Table 1.4

*Data for 2010–2050 are projected.
Source: U.S., Bureau of the Census, *Statistical Abstract of the United States: 1,* 2006.

S O L U T I O N

In Table 1.4 we can think of a year and its associated median age as an ordered pair of the form (year, median age). For example, the first row corresponds to the ordered pair (1850, 18.9) and the second row corresponds to (1860, 19.4). Figure 1.7 shows a scatter plot of the data. The graph is called a *time series* because it shows changes over time. In newspapers and magazines, the time series is the most frequently used form of data graphic.[7]

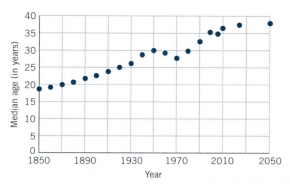

Figure 1.7 Median age of U.S. population over time.

[7]Edward Tufte in *The Visual Display of Information* (Cheshire, Conn.: Graphics Press, 2001, p. 28) reported on a study that found that more than 75% of all graphics published were time series.

? SOMETHING TO THINK ABOUT

What are some of the trade-offs in using the median instead of the mean age to describe changes over time?

Our graph shows that the median age of the U.S. population grew quite steadily for one hundred years, from 1850 to 1950. Although the median age decreased between 1950 and 1970, since 1970 it has continued to increase. From 1850 to the present, the median age nearly doubled, and projections for 2025 and 2050 indicate continued increases, though at a slower pace.

Constructing a "60-Second Summary"

To communicate effectively, you need to describe your ideas succinctly and clearly. One tool for doing this is a "60-second summary"—a brief synthesis of your thoughts that could be presented in one minute. Quantitative summaries strive to be straightforward and concise. They often start with a topic sentence that summarizes the key idea, followed by supporting quantitative evidence.

After you have identified a topic you wish to write about or present orally, some recommended steps for constructing a 60-second summary are:

- Collect relevant information (possibly from multiple sources, including the Internet).
- Search for patterns, taking notes.
- Identify a key idea (out of possibly many) that could provide a topic sentence.
- Select evidence and arguments that support your key idea.
- Examine counterevidence and arguments and decide if they should be included.
- Construct a 60-second summary, starting with your topic sentence.

You will probably weave back and forth among the steps in order to refine or modify your ideas. You can help your ideas take shape by putting them down on paper. Quantitative reports should not be written in the first person. For example, you might say something like "The data suggest that . . ." rather than "I found that the data . . ."

EXAMPLE 2 **A 60-Second Summary**

The annual federal surplus (+) or deficit (−) since World War II is shown in Table 1.5 and Figure 1.8 (a scatter plot where the points have been connected). Construct a 60-second summary describing the changes over time.

Federal Budget: Surplus (+) or Deficit (–)

Year	Billions of Dollars	Year	Billions of Dollars	Year	Billions of Dollars
1945	−$48	1979	−$41	1993	−$255
1950	−$3	1980	−$74	1994	−$203
1955	−$3	1981	−$79	1995	−$164
1960	$0	1982	−$128	1996	−$108
1965	−$1	1983	−$208	1997	−$22
1970	−$3	1984	−$185	1998	+$69
1971	−$23	1985	−$212	1999	+$126
1972	−$23	1986	−$221	2000	+$236
1973	−$15	1987	−$150	2001	+$127
1974	−$6	1988	−$155	2002	−$159
1975	−$53	1989	−$152	2003	−$378
1976	−$74	1990	−$221	2004	−$412
1977	−$54	1991	−$269	2005	−$427
1978	−$59	1992	−$290		

Table 1.5
Source: U.S. office of Management and Budget.

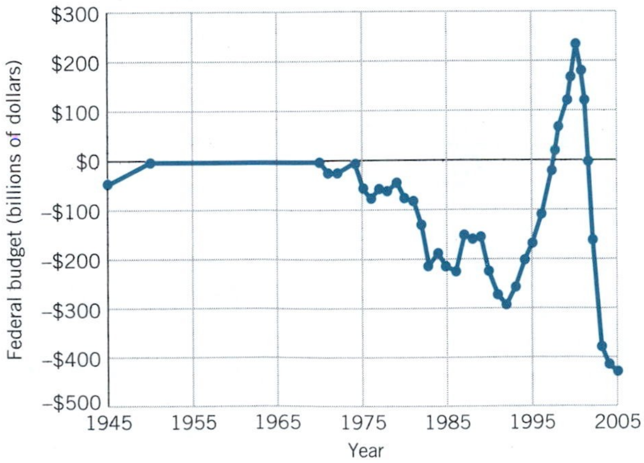

Figure 1.8 Annual federal budget surplus or deficit in billions of dollars.

SOLUTION Between 1945 and 2005 the annual U.S. federal deficit moved from a 30-year stable period, with as little as $0 deficit, to a period of oscillations, leading in 2005 to the largest deficit ever recorded. From 1971 to 1992, the federal budget ran an annual deficit, which generally was getting larger until it reached almost $300 billion in 1992. From 1992 to 1997, the deficit steadily decreased, and from 1998 to 2001 there were relatively large surpluses. The maximum surplus occurred in 2000, when it reached $236 billion. But by 2002 the federal government was again running large deficits. In 2005 the deficit reached $427 billion, the largest recorded up to that time.

Algebra Aerobics 1.2a

1. The net worth of a household at any given time is the difference between assets (what you *own*) and liabilities (what you *owe*). Table 1.6 and Figure 1.9 show the median net worth of U.S. households, adjusted for inflation.[8]

    Median Net Worth of Households (adjusted for inflation using year 2000 dollars)

Year	Median Net Worth ($)
1984	50,018
1988	49,855
1991	44,615
1993	43,567
1995	44,578
1998	49,932
2000	55,000

    Table 1.6
    Source: U.S. Bureau of the Census, *www.census.gov.*

    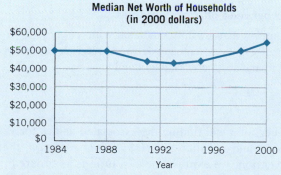

    Median Net Worth of Households (in 2000 dollars)

    Figure 1.9

    a. Write a few sentences about the trend in U.S. median household net worth.

    b. What additional information might be useful in describing the trend in median net worth?

2. Use Figure 1.10 to estimate:

    a. The year when the world population reached 4 billion,

    b. The year that it is projected to reach 8 billion.

    c. The number of years it will take to grow from 4 to 8 billion.

3. Use Figure 1.10 to estimate the following projections for the year 2150.

    a. The total world population.

    b. The total populations of all the more developed countries.

    c. The total populations of all the less developed countries.

    d. Write a topic sentence about the estimated world population in 2150.

4. Use Figure 1.10 to answer the following:

    a. The world population in 2000 was how many times greater than the world population in 1900? What was the difference in population size?

    b. The world population in 2100 is projected to be how many times greater than the world population in 2000? What is the difference in population size?

    c. Describe the difference in the growth in world population in the twentieth century (1900–2000) versus the projected growth in the twenty-first century (2000–2150).

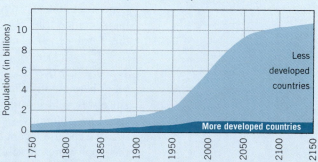
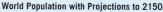

Figure 1.10 World population growth, 1750–2150 (est.).
Source: Population Reference Bureau, *www.prb.org.*

---

[8]"Constant dollars" is a measure used by economists to compare incomes and other variables in terms of purchasing power, eliminating the effects of inflation. To say the median income in 1986 was $37,546 in "constant 2000 dollars" means that the median income in 1986 could buy an amount of goods and services that would cost $37,546 to buy in 2000. The actual median income in 1986 (measured in what economists call "current dollars") was much lower. Income corrected for inflation is sometimes called "real" income.

Using Equations to Describe Change

The relationship between two variables can also be described with an equation. An equation gives a rule on how change in the value of one variable affects change in the value of the other. If the variable n represents the number of years of education beyond grammar school and e represents yearly median earnings (in dollars) for people living in the United States, then the following equation models the relationship between e and n:

$$e = 3780 + 4320n$$

This equation provides a powerful tool for describing how earnings and education are linked and for making predictions.[9] For example, to predict the median earnings, e, for those with a high school education, we replace n with 4 (representing 4 years beyond grammar school, or a high school education) in our equation to get

$$e = 3780 + 4320 \cdot 4$$
$$= \$21{,}060$$

Thus our equation predicts that for those with a high school education, median earnings will be about $21,060.

An equation that is used to describe a real-world situation is called a *mathematical model*. Such models offer compact, often simplified descriptions of what may be a complex situation. Thus, the accuracy of the predictions made with such models can be questioned and disciplines outside of mathematics may be needed to help answer such questions. Yet these models are valuable guides in our quest to understand social and physical phenomena in our world.

Describing the relationship between abstract variables

Variables can represent quantities that are not associated with real objects or events. The following equation or mathematical sentence defines a relationship between two quantities, which are named by the abstract variables x and y:

$$y = x^2 + 2x - 3$$

By substituting various values for x and finding the associated values for y, we can generate pairs of values for x and y, called *solutions to the equation*, that make the sentence true. By convention, we express these solutions as ordered pairs of the form (x, y). Thus, $(1, 0)$ would be a solution to $y = x^2 + 2x - 3$, since $0 = 1^2 + 2(1) - 3$, whereas $(0, 1)$ would not be a solution, since $1 \neq 0^2 + 2(0) - 3$.

There are infinitely many possible solutions to the equation $y = x^2 + 2x - 3$, since we could substitute any real number for x and find a corresponding y. Table 1.7 lists a few solutions.

We can use technology to graph the equation (see Figure 1.11). All the points on the graph represent solutions to the equation, and every solution is a point on the graph of the equation.

---

[9]In "Extended Exploration: Looking for Links between Education and Earnings," which follows Chapter 2, we show how such equations are derived and how they are used to analyze the relationship between education and earnings.

x	y
-4	5
-3	0
-2	-3
-1	-4
0	-3
1	0
2	5
3	12

Table 1.7

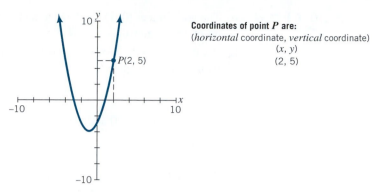

Figure 1.11 Graph of $y = x^2 + 2x - 3$ where one solution of infinitely many solutions is labeled.

Note that sometimes an arrow is used to show that a graph extends indefinitely in the indicated direction. In Figure 1.11, the arrows show that both arms of the graph extend indefinitely upward.

Solutions of an Equation

The *solutions* of an equation in two variables x and y are the ordered pairs (x, y) that make the equation a true statement.

Graph of an Equation

The *graph* of an equation in two variables displays the set of points that are solutions to the equation.

EXAMPLE 3   **Solutions for equations in one or two variables**

Describe how the solutions for the following equations are similar and how they differ.

$$3x + 5 = 11$$

$$x + 2 = x + 2$$

$$3 + x = y + 5$$

SOLUTION   The solutions are similar in the sense that each solution for each particular statement makes the statement true. They are different because:

There is only one solution ($x = 2$) of the single-variable equation $3x + 5 = 11$.

There are an infinite number of solutions for x of the single-variable equation $x + 2 = x + 2$, since any real number will make the statement a true statement.

There are infinitely many solutions, in the form of ordered pairs (x, y), of the two-variable equation $3 + x = y + 5$.

EXAMPLE 4   **Estimating solutions from a graph**

The graph of the equation $x^2 + 4y^2 = 4$ is shown in Figure 1.12.

a. From the graph, estimate three solutions of the equation.

b. Check your solutions using the equation.

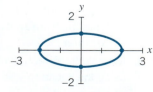

Figure 1.12

S O L U T I O N   **a.** The coordinates $(0, 1)$, $(-2, 0)$, and $(1, 0.8)$ appear to lie on the ellipse, which is the graph of the equation $x^2 + 4y^2 = 4$.

b. If substituting the ordered pair $(0, 1)$ into the equation makes it a true statement, then $(0, 1)$ is a solution.

Given	$x^2 + 4y^2 = 4$
substitute $x = 0$ and $y = 1$	$(0)^2 + 4(1)^2 = 4$
evaluate	$4 = 4$

We get a true statement, so $(0, 1)$ is a solution to the equation.
For the ordered pair $(-2, 0)$:

Given	$x^2 + 4y^2 = 4$
substitute $x = -2$ and $y = 0$	$(-2)^2 + 4(0)^2 = 4$
evaluate	$4 + 0 = 4$
	$4 = 4$

Again we get a true statement, so $(-2, 0)$ is a solution to the equation.
For the ordered pair $(1, 0.8)$:

Given	$x^2 + 4y^2 = 4$
substitute $x = 1$ and $y = 0.8$	$(1)^2 + 4(0.8)^2 = 4$
evaluate	$1 + 4(0.64) = 4$
	$3.56 \neq 4$

We get a false statement, so $(1, 0.8)$ is not a solution, although it is close to a solution.

Algebra Aerobics 1.2b

Problem 4(c) requires a graphing program.

1. a. Describe in your own words how to compute the value for y, given a value for x, using the following equation:

$$y = 3x^2 - x + 1$$

b. Which of the following ordered pairs represent solutions to the equation?

$(0, 0)$, $(0, 1)$, $(1, 0)$, $(-1, 2)$, $(-2, 3)$, $(-1, 0)$

c. Use $x = 0$, ± 1, ± 2, ± 3 to generate a small table of values that represent solutions to the equation.

2. Repeat the directions in Problem 1(a), (b), and (c) using the equation $y = (x - 1)^2$.

3. Given the equations $y_1 = 4 - 3x$ and $y_2 = -2x^2 - 3x + 5$, fill in Table 1.8.

x	-4	-2	-1	0	1	2	4
y_1							
y_2							

Table 1.8

a. Use the table to create two scatter plots, one for the ordered pairs (x, y_1) and the other for (x, y_2).

b. Draw a smooth curve through the points on each graph.

c. Is (1, 1) a solution for equation y_1? For y_2?

d. Is $(-1, 6)$ a solution for equation y_1? For y_2?

e. Look at the graphs. Is the ordered pair $(-3, 2)$ a solution for either equation? Verify your answer by substituting the values into each equation.

4. Given the equation $y = x^2 - 3x + 2$,

  a. If $x = 3/2$, find y.

  b. Find two points that are *not* solutions to this equation.

  c. If available, use technology to graph the equation and then confirm your results for parts (a) and (b).

Exercises for Section 1.2

Course software recommended for Exercise 20.

1. Assume you work for a newspaper and are asked to report on the following data.

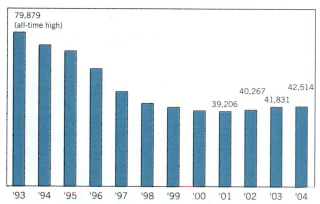

AIDS Cases in U.S.

79,879 (all-time high)

40,267   42,514

39,206   41,831

'93 '94 '95 '96 '97 '98 '99 '00 '01 '02 '03 '04

Source: Centers for Disease Control and Prevention, *www.cdc.gov.*

  a. What are three important facts that emerge from this graph?

  b. Construct a 60-second summary that could accompany the graph in the newspaper article.

2. The following graph shows changes in the Dow Jones Industrial Average, which is based on 30 stocks that trade on the New York Stock Exchange and is the best-known index of U.S. stocks.

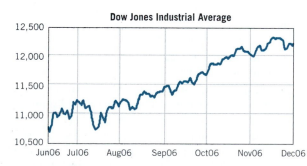

Dow Jones Industrial Average

12,500

12,000

11,500

11,000

10,500

Jun06 Jul06 Aug06 Sep06 Oct06 Nov06 Dec06

Source: http://finance.yahoo.com.

  a. What time period does the graph cover?

  b. Estimate the lowest Dow Jones Industrial Average. During what month did it occur?

  c. Estimate the highest value for the Dow Jones during that period. When did it occur?

  d. Write a topic sentence describing the change in the Dow Jones over the given time period.

3. The accompanying table shows the number of personal and property crimes in the United States from 1995 to 2003.

Year	Personal Crimes (in thousands)	Property Crimes (in thousands)
1995	1,799	12,064
1998	1,534	10,952
2000	1,425	10,183
2003	1,381	10,436

Source: U.S. Bureau of the Census, Statistical *Abstract, 2006.*

  a. Create a scatter plot of the personal crimes over time. Connect the points with line segments.

  b. Approximately how many *times* more property crimes than personal crimes were committed in 1995? In 2003?

  c. Write a topic sentence that compares property and personal crime from 1995 to 2003.

4. The National Cancer Institute now estimates that after 70 years of age, 1 woman in 8 will have gotten breast cancer. Fortunately, they also estimate that 95% of breast cancer can be cured, especially if caught early. The data in the accompanying table show how many women in different age groups are likely to get breast cancer.

Lifetime Risk of Developing Breast Cancer

Age Group	Chance of Developing Cancer	Chance in 1000 Women
30–39	1 in 229	4 per 1000
40–49	1 in 68	15 per 1000
50–59	1 in 37	27 per 1000
60–69	1 in 26	38 per 1000
70+	1 in 8	125 per 1000

(*Note:* Men may get breast cancer too, but less than 1% of all breast cancer cases occur in men.)

  a. What is the overall relationship between age and breast cancer?

b. Make a bar chart using the chance of breast cancer in 1000 women for the age groups given.

c. Using the "chance in 1000 women" data, estimate how much more likely that women in their 40s would have had breast cancer than women in their 30s. How much more likely for women in their 50s than women in their 40s?

d. It is common for women to have yearly mammograms to detect breast cancer after they turn 50, and health insurance companies routinely pay for them. Looking at these data, would you recommend an earlier start for yearly mammograms? Explain your answer in terms of the interests of the patient and the insurance company. (*Note:* Some research says that mammograms are not that good at detection.)

5. The National Center for Chronic Disease Prevention and Health Promotion published the following data on the chances that a man has had prostate cancer at different ages.

Lifetime Risk of Developing Prostate Cancer

Age	Risk	Percent Risk
45	1 in 25,000	0.004%
50	1 in 476	0.21%
55	1 in 120	0.83%
60	1 in 43	2.3%
65	1 in 21	4.8%
70	1 in 13	7.7%
75	1 in 9	11.1%
80	1 in 6	16.7%

a. What is the relationship between age and getting prostate cancer?

b. Make a scatter plot of the percent risk for men of the ages given.

c. Using the "percent risk" data, how much more likely are men 50 years old to have had prostate cancer than men who are 45? How much more likely are men 55 years old to have had prostate cancer than men who are 50?

d. Looking at these data, when would you recommend annual prostate checkups to begin for men? Explain your answer in terms of the interests of the patient and the insurance company.

6. Birth rate data in the United States are given as the number of live births per 1000 women in each age category.

Age	1950	2000
10–14	1.0	0.9
15–19	81.6	48.5
20–24	196.6	112.3
25–29	166.1	121.4
30–34	103.7	94.1
35–39	52.9	40.4
40–44	15.1	7.9
45–49	1.2	0.5

Source: National Center for Health Statistics, U.S. Dept. of Health and Human Services.

a. Construct a bar chart showing the birth rates for the year 1950. Which mother's age category had the highest rate of live births? What percentage of women in that category delivered live babies? In which age category was the lowest rate of babies born? What percentage of women in that category delivered live babies?

b. Construct a bar chart showing the birth rates for the year 2000. Which mother's age category had the highest rate of live births? What percentage of women in that category delivered live babies? In which age category was the lowest rate of babies born? What percentage of women in that category delivered live babies?

c. Write a paragraph comparing and contrasting the birth rates in 1950 and in 2000. Bear in mind that since 1950 there have been considerable medical advances in saving premature babies and in increasing the fertility of couples.

7. The National Center for Health Statistics published the accompanying chart on childhood obesity.

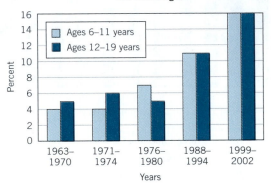

Source: National Center for Health Statistics, *www.cdc.gov/nchs.*

a. How would you describe the overall trend in the weights of American children?

b. Over which years did the percentage of overweight children age 6 to 11 increase?

c. Over which time period was there no change in the percentage of overweight children age 6 to 11?

d. During which time period were there relatively more overweight 6- to 11-year-olds than 12- to 19-year-olds?

e. One of the national health objectives for the year 2010 is to reduce the prevalence of obesity among children to less than 15%. Does this seem like a reasonable goal?

8. Some years are more severe for influenza- and pneumonia-related deaths than others. The table at the top of the next page shows data from Centers for Disease Control figures for the U.S. for selected years from 1950 to 2000.

Age-Adjusted Death Rate for Influenza and Pneumonia

	Death Rate per 100,000	
Year	Males	Females
1950	55	42
1960	66	44
1970	54	33
1980	42	25
1990	48	31
2000	29	21

Source: Centers for Disease Control and Prevention, *www.cdc.gov.*

a. Create a double bar chart showing the death rates both for men and for women who died of influenza and pneumonia between 1950 and 2000.

b. In which year were death rates highest for both men and women?

c. Were there any decades in which there was an increase in male deaths but a decrease for women?

d. Write a 60-second summary about deaths due to influenza and pneumonia over the years 1950 to 2000.

9. The following three graphs describe two cars, A and B.

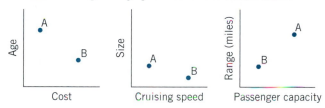

For parts (a)–(d), decide whether the statement is true or false. Explain your reasoning.

a. The newer car is more expensive.

b. The slower car is larger.

c. The larger car is newer.

d. The less expensive car carries more passengers.

e. State two other facts you can derive from the graphs.

f. Which car would you buy? Why?

10. a. Which (if any) of the following ordered pairs (x, y) is a solution to the equation $y = x^2 - 2x + 1$? Show how you came to your conclusion.

$$(-2, 7), \quad (1, 0), \quad (2, 1)$$

b. Find one additional ordered pair that is a solution to the equation above. Show how you found your solution.

11. Consider the equation $R = 2 - 5T$.

a. Determine which, if any, of the following points (T, R) satisfy this equation.

$$(0, 4), \quad (1, -3), \quad (2, 0)$$

b. Find two additional ordered pairs that are solutions to the equation.

c. Make a scatter plot of the solution points found.

d. What does the scatter plot suggest about where more solutions could be found? Check your predictions.

12. Use the accompanying graph to estimate the missing values for x or y in the table.

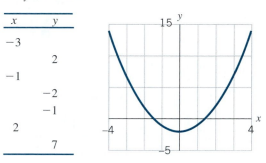

x	y
-3	
	2
-1	
	-2
	-1
2	
	7

13. For parts (a)–(d) use the following equation: $y = \dfrac{x + 1}{x - 1}$.

a. Describe in words how to find the value for y given a value for x.

b. Find the ordered pair that represents a solution to the equation when the value of x is 5.

c. Find the ordered pair that represents a solution to the equation when the value of y is 3.

d. Is there an ordered-pair solution to the equation when the value of x is 1? If so, find it; if not, explain why.

14. For parts (a)–(d) use the following equation: $y = \dfrac{1}{x + 1}$.

a. Describe in words how to find the value for y given a value for x.

b. Find the ordered pair that represents a solution to the equation when the value of x is 0.

c. Find the ordered pair that represents a solution to the equation when the value of y is 4.

d. Is there an ordered-pair solution to the equation when the value of x is -1? If so, find it; if not, explain why.

15. For parts (a)–(d) use the following equation: $y = -2x^2$.

a. If $x = 0$, find the value of y.

b. If x is greater than zero, what can you say about the value of y?

c. If x is a negative number, what can you say about the value of y?

d. Can you find an ordered pair that represents a solution to the equation when y is greater than zero? If so, find it; if not, explain why.

16. Find the ordered pairs that represent solutions to each of the following equations when $x = 0$, when $x = 3$, and when $x = -2$.

a. $y = 2x^2 + 5x$

b. $y = -x^2 + 1$

c. $y = x^3 + x^2$

d. $y = 3(x - 2)(x - 1)$

17. Given the four ordered pairs $(-1, 3)$, $(1, 0)$, $(2, 3)$, and $(1, 2)$, for each of the following equations, identify which points (if any) are solutions for that equation.

a. $y = 2x + 5$

b. $y = x^2 - 1$

c. $y = x^2 - x + 1$

d. $y = \dfrac{4}{x + 1}$

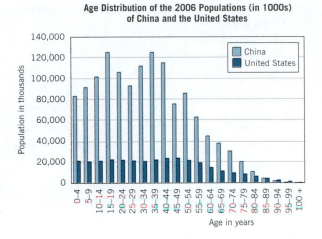

Age Distribution of the 2006 Populations (in 1000s) of China and the United States

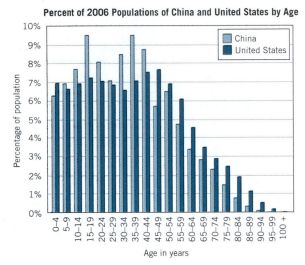

Percent of 2006 Populations of China and United States by Age

Source: U.S. Bureau of the Census, *www.census.gov.*

18. The accompanying graphs (and related Excel and graph link file USCHINA) contain information about the populations of the United States and China. Write a 60-second summary comparing the two populations.

19. Read *The New York Times* op-ed article "A Fragmented War on Cancer" by Hamilton Jordan, who was President Jimmy Carter's chief of staff. Jordan claims that we are on the verge of a cancer epidemic.

   a. Use what you know about the distribution of ages over time in the United States to refute his claim. (See Sec.1.1 Exercise 28 for some ideas.) Are there other arguments that refute his claim?

   b. Read *The New York Times* letter to the editor by William M. London, Director of Public Health, American Council on Science and Health. London argues that Hamilton Jordan's assertions are misleading. What questions are raised by the arguments of William London? What additional data would you need to evaluate his arguments?

   c. Write a paragraph refuting Hamilton Jordan's claim that we are on the verge of a cancer epidemic.

20. (Computer required.) Make a prediction about the distribution of income for males and females in the United States. Check your predictions using the course software "F1: Histograms" in *FAM1000 Census Graphs* and/or using data from the U.S. Census Bureau at *www.census.gov.* Write a 60-second summary describing your results.

1.3 *An Introduction to Functions*

What Is a Function?

When we speak informally of one quantity being a function of some other quantity, we mean that one depends on the other. For example, someone may say that what they wear is a function of where they are going, or what they weigh is a function of what they eat, or how well a car runs is a function of how well it is maintained.

In mathematics, the word "function" has a precise meaning. A function is a special relationship between two quantities. If the value of one quantity uniquely determines the value of a second quantity, then the second quantity is a *function* of the first.

Median age and the federal deficit are functions of time since each year determines a unique (one and only one) value of median age or the federal deficit. The equation $y = x^2 + 2x - 3$ defines y as a function of x since each value of x we substitute in the equation determines a unique value of y.

> **Definition of a Function**
>
> A variable y is a *function* of a variable x if each value of x determines a unique value of y.

Representing Functions in Multiple Ways

We can think of a function as a "rule" that takes certain inputs and assigns to each input value exactly one output value. The rule can be described using words, data tables, graphs, or equations.

EXAMPLE 1   **Sales tax**

Eleven states have a sales tax of 6%; that is, for each dollar spent in a store in these states, the law says that you must pay a tax of 6 cents, or $0.06. Represent the sales tax as a function of purchase price using an equation, table, and graph.[10]

SOLUTION   **Using an Equation**

We can write this relationship as an equation where T represents the amount of sales tax and P represents the price of the purchase (both measured in dollars):

$$\text{Amount of sales tax} = 0.06 \cdot \text{price of purchase}$$

$$T = 0.06P$$

Our function rule says: "Take the given value of P and multiply it by 0.06; the result is the corresponding value of T." The equation represents T as a function of P, since for each value of P the equation determines a unique (one and only one) value of T. The purchase price, P, is restricted to dollar amounts greater than or equal to zero.

Using a Table

We can use this formula to make a table of values for T determined by the different values of P (see Table 1.9). Such tables were once posted next to many cash registers.

P (purchase price in $)	0	1	2	3	4	5	6	7	8	9	10
T (sales tax in $)	0.00	0.06	0.12	0.18	0.24	0.30	0.36	0.42	0.48	0.54	0.60

Table 1.9

Using a Graph

The points in Table 1.9 were used to create a graph of the function (Figure 1.13). The table shows the sales tax only for selected purchase prices, but we could have used any positive dollar amount for P. We connected the points on the scatter plot to suggest the many possible intermediate values for price. For example, if $P = \$2.50$, then $T = \$0.15$.

Figure 1.13 Graph of 6% sales tax.

---

[10]In 2007, a sales tax of 6% was the most common rate for a sales tax in the United States. See *www.taxadmin.org* for a listing of the sales tax rates for all of the states.

Independent and Dependent Variables

Since a function is a rule that assigns to each input a unique output, we think of the output as being dependent on the input. We call the input of a function the *independent variable* and the output the *dependent variable*. When a set of ordered pairs represents a function, then each ordered pair is written in the form

(independent variable, dependent variable)

or equivalently,   (input, output)

If x is the independent and y the dependent variable, then the ordered pairs would be of the form

$$(x, y)$$

The mathematical convention is for the first variable, or input of a function, to be represented on the horizontal axis and the second variable, or output, on the vertical axis.

Sometimes the choice of the independent variable is arbitrary or not obvious. For example, economists argue as to whether wealth is a function of education or education is a function of wealth. As seen in the next example, there may be more than one correct choice.

E X A M P L E   2

Identifying Independent and dependent variables

In the sales tax example, the equation $T = 0.06P$ gives the sales tax, T, as a function of purchase price, P. In this case T is the dependent variable, or output, and P is the independent variable, or input. But for this equation we can see that P is also a function of T; that is, each value of T corresponds to one and only one value of P. It is easier to see the relationship if we solve for P in terms of T, to get

$$P = \frac{T}{0.06}$$

Now we are thinking of the purchase price, P, as the dependent variable, or output, and the sales tax, T, as the independent variable, or input. So, if you tell me how much tax you paid, I can find the purchase price.

When Is a Relationship Not a Function?

Not all relationships define functions. A function is a special type of relationship, one where for each input, the rule specifies one and only one output. Examine the following examples.

Function		Not a Function		Function	
Input	Output	Input	Output	Input	Output
1 ⟶ 6		1 ⟶ 6		1 ⟶	
2 ⟶ 7		⟶ 7		2 ⟶ 6	
3 ⟶ 8		2 ⟶ 8		3 ⟶	
		3 ⟶ 9			
Each input has only one output.		The input of 1 gives *two different outputs*, 6 and 7, so this relationship is *not* a function.		Each input has only one output. Note that a function may have identical outputs for different inputs.	

E X A M P L E   3

Does the table represent a function?

Consider the set of data in Table 1.10. The first column shows the year, T, of the Olympics. The second column shows the winning distance, D (in feet), for that year for the men's Olympic 16-pound shot put.

a. Is D a function of T?

b. Is T a function of D?

c. What should be your choice for the dependent and independent variables?

Olympic Shot Put

Year, *T*	Winning Distance in Feet Thrown, *D*
1960	65
1964	67
1968	67
1972	70
1976	70
1980	70
1984	70
1988	74
1992	71
1996	71
2000	70
2004	69

Table 1.10

Source: The World Almanac and Book of Facts, 2006.

SOLUTION

a. D **is a function of** T**.**

To determine if D is a function of T, we need to find out if each value of T (the input) determines one and only one value for D (the output). So, the ordered pair representing the relationship would be of the form (T, D). Using Table 1.10, we can verify that for each T, there is one and only one D. So D, the winning shot put distance, is a function of the year of the Olympics, T. Note that different inputs (such as 1964 and 1968) can have the same output (67 feet), and the relationship can still be a function. There are even 5 different years that have 70 feet as their output.

b. T **is** *not* **a function of** D**.**

To determine if T is a function of D, we need to find out if each value of D (now the input) determines one and only one value for T (the output). Now the ordered pairs representing the relationship would be of the form (D, T). Table 1.11 shows this new pairing, where D is thought of as the input and T as the output.

The year, T, is *not* a function of D, the winning distance, since some values of D give more than one value for T. For example, when $D = 67$ there are two corresponding values for T, 1964 and 1968, and this violates the condition of a unique (one and only one) output for each input.

Olympic Shot Put

Winning Distance in Feet Thrown, *D*	Year, *T*
65	1960
67	1964
67	1968
70	1972
70	1976
70	1980
70	1984
74	1988
71	1992
71	1996
70	2000
69	2004

Table 1.11

c. *D*, the winning distance, is a function of the year, *T*, but *T* is *not* a function of *D*. The distance, *D*, depends on the year, *T*, but *T* does not depend on *D*. To construct a function relating *T* and *D*, we must choose *T* as the independent variable and *D* as the dependent variable. The ordered pairs that represent the function would be written as (T, D).

EXAMPLE 4 How would the axes be labeled for each graph of the following functions?

a. Density of water is a function of temperature.

b. Radiation intensity is a function of wavelength.

c. A quantity *Q* is a function of time *t*.

SOLUTION

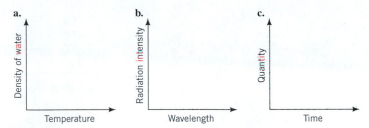

Figure 1.14 Various labels for axes, depending on the context.

How to tell if a graph represents a function: The vertical line test

For a graph to represent a function, each value of the input on the horizontal axis must be associated with one and only one value of the output on the vertical axis. If you can draw a vertical line that intersects a graph in more than one point, then at least one input is associated with two or more outputs, and the graph does not represent a function.

The graph in Figure 1.15 represents *y* as a function of *x*. For each value of *x*, there is only one corresponding value of *y*. No vertical line intersects the curve in more than one point. The graph in Figure 1.16 does *not* represent a function. One can draw a vertical line (an infinite number, in fact) that intersects the graph in more than one point. Figure 1.16 shows a vertical line that intersects the graph at both $(4, 2)$ and $(4, -2)$. That means that the value $x = 4$ does not determine one and only one value of *y*. It corresponds to *y* values of both 2 and -2.

 Exploration 1.2 will help you develop an intuitive sense of functions.

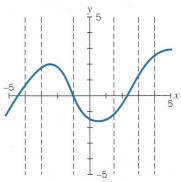

Figure 1.15 The graph represents *y* as a function of *x* since there is no vertical line that intersects the curve at more than one point.

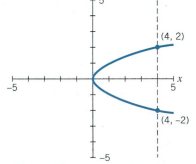

Figure 1.16 The graph does not represent *y* as a function of *x* since there is at least one vertical line that intersects this curve at more than one point.

Vertical Line Test

If there is a vertical line that intersects a graph more than once, the graph does not represent a function.

Algebra Aerobics 1.3

1. Which of the following tables represent functions? Justify your answer.

Input	Output
1	5
2	8
3	8
4	10

Table A

Input	Output
1	5
2	7
2	8
4	10

Table B

2. Does the following table represent a function? If so, why? If not, how could you change the values in the table so it represents a function?

Input	Output
1	5
1	7
3	8
4	10

3. Refer to the graph in Figure 1.17. Is y a function of x?

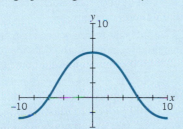

Figure 1.17 Graph of an abstract relationship between x and y.

4. Which of the graphs in Figure 1.18 represent functions and which do not? Why?

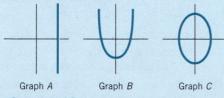

Graph *A*   Graph *B*   Graph *C*

Figure 1.18

5. Consider the scatter plot in Figure 1.19. Is weight a function of height? Is height a function of weight? Explain your answer.

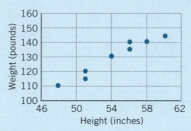

Figure 1.19 Graph of weight versus height.

6. Consider Table 1.12.
   a. Is D a function of Y?
   b. Is Y a function of D?

Y	1992	1993	1994	1995	1996	1997
D	$2.50	$2.70	$2.40	−$0.50	$0.70	$2.70

Table 1.12

7. Plot the following points with x on the horizontal and y on the vertical axis. Draw a line through the points and determine if the line represents a function.

a.

x	y
−1	3
0	3
2	3
3	3

b.

x	y
−2	−1
−2	0
−2	1
−2	4

8. a. Write an equation for computing a 15% tip in a restaurant. Does your equation represent a function? If so, what are your choices for the independent and dependent variables?
   b. How much would the equation suggest you tip for an $8 meal?
   c. Compute a 15% tip on a total check of $26.42.

Exercises for Section 1.3

A graphing program is required for Exercise 12.

1. The following table gives the high temperature in Rome, Italy, for each of five days in October 2006.
   a. Is the temperature a function of the date?
   b. Is the date a function of the temperature?

Date	Rome High Temperature
Oct. 26	27°C
Oct. 27	27°C
Oct. 28	25°C
Oct. 29	26°C
Oct. 30	22°C

2. Which of the following tables describe functions? Explain your answer.

a.

Input value	−2	−1	0	1	2
Output value	−8	−1	0	1	8

b.

Input value	0	1	2	1	0
Output value	−4	−2	0	2	4

c.

Input value	10	7	4	7	10
Output value	3	6	9	12	15

d.

Input value	0	3	9	12	15
Output value	3	3	3	3	3

3. Determine whether each set of points represents a function. (*Hint:* It may be helpful to plot the points.)

a. $(2, 6), (−4, 6), (1, −3), (4, −3)$

b. $(2, −2), (3, −2), (4, −2), (6, −2)$

c. $(2, −3), (2, 3), (−2, −3), (−2, 3)$

d. $(−1, 2), (−1, 0), (−1, −1), (−1, −2)$

4. Which of the accompanying graphs describe functions? Explain your answer.

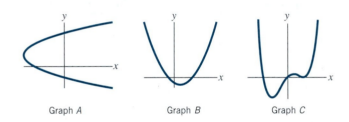

Graph *A*   Graph *B*   Graph *C*

5. Which of the accompanying graphs describe functions? Explain your answer.

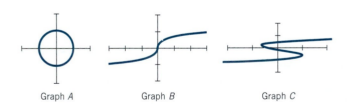

Graph *A*   Graph *B*   Graph *C*

6. Consider the accompanying table, listing the weights (W) and heights (H) of five individuals. Based on this table, is height a function of weight? Is weight a function of height? Justify your answers.

Weight W (lb)	Height H (in)
120	54
120	55
125	58
130	60
135	56

7. a. Find an equation that represents the relationship between x and y in each of the accompanying tables.

i.

x	y
0	5
1	6
2	7
3	8
4	9

ii.

x	y
0	1
1	2
2	5
3	10
4	17

iii.

x	y
0	3
1	3
2	3
3	3
4	3

b. Which of your equations represents y as function of x? Justify your answer.

8. For each of the accompanying tables find a function formula that takes the x values and produces the given y values.

a.

x	y
0	0
1	3
2	6
3	9
4	12

b.

x	y
0	−2
1	1
2	4
3	7
4	10

c.

x	y
0	0
1	−1
2	−4
3	−9
4	−16

9. The basement of a large department store features discounted merchandise. Their policy is to reduce the previous month's price of the item by 10% each month for 5 months, and then give the unsold items to charity.

a. Let S_1 be the sale price for the first month and P the original price. Express S_1 as a function of P. What is the price of a $100 garment on sale for the first month?

b. Let S_2 be the sale price for the second month and P the original price. Express S_2 as a function of P. What is the price of a $100 garment on sale for the second month?

c. Let S_3 be the sale price for the third month and P the original price. Express S_3 as a function of P. What is the price of a $100 garment on sale for the third month?

d. Let S_5 be the sale price for the fifth month and P the original price. Express S_5 as a function of P. What is the final price of a $100 garment on sale for the fifth month? By what total percentage has the garment now been reduced from its original price?

10. Write a formula to express each of the following sentences:

a. The sale price is 20% off the original price. Use S for sale price and P for original price to express S as a function of P.

b. The time in Paris is 6 hours ahead of New York. Use P for Paris time and N for New York time to express P as a function of N. (Represent your answer in terms of a 12-hour clock.) How would you adjust your formula if P comes out greater than 12?

c. For temperatures above 0°F the wind chill effect can be estimated by subtracting two-thirds of the wind speed (in miles per hour) from the outdoor temperature. Use C for the effective wind chill temperature, W for wind speed, and T for the actual outdoor temperature to write an equation expressing C in terms of W and T.

11. Determine whether y is a function of x in each of the following equations. If the equation does not define a function, find a value of x that is associated with two different y values.

a. $y = x^2 + 1$ **c.** $y = 5$

b. $y = 3x - 2$ **d.** $y^2 = x$

12. (Graphing program required.) For each equation below, write an equivalent equation that expresses z in terms of t. Use technology to sketch the graph of each equation. Is z a function of t? Why or why not?

   a. $3t - 5z = 10$ **c.** $2(t - 4) - (z + 1) = 0$

   b. $12t^2 - 4z = 0$

13. If we let D stand for ampicillin dosage expressed in milligrams and W stand for a child's weight in kilograms, then the equation

$$D = 50W$$

gives a rule for finding the safe maximum daily drug dosage of ampicillin (used to treat respiratory infections) for children who weigh less than 10 kilograms (about 22 pounds).[11]

   a. What are logical choices for the independent and dependent variables?

   b. Does the equation represent a function? Why?

   c. Generate a small table and graph of the function.

   d. Think of the function $D = 50W$ for ampicillin dosage as an abstract mathematical equation. How will the table and graph change?

1.4 *The Language of Functions*

As we have seen, not all equations represent functions. But functions have important qualities, so it is useful to have a way to indicate when a relationship is a function.

Function Notation

When a quantity y is a function of x, we can write

$$y \text{ is a function of } x$$

or in abbreviated form,

$$y \text{ equals "} f \text{ of } x\text{"}$$

or using function notation,

$$y = f(x)$$

The expression $y = f(x)$ means that the rule f is applied to the input value x to give the output value, $f(x)$:

$$\text{output} = f(\text{input})$$

or

$$\text{dependent variable} = f(\text{independent variable})$$

The letter f is often used to denote the function, but we could use any letter, not just f.

Understanding the symbols

Suppose we have a function that triples the input. We could write this function as

$$y = 3x \qquad (1)$$

or with function notation as

$$T(x) = 3x \qquad \text{where } y = T(x) \qquad (2)$$

Equations (1) and (2) represent the same function, but with function notation we name the function—in this case T—and identify the input, x, and output, $3x$.

Function notation can provide considerable economy in writing and reading. For example, throughout a discussion we can use $T(x)$ instead of the full expression to represent the function.

---

[11]Information extracted from Anna M. Curren and Laurie D. Muntlay, *Math for Meds; Dosages and Solutions*, 6th ed. (San Diego: W. I. Publications, 1990), p. 198.

The Language of Functions

If y is a function, f, of x, then

$$y = f(x)$$

where f is the name of the function,
                y is the output or *dependent* variable,
                x is the input or *independent* variable.

                output $= f$(input)
                dependent $= f$(independent)

Finding output values: Evaluating a function

Function notation is particularly useful when a function is being evaluated at a specific input value. Suppose we want to find the value of the previous function $T(x)$ when our input value is 10. Using equation (1) we would say, "find the value of y when $x = 10$." With function notation, we simply write $T(10)$. To evaluate the function T at 10 means calculating the value of the output when the value of the input is 10:

Given $T(x) = 3x$

Substitute 10 for x $T(10) = 3(10)$
                                    $= 30$

So, applying the function rule T to the input value of 10 gives an output value of 30.

Common Error

The expression $f(x)$ does not mean "f times x." It means the function f evaluated at x.

E X A M P L E   1   **Using function notation with equations**

a. Given $f(x) = 2x^2 + 3$, evaluate $f(5)$, $f(0)$, and $f(-2)$

b. Evaluate $g(0)$, $g(2)$, and $g(-2)$ for the function

$$g(x) = \frac{1}{x - 1}$$

S O L U T I O N   a. To evaluate $f(5)$, we replace every x in the formula with 5.

Given $f(x) = 2x^2 + 3$
Substitute 5 for x $f(5) = 2(5)^2 + 3$
                            $= 2(5)(5) + 3$
                            $= 53$

Similarly,

$$f(0) = 2(0)^2 + 3 = 3$$
$$f(-2) = 2(-2)^2 + 3$$
$$= (2 \cdot 4) + 3 = 11$$

b. $g(0) = \dfrac{1}{0 - 1} = -1, \quad g(2) = \dfrac{1}{2 - 1} = 1, \quad g(-2) = \dfrac{1}{-2 - 1} = \dfrac{1}{-3} = -\dfrac{1}{3}$

EXAMPLE   2   **Using function notation with data tables**

Use Table 1.13 to fill in the missing values:

a. $S(0) = ?$

b. $S(-1) = ?$

c. $S(?) = 4$

Input, x	-1	-2	0	1	2
Output, $S(x)$	1	4	0	1	4

Table 1.13

SOLUTION   **a.** $S(0)$ means to evaluate S when the input $x = 0$. The table says that the corresponding output is also 0, so $S(0) = 0$.

b. $S(-1) = 1$.

c. $S(?) = 4$ means to find the input when the output is 4. When the output is 4, the input is -2 or 2, so $S(-2) = 4$ and $S(2) = 4$.

EXAMPLE   3   **Using function notation with graphs**

Use the graph in Figure 1.20 to estimate the missing values:

a. $f(0) = ?$

b. $f(-5) = ?$

c. $f(?) = 0$

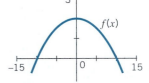

Figure 1.20 Graph of a function.

SOLUTION   Remember that by convention the horizontal axis represents the input (or independent variable) and the vertical axis represents the output (or dependent variable).

a. $f(0) = 2$

b. $f(-5) \doteq 1.5$

c. $f(10) = 0$ and $f(-10) = 0$

Rewriting equations using function notation

In order to use function notation, an equation needs to be in the form

$$\text{output} = \text{some rule applied to input}$$

or equivalently

$$\text{dependent variable} = \text{some rule applied to independent variable}$$

Translating an equation into this format is called putting the equation in *function form*. Many graphing calculators and computer graphing programs accept only equations in function form as input.

To put an equation into function form, we first need to identify the independent and the dependent variables. The choice is sometimes obvious, at other times arbitrary. If we use the mathematical convention that x represents the input or independent variable and y the output or dependent variable, when we put equations into function form, we want

$$y = \text{some rule applied to } x$$

EXAMPLE   4   Analyze the equation $4x - 3y = 6$. Decide whether or not the equation represents a function. If it does, write the relationship using function notation.

SOLUTION   First, put the equation into function form. Assume y is the output.

Given the equation $\qquad\qquad 4x - 3y = 6$

subtract $4x$ from both sides $\qquad -3y = 6 - 4x$

divide both sides by -3 $\qquad \dfrac{-3y}{-3} = \dfrac{6 - 4x}{-3}$

simplify
$$y = \frac{6}{-3} + \frac{-4x}{-3}$$

simplify and rearrange terms
$$y = \frac{4}{3}x - 2$$

We now have an expression for y in terms of x.

Using technology or by hand, we can generate a graph of the equation (see Figure 1.21). Since the graph passes the vertical line test, y is a function of x.

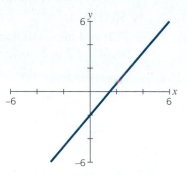

Figure 1.21 Graph of $y = \frac{4}{3}x - 2$.

If we name our function f, then using function notation, we have

$$y = f(x) \qquad \text{where } f(x) = \frac{4}{3}x - 2$$

E X A M P L E 5 Analyze the equation $y^2 - x = 0$. Generate a graph of the equation. Decide whether or not the equation represents a function. If the equation represents a function, write the relationship using function notation. Assume y is the output.

S O L U T I O N Put the equation in function form:

Given the equation
$$y^2 - x = 0$$

add x to both sides
$$y^2 = x$$

To solve this equation, we take the *square root* of both sides of the equation and we get

$$y = \pm\sqrt{x}$$

This gives us two solutions for any value of $x > 0$ as shown in Table 1.14. For example, if $x = 4$, then y can either be 2 or -2 since both $2^2 = 4$ and $(-2)^2 = 4$.

x	y
0	0
1	1 or -1
2	$\sqrt{2}$ or $-\sqrt{2}$
4	2 or -2
9	3 or -3

Table 1.14

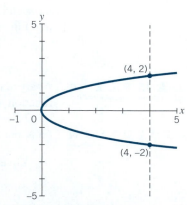

Figure 1.22 Graph of the equation $y^2 = x$.

The graph of the equation in Figure 1.22 does not pass the vertical line test. In particular, the solutions $(4, -2)$ and $(4, 2)$ lie on the same vertical line. So y is not a function of x and we cannot use function notation to represent this relationship.

Algebra Aerobics 1.4a

1. Given $g(x) = 3x$, evaluate $g(0)$, $g(-1)$, $g(1)$, $g(20)$, and $g(100)$.

2. Consider the function $f(x) = x^2 - 5x + 6$. Find $f(0), f(1)$, and $f(-3)$.

3. Given the function $f(x) = \dfrac{2}{x - 1}$, evaluate $f(0)$, $f(-1), f(1)$, and $f(-3)$.

4. Determine the value of t for which each of the functions has a value of 3.

   $r(t) = 5 - 2t \qquad p(t) = 3t - 9 \qquad m(t) = 5t - 12$

In Problems 5–7 solve for y in terms of x. Determine if y is a function of x. If it is, rewrite using $f(x)$ notation.

5. $2(x - 1) - 3(y + 5) = 10$
6. $x^2 + 2x - y + 4 = 0$
7. $7x - 2y = 5$
8. From the graph in Figure 1.23, estimate $f(-4), f(-1)$, $f(0)$, and $f(3)$. Find two approximate values for x such that $f(x) = 0$.

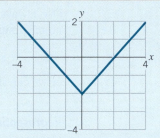

Figure 1.23 Graph of $f(x)$.

9. From Table 1.15 find $f(0)$ and $f(20)$. Find two values of x for which $f(x) = 10$. Explain why $f(x)$ is a function.

x	$f(x)$
0	20
10	10
20	0
30	10
40	20

Table 1.15

Domain and Range

A function is often defined only for certain values of the input (or independent variable). The set of all possible values for the input is called the *domain* of the function. The set of corresponding values of the output (or dependent variable) is called the *range* of the function.

> **Domain and Range of a Function**
>
> The *domain* of a function is the set of possible values of the input.
> The *range* is the set of corresponding values of the output.

E X A M P L E   6   **Finding a reasonable domain and range**

In the sales tax example at the beginning of this section, we used the equation

$$T = 0.06P$$

to represent the sales tax, T, as a function of the purchase price, P (where all units are in dollars). What are the domain and range of this function?

S O L U T I O N   Since negative values for P are meaningless, P is restricted to dollar amounts greater than or equal to zero. In theory there is no upper limit on prices, so we assume P has no maximum amount. In this example,

the domain is all dollar values of P greater than or equal to 0

We can express this more compactly with the symbol ≥, which means "greater than or equal to." So,

the domain is all dollar values of P such that $P \geq 0$

or abbreviated to   the domain is $P \geq 0$

What are the corresponding values for the tax T? The values for T in our model will also always be nonnegative. As long as there is no maximum value for P, there will be no maximum value for T. So,

the range is all dollar values of T greater than or equal to 0

or we can shorten this to   the range is $T \geq 0$

Representing the Domain and Range with Interval Notation

Interval notation is often used to represent the domain and range of a function.

Interval Notation

A *closed interval* $[a, b]$ indicates all real numbers x for which $a \leq x \leq b$. Closed intervals include their endpoints.

An *open interval* (a, b) indicates all real numbers x for which $a < x < b$. Open intervals exclude their endpoints.

Half-open (or equivalently half-closed) intervals are represented by $[a, b)$ which indicates all real numbers x for which $a \leq x < b$ or $(a, b]$ which indicates all real numbers x for which $a < x \leq b$.

For example, if the domain is values of n greater than or equal to 50 and less than or equal to 100, then

$$\text{domain} = \text{all } n \text{ values with } 50 \leq n \leq 100$$
$$= [50, 100]$$

If the domain is values of n greater than 50 and less than 100, then

$$\text{domain} = \text{all } n \text{ values with } 50 < n < 100$$
$$= \text{interval } (50, 100)$$

If we want to exclude 50 but include 100 as part of the domain, we would represent the interval as $(50, 100]$. The interval can be displayed on the real number line as:

In general, a hollow dot indicates exclusion and a solid dot inclusion.
Note: Since the notation (a, b) can also mean the coordinates of a point, we will say the *interval* (a, b) when we want to refer to an interval.

EXAMPLE 7 **Finding the domain and range from a graph**
The graph in Figure 1.24 shows the water level of the tides in Pensacola, Florida, over a 24-hour period. Are the Pensacola tides a function of the time of day? If so, identify the independent and dependent variables. Use interval notation to describe the domain and range of this function.

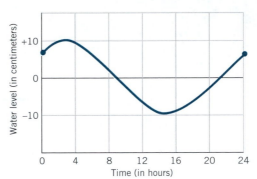

Figure 1.24 Diurnal tides in a 24-hour period in Pensacola, Florida.
Source: Adapted from Fig. 8.2 in *Oceanography: An Introduction to the Planet Oceanus,* by Paul R. Pinet. Copyright © 1992 by West Publishing Company, St. Paul, MN. All rights reserved.

SOLUTION   The Pensacola tides are a function of the time of day since the graph passes the vertical line test. The independent variable is time, and the dependent variable is water level. The domain is from 0 to 24 hours, and the range is from about −10 to +10 centimeters. Using interval notation:

$$\text{domain} = [0, 24]$$

$$\text{range} = [-10, 10]$$

When are there restrictions on the domain and range?

When specifying the domain and range of a function, we need to consider whether the function is undefined for any values. For example, for the function

$$y = \frac{1}{x}$$

the expression $1/x$ is undefined when $x = 0$. For any other value for x, the function is defined. Thus,

the domain is all real numbers except 0

To find the range, we need to determine the possible output values for y. Sometimes it is easier to find the y values that are not possible. In this case, y can't equal zero. Why? Our rule says to take 1 and divide by x, but it is impossible to divide 1 by a real number in our domain and get zero as a result. Thus,

the range is all real numbers except 0

The interval (−infinity, +infinity) or $(-\infty, \infty)$ represents all real numbers. To represent all real numbers except 0 using interval notation, we use the union symbol, $\cup$, of two intervals:

$$(-\infty, 0) \cup (0, \infty)$$

Using the graph of the function in Figure 1.25 we can check to see if our domain and range are reasonable. The graph suggests that x comes very close to 0, but does not equal 0. The same is true for y.

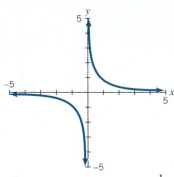

Figure 1.25 Graph of $y = \dfrac{1}{x}$.

Another expression that is undefined is $\sqrt{\text{negative number}}$, if we restrict ourselves to the real number system.

EXAMPLE 8 **Finding restrictions on the domain and range**

Specify the domain and range for each of the following functions:

a. $y = \sqrt{x - 4}$ **b.** $y = \dfrac{1}{x - 2}$

SOLUTION **a.** First, we ask ourselves, "Is the domain restricted?" In this case, the answer is yes. The expression $\sqrt{x - 4}$ is not defined for negative values, so we must have the expression $x - 4 \geq 0 \Rightarrow x \geq 4$. Corresponding y values must be ≥ 0. So,

Domain:	all real numbers ≥ 4
Range:	all real numbers ≥ 0.

Using interval notation, we get

Domain:	$[4, +\infty)$
Range:	$[0, +\infty)$

b. The domain of $y = \dfrac{1}{x - 2}$ is also restricted. In this case, the denominator cannot equal 0 since $\dfrac{1}{0}$ is not defined. This means the expression $x - 2 \neq 0 \Rightarrow x \neq 2$. The range is also restricted. It is not possible for $y = 0$, since 1 divided by any real number is never 0. So,

Domain:	all real numbers except 2
Range:	all real numbers except 0

Using interval notation, we get

Domain:	$(-\infty, 2) \cup (2, +\infty)$
Range:	$(-\infty, 0) \cup (0, +\infty)$

Two Cases Where the Domain and Range Are Restricted

$y = \dfrac{1}{x}$ Domain: all real numbers except 0.
Range: all real numbers except 0

$y = \sqrt{x}$ Domain: all real numbers ≥ 0
Range: all real numbers ≥ 0

EXAMPLE 9 **When is a function undefined?**

Match each function graph in Figure 1.26 with the appropriate domain and range listed in parts (a) to (e). [*Note:* The dotted line in the graph of $f(x)$ is not part of the function.]

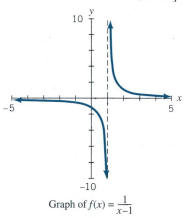

Graph of $f(x) = \dfrac{1}{x-1}$

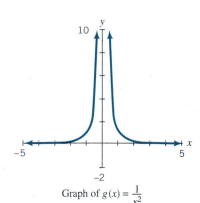

Graph of $g(x) = \dfrac{1}{x^2}$

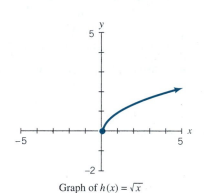

Graph of $h(x) = \sqrt{x}$

Figure 1.26 Graphs for three functions.

   a. Domain: $[0, \infty)$ Range: $[0, \infty)$

   b. Domain: $(0, \infty)$ Range: $(0, \infty)$

   c. Domain: $(-\infty, 1) \cup (1, \infty)$ Range: $(-\infty, 1) \cup (1, \infty)$

   d. Domain: $(-\infty, 1) \cup (1, \infty)$ Range: $(-\infty, 0) \cup (0, \infty)$

   e. Domain: $(-\infty, 0) \cup (0, \infty)$ Range: $(0, \infty)$

SOLUTION $f(x) = \dfrac{1}{x-1}$ matches with (d)

                     $g(x) = \dfrac{1}{x^2}$ matches with (e)

                     $h(x) = \sqrt{x}$ matches with (a)

Algebra Aerobics 1.4b

1. Express each of the following using interval notation.

   a. $x > 2$

   b. $4 \le x < 20$

   c. $t \le 0$ or $t > 500$

2. Express the given interval as an inequality.

   a. $[-3, 10)$

   b. $(-2.5, 6.8]$

   c. $(-\infty, 5] \cup [12, \infty)$

3. Express each of the following statements in interval notation.

   a. Harry's GPA is at least 2.5 but at most 3.6.

   b. A good hitter has a batting average of at least 0.333.

   c. Starting annual salary at a position is anything from $35,000 to $50,000 depending upon experience.

In Problems 4–8 solve for y in terms of x. Determine if y is a function of x. If it is, rewrite using $f(x)$ notation and determine the domain and range.

4. $2(x + 1) + 3y = 5$

5. $x + 2y = 3x - 4$

6. $y = \sqrt{x}$

7. $2xy = 6$

8. $\dfrac{x}{2} + \dfrac{y}{3} = 1$

9. Find values of x for which the function is undefined, and determine the domain and range.

$$f(x) = \frac{x+1}{x+5}, \quad g(x) = \frac{1}{x+1}, \quad h(x) = \sqrt{x-10}$$

Exercises for Section 1.4

A graphing program is required for Exercise 6.

1. Given $T(x) = x^2 - 3x + 2$, evaluate $T(0)$, $T(-1)$, $T(1)$, and $T(-5)$.

2. Given $f(x) = \dfrac{x}{x-1}$, evaluate $f(0)$, $f(-1)$, $f(1)$, $f(20)$, and $f(100)$.

3. Assume that for persons who earn less than $20,000 a year, income tax is 16% of their income.

   a. Generate a formula that describes income tax in terms of income for people earning less than $20,000 a year.

   b. What are you treating as the independent variable? The dependent variable?

   c. Does your formula represent a function? Explain.

   d. If it is a function, what is the domain? The range?

4. Suppose that the price of gasoline is $3.09 per gallon.

   a. Generate a formula that describes the cost, C, of buying gas as a function of the number of gallons of gasoline, G, purchased.

   b. What is the independent variable? The dependent variable?

   c. Does your formula represent a function? Explain.

   d. If it is a function, what is the domain? The range?

   e. Generate a small table of values and a graph.

5. The cost of driving a car to work is estimated to be $2.00 in tolls plus 32 cents per mile. Write an equation for computing the total cost C of driving M miles to work. Does your equation represent a function? What is the independent variable? What is the dependent variable? Generate a table of values and then graph the equation.

6. (Graphing program required.) For each equation, write the equivalent equation that expresses y in terms of x. Use technology to graph each function and then estimate its domain and range.

   a. $3x + 5x - y = 3y$ c. $x(x - 1) + y = 2x - 5$

   b. $3x(5 - x) = x - y$ d. $2(y - 1) = y + 5x(x + 1)$

7. If $f(x) = x^2 - x + 2$, find:

   a. $f(2)$ b. $f(-1)$ c. $f(0)$ d. $f(-5)$

8. If $g(x) = 2x + 3$, evaluate $g(0)$, $g(1)$, and $g(-1)$.

9. Look at the accompanying table.

   a. Find $p(-4)$, $p(5)$, and $p(1)$.

   b. For what value(s) of n does $p(n) = 2$?

n	-4	-3	-2	-1	0	1	2	3	4	5
$p(n)$	0.063	0.125	0.25	0.5	1	2	4	8	16	32

10. Consider the function $y = f(x)$ graphed in the accompanying figure.

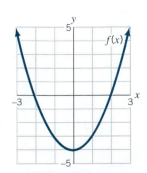

a. Find $f(-3)$, $f(0)$, $f(1)$, and $f(2.5)$.

b. Find two values of x such that $f(x) = 0$.

11. From the accompanying graph of $y = f(x)$:

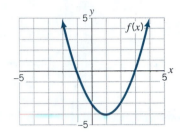

a. Find $f(-2)$, $f(-1)$, $f(0)$, and $f(1)$.

b. Find two values of x for which $f(x) = -3$.

c. Estimate the range of f. Assume that the arms of the graph extend upward indefinitely.

12. Find $f(3)$, if it exists, for each of the following functions:

   a. $f(x) = (x - 3)^2$ c. $f(x) = \dfrac{x + 1}{x - 3}$

   b. $f(x) = \dfrac{1}{x}$ d. $f(x) = \dfrac{2x}{x - 1}$

   Determine the domain for each function.

13. If $f(x) = (2x - 1)^2$, evaluate $f(0)$, $f(1)$, and $f(-2)$.

14. Find the domain for each of the following functions:

   a. $f(x) = 300.4 + 3.2x$ d. $k(x) = 3$

   b. $g(x) = \dfrac{5 - 2x}{2}$ e. $f(x) = x^2 + 3$

   c. $j(x) = \dfrac{1}{x + 1}$

15. Given $f(x) = 1 - 0.5x$ and $g(x) = x^2 + 1$, evaluate:

   a. $f(0)$, $g(0)$ c. $f(2)$, $f(1)$

   b. $f(-2)$, $g(-3)$ d. $f(3)$, $g(3)$

16. Each of the following functions has a restricted domain and range. Find the domain and range for each function and explain why the restrictions occur.

   a. $f(x) = \dfrac{3}{x + 2}$

   b. $g(x) = \sqrt{x - 5}$

   c. $h(x) = -\dfrac{1}{2x - 3}$

   d. $k(x) = \dfrac{1}{x^2 - 4}$

   e. $l(x) = \sqrt{x + 3}$

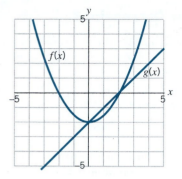

17. For the functions $f(x)$ and $g(x)$ shown on the accompanying graph, find the values of x that make the following true.

   a. $f(x) = 0$ **b.** $g(x) = 0$ **c.** $f(x) = g(x)$

18. Determine the domain of each of the following functions. Explain your answers.

$$f(x) = 4 \qquad\qquad g(x) = 3x + 5 \qquad h(x) = \frac{x - 1}{x - 2}$$

$$F(x) = x^2 - 4 \qquad G(x) = \frac{x - 1}{x^2 - 4} \qquad H(x) = \sqrt{x - 2}$$

1.5 *Visualizing Functions*

In this section we return to the question: How does change in one variable affect change in another variable? Graphs are one of the easiest ways to recognize change. We start with three basic questions:

Is There a Maximum or Minimum Value?

If a function has a *maximum* (or *minimum*) value, then it appears as the highest point (or lowest point) on its graph.

EXAMPLE 1   Determine if each function in Figure 1.27 has a maximum or minimum, then estimate its value.

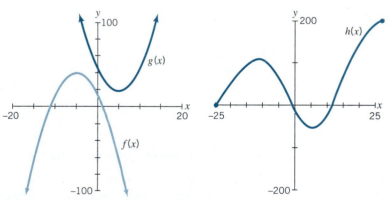

Figure 1.27 Graphs of $f(x)$, $g(x)$, and $h(x)$.

SOLUTION   The function $f(x)$ in Figure 1.27 appears to have a maximum value of 40 when $x = -5$ but has no minimum value since both arms of the function extend indefinitely downward.

The function $g(x)$ appears to have a minimum value of 20 when $x = 5$, but no maximum value since both arms of the function extend indefinitely upward.

The function $h(x)$ appears to have a maximum value of 200, which occurs when $x = 25$, and a minimum value of -50, when $x = 5$.

Is the Function Increasing or Decreasing?

Decreasing   Increasing

A function *f* is *decreasing* over a specified interval if the values of *f(x)* decrease as *x* increases over the interval. A function *f* is *increasing* over a specified interval if the values of *f(x)* increase as *x* increases over the interval.

The graph of an increasing function climbs as we move from left to right. The graph of a decreasing function falls as we move from left to right.

E X A M P L E   2   **Increasing and decreasing production**

Figure 1.28 shows a 75-year history of annual natural gas production in the United States. Create a 60-second summary about gas production from 1930 to 2005 in the U.S.

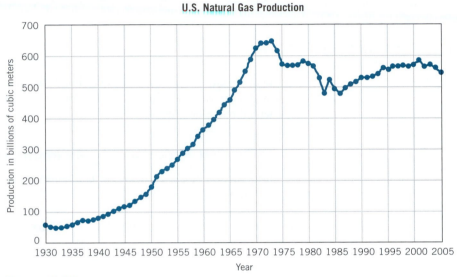

U.S. Natural Gas Production

Production in billions of cubic meters (y-axis: 0, 100, 200, 300, 400, 500, 600, 700)

Year (x-axis: 1930 1935 1940 1945 1950 1955 1960 1965 1970 1975 1980 1985 1990 1995 2000 2005)

Figure 1.28
Source: Wikipedia. *http://www.answers.com/topic/natural-gas-prices.*

S O L U T I O N   Natural gas production in the United States increased from approximately 50 billion cubic meters in 1930 to a high of approximately 650 billion cubic meters in 1973. For the next 10 years production generally decreased to under 500 billion cubic meters in 1983. Between 1983 and 2005, annual production oscillated between 500 and 600 billion cubic meters, with production at approximately 550 billion in 2005.

Is the Graph Concave Up or Concave Down?

What does the concavity of a graph mean? The graph of a function is *concave up* if it bends upward and it is *concave down* if it bends downward.

Concave up   Concave down

Concavity is independent of whether the function is increasing or decreasing.

Increasing   Decreasing

Concave down   Concave up   Concave down   Concave up

EXAMPLE 3 **Graphs are not necessarily pictures of events**[12]
The graph in Figure 1.29 shows the speed of a roller coaster car as a function of time.

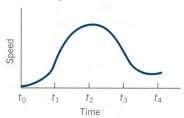

Figure 1.29

a. Describe how the speed of the roller coaster car changes over time. Describe the changes in the graph as the speed changes over time.

b. Draw a picture of a possible track for this roller coaster.

SOLUTION **a.** The speed of the roller coaster car increases from t_0 to t_2, reaching a maximum for this part of the ride at t_2. The speed decreases from t_2 to t_4.

The graph of speed versus time is concave up and increasing from t_0 to t_1 and then concave down and increasing from t_1 to t_2. From t_2 to t_3 the graph is concave down and decreasing, and from t_3 to t_4 it is concave up and decreasing.

b. A picture for a possible track of the roller coaster is shown in Figure 1.30. Notice how the track is an upside-down picture of the graph of speed versus time. (When the roller coaster car goes down the speed increases, and when the roller coaster car goes up, the speed decreases.)

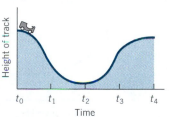

Figure 1.30

EXAMPLE 4 **Growth patterns**
Figure 1.31 shows the growth patterns for three areas in Virginia. Compare the differences in growth for these areas between 1900 and 2005.

Figure 1.31
Source: www.savethebay.org/land/images.

[12]Example 3 is adapted from Shell Centre for Mathematical Education, *The Language of Functions and Graphs.* Manchester, England: University of Nottingham, 1985.

SOLUTION

From 1900 to about 1988, Richmond City consistently had the largest population; however, after about 1950 Henrico and Chesterfield counties were growing faster than Richmond City. By 1990 all three areas had populations of about 200,000. After 1990, Henrico's and Chesterfield's populations continued to grow and Richmond's continued to decline.

Getting the Big Idea

We now have the basic vocabulary for describing a function's behavior. Think of a function as telling a story. We want to decipher not just the individual words, but the overall plot. In each situation we should ask, What is really happening here? What do the words tell us about the shape of the graph? What does the graph tell us about the underlying phenomenon?

EXAMPLE 5

Generate a rough sketch of each of the following situations.

a. A cup of hot coffee cooling.

b. U.S. venture capital (money provided by investment companies to business start-ups) increased modestly but steadily in the mid-1990s, soared during the "dotcom bubble" (in the late 1990s), with a high in 2000, and then suffered a drastic decrease back to pre-dotcom levels.

c. Using a simple predator-prey model: initially as the number of lions (the predators) increases, the number of gazelles (their prey) decreases. When there are not enough gazelles to feed all the lions, the number of lions decreases and the number of gazelles starts to increase.

SOLUTION

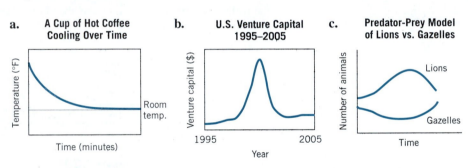

EXAMPLE 6

You are a TV journalist. Summarize for your viewers the essence of each of the graphs in Figures 1.32 and 1.33.

a. *Note:* The vertical scale is used in two different ways on this graph from the U.S. Bureau of the Census.

Figure 1.32

Source: U.S. Bureau of the Census

www.census.gov/compendia/statab/incom_expenditures_wealth/household_incom.

b.

Data on the top 400 taxpayers in the United States

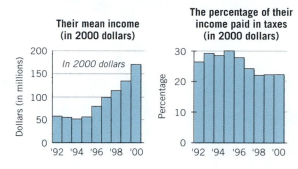

Their mean income
(in 2000 dollars)

The percentage of their
income paid in taxes
(in 2000 dollars)

Figure 1.33
Source: The New York Times, June 26, 2003.

S O L U T I O N   **a.** In 1959, according to the U.S. Bureau of the Census, there were about 40 million Americans in poverty, representing almost 23% of the population. Between 1959 and 1974 both the number and the percentage of Americans in poverty decreased to a low of about 23 million and 12%, respectively. From 1974 to 2004, the percentage in poverty continued to hover between 12% and 15%. During the same time period the number in poverty vacillated, but the overall trend was an increase to about 37 million in 2004.

b. From 1994 to 2000 the mean income of the top 400 taxpayers grew steadily, but overall the percentage of their income that went toward taxes decreased.

We will spend the rest of this text examining patterns in functions and their graphs. We'll study "families of functions"—linear, exponential, logarithmic, power, and polynomial—that will provide mathematical tools for describing the world around us.

Algebra Aerobics 1.5

1. Create a title for each of the following graphs.

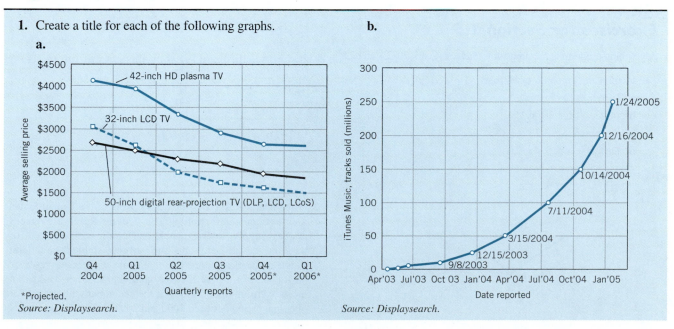

a.

*Projected.
Source: Displaysearch.*

b.

Source: Displaysearch.

2. Use Figure 1.34 to answer the following questions.

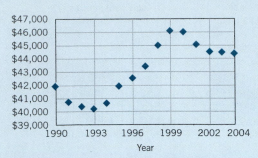

Figure 1.34 Median household income adjusted for inflation (in constant 2004 dollars).

a. Estimate the maximum value for median household income during the time period represented on the graph. In what year does the maximum occur? What are the approximate coordinates at the maximum point?

b. What is the minimum value for median household income? In what year does this occur? What are the coordinates of this point?

c. Describe the changes in median household income from 1990 to 2004.

3. Choose the "best" graph in Figure 1.35 to describe the following situation. Speed (S) is on the vertical axis and time (t) is on the horizontal axis.

A child in a playground tentatively climbs the steps of a large slide, first at a steady pace, then gradually slowing down until she reaches the top, where she stops to rest before sliding down.

Graph A

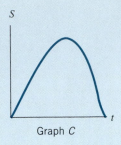

Graph C

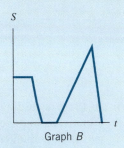

Graph B

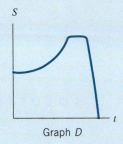

Graph D

Figure 1.35

4. Generate a rough sketch of the following situation. U.S. AIDS cases increased dramatically, reaching an all-time high for a relatively short period, and then consistently decreased, until a recent small increase.

5. Sketch a graph for each of the following characteristics, and then indicate with an arrow which arm of the graph is increasing and which is decreasing.

a. Concave up with a minimum point at $(-2, 1)$.

b. Concave down with a maximum point at $(3, -2)$.

Exercises for Section 1.5

A graphing program is required for Exercises 12, 13, 20, 27, and 29.

1. Identify the graph (A or B) that

a. Increases for $1 < x < 3$

b. Increases for $2 < x < 5$

c. Decreases for $-2 < x < 2$

d. Decreases for $3 < x < 5$

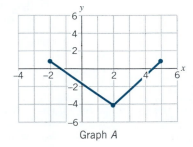

Graph A

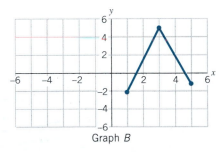

Graph B

2. The Federal Reserve is the central bank of the United States that sets monetary policy. The Federal Reserve oversees money supply, interest rates, and credit with the goal of keeping the U.S. economy and currency stable. The federal funds rate is the interest rate that banks with excess reserves at a Federal Reserve district bank charge other banks that need overnight loans. Look at the accompanying graphs for the time period between 2000 and 2003.

a. Describe the overall trends for this time period for the federal funds rate.

b. Describe the overall trends for this time period for credit card rates.

c. Describe the overall trends for this time period for the 30-year mortgage rates.

d. Estimate the maximum federal funds rate for this time period. When did it occur?

e. Approximate the minimum federal funds rate for this time period. When did it occur?

f. Write a topic sentence that compares the federal funds rate with the consumer loan rates for credit cards and mortgages for this time period.

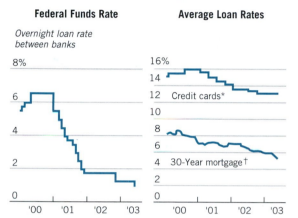

Federal Funds Rate

Overnight loan rate between banks

Average Loan Rates

Credit cards*

30-Year mortgage†

* Average annual percentage rate for all credit card accounts in banks in the United States that respond to the Federal Reserve's Survey.

† National average for a 30-year fixed rate on a single-family home.

Source: www.federalreserve.gov

3. The accompanying graphs show the price of shares of stock of two companies over the one-week period January 9 to 16, 2004. Describe the changes in price over this time period.

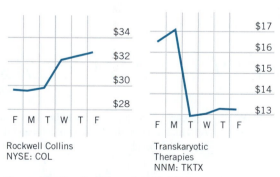

Rockwell Collins
NYSE: COL

Transkaryotic
Therapies
NNM: TKTX

Source: The New York Times, Jan. 17, 2004.

4. For each of the following functions,

a. Over which interval is the function decreasing?

b. Over which interval is the function increasing?

c. Does the function appear to have a minimum? If so, where?

d. Does the function appear to have a maximum? If so, where?

e. Describe the concavity.

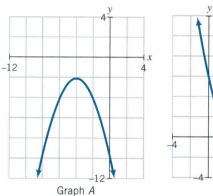
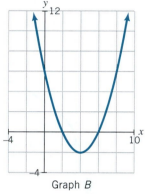

Graph *A*

Graph *B*

5. For the following function,

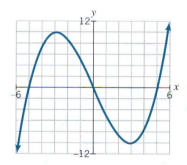

a. Over which interval(s) is the function positive?

b. Over which interval(s) is the function negative?

c. Over which interval(s) is the function decreasing?

d. Over which interval(s) is the function increasing?

e. Does the function appear to have a minimum? If so, where?

f. Does the function appear to have a maximum? If so, where?

6. Choose which graph(s) at the top of the next page match the description: As *x* increases, the graph is:

a. Increasing and concave up

b. Increasing and concave down

c. Concave up and appears to have a minimum value at $(-3, 2)$

d. Concave down and appears to have a maximum value at $(-3, 2)$

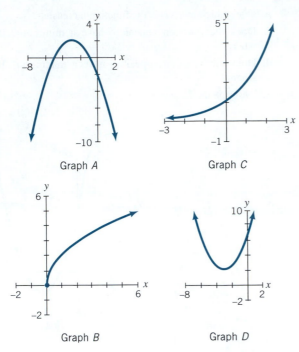

Graph A

Graph C

Graph B

Graph D

7. Examine each of the graphs in Exercise 6. Assume each graph describes a function $f(x)$. The arrows indicate that the graph extends indefinitely in the direction shown.

a. For each function estimate the domain and range.

b. For each function estimate the x interval(s) where $f(x) > 0$.

c. For each function estimate the x interval(s) where $f(x) < 0$.

8. Look at the graph of $y = f(x)$ in the accompanying figure.

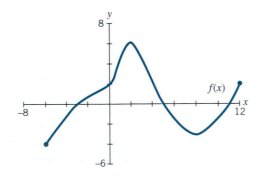

$f(x)$

a. Find $f(-6)$, $f(2)$, and $f(12)$.

b. Find $f(0)$.

c. For what values of x is $f(x) = 0$?

d. Is $f(8) > 0$ or is $f(8) < 0$?

e. How many times would the line $y = 1$ intersect the graph of $f(x)$?

f. What are the domain and range of $f(x)$?

g. What is the maximum? The minimum?

9. Use the graph of Exercise 8 to answer the following questions about $f(x)$.

a. Over which interval(s) is $f(x) < 0$?

b. Over which interval(s) is $f(x) > 0$?

c. Over which interval(s) is $f(x)$ increasing?

d. Over which interval(s) is $f(x)$ decreasing?

e. How would you describe the concavity of $f(x)$ over the interval $(0, 5)$ for x? Over $(5, 8)$ for x?

f. Find a value for x when $f(x) = 4$.

g. $f(-8) = $?

10. Match each graph with the best description of the function. Assume that the horizontal axis represents time, t.

i. The height of a ball thrown straight up is a function of time.

ii. The distance a truck travels at a constant speed is a function of time.

iii. The number of daylight hours is a function of the day of the year.

iv. The temperature of a pie baking in an oven is a function of time.

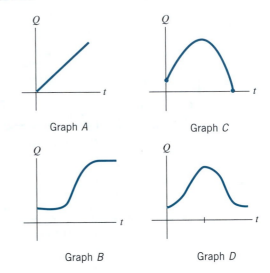

Graph A

Graph C

Graph B

Graph D

11. Choose the best graph to describe the situation.

a. A student in a large urban area takes a local bus whose route ends at the college. Time, t, is on the horizontal axis and speed, s, is on the vertical axis.

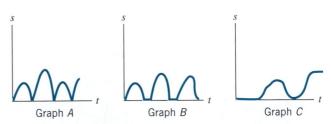

Graph A

Graph B

Graph C

b. What graph depicts the total distance the student traveled in the bus? Time, t, is on the horizontal axis and distance, d, is on the vertical axis.

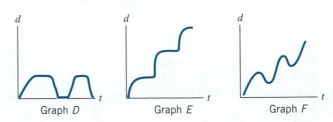

Graph D

Graph E

Graph F

12. (Graphing program required.) Use technology to graph each function. Then approximate the x intervals where the function is concave up, and then where it is concave down

　a. $f(x) = x^3$　　**b.** $g(x) = x^3 - 4x$

13. (Graphing program required.) Use technology to graph each function. Then approximate the x intervals where the function is concave up, and then where it is concave down.

　a. $h(x) = x^4$

　b. $k(x) = x^4 - 24x + 50$ (*Hint:* Use an interval of $[-5, 5]$ for x and $[0, 200]$ for y.)

14. a. In the accompanying graphs, estimate the coordinates of the maximum and minimum points (if any) of the function.

　b. Specify the interval(s) over which each function is increasing.

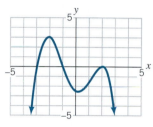
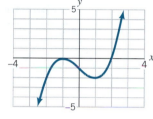

Graph *A*　　　　　　　　　Graph *B*

15. Consider the accompanying graph of U.S. military sales to foreign governments from 1995 to 2003.

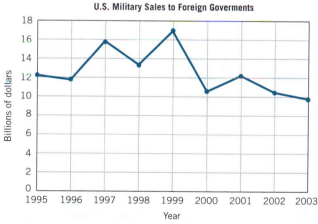

U.S. Military Sales to Foreign Goverments

Source: U.S. Department of Defense, Defense Security Cooperation Agency, *DSCA Data and Statistics.*

　a. Between what years did sales increase?

　b. Between what years did sales decrease?

　c. Estimate the maximum value for sales.

　d. Estimate the minimum value for sales.

16. Sketch a plausible graph for each of the following and label the axes.

　a. The amount of snow in your backyard each day from December 1 to March 1.

　b. The temperature during a 24-hour period in your home town during one day in July.

　c. The amount of water inside your fishing boat if your boat leaks a little and your fishing partner bails out water every once in a while.

　d. The total hours of daylight each day of the year.

　e. The temperature of an ice-cold drink left to stand.

17. Examine the accompanying graph, which shows the populations of two towns.

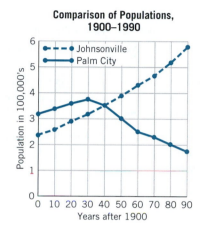

Comparison of Populations, 1900–1990

　a. What is the range of population size for Johnsonville? For Palm City?

　b. During what years did the population of Palm City increase?

　c. During what years did the population of Palm City decrease?

　d. When were the populations equal?

18. In Section 1.2 we examined the annual federal budget *surplus* or *deficit*. The federal *debt* takes into account the cumulative effect of all the deficits and surpluses for each year together with any interest or payback of principal. The accompanying graph shows the accumulated gross federal debt from 1950 to 2005. (See related Excel or graph link file FEDDEBT.) Create a topic sentence for this graph for a newspaper article.

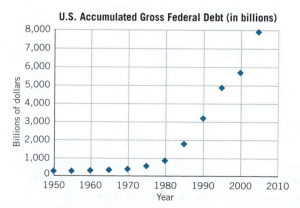

U.S. Accumulated Gross Federal Debt (in billions)

Source: U.S. Dept. of Treasury, www.ustreas.gov

19. Consider the accompanying table.

Y	P
1990	$1.4
1991	$2.3
1992	$0
1993	−$0.5
1994	$1.4
1995	$1.2

a. Is *P* a function of *Y*?

b. What is the domain? What is the range?

c. What is the maximum value of *P*? In what year did this occur?

d. During what intervals was *P* increasing? Decreasing?

e. Now consider *P* as the independent variable and *Y* as the dependent variable. Is *Y* a function of *P*?

20. (Graphing program required.) Using technology, graph each function over the intervals $[-6, 6]$ for *x* and $[-20, 20]$ for *y*.

$$y_1 = x^2 - 3x + 2 \qquad y_2 = 0.5x^3 - 2x - 1$$

For each function,

a. Estimate the maximum value of *y* on each interval.

b. Estimate the minimum value of *y* on each interval.

21. The accompanying graph shows the world population over time and future predictions.

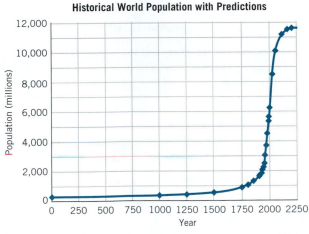

Historical World Population with Predictions

Source: www.worldonline.n1/invd/world/whist2.html

a. Over what interval does the world population show dramatic growth?

b. Does the dramatic growth slow down?

c. Write a topic sentence for a report for the United Nations.

22. Make a graph showing what you expect would be the relative ups and downs throughout the year of sales (in dollars) of

a. Turkeys **c.** Bathing suits in your state

b. Candy **d.** Textbooks at your school bookstore

23. Examine the graphs of military reserve enlisted personnel over the years 1990 to 2004.

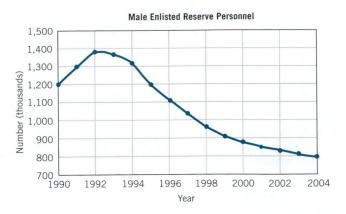

Male Enlisted Reserve Personnel

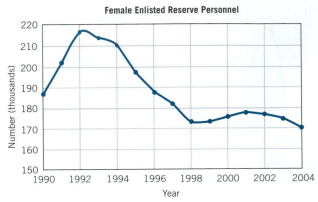

Female Enlisted Reserve Personnel

Source: U.S. Dept. of Defense, *Official Guard and Reserve Manpower Strengths and Statistics*, annual.

a. In what year(s) did female and male enlisted personnel reach a maximum?

b. What was the maximum and minimum for both men and women over the time interval [1990, 2004]?

c. Describe the trends in number of female and male reserve enlisted personnel.

d. Given the current state of world affairs, what would you expect would happen to the enlistment numbers for 2004 to the present?

24. Generate a rough sketch of a graph of internal pressure vs. time for the following situation: When a soda is removed from the fridge, the internal pressure is slightly above the surrounding air pressure. With the can unopened, the internal pressure soon more than doubles, stabilizing at a level three times the surrounding air pressure.

25. A student breaks her ankle and is taken to a doctor, who puts a cast on her leg and tells her to keep the foot off the ground altogether. After 2 weeks she is given crutches and can begin to walk around more freely, but at the beginning of the third week she falls and is resigned to keeping stationary again for a while. After 6 weeks from her first fall, she is given a walking cast in which she can begin to put her foot on the ground again. She is now able to limp around using crutches. Her walking speed slowly progresses. At 12 weeks she hits a plateau and, seeing no increase in her mobility, starts physical therapy. She rapidly improves. At 16 weeks the cast is removed and she can walk freely. Make a graph of the student's mobility level during her recovery.

26. Every January 1 a hardy group called the L Street Brownies celebrates the New Year by going for a swim at the L Street Beach in South Boston. The water is always very cold, and swimmers adopt a variety of strategies for getting into it. The graph shows the progress of three different friends who join in the event, with percentage of body submerged on the vertical axis and time on the horizontal axis. Match the graphs to the descriptions below of how each of the friends manages to get completely submerged in the icy ocean.

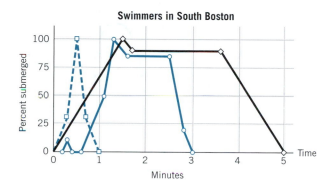

a. Ali has done this before and confidently walks in until his head is underwater; then he puts his head out and swims around a few minutes; then he walks out.

b. Ben dashes in until the water is up to his knee, trips over a hidden rock, and falls in completely; he stands up and, since he is now totally wet, runs back out of the water.

c. Cat puts one foot in, takes it out again, and shivers. She makes up her mind to get it over with, runs until she is up to her waist, dives in, swims back as close to the water's edge as she can get, stands up, and steps out of the water.

27. (Graphing program required.) Use technology to graph the following functions and then complete both sentences for each function.

$$y_1 = x^3, \qquad y_2 = x^2, \qquad y_3 = \frac{1}{x+3}, \qquad y_4 = \frac{1}{x} + 2$$

   a. As x approaches positive infinity, y approaches _____.

   b. As x approaches negative infinity, y approaches _____.

28. Describe the behavior of $f(x)$ in the accompanying figure over the interval $(-\infty, +\infty)$ for x, using such words as "increases," "decreases," "concavity," "maximum/minimum," and "approaches infinity."

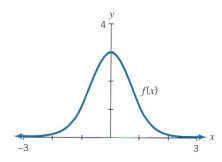

29. (Graphing program required.) This exercise is to be done with a partner. Name the partners person #1 and person #2.

   a. Person #1, using technology, graphs the function $f(x) = 0.5(x - 3)(x + 2)^2$, but does not show the graph to person #2.

   b. Person #1 describes to person #2 the behavior of the graph of $f(x)$ so that he/she can sketch it on a piece of paper.

   c. Switch roles; now person #2, using technology, graphs $g(x) = -0.5(x - 3)(x + 2)^2$, but does not show the graph to person #1.

   d. Person #2 describes to person #1 the behavior of the graph of $g(x)$ so that he/she can sketch it on a piece of paper.

   e. Compare the accuracy of the graphs and compare the shapes of the two graphs. What do $f(x)$ and $g(x)$ have in common? How do they differ?

CHAPTER SUMMARY

Single-Variable Data

Single-variable data are often represented with bar charts, pie charts, and histograms.

Two-Variable Data

Graphs of two-variable data show how change in one variable affects change in the other. The accompanying graph is called a *time series* because it shows changes over time.

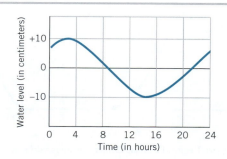

Equations in Two Variables

The *solutions of an equation* in two variables x and y are the ordered pairs (x, y) that make the equation a true statement. For example, $(3, -5)$ is one solution of the equation $2x + y = 1$.

The *graph of an equation* in two variables displays the set of points that are solutions to the equation.

Functions

A variable y is a *function* of a variable x if each value of x determines a unique (one and only one) value of y. Functions can be represented with words, graphs, equations, and tables.

When a set of ordered pairs represents a function, then each ordered pair is written in the form

(independent variable, dependent variable)

(input, output)

(x, y)

By convention, on the graph of a function, the independent variable is represented on the horizontal axis and the dependent variable on the vertical axis.

A graph does not represent a function if it fails the *vertical line test*. If you can draw a vertical line that crosses the graph two or more times, the graph does not represent a function.

The *domain* of a function is the set of all possible values of the independent variable.

The *range* is the set of corresponding values of the dependent variable.

Function Notation

The expression $f(x)$ means the rule f is applied to the input value x to give the output value, $f(x)$:

$$f(\text{input}) = \text{output}$$

For example, $f(x) = 2x + 5$ tells us f is the name of a function where the input is x, the output is $f(x)$, and the rule is to multiply the input by 2 and add 5. To evaluate this function when the input is 4, we write

$$f(4) = (2 \cdot 4) + 5 \Rightarrow f(4) = 13$$

Visualizing Functions

Graphs of functions may demonstrate the following properties:

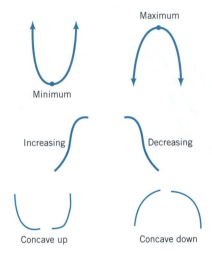

CHECK YOUR UNDERSTANDING

I. Is each of the statements in Problems 1–27 true or false?

1. Histograms, bar charts, and pie charts are used to graph single-variable data.

2. Means and medians, both measures of central tendency, can be used interchangeably.

3. A scatter plot is a plot of data points (x, y) for two-variable data.

Problems 4 and 5 refer to the accompanying figure.

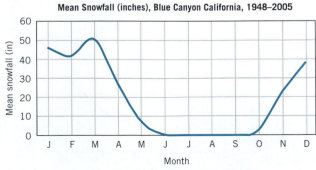

Source: Western Regional Climate Center, *www.wrcc.dri.edu.*

4. The graph shows the changes in mean snowfall for each month of the year over the years 1948–2005.

5. The maximum snowfall occurs in March, with mean snowfall of about 52 inches.

6. The accompanying figure, of the wolf population in Yellowstone National Park (YNP) and the Northern

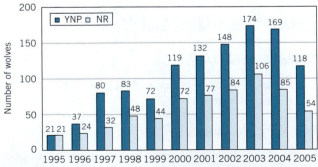

Source: Yellowstone Wolf Project, Annual Report 2005, U.S. Department of the Interior.

Region (NR) of Yellowstone, shows that the wolf population in the Greater Yellowstone area is increasing.

7. The independent variable of a function is also called the output of the function.

8. The graph of the equation describing the sales revenue R ($ million) as a function of the amount spent on advertising A ($ thousand) is the set of ordered pairs (R, A) that satisfy the equation.

9. If sales revenue R is a function of advertising A, then R is the dependent variable and A is the independent variable.

10. If $M = F(q)$, then F is a function of q.

Problems 11 and 12 refer to the function $F(q) = \dfrac{2}{3q}$.

11. $F(-1) = \dfrac{-2}{3}$.

12. The domain of $F(q)$ is the set of all real numbers.

13. In the accompanying graph y is a function of x.

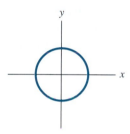

Problems 14–16 refer to the accompanying graph of $h(x)$.

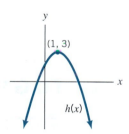

14. The function $h(x)$ has a maximum value at $(1, 3)$.

15. The function $h(x)$ is concave down.

16. The function $h(x)$ increases to the right of $x = 1$ and increases to the left of $x = 1$.

Problems 17 and 18 refer to the following table.

R	-1	2	0	3
S	10	8	6	4

17. The variable S is a function of R.

18. The variable S is decreasing over the domain for R.

19. A set of ordered pairs of the form (M, C) implies that M is a function of C.

Problems 20–22 refer to the accompanying graph of $f(x)$.

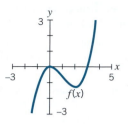

20. The function $f(x)$ has both a maximum and a minimum over the interval $[-1, 3)$.

21. The function $f(x)$ decreases over the interval $(0, 2)$.

22. The function $f(x)$ is concave up for $x > 1$.

Problems 23 and 24 refer to the accompanying figure.

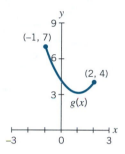

23. The domain of the function $g(x)$ is the interval $[-1, 2]$.

24. The range of the function $g(x)$ is the interval $[4, 7]$.

Problems 25–27 refer to the accompanying figure.

U.S. Bicycle Use by Age in Year 2000

Source: National Survey on Recreation and the Environment, *www.bts.gov*

25. The histogram describes the number of bicycle riders by age.

26. There are about twice as many bicycle riders who are 18 to 44 years old as there are riders who are 45 years or older.

27. The total number of bicycle riders is over 30 million.

II. In Problems 28–34, give an example of a graph, relationship, function, or functions with the specified properties.

28. A relationship between two variables w and z described with an equation where z is not a function of w.

29. A relationship between two variables w and z described as a table where w is a function of z but z is not a function of w.

30. A graph of a function that is increasing and concave down.

31. A graph of a function that is decreasing and concave up.

32. A graph of a function that is concave up and has a minimum value at the point $(-2, 0)$.

33. A graph of a function where the domain is the set of real numbers and that has no maximum or minimum values.

34. A topic sentence describing the wolf pups that were born and survived in Yellowstone Park as shown in the accompanying graph.

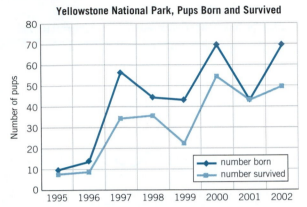

Yellowstone National Park, Pups Born and Survived

Source: Yellowstone Wolf Project, Annual Report 2002, U.S. Department of the Interior.

III. Is each of the statements in Problems 35–42 true or false? If a statement is true, explain how you know. If a statement is false, give a counterexample or explain why it is false.

35. A function can have either a maximum or a minimum but not both.

36. Neither horizontal nor vertical lines are functions.

37. A function is any relationship between two quantities.

38. The graph in the accompanying figure is decreasing and concave down.

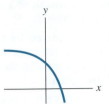

39. The graph in the accompanying figure is increasing and concave up.

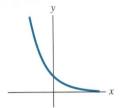

40. All functions have at least one minimum value.

41. Sometimes the choice as to which variable will be the independent variable for a function is arbitrary.

42. The graph in the accompanying figure represents P as a function of Q.

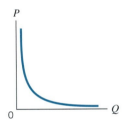

CHAPTER 1 REVIEW: PUTTING IT ALL TOGETHER

1. A man weighed 160 pounds at age 20. Now, at age 60, he weighs 200 pounds.

   a. What percent of his age 20 weight is his weight gain?

   b. If he went on a diet and got back to 160 pounds, what percent of his age 60 weight would be his weight loss?

   c. Explain why the weight gain is a different percentage than the weight loss even though it is the same number of pounds.

2. Data from the *World Health Organization Report* 2005 shows vast differences in average spending per person on health for different countries. Data on health spending from selected countries is shown here.

Country	As % of Gross Domestic Product (GDP)	$ per Capita
China	5.8	63
India	6.1	30
Iraq	1.5	11
North Korea	4.6	0.3
South Korea	5.0	532
United States	14.6	5274

   a. North and South Korea spent close to the same percentage of their gross domestic product on health, but a very different amount per person (per capita) in 2005. How can this be?

b. In 2005, how many Iraqis could receive health care (at the Iraqi level of spending) for the amount the United States spent on one person? If Iraq spent the same percent of its GDP on health care as the United States, how much would they be spending per person?

c. China and India had the two largest populations in 2005, with China at 1.31 billion and India at 1.08 billion. Estimate how much each of these countries is spending on health care.

3. In April 2005 there was considerable debate in the media about whether "average" incomes had gone up or down in the United Kingdom (UK). The Institute for Fiscal Studies produced a report in which they stated that the mean household income in the UK in 2004 fell by 0.2% over the previous year and that median household income in the UK rose for the same period by 0.5%. Explain how the mean income could go down while the median income rose.

4. In 2001 there were 130,651 thousand metric tons of commercial fish caught worldwide. The chart below lists the world's 10 leading commercial fishing nations.

The 10 Leading Commercial Fishing Nations in 2001

	Fish caught (in thousands of metric tons)
China	42,579
Peru	7,996
India	5,897
Japan	5,515
United States	5,424
Indonesia	5,137
Chile	4,363
Russia	3,718
Thailand	3,657
Norway	3,198

Source: U.S. Bureau of the Census, *Statistical Abstract of the United States:* 2006.

a. What are the mean and median for this set of data?

b. China caught what percent of the commercial fish among the leading 10 nations? Of the world?

c. Will the mean and median for the 10 leading nations change if China substantially increases the amount of commercial fish it catches while other nations remain at the same level?

5. The following table shows the four leading causes of death around the world.

The Four Leading Causes of Death Worldwide

Cause	% of All Deaths
Heart disease	12.6
Stroke	9.7
Lower respiratory infections (e.g., pneumonia, emphysema, bronchitis)	6.8
HIV/AIDS	4.9

Source: World Health Organization (WHO), *The World Health Report*, 2003.

a. What is the leading cause of death worldwide?

b. Create a bar chart using these data.

c. What percentage of worldwide deaths is not accounted for in the table? List at least two other diseases or conditions that can cause death. Could either of them account for more that 5% of all deaths worldwide?

6. Before doctors transfuse blood they must know the patient's blood type and which types of blood are compatible with that type.

a. On the following graph, types O+ and O− make up approximately 37% and 6%, respectively, of all U.S. blood types. Estimate the percentage of the population in the United States for each of the other blood types.

b. How could you determine the number of people with each blood type?

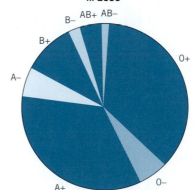

% U.S. Population by Blood Type in 2006

Source: http://bloodcenter.stanford.edu/about_blood/blood_types.html.

7. In a certain state, car buyers pay an excise tax of 2.5%. This means that someone who buys a car must pay the state a one-time fee of 2.5% of the car's value.

Use P to represent the price of the car and E to represent excise tax, then express E as a function of P.

8. A waitress makes $2.50 an hour in wages and receives tips totaling 18% of the price of each meal she serves. Let *H* represent the hours she works, *P* represent the total cost of the meals she served, and *E* represent what she earns in a week before taxes. Write an equation expressing *E* in terms of *H* and *P*.

9. The following graph shows the number of babies per million babies that were named Emma in the United States between 1880 and 2004. Create a title for the graph and explain in a few sentences what is happening over time to the name Emma.

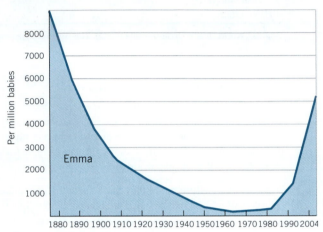

Source: thebabywizard.ivillage.com.

10. During the 1990s natural disasters worldwide took more than 666,000 lives and cost over $683 billion (US $). Using the accompanying chart, write a 60-second summary describing the worldwide financial costs of natural disasters during this period.

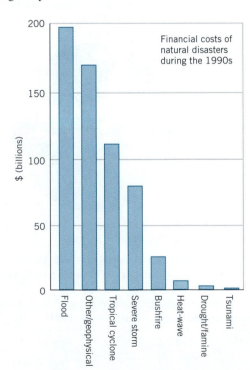

Source: Australian government, Department of Meteorology.

11. a. Determine whether each set of points represents a function.

  i. $(6, -1)$, $(5, -1)$, $(8, 3)$, $(-3, 2)$

  ii. $(5, 3)$, $(7, -1)$, $(5, -2)$, $(4, -3)$

  iii. $(7, -3)$, $(8, -3)$, $(-2, -3)$, $(9, -3)$

  iv. $(4, 0)$, $(4, -3)$, $(4, 7)$, $(4 - 1)$

 b. Create a table of values for the variables *A* and *B*, where *B* is a function of *A* but *A* is not a function of *B*.

 c. Create two graphs, one that represents a function and one that does not represent a function.

12. Find an equation that represents the relationship between *x* and *y* in each of the following tables. Specify in each case if *y* is a function of *x*.

a.

x	y
0	2
1	4
2	6
3	8
4	10

b.

x	y
0	-1
1	0
2	3
3	8
4	15

c.

x	y
0	1
1	2
2	9
3	28
4	65

13. The following graph shows changes in U.S. opinion regarding the war in Iraq over the period between October 2003 and June 2005.

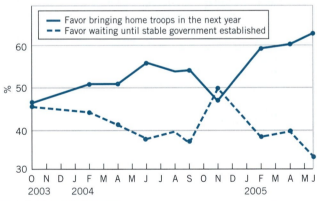

Source: Harris Interactive.

 a. In February 2004, what was the approximate percentage of Americans who favored bringing troops home in the next year?

 b. In the period covered by this graph, when did more people favor waiting for the establishment of a stable government than favored a withdrawal in the next year?

 c. During which interval(s) did the amount of people in favor of waiting for a stable government increase?

 d. On which date was the difference in the popularity of the two opinions the greatest? Estimate this difference.

14. The following graph shows fuel economy in relation to speed.

 a. Does the graph represent a function? If not, why not? If so, what are the domain and range?

b. Describe in words how speed relates to fuel economy.

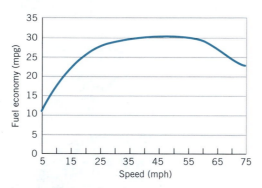

(mpg = miles per gallon of gas, mph = miles per hour)
Source: www.fueleconomy.gov.

15. In 2006 the exchange rate for the Chinese yuan and U.S. dollar was about 8 to 1; that is, you could exchange 8 yuan for 1 U.S. dollar.

a. Let y = number of Chinese yuan and d = number of U.S. dollars. Complete the following table.

d	1	2	3	4	10	20	100
y							

b. Find an equation that expresses y as a function of d. What is the independent variable? The dependent variable?

c. Complete the following table:

y	8	12	16	20	50	80	100
d							

d. Find an equation that expresses d as a function of y. What is the independent variable? The dependent variable?

16. The average number of calories burned with exercising varies by the weight of the person. The accompanying table gives the approximate calories burned per hour from disco dancing.

Number of Calories Burned/hr from Disco Dancing

Weight	100 lb. person	125 lb. person	150 lb. person	175 lb. person	200 lb. person
calories burned per hour	264	330	396	462	528

Source: http://www.fitresource.com/Fitness.

Let t be the number of hours spent exercising and C the total number of calories burned.

a. Write an equation expressing C_1 in terms of t for a 125 lb. person disco dancing and an equation for C_2 for a 175 lb. person disco dancing.

b. Do your equations in part (a) represent functions? If not, why not?

c. For the equations in part (a) that represent functions, what is the independent variable? The dependent variable? What is a reasonable domain? What is a reasonable range?

d. Generate small tables of values and graphs for C_1 and C_2. How do the two graphs compare?

e. How do you think these graphs would compare to the corresponding graphs for a 200 lb. person? Explain.

17. Sketch the graph of a function $f(t)$ with domain = [0, 10] and range = [0, 100] that satisfies all of the given conditions:

$$f(0) = 100, \ f(5) = 60, \ f(10) = 0$$
$f(t)$ is decreasing and concave down for interval $(0, 5)$
$f(t)$ is decreasing and concave up for interval $(5, 10)$

18. Below is the graph of $f(x)$.

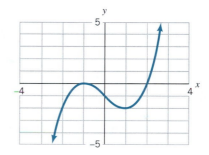

a. Determine the values for $f(-1)$, $f(0)$, $f(2)$.

b. For what value(s) of x does $f(x) = -2$?

c. Specify the interval (s) over which the function is
  i. Increasing
  ii. Decreasing
  iii. Concave up
  iv. Concave down

19. The following graph shows the life expectancy in Botswana compared with world life expectancy. The dotted lines on the graph represent projections.

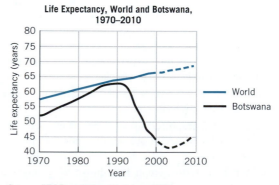

Source: UN Population Division, 1999.

a. Does the graph for Botswana represent a function? If not, why not? If so, what are the domain and range?

b. Over the given time period, when did life expectancy in Botswana reach its maximum? Estimate this maximum value.

c. When was life expectancy in Botswana projected to reach its minimum? Estimate this minimum value. What might have caused the sudden drop in life expectancy?

d. Over what interval(s) was life expectancy (actual and projected) in Botswana increasing? Decreasing?

e. Over what interval(s) is Botswana's graph concave down? Concave up?

f. Write a short description of life expectancy in Botswana compared with world life expectancy.

20. Below is a graph of data collected by the FBI.

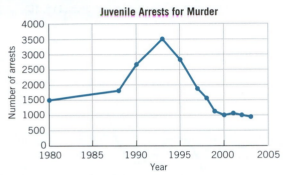

Juvenile Arrests for Murder

Source: U.S. Bureau of The Census, *Statistical Abstract of the United States: 2006.*

If we let $N(x)$ = number of juvenile arrests for murder in year x,

a. Estimate $N(1993)$. What are the coordinates of this point? What does this point represent?

b. Estimate the coordinates of the minimum point. What does this point represent?

c. Over what interval is the function increasing? Over what interval is the function decreasing?

d. What is the domain of $N(x)$? What is the range?

e. Write a brief summary about juvenile arrests between 1980 and 2003.

21. UNICEF has made children's education worldwide a top priority. According to UNICEF, 25 years ago only half of the world's children received a primary school education. Today, 86% receive a primary education. The following pie chart shows for each region the percentage of the *total* number of worldwide children out of school.

a. Explain what the label 3% means next to the region Latin America/Caribbean.

b. Does this graph tell us what percentage of children in Sub-Saharan Africa are out of school? Explain why or why not.

c. What does this graph not show? Why might this be confusing?

d. What is the region with the largest percentage of the world's children who are out of school? What is the total percentage of the two regions with the most children out of school? What is the region with the smallest percentage?

e. Write a 60-second summary about where you would suggest UNICEF focus its efforts in childhood education.

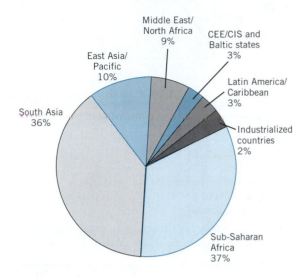

Where Children Are Out of School

22. a. Given $g(x) = 2x^2 + 3x - 1$, evaluate $g(0)$ and $g(-1)$.

b. Find the domain of $g(x)$.

23. a. Given $f(x) = \dfrac{1}{x - 2}$, evaluate $f(0), f(-1),$ and $f(2)$.

b. Find the domain of $f(x)$.

24. Find the domain and range of the following functions:

a. $h(x) = \sqrt{x}$ **b.** $f(x) = \sqrt{x - 6}$

25. The following table shows the cost each month in a small condominium in Maine for natural gas, which is used for cooking, heating, and hot water.

	Jan	Feb	Mar	Apr	May	Jun	Jul	Aug	Sep	Oct	Nov	Dec
$ gas	220	234	172	83	51	42	30	32	51	89	102	150

a. What is the month with the maximum cost for gas? The month with the minimum? Do these numbers seem reasonable? Why or why not?

b. What is the mean monthly cost of gas for the year? What is the median monthly cost of gas? Describe in words what these descriptors tell you about the cost of gas in this condominium.

c. Create a table with intervals $0–49, $50–99, $100–149, and so on, and next to each interval show the number of months that fall within that cost interval. Use the table to construct a histogram of these data (with the cost intervals on the horizontal axis).

d. Describe two patterns in the cost of gas for this condominium.

26. The following graphs are population pyramids for Yemen (a developing country) and Japan (a highly developed country).

Write a 60-second summary comparing and contrasting the populations in the year 2000 in these two countries.

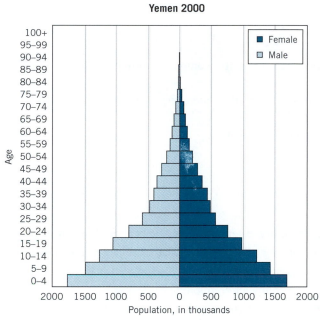

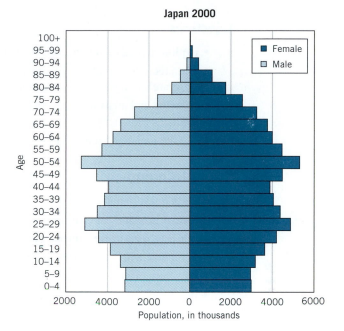

Source. WHO World Population Prospects, 2002.

Collecting, Representing, and Analyzing Data

Objectives

- explore issues related to collecting data.
- learn techniques for organizing and graphing data using a computer (with a spreadsheet program) or a graphing calculator.
- describe and analyze the overall shape of single-variable data using frequency and relative frequency histograms.
- use the mean and median to represent single-variable data.

Material/Equipment

- class questionnaire
- measuring tapes in centimeters and inches
- optional measuring devices: eye chart, flexibility tester, measuring device for blood pressure
- computer with spreadsheet program or graphing calculator with statistical plotting features
- data from class questionnaire or other small data set formatted either as spreadsheet or graph link file
- overhead projector and projection panel for computer or graphing calculator
- transparencies for printing or drawing graphs for overhead projector (optional)

Related Readings

(On the web at *www.wiley.com/college/kimeclark*)
"U.S. Government Definitions of Census Terms"
"Health Measurements"

Related Software

"F1: Histograms," in *FAM 1000 Census Graphs*

Procedure

This exploration may take two class periods.

Day One

In a Small Group or with a Partner

1. Pick (or your instructor will assign you) one of the undefined variables on the questionnaire. Spend about 15 minutes coming up with a workable definition and a strategy for measuring that variable. Be sure there is a way in which a number or single letter can be used to record each individual's response on the questionnaire.
2. Consult the reading "Health Measurements" if you decide to collect health data.

Class Discussion

After all of the definitions are recorded on the board, discuss your definition with the class. Is it clear? Does everyone in the class fall into one of the categories of your definition? Can anyone think of someone who might not fit into any of the categories? Modify the definition until all can agree on some wording. As a class, decide on the final version of the questionnaire and record it in your class notebook.

In a Small Group or with a Partner

Help each other when necessary to take measurements and fill out the entire questionnaire. Questionnaires remain anonymous, and you can leave blank any question you can't or don't want to answer. Hand in your questionnaire to your instructor by the end of class.

Exploration-Linked Homework

Read "U.S. Government Definitions of Census Terms" for a glimpse into the federal government's definitions of the variables you defined in class. How do the "class" definitions and the "official" ones differ?

Day Two

Class Demonstration

1. If you haven't used a spreadsheet or graphing calculator before, you'll need a basic technical introduction. (*Note:* If you are using a TI-83 or TI-84 graphing calculator, there are basic instructions in the Graphing Calculator Manual on *www.wiley.com/college/kimeclark.*) Then you'll need an electronic version of the data set from which you will choose one variable for the whole class to study (e.g., age from the class data).

   a. If you're using a spreadsheet:

   - Copy the column with the data onto a new spreadsheet and graph the data. What does this graph tell you about the data?
   - Sort and replot the data. Is this graph any better at conveying information about the data?

   b. If you're using a graphing calculator:

   - Discuss window sizes, changing interval sizes, and statistical plot procedures.

2. Select an interval size and then construct a frequency histogram and a relative frequency histogram. If possible, label one of these carefully and print it out. If you have access to a laser printer, you can print onto an overhead transparency.

3. Calculate the mean and median using spread sheet or graphing calculator functions.

In a Small Group or with a Partner

Choose another variable from your data. Pick an interval size, and then generate both a frequency histogram and a relative frequency histogram. If possible, make copies of the histograms for both your partner and yourself. Calculate the mean and median.

Discussion/Analysis

With your partner(s), analyze and jot down patterns that emerge from the data. How could you describe your results? What are some limitations of the data? What other questions are raised and how might they be resolved? Record your ideas for a 60-second summary.

Exploration-Linked Homework

Prepare a verbal 60-second summary to give to the class. If possible, use an overhead projector with a transparency of your histogram or a projector linked to your graphing calculator. If not, bring in a paper copy of your histogram. Construct a written 60-second summary. (See Section 1.2 for some writing suggestions.)

ALGEBRA CLASS QUESTIONNAIRE
(You may leave any category blank)

1. Age (in years)

2. Sex (female = 1, male = 2)

3. Your height (in inches)

4. Distance from your navel to the floor (in centimeters)

5. Estimate your average travel time to school (in minutes)

6. What is your most frequent mode of transportation to school?
   (F = by foot, C = car, P = public transportation, B = bike, O = other)

The following variables will be defined in class. We will discuss ways of coding possible responses and then use the results to record our personal data.

7. The number of people in your household

8. Your employment status

9. Your ethnic classification

10. Your attitude toward mathematics

Health Data

11. Your pulse rate before jumping (beats per minute)

12. Your pulse rate after jumping for 1 minute (beats per minute)

13. Blood pressure: systolic (mm Hg)

14. Blood pressure: diastolic (mm Hg)

15. Flexibility (in inches)

16. Vision, left eye

17. Vision, right eye

Other Data

Picturing Functions

Objective

- develop an intuitive understanding of functions.

Material/Equipment

None required

Procedure

Part I

Class Discussion

Bridget, the 6-year-old daughter of a professor at the University of Pittsburgh, loves playing with her rubber duckie in the bath at night. Her mother drew the accompanying graph for her math class. It shows the water level (measured directly over the drain) in Bridget's tub as a function of time.

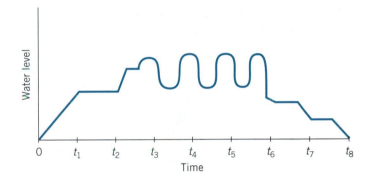

Pick out the time period during which:

- The tub is being filled
- Bridget is entering the tub
- She is playing with her rubber duckie
- She leaves the tub
- The tub is being drained

With a Partner

Create your own graph of a function that tells a story. Be as inventive as possible. Some students have drawn functions that showed the decibel levels during a phone conversation of a boyfriend and girlfriend, number of hours spent doing homework during one week, and amount of money in one's pocket during the week.

Class Discussion

Draw your graph on the blackboard and tell its story to the class.

Part II

With a Partner

Generate a plausible graph for each of the following:

1. Time spent driving to work as a function of the amount of snow on the road. (*Note:* The first inch or so may not make any difference; the domain may be only up to about a foot of snow since after that you may not be able to get to work.)
2. The hours of nighttime as a function of the time of year.
3. The temperature of ice cream taken out of the freezer and left to stand.
4. The distance that a cannonball (or javelin or baseball) travels as a function of the angle of elevation at which is it is launched. (The maximum distance is attained for angles of around 45° from the horizontal.)
5. Assume that you leave your home walking at a normal pace, realize you have forgotten your homework and run back home, and then run even faster to school. You sit for a while in a classroom and then walk leisurely home. Now plot your distance from home as a function of time.

Bonus Question

Assume that water is pouring at a constant rate into each of the containers shown. The height of water in the container is a function of the volume of liquid. Sketch a graph of this function for each container.

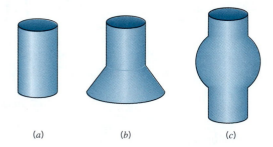

(a)   (b)   (c)

Discussion/Analysis

Are your graphs similar to those generated by the rest of the class? Can you agree as a class as to the basic shape of each of the graphs? Are there instances in which the graphs could look quite different?

Deducing Formulas to Describe Data

Objective

- find and describe patterns in data.
- deduce functional formulas from data tables.
- extend patterns using functional formulas.

Material/Equipment

None required

Procedure

Class Discussion

1. Examine data tables (a) and (b). In each table, look for a pattern in terms of how y changes when x changes. Explain in your own words how to find y in terms of x.

a.

x	y
0	0.0
1	0.5
2	1.0
3	1.5
4	2.0

b.

x	y
0	5
1	8
2	11
3	14
4	17

2. Assuming that the pattern continues indefinitely, use the rule you have found to extend the data table to include negative numbers for x.

3. Check your extended data tables. Did you find only one value for y given a particular value for x?

4. Use a formula to describe the pattern that you have found. Do you think this formula describes a function? Explain.

On Your Own

1. For each of the data tables (c) to (h), explain in your own words how to find y in terms of x. Then extend each table using the rules you have found.

c.

x	y
0	0
1	1
2	4
3	9
4	16

d.

x	y
0	0
1	1
2	8
3	27
4	64

e.

x	y
0	0
1	2
2	12
3	36
4	80

[*Hint:* For table (e) think about some combination of data in tables (c) and (d).]

f.

x	y
−2	0
0	10
5	35
10	60
100	510

g.

x	y
0	−1
1	0
2	3
3	8
4	15

h.

x	y
0	3
10	8
20	13
30	18
100	53

2. For each table, construct a formula to describe the pattern you have found.

Discussion/Analysis

With a Partner

Compare your results. Do the formulas that you have found describe functions? Explain.

Class Discussion

Does the rest of the class agree with your results? Remember that formulas that look different may give the same results.

Exploration-Linked Homework

1. a. For data tables (i) and (j), explain in your own words how to find *y* in terms of *x*. Using the rules you have found, extend the data tables to include negative numbers.

i.

x	y
−10	10.0
0	0.0
3	0.9
8	6.4
10	10.0

j.

x	y
0	−3
1	1
2	5
3	9
4	13

b. For each table, find a formula to describe the pattern you have found. Does your formula describe a function? Explain.

2. Make up a functional formula, generate a data table, and bring the data table on a separate piece of paper to class. The class will be asked to find your rule and express it as a formula.

k.

x	y

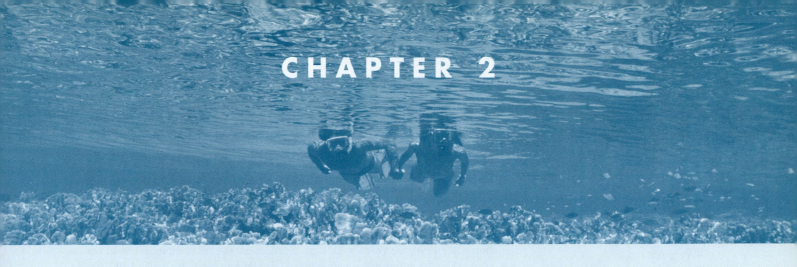

CHAPTER 2

RATES OF CHANGE AND LINEAR FUNCTIONS

OVERVIEW

How does the U.S. population change over time? How do children's heights change as they age? Average rates of change provide a tool for measuring how change in one variable affects a second variable. When average rates of change are constant, the relationship is linear.

After reading this chapter you should be able to

- calculate and interpret average rates of change

- understand how representations of data can be biased

- recognize that a constant rate of change denotes a linear relationship

- construct a linear equation given a table, graph, or description

- derive by hand a linear model for a set of data

2.1 *Average Rates of Change*

In Chapter 1 we looked at how change in one variable could affect change in a second variable. In this section we'll examine how to measure that change.

Describing Change in the U.S. Population over Time

We can think of the U.S. population as a function of time. Table 2.1 and Figure 2.1 are two representations of that function. They show the changes in the size of the U.S. population since 1790, the year the U.S. government conducted its first decennial census. Time, as usual, is the independent variable and population size is the dependent variable.

Change in population

Figure 2.1 clearly shows that the size of the U.S. population has been growing over the last two centuries, and growing at what looks like an increasingly rapid rate. How can the change in population over time be described quantitatively? One way is to pick two points on the graph of the data and calculate how much the population has changed during the time period between them.

Suppose we look at the change in the population between 1900 and 1990. In 1900 the population was 76.2 million; by 1990 the population had grown to 248.7 million. How much did the population increase?

$$\text{change in population} = (248.7 - 76.2) \text{ million people}$$
$$= 172.5 \text{ million people}$$

This difference is portrayed graphically in Figure 2.2, at the top of the next page.

Change in time

Knowing that the population increased by 172.5 million tells us nothing about how rapid the change was; this change clearly represents much more dramatic growth if it happened over 20 years than if it happened over 200 years. In this case, the length of time over which the change in population occurred is

$$\text{change in years} = (1990 - 1900) \text{ years}$$
$$= 90 \text{ years}$$

Population of the United States: 1790–2000

Year	Population in Millions
1790	3.9
1800	5.3
1810	7.2
1820	9.6
1830	12.9
1840	17.1
1850	23.2
1860	31.4
1870	39.8
1880	50.2
1890	63.0
1900	76.2
1910	92.2
1920	106.0
1930	123.2
1940	132.2
1950	151.3
1960	179.3
1970	203.3
1980	226.5
1990	248.7
2000	281.4

Table 2.1

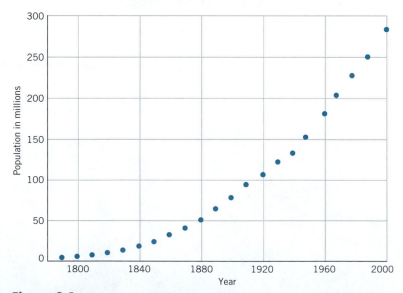

Figure 2.1 Population of the United States.
Source: U.S. Bureau of the Census, *Statistical Abstract of the United States: 2002.*

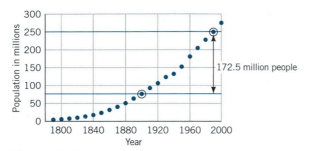

Figure 2.2 Population change: 172.5 million people.

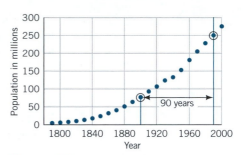

Figure 2.3 Time change: 90 years.

This interval is indicated in Figure 2.3 above.

Average rate of change

To find the *average rate of change* in population per year from 1900 to 1990, divide the change in the population by the change in years:

$$\text{average rate of change} = \frac{\text{change in population}}{\text{change in years}}$$

$$= \frac{172.5 \text{ million people}}{90 \text{ years}}$$

$$\approx 1.92 \text{ million people/year}$$

In the phrase "million people/year" the slash represents division and is read as "per." So our calculation shows that "on average," the population grew at a rate of 1.92 million people per year from 1900 to 1990. Figure 2.4 depicts the relationship between time and population increase.

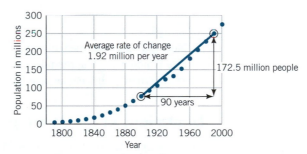

Figure 2.4 Average rate of change: 1900–1990.

Defining the Average Rate of Change

The notion of average rate of change can be used to describe the change in any variable with respect to another. If you have a graph that represents a plot of data points of the form (x, y), then the average rate of change between any two points is the change in the y value divided by the change in the x value.

> The *average rate of change* of y with respect to $x = \dfrac{\text{change in } y}{\text{change in } x}$

If the variables represent real-world quantities that have units of measure (e.g., millions of people or years), then the average rate of change should be represented in terms of the appropriate units:

$$\text{units of the average rate of change} = \frac{\text{units of } y}{\text{units of } x}$$

For example, the units might be dollars/year (read as "dollars per year") or pounds/person (read as "pounds per person").

EXAMPLE 1   Between 1850 and 1950 the median age in the United States rose from 18.9 to 30.2, but by 1970 it had dropped to 28.0.

a. Calculate the average rate of change in the median age between 1850 and 1950.

b. Compare your answer in part (a) to the average rate of change between 1950 and 1970.

SOLUTION   **a.** Between 1850 and 1950,

$$\text{average rate of change} = \frac{\text{change in median age}}{\text{change in years}}$$

$$= \frac{(30.2 - 18.9) \text{ years}}{(1950 - 1850) \text{ years}} = \frac{11.3 \text{ years}}{100 \text{ years}}$$

$$= 0.113 \text{ years/year}$$

The units are a little confusing. But the results mean that between 1850 and 1950 the median age increased an average of 0.113 years each calendar year.

b. Between 1950 and 1970,

$$\text{average rate of change} = \frac{\text{change in median age}}{\text{change in years}}$$

$$= \frac{(28.0 - 30.2) \text{ years}}{(1970 - 1950) \text{ years}} = \frac{-2.2 \text{ years}}{20 \text{ years}}$$

$$= -0.110 \text{ years/year}$$

Note that since the median age dropped in value between 1950 and 1970, the average rate is negative. The median age decreased by 0.110 years/year between 1950 and 1970, whereas the median age increased by 0.113 years/year beween 1850 and 1950.

Limitations of the Average Rate of Change

The average rate of change is an average. Average rates of change have the same limitations as any average. Although the average rate of change of the U.S. population from 1900 to 1990 was 1.92 million people/year, it is highly unlikely that in each year the population grew by exactly 1.92 million. The number 1.92 million people/year is, as the name states, an average. Similarly, if the arithmetic average, or *mean*, height of students in your class is 67 inches, you wouldn't expect every student to be 67 inches tall. In fact, it may be the case that not even one student is 67 inches tall.

The average rate of change depends on the end points. If the data points do not all lie on a straight line, the average rate of change varies for different intervals. For instance, the average rate of change in population for the time interval 1840 to 1940 is 1.15 million people/year and from 1880 to 1980 is 1.76 million people/year. (See Table 2.2. *Note:* Here we abbreviate "million people" as "million.") You can see on the graphs that the line segment is much steeper from 1880 to 1980 than from 1840 to 1940 (Figures 2.5 and 2.6). Different intervals give different impressions of the rate of change in the U.S. population, so it is important to state which end points are used.

Time Interval	Change in Time	Change in Population	Average Rate of Change
1840–1940	100 yr	$132.2 - 17.1 = 115.1$ million	$\dfrac{115.1 \text{ million}}{100 \text{ yr}} \approx 1.15$ million/yr
1880–1980	100 yr	$226.5 - 50.2 = 176.3$ million	$\dfrac{176.3 \text{ million}}{100 \text{ yr}} \approx 1.76$ million/yr

Table 2.2

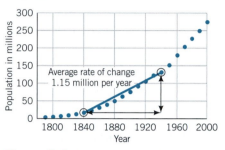

Figure 2.5 Average rate of change: 1840–1940.

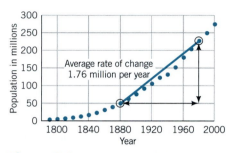

Figure 2.6 Average rate of change: 1880–1980.

The average rate of change does not reflect all the fluctuations in population size that may occur between the end points. For more specific information, the average rate of change can be calculated for smaller intervals.

Algebra Aerobics 2.1

1. Suppose your weight five years ago was 135 pounds and your weight today is 143 pounds. Find the average rate of change in your weight with respect to time.

2. Table 2.3 shows data on U.S. international trade as reported by the U.S. Bureau of the Census.

Year	U.S. Exports (billions of $)	U.S. Imports (billions of $)	U.S. Trade Balance = Exports − Imports (billions of $)
1990	537.2	618.4	−81.2
2006	820.2	1273.2	−453.0

Table 2.3

   a. What is the average rate of change between 1990 and 2006 for:

      i. Exports?

      ii. Imports?

      iii. The trade balance, the difference between what we sell abroad (exports) and buy from abroad (imports)?

   b. What do these numbers tell us?

3. Table 2.4 indicates the number of deaths in motor vehicle accidents in the United States as listed by the U.S. Bureau of the Census.

Annual Deaths in Motor Vehicle Accidents (thousands)

1980	1990	2000	2004
52.1	44.6	41.8	42.6

Table 2.4

Find the average rate of change:

a. From 1980 to 2000

b. From 2000 to 2004

Be sure to include units.

4. A car is advertised to go from 0 to 60 mph in 5 seconds. Find the average rate of change (i.e., the average acceleration) over that time.

5. According to the National Association of Insurance Commissioners, the average cost for automobile insurance has gone from $689 in 2000 to $867 in 2006. What is the average rate of change?

6. A football player runs for 1056 yards in 2002 and for 978 yards in 2006. Find the average rate of change in his performance.

7. The African elephant is an endangered species, largely because poachers (people who illegally hunt elephants) kill elephants to sell the ivory from their tusks. In the African country of Kenya in the last 10 years, the elephant population has dropped from 150,000 to 30,000. Calculate the average rate of change and describe what it means.

Exercises for Section 2.1

1. If r is measured in inches, s in pounds, and t in minutes, identify the units for the following average rates of change:

a. $\dfrac{\text{change in } r}{\text{change in } s}$ b. $\dfrac{\text{change in } t}{\text{change in } r}$ c. $\dfrac{\text{change in } s}{\text{change in } r}$

2. Assume that R is measured in dollars, S in ounces, T in dollars per ounce, and V in ounces per dollar. Write a product of two of these terms whose resulting units will be:

a. Dollars b. Ounces

3. Your car's gas tank is full and you take a trip. You travel 212 miles, then you fill your gas tank up again and it takes 10.8 gallons. If you designate your change in distance as 212 miles and your change in gallons as 10.8, what is the average rate of change of gasoline used, measured in miles per gallon?

4. The gas gauge on your car is broken, but you know that the car averages 22 miles per gallon. You fill your 15.5-gallon gas tank and tell your friend, "I can travel 300 miles before I need to fill up the tank again." Explain why this is true.

5. The consumption of margarine (in pounds per person) decreased from 9.4 in 1960 to 7.5 in 2000. What was the annual average rate of change? (*Source: www.census.gov*)

6. The percentage of people who own homes in the United States has gone from 65.5% in 1980 to 69.0% in 2005. What is the average rate of change in percentage points per year?

7. The accompanying table shows females' SAT scores in 2000 and 2005.

Year	Average Female Verbal SAT	Average Female Math SAT
2000	504	498
2005	505	504

Source: www.collegeboard.com.

Find the average rate of change:
a. In the math scores from 2000 to 2005
b. In the verbal scores from 2000 to 2005

8. a. In 1992 the aerospace industry showed a net loss (negative profit) of $1.84 billion. In 2002 the industry had a net profit of $8.97 billion. Find the average annual rate of change in net profits from 1992 to 2002.

b. In 2005, aerospace industry net profits were $2.20 billion. Find the average rate of change in net profits:
  i. From 1992 to 2005
  ii. From 2002 to 2005

9. According to the U.S. Bureau of the Census, in elementary and secondary schools, in the academic year ending in 1985 there were about 630,000 computers being used for student instruction, or about 84.1 students per computer. In the academic year ending in 2005, there were about 13,600,000 computers being used, or about 4.0 students per computer. Find the average rate of change from 1985 to 2005 in:

a. The number of computers being used
b. The number of students per computer

10. According to the U.S. Bureau of the Census, the percentage of persons 25 years old and over completing 4 or more years of college was 4.6 in 1940 and 27.6 in 2005.

a. Plot the data, labeling both axes and the coordinates of the points.
b. Calculate the average rate of change in percentage points per year.
c. Write a topic sentence summarizing what you think is the central idea to be drawn from these data.

11. Though reliable data about the number of African elephants are hard to come by, it is estimated that there were about 4,000,000 in 1930 and only 500,000 in 2000.

a. What is the average annual rate of change in elephants over time? Interpret your result.
b. During the 1980s it was estimated that 100,000 elephants were being killed each year due to systematic poaching for ivory. How does this compare with your answer in part (a)? What does this tell you about what was happening before or after the 1980s? (*Source: www.panda.org*)

12. a. According to the U.S. Bureau of the Census, between 1980 and 2004 domestic new car sales declined from 6581 thousand cars to 5357 thousand. (*Note:* This does not include trucks, vans, or SUVs.) Calculate the annual average rate of change.

b. During the same period Japanese car sales in the United States dropped from 1906 thousand to 798 thousand. Calculate the average rate of change.

c. What do the two rates suggest about car sales in the United States?

13. Use the information in the accompanying table to answer the following questions.

Percentage of Persons 25 Years Old and Over Who Have Completed 4 Years of High School or More

	1940	2004
All	24.5	85.2
White	26.1	85.8
Black	7.3	80.6
Asian/Pacific Islander	22.6	85.0

Source: U.S. Bureau of the Census, Statistical Abstract of the United States: 2006.

a. What was the average rate of change (in percentage points per year) of completion of 4 years of high school from 1940 to 2004 for whites? For blacks? For Asian/Pacific Islanders? For all?

b. If these rates continue, what percentages of whites, of blacks, of Asian/Pacific Islanders, and of all will have finished 4 years of high school in the year 2007? Check the Internet to see if your predictions are accurate.

c. Write a 60-second summary describing the key elements in the high school completion data. Include rates of change and possible projections for the near future.

d. If these rates continue, in what year will 100% of whites have completed 4 years of high school or more? In what year 100% of blacks? In what year 100% of Asian/Pacific Islanders? Do these projections make sense?

14. The accompanying data show U.S. consumption and exports of cigarettes.

Year	U.S. Consumption (billions)	Exports (billions)
1960	484	20
1980	631	82
2000	430	148
2005	378	113

Source: U.S. Department of Agriculture.

a. Calculate the average rates of change in U.S. cigarette consumption from 1960 to 1980, from 1980 to 2005, and from 1960 to 2005.

b. Compute the average rate of change for cigarette exports from 1960 to 2005. Does this give an accurate image of cigarette exports?

c. The total number of cigarettes consumed in the United States in 1960 was 484 billion, very close to the number consumed in 1995, 487 billion. Does that mean smoking was as popular in 1995 as it was in 1960? Explain your answer.

d. Write a paragraph summarizing what the data tell you about the consumption and exports of cigarettes since 1960, including average rates of change.

15. Use the accompanying table on life expectancy to answer the following questions.

Average Number of Years of Life Expectancy in the United States by Race and Sex Since 1900

Life Expectancy at Birth by Year	White Males	White Females	Black Males	Black Females
1900	46.6	48.7	32.5	33.5
1950	66.5	72.2	58.9	62.7
2000	74.8	80.0	68.2	74.9
2005	75.4	81.1	69.9	76.8

Source: U.S. National Center for Health Statistics, *Statistical Abstract of the United States*, 2007.

a. What group had the highest life expectancy in 1900? In 2005? What group had the lowest life expectancy in 1900? In 2005?

b. Which group had the largest average rate of change in life expectancy between 1900 and 2005?

c. Write a short summary of the patterns in U.S. life expectancy from 1900 to 2005 using average rates of change to support your points.

16. The accompanying table gives the number of unmarried males and females over age 15 in the United States.

Marital Status of Population 15 Years Old and Older

Year	Number of unmarried males (in thousands)	Number of unmarried females (in thousands)
1950	17,735	19,525
1960	18,492	22,024
1970	23,450	29,618
1980	30,134	36,950
1990	36,121	43,040
2000	43,429	50,133
2004	41,214	47,616

Source: U.S. Bureau of the Census, *www.census.gov.*

a. Calculate the average rate of change in the number of unmarried males between 1950 and 2004. Interpret your results.

b. Calculate the average rate of change in the number of unmarried females between 1950 and 2004. Interpret your results.

c. Compare the two results.

d. What does this tell you, if anything, about the *percentages* of unmarried males and females?

2.2 *Change in the Average Rate of Change*

We can obtain an even better sense of patterns in the U.S. population if we look at how the average rate of change varies over time. One way to do this is to pick a fixed interval period for time and then calculate the average rate of change for each successive time period. Since we have the U.S. population data in 10-year intervals, we can calculate the average rate of change for each successive decade. The third column in Table 2.5 shows the results of these calculations. Each entry represents the average population growth *per year* (the average annual rate of change) during the previous decade. A few of these calculations are worked out in the last column of the table.

Average Annual Rates of Change of U.S. Population: 1790–2000

Year	Population (millions)	Average Annual Rate for Prior Decade (millions/yr)	Sample Calculations
1790	3.9	Data not available	
1800	5.3	0.14	$0.14 = (5.3 - 3.9)/(1800 - 1790)$
1810	7.2	0.19	
1820	9.6	0.24	
1830	12.9	0.33	
1840	17.1	0.42	$0.42 = (17.1 - 12.9)/(1840 - 1830)$
1850	23.2	0.61	
1860	31.4	0.82	
1870	39.8	0.84	
1880	50.2	1.04	
1890	63.0	1.28	
1900	76.2	1.32	
1910	92.2	1.60	
1920	106.0	1.38	
1930	123.2	1.72	
1940	132.2	0.90	$0.90 = (132.2 - 123.2)/(1940 - 1930)$
1950	151.3	1.91	
1960	179.3	2.80	
1970	203.3	2.40	
1980	226.5	2.32	
1990	248.7	2.22	
2000	281.4	3.27	

Table 2.5

Source: U.S. Bureau of the Census, *Statistical Abstract of the United States: 2002.*

What is happening to the average rate of change over time?

Start at the top of the third column and scan down the numbers. Notice that until 1910 the average rate of change increases every year. Not only is the population growing every decade until 1910, but it is growing at an increasing rate. It's like a car that is not only moving forward but also accelerating. A feature that was not so obvious in the original data is now evident: In the intervals 1910 to 1920, 1930 to 1940, and 1960 to 1990 we see an increasing population but a decreasing rate of growth. It's like a car decelerating—it is still moving forward but it is slowing down.

The graph in Figure 2.7, with years on the horizontal axis and average rates of change on the vertical axis, shows more clearly how the average rate of change

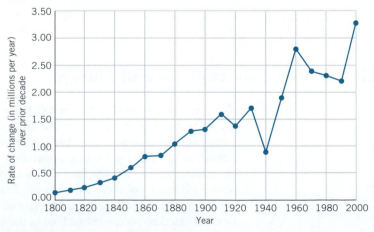

Figure 2.7 Average rates of change in the U.S. population by decade.

fluctuates over time. The first point, corresponding to the year 1800, shows an average rate of change of 0.14 million people/year for the decade 1790 to 1800. The rate 1.72, corresponding to the year 1930, means that from 1920 to 1930 the population was increasing at a rate of 1.72 million people/year.

What does this tell us about the U.S. population?

The pattern of growth was fairly steady up until about 1910. Why did it change? A possible explanation for the slowdown in the decade prior to 1920 might be World War I and the 1918 flu epidemic, which by 1920 had killed nearly 20,000,000 people, including about 500,000 Americans.

In Figure 2.7, the steepest decline in the average rate of change is between 1930 and 1940. One obvious suspect for the big slowdown in population growth in the 1930s is the Great Depression. Look back at Figure 2.1, the original graph that shows the overall growth in the U.S. population. The decrease in the average rate of change in the 1930s is large enough to show up in our original graph as a visible slowdown in population growth.

The average rate of change increases again between 1940 and 1960, then drops off from the 1960s through the 1980s. The rate increases once more in the 1990s. This latest surge in the growth rate is attributed partially to the "baby boom echo" (the result of baby boomers having children) and to a rise in birth rates and immigration.

Algebra Aerobics 2.2

1. Table 2.6 and Figure 2.8 show estimates for world population between 1800 and 2050.
   a. Fill in the third column of the table by calculating the average annual rate of change.
   b. Graph the annual average rate of change versus time.

World Population

Year	Total Population (millions)	Annual Average Rate of Change (over prior 50 years)
1800	980	n.a.
1850	1260	
1900	1650	
1950	2520	
2000	6090	
2050	9076 (est.)	

Table 2.6

Source: Population Division of the United Nations, *www.un.org/popin.*

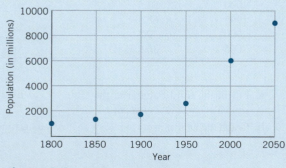

Figure 2.8 World population.

c. During what 50-year period was the average annual rate of change the largest?
   d. Describe in general terms what happened (and is predicted to happen) to the world population and its average rate of change between 1800 and 2050.

2. A graph illustrating a corporation's profits indicates a positive average rate of change between 2003 and 2004, another positive rate of change between 2004 and 2005, a zero rate of change between 2005 and 2006, and a negative rate of change between 2006 and 2007. Describe the graph and the company's financial situation over the years 2003–2007.

3. Table 2.7 shows educational data collected on 18- to 24-year-olds between 1960 and 2004 by the National Center for Educational Statistics. The table shows the number of students who graduated from high school or completed a GED (a high school equivalency exam) during the indicated year.

High School Completers

Year	Number (thousands)	Average Rate of Change (thousands per year)
1960	1679	n.a.
1970	2757	
1980	3089	
1990	2355	
2000	2756	
2004	2752	

Table 2.7

a. Fill in the blank cells with the appropriate average rates of change for high school completers.

b. Describe the pattern in the number of high school completers between 1960 and 2004.

c. What does it mean here when the average rate of change is positive? Give a specific example from your data.

d. What does it mean here when the average rate of change is negative? Give another specific example.

e. What does it mean when two adjacent average rates of change are positive, but the second one is smaller than the first?

Exercises for Section 2.2

Technology is optional for Exercises 4 and 11.

1. Calculate the average rate of change between adjacent points for the following function. (The first few are done for you.)

x	$f(x)$	Average Rate of Change
0	0	n.a.
1	1	1
2	8	7
3	27	
4	64	
5	125	

a. Is the function $f(x)$ increasing, decreasing, or constant throughout?

b. Is the average rate of change increasing, decreasing, or constant throughout?

2. Calculate the average rate of change between adjacent points for the following function. The first one is done for you.

x	$f(x)$	Average Rate of Change
0	0	n.a.
1	1	1
2	16	
3	81	
4	256	
5	625	

a. Is the function $f(x)$ increasing, decreasing, or constant throughout?

b. Is the average rate of change increasing, decreasing, or constant throughout?

3. The accompanying table shows the number of registered motor vehicles in the United States.

Year	Registered Motor Vehicles (millions)	Annual Average Rate of Change (over prior decade)
1960	74	n.a.
1970	108	
1980	156	
1990	189	
2000	218	

a. Fill in the third column in the table.

b. During which decade was the average rate of change the smallest?

c. During which decade was the average rate of change the largest?

d. Write a paragraph describing the change in registered motor vehicles between 1960 and 2000.

4. (Graphing program optional.) The accompanying table indicates the number of juvenile arrests (in thousands) in the United States for aggravated assault.

Year	Juvenile Arrests (thousands)	Annual Average Rate of Change over Prior 5 Years
1985	36.8	n.a.
1990	54.5	
1995	68.5	
2000	49.8	
2005	36.9	

a. Fill in the third column in the table by calculating the annual average rate of change.

b. Graph the annual average rate of change versus time.

c. During what 5-year period was the annual average rate of change the largest?

d. Describe the change in aggravated assault cases during these years by referring both to the number and to the annual average rate of change.

5. Calculate the average rate of change between adjacent points for the following functions and place the values in a third column in each table. (The first entry is "n.a.")

x	$f(x)$
0	5
10	25
20	45
30	65
40	85
50	105

x	$g(x)$
0	270
10	240
20	210
30	180
40	150
50	120

a. Are the functions $f(x)$ and $g(x)$ increasing, decreasing, or constant throughout?

b. Is the average rate of change of each function increasing, decreasing, or constant throughout?

6. Calculate the average rate of change between adjacent points for each of the functions in Tables A–D and place the values in a third column in each table. (The first entry is "n.a.") Then for each function decide which statement best describes it.

Table A

x	$f(x)$
0	1
1	3
2	9
3	27
4	81
5	243

Table C

x	$h(x)$
0	50
10	55
20	60
30	65
40	70
50	75

Table B

x	$g(x)$
0	200
15	155
30	110
45	65
60	20
75	−25

Table D

x	$k(x)$
0	40
1	31
2	24
3	19
4	16
5	15

a. As x increases, the function increases at a constant rate.

b. As x increases, the function increases at an increasing rate.

c. As x increases, the function decreases at a constant rate.

d. As x increases, the function decreases at an increasing rate.

7. Each of the following functions has a graph that is increasing. If you calculated the average rate of change between sequential equal-size intervals, which function can be said to have an average rate of change that is:

a. Constant?   **b.** Increasing?   **c.** Decreasing?

Graph *A*

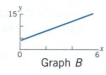
Graph *B*

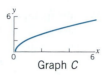
Graph *C*

8. Match the data table with its graph.

Table A

x	y
0	2
1	5
2	8
3	11
4	14
5	17
6	20

Table B

x	y
0	0
1	0.5
2	2
3	4.5
4	8
5	12.5
6	18

Table C

x	y
0	0
1	2
4	4
9	6
16	8
25	10
36	12

Graph *D*

Graph *E*

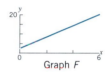
Graph *F*

9. Refer to the first two data tables (A and B) in Exercise 8. Insert a third column in each table and label the column "average rate of change."

a. Calculate the average rate of change over adjacent data points.

b. Identify whether the table represents an average rate of change that is constant, increasing, or decreasing.

c. Explain how you could tell this by looking at the corresponding graph in Exercise 8.

10. Following are data on the U.S. population over the time period 1830–1930 (extracted from Table 2.1).

U.S. Population

Year	Population (in millions)	Average Rate of Change (millions/yr)
1830	12.9	n.a.
1850	23.2	
1870		0.83
1890	63.0	1.16
1910	92.2	
1930		1.55

Source: U.S. Bureau of the Census, *www.census.gov.*

a. Fill in the missing parts of the chart.

b. Which 20-year interval experienced the largest average rate of change in population?

c. Which 20-year interval experienced the smallest average rate of change in population?

11. (Technology recommended.) The accompanying data give a picture of the two major methods of news communication in the United States. (See also Excel or graph link files NEWPRINT and ONAIRTV.)

a. Use the U.S. population numbers from Table 2.1 (at the beginning of this chapter) to calculate and compare the number of copies of newspapers *per person* in 1920 and in 2000.

b. Create a table that displays the annual average rate of change in TV stations for each decade since 1950. Create a similar table that displays the annual average rate of change in newspapers published for the same period. Graph the results.

c. If new TV stations continue to come into existence at the same rate as from 1990 to 2000, how many will there be by the year 2010? Do you think this is likely to be a reasonable projection, or is it overly large or small judging from past rates of growth? Explain.

d. What trends do you see in the dissemination of news as reflected in these data?

Number of U.S. Newspapers

Year	Newspapers (thousands of copies printed)	Number of Newspapers Published
1920	27,791	2042
1930	39,589	1942
1940	41,132	1878
1950	53,829	1772
1960	58,882	1763
1970	62,108	1748
1980	62,202	1745
1990	62,324	1611
2000	55,800	1480

Source: U.S. Bureau of the Census, *www.census.gov.*

Number of U.S. Commercial TV Stations

Year	Number of Commercial TV Stations
1950	98
1960	515
1970	677
1980	734
1990	1092
2000	1248

Source: U.S. Bureau of the Census, *www.census.gov.*

2.3 The Average Rate of Change Is a Slope

Calculating Slopes

The reading "Slopes" describes many of the practical applications of slopes, from cowboy boots to handicap ramps.

On a graph, the average rate of change is the *slope* of the line connecting two points. The slope is an indicator of the steepness of the line.

If (x_1, y_1) and (x_2, y_2) are two points, then the change in y equals $y_2 - y_1$ (see Figure 2.9). This difference is often denoted by Δy, read as "delta y," where Δ is the Greek letter capital D (think of D as representing difference): $\Delta y = y_2 - y_1$. Similarly, the change in x (delta x) can be represented by $\Delta x = x_2 - x_1$. Then

$$\text{slope} = \frac{y_2 - y_1}{x_2 - x_1} = \frac{\Delta y}{\Delta x} = \frac{\text{change in } y}{\text{change in } x}$$

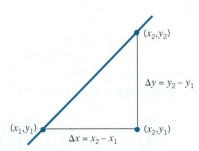

Figure 2.9 Slope $= \Delta y/\Delta x$.

> The average rate of change represents a *slope*. Given two points (x_1, y_1) and (x_2, y_2),
>
> $$\text{average rate of change} = \frac{\text{change in } y}{\text{change in } x} = \frac{\Delta y}{\Delta x} = \frac{y_2 - y_1}{x_2 - x_1} = \text{slope}$$

When calculating a slope, it doesn't matter which point is first

Given two points, (x_1, y_1) and (x_2, y_2), it doesn't matter which one we use as the first point when we calculate the slope. In other words, we can calculate the slope between (x_1, y_1) and (x_2, y_2) as

$$\frac{y_2 - y_1}{x_2 - x_1} \quad \text{or as} \quad \frac{y_1 - y_2}{x_1 - x_2}$$

The two calculations result in the same value.
We can show that the two forms are equivalent.

$$
\begin{aligned}
\text{Given} && \text{slope} &= \frac{y_2 - y_1}{x_2 - x_1} \\[2mm]
\text{multiply by } \frac{-1}{-1} && &= \frac{-1}{-1} \cdot \frac{y_2 - y_1}{x_2 - x_1} \\[2mm]
\text{simplify} && &= \frac{-y_2 + y_1}{-x_2 + x_1} \\[2mm]
\text{rearrange terms} && &= \frac{y_1 - y_2}{x_1 - x_2}
\end{aligned}
$$

In calculating the slope, we do need to be consistent in the order in which the coordinates appear in the numerator and the denominator. If y_1 is the first term in the numerator, then x_1 must be the first term in the denominator.

EXAMPLE 1   Plot the two points $(-2, -6)$ and $(7, 12)$ and calculate the slope of the line passing through them.

SOLUTION   Treating $(-2, -6)$ as (x_1, y_1) and $(7, 12)$ as (x_2, y_2) (Figure 2.10), then

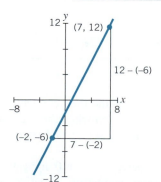

Figure 2.10

$$\text{slope} = \frac{y_2 - y_1}{x_2 - x_1} = \frac{12 - (-6)}{7 - (-2)} = \frac{18}{9} = 2$$

We could also have used -6 and -2 as the first terms in the numerator and denominator, respectively:

$$\text{slope} = \frac{y_1 - y_2}{x_1 - x_2} = \frac{-6 - 12}{-2 - 7} = \frac{-18}{-9} = 2$$

Either way we obtain the same answer.

E X A M P L E   2

The percentage of the U.S. population living in rural areas decreased from 84.7% in 1850 to 21.0% in 2000. Plot the data, then calculate and interpret the average rate of change in the rural population over time.

S O L U T I O N

If we treat year as the independent and percentage as the dependent variable, our given data can be represented by the points (1850, 84.7) and (2000, 21.0). See Figure 2.11.

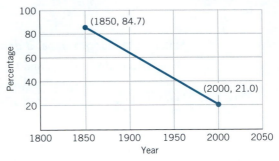

Figure 2.11 Percentage of the U.S. population living in rural areas.
Source: U.S. Bureau of the Census, *www.census.gov.*

$$\text{The average rate of change} = \frac{\text{change in the percentage of rural population}}{\text{change in time}}$$

$$= \frac{(21.0 - 84.7)\%}{(2000 - 1850)\text{ years}}$$

$$= \frac{-63.7\%}{150 \text{ years}}$$

$$\approx -0.42\% \text{ per year}$$

The sign of the average rate of change is negative since the percentage of people living in rural areas was decreasing. (The negative slope of the graph in Figure 2.11 confirms this.) The value tells us that, on average, the percentage living in rural areas decreased by 0.42 percentage points (or about one-half of 1%) each year between 1850 and 2000. The change per year may seem small, but in a century and a half the rural population went from being the overwhelming majority (84.7%) to about one-fifth (21%) of the population.

If the slope between any two points (the average rate of change of *y* with respect to *x*) is *positive*, then the graph of the relationship rises when read from left to right. This means that as *x* increases in value, *y* also increases in value.

If the slope is *negative*, the graph falls when read from left to right. As *x* increases, *y* decreases.

If the slope is *zero*, the graph is flat. As *x* increases, there is no change in *y*.

E X A M P L E   3

Given Table 2.8, a listing of civil disturbances over time, plot and then connect the points, and (without doing any calculations) indicate on the graph when the average rate of change between adjacent data points is positive (+), negative (−), and zero (0).

S O L U T I O N

The data are plotted in Figure 2.12. Each line segment is labeled +, −, or 0, indicating whether the average rate of change between adjacent points is positive, negative, or zero. The largest positive average rate of change, or steepest upward slope, seems to be

Civil Disturbances in U.S. Cities

Year	Period	Number of Disturbances
1968	Jan.–Mar.	6
	Apr.–June	46
	July–Sept.	25
	Oct.–Dec.	3
1969	Jan.–Mar.	5
	Apr.–June	27
	July–Sept.	19
	Oct.–Dec.	6
1970	Jan.–Mar.	26
	Apr.–June	24
	July–Sept.	20
	Oct.–Dec.	6
1971	Jan.–Mar.	12
	Apr.–June	21
	July–Sept.	5
	Oct.–Dec.	1
1972	Jan.–Mar.	3
	Apr.–June	8
	July–Sept.	5
	Oct.–Dec.	5

Table 2.8

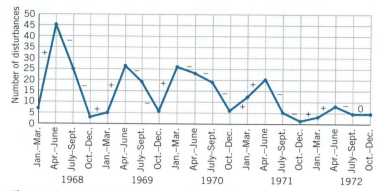

Figure 2.12 Civil disturbances: 1968–1972.
Source: D. S. Moore and G. P. McCabe, *Introduction to the Practice of Statistics.*
Copyright © 1989 by W.H. Freeman and Company. Used with permission.

between the January-to-March and April-to-June counts in 1968. The largest negative average rate of change, or steepest downward slope, appears later in the same year (1968) between the July-to-September and October-to-December counts.

Civil disturbances between 1968 and 1972 occurred in cycles: The largest numbers generally occurred in the spring months and the smallest in the winter months. The peaks decrease over time. What was happening in America that might correlate with the peaks? This was a tumultuous period in our history. Many previously silent factions of society were finding their voices. Recall that in April 1968 Martin Luther King was assassinated and in January 1973 the last American troops were withdrawn from Vietnam.

Algebra Aerobics 2.3

1. a. Plot each pair of points and then calculate the slope of the line that passes through them.

  i. (4, 1) and (8, 11)

  ii. (−3, 6) and (2, 6)

  iii. (0, −3) and (−5, −1)

  b. Recalculate the slopes in part (a), reversing the order of the points. Check that your answers are the same.

2. Specify the intervals on the graph in Figure 2.13 for which the average rate of change between adjacent data points is approximately zero.

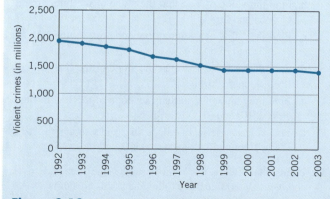

Figure 2.13 Violent crimes in the United States.

3. Specify the intervals on the graph in Figure 2.14 for which the average rate of change between adjacent data points appears positive, negative, or zero.

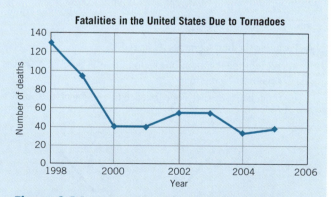

Figure 2.14 Deaths due to tornadoes: 1998–2005.

4. What is the missing y-coordinate that would produce a slope of 4, if a line were drawn through the points (3, −2) and (5, y)?

5. Find the slope of the line through the points (2, 9) and (2 + h, 9 + 2h).

6. Consider points $P_1 = (0, 0)$, $P_2 = (1, 1)$, $P_3 = (2, 4)$, and $P_4 = (3, 9)$.

   a. Verify that these four points lie on the graph of $y = x^2$.

   b. Find the slope of the line segments connecting P_1 and P_2, P_2 and P_3, and P_3 and P_4.

   c. What do these slopes suggest about the graph of the function within those intervals?

Exercises for Section 2.3

1. Find the slope of a straight line that goes through:

   a. $(-5, -6)$ and $(2, 3)$

   b. $(-5, 6)$ and $(2, -3)$

2. Find the slope of each line using the points where the graph intersects the x and y axes.

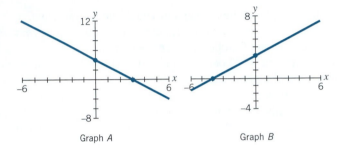

Graph A   Graph B

3. Find the slope of each line using the points where the line crosses the x- or y-axis.

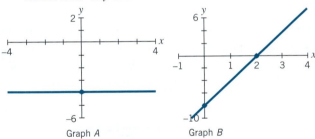

Graph A   Graph B

4. Examine the line segments A, B, and C.

   a. Which line segment has a slope that is positive? That is negative? That is zero?

   b. Calculate the exact slope for each line segment A, B, and C.

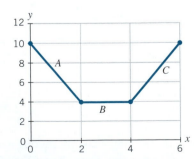

5. Given the graph at the top of the next column:

   a. Estimate the slope for each line segment A–F.

   b. Which line segment is the steepest?

   c. Which line segment has a slope of zero?

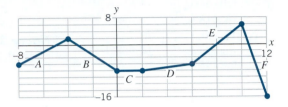

6. Plot each pair of points and calculate the slope of the line that passes through them.

   a. $(3, 5)$ and $(8, 15)$ d. $(-2, 6)$ and $(2, -6)$

   b. $(-1, 4)$ and $(7, 0)$ e. $(-4, -3)$ and $(2, -3)$

   c. $(5, 9)$ and $(-5, 9)$

7. The following problems represent calculations of the slopes of different lines. Solve for the variable in each equation.

   a. $\dfrac{150 - 75}{20 - 10} = m$ c. $\dfrac{182 - 150}{28 - x} = 4$

   b. $\dfrac{70 - y}{0 - 8} = 0.5$ d. $\dfrac{6 - 0}{x - 10} = 0.6$

8. Find the slope m of the line through the points $(0, b)$ and (x, y), then solve the equation for y.

9. Find the value of t if m is the slope of the line that passes through the given points.

   a. $(3, t)$ and $(-2, 1)$, $m = -4$

   b. $(5, 6)$ and $(t, 9)$, $m = \frac{2}{3}$

10. a. Find the value of x so that the slope of the line through $(x, 5)$ and $(4, 2)$ is $\frac{1}{3}$.

    b. Find the value of y so that the slope of the line through $(1, -3)$ and $(-4, y)$ is -2.

    c. Find the value of y so that the slope of the line through $(-2, 3)$ and $(5, y)$ is 0.

    d. Find the value of x so that the slope of the line through $(-2, 2)$ and $(x, 10)$ is 2.

    e. Find the value of y so that the slope of the line through $(-100, 10)$ and $(0, y)$ is $-\frac{1}{10}$.

    f. Find at least one set of values for x and y so that the slope of line through $(5, 8)$ and (x, y) is 0.

11. Points that lie on the same straight line are said to be *collinear*. Determine if the following points are collinear.

    a. $(2, 3)$, $(4, 7)$, and $(8, 15)$

    b. $(-3, 1)$, $(2, 4)$, and $(7, 8)$

12. a. Find the slope of the line through each of the following pairs of points.

      i. $(-1, 4)$ and $(-2, 4)$

      ii. $(7, -3)$ and $(-7, -3)$

      iii. $(-2, -6)$ and $(5, -6)$

  b. Summarize your findings.

13. Graph a line through each pair of points and then calculate its slope.

  a. The origin and $(6, -2)$ **b.** The origin and $(-4, 7)$

14. Find some possible values of the y-coordinates for the points $(-3, y_1)$ and $(6, y_2)$ such that the slope $m = 0$.

15. Calculate the slope of the line passing through each of the following pairs of points.

  a. $(0, \sqrt{2})$ and $(\sqrt{2}, 0)$

  b. $\left(0, -\frac{3}{2}\right)$ and $\left(-\frac{3}{2}, 0\right)$

  c. $(0, b)$ and $(b, 0)$

  d. What do these pairs of points and slopes all have in common?

16. Which pairs of points produce a line with a negative slope?

  a. $(-5, -5)$ and $(-3, -3)$ **d.** $(4, 3)$ and $(12, 0)$

  b. $(-2, 6)$ and $(-1, 4)$ **e.** $(0, 3)$ and $(4, -10)$

  c. $(3, 7)$ and $(-3, -7)$ **f.** $(4, 2)$ and $(6, 2)$

17. In the previous exercise, which pairs of points produce a line with a positive slope?

18. A study on numerous streams examined the effects of a warming climate. It found an increase in water temperature of about 0.7°C for every 1°C increase in air temperature.

  a. Find the rate of change in water temperature with respect to air temperature. What are the units?

  b. If the air temperature increased by 5°C, by how much would you expect the stream temperature to increase?

19. Handicapped Vietnam veterans successfully lobbied for improvements in the architectural standards for wheelchair access to public buildings.

  a. The old standard was a 1-foot rise for every 10 horizontal feet. What would the slope be for a ramp built under this standard?

  b. The new standard is a 1-foot rise for every 12 horizontal feet. What would the slope of a ramp be under this standard?

  c. If the front door is 3 feet above the ground, how long would the handicapped ramp be using the old standard? Using the new?

20. a. The accompanying graph gives information about the unemployment rate. Specify the intervals on the graph for which the slope of the line segment between adjacent data points appears positive. For which does it appear negative? For which zero?

  b. What might have caused the increase in the unemployment rate just after 2000?

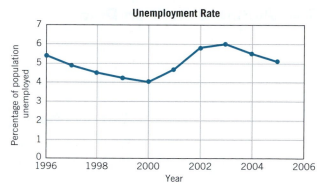

Source: U.S. Department of Labor, Department of Labor Statistics, *www.bls.gov*

21. Read the Anthology article "Slopes" on the course website and then describe two practical applications of slopes, one of which is from your own experience.

22. The following graph shows the relationship between education and health in the United States.

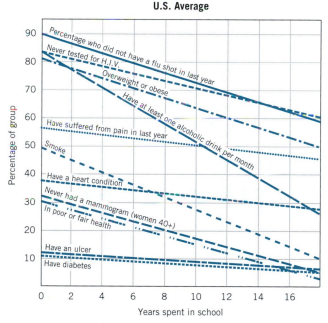

Source: "A Surprising Secret to a Long Life, According to Studies: Stay in School," *New York Times,* January 3, 2007.

  a. Estimate the percentages for those with 8 years of education and for those with 16 years of education (college graduates) who:

    i. Are overweight or obese

    ii. Have at least one alcoholic drink per month

    iii. Smoke

  b. For each category in part (a), calculate the average rate of change with respect to years of education. What do they all have in common?

  c. What does the graph suggest about the link between education and health? Could there be other factors at play?

2.4 *Putting a Slant on Data*

Whenever anyone summarizes a set of data, choices are being made. One choice may not be more "correct" than another. But these choices can convey, either accidentally or on purpose, very different impressions.

Slanting the Slope: Choosing Different End Points

Within the same data set, one choice of end points may paint a rosy picture, while another choice may portray a more pessimistic outcome.

EXAMPLE 1   The data in Table 2.9 and the scatter plot in Figure 2.15 show the number of people below the poverty level in the United States from 1960 to 2005. How could we use the information to make the case that the poverty level has decreased? Has increased?

People in Poverty in the United States

Year	Number of People in Poverty (in thousands)
1960	39,851
1970	25,420
1980	29,272
1990	33,585
2000	31,581
2005	36,950

Table 2.9
Source: U.S. Bureau of the Census, *www. census.gov.*

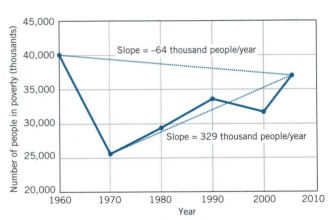

Figure 2.15 Number of people in poverty in the United States between 1960 and 2005.

SOLUTION   *Optimistic case:* To make an upbeat case that poverty numbers have decreased, we could choose as end points (1960, 39851) and (2005, 36950). Then

$$\text{average rate of change} = \frac{\text{change in no. of people in poverty (000s)}}{\text{change in years}}$$

$$= \frac{36{,}950 - 39{,}851}{2005 - 1960}$$

$$= \frac{-2901}{45}$$

$$\approx -64 \text{ thousand people/year}$$

So between 1960 and 2005 the number of impoverished individuals *decreased* on average by 64 thousand (or 64,000) each year. We can see this reflected in Figure 2.15 in the negative slope of the line connecting (1960, 39851) and (2005, 36950).

Pessimistic case: To make a depressing case that poverty numbers have risen, we could choose (1970, 25420) and (2005, 36950) as end points. Then

$$\text{average rate change of change} = \frac{\text{change in no. of people in poverty (000s)}}{\text{change in years}}$$

$$= \frac{36{,}950 - 25{,}420}{2005 - 1970}$$

$$= \frac{11{,}530}{35}$$

$$\approx 329 \text{ thousand people/year}$$

So between 1970 and 2005 the number of impoverished individuals *increased* on average by 329 thousand (or 329,000) per year. This is reflected in Figure 2.15 in the positive slope of the line connecting (1970, 25420) and (2005, 36950). Both average rates of change are correct, but they give very different impressions of the changing number of people living in poverty in America.

Slanting the Data with Words and Graphs

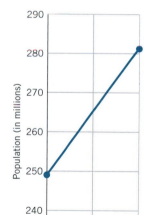

See "C4: Distortion by Clipping and Scaling" in *Rates of Change*.

If we wrap data in suggestive vocabulary and shape graphs to support a particular viewpoint, we can influence the interpretation of information. In Washington, D.C., this would be referred to as "putting a spin on the data."

Take a close look at the following three examples. Each contains exactly the same underlying facts: the same average rate of change calculation and a graph with a plot of the same two data points (1990, 248.7) and (2000, 281.4), representing the U.S. population (in millions) in 1990 and in 2000.

The U.S. population increased by only 3.27 million/year between 1990 and 2000

Stretching the scale of the horizontal axis relative to the vertical axis makes the slope of the line look almost flat and hence minimizes the impression of change (Figure 2.16).

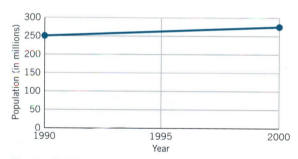

Figure 2.16 "Modest" growth in the U.S. population.

The U.S. population had an explosive growth of over 3.27 million/year between 1990 and 2000

Cropping the vertical axis (which now starts at 240 instead of 0) and stretching the scale of the vertical axis relative to the horizontal axis makes the slope of the line look steeper and strengthens the impression of dramatic change (Figure 2.17).

The U.S. population grew at a reasonable rate of 3.27 million/year during the 1990s

Visually, the steepness of the line in Figure 2.18 seems to lie roughly halfway between the previous two graphs. *In fact, the slope of 3.27 million/year is precisely the same for all three graphs.*

How could you decide upon a "fair" interpretation of the data? You might try to put the data in context by asking, How does the growth between 1990 and 2000 compare with other decades in the history of the United States? How does it compare with growth in other countries at the same time? Was this rate of growth easily accommodated, or did it strain national resources and overload the infrastructure?

Figure 2.17

"Explosive" growth in the U.S. population.

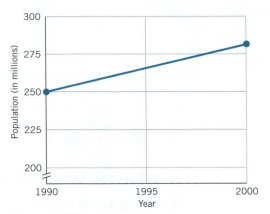

Figure 2.18 "Reasonable" growth in the U.S. population.

Exploration 2.1 gives you a chance to put your own "spin" on data.

A statistical claim is never completely free of bias. For every statistic that is quoted, others have been left out. This does not mean that you should discount all statistics. However, you will be best served by a thoughtful approach when interpreting the statistics to which you are exposed on a daily basis. By getting in the habit of asking questions and then coming to your own conclusions, you will develop good sense about the data you encounter.

Algebra Aerobics 2.4

1. Table 2.10 shows the *percentage* of the U.S. population in poverty between 1960 and 2005. In each case identify two end points you could use to make the case that poverty:

Year	% of Population in Poverty
1960	22.2
1970	12.6
1980	13.0
1990	13.5
2000	11.3
2005	12.6

Table 2.10

Source: U.S. Bureau of the Census, *www.census.gov.*

a. Has declined dramatically

b. Has remained stable

c. Has increased substantially

2. Assume you are the financial officer of a corporation whose stock earnings were $1.02 per share in 2005, and $1.12 per share in 2006, and $1.08 per share in 2007. How could you make a case for:

a. Dramatic growth?

b. Dramatic decline?

3. Sketch a graph and compose a few sentences to forcefully convey the views of the following persons.

a. You are an antiwar journalist reporting on American casualties during a war. In week 1 there were 17, in week 2 there were 29, and in week 3 there were 26.

b. You are the president's press secretary in charge of reporting war casualties [listed in part (a)] to the public.

4. The graph in Figure 2.19 appeared as part of an advertisement in the *Boston Globe* on June 27, 2003. Identify at least three strategies used to persuade you to buy gold.

Figure 2.19 Prices for an ounce of gold.

Exercises for Section 2.4

Graphing program optional for Exercise 10. Course software is required for Exercise 12.

1. Examine the three graphs *A*, *B*, and *C*. All report the same data on 30-year mortgage interest rates for the first six months in 2006. Create a different title for each graph exaggerating their differences.

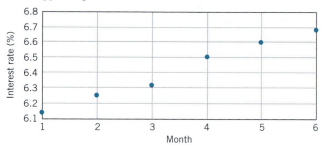

Graph *A*

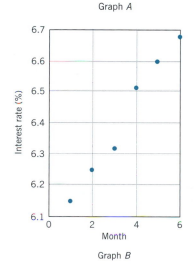

Graph *B*

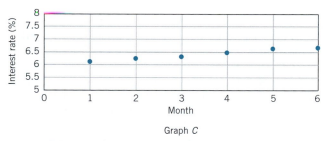

Graph *C*

2. Compare the accompanying graphs.
   a. Which line appears to have the steeper slope?
   b. Use the intercepts to calculate the slope. Which graph actually does have the steeper slope?

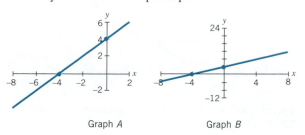

Graph *A*   Graph *B*

3. Compare the accompanying graphs.
   a. Which line appears to have the steeper slope?
   b. Use the intercepts to calculate the slope. Which graph has the more negative (and hence steeper) slope?

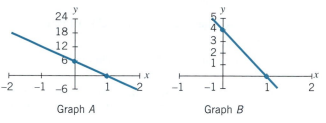

Graph *A*   Graph *B*

4. The following graphs show the same function graphed on different scales.
   a. In which graph does *Q* appear to be growing at a faster rate?
   b. In which graph does *Q* appear to be growing at a near zero rate?
   c. Explain why the graphs give different impressions.

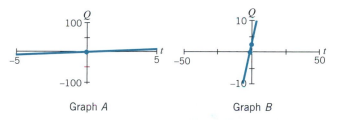

Graph *A*   Graph *B*

5. The following graph shows both federal spending on K–12 education under the Education and Secondary Education Act (ESEA) and National Assessment of Educational Progress (NAEP) reading scores (for age 9).

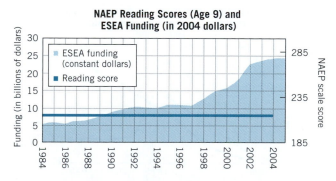

   a. Summarize the message that this chart conveys concerning NAEP reading scores and federal spending on ESEA.
   b. What strategy is used to make the reading scores seem lower?

6. Generate two graphs and on each draw a line through the points (0, 3) and (4, 6), choosing *x* and *y* scales such that:
   a. The first line appears to have a slope of almost zero.
   b. The second line appears to have a very large positive slope.

7. a. Generate a graph of a line through the points $(0, -2)$ and $(5, -2)$.

b. On a new grid, choose different scales so that the line through the same points appears to have a large positive slope.

c. What have you discovered?

8. What are three adjectives (like "explosive") that would imply rapid growth?

9. What are three adjectives (like "severe") that would imply rapid decline?

10. (Graphing program optional.) Examine the data given on women in the U.S. military forces.

Women in Uniform: Female Active-Duty Military Personnel

Year	Total	Army	Navy	Marine Corps	Air Force
1970	41,479	16,724	8,683	2,418	13,654
1980	171,418	69,338	34,980	6,706	60,394
1990	227,018	83,621	59,907	9,356	74,134
2000	204,498	72,021	53,920	9,742	68,815
2005	189,465	71,400	54,800	8,498	54,767

Source: U.S. Defense Department.

a. Make the case with graphs and numbers that women are a growing presence in the U.S. military.

b. Make the case with graphs and numbers that women are a declining presence in the U.S. military.

c. Write a paragraph that gives a balanced picture of the changing presence of women in the military using appropriate statistics to make your points. What additional data would be helpful?

11. The first case in the United States of what later came to be called AIDS appeared in June 1981. The accompanying graph shows the progress of AIDS cases in the United States as reported by the Centers for Disease Control and Prevention (CDC).

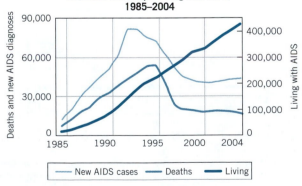

Estimated New AIDS Cases, Deaths Among Persons with AIDS and People Living with AIDS, 1985–2004

— New AIDS cases — Deaths — Living

a. Find something encouraging to say about these data by using numerical evidence, including estimated average rates of change.

b. Find numerical support for something discouraging to say about the data.

c. How might we explain the enormous increase in new AIDS cases reported from 1985 to a peak in 1992–1993, and the drop-off the following year?

12. (Course software required.) Open "L6: Changing Axis Scales" in *Linear Functions*. Generate a line in the upper left-hand box. The same line will appear graphed in the three other boxes but with the axes scaled differently. Describe how the axes are rescaled in order to create such different impressions.

13. The accompanying graphs show the same data on the median income of a household headed by a black person as a percentage of the median income of a household headed by a white person. Describe the impression each graph gives and how that was achieved. (Data from the U.S. Bureau of the Census.)

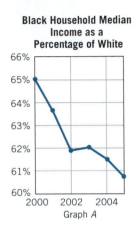

Black Household Median Income as a Percentage of White

Graph *A*

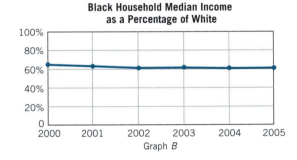

Black Household Median Income as a Percentage of White

Graph *B*

14. The accompanying graph shows the number of Nobel Prizes awarded in science for various countries between 1901 and 1974. It contains accurate information but gives the impression that the number of prize winners declined drastically in the 1970s, which was not the case. What flaw in the construction of the graph leads to this impression?

Nobel Prizes Awarded in Science, for Selected Countries, 1901–1974

Source: E. R. Tufte, *The Visual Display of Quantitative Information* (Cheshire, Conn.: Graphics Press, 1983).

2.5   *Linear Functions: When Rates of Change Are Constant*

In many of our examples so far, the average rate of change has varied depending on the choice of end points. Now we will examine the special case when the average rate of change remains constant.

What If the U.S. Population Had Grown at a Constant Rate? A Hypothetical Example

In Section 2.2 we calculated the average rate of change in the population between 1790 and 1800 as 0.14 million people/year. We saw that the average rate of change was different for different decades. What if the average rate of change had remained constant? What if in every decade after 1790 the U.S. population had continued to grow at the same rate of 0.14 million people/year? That would mean that starting with a population estimated at 3.9 million in 1790, the population would have grown by 0.14 million each year. The slopes of all the little line segments connecting adjacent population data points would be identical, namely 0.14 million people/year. The graph would be a straight line, indicating a constant average rate of change.

On the graph of actual U.S. population data, the slopes of the line segments connecting adjacent points are increasing, so the graph curves upward. Figure 2.20 compares the actual and hypothetical results.

Experiment with varying the average velocities and then setting them all constant in "C3: Average Velocity and Distance" in Rates of Change.

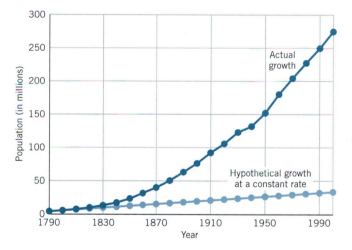

Figure 2.20 U.S. population: A hypothetical example.

Any function that has the same average rate of change on every interval has a graph that is a straight line. The function is called linear. This hypothetical example represents a linear function. When the average rate of change is constant, we can drop the word "average" and just say "rate of change."

> A linear function has a constant rate of change. Its graph is a straight line.

Real Examples of a Constant Rate of Change

EXAMPLE 1   According to the standardized growth and development charts used by many American pediatricians, the median weight for girls during their first six months of life increases at an almost constant rate. Starting at 7.0 pounds at birth, female median weight increases by approximately 1.5 pounds per month. If we assume that the median weight for females, W, is increasing at a constant rate of 1.5 pounds per month, then W is a linear function of age in months, A.

a. Generate a table that gives the median weight for females, W, for the first six months of life and create a graph of W as a function of A.

b. Find an equation for W as a function of A. What is an appropriate domain for this function?

c. Express the equation for part (b) using only units of measure.

SOLUTION
a. For female infants at birth ($A = 0$ months), the median weight is 7.0 lb ($W = 7.0$ lb). The rate of change, 1.5 pounds/month, means that as age increases by 1 month, weight increases by 1.5 pounds. See Table 2.11.

The dotted line in Figure 2.21 shows the trend in the data.

Median Weight for Girls

A, Age (months)	W, Weight (lb)
0	7.0
1	8.5
2	10.0
3	11.5
4	13.0
5	14.5
6	16.0

Table 2.11
Source: Data derived from the Ross Growth and Development Program, Ross Laboratories, Columbus, OH.

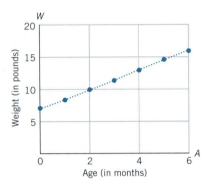
Figure 2.21 Median weight for girls.

b. To find a linear equation for W (median weight in pounds) as a function of A (age in months), we can study the table of values in Table 2.11.

$$W = \text{initial weight} + \text{weight gained}$$
$$= \text{initial weight} + \text{rate of growth} \cdot \text{number of months}$$
$$= 7.0 \text{ lb} + 1.5 \text{ lb/month} \cdot \text{number of months}$$

The equation would be

$$W = 7.0 + 1.5A$$

An appropriate domain for this function would be $0 \le A \le 6$.

c. Since our equation represents quantities in the real world, each term in the equation has units attached to it. The median weight W is in pounds (lb), rate of change is in lb/month, and age, A, is in months. If we display the equation

$$W = 7.0 + 1.5A$$

showing only the units, we have

$$\text{lb} = \text{lb} + \left(\frac{\text{lb}}{\text{month}}\right)\text{month}$$

The rules for canceling units are the same as the rules for canceling numbers in fractions. So,

$$\text{lb} = \text{lb} + \left(\frac{\text{lb}}{\cancel{\text{month}}}\right)\cancel{\text{month}}$$
$$\text{lb} = \text{lb} + \text{lb}$$

SOMETHING TO THINK ABOUT

If the median birth weight for baby boys is the same as for baby girls, but boys put on weight at a faster rate, which numbers in the model would change and which would stay the same? What would you expect to be different about the graph?

This equation makes sense in terms of the original problem since pounds (lb) added to pounds (lb) should give us pounds (lb).

E X A M P L E   2   You spend $1200 on a computer and for tax purposes choose to depreciate it (or assume it decreases in value) to $0 at a constant rate over a 5-year period.

    a. Calculate the rate of change of the assumed value of the equipment over 5 years. What are the units?

    b. Create a table and graph showing the value of the equipment over 5 years.

    c. Find an equation for the value of the computer as a function of time in years. Why is this a linear function?

    d. What is an appropriate domain for this function? What is the range?

S O L U T I O N   **a.** After 5 years, your computer is worth $0. If V is the value of your computer in dollars and t is the number of years you own the computer, then the rate of change of V from $t = 0$ to $t = 5$ is

$$\text{rate of change} = \frac{\text{change in value}}{\text{change in time}} = \frac{\Delta V}{\Delta t}$$

$$= -\frac{\$1200}{5 \text{ years}} = -\$240/\text{year}$$

Thus, the worth of your computer drops at a constant rate of $240 per year. The rate of change in V is negative because the worth of the computer decreases over time. The units for the rate of change are dollars per year.

Value of Computer Depreciated over 5 Years

Numbers of Years	Value of Computer ($)
0	1200
1	960
2	720
3	480
4	240
5	0

Table 2.12

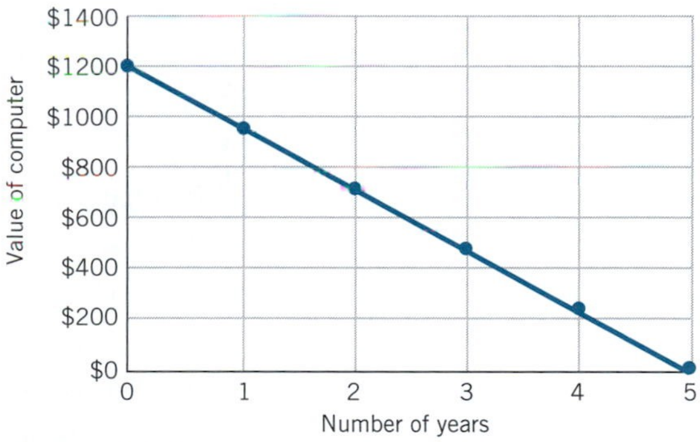

Figure 2.22 Value of computer over 5 years.

    b. Table 2.12 and Figure 2.22 show the depreciated value of the computer.

    c. To find a linear equation for V as a function of t, think about how we found the table of values.

$$\text{value of computer} = \text{initial value} + (\text{rate of decline}) \cdot (\text{number of years})$$
$$V = \$1200 + (-\$240/\text{year}) \cdot t$$
$$V = 1200 - 240t$$

This equation describes V as a function of t because for every value of t, there is one and only one value of V. It is a linear function because the rate of change is constant.

    d. The domain is $0 \le t \le 5$ and the range is $0 \le V \le 1200$.

The General Equation for a Linear Function

Explorations 2.2A and 2.2B (along with "L1: m & b Sliders" in Linear Functions) allow you to examine the effects of m and b on the graph of a linear function.

The equations in Examples 1 and 2 can be rewritten in terms of the output (dependent variable) and the input (independent variable).

$$\text{weight} = \text{initial weight} + \text{rate of growth} \cdot \text{number of months}$$

$$\text{value of computer} = \text{initial value} + \text{rate of decline} \cdot \text{number of years}$$

$$\underbrace{\text{output}}_{y} = \underbrace{\text{initial value}}_{b} + \underbrace{\text{rate of change}}_{m} \cdot \underbrace{\text{input}}_{x}$$

Thus, the general linear equation can be written in the form

$$y = b + mx$$

where we use the traditional mathematical choices of y for the output (dependent variable) and x for the input (independent variable). We let m stand for the rate of change; thus

$$m = \frac{\text{change in } y}{\text{change in } x} = \frac{\Delta y}{\Delta x}$$
$$= \text{slope of the graph of the line}$$

In our general equation, we let b stand for the initial value. Why is b called the initial value? The number b is the value of y when $x = 0$. If we let $x = 0$, then

given
$$y = b + mx$$
$$y = b + m \cdot 0$$
$$y = b$$

The point $(0, b)$ satisfies the equation and lies on the y-axis. The point $(0, b)$ is technically the vertical intercept. However, since the coordinate b tells us where the line crosses the y-axis, we often just refer to b as the *vertical* or *y-intercept* (see Figure 2.23).

Figure 2.23 Graph of $y = b + mx$.

A Linear Function

A function $y = f(x)$ is called *linear* if it can be represented by an equation of the form

$$y = b + mx$$

Its graph is a straight line where m is the *slope*, the rate of change of y with respect to x, so

$$m = \frac{\Delta y}{\Delta x}$$

b is the *vertical* or *y-intercept* and is the value of y when $x = 0$.

The equation $y = b + mx$ could, of course, be written in the equivalent form $y = mx + b$. In linear mathematical models, b is often the initial or starting value of the output, so it is useful to place it first in the equation.

EXAMPLE 3   For each of the following equations, identify the value of b and the value of m.

a. $y = -4 + 3.25x$ **c.** $y = -4x + 3.25$

b. $y = 3.25x - 4$ **d.** $y = 3.25 - 4x$

SOLUTION   **a.** $b = -4$ and $m = 3.25$

b. $b = -4$ and $m = 3.25$

c. $b = 3.25$ and $m = -4$

d. $b = 3.25$ and $m = -4$

EXAMPLE 4   In the following equations, L represents the legal fees (in dollars) charged by four different law firms and h represents the number of hours of legal advice.

$$L_1 = 500 + 200h \qquad\qquad L_3 = 800 + 350h$$
$$L_2 = 1000 + 150h \qquad\qquad L_4 = 500h$$

a. Which initial fee is the highest?

b. Which rate per hour is the highest?

c. If you need 5 hours of legal advice, which legal fee will be the highest?

SOLUTION   a. L_2 has the highest initial fee of $1000.

b. L_4 has the highest rate of $500 per hour.

c. Evaluate each equation for $h = 5$ hours.

$$
\begin{aligned}
L_1 &= 500 + 200h & L_3 &= 800 + 350h \\
&= 500 + 200(5) & &= 800 + 350(5) \\
&= \$1500 & &= \$2550
\end{aligned}
$$

$$
\begin{aligned}
L_2 &= 1000 + 150h & L_4 &= 500h \\
&= 1000 + 150(5) & &= 500(5) \\
&= \$1750 & &= \$2500
\end{aligned}
$$

For 5 hours of legal advice, L_3 has the highest legal fee.

Algebra Aerobics 2.5

1. From Figure 2.21, estimate the weight W of a baby girl who is 4.5 months old. Then use the equation $W = 7.0 + 1.5A$ to calculate the corresponding value for W. How close is your estimate?

2. From the same graph, estimate the age of a baby girl who weighs 11 pounds. Then use the equation to calculate the value for A.

3. a. If $C = 15P + 10$ describes the relationship between the number of persons (P) in a dining party and the total cost in dollars (C) of their meals, what is the unit of measure for 15? For 10?

   b. The equation $W = 7.0 + 1.5A$ (modeling weight as a function of age) expressed in units of measure only is

   $$\text{lb} = \text{lb} + \left(\frac{\text{lb}}{\text{month}}\right)\text{months}$$

   Express $C = 15P + 10$ from part (a) in units of measure only.

4. (True story.) A teenager travels to Alaska with his parents and wins $1200 in a rubber ducky race. (The race releases 5000 yellow rubber ducks marked with successive integers from one bridge over a river and collects them at the next bridge.) Upon returning home he opens up a "Rubber Ducky Savings Account," deposits his winnings, and continues to deposit $50 each month. If D = amount in the account and M = months since the creation of the account,

then $D = 1200 + 50M$ describes the amount in the account after M months.

   a. What are the units for 1200? For 50?

   b. Express the equation using only units of measure.

5. Assume $S = 0.8Y + 19$ describes the projected relationship between S, sales of a company (in millions of dollars), for Y years from today.

   a. What are the units of 0.8 and what does it represent?

   b. What are the units of 19 and what does it represent?

   c. What would be the projected company sales in three years?

   d. Express the equation using only units of measure.

6. Assume $C = 0.45N$ represents the total cost C (in dollars) of operating a car for N miles.

   a. What does 0.45 represent and what are its units?

   b. Find the total cost to operate a car that has been driven 25,000 miles.

   c. Express the equation using only units of measure.

7. The relationship between the balance B (in dollars) left on a mortgage loan and N, the number of monthly payments, is given by $B = 302{,}400 - 840N$.

   a. What is the monthly mortgage payment?

   b. What does 302,400 represent?

   c. What is the balance on the mortgage after 10 years? 20 years? 30 years? (*Hint:* Remember there are 12 months in a year.)

8. Identify the slope, m, and the vertical intercept, b, of the line with the given equation.

   a. $y = 5x + 3$ **e.** $f(x) = 7.0 - x$

   b. $y = 5 + 3x$ **f.** $h(x) = -11x + 10$

   c. $y = 5x$ **g.** $y = 1 - \frac{2}{3}x$

   d. $y = 3$ **h.** $2y + 6 = 10x$

9. If $f(x) = 50 - 25x$:

   a. Why does $f(x)$ describe a linear function?

   b. Evaluate $f(0)$ and $f(2)$.

   c. Use your answers in part (b) to verify that the slope is -25.

10. Identify the functions that are linear. For each linear function, identify the slope and the vertical intercept.

   a. $f(x) = 3x + 5$ **c.** $f(x) = 3x^2 + 2$

   b. $f(x) = x$ **d.** $f(x) = 4 - \frac{2}{3}x$

11. Write an equation for the line in the form $y = mx + b$ for the indicated values.

   a. $m = 3$ and $b = 4$

   b. $m = -1$ and passes through the origin

   c. $m = 0$ and $b = -3$

12. Write the equation of the graph of each line in Figure 2.24 in $y = mx + b$ form. Use the y-intercept and the slope indicated on each graph.

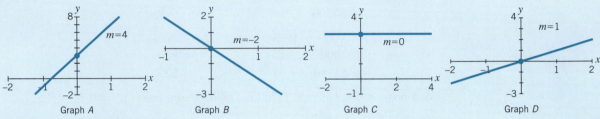

Figure 2.24 Four linear graphs.

Exercises for Section 2.5

You might wish to hone your algebraic mechanical skills with three programs in the course software; *Linear Functions:* "L1: Finding m & b", "L3: Finding a Line through 2 Points" and "L4: Finding 2 Points on a Line." They offer practice in predicting values for m and b, generating linear equations, and finding corresponding solutions.

1. Consider the equation $E = 5000 + 100n$.

   a. Find the value of E for $n = 0, 1, 20$.

   b. Express your answers to part (a) as points with coordinates (n, E).

2. Consider the equation $G = 12,000 + 800n$.

   a. Find the value of G for $n = 0, 1, 20$.

   b. Express your answers to part (a) as points with coordinates (n, G).

3. Determine if any of the following points satisfy one or both of the equations in Exercises 1 and 2.

   a. $(5000, 0)$ **b.** $(15, 24000)$ **c.** $(35, 40000)$

4. Suppose during a 5-year period the profit P (in billions of dollars) for a large corporation was given by $P = 7 + 2Y$, where Y represents the year.

   a. Fill in the chart.

Y	0	1	2	3	4
P					

   b. What are the units of P?

   c. What does the 2 in the equation represent, and what are its units?

   d. What was the initial profit?

5. Consider the equation $D = 3.40 + 0.11n$

   a. Find the values of D for $n = 0, 1, 2, 3, 4$.

   b. If D represents the average consumer debt, in thousands of dollars, over n years, what does 0.11 represent? What are its units?

   c. What does 3.40 represent? What are its units?

6. Suppose the equations $E = 5000 + 1000n$ and $G = 12,000 + 800n$ give the total cost of operating an electrical (E) versus a gas (G) heating/cooling system in a home for n years.

   a. Find the cost of heating a home using electricity for 10 years.

   b. Find the cost of heating a home using gas for 10 years.

   c. Find the initial (or installation) cost for each system.

   d. Determine how many years it will take before $40,000 has been spent in heating/cooling a home that uses:

      i. Electricity   **ii.** Gas

7. If the equation $E = 5000 + 1000n$ gives the total cost of heating/cooling a home after n years, rewrite the equation using only units of measure.

8. Over a 5-month period at Acadia National Park in Maine, the average night temperature increased on average 5 degrees Fahrenheit per month. If the initial temperature is 25 degrees, create a formula for the night temperature N for month t, where $0 \le t \le 4$.

9. A residential customer in the Midwest purchases gas from a utility company that charges according to the formula $C(g) = 11 + 10.50(g)$, where $C(g)$ is the cost, in dollars, for g thousand cubic feet of gas.

   a. Find $C(0)$, $C(5)$, and $C(10)$.

   b. What is the cost if the customer uses no gas?

   c. What is the rate per thousand cubic feet charged for using the gas?

   d. How much would it cost if the customer uses 96 thousand cubic feet of gas (the amount an average Midwest household consumes during the winter months)?

10. Create the formula for converting degrees centigrade, C, to degrees Fahrenheit, F, if for every increase of 5 degrees centigrade the Fahrenheit temperature increases by 9 degrees, with an initial point of $(C, F) = (0, 32)$.

11. Determine the vertical intercept and the rate of change for each of these formulas:

   a. $P = 4s$ **c.** $C = 2\pi r$

   b. $C = \pi d$ **d.** $C = \frac{5}{9}F - 17.78$

12. A hiker can walk 2 miles in 45 minutes.

   a. What is his average speed in miles per hour?

   b. What formula can be used to find the distance traveled, d, in t hours?

13. The cost $C(x)$ of producing x items is determined by adding fixed costs to the product of the cost per item and the number of items produced, x. Below are three possible cost functions $C(x)$, measured in dollars. Match each description [in parts (e) − (g)] with the most likely cost function and explain why.

   a. $C(x) = 125,000 + 42.50x$

   b. $C(x) = 400,000 + 0.30x$

   c. $C(x) = 250,000 + 800x$

   e. The cost of producing a computer

   f. The cost of producing a college algebra text

   g. The cost of producing a CD

14. A new \$25,000 car depreciates in value by \$5000 per year. Construct a linear function for the value V of the car after t years.

15. The state of Pennsylvania has a 6% sales tax on taxable items. (*Note:* Clothes, food, and certain pharmaceuticals are not taxed in Pennsylvania.)

   a. Create a formula for the total cost (in dollars) of an item $C(p)$ with a price tag p. (Be sure to include the sales tax.)

   b. Find $C(9.50)$, $C(115.25)$, and $C(1899)$. What are the units?

16. a. If $S(x) = 20,000 + 1000x$ describes the annual salary in dollars for a person who has worked for x years for the Acme Corporation, what is the unit of measure for 20,000? For 1000?

   b. Rewrite $S(x)$ as an equation using only units of measure.

   c. Evaluate $S(x)$ for x values of 0, 5, and 10 years.

   d. How many years will it take for a person to earn an annual salary of \$43,000?

17. The following represent linear equations written using only units of measure. In each case supply the missing unit.

   a. inches = inches + (inches/hour) · (?)

   b. miles = miles + (?) · (gallons)

   c. calories = calories + (?) · (grams of fat)

18. Identify the slopes and the vertical intercepts of the lines with the given equations.

   a. $y = 3 + 5x$ **d.** $Q = 35t - 10$

   b. $f(t) = -t$ **e.** $f(E) = 10,000 + 3000E$

   c. $y = 4$

19. For each of the following, find the slope and the vertical intercept, then sketch the graph. (*Hint*: Find two points on the line.)

   a. $y = 0.4x - 20$ **b.** $P = 4000 - 200C$

20. Construct an equation and sketch the graph of its line with the given slope, m, and vertical intercept, b. (*Hint*: Find two points on the line.)

   a. $m = 2$, $b = -3$ **c.** $m = 0$, $b = 50$

   b. $m = -\frac{3}{4}$, $b = 1$

In Exercises 21 to 23 find an equation, generate a small table of solutions, and sketch the graph of a line with the indicated attributes.

21. A line that has a vertical intercept of −2 and a slope of 3.

22. A line that crosses the vertical axis at 3.0 and has a rate of change of −2.5.

23. A line that has a vertical intercept of 1.5 and a slope of 0.

24. Estimate b (the y-intercept) and m (the slope) for each of the accompanying graphs. Then, for each graph, write the corresponding linear function.

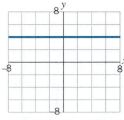

Graph A

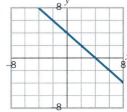

Graph C

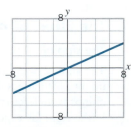

Graph B

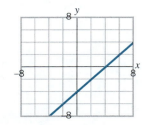

Graph D

2.6 *Visualizing Linear Functions*

The values for b and m in the general form of the linear equation, $y = b + mx$, tell us about the graph of the function. The value for b tells us where to anchor the line on the y-axis. The value for m tells us whether the line climbs or falls and how steep it is.

The Effect of b

In the equation $y = b + mx$, the value b is the vertical intercept, so it anchors the line at the point $(0, b)$ (see Figure 2.25).

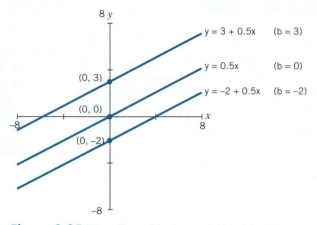

Figure 2.25 The effect of b, the vertical intercept.

EXAMPLE 1

Explain how the graph of $y = 4 + 5x$ differs from the graphs of the following functions:

a. $y = 8 + 5x$

b. $y = 2 + 5x$

c. $y = 5x + 4$

SOLUTION

Although all of the graphs are straight lines with a slope of 5, they each have a different vertical intercept. The graph of $y = 4 + 5x$ has a vertical intercept at 4.

a. The graph of $y = 8 + 5x$ intersects the y-axis at 8, four units above the graph of $y = 4 + 5x$.

b. The graph of $y = 2 + 5x$ intersects the y-axis at 2, two units below the graph of $y = 4 + 5x$.

c. Since $y = 5x + 4$ and $y = 4 + 5x$ are equivalent equations, they have the same graph.

The Effect of m

The sign of m

The sign of m in the equation $y = b + mx$ determines whether the line climbs (slopes up) or falls (slopes down) as we move left to right on the graph. If m is positive, the line climbs from left to right (as x increases, y increases). If m is negative, the line falls from left to right (as x increases, y decreases) (see Figure 2.26).

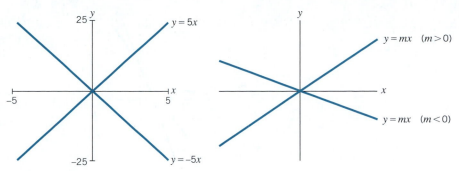

Figure 2.26 The effect of the sign of *m*.

EXAMPLE 2 Match the following functions to the lines in Figure 2.27.

$$f(x) = 3 - 2x$$
$$g(x) = 3 + 2x$$
$$h(x) = -5 - 2x$$
$$j(x) = -5 + 2x$$

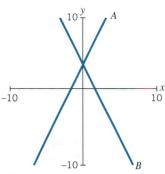

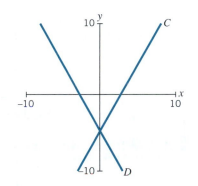

Figure 2.27 Matching graphs.

SOLUTION *A* is the graph of $g(x)$.

B is the graph of $f(x)$.

C is the graph of $j(x)$.

D is the graph of $h(x)$.

The magnitude of m

The magnitude (absolute value) of *m* determines the steepness of the line. Recall that the absolute value of *m* is the value of *m* stripped of its sign; for example, $|-3| = 3$. The greater the magnitude $|m|$, the steeper the line. This makes sense since *m* is the slope or the rate of change of *y* with respect to *x*. Notice how the steepness of each line in Figure 2.28 (next page) increases as the magnitude of *m* increases. For example, the slope ($m = 5$) of $y = 7 + 5x$ is steeper than the slope ($m = 3$) of $y = 7 + 3x$.

In Figure 2.29 we can see that the slope ($m = -5$) of $y = 7 - 5x$ is steeper than the slope ($m = -3$) of $y = 7 - 3x$ since $|-5| = 5 > |-3| = 3$. The lines $y = 7 - 5x$ and $y = 7 + 5x$ have the same steepness of 5 since $|-5| = |5| = 5$.

SOMETHING TO THINK ABOUT

Describe the graph of a line where m = 0.

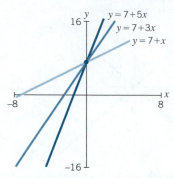

Figure 2.28 Graphs with positive values for *m*.

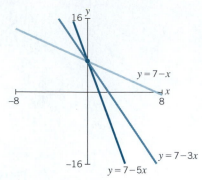

Figure 2.29 Graphs with negative values for *m*.

EXAMPLE 3

Pair each graph in Figure 2.30 with a matching equation.

$$f(x) = 9 - 0.4x$$
$$g(x) = -4 + x$$
$$h(x) = 9 - 0.2x$$
$$i(x) = -4 + 0.25x$$
$$j(x) = 9 + 0.125x$$
$$k(x) = 4 - 0.25x$$

SOLUTION

A is the graph of $j(x)$.
B is the graph of $h(x)$.
C is the graph of $f(x)$.
D is the graph of $k(x)$.
E is the graph of $g(x)$.
F is the graph of $i(x)$.

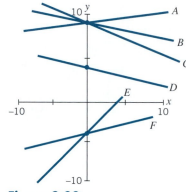

Figure 2.30 Graphs of multiple linear functions.

EXAMPLE 4

Without graphing the following functions, how can you tell which graph will have the steepest slope?

a. $f(x) = 5 - 2x$ **b.** $g(x) = 5 + 4x$ **c.** $h(x) = 3 - 6x$

SOLUTION

The graph of the function *h* will be steeper than the graphs of the functions *f* and *g* since the magnitude of *m* is greater for *h* than for *f* or *g*. The greater the magnitude of *m*, the steeper the graph of the line. For $f(x)$, the magnitude of the slope is $|-2| = 2$. For $g(x)$, the magnitude of the slope is $|4| = 4$. For $h(x)$, the magnitude of the slope is $|-6| = 6$.

EXAMPLE 5

Which of the graphs in Figure 2.31 has the steeper slope?

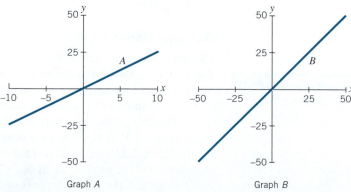

Graph *A*   Graph *B*

Figure 2.31 Comparing slopes.

SOLUTION Remember from Section 2.4 that the steepness of a linear graph is not related to its visual impression, but to the numerical magnitude of the slope. The scales of the horizontal axes are different for the two graphs in Figure 2.31, so the impression of relative steepness is deceiving. Line A passes through $(0, 0)$ and $(10, 25)$, so its slope is $(25 - 0)/(10 - 0) = 2.5$. Line B passes through $(0, 0)$ and $(50, 50)$ so its slope is $(50 - 0)/(50 - 0) = 1$. So line A has a steeper slope than line B.

The Graph of a Linear Equation

Given the general linear equation, $y = b + mx$, whose graph is a straight line:

The y-intercept, b, tells us where the line crosses the y-axis.

The slope, m, tells us how fast the line is climbing or falling. The larger the magnitude (or absolute value) of m, the steeper the graph.

If the slope, m, is positive, the line climbs from left to right. If m is negative, the line falls from left to right.

Algebra Aerobics 2.6

1. Place these numbers in order from smallest to largest.

$$|-12|, |-7|, |-3|, |-1|, 0, 4, 9$$

2. Without graphing the function, explain how the graph $y = 6x - 2$ differs from the graph of
   a. $y = 6x$ c. $y = -2 + 3x$
   b. $y = 2 + 6x$ d. $y = -2 - 2x$

3. Without graphing, order the graphs of the functions from least steep to steepest.
   a. $y = 100 - 2x$ c. $y = -3x - 5$
   b. $y = 1 - x$ d. $y = 3 - 5x$

4. On an x-y coordinate system, draw a line with a positive slope and label it $f(x)$.
   a. Draw a line $g(x)$ with the same slope but a y-intercept three units above $f(x)$.
   b. Draw a line $h(x)$ with the same slope but a y-intercept four units below $f(x)$.
   c. Draw a line $k(x)$ with the same steepness as $f(x)$ but with a negative slope.

5. Which function has the steepest slope? Create a table of values for each function and graph the function to show that this is true.
   a. $f(x) = 3x - 5$ b. $g(x) = 7 - 8x$

6. Create three functions with a y-intercept of 4 and three different negative slopes. Indicate which function has the steepest slope.

7. Match the four graphs in Figure 2.32 with the given functions.
   a. $f(x) = 2 + 3x$ c. $h(x) = \frac{1}{2}x - 2$
   b. $g(x) = -2 - 3x$ d. $k(x) = 2 - 3x$

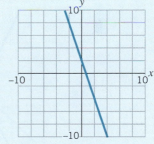

Graph A

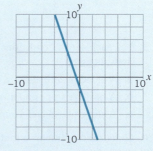

Graph C

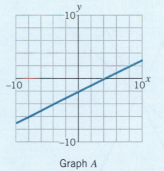

Graph B

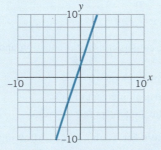

Graph D

Figure 2.32 Four graphs of linear functions.

Exercises for Section 2.6

1. Assuming m is the slope, identify the graph(s) where:

    a. $m = 3$ **b.** $|m| = 3$ **c.** $m = 0$ **d.** $m = -3$

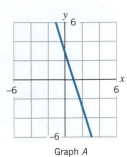

Graph A

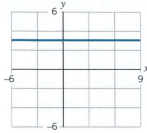

Graph C

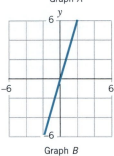

Graph B

Graph D

2. Which line(s), if any, have a slope m such that:

    a. $|m| = 2$?   **b.** $m = \frac{1}{2}$?

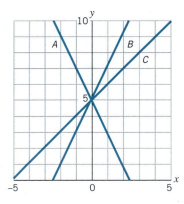

3. For each set of conditions, construct a linear equation and draw its graph.

    a. A slope of zero and a y-intercept of -3

    b. A positive slope and a vertical intercept of -3

    c. A slope of -3 and a vertical intercept that is positive

4. Construct a linear equation for each of the following conditions.

    a. A negative slope and a positive y-intercept

    b. A positive slope and a vertical intercept of -10.3

    c. A constant rate of change of \$1300/year

5. Match the graph with the correct equation.

    a. $y = x$ **b.** $y = 2x$ **c.** $y = \frac{x}{2}$ **d.** $y = 4x$

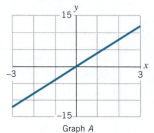

Graph A

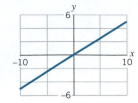

Graph C

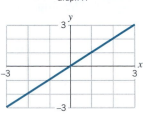

Graph B

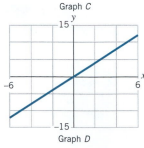

Graph D

6. Match the function with its graph. (*Note:* There is one graph that has no match.)

    a. $y = -4 + 3x$ **b.** $y = -3x + 4$ **c.** $y = 4 + 3x$

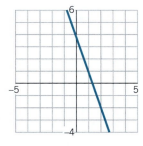

Graph A

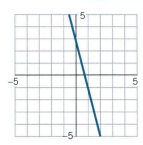

Graph C

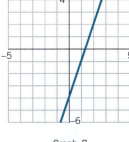

Graph B

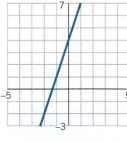

Graph D

7. In each part construct three different linear equations that all have:

    a. The same slope

    b. The same vertical intercept

8. Which equation has the steepest slope?

    a. $y = 2 - 7x$ **b.** $y = 2x + 7$ **c.** $y = -2 + 7x$

9. Given the function $Q(t) = 13 - 5t$, construct a related function whose graph:

    a. Lies five units above the graph of $Q(t)$

    b. Lies three units below the graph of $Q(t)$

    c. Has the same vertical intercept

    d. Has the same slope

    e. Has the same steepness, but the slope is positive

10. Given the equation $C(n) = 30 + 15n$, construct a related equation whose graph:

    a. Is steeper

    b. Is flatter (less steep)

    c. Has the same steepness, but the slope is negative

11. On the same axes, graph (and label with the correct equation) three lines that go through the point (0, 2) and have the following slopes:

    a. $m = \frac{1}{2}$ **b.** $m = 2$ **c.** $m = \frac{5}{6}$

12. On the same axes, graph (and label with the correct equation) three lines that go through the point (0, 2) and have the following slopes:

    a. $m = -\frac{1}{2}$ **b.** $m = -2$ **c.** $m = -\frac{5}{6}$

13. Given the following graphs of three straight lines:

    a. Which has the steepest slope?

    b. Which has the flattest slope?

    c. If the slope of the lines A, B, and C are m_1, m_2, and m_3, respectively, list them in increasing numerical order.

    d. In this example, why does the line with the steepest slope have the smallest numerical value?

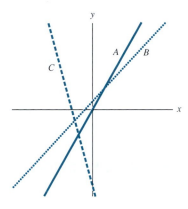

2.7 *Finding Graphs and Equations of Linear Functions*

Finding the Graph

EXAMPLE 1

Given the equation

A forestry study measured the diameter of the trunk of a red oak tree over 5 years. The scientists created a linear model $D = 1 + 0.13Y$, where D = diameter in inches and Y = number of years from the beginning of the study.

    a. What do the numbers 1 and 0.13 represent in this context?

    b. Sketch a graph of the function model.

SOLUTION

    a. The number 1 represents a starting diameter of 1 inch. The number 0.13 represents the annual growth rate of the oak's diameter (change in diameter/change in time), 0.13 inches per year.

    b. The linear equation $D = 1 + 0.13Y$ tells us that 1 is the vertical intercept, so the point (0, 1) lies on the graph. The graph represents solutions to the equation. So to find a second point, we can evaluate D for any other value of Y. If we set $Y = 1$, then $D = 1 + (0.13 \cdot 1) = 1 + 0.13 = 1.13$. So (1, 1.13) is another point on the line. Since two points determine a line, we can sketch our line through (0, 1) and (1, 1.13) (see Figure 2.33).

> The program "L4: Finding 2 Points on a Line" in *Linear Functions* will give you practice in finding solutions to equations.

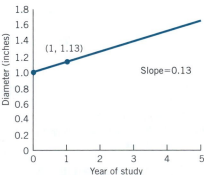

Figure 2.33 The diameter of a red oak over time.

EXAMPLE 2

Given a point off the *y*-axis and the slope

Given a point (2, 3) and a slope of $m = -5/4$, describe at least two ways you could find a second point to plot a line with these characteristics without constructing the equation.

SOLUTION Plot the point (2, 3).

a. If we write the slope as (−5)/4, then a change of 4 in x corresponds to a change of −5 in y. So starting at (2, 3), moving horizontally four units to the right (adding 4 to the x-coordinate), and then moving vertically five units down (subtracting 5 from the y-coordinate) gives us a second point on the line at (2 + 4, 3 − 5) = (6, −2). Now we can plot our second point (6, −2) and draw the line through it and our original point, (2, 3) (see Figure 2.34).

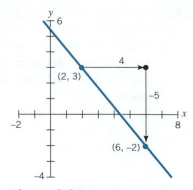

Figure 2.34 Graph of the line through (2, 3) with slope −5/4.

b. If we are modeling real data, we are more likely to convert the slope to decimal form. In this case $m = (-5)/4 = -1.25$. We can generate a new point on the line by starting at (2, 3) and moving 1 unit to the right and down (since m is negative) −1.25 units to get the point (2 + 1, 3 − 1.25) = (3, 1.75).

EXAMPLE 3 **Given a general description**

A recent study reporting on the number of smokers showed:

a. A linear increase in Georgia

b. A linear decrease in Utah

c. A nonlinear decrease in Hawaii

d. A nonlinear increase in Oklahoma

Generate four rough sketches that could represent these situations.

SOLUTION See Figure 2.35.

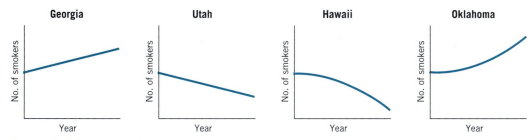

Figure 2.35 The change in the number of smokers over time in four states.

Finding the Equation

The program "L.3: Finding a Line Through 2 Points" in *Linear Functions* will give you practice in this skill.

To determine the equation of any particular linear function $y = b + mx$, we only need to find the specific values for m and b.

EXAMPLE 4 **Given two points**

Find the equation of the line through the two points (−3, −5) and (4, 9).

SOLUTION If we think of these points as being in the form (x, y), then the slope m of the line connecting them is

$$\frac{\text{change in } y}{\text{change in } x} = \frac{9 - (-5)}{4 - (-3)} = \frac{14}{7} = 2$$

So the equation is of the form $y = b + 2x$. To find b, the y-intercept, we can substitute either point into the equation. Substituting in $(4, 9)$ we get $9 = b + (2 \cdot 4) \Rightarrow 9 = b + 8 \Rightarrow b = 1$. So the equation is $y = 1 + 2x$.

EXAMPLE 5   **From a graph**

Find the equation of the linear function graphed in Figure 2.36.

SOLUTION   We can use any two points on the graph to calculate m, the slope. If, for example, we take $(-3, 0)$ and $(3, -2)$, then

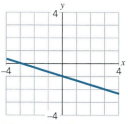

Figure 2.36 Graph of a linear function.

$$\text{slope} = \frac{\text{change in } y}{\text{change in } x} = \frac{-2 - 0}{3 - (-3)} = \frac{-2}{6} = \frac{-1}{3}$$

From the graph we can estimate the y-intercept as -1. So $b = -1$.

Hence the equation is $y = -1 - \frac{1}{3}x$.

EXAMPLE 6   **From a verbal description**

In 2006 AT&T's One Rate plan charged a monthly base fee of $3.95 plus $0.07 per minute for long-distance calls. Construct an equation to model a monthly phone bill.

SOLUTION   In making the transition from words to an equation, it's important to first identify which is the independent and which the dependent variable. We usually think of the phone bill, B, as a function of the number of minutes you talk, N. If you haven't used any phone minutes, then $N = 0$ and your bill $B = \$3.95$. So $3.95 is the vertical intercept. The number $0.07 is the rate of change of the phone bill with respect to number of minutes talked. The rate of change is constant, making the relationship linear. So the slope is $0.07/minute and the equation is

$$B = 3.95 + 0.07N$$

EXAMPLE 7   The top speed a snowplow can travel on dry pavement is 40 miles per hour, which decreases by 0.8 miles per hour with each inch of snow on the highway.

a. Construct an equation describing the relationship between snowplow speed and snow depth.

b. Determine a reasonable domain and then graph the function.

SOLUTION   a. If we think of the snow depth, D, as determining the snowplow speed, P, then we need an equation of the form $P = b + mD$. If there is no snow, then the snowplow can travel at its maximum speed of 40 mph; that is, when $D = 0$, then $P = 40$. So the point $(0, 40)$ lies on the line, making the vertical intercept $b = 40$. The rate of change, m, (change in snowplow speed)/(change in snow depth) $= -0.8$ mph per inch of snow. So the desired equation is

$$P = 40 - 0.8D$$

b. Consider the snowplow as only going forward (i.e., not backing up). Then the snowplow speed does not go below 0 mph. So if we let $P = 0$ and solve for D, we have

$$0 = 40 - 0.8D$$
$$0.8D = 40$$
$$D = 50$$

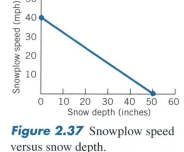

Figure 2.37 Snowplow speed versus snow depth.

So when the snow depth reaches 50 inches, the plow is no longer able to move. A reasonable domain then would be $0 \leq D \leq 50$. (See Figure 2.37.)

EXAMPLE 8

Pediatric growth charts suggest a linear relationship between age (in years) and median height (in inches) for children between 2 and 12 years. Two-year-olds have a median height of 35 inches, and 12-year-olds have a median height of 60 inches.

a. Generate the average rate of change of height with respect to age. (Be sure to include units.) Interpret your result in context.

b. Generate an equation to describe height as a function of age. What is an appropriate domain?

c. What would this model predict as the median height of 8-year-olds?

SOLUTION

a. Average rate of change $= \dfrac{\text{change in height}}{\text{change in age}} = \dfrac{60 - 35}{12 - 2} = \dfrac{25}{10} = 2.5$ inches/year

The chart suggests that, on average, children between the ages of 2 and 12 grow 2.5 inches each year.

b. If we think of height, H (in inches), depending on age, A (in years), then we want an equation of the form $H = b + m \cdot A$. From part (a) we know $m = 2.5$, so our equation is $H = b + 2.5A$. To find b, we can substitute the values for any known point into the equation. When $A = 2$, then $H = 35$. Substituting in, we get

$$H = b + 2.5A$$
$$35 = b + (2.5 \cdot 2)$$
$$35 = b + 5$$
$$30 = b$$

So the final form of our equation is

$$H = 30 + 2.5A$$

where the domain is $2 \le A \le 12$.

c. When $A = 8$ years, our model predicts that the median height is
$H = 30 + (2.5 \cdot 8) = 30 + 20 = 50$ inches.

EXAMPLE 9

From a table

a. Determine if the data in Table 2.13 represent a linear relationship between values of blood alcohol concentration and number of drinks consumed for a 160-pound person. (One drink is defined as 5 oz of wine, 1.25 oz of 80-proof liquor, or 12 oz of beer.)

D, Number of Drinks	A, Blood Alcohol Concentration
2	0.047
4	0.094
6	0.141
10	0.235

Table 2.13

Note that federal law requires states to have 0.08 as the legal BAC limit for driving drunk.

b. If the relationship is linear, determine the corresponding equation.

SOLUTION

a. We can generate a third column in the table that represents the average rate of change between consecutive points (see Table 2.14). Since the average rate of change of A with respect to D remains constant at 0.0235, these data represent a linear relationship.

D	A	Average Rate of Change
2	0.047	n.a.
4	0.094	$\frac{0.094 - 0.047}{4 - 2} = \frac{0.047}{2} = 0.0235$
6	0.141	$\frac{0.141 - 0.094}{6 - 4} = \frac{0.047}{2} = 0.0235$
10	0.235	$\frac{0.235 - 0.141}{10 - 6} = \frac{0.094}{4} = 0.0235$

Table 2.14

b. The rate of change is the slope, so the corresponding linear equation will be of the form

$$A = b + 0.0235D \qquad (1)$$

To find b, we can substitute any of the original paired values for D and A, for example, $(4, 0.094)$, into Equation (1) to get

$$0.094 = b + (0.0235 \cdot 4)$$
$$0.094 = b + 0.094$$
$$0.094 - 0.094 = b$$
$$0 = b$$

So the final equation is

$$A = 0 + 0.0235D$$
or just $$A = 0.0235D$$

So when D, the number of drinks, is 0, A, the blood alcohol concentration, is 0, which makes sense.

Algebra Aerobics 2.7

For Problems 1 and 2, find an equation, make an appropriate table, and sketch the graph of:

1. A line with slope 1.2 and vertical intercept -4.

2. A line with slope -400 and vertical intercept 300 (be sure to think about scales on both axes).

3. Write an equation for the line graphed in Figure 2.38.

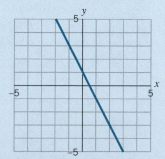

Figure 2.38 Graph of a linear function.

4. a. Find an equation to represent the current salary after x years of employment if the starting salary is $12,000 with annual increases of $3000.

b. Create a small table of values and sketch a graph.

5. a. Plot the data in Table 2.15.

Years of Education	Hourly Wage
8	$5.30
10	$8.50
13	$13.30

Table 2.15

b. Is the relationship between hourly wage and years of education linear? Why or why not?

c. If it is linear, construct a linear equation to model it.

6. Complete this statement regarding the graph of the line with equation $y = 6.2 + 3x$: Beginning with any point on the graph of the line, we could find another point by moving up ___ units for each unit that we move horizontally to the right.

7. Given the equation $y = 8 - 4x$, complete the following statements.

   a. Beginning with the vertical intercept, if we move one unit horizontally to the right, then we need to move down ___ units vertically to stay on the line and arrive at point (___, ___).

   b. Beginning with the vertical intercept, if we move one unit horizontally to the left, then we need to move up ___ units vertically to stay on the line and arrive at point (___, ___).

8. The relationship between the number of payments P made and the balance B (in dollars) of a $10,800 car loan can be represented by Table 2.16.

P, Number of Monthly Payments	B, Amount of Loan Balance ($)
0	10,800
1	10,500
2	10,200
3	9,900
4	9,600
5	9,300
6	9,000

Table 2.16

   a. Based on the table, develop a linear equation for the amount of the car loan balance B as a function of the number of monthly payments P.

   b. What is the monthly car payment?

   c. What is the balance after 24 payments?

   d. How many months are needed to produce a balance of zero?

9. The relationship between the number of tickets purchased for a movie and the revenue generated from that movie is indicated in Table 2.17.

Number of Tickets Purchased, T	Revenue, R ($)
0	0
10	75
20	150
30	225
40	300

Table 2.17

   a. Based on this table, construct a linear equation for the relationship between revenue, R, and the number of tickets purchased, T.

   b. What is the cost per ticket?

   c. Find the revenue generated by 120 ticket purchases.

10. For each graph in Figure 2.39, identify two points on each line, determine the slope, then write an equation for the line.

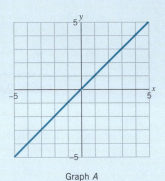

Graph A

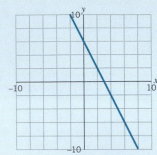

Graph C

Graph B

Figure 2.39

11. The revenue from one season of a college baseball team is the sum of allotted funds from alumni gifts and ticket sales from home games. In 2007 a college baseball team began the season with $15,000 in allotted funds. Tickets are sold at an average price of $12 each.

   a. Write an equation for the relationship between tickets sold at home games, T, and revenue, R (in dollars).

   b. Find the revenue for the team if 40,000 home game tickets are sold over the entire season.

12. Write the equations of three lines each with vertical intercept of 6.

13. Write the equations of three lines with slope -3.

14. Write an equation for the line that passes through:

   a. The point $(-2, 1)$ and has a slope of 4

   b. The point $(3, 5)$ and has a slope of $-2/3$

   c. The point $(-1, 3)$ and has a slope of -10

   d. The point $(1.2, 4.5)$ and has a slope of 2

15. Write an equation for the line:
  a. Through $(2, 5)$ and $(4, 11)$
  b. Through $(-3, 2)$ and $(6, 1)$
  c. Through $(4, -1)$ and $(-2, -7)$

16. Write each of the following in the form $y = mx + b$.
  a. $3x + 4y = -12$
  b. $7x - y = 5$
  c. $2x + 8y = 1$
  d. $x - 2y = 0$

  e. $y - 2 = 3(x + 1)$
  f. $y + 4 = -5(x - 2)$

17. Sketch the graph of the line:
  a. With slope 2 and vertical intercept 5
  b. With slope $-1/2$ and vertical intercept 6
  c. With slope $-3/4$ and passing through the point $(-4, 2)$
  d. With slope 3 and passing through the point $(-5, -6)$

Exercises for Section 2.7

Graphing program optional in Exercises 8, 10 and 23.

1. Write an equation for the line through $(-2, 3)$ that has slope:
  a. 5   **b.** $-\frac{3}{4}$ **c.** 0

2. Write an equation for the line through $(0, 50)$ that has slope:
  a. -20 **b.** 5.1   **c.** 0

3. Calculate the slope and write an equation for the linear function represented by each of the given tables.

a.

x	y
2	7.6
4	5.1

b.

A	W
5	12
7	16

4. Determine which of the following tables represents a linear function. If it is linear, write the equation for the linear function.

a.

x	y
0	3
1	8
2	13
3	18
4	23

c.

x	$g(x)$
0	0
1	1
2	4
3	9
4	16

e.

x	$h(x)$
20	20
40	-60
60	-140
80	-220
100	-300

b.

q	R
0	0.0
1	2.5
2	5.0
3	7.5
4	10.0

d.

t	r
10	5.00
20	2.50
30	1.67
40	1.25
50	1.00

f.

p	T
5	0.25
10	0.50
15	0.75
20	1.00
25	1.25

5. Plot each pair of points, then determine the equation of the line that goes through the points.
  a. $(2, 3), (4, 0)$ **c.** $(2, 0), (0, 2)$
  b. $(-2, 3), (2, 1)$ **d.** $(4, 2), (-5, 2)$

6. Find the equation for each of the lines A–C on the accompanying graph.

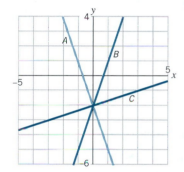

7. Put the following equations in $y = mx + b$ form, then identify the slope and the vertical intercept.
  a. $2x - 3y = 6$ **d.** $2y - 3x = 0$
  b. $3x + 2y = 6$ **e.** $6y - 9x = 0$
  c. $\frac{1}{3}x + \frac{1}{2}y = 6$ **f.** $\frac{1}{2}x - \frac{2}{3}y = -\frac{1}{6}$

8. (Graphing program optional.) Solve each equation for y in terms of x, then identify the slope and the y-intercept. Graph each line by hand. Verify your answers with a graphing utility if available.
  a. $-4y - x - 8 = 0$ **c.** $-4x - 3y = 9$
  b. $\frac{1}{2}x - \frac{1}{4}y = 3$ **d.** $6x - 5y = 15$

9. Complete the table for each of the linear functions, and then sketch a graph of each function. Make sure to choose an appropriate scale and label the axes.

a.

x	$f(x) = 0.10x + 10$
-100	
0	
100	

b.

x	$h(x) = 50x + 100$
-0.5	
0	
0.5	

10. (Graphing program optional.) The equation $K = 4F - 160$ models the relationship between F, the temperature in degrees Fahrenheit, and K, the number of chirps per minute for the snow tree cricket.

   a. Assuming F is the independent variable and K is the dependent variable, identify the slope and vertical intercept in the given equation.

   b. Identify the units for K, 4, F, and -160.

   c. What is a reasonable domain for this model?

   d. Generate a small table of points that satisfy the equation. Be sure to choose realistic values for F from the domain of your model.

   e. Calculate the slope directly from two data points. Is this value what you expected? Why?

   f. Graph the equation, indicating the domain.

11. Find an equation to represent the cost of attending college classes if application and registration fees are $150 and classes cost $120 per credit.

12. a. Write an equation that describes the total cost to produce x items if the startup cost is $200,000 and the production cost per item is $15.

   b. Why is the total average cost per item less if the item is produced in large quantities?

13. Your bank charges you a $2.50 monthly maintenance fee on your checking account and an additional $0.10 for each check you cash. Write an equation to describe your monthly checking account costs.

14. If a town starts with a population of 63,500 that declines by 700 people each year, construct an equation to model its population size over time. How long would it take for the population to drop to 53,000?

15. A teacher's union has negotiated a uniform salary increase for each year of service up to 20 years. If a teacher started at $26,000 and 4 years later had a salary of $32,000:

   a. What was the annual increase?

   b. What function would describe the teacher's salary over time?

   c. What would be the domain for the function?

16. Your favorite aunt put money in a savings account for you. The account earns simple interest; that is, it increases by a fixed amount each year. After 2 years your account has $8250 in it and after 5 years it has $9375.

   a. Construct an equation to model the amount of money in your account.

   b. How much did your aunt put in initially?

   c. How much will your account have after 10 years?

17. You read in the newspaper that the river is polluted with 285 parts per million (ppm) of a toxic substance, and local officials estimate they can reduce the pollution by 15 ppm each year.

   a. Derive an equation that represents the amount of pollution, P, as a function of time, t.

   b. The article states the river will not be safe for swimming until pollution is reduced to 40 ppm. If the cleanup proceeds as estimated, in how many years will it be safe to swim in the river?

18. The women's recommended weight formula from Harvard Pilgrim Healthcare says, "Give yourself 100 lb for the first 5 ft plus 5 lb for every inch over 5 ft tall."

   a. Find a mathematical model for this relationship. Be sure you clearly identify your variables.

   b. Specify a reasonable domain for the function and then graph the function.

   c. Use your model to calculate the recommended weight for a woman 5 feet, 4 inches tall; and for one 5 feet, 8 inches tall.

19. In 1977 a math professor bought her condominium in Cambridge, Massachusetts, for $70,000. The value of the condo has risen steadily so that in 2007 real estate agents tell her the condo is now worth $850,000.

   a. Find a formula to represent these facts about the value of the condo $V(t)$, as a function of time, t.

   b. If she retires in 2010, what does your formula predict her condo will be worth then?

20. The y-axis, the x-axis, the line $x = 6$, and the line $y = 12$ determine the four sides of a 6-by-12 rectangle in the first quadrant (where $x > 0$ and $y > 0$) of the xy plane. Imagine that this rectangle is a pool table. There are pockets at the four corners and at the points $(0, 6)$ and $(6, 6)$ in the middle of each of the longer sides. When a ball bounces off one of the sides of the table, it obeys the "pool rule": The slope of the path after the bounce is the negative of the slope before the bounce. (*Hint:* It helps to sketch the pool table on a piece of graph paper first.)

   a. Your pool ball is at $(3, 8)$. You hit it toward the y-axis, along the line with slope 2.

      i. Where does it hit the y-axis?

      ii. If the ball is hit hard enough, where does it hit the side of the table next? And after that? And after that?

      iii. Show that the ball ultimately returns to $(3, 8)$. Would it do this if the slope had been different from 2? What is special about the slope 2 for this table?

   b. A ball at $(3, 8)$ is hit toward the y-axis and bounces off it at $\left(0, \frac{16}{3}\right)$. Does it end up in one of the pockets? If so, what are the coordinates of that pocket?

   c. Your pool ball is at $(2, 9)$. You want to shoot it into the pocket at $(6, 0)$. Unfortunately, there is another ball at $(4, 4.5)$ that may be in the way.

      i. Can you shoot directly into the pocket at $(6, 0)$?

      ii. You want to get around the other ball by bouncing yours off the y-axis. If you hit the y-axis at $(0, 7)$, do you end up in the pocket? Where do you hit the line $x = 6$?

      iii. If bouncing off the y-axis at $(0, 7)$ didn't work, perhaps there is some point $(0, b)$ on the y-axis from which the ball would bounce into the pocket at $(6, 0)$. Try to find that point.

21. Find the equation of the line shown on the accompanying graph. Use this equation to create two new graphs, taking care to label the scales on your new axes. For one of your graphs, choose scales that make the line appear steeper than in the original graph. For your second graph, choose scales that make the line appear less steep than in the original graph.

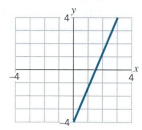

22. The exchange rate that a bank gave for euros in October 2006 was 0.79 euros for $1 U.S. They also charged a constant fee of $5 per transaction. The bank's exchange rate from euros to British pounds was 0.66 pounds for 1 euro, with a transaction fee of 4.1 euros.

    a. Write a general equation for how many euros you got when changing dollars. Use E for euros and D for dollars being exchanged. Draw a graph of E versus D.

    b. Would it have made any sense to exchange $10 for euros?

    c. Find a general expression for the percentage of the total euros converted from dollars that the bank kept for the transaction fee.

    d. Write a general equation for how many pounds you would get when changing euros. Use P for British pounds and E for the euros being exchanged. Draw a graph of P versus E.

23. (Graphing program optional.) Suppose that:

    • For 8 years of education, the mean annual earnings for women working full-time are approximately $19,190.

    • For 12 years of education, the mean annual earnings for women working full-time are approximately $31,190.

    • For 16 years of education, the mean annual earnings for women working full-time are approximately $43,190.

    a. Plot this information on a graph.

    b. What sort of relationship does this information suggest between earnings and education for women? Justify your answer.

    c. Generate an equation that could be used to model the data from the limited information given (letting E = years of education and M = mean earnings). Show your work.

24. a. Create a third column in Tables A and B, and insert values for the average rate of change. (The first entry will be "n.a.".)

t	d
0	400
1	370
2	340
3	310
4	280
5	250

Table A

t	d
0	1.2
1	2.1
2	3.2
3	4.1
4	5.2
5	6.1

Table B

b. In either table, is d a linear function of t? If so, construct a linear equation relating d and t for that table.

25. Adding minerals or organic compounds to water lowers its freezing point. Antifreeze for car radiators contains glycol (an organic compound) for this purpose. The accompanying table shows the effect of salinity (dissolved salts) on the freezing point of water. Salinity is measured in the number of grams of salts dissolved in 1000 grams of water. So our units for salinity are in parts per thousand, abbreviated ppt. Is the relationship between the freezing point and salinity linear? If so, construct an equation that models the relationship. If not, explain why.

Relationship between Salinity and Freezing Point

Salinity (ppt)	Freezing Point (°C)
0	0.00
5	−0.27
10	−0.54
15	−0.81
20	−1.08
25	−1.35

Source: Data adapted from P.R. Pinel, *Oceanography: An Introduction to the Planet Oceanus* (St. Paul, MN: West, 1992), p. 522.

26. The accompanying data show rounded average values for blood alcohol concentration (BAC) for people of different weights, according to how many drinks (5 oz wine, 1.25 oz 80-proof liquor, or 12 oz beer) they have consumed.

Blood Alcohol Concentration for Selected Weights

Number of Drinks	100 lb	140 lb	180 lb
2	0.075	0.054	0.042
4	0.150	0.107	0.083
6	0.225	0.161	0.125
8	0.300	0.214	0.167
10	0.375	0.268	0.208

a. Examine the data on BAC for a 100-pound person. Are the data linear? If so, find a formula to express blood alcohol concentration, A, as a function of the number of drinks, D, for a 100-pound person.

b. Examine the data on BAC for a 140-pound person. Are the data linear? If they're not precisely linear, what might be a reasonable estimate for the average rate of change of blood alcohol concentration, A, with respect to number of drinks, D? Find a formula to estimate blood alcohol concentration, A, as a function of number of drinks, D, for a 140-pound person. Can you make any general conclusions about BAC as a function of number of drinks for all of the weight categories?

c. Examine the data on BAC for people who consume two drinks. Are the data linear? If so, find a formula to express blood alcohol concentration, A, as a function of weight, W, for people who consume two drinks. Can you make any general conclusions about BAC as a function of weight for any particular number of drinks?

2.8 *Special Cases*

Direct Proportionality

The simplest relationship between two variables is when one variable is equal to a constant multiple of the other. For instance, in the previous example $A = 0.0235D$; blood alcohol concentration A equals a constant, 0.0235, times D, the number of drinks. We say that A *is directly proportional to D.*

How to recognize direct proportionality

Linear functions of the form

$$y = mx \qquad (m \neq 0)$$

describe a relationship where y is directly proportional to x. If two variables are directly proportional to each other, the graph will be a straight line that passes through the point $(0, 0)$, the origin. Figure 2.40 shows the graphs of two relationships in which y is directly proportional to x, namely $y = 2x$ and $y = -x$.

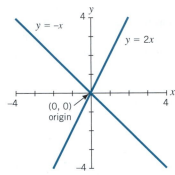

Figure 2.40 Graphs of two relationships in which y is directly proportional to x. Note that both graphs are lines that go through the origin.

SOMETHING TO THINK ABOUT

If y is directly proportional to x, is x directly proportional to y?

Direct Proportionality

In a linear equation of the form

$$y = mx \qquad (m \neq 0)$$

we say that

$$y \text{ is } directly \text{ } proportional \text{ } to \text{ } x.$$

Its graph will go through the origin.

EXAMPLE 1   Braille is a code, based on six-dot cells, that allows blind people to read. One page of regular print translates into 2.5 Braille pages. Construct a function describing this relationship. Does it represent direct proportionality? What happens if the number of print pages doubles? Triples?

SOLUTION   If $P =$ number of regular pages and $B =$ number of Braille pages, then $B = 2.5P$. So B is directly proportional to P. If the number of print pages doubles, the number of Braille pages doubles. If the number of print pages triples, the number of Braille pages will triple.

EXAMPLE 2 You are traveling to Canada and need to exchange American dollars for Canadian dollars. On that day the exchange rate is approximately 1 American dollar for 1.13 Canadian dollars.

a. Construct an equation for converting American to Canadian dollars. Does it represent direct proportionality?

b. Suppose the Exchange Bureau charges a $2 flat fee to change money. Alter your equation from part (a) to include the service fee. Does the new equation represent direct proportionality?

SOLUTION **a.** If we let A = American dollars and C = Canadian dollars, then the equation

$$C = 1.13A$$

describes the conversion from American (the input) to Canadian (the output). The amount of Canadian money you receive is directly proportional to the amount of American money you exchange.

b. If there is a $2 service fee, you would have to subtract $2 from the American money you have before converting to Canadian. The new equation is

$$C = 1.13(A - 2)$$

or equivalently,

$$C = 1.13A - 2.26$$

where 2.26 is the service fee in Canadian dollars. Then C is no longer directly proportional to A.

EXAMPLE 3 A prominent midwestern university decided to change its tuition cost. Previously there was a ceiling on tuition (which included fees). Currently the university uses what it calls a linear model, charging $106 per credit hour for in-state students and $369 per credit hour for out-of-state students.

a. Is it correct to call this pricing scheme a linear model for in-state students? For out-of-state students? Why?

b. Generate equations and graphs for the cost of tuition for both in-state and out-of-state students. If we limit costs to one semester during which the usual maximum credit hours is 15, what would be a reasonable domain?

c. In each case is the tuition directly proportional to the number of credit hours?

SOLUTION **a.** Yes, both relationships are linear since the rate of change is constant in each case: $106 per credit hour for in-state students and $369 per credit hour for out-of-state students.

b. Let N = number of credit hours, C_i = cost for an in-state student, and C_o = cost for an out-of-state student. In each case if the number of credit hours is zero ($N = 0$), then the cost would be zero ($C_i = 0 = C_o$). Hence both lines would pass through the origin $(0, 0)$, making the vertical intercept 0 for both equations. So the results would be of the form

$$C_i = 106N \quad \text{and} \quad C_o = 369N$$

which are graphed in Figure 2.41. A reasonable domain would be $0 \leq N \leq 15$.

c. In both cases the tuition is directly proportional to the number of credit hours. The graphs verify this since both are straight lines going through the origin.

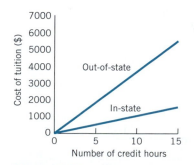

Figure 2.41 Tuition for in-state and out-of-state students at a midwestern university.

Algebra Aerobics 2.8a

1. Construct an equation and draw the graph of the line that passes through the origin and has the given slope.

   a. $m = -1$ **b.** $m = 0.5$

2. For each of Tables 2.18 and 2.19, determine whether x and y are directly proportional to each other. Represent each relationship with an equation.

   a.

x	y
-2	6
-1	3
0	0
1	-3
2	-6

   Table 2.18

   b.

x	y
0	5
1	8
2	11
3	14
4	17

   Table 2.19

3. In September 2006 the exchange rate was $1.00 U.S. to 0.79 euros, the common European currency.

   a. Find a linear function that converts U.S. dollars to euros.

   b. Find a linear function that converts U.S. dollars to euros with a service fee of $2.50.

   c. Which function represents a directly proportional relationship and why?

4. Suppose you go on a road trip, driving at a constant speed of 60 miles per hour. Create an equation relating distance d in miles and time traveled t in hours. Does it represent direct proportionality? What happens to d if the value for t doubles? If t triples?

5. The total cost C for football tickets is directly proportional to the number of tickets purchased, N. If two tickets cost $50, construct the formula relating C and N. What would the total cost of 10 tickets be?

6. Write a formula to describe each situation.

   a. y is directly proportional to x, and y is 4 when x is 12.

   b. d is directly proportional to t, and d is 300 when t is 50.

7. Write a formula to describe the following:

   a. The diameter, d, of a circle is directly proportional to the circumference, C.

   b. The amount of income tax paid, T, is directly proportional to income, I.

   c. The tip amount t, is directly proportional to the cost of the meal, c.

8. Assume that a is directly proportional to b. When $a = 10$, $b = 15$.

   a. Find a if b is 6.

   b. Find b if a is 4.

Horizontal and Vertical Lines

The slope, m, of any horizontal line is 0. So the general form for the equation of a horizontal line is

$$y = b + 0x$$

or just

$$y = b$$

For example, Table 2.20 and Figure 2.42 show points that satisfy the equation of the horizontal line $y = 1$. If we calculate the slope between any two points in the table—for example, $(-2, 1)$ and $(2, 1)$—we get

$$\text{slope} = \frac{1 - 1}{-2 - 2} = \frac{0}{-4} = 0$$

x	y
-4	1
-2	1
0	1
2	1
4	1

Table 2.20

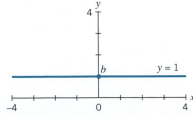

Figure 2.42 Graph of the horizontal line $y = 1$.

For a vertical line the slope, m, is undefined, so we can't use the standard $y = b + mx$ format. The graph of a vertical line (as in Figure 2.43) fails the vertical line test, so y is not a function of x. However, every point on a vertical line does have the same horizontal coordinate, which equals the coordinate of the horizontal intercept. Therefore, the general equation for a vertical line is of the form

$$x = c \quad \text{where } c \text{ is a constant (the horizontal intercept)}$$

For example, Table 2.21 and Figure 2.43 show points that satisfy the equation of the vertical line $x = 1$.

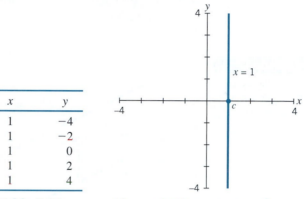

x	y
1	−4
1	−2
1	0
1	2
1	4

Table 2.21

Figure 2.43 Graph of the vertical line $x = 1$.

Note that if we tried to calculate the slope between two points, say $(1, -4)$ and $(1, 2)$, on the vertical line $x = 1$ we would get

$$\text{slope} = \frac{-4 - 2}{1 - 1} = \frac{-6}{0} \quad \text{which is undefined.}$$

> The general equation of a *horizontal line* is
>
> $$y = b$$
>
> where b is a constant (the vertical intercept) and the slope is 0.
>
> The general equation of a *vertical line* is
>
> $$x = c$$
>
> where c is a constant (the horizontal intercept) and the slope is undefined.

EXAMPLE 4 Find the equation for each line in Figure 2.44.

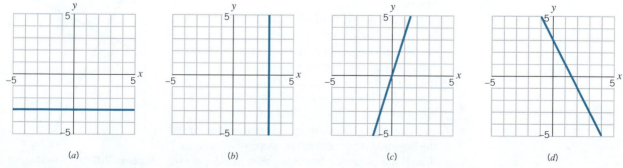

(a) (b) (c) (d)

Figure 2.44 Four linear graphs.

SOLUTION

a. $y = -3$, a horizontal line

b. $x = 3$, a vertical line

c. $y = 3x$, a direct proportion, slope $= 3$

d. $y = -2x + 3$, a line with slope -2 and y-intercept 3

Parallel and Perpendicular Lines

Parallel lines have the same slope. So if the two equations $y = b_1 + m_1x$ and $y = b_2 + m_2x$ describe two parallel lines, then $m_1 = m_2$. For example, the two lines $y = 2.0 - 0.5x$ and $y = -1.0 - 0.5x$ each have a slope of -0.5 and thus are parallel (see Figure 2.45).

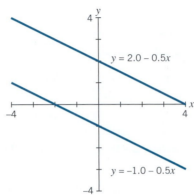

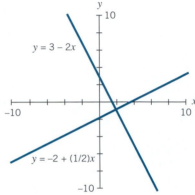

Figure 2.45 Two parallel lines have the same slope.

Figure 2.46 Two perpendicular lines have slopes that are negative reciprocals.

> **SOMETHING TO THINK ABOUT**
>
> Describe the equation for any line perpendicular to the horizontal line $y = b$.

Two lines are perpendicular if their slopes are negative reciprocals. If $y = b_1 + m_1x$ and $y = b_2 + m_2x$ describe two perpendicular lines, then $m_1 = -1/m_2$. For example, in Figure 2.46 the two lines $y = 3 - 2x$ and $y = -2 + \frac{1}{2}x$ have slopes of -2 and $\frac{1}{2}$, respectively. Since -2 is the negative reciprocal of $\frac{1}{2}$ (i.e., $-\frac{1}{\left(\frac{1}{2}\right)} = -1 \div \frac{1}{2} = -1 \cdot \frac{2}{1} = -2$), the two lines are perpendicular.

Why does this relationship hold for perpendicular lines?

Consider a line whose slope is given by v/h. Now imagine rotating the line 90 degrees clockwise to generate a second line perpendicular to the first (Figure 2.47). What would the slope of this new line be?

The positive vertical change, v, becomes a positive horizontal change. The positive horizontal change, h, becomes a negative vertical change. The slope of the original line is v/h, and the slope of the line rotated 90 degrees clockwise is $-h/v$. Note that $-h/v = -1/(v/h)$, which is the original slope inverted and multiplied by -1.

In general, the slope of a perpendicular line is the negative reciprocal of the slope of the original line. If the slope of a line is m_1, then the slope, m_2, of a line perpendicular to it is $-1/m_1$.

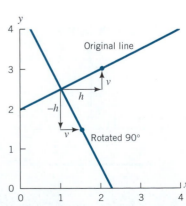

Figure 2.47 Perpendicular lines $m_2 = -1/m_1$.

This is true for any pair of perpendicular lines for which slopes exist. It does not work for horizontal and vertical lines since vertical lines have undefined slopes.

> Parallel lines have the same slope.
>
> Perpendicular lines have slopes that are negative reciprocals of each other.

EXAMPLE 5 Determine from the equations which pairs of lines are parallel, perpendicular, or neither.

a. $y = 2 + 7x$ and $y = 7x + 3$

b. $y = 6 - x$ and $y = 6 + x$

c. $y = 5 + 3x$ and $y = 5 - 3x$

d. $y = 3x + 13$ and $3y + x = 2$

SOLUTION a. The two lines are parallel since they share the same slope, 7.

b. The two lines are perpendicular since the negative reciprocal of -1 (the slope of first line) equals $-(1/(-1)) = -(-1) = 1$, the slope of the second line.

c. The lines are neither parallel nor perpendicular.

d. The lines are perpendicular. The slope of the first line is 3. If we solve the second equation for y, we get

$$3y + x = 2$$
$$3y = 2 - x$$
$$y = 2/3 - (1/3)x$$

So the slope of the second line is $-(1/3)$, the negative reciprocal of 3.

Algebra Aerobics 2.8b

1. In each case, find an equation for the horizontal line that passes through the given point.
   a. $(3, -5)$ b. $(5, -3)$ c. $(-3, 5)$

2. In each case, find an equation for the vertical line that passes through the given point.
   a. $(3, -5)$ b. $(5, -3)$ c. $(-3, 5)$

3. Construct the equation of the line that passes through the points.
   a. $(0, -7), (3, -7),$ and $(350, -7)$
   b. $(-4.3, 0), (-4.3, 8)$ and $(-4.3, -1000)$

4. Find the equation of the line that is parallel to $y = 4 - x$ and that passes through the origin.

5. Find the equation of the line that is parallel to $W = 360C + 2500$ and passes through the point where $C = 4$ and $W = 1000$.

6. Find the slope of a line perpendicular to each of the following.
   a. $y = 4 - 3x$ c. $y = 3.1x - 5.8$
   b. $y = x$ d. $y = -\frac{3}{5}x + 1$

7. a. Find an equation for the line that is perpendicular to $y = 2x - 4$ and passes through $(3, -5)$.
   b. Find the equations of two other lines that are perpendicular to $y = 2x - 4$ but do not pass through the point $(3, -5)$.
   c. How do the three lines from parts (a) and (b) that are perpendicular to $y = 2x - 4$ relate to each other?
   d. Check your answers by graphing the equations.

8. Find the slope of the line $Ax + By = C$ assuming that y is a function of x. (*Hint:* Solve the equation for y.)

9. Use the result of the previous exercise to determine the slope of each line described by the following linear equations (again assuming y is a function of x).
   a. $2x + 3y = 5$ d. $x = -5$
   b. $3x - 4y = 12$ e. $x - 3y = 5$
   c. $2x - y = 4$ f. $y = 4$

10. Solve the equation $2x + 3y = 5$ for y, identify the slope, then find an equation for the line that is parallel to the line $2x + 3y = 5$ and passes through the point $(0, 4)$.

11. Solve the equation $3x + 4y = -7$ for y, identify the slope, then find an equation for the line that is perpendicular to the line of $3x + 4y = -7$ and passes through $(0, 3)$.

12. Solve the equation $4x - y = 6$ for y, identify the slope, then find an equation for the line that is perpendicular to the line $4x - y = 6$ and passes through $(2, -3)$.

13. Determine whether each equation could represent a vertical line, a horizontal line, or neither.

    a. $x + 1.5 = 0$ **b.** $2x + 3y = 0$ **c.** $y - 5 = 0$

14. Write an equation for the line perpendicular to $2x + 3y = 6$:

    a. That has vertical intercept of 5

    b. That passes through the point $(-6, 1)$

15. Write an equation for the line parallel to $2x - y = 7$:

    a. That has a vertical intercept of 9

    b. That passes through $(4, 3)$

Piecewise Linear Functions

Some functions are not linear throughout but are made up of linear segments. They are called *piecewise linear functions*. For example, we could define a function $f(x)$ where:

$$f(x) = 2 + x \quad \text{for } x \le 1$$
$$f(x) = 3 \quad \text{for } x > 1 \quad \text{or, more compactly,} \quad f(x) = \begin{cases} 2 + x & \text{for } x \le 1 \\ 3 & \text{for } x > 1 \end{cases}$$

The graph of $f(x)$ in Figure 2.48 clearly shows the two distinct linear segments.

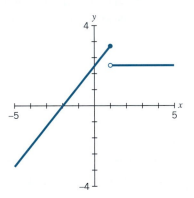

Figure 2.48 Graph of a piecewise linear function.

EXAMPLE 6 Consider the amount of gas in your car during a road trip. You start out with 20 gallons and drive for 3 hours, leaving you with 14 gallons in the tank. You stop for lunch for an hour and then drive for 4 more hours, leaving you with 6 gallons.

    a. Construct a piecewise linear function for the amount of gas in the tank as a function of time in hours.

    b. Graph the results.

SOLUTION **a.** Let t = time (in hours). For $0 \le t < 3$, the average rate of change in gasoline over time is $(14-20)$ gallons/$(3 \text{ hr}) = (-6 \text{ gallons})/(3 \text{ hr}) = -2$ gallons/hr; that is, you are consuming 2 gallons per hour. The initial amount of gas is 20 gallons, so $G(t)$, the amount of gas in the tank at time t, is given by

$$G(t) = 20 - 2t \quad \text{for } 0 \le t < 3$$

While you are at lunch for an hour, you consume no gasoline, so the amount of gas stays constant at 14 gallons. So

$$G(t) = 14 \quad \text{for } 3 \le t < 4$$

At the end of lunch, $t = 4$ and $G(t) = 14$. You continue to drive for 4 more hours, ending up with 6 gallons. You are still consuming 2 gallons per hour since (6–14) gallons/4 hr $= -2$ gallons/hr. So your equation will be of the form $G(t) = b - 2t$. Substituting in $t = 4$ and $G(t) = 14$, we have $14 = b - 2 \cdot 4 \Rightarrow$ $b = 22$. So

$$G(t) = 22 - 2t \quad \text{for } 4 \le t < 8$$

Writing $G(t)$ more compactly, we have

$$G(t) = \begin{cases} 20 - 2t & \text{if } 0 \le t < 3 \\ 14 & \text{if } 3 \le t < 4 \\ 22 - 2t & \text{if } 4 \le t < 8 \end{cases}$$

b. The graph of $G(t)$ is shown in Figure 2.49.

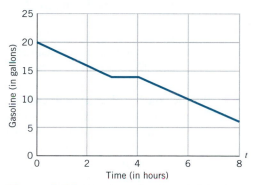

Figure 2.49 Number of gallons left in the car's tank.

The absolute value function

The absolute value of x, written as $|x|$, strips x of its sign. That means we consider only the magnitude of x, so $|x|$ is never negative.[1]

- If x is positive (or 0), then $|x| = x$.
- If x is negative, then $|x| = -x$.

For example, $|-5| = -(-5) = 5$.

We can construct the absolute value function using piecewise notation.

> **The absolute value function**
>
> If $f(x) = |x|$, then $f(x) = \begin{cases} x & \text{for } x \ge 0 \\ -x & \text{for } x < 0 \end{cases}$

EXAMPLE 7   If $f(x) = |x|$, then:

a. What is $f(6)$? $f(0)$? $f(-6)$?

b. Graph the function $f(x)$ for x between -6 and 6.

c. What is the slope of the line segment when $x > 0$? When $x < 0$?

---

[1] Graphing calculators and spreadsheet programs usually have an absolute value function. It is often named *abs* where *abs* $(x) = |x|$. So *abs*$(-3) = |3| = 3$.

SOLUTION   **a.** $f(6) = 6; f(0) = 0; f(-6) = |-6| = 6$

b. See Figure 2.50.

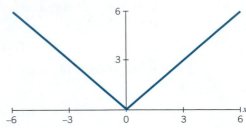

Figure 2.50 Graph of the absolute value function $f(x) = |x|$.

c. When $x > 0$, the slope is 1; when $x < 0$, the slope is -1.

Absolute value inequalities (in one variable) are frequently used in describing an allowable range above or below a certain amount.

EXAMPLE 8   **Range in values of poll results**

Poll figures are often given with a margin of error. For example, in January 2007 a CNN poll said that 63% of Americans felt that the economy was in good condition, with a sample error of ± 3 points. Construct an absolute value inequality that describes the range of percentages P that are possible within this poll. Restate this condition without using absolute values, and display it on a number line.

SOLUTION $|P - 63| \le 3$; that is, the poll takers are confident that the difference between the estimated percentage, 63%, and the actual percentage, P, is less than or equal to 3%.

Equivalently we could write $60 \le P \le 66$; that is, the actual percentage P is somewhere between 60% and 66% (see Figure 2.51).

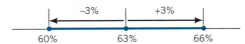

Figure 2.51 Range of error around 63% is ± 3 percentage points.

Absolute value functions are useful in describing distances between objects.

EXAMPLE 9   **Distance between a cell phone and cell tower**

You are a passenger in a car, talking on a cell phone. The car is traveling at 60 mph along a straight highway and the nearest cell phone tower is 6 miles away.

a. How long will it take you to reach the cell phone tower (assuming it is right by the road)?

b. Construct a linear function $D(t)$ that describes your distance *to* the cell tower (in miles ≥ 0) or *from* the cell tower (in miles < 0), where t is the number of hours traveled.

c. Graph your function using a reasonable domain for t.

d. What does $|D(t)|$ represent? Graph $|D(t)|$ on a separate grid and compare it with the graph of $D(t)$.

SOLUTION **a.** Traveling at 60 miles per hour is equivalent traveling at 1 mile per minute. So traveling 6 miles from the start will take you 6 minutes or 0.1 hr to reach the cell tower.

b. At the starting time $t = 0$ hours, the distance to the nearest cell phone tower is 6 miles, so $D(0) = 6$ miles. Thus, the vertical intercept is at $(0, 6)$. After $t = 0.1$ hr you are at the cell tower, so the distance between you and the cell tower is 0. Thus, $D(0.1) = 0$ miles and hence the horizontal intercept is $(0.1, 0)$. The slope of the line is $(0 - 6)/(0.1 - 0) = -60$. So the distance function is $D(t) = 6 - 60t$, where t is in hours and $D(t)$ is in miles (to or from the cell tower).

c. A reasonable domain might be $0 \leq t \leq 0.2$ hours. See Figure 2.52.

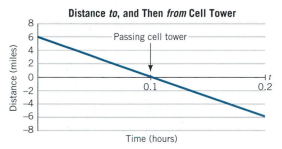

Figure 2.52 Graph of $D(t) = 6 - 60t$.

d. $|D(t)|$ describes the absolute value of the distance between you and the cell tower, indicating that the direction of travel no longer matters. Whether you are driving toward or away from the tower, the absolute value of the distance is always positive (or 0). For example, $|D(0.2)| = |6 - 60 \cdot 0.2| = |6 - 12| = |-6| = 6$ miles, which means that after 0.2 hours (or 12 minutes) you are 6 miles away from the tower. See Figure 2.53.

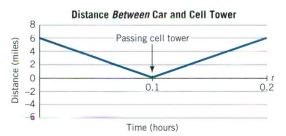

Figure 2.53 Graph of $|D(t)| = |5 - 25t|$.

Step functions

Some piecewise linear functions are called *step functions* because their graphs look like the steps of a staircase. Each "step" is part of a horizontal line.

EXAMPLE 10 **A step function: Minimum wages**
The federal government establishes a national minimum wage per hour. Table 2.22 shows the value of the minimum wage over the years 1990 to 2007.

Federal Minimum Wage for 1990–2007

Year New Minimum Wage Set	Minimum Wage (per hour)
1990	$3.80
1991	$4.25
1996	$4.75
1997	$5.15
2007	$5.85

Table 2.22

a. Construct a step function $M(x)$, where $M(x)$ is the minimum wage at year x. What is the domain?

b. What is $M(1992)$? $M(2006)$?

c. Graph the step function.

d. Why do you think there was a lot of discussion in 2006 about raising the minimum wage? (*Note:* Individual states can set a higher minimum for their workers.)

SOLUTION

a. $M(x) = \begin{cases} 3.80 & \text{for } 1990 \le x < 1991 \\ 4.25 & \text{for } 1991 \le x < 1996 \\ 4.75 & \text{for } 1996 \le x < 1997 \\ 5.15 & \text{for } 1997 \le x < 2007 \\ 5.85 & \text{for } 2007 \le x < 2008 \end{cases}$

The domain is $1990 \le x < 2008$.

b. $M(1992) = \$4.25/\text{hr}; M(2006) = \$5.15/\text{hr}$

c. See Figure 2.54.

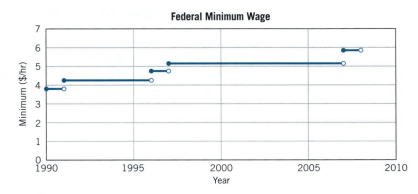

Federal Minimum Wage

Figure 2.54 Step function for federal minimum wage between 1990 and 2007.

d. By the end of 2006 the federal minimum wage had stayed the same (\$5.15/hr) for 9 years. Inflation always erodes the purchasing power of the dollar, so \$5.15 in 2006 bought a lot less than \$5.15 in 1997. So many felt an increase in the minimum wage was way overdue.

Algebra Aerobics 2.8c

1. Construct the graphs of the following piecewise linear functions. (Be sure to indicate whether each endpoint is included on or excluded from the graph.)

a. $f(x) = \begin{cases} 1 & \text{for } 0 < x \le 1 \\ 0 & \text{for } 1 < x \le 2 \\ -1 & \text{for } 2 < x \le 3 \end{cases}$

b. $g(x) = \begin{cases} x + 3 & \text{for } -4 \le x < 0 \\ 2 - x & \text{for } 0 \le x \le 4 \end{cases}$

2. Construct piecewise linear functions $Q(t)$ and $C(r)$ to describe the graphs in Figures 2.55 and 2.56.

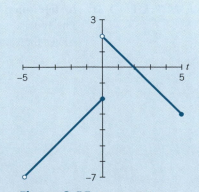

Figure 2.55 Graph of $Q(t)$.

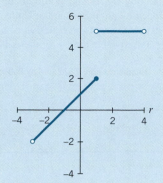

Figure 2.56 Graph of $C(r)$.

3. Evaluate the following:
   a. $|-2|$ **c.** $|3-5|$ **e.** $-|3| \cdot |-5|$
   b. $|6|$ **d.** $|3| - |5|$

4. Given the function $g(x) = |x - 3|$:
   a. What is $g(-3)$? $g(0)$? $g(3)$? $g(6)$?
   b. Sketch the graph of $g(x) = |x - 3|$ for $-6 \le x \le 6$.
   c. Compare the graph of $g(x) = |x - 3|$ with the graph of $f(x) = |x|$.
   d. Write $g(x)$ using piecewise linear notation.

5. Rewrite the following expressions without using an absolute value sign, and then describe in words the result.
   a. $|t - 5| \le 2$ **b.** $|Q - 75| < 6$

6. The optimal water temperature for trout is 55°F, but they can survive water temperatures that are 20° above or below that. Write an absolute value inequality that describes the temperature values, T, that lie within the trout survival temperature range. Then write an equivalent expression without the absolute value sign.

7. **a.** Sketch a piecewise linear graph of the *total* distance you would travel if you walked at a constant speed from home to a coffee shop, stopped for a cup, and then walked home at a faster pace.
   b. For the scenario in part (a), sketch a piecewise linear graph that shows the distance *between* you and the coffee shop over time.

8. The federal funds rate is the short-term interest rate charged by the Federal Reserve for overnight loans to other federal banks. This is one of the major tools the Federal Reserve Board uses to stimulate the economy (with a rate decrease) or control inflation (with a rate increase). Table 2.23 shows the week of each rate change during 2006.

Federal Funds Rates During 2006

Week in 2006 When Rate Was Changed	Federal Funds Rate (%)
Week 1	4.25
Week 5	4.50
Week 13	4.75
Week 19	5.00
Week 26	5.25

Table 2.23

   a. What was the federal funds rate in week 4? In week 52? What was the longest period in 2006 over which the federal funds rate remained the same?
   b. What appears to be the typical percentage increase used by the Board? Do the increases occur at regular intervals?
   c. During 2006 was the Federal Reserve Board more concerned about stimulating economic growth or curbing inflation?
   d. Construct a piecewise linear function to describe the federal funds rate during 2006.
   e. Sketch a graph of your function.

Exercises for Section 2.8

1. Using the general formula $y = mx$ that describes direct proportionality, find the value of m if:
   a. y is directly proportional to x and $y = 2$ when $x = 10$.
   b. y is directly proportional to x and $y = 0.1$ when $x = 0.2$.
   c. y is directly proportional to x and $y = 1$ when $x = \frac{1}{4}$.

2. For each part, construct an equation and then use it to solve the problem.
   a. Pressure P is directly proportional to temperature T, and P is 20 lb per square inch when T is 60 degrees Kelvin. What is the pressure when the temperature is 80 degrees Kelvin?
   b. Earnings E are directly proportional to the time T worked, and E is \$46 when T is 2 hours. How long has a person worked if she earned \$471.50?

   c. The number of centimeters of water depth W produced by melting snow is directly proportional to the number of centimeters of snow depth S. If W is 15.9 cm when S is 150 cm, then how many centimeters of water depth are produced by a 100-cm depth of melting snow?

3. In the accompanying table y is directly proportional to x.

Number of CDs purchased (x)	3	4	5
Cost of CDs (y)	42.69	56.92	

   a. Find the formula relating y and x, then determine the missing value in the table.
   b. Interpret the coefficient of x in this situation.

4. The electrical resistance R (in ohms) of a wire is directly proportional to its length l (in feet).

   a. If 250 feet of wire has a resistance of 1.2 ohms, find the resistance for 150 ft of wire.

   b. Interpret the coefficient of l in this context.

5. For each of the following linear functions, determine the independent and dependent variables and then construct an equation for each function.

   a. Sales tax is 6.5% of the purchase price.

   b. The height of a tree is directly proportional to the amount of sunlight it receives.

   c. The average salary for full-time employees of American domestic industries has been growing at an annual rate of $1300/year since 1985, when the average salary was $25,000.

6. On the scale of a map 1 inch represents a distance of 35 miles.

   a. What is the distance between two places that are 4.5 inches apart on the map?

   b. Construct an equation that converts inches on the map to miles in the real world.

7. Find a function that represents the relationship between distance, d, and time, t, of a moving object using the data in the accompanying table. Is d directly proportional to t? Which is a more likely choice for the object, a person jogging or a moving car?

t (hours)	d (miles)
0	0
1	5
2	10
3	15
4	20

8. Determine which (if any) of the following variables (w, y, or z) is directly proportional to x:

x	w	y	z
0	1	0.0	0
1	2	2.5	$-\frac{1}{3}$
2	5	5.0	$-\frac{2}{3}$
3	10	7.5	-1
4	17	10.0	$-\frac{4}{3}$

9. Find the slope of the line through the pair of points, then determine the equation.

   a. $(2, 3)$ and $(5, 3)$ c. $(-3, 8)$ and $(-3, 4)$

   b. $(-4, -7)$ and $(12, -7)$ d. $(2, -3)$ and $(2, -1)$

10. Describe the graphs of the following equations.

   a. $y = -2$ c. $x = \frac{2}{5}$ e. $y = 324$

   b. $x = -2$ d. $y = \frac{x}{4}$ f. $y = \frac{2}{3}$

11. The accompanying figure shows the quantity of books (in millions) shipped by publishers in the United States between 2001 and 2005. Construct the equation of a horizontal line that would be a reasonable model for these data.

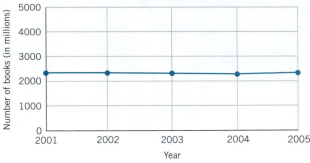

Number of Books Shipped by Publishers

Source: U.S. Bureau of the Census. *Statistical Abstract of the United States*, 2006.

12. An employee for an aeronautical corporation had a starting salary of $25,000/year. After working there for 10 years and not receiving any raises, he decides to seek employment elsewhere. Graph the employee's salary as a function of time for the time he was employed with this corporation. What is the domain? What is the range?

13. For each of the given points write equations for three lines that all pass through the point such that one of the three lines is horizontal, one is vertical, and one has slope 2.

   a. $(1, -4)$ b. $(2, 0)$

14. Consider the function $f(x) = 4$.

   a. What is $f(0)$? $f(30)$? $f(-12.6)$?

   b. Describe the graph of this function.

   c. Describe the slope of this function's graph.

15. A football player who weighs 175 pounds is instructed at the end of spring training that he has to put on 30 pounds before reporting for fall training.

   a. If fall training begins 3 months later, at what (monthly) rate must he gain weight?

   b. Suppose that he eats a lot and takes several nutritional supplements to gain weight, but due to his metabolism he still weighs 175 pounds throughout the summer and at the beginning of fall training. Sketch a graph of his weight versus time for those 3 months.

16. a. Write an equation for the line parallel to $y = 2 + 4x$ that passes through the point $(3, 7)$.

   b. Find an equation for the line perpendicular to $y = 2 + 4x$ that passes through the point $(3, 7)$.

17. a. Write an equation for the line parallel to $y = 4 - x$ that passes through the point $(3, 7)$.

   b. Find an equation for the line perpendicular to $y = 4 - x$ that passes through the point $(3, 7)$.

18. Construct the equation of a line that goes through the origin and is parallel to the graph of given equation.

   a. $y = 6$ b. $x = -3$ c. $y = -x + 3$

19. Construct the equation of a line that goes through the origin and is perpendicular to the given equation.

    a. $y = 6$ **b.** $x = -3$ **c.** $y = -x + 3$

20. Which lines are parallel to each other? Which lines are perpendicular to each other?

    a. $y = \frac{1}{3}x + 2$ **c.** $y = -2x + 10$ **e.** $2y + 4x = -12$

    b. $y = 3x - 4$ **d.** $y = -3x - 2$ **f.** $y - 3x = 7$

21. Because different scales may be used on the horizontal and vertical axes, it is often difficult to tell if two lines are perpendicular to each other. In parts (a) and (b), determine the equations of each pair of lines and show whether or not the paired lines are perpendicular to each other.

    a.   **b.**

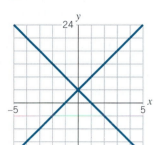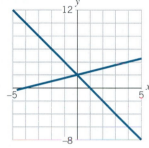

22. In each part construct the equations of two lines that:

    a. Are parallel to each other

    b. Intersect at the same point on the y-axis

    c. Both go through the origin

    d. Are perpendicular to each other

23. For each of the accompanying graphs you don't need to do any calculations or determine the actual equations. Rather, using just the graphs, determine if the slopes for each pair of lines are the same. Are the slopes both positive or both negative, or is one negative and one positive? Do the lines have the same y-intercept?

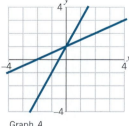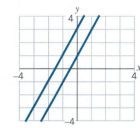

    Graph *A*   Graph *C*

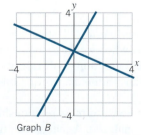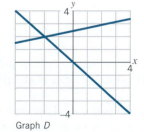

    Graph *B*   Graph *D*

24. Find the equation of the line in the form $y = mx + b$ for each of the following sets of conditions. Show your work.

    a. Slope is \$1400/year and line passes through the point (10 yr, \$12,000).

    b. Line is parallel to $2y - 7x = y + 4$ and passes through the point $(-1, 2)$.

    c. Equation is $1.48x - 2.00y + 4.36 = 0$.

    d. Line is horizontal and passes through $(1.0, 7.2)$.

    e. Line is vertical and passes through $(275, 1029)$.

    f. Line is perpendicular to $y = -2x + 7$ and passes through $(5, 2)$.

25. In the equation $Ax + By = C$:

    a. Solve for y so as to rewrite the equation in the form $y = mx + b$.

    b. Identify the slope.

    c. What is the slope of any line parallel to $Ax + By = C$?

    d. What is the slope of any line perpendicular to $Ax + By = C$?

26. Use the results of Exercise 25, parts (c) and (d), to find the slope of any line that is parallel and then one that is perpendicular to the given lines.

    a. $5x + 8y = 37$ **b.** $7x + 16y = -14$ **c.** $30x + 47y = 0$

27. Construct the graphs of the following piecewise linear functions. Be sure to indicate whether an endpoint is included in or excluded from the graph.

    a. $f(x) = \begin{cases} 2 & \text{for } -3 < x \le 0 \\ 1 & \text{for } 0 < x \le 3 \end{cases}$

    b. $g(x) = \begin{cases} x + 3 & \text{for } -4 < x \le 0 \\ 2 - x & \text{for } 0 < x \le 4 \end{cases}$

28. Construct piecewise linear functions for the following graphs.

    a. Graph of $f(x)$

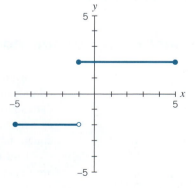

    b. Graph of $g(x)$

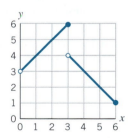

29. Given the following graph of $g(x)$:

   a. Construct a piecewise linear description of $g(x)$.

   b. Construct another description using absolute values.

   c. Describe the relationship between this graph and the graph of $f(x) = |x|$.

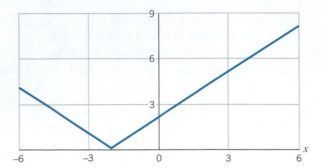

30. a. Normal human body temperature is often cited as 98.6°F. However, any temperature that is within 1°F more or less than that is still considered normal. Construct an absolute value inequality that describes normal body temperatures T that lie within that range. Then rewrite the expression without the absolute value sign.

   b. The speed limit is set at 65 mph on a highway, but police do not normally ticket you if you go less than 5 miles above or below that limit. Construct an absolute value inequality that describes the speeds S at which you can safely travel without getting a ticket. Rewrite the expression without using the absolute value sign.

31. Assume two individuals, A and B, are traveling by car and initially are 400 miles apart. They travel toward each other, pass and then continue on.

   a. If A is traveling at 60 miles per hour, and B is traveling at 40 mph, write two functions, $d_A(t)$ and $d_B(t)$, that describe the distance (in miles) that A and B each has traveled over time t (in hours).

   b. Now construct a function for the distance $D_{AB}(t)$ *between* A and B at time t (in hours). Graph the function for $0 \le t \le 8$ hours.

   c. At what time will A and B cross paths? At that point, how many miles has each traveled?

d. What is the distance between them one hour before they meet? An hour after they meet? Interpret both values in context. (*Hint:* If they are traveling toward each other, the distance between them is considered positive. Once they have met and are traveling away from each other, the distance between them is considered negative.)

e. Now construct an absolute value function that describes the (positive) distance between A and B at any point, and graph your result for $0 \le t \le 8$ hours.

32. The greatest integer function $y = [x]$ is defined as the greatest integer $\le x$ (i.e., it rounds x down to the nearest integer below it).

   a. What is [2] ? [2.5] ? [2.9999999] ?

   b. Sketch a graph of the greatest integer function for $0 \le x < 5$. Be sure to indicate whether each endpoint is included or excluded.

[*Note:* A bank employee embezzled hundreds of thousands of dollars by inserting software to round down transactions (such as generating interest on an account) to the nearest cent, and siphoning the round-off differences into his account. He was eventually caught.]

33. The following table shows U.S. first-class stamp prices (per ounce) over time.

Year	Price for First-Class Stamp
2001	34 cents
2002	37 cents
2006	39 cents

   a. Construct a step function describing stamp prices for 2001–2006.

   b. Graph the function. Be sure to specify whether each of the endpoints is included or excluded.

   c. In 2007 the price of a first-class stamp was raised to 41 cents. How would the function domain and the graph change?

34. Sketch a graph for each of the following situations.

   a. The amount in your savings account over a month, where you direct-deposit your paycheck each week and make one withdrawal during the month.

   b. The amount of money in an ATM machine over one day, where the ATM is stocked with dollars at the beginning of the day, and then ATM withdrawals of various sizes are made.

2.9 *Constructing Linear Models of Data*

According to Edward Tufte in *Data Analysis of Politics and Policy,* "Fitting lines to relationships is the major tool of data analysis." Of course, when we work with actual data searching for an underlying linear relationship, the data points will rarely fall exactly in a straight line. However, we can model the trends in the data with a linear equation.

Linear relationships are of particular importance not because most relationships are linear, but because straight lines are easily drawn and analyzed. A human can fit a straight line by eye to a scatter plot almost as well as a computer. This paramount convenience of linear equations as well as their relative ease of manipulation and interpretation means that lines are often used as first approximations to patterns in data.

Fitting a Line to Data: The Kalama Study

Children's heights were measured monthly over several years as a part of a study of nutrition in developing countries. Table 2.24 and Figure 2.57 show data collected on the mean heights of 161 children in Kalama, Egypt.

Mean Heights of Kalama Children

Age (months)	Height (cm)
18	76.1
19	77.0
20	78.1
21	78.2
22	78.8
23	79.7
24	79.9
25	81.1
26	81.2
27	81.8
28	82.8
29	83.5

Table 2.24

Source: D. S. Moore and G. P. McCabe, *Introduction to the Practice of Statistics.* Copyright © 1989 by W. H. Freeman and Company. Used with permission.

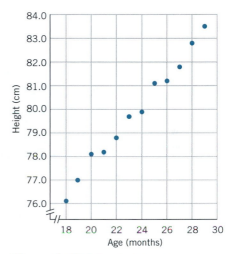

Figure 2.57 Mean heights of children in Kalama, Egypt.

Sketching a line through the data

Although the data points do not lie exactly on a straight line, the overall pattern seems clearly linear. Rather than generating a line through two of the data points, try eyeballing a line that approximates all the points. A ruler or a piece of black thread laid down through the dots will give you a pretty accurate fit.

In the Extended Exploration on education and earnings following this chapter, we will use technology to find a "best-fit" line. Figure 2.58 on the next page shows a line sketched that approximates the data points. This line does not necessarily pass through any of the original points.

Finding the slope

Estimating the coordinates of two points *on the line,* say (20, 77.5) and (26, 81.5), we can calculate the slope, *m*, or rate of change, as

$$m = \frac{(81.5 - 77.5) \text{ cm}}{(26 - 20) \text{ months}}$$
$$= \frac{4.0 \text{ cm}}{6 \text{ months}}$$
$$\approx 0.67 \text{ cm/month}$$

So our model predicts that for each additional month an "average" Kalama child will grow about 0.67 centimeter.

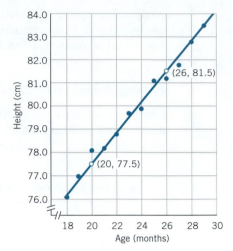

Figure 2.58 Estimated coordinates of two points on the line.

Constructing the equation

Since the slope of our linear model is 0.67 cm/month, then our equation is of the form

$$H = b + 0.67A \qquad (1)$$

where A = age in months and H = mean height in centimeters.

How can we find b, the vertical intercept? We have to resist the temptation to estimate b directly from the graph. As is frequently the case in social science graphs, both the horizontal and the vertical axes are cropped. Because the horizontal axis is cropped, we can't read the vertical intercept off the graph. We'll have to calculate it.

Since the line passes through (20, 77.5) we can

substitute (20, 77.5) in Equation (1)	$77.5 = b + (0.67)(20)$
simplify	$77.5 = b + 13.4$
solve for b	$b = 64.1$

Having found b, we complete the linear model:

$$H = 64.1 + 0.67A$$

where A = age in months and H = height in centimeters. It offers a compact summary of the data.

What is the domain of this model? In other words, for what inputs does our model apply? The data were collected on children age 18 to 29 months. We don't know its predictive value outside these ages, so

the domain consists of all values of A for which $18 \leq A \leq 29$

The vertical intercept may not be in the domain

Although the H-intercept is necessary to write the equation for the line, it lies outside of the domain.

Compare Figure 2.58 with Figure 2.59. They show graphs of the same equation, $H = 64.1 + 0.67A$. In Figure 2.58 both axes are cropped, while Figure 2.59 includes the origin (0, 0). In Figure 2.59 the vertical intercept is visible, and the shaded area between the dotted lines indicates the region that applies to our model. So a word of warning when reading graphs: Always look carefully to see if the axes have been cropped.

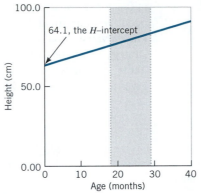

Figure 2.59 Graph of $H = 64.1 + 0.67A$ that includes the origin $(0, 0)$. Shaded area shows the region that models the Kalama data.

Reinitializing the Independent Variable

When we model real data, it often makes sense to reinitialize the independent variable in order to have a reasonable vertical intercept. This is especially true for time series, as shown in the following example, where the independent variable is the year.

EXAMPLE 1   **Time series**
How can we find an equation that models the trend in smoking in the United States?

SOLUTION   The American Lung Association website provided the data reproduced in Table 2.25 and graphed in Figure 2.60. Although in some states smoking has increased, the overall trend is a steady decline in the percentage of adult smokers in the United States between 1965 and 2005.

Year	Percentage of Adults Who Smoke
1965	41.9
1974	37.0
1979	33.3
1983	31.9
1985	29.9
1990	25.3
1995	24.6
2000	23.1
2005	20.9

Table 2.25

Figure 2.60 Percentage of adults who smoke.

The relationship appears reasonably linear. So the equation of a best-fit line could provide a fairly accurate description of the data. Since the horizontal axis is cropped, starting at the year 1965, the real vertical intercept would occur 1965 units to the left, at 0 A.D.! If you drew a big enough graph, you'd find that the vertical intercept would occur at approximately (0, 1220). This nonsensical extension of the model outside its known values would say that in 0 A.D., 1220% of the adult population smoked. A better strategy would be to define the independent variable as the number of years *since* 1965. Table 2.26 shows the reinitialized values for the independent variable, and Figure 2.61 gives a sketched-in best-fit line.

Year	No. of Years Since 1965	% of Adults Who Smoke
1965	0	41.9
1974	9	37.0
1979	14	33.3
1983	18	31.9
1985	20	29.9
1990	25	25.3
1995	30	24.6
2000	35	23.1
2005	40	20.9

Table 2.26

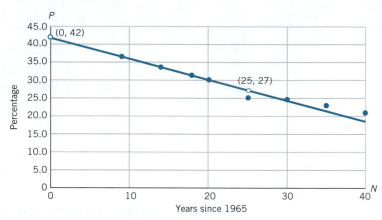

Figure 2.61 Percentage of adult smokers since 1965 with estimated best-fit line.

We can estimate the coordinates of two points, (0, 42) and (25, 27), on our best-fit line. Using them, we have

$$\text{vertical intercept} = 42 \quad \text{and} \quad \text{slope} = \frac{42 - 27}{0 - 25} = \frac{15}{-25} = -0.6$$

If we let N = the number of years since 1965 and P = percentage of adult smokers, then the equation for our best-fit line is

$$P = 42 - 0.6N$$

where the domain is $0 \leq N \leq 35$ (see Figure 2.61). This model says that starting in 1965, when about 42% of U.S. adults smoked, the percentage of the adult smokers has declined on average by 0.6 percentage points a year for 35 years.

What this model doesn't tell us is that (according to the U.S. Bureau of the Census) the total number of smokers during this time has remained fairly constant, at 50 million.

Interpolation and Extrapolation: Making Predictions

We can use this linear model on smokers to make predictions. We can *interpolate* or estimate new values between known ones. For example, in our smoking example we have no data for the year 1970. Using our equation we can estimate that in 1970 (when $N = 5$), $P = 42 - (0.6 \cdot 5) = 39\%$ of adults smoked. Like any other point on the best-fit line, this prediction is only an estimate and may, of course, be different from the actual percentage of smokers (see Figure 2.62).

We can also use our model to *extrapolate* or to predict beyond known values. For example, our model predicts that in 2010 (when $N = 45$), $P = 42 - (0.6 \cdot 45) = 15\%$ of adults will smoke. Extrapolation much beyond known values is risky. For 2035 (where $N = 70$) our model predicts that 0% will smoke, which seems unlikely. After 2035 our model would give the impossible answer that a negative percentage of adults will smoke.

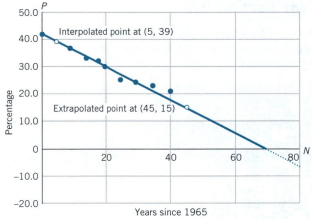

Figure 2.62 Interpolation and extrapolation of percentage of smokers.

Algebra Aerobics 2.9

1. Figure 2.63 shows the total number of U.S. college graduates (age 25 or older) between 1960 and 2005.

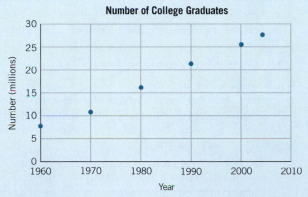

Figure 2.63 Total number of U.S. college graduates over time.
Source: U.S. National Center for Education Statistics, *Digest of Education Statistics,* annual.

a. Estimate from the scatter plot the number of college graduates in 1960 and in 2005.

b. Since the data look fairly linear, sketch a line that would model the growth in U.S. college graduates.

c. Estimate the coordinates of two points *on your line* and use them to calculate the slope.

d. If x = number of years *since* 1960 and y = total number (in millions) of U.S. college graduates, what would the coordinates of your two points in part (c) be in terms of x and y?

e. Construct a linear equation using the x and y values defined in part (d).

f. What does your model tell you about the number of college graduates in the United States?

2. Figure 2.64 shows the percentage of adults who (according to the U.S. Census Bureau) had access to the Internet either at home or at work between 1997 and 2003.

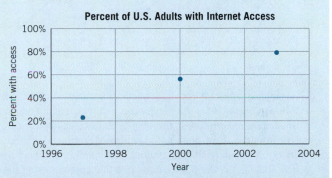

Figure 2.64 U.S. adults with Internet access.

a. Since the data appear roughly linear, sketch a best-fit line. (This line need not pass through any of the three data points.)

b. Reinitialize the years so that 1996 becomes year 0. Then identify the coordinates of any two points that lie on the line that you drew. Use these coordinates to find the slope of the line. What does this tell you about the percentage of adults with Internet access?

c. Give the approximate vertical intercept of the line that you drew (using the reinitialized value for the year).

d. Write an equation for your line.

e. Use your equation to estimate the number of adults with Internet access in 1998 and in 2002.

f. What would you expect to happen to the percentage after 2003? Do you think your linear model will be a good predictor after 2003?

Exercises for Section 2.9

"Extended Exploration: Looking for Links between Education and Earnings," which follows this chapter, has many additional exercises that involve finding best-fit lines using technology.

Graphing program recommended (or optional) for Exercises 3, 4, 13, and 15.

1. Match each equation with the appropriate table.

   a. $y = 3x + 2$ **b.** $y = \frac{1}{2}x + 2$ **c.** $y = 1.5x + 2$

   A.

x	y
0	2
2	3
4	4
6	5
8	6

   B.

x	y
0	2
2	5
4	8
6	11
8	14

   C.

x	y
0	2
2	8
4	14
6	20
8	26

2. Match each of the equations with the appropriate graph.

   a. $y = 10 - 2x$ **b.** $y = 10 - 5x$ **c.** $y = 10 - 0.5x$

   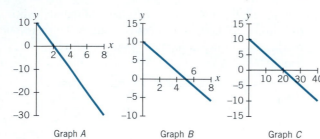

   Graph *A*   Graph *B*   Graph *C*

3. (Graphing program recommended.) Identify which of the following data tables represent exact and which approximate linear relationships. For the one(s) that are exactly linear, construct the corresponding equation(s). For the one(s) that are approximately linear, generate the equation of a best-fit line; that is, plot the points, draw in a line approximating the data, pick two points on the line (not necessarily from your data) to generate the slope, and then construct the equation.

   a.

x	-2	-1	0	1	2	3
y	-6.5	-5.0	-3.5	-2.0	-0.5	1.0

   b.

t	-1	0	1	2	3	4
Q	8.5	6.5	3.0	1.2	-1.5	-2

   c.

N	0	15	23	45	56	79
P	35	80	104	170	203	272

4. (Graphing program recommended.) Plot the data in each of the following data tables. Determine which data are exactly linear and which are approximately linear. For those that are approximately linear, sketch a line that looks like a best fit to the data. In each case generate the equation of a line that you think would best model the data.

   a. The number of cocaine-related emergency room episodes.

Year	1994	1996	1998	2000
Cocaine-related emergency room episodes (in thousands)	143	152	172	174

   Source: Centers for Disease Control and Prevention, National Center for Health-Related Statistics, 2005.

   b. The amount of tax owed on a purchase price.

Price	$2.00	$5.00	$10.00	$12.00
Tax	$0.12	$0.30	$0.60	$0.72

   c. The number of pounds in a given number of kilograms.

Kilograms	1	5	10	20
Pounds	2.2	11	22	44

   Use the linear equations found in parts (a), (b), and (c) to approximate the values for:

   d. The number of cocaine-related emergency room episodes predicted for 2004. How does your prediction compare with the actual value in 2004 of 131 thousand cases?

   e. The amount of tax owed on $7.79 and $25.75 purchase prices.

   f. The number of pounds in 15 kg and 150 kg.

5. Determine which data represent exactly linear and which approximately linear relationships. For the approximately linear data, sketch a line that looks like a best fit to the data. In each case generate the equation of a line that you think would best model the data.

   a. The number of solar energy units consumed (in quadrillions of British thermal units, called Btus)

Year	1998	1999	2000	2001
Solar consumption (in quadrillions of Btus)	0.07	0.07	0.07	0.07

   Source: U.S. Bureau of the Census, *Statistical Abstract.*

   b. Gross farm output (in billions of dollars)

   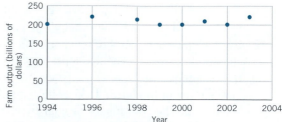

   Source: U.S. Census Bureau, *Statistical Abstract.*

6. In 1995 the United States consumed 464 million gallons of wine and in 2004, about 667 million gallons.

   a. Assuming the growth was linear, create a function that could model the trend in wine consumption.

b. Estimate the amount of wine consumed in 2005. The actual amount was about 703 million gallons worth of wine. How accurate was your approximation?

(*Source*: U.S. Department of Agriculture, Economic Research Service, 2007.)

7. The percentage of medical degrees awarded to women in the United States between 1970 and 2002 is shown in the accompanying graph.

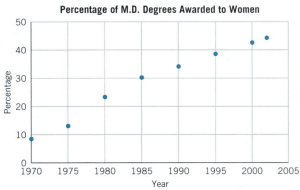

Percentage of M.D. Degrees Awarded to Women

Source: U.S. National Center for Education Statistics, "Digest of Education Statistics," annual; *Statistical Abstract of the United States.*

The data show that since 1970 the percentage of female doctors has been rising.

a. Sketch a line that best represents the data points. Use your line to estimate the rate of change of the percentage of M.D. degrees awarded to women.

b. If you extrapolate your line, estimate when 100% of doctors' degrees will be awarded to women.

c. It seems extremely unlikely that 100% of medical degrees will ever be granted to women. Comment on what is likely to happen to the rate of growth of women's degrees in medicine; sketch a likely graph for the continuation of the data into this century.

8. The accompanying graph shows the mortality rates (in deaths per 1000) for male and female infants in the United States from 1980 to 2000.

U.S. Infant Mortality Rates

Source: Centers for Disease Control and Prevention, *www.cdc.gov.*

a. Sketch a line through the graph of the data that best represents female infant mortality rates. Does the line seem to be a reasonable model for the data? What is the approximate slope of the line through these points? Show your work.

b. Sketch a similar line through the male mortality rates. Is it a reasonable approximation? Estimate its slope. Show your work.

c. List at least two important conclusions from the graph.

9. The accompanying scatter plot shows the relationship between literacy rate (the percentage of the population who can read and write) and infant mortality rate (infant deaths per 1000 live births) for 91 countries. The raw data are contained in the Excel or graph link file NATIONS and are described at the end of the Excel file. (You might wish to identify the outlier, the country with about a 20% literacy rate and a low infant mortality rate of about 40 per 1000 live births.) Construct a linear model. Show all your work and clearly identify the variables and units. Interpret your results.

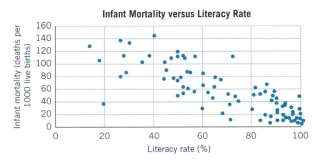

Infant Mortality versus Literacy Rate

10. The accompanying graph shows data for the men's Olympic 16-pound shot put.

Men's Olympic 16-lb Shot Put Winning Results

Source: 2006 World Almanac and Book of Facts.
Note: There were no Olympics in 1940 and 1944 due to World War II.

a. The shot put results are roughly linear between 1920 and 1972. Sketch a best-fit line for those years. Estimate the coordinates of two points on the line to calculate the slope. Interpret the slope in this context.

b. What is happening to the winning shot put results after 1972? Estimate the slope of the best-fit line for the years after 1972.

c. Letting x = years since 1920, construct a piecewise linear function $S(x)$ that describes the winning Olympic shot put results (in feet thrown) between 1920 and 2004.

11. The accompanying chart shows the percentage of U.S. households that have at least one cell phone.

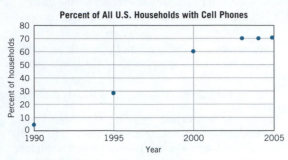

Percent of All U.S. Households with Cell Phones

Source: 2006 World Almanac and Book of Facts Forrester Reports.

a. A linear model seems reasonable between 1990 and 2003. Sketch a best-fit line for those dates, then pick two points on your line and calculate the slope. Interpret the slope in this context.

b. What appears seems to be happening after 2003? What might be the explanation?

c. Letting x = years since 1990, construct a piecewise linear function P(x) that would describe the percentage of households with cell phones between 1990 and 2005.

12. The accompanying graph shows the relationship between the age of a woman when she has her first child and her lifetime risk of getting breast cancer relative to a childless woman.

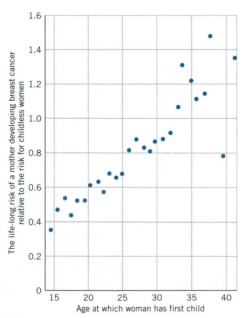

Source: J. Cairns, *Cancer: Science and Society* (San Francisco: W. H. Freeman, 1978), p. 49.

a. If a woman has her first child at age 18, approximately what is her risk of developing cancer relative to a woman who has never borne a child?

b. At roughly what age are the chances the same that a woman will develop breast cancer whether or not she has a child?

c. If a first-time mother is beyond the age you specified in part (b), is she more or less likely to develop breast cancer than a childless woman?

d. Sketch a line that looks like a best fit to the data, estimate the coordinates of two points on the line, and use them to calculate the slope.

e. Interpret the slope in this context.

f. Construct a linear model for these data, identifying your independent and dependent variables.

13. (Graphing program recommended.) The given data show that health care is becoming more expensive and is taking a bigger share of the U.S. gross domestic product (GDP). The GDP is the market value of all goods and services that have been bought for final use.

Year	1960	1970	1980	1990	2000	2003
U.S. health care costs as a percentage of GDP	5.1	7.1	8.9	12.2	13.3	15.3
Cost per person, $	141	341	1052	2691	4675	5671

Source: U.S. Health Care Financing Administration and Centers for Medicare and Medicaid Services.

a. Graph health care costs as a percentage of GDP versus year, with time on the horizontal axis. Measure time in years since 1960. Draw a straight line by eye that appears to be the closest fit to the data. Figure out the slope of your line and create a function H(t) for health care's percentage of the GDP as a function of t, years since 1960.

b. What does your formula predict for health care as a percentage of GDP for the year 2010?

c. Why do you think the health care cost per person has gone up so much more dramatically than the health care percentage of the GDP?

14. a. From the accompanying chart showing sport utility vehicle (SUV) sales, estimate what the rate of increase in sales has been from 1990 to 2004. (*Hint:* Convert the chart into an equivalent scatter plot.)

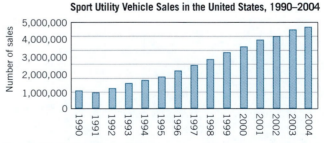

Sport Utility Vehicle Sales in the United States, 1990–2004

Source: Ward's Communications.

b. Estimate a linear formula to represent sales in years since 1990.

c. Do you think the popularity of SUVs will continue to grow at the same rate? Why or why not?

15. (Graphing program optional.) The Gas Guzzler Tax is imposed on manufacturers on the sale of new-model cars (*not* minivans, sport utility vehicles, or pickup trucks) whose fuel economy fails to meet certain statutory regulations, to discourage the production of fuel-inefficient vehicles. The tax is collected by the IRS and paid by the manufacturer. The table shows the amount of tax that the manufacturer must pay for a vehicle's miles per gallon fuel efficiency.

Gas Guzzler Tax

MPG	Tax per Car
12.5	$6400
13.5	$5400
14.5	$4500
15.5	$3700
16.5	$3700
17.5	$2600
18.5	$2100
19.5	$1700
20.5	$1300
21.5	$1000
22.5	$0

Source: http://www.epa.gov/otaq/ cert/factshts/fefact 0.1.pdf.

a. Plot the data, verify that they are roughly linear, and add a line of best fit.

b. Choose two points on the line, find the slope, and then form a linear equation with *x* as the fuel efficiency in mpg and *y* as the tax in dollars.

c. What is the rate of change of the amount of tax imposed on fuel-inefficient vehicles? Interpret the units.

16. A veterinarian's office displayed the following table comparing dog age (in dog years) to human age (in human years). The chart shows that the relationship is fairly linear.

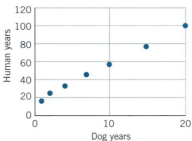

Comparative Ages of Dogs and Humans

a. Draw a line that looks like a best fit to the data.

b. Estimate the coordinates and label two points on the line. Use them to find the slope. Interpret the slope in this context.

c. Using *H* for human age and *D* for dog age, identify which you are using as the independent and which you are using as the dependent variable.

d. Generate the equation of your line.

e. Use the linear model to determine the "human age" of a dog that is 17 dog years old.

f. Middle age in humans is 45–59 years. Use your model equation to find the corresponding middle age in dog years.

g. What is the domain for your model?

CHAPTER SUMMARY

The Average Rate of Change

The average rate of change of *y* with respect to *x* is

$$\frac{\text{change in } y}{\text{change in } x}$$

The units of the average rate of change $= \dfrac{\text{units of } y}{\text{units of } x}$

For example, the units might be dollars/year (read as "dollars per year") or pounds/person (read as "pounds per person").

The average rate of change between two points is the slope of the straight line connecting the points. Given two points (x_1, y_1) and (x_2, y_2),

$$\text{average rate of change} = \frac{\text{change in } y}{\text{change in } x}$$

$$= \frac{\Delta y}{\Delta x} = \frac{y_2 - y_1}{x_2 - x_1} = \text{slope}$$

If the slope, or average rate of change, of *y* with respect to *x* is *positive,* then the graph of the relationship rises when read from left to right. This means that as *x* increases in value, *y* increases in value.

If the slope is *negative,* the graph falls when read from left to right. As *x* increases, *y* decreases.

If the slope is *zero,* the graph is flat. As *x* increases, there is no change in *y*.

Linear Functions

A *linear function* has a constant average rate of change. It can be described by an equation of the form

$$\underbrace{\text{Output}}_{y} = \underbrace{\text{initial value}}_{b} + \underbrace{\text{rate of change}}_{m} \cdot \underbrace{\text{input}}_{x}$$

or

$$y = b + mx$$

where *b* is the vertical intercept and *m* is the slope, or rate of change of *y* with respect to *x*.

The Graph of a Linear Function

The graph of the linear function $y = b + mx$ is a straight line.

The *y*-intercept, *b*, tells us where the line crosses the *y*-axis.

The slope, m, tells us how fast the line is climbing or falling. The larger the magnitude (or absolute value) of m, the steeper the graph.

If the slope, m, is positive, then the line climbs from left to right. If m is negative, the line falls from left to right.

Special Cases of Linear Functions

Direct proportionality: y is *directly proportional to (or varies directly with)* x if

$$y = mx \quad \text{where the constant } m \neq 0$$

This equation represents a linear function in which the y-intercept is 0, so the graph passes through $(0, 0)$, the origin.

Horizontal line: A line of the form $y = b$, with slope 0.

Vertical line: A line of the form $x = c$, with slope undefined.

Parallel lines: Two lines that have the same slope.

Perpendicular lines: Two lines whose slopes are negative reciprocals.

Piecewise linear function: A function constructed from different linear segments. Some examples are the *absolute value function,* $f(x) = |x|$, and *step functions,* whose linear segments are all horizontal.

Fitting Lines to Data

We can visually position a line to fit data whose graph exhibits a linear pattern. The equation of a best-fit line offers an approximate but compact description of the data. Lines are of special importance in describing patterns in data because they are easily drawn and manipulated and give a quick first approximation of trends.

CHECK YOUR UNDERSTANDING

I. Is each of the statements in Problems 1–30 true or false? Give an explanation for your answer.

In Problems 1–4 assume that y is a function of x.

1. The graph of $y - 5x = 5$ is decreasing.

2. The graph of $2x + 3y = -12$ has a negative slope and negative vertical intercept.

3. The graph of $x - 3y + 9 = 0$ is steeper than the graph of $3x - y + 9 = 0$.

4. The accompanying figure is the graph of $5x - 3y = -2$.

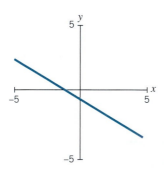

5. Health care costs between the years 1990 and 2007 would most likely show a positive average rate of change.

6. If we choose any two points on the graph in the accompanying figure, the average rate of change between them would be positive.

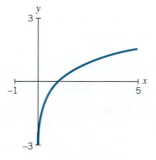

7. The average rate of change between two points (t_1, Q_1) and (t_2, Q_2) is the same as the slope of the line joining these two points.

8. The average rate of change of a variable M between the years 1990 and 2000 is the slope of the line joining two points of the form $(M_1, 1990)$ and $(M_2, 2000)$.

9. To calculate the average rate of change of a variable over an interval, you must have two distinct data points.

10. A set of data points of the form (x, y) that do not fall on a straight line will generate varying average rates of change depending on the choice of endpoints.

11. If the average rate of change of women's salaries from 2003 to 2007 is \$1000/year, then women's salaries increased by exactly \$1000 between 2003 and 2007.

12. If the average rate of change is positive, the acceleration (or rate of change of the average rate of change) may be positive, negative, or zero.

13. If the average rate of change is constant, then the acceleration (or rate of change of the average rate of change) is zero.

14. The average rate of change between (W_1, D_1) and (W_2, D_2) can be written as either $(W_1 - W_2)/(D_1 - D_2)$ or $(W_2 - W_1)/(D_2 - D_1)$.

15. If we choose any two distinct points on the line in the accompanying figure, the average rate of change between them would be the same negative number.

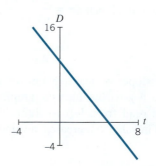

16. The average rate of change between (t_1, Q_1) and (t_2, Q_2) can be written as either $(Q_1 - Q_2)/(t_1 - t_2)$ or $(Q_2 - Q_1)/(t_2 - t_1)$.

17. On a linear graph, it does not matter which two distinct points on the line you use to calculate the slope.

18. If the distance a sprinter runs (measured in meters) is a function of the time (measured in minutes), then the units of the average rate of change are minutes per meter.

19. Every linear function crosses the horizontal axis exactly one time.

20. If a linear function in the form $y = mx + b$ has slope m, then increasing the x value by one unit changes the y value by m units.

21. If the units of the dependent variable y are pounds and the units of the independent variable x are square feet, then the units of the slope are pounds per square foot.

22. If the average rate of change between any two data points is increasing as you move from left to right, then the function describing the data is linear and is increasing.

23. The function in the accompanying figure has a slope that is increasing as you move from left to right.

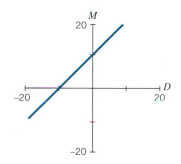

24. The slope of the function $f(x)$ is of greater magnitude than the slope of the function $g(x)$ in the accompanying figures.

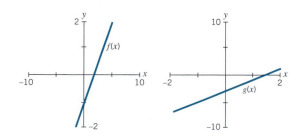

25. You can calculate the slope of a line that goes through any two points on the plane.

26. If $f(x) = y$ is decreasing throughout, then the y values decrease as the x values decrease.

27. Having a slope of zero is the same as having an undefined slope.

28. Vertical lines are linear functions.

29. Two nonvertical lines that are perpendicular must have slopes of opposite sign.

30. All linear functions can be written in the form $y = b + mx$.

II. In Problems 31–40, give an example of a function or functions with the specified properties. Express your answer using equations, and specify the independent and dependent variables.

31. Linear and decreasing with positive vertical intercept

32. Linear and horizontal with vertical intercept 0

33. Linear and with positive horizontal intercept and negative vertical intercept

34. Linear and does not pass through the first quadrant (where $x > 0$ and $y > 0$)

35. Linear with average rate of change of 37 minutes/lap

36. Linear describing the value of a stock that is currently at $19.25 per share and is increasing exactly $0.25 per quarter

37. Two linear functions that are parallel, such that moving one of the functions horizontally to the right two units gives the graph of the other

38. Four distinct linear functions all passing through the point $(0, 4)$

39. Two linear functions where the slope of one is m and the slope of the second is $-1/m$, where m is a negative number

40. Five data points for which the average rates of change between consecutive points are positive and are increasing at a decreasing rate

III. Is each of the statements in Problems 41–53 true or false? If a statement is true, explain how you know. If a statement is false, give a counterexample.

41. If the average rate of change between any two points of a data set is constant, then the data are linear.

42. If the slope of a linear function is negative, then the average rate of change decreases.

43. For any two distinct points, there is a linear function whose graph passes through them.

44. To write the equation of a specific linear function, one needs to know only the slope.

45. Function A in the accompanying figure is increasing at a faster rate than function B.

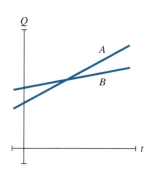

46. The graph of a linear function is a straight line.

47. Vertical lines are not linear functions because they cannot be written in the form $y = b + mx$, as they have undefined slope.

48. There exist linear functions that slant upward moving from left to right but have negative slope.

49. A constant average rate of change means that the slope of the graph of a function is zero.

50. All linear functions in x and y describe a relationship where y is directly proportional to x.

51. The function $h(t) = |t - 2|$ is always positive.

52. The function $h(t) = |t - 2|$ can be written as a piecewise linear function.

53. The graph to the right shows a step function.

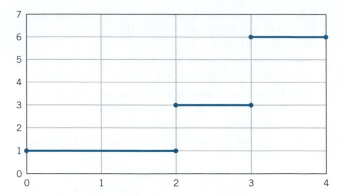

CHAPTER 2 REVIEW: PUTTING IT ALL TOGETHER

1. According to the National Association of Realtors, the median price for a single-family home rose from $147,300 in 2000 to $217,300 in 2006. Describe the change in the following ways.

  a. The absolute change in dollars

  b. The percent increase

  c. The annual average rate of change of the price with respect to year

(*Note:* Be sure to include units in all your answers.)

2. a. According to Apple Computer, sales of its iPod (the world's best-selling digital audio player) soared from 304,000 in the third fiscal quarter of 2003 to 8,111,000 in the third fiscal quarter of 2006. What was the average rate of change in iPods sold per quarter? What was the average rate of change per month?

  b. iPod sales reached an all-time high of 14,043,000 in Apple's first 2006 fiscal quarter (which included December 2005). What was the average rate of change in the number of iPods sold between the first and third fiscal quarters of 2006? What might be the reason for the decline?

3. Given the following graph of the function $h(t)$, identify any interval(s) over which:

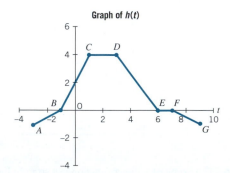

Graph of $h(t)$

  a. The function is positive; is negative; is zero

  b. The slope is positive; is negative; is zero

4. Which line has the steeper slope? Explain why.

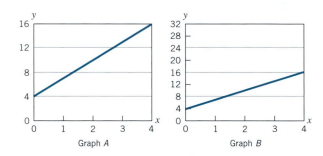

Graph A   Graph B

5. You are traveling abroad and realize that American, British, and French clothing sizes are different. The accompanying table shows the correspondence among female clothing sizes in these different countries.

Women's Clothing Sizes

U.S.	Britain	France
4	10	38
6	12	40
8	14	42
10	16	44
12	18	46
14	20	48

  a. Are the British and French sizes both linear functions of the U.S. sizes? Why or why not?

  b. Write a sentence describing the relationship between British sizes and U.S. sizes.

  c. Construct an equation that describes the relationship between French and U.S. sizes.

6. A bathtub that initially holds 50 gallons of water starts draining at 10 gallons per minute.

  a. Construct a function $W(t)$ for the volume of water (in gallons) in the tub after t minutes.

b. Graph the function and label the axes. Evaluate $W(0)$ and $W(5)$ and describe what they represent in this context.

c. How would the original function and its graph change if the initial volume were 60 gallons? Call the new function $U(t)$.

d. How would the original function and graph change if the drain rate were 12 gallons per minute? Call the new function $V(t)$.

e. Add the graphs of $U(t)$ and $V(t)$ to the original graph of $W(t)$.

7. Cell phones have produced a seismic cultural shift. No other recent invention has incited so much praise—and criticism. In 2000 there were 109.4 million cell phone subscriptions in the United States; since then subscriptions steadily increased to reach 207.9 million in 2005. (*Note:* Some people had more than one subscription.) In 2005 about 66% of the U.S. population had a cell phone.

a. How many Americans were *without* a cell phone in 2005 (when the U.S. population was 296.4 million)?

b. What was the average rate of change in millions of cell phone subscriptions per year between 2000 and 2005?

c. Construct a linear function $C(t)$ for cell phone subscriptions (in millions) for t = years from 2000.

d. If U.S. cell phone subscriptions continue to increase at the same rate, how many will there be in 2010? Does your result sound plausible? (See Section 2.9, Exercise 11.)

8. The following graph shows average salaries for major league baseball players from 1990 to 2006.

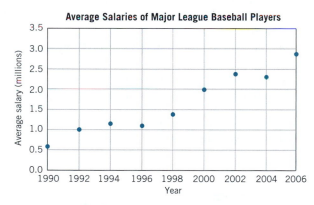

Average Salaries of Major League Baseball Players

a. On the graph, sketch a straight line that is an approximate mathematical model for the data.

b. Specify the coordinates of two points *on your line* and describe what they represent in terms of your model.

c. Calculate the slope and then interpret it in terms of the salary in millions of dollars and the time interval.

d. Let $S(x)$ be a linear approximation of the salaries where x = years since 1990. What is the slope of the graph of $S(x)$? The vertical intercept? What is the equation for $S(x)$?

9. Consider the following linear equations.

   i. $3y + 2x - 15 = 0$

   ii. $y + 4x = 1$

   iii. $3x - y = 0$

   iv. $8x - 5 = 2y$

Assuming x is graphed on the horizontal axis, which equation(s) have graphs that:

a. Are decreasing?

b. Have a positive vertical intercept?

c. Pass through the origin?

d. Is (are) the steepest?

10. Match one or more of the following graphs with the following descriptions. Be sure to note the scale on each axis.

a. Represents a constant rate of change of 3

b. Has a vertical intercept of 10

c. Is parallel to the line $y = 5 - 2x$

d. Has a steeper slope than the line C

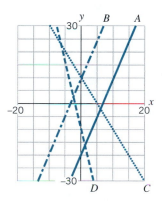

11. Gasoline prices spiked during the summer of 2006. The following table gives the weekly national average prices for regular gasoline as reported by the Department of Energy for three different weeks.

Regular Gas Prices in 2006

Week 1 (in January)	Week 31 (in August)	Week 43 (in October)
$2.22/gal	$3.04/gal	$2.22/gal

a. Calculate the average rates of change for gasoline prices (in $/gal/week) between weeks 1 and 31; between weeks 31 and 43; between weeks 1 and 43.

b. Write a 60-second summary about the price of gas during 2006.

(*Note:* Unfortunately in the summer of 2007, gas prices once again went over $3.00 per gallon.)

12. You plan a trip to Toronto and find that on July 5, 2006, the exchange rate for Canadian money is CA\$1 = US\$0.899. Although this exchange rate favors the U.S. visitor to Canada, you are not delighted to discover that the Canadians have a 14% sales tax.

a. If a taxable item costs CA\$50, how much do you actually have to pay in Canadian dollars?

b. What is this worth in U.S. dollars?

c. Construct a function that describes the total cost C_{us} (in U.S.) of purchasing a taxable item with a price of P_{ca} (in Canadian dollars).

d. Is the cost in U.S. dollars, C_{us}, directly proportional to the Canadian price, P_{ca}?

13. Create a linear function for each condition listed below, using (at most) the numbers -2 and 3 and the variables Q and t. Assume t is the independent variable. The function's graph is:

a. Increasing

b. Decreasing

c. Horizontal

d. Steeper than that of $Q = 5 - t$

e. Parallel to $Q = 3t - 4$

f. Perpendicular to $Q = (1/2)t + 6$

14. Given the following graph:

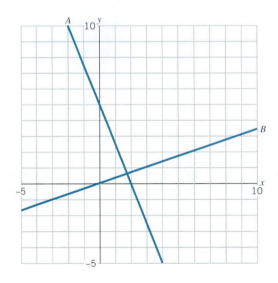

a. Find the equations of line A and line B.

b. Show that line A is (or is not) perpendicular to line B.

15. According to the Environmental Protection Agency's "National Coastal Conditions Report II," more than half of the U.S. population live in the narrow coastal fringes. Increasing population in these areas contributes to degradation (by runoff, sewage spills, construction, and overfishing) of the same resources that make the coasts desirable. In 2003 about 153 million people lived in coastal counties. This coastal population is currently increasing by an average of 3600 people per day.

a. What is the average rate of change in millions of people per *year*?

b. Let x = years since 2003 and construct a linear function $P(x)$ for the coast population (in millions).

c. Graph the function $P(x)$ over a reasonable domain.

d. What does your model give for the projected population in 2008?

16. Generate the equation for a line under each of the following conditions.

a. The line goes through the points $(-1, -4)$ and $(4, 6)$.

b. The line is parallel to the line in part (a) and goes through the point $(0, 5)$.

c. The line is perpendicular to the line in part (a) and goes through the origin.

17. In 2005 a record 51.5% of paper consumed in the United States (51.3 million tons) was recycled. The American Forest and Paper Association states that its goal is to have 55% recovery by 2012.

a. Plot two data points that represent the percentage of paper recycled at t number of years since 2005. (You may want to crop the vertical axis to start at 50%.) Connect the two points with a line and calculate its slope.

b. Assuming linear growth, the line represents a model, $R(t)$, for the percentage of paper recycled at t years since 2005. Find the equation for $R(t)$.

c. Find $R(0)$, $R(5)$, and $R(20)$. What does $R(20)$ represent?

18. Sketch a graph that could represent each of the following situations. Be sure to label your axes.

a. The volume of water in a kettle being filled at a constant rate.

b. The height of an airplane flying at 30,000 feet for 3 hours.

c. The unlimited demand for a certain product, such that if the company produces 10,000 of them, they can sell them for virtually any price. (Use the convention of economists, placing price on the vertical axis and quantity on the horizontal.)

d. The price of subway tokens that periodically increase over time.

e. The distance from your destination over time when you travel three subway stops to get to your destination.

19. Generate an equation for each line A, B, C, D, and E shown in the accompanying graph.

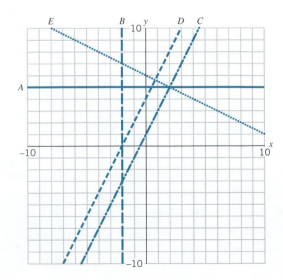

20. U.S. life expectancy has been steadily increasing (see Exercise 15 in Section 2.1 and data file for LIFEXEC). The following graph shows the roughly linear trends for both males and females over more than three decades.

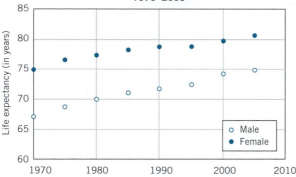

Trends in U.S. Life Expectancy by Sex, 1970–2005

Source: U.S. Bureau of the Census.

a. According to these data, what is the life expectancy of a female born in 1970? Of a male born in 2005?

b. Sketch a straight line to create a linear approximation of the data for your gender. Males and females will have different lines.

c. Identify two points *on your line* and calculate the slope. What does your slope mean in this context? The male slope is greater than the female slope. What does that imply?

d. Generate a linear function for your line. (*Hint:* Let the independent variable be the number of years since 1970.)

e. What would your model predict for your sex's life expectancy in 2010? How close does your model come to Census Bureau current predictions of life expectancies in 2010 of 75.6 years for males and 81.4 for females?

21. For the function $g(x) = |x| + 2$:

a. Generate a table of values for $g(x)$ using integer values of x between -3 and 3.

b. Use the table to sketch a graph of $g(x)$.

c. How does this graph compare with the graph of $f(x) = |x|$?

22. In Exercise 32 in Section 2.8 we met the greatest integer function $f(x) = [x]$, where $[x]$ is the greatest integer $\le x$; (i.e., you round down to the nearest integer). So $[-1.5] = -2$ and $[3.99] = 3$. This function is sometimes written using the notation $\lfloor x \rfloor$ and called the *floor* function. There is a similar *ceiling* function, $g(x) = \lceil x \rceil$, where $\lceil x \rceil$ is the smallest integer $\ge x$; (i.e., here you round up to the nearest integer). So $\lceil -1.5 \rceil = -1$ and $\lceil 3.000001 \rceil = 4$.

a. Complete the following table.

x	-2	-1.5	-1	-0.5	0	0.5	1	1.5	2
Floor $f(x) = \lfloor x \rfloor$									
Ceiling $g(x) = \lceil x \rceil$									

b. Plot the ceiling function and the floor function on two separate graphs, being sure to specify whether the endpoints of each line segment are included or excluded.

c. Why might a telephone company be interested in a ceiling function?

23. The table below gives historical data on voter turnout as a percentage of voting-age population (18 years and older) in U.S. presidential elections since 1960.

Year	Voting-Age Population	Turnout	%Turnout of Voting-Age Pop.
2004	221,256,931	122,294,978	55.3%
2000	205,815,000	105,586,274	51.3%
1996	196,511,000	96,456,345	49.1%
1992	189,529,000	104,405,155	55.1%
1988	182,778,000	91,594,693	50.1%
1984	174,466,000	92,652,680	53.1%
1980	164,597,000	86,515,210	52.6%
1976	152,309,190	81,555,789	53.5%
1972	140,776,000	77,718,554	55.2%
1968	120,328,186	73,211,875	60.8%
1964	114,090,000	70,644,592	61.9%
1960	109,159,000	68,838,204	63.1%

Using selected data from the table, graphs, and dramatically persuasive language, provide convincing arguments in a 60-second summary that during the years 1960 to 2004:

i. Voter turnout has plummeted.

ii. Voter turnout has soared.

24. a. Construct the equation of a line y_1 whose graph is steeper than that of $8x + 2y = 6$ and does not pass through quadrant I (where both $x > 0$ and $y > 0$). Plot the lines for both equations on the same graph.

b. Construct another equation of a line y_2 that is perpendicular to the original equation, $8x + 2y = 6$, and *does* go through quadrant I. Add the plot of this line to the graph in part (a).

25. Sweden has kept meticulous records of its population for many years. The following graph shows the mortality rate (as a percentage) for female and male children in Sweden over a 250-year period.

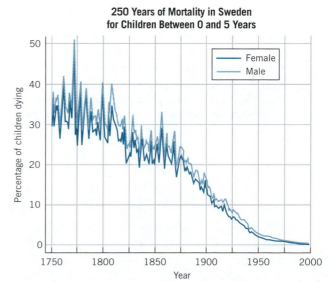

250 Years of Mortality in Sweden for Children Between 0 and 5 Years

Source: Statistics Sweden.

a. Describe the overall trend, and the similarities and differences between the female and male graphs.

b. Assuming the graphs are roughly linear between 1800 and 1950, draw a line approximating the female childhood mortality rate during this time. What is the slope of the line and what does it tell you in this context?

c. Construct a linear function to model the deaths over this period.

d. What is happening in the years after 1950?

Note: The large spike between 1750 and 1800 represents the more than 300,000 Swedish children (out of a total population of about 2 million) who died from smallpox, the most feared disease of that century.

26. According to the U.S. Environmental Protection Agency, carbon dioxide makes up 84.6% of the total U.S. greenhouse gas emissions. Carbon dioxide, or CO_2, arises from the combustion of coal, oil, and gas. Greenhouse gases trap heat on our planet, causing it to become warmer. Effects of this phenomenon are already being seen in increased melting of glaciers and permafrost. If this continues unchecked, water levels will rise, causing flooding in coastal communities, where the majority of the U.S. population lives. (See Exercise 15.)

The following graph shows the average rate of change in the concentration of atmospheric CO_2 over time.

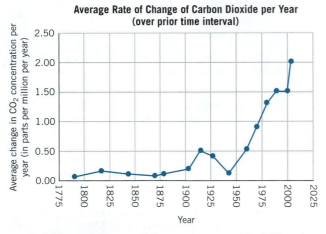

Average Rate of Change of Carbon Dioxide per Year (over prior time interval)

a. Is the average rate of change always positive? If so, does that imply that the amount of atmospheric carbon dioxide is always increasing? Explain your answer.

b. What does it mean when the average rate of change was decreasing (though not negative) between about 1915 and 1943? What was happening in the world during this period

that might have caused carbon dioxide discharges to increase at a lower rate?

c. After 1943 the average rate of change steadily increases. What does that tell you about the increase of carbon dioxide?

27. Examine the following graph of a function $y = f(x)$.

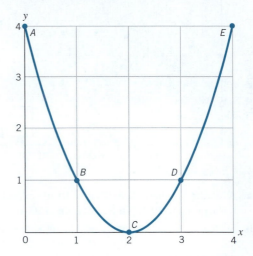

a. Using the points A, B, C, D, and E labeled on the graph, fill in the rest of the accompanying table.

Point	x-Coordinate of Point	y-Coordinate of Point	Average Rate of Change Between Two Adjacent Points
A	0	4	n.a.
B	1	1	-3
C			
D			
E			

b. Over what x interval is the function increasing? What is happening to the average rate of change between the points in that interval?

c. Over what x interval is the function decreasing? What is happening to the absolute value of the average rate of change of the points in that interval? What does this mean? (Recall that the absolute value of a slope gives the steepness of the line.)

d. What is the concavity of the entire function? How do the changes in the steepness of the curve help confirm the concavity of the function graph?

EXPLORATION 2.1

Having It Your Way

Objective

- construct arguments supporting opposing points of view from the same data

Material/Equipment

- excerpts from the *Student Statistical Portrait* of the University of Massachusetts, Boston (in the Appendix) or from the equivalent for the student body at your institution
- computer with spreadsheet program and printer or graphing calculator with projection system (all optional)
- graph paper and/or overhead transparencies (and overhead projector)

Procedure

Working in Small Groups

Examine the data and graphs from the *Student Statistical Portrait* of the University of Massachusetts, Boston, or from your own institution. Explore how you would use the data to construct arguments that support at least two different points of view. Decide on the arguments you are going to make and divide up tasks among your team members.

Rules of the Game

- Your arguments needn't be lengthy, but you need to use graphs and numbers to support your position. You may use only legitimate numbers, but you are free to pick and choose those that support your case. If you construct your own graphs, you may, of course, use whatever scaling you wish on the axes.
- For any data that represent a time series, as part of your argument pick two appropriate endpoints and calculate the associated average rate of change.
- Use "loaded" vocabulary (e.g., "surged ahead," "declined drastically"). This is your chance to be outrageously biased, write absurdly flamboyant prose, and commit egregious sins of omission.
- Decide as a group how to present your results to the class. Some students enjoy realistic "role playing" in their presentations and have added creative touches such as mock protesters complete with picket signs.

Suggested Topics

Your instructor might ask your group to construct one or both sides of the arguments on one topic. If you're using data from your own institution, answer the questions provided by your instructor.

Using the *Student Statistical Portrait* from the University of Massachusetts, Boston (data located in the Appendix)

1. Use the table "Undergraduate Admissions Summary" to construct a persuasive case for each situation.
   a. You are the Provost, the chief academic officer of the university, arguing in front of the Board of Trustees that the university is becoming more appealing to students.
   b. You are a student activist arguing that the university is becoming less appealing to students.

2. Use the table "Trends in New Student Race/Ethnicity in the College of Liberal Arts" to make a convincing case for each of the following.

   a. You are the Affirmative Action Officer arguing that her office has done a terrific job.

   b. You are a reporter for the student newspaper criticizing the university for its neglect of minority students.

3. Use the table "SAT Scores of New Freshmen by College/Program" to "prove" each of the contradictory viewpoints.

   a. You are the Dean of the College of Liberal Arts arguing that you have brighter freshmen than those in the College of Management.

   b. You are the Dean of the College of Management arguing that your freshmen are superior.

Exploration-Linked Homework

With your partner or group prepare a short class presentation of your arguments, using, if possible, overhead transparencies or a projection panel. Then write individual 60-second summaries to hand in.

EXPLORATION 2.2A

Looking at Lines with the Course Software

Objective

- find patterns in the graphs of linear equations of the form $y = mx + b$

Equipment

- computer with course software "L1: m & b Sliders" in *Linear Functions*

Procedure

In each part try working first in pairs, comparing your observations and taking notes. Your instructor may then wish to bring the whole class back together to discuss everyone's results.

Part I: Exploring the Effect of m and b on the Graph of $y = mx + b$

Open the program *Linear Functions* and click on the button "L1: m & b Sliders."

1. What is the effect of m on the graph of the equation?

Fix a value for b. Construct four graphs with the same value for b but with different values for m. Continue to vary m, jotting down your observations about the effect on the line when m is positive, negative, or equal to zero. Do you think your conclusions work for values of m that are not on the slider?

Choose a new value for b and repeat your experiment. Are your observations still valid? Compare your observations with those of your partner.

2. What is the effect of b on the graph of the equation?

Fix a value for m. Construct four graphs with the same value for m but with different values for b. What is the effect on the graph of changing b? Record your observations. Would your conclusions still hold for values of b that are not on the slider?

Choose a new value for m and repeat your experiment. Are your observations still valid? Compare your observations with those of your partner.

3. Write a 60-second summary on the effect of m and b on the graph of $y = mx + b$.

Part II: Constructing Lines under Certain Constraints

1. Construct the following sets of lines still using "L1: m & b Sliders." Be sure to write down the equations for the lines you construct. What generalizations can you make about the lines in each case? Are the slopes of the lines related in some way? Are the vertical intercepts of the lines related?

Construct any line. Then construct another line that has a steeper slope, and then construct one that has a shallower slope.

Construct three *parallel lines*.

Construct three lines with the *same y-intercept*, the point where the line crosses the y-axis.

Construct a pair of lines that are *horizontal*.

Construct a pair of lines that go *through the origin*.

Construct a pair of lines that are *perpendicular* to each other.

2. Write a 60-second summary of what you have learned about the equations of lines.

EXPLORATION 2.2B

Looking at Lines with a Graphing Calculator

Objective

- find patterns in the graphs of linear equations of the form $y = mx + b$

Material/Equipment

- graphing calculator (instructions for the TI-83 and TI-84 families of calculators are available in the Graphing Calculator Manual)

Procedure

Getting Started

Set your calculator to the integer window setting. For the TI-83 or TI-84 do the following:

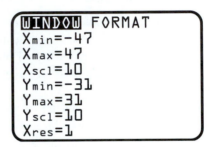

1. Press ZOOM and select [6:ZStandard].
2. Press ZOOM and select [8:ZInteger], ENTER.
3. Press WINDOW to see whether the settings are the same as the duplicated screen image.

Working in Pairs

In each part try working first in pairs, comparing your observations and taking notes. Your instructor may then wish to bring the whole class back together to discuss everyone's results.

Part I: Exploring the Effect of m and b on the Graph of $y = mx + b$

1. What is the effect of m on the graph of the equation $y = mx$?

   a. Case 1: $m > 0$

   Enter the following functions into your calculator and then plot the graphs. To get started, try $m = 1, 2, 5$. Try a few other values of m where $m > 0$.

   $Y1 = x$
   $Y2 = 2x$
   $Y3 = 5x$
   $Y4 = \cdots$
   $Y5 =$
   $Y6 =$

   Compare your observations with those of your partner. In your notebook describe the effect of multiplying x by a positive value for m in the equation $y = mx$.

b. Case 2: $m < 0$

Begin by comparing the graphs of the lines when $m = 1$ and $m = -1$. Then experiment with other negative values for m and compare the graphs of the equations.

$Y1 = x$
$Y2 = -x$
$Y3 = \cdots$
$Y4 =$
$Y5 =$
$Y6 =$

Alter your description in part 1(a) to describe the effect of multiplying x by any real number m for $y = mx$ (remember to also explore what happens when $m = 0$).

2. What is the effect of b on the graph of an equation $y = mx + b$?

   a. Enter the following into your calculator and then plot the graphs. To get started, try $m = 1$ and $b = 0, 20, -20$. Try other values for b as well.

   $Y1 = x$
   $Y2 = x + 20$
   $Y3 = x - 20$
   $Y4 = \cdots$
   $Y5 =$
   $Y6 =$

   b. Discuss with your partner the effect of adding any number b to x for $y = x + b$. (*Hint:* Use "trace" to find where the graph crosses the y-axis.) Record your comments in your notebook.

   c. Choose another value for m and repeat the exercise. Are your observations still valid?

3. Write a 60-second summary on the effect of m and b on the graph of $y = mx + b$.

Part II: Constructing Lines under Certain Constraints

1. Construct the following sets of lines using your graphing calculator. Be sure to write down the equations for the lines you construct. What generalizations can you make about the lines in each case? Are the slopes of the lines related in any way? Are the vertical intercepts related? Which graph is the steepest?

   Construct three *parallel lines.*

   Construct three lines with the *same y-intercept.*

   Construct a pair of lines that are *horizontal.*

   Construct a pair of lines that go *through the origin.*

   Construct a pair of lines that are *perpendicular* to each other.

2. Write a 60-second summary of what you have learned about the equations of lines.

LOOKING FOR LINKS BETWEEN EDUCATION AND EARNINGS

OVERVIEW

Is learning the key to earning? Does going to school pay off? In this extended exploration, you use a large data set from the U.S. Bureau of the Census to examine ways in which education and earnings may be related. Technology is used to fit lines to data, and you learn how to interpret the resulting linear models, called regression lines. You can explore further by finding evidence to support or disprove conjectures, examining questions raised by the analysis, and posing your own questions.

In this exploration, you will

- analyze U.S. Census data

- use regression lines to summarize data

- make conjectures about the relationship between education and earnings in the United States

- find evidence to support or refute your conjectures

- examine the distinction between correlation and causation

Using U.S. Census Data

Does more education mean higher earnings? The answer may seem obvious. We may reasonably expect that having more education gives access to higher-paying jobs. Is this indeed the case? We explore how a social scientist might start to answer these questions using a random sample from U.S. Census data. Our data set, called FAM1000, provides information on 1000 individuals and their families.

The Bureau of the Census, as mandated by the Constitution, conducts a nationwide census every 10 years. To collect more up-to-date information, the Census Bureau also conducts a monthly survey of American households for the Bureau of Labor Statistics called the Current Population Survey, or CPS. The CPS is the largest survey taken between census years. The CPS is based on data collected each month from approximately 50,000 households. Questions are asked about race, education, housing, number of people in the household, earnings, and employment status.[1] The March surveys are the most extensive. Our sample of census data, FAM1000, was extracted from the March 2006 Current Population Survey. It contains information about 1000 individuals randomly chosen from those 16 and older who worked at least 1 week in 2005.

You can use the FAM1000 data and the related software, called *FAM1000 Census Graphs,* to follow the discussion in the text and/or conduct your own case study. The full FAM1000 data set is in the Excel file FAM1000, and condensed versions are in the graph link files FAM1000 A to H. These data files and related software can be downloaded from the web at *www.wiley.com/college/kimeclark* or on your class Wiley Plus site. The software provides easy-to-use interactive tools for analyzing the FAM1000 data.

Table 1 on page 147 is a *data dictionary* with short definitions for each data category. Think of the data as a large array of rows and columns of facts. Each row represents all the information obtained from one particular respondent about his or her family. Each column contains the coded answers of all the respondents to one particular question. Try deciphering the information contained in the first row of the 1000 rows in the FAM1000 data set in Table 2.

age	sex	region	cencity	marstat	famsize	edu	occup	hrswork	wkswork	yrft	pearnings	ptotinc	faminc	race	hispanic
42	2	4	1	0	3	12	4	40	52	1	$25,000	$25,559	$46,814	5	0

Table 2

Referring to the data dictionary, we learn that the respondent is a 42-year-old female who lives in a city somewhere in the West. She is married with two other people in her family and she has a high school degree. In 2005 she worked in sales or a related occupation for 40 hours a week, 52 weeks of the year. Her personal earnings from work were $25,000, but her personal total income, was $25,559. That means she had an additional source of unearned income, such as dividends from stocks or bonds, or interest on a savings account. Her family income totaled $46,814. She did not identify with any of the racial or Hispanic categories listed on the census form.

[1]The results of the survey are used to estimate numerous economic and demographic variables, such as the size of the labor force, the employment rate, earnings, and education levels. The results are widely quoted in the popular press and are published monthly in *The Monthly Labor Review* and *Employment and Earnings,* irregularly in *Current Population Reports* and *Special Labor Force Reports,* and yearly in *Statistical Abstract of the United States* and *The Economic Report of the President.*

Data Dictionary for March 2006 Current Population Survey

Variable	Definition	Unit of Measurement (code and allowable range)	Variable	Definition	Unit of Measurement (code and allowable range)
age	Age	Range is 16 to 85	occup	Occupation group of respondent	7 = Construction and extraction occupations
sex	Sex	1 = Male			8 = Installation, maintenance, and repair occupations
		2 = Female			9 = Production occupations
region	Census region	1 = Northeast			10 = Transportation and material moving occupations
		2 = Midwest			11 = Armed Forces
		3 = South			
		4 = West	hrswork	Usual hours worked per week	Range is 1 to 99
cencity	Residence location	1 = Metropolitan			
		2 = Nonmetropolitan	wkswork	Weeks worked in 2005	Range is 1 to 52
		3 = Not Identified	yrft	Employed full-time year-round	1 = full-time year-round
marstat	Marital status	0 = Presently married			2 = part-time year-round
		1 = Presently not married			3 = full-time part of the year
famsize	Family size	Range is 1 to 39			4 = part-time part of the year
educ	Years of education	8 = 8 or fewer years of education			
		10 = No high school degree, 9–12 years of education	pearnings	Total personal earnings from work	Range is $0 to $650,000
		12 = High school degree	ptotinc	Personal total income	Range is negative $1,000,000 to $10,000,000
		13 = Some college			
		14 = Associate's degree	faminc	Family income	Range is negative $400,000 to $24,000,000
		16 = Bachelor's degree			
		18 = Master's degree	race	Race of respondent	1 = White
		20 = Doctorate			2 = Black
		22 = Professional degree (e.g., MD)			3 = American Indian, Alaskan Native Only
occup	Occupation group of respondent	0 = Not in universe, or children			4 = Asian or Pacific Islander
		1 = Management, business, and financial occupations			5 = Other
		2 = Professional and related occupations	hispanic	Hispanic heritage	0 = Not in universe
		3 = Service occupations			1 = Mexican
		4 = Sales and related occupations			2 = Puerto Rican
		5 = Office and administrative support occupations			3 = Cuban
		6 = Farming, fishing, and forestry occupations			4 = Central/South American
					5 = Other Spanish

Table 1

Source: Adapted from the U.S. Bureau of the Census, Current Population Survey, March 2006, by Jie Chen, Computing Services, University of Massachusetts, Boston.

Summarizing the Data: Regression Lines

Is There a Relationship between Education and Earnings?

In the physical sciences, the relationship among variables is often quite direct; if you hang a weight on a spring, it is clear, even if the exact relationship is not known, that the amount the spring stretches is definitely dependent on the heaviness of the weight. Further, it is reasonably clear that the weight is the *only* important variable; the temperature and the phase of the moon, for example, can safely be neglected.

In the social and life sciences it is usually difficult to tell whether one variable truly depends on another. For example, it is certainly plausible that a person's earnings depend in part on how much formal education he or she has had, since we may suspect that having more education gives access to higher-paying jobs, but many other factors also play a role. Some of these factors, such as the person's age or type of work, are measured in the FAM1000 data set; others, such as family background or good luck, may not have been measured or even be measurable. Despite this complexity, we attempt to determine as much as we can by first looking at the relationship between earnings and education alone.

We start with a scatter plot of education and personal earnings from the FAM1000 data set. If we hypothesize that earnings depend on education, then the convention is to graph education on the horizontal axis. Each ordered pair of data values gives a point with coordinates of the form

(education, personal earnings)

Take a moment to examine the scatter plot in Figure 1. Each point refers to respondents with a particular level of education and income and has two coordinates. The first coordinate gives the years of education past grade 8 (So zero represents an

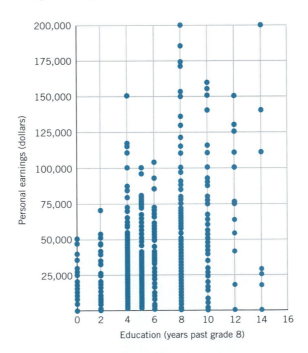

Figure 1 Personal earnings vs. education. In attempting to pick a reasonable scale to display personal earnings on this graph, the vertical axis was cropped at $200,000, which meant excluding from the display the points for three individuals who each earned more than $200,000 in wages.

eighth-grade education or less) and the second gives the personal earnings. For example, the points at the very top of the graph represent individuals who make $200,000 in personal wages. One of these points represents those with 16 years (8 years past eighth grade) of education and the other, those with 22 years (14 years past eighth grade) of education. The coordinates for these points are (8, $200,000) and (14, $200,000). We can refer back to the original data set to find out more information about these points, as well as the outliers that are not shown on the graph. Note that some dots represent more than one individual (i.e., for a particular level of education there might be many people with the same income).

How might we think of the relationship between these two variables? Clearly, personal earnings are not a function of education in the mathematical sense since people who have the same amount of education earn widely different amounts. The scatter plot obviously fails the vertical line test.

But suppose that, to form a simple description of these data, we were to insist on finding a simple functional description. And suppose we insist that this simple relationship be a linear function. In Chapter 2, we informally fit linear functions to data. A formal mathematical procedure called *regression analysis* lets us determine what linear function is the "best" approximation to the data; the resulting "best-fit" line is called a *regression line* and is similar in spirit to reporting only the mean of a set of single-variable data, rather than the entire data set. It can be a useful and powerful method of summarizing a set of data.

We can measure how well a line represents a data set by summing the vertical distances squared between the line and the data points. The regression line is the line that makes this sum as small as possible. The calculations necessary to compute this line are tedious, although not difficult, and are easily carried out by computer software and graphing calculators. We can use the *FAM1000 Census Graphs* or Excel program to find regression lines. The techniques for determining regression lines are beyond the scope of this course.

In Figure 2, we show the FAM1000 data set along with a regression line determined from the data points. The equation of the line is

$$\text{personal earnings} = 4188 + 5693 \cdot \text{yrs. of educ. past grade 8}$$

If you are interested in a standard technique for generating regression lines, a summary of the method used in the course software is provided in the reading "Linear Regression Summary."

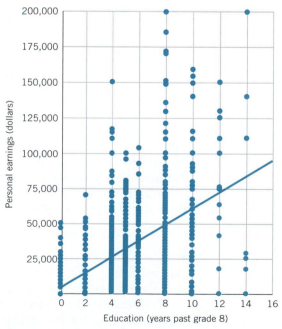

Figure 2 Regression line for personal earnings vs. education.

Since personal earnings are in dollars, the units for the term 4188 must be dollars, and the units for 5693 must be dollars per year of education.

This line is certainly more concise than the original set of data points. Looking at the graph, you may judge with your eyes to what extent the line is a good description of the original data set. From the equation, the vertical intercept is 4188. Thus, this model predicts that individual personal earnings for those with an eighth-grade education or less will be $4188. The number 5693 represents the slope of the regression line, or the average rate of change of personal earnings with respect to years of education. Thus, this model predicts that for each additional year of education, individual personal earnings increase by $5693.

We emphasize that, although we can construct an approximate linear model for any data set, this does not mean that we really believe that the data are truly represented by a linear relationship. In the same way, we may report the median to summarize a set of data, without believing that the data values are at the median. In both cases, there are features of the original data set that may or may not be important and that we do not report.

The data points are widely scattered about the line, for reasons that are clearly not captured by the linear model. We can eliminate the clutter by grouping together all people with the same years of education and plotting the median personal earnings of each group. The result is the graph in Figure 3, which includes a regression line for the new data.

For each year of education, only a single median earnings point has been graphed. For instance, the point corresponding to 12 years of education past grade 8 has a vertical value of approximately $75,000; hence the median of the personal earnings of everyone in the FAM1000 data with 20 years of education is about $75,000. The pattern is now clearer: An upward trend to the right is more obvious in this graph.

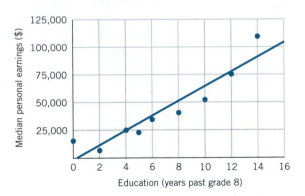

Figure 3 Median personal earnings vs. education.

Every time we construct a simplified representation of an original data set, we should ask ourselves what information has been suppressed. In Figure 3 we have suppressed the spread of data in the vertical direction. For example, there are only 36 people with an eighth-grade education or less but 175 with 16 years of education. Yet each of these sets is represented by a single point.

We can fit a line to the graph in Figure 3 using the same method of linear regression. The equation of this straight line is

median personal earnings = −2237 + 6592 · yrs. of educ. past grade 8

Here, −2237 is the vertical intercept and 6592 is the slope or rate of change of median personal earnings with respect to education. This model predicts that for each additional year of education, median personal earnings increase by $6592.

Note that this linear model predicts *median personal earnings for the group*, not personal earnings for an individual. The vertical intercept of this line is negative even though all earnings in the original data set are positive. The linear model is clearly inaccurate for those with an eighth-grade education or less.

Figure 4 shows two regression lines: One represents the fit to the medians and the other represents the fit to all of the data. Both of these straight lines are reasonable answers to the question "What straight line best describes the relationship between education and earnings?" and the difference between them indicates the uncertainty in answering such a question. We may argue that the benefit in earnings for each year of education is $5693, or $6592, or something between these values.

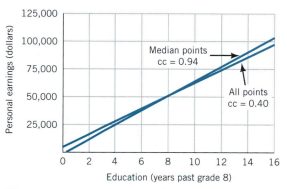

Figure 4 Regression lines for personal earnings vs. education.

Regression Line: How Good a Fit?

Once we have determined a line that approximates our data, we must ask, "How good a fit is our regression line?" To help answer this, statisticians calculate a quantity called the *correlation coefficient*. This number can be computed by statistical software, and we have included it on our graphs and labeled it "cc."[2] The correlation coefficient is always between −1 (negative association with no scatter; the data points fit exactly on a line with a negative slope) and 1 (positive association with no scatter; the data points fit exactly on a line with a positive slope). The closer the absolute value of the correlation coefficient is to 1, the better the fit and the stronger the linear association between the variables.

A small correlation coefficient (with absolute value close to zero) indicates that the variables do not depend linearly on each other. This may be because there is no relationship between them, or because there is a relationship that is something more complicated than linear. In future chapters we discuss many possible nonlinear functional relationships.

There is no definitive answer to the question of when a correlation coefficient is "good enough" to say that the linear regression line is a good fit to the data. A fit to the graph of medians (or means) generally gives a higher correlation coefficient than a fit to the original data set because the scatter has been smoothed out. (See the cc's in Figure 4.) When in doubt, plot all the data along with the linear model and use your best judgment. The correlation coefficient is only a tool that may help you decide among different possible models or interpretations.

The programs R1–R4 and R7 in *Linear Regression* can help you visualize the links among scatter plots, best-fit lines, and correlation coefficients.

The reading "The Correlation Coefficient" explains how to calculate and interpret the correlation coefficient.

---

[2]We use the label "cc" for the correlation coefficient in the text and software to minimize confusion. In a statistics course the correlation coefficient is usually referred to as *Pearson's r* or just *r*.

EXAMPLE 1

Interpreting the correlation coefficient

Figure 5 contains the data and regression line for *mean* personal earnings vs. education.

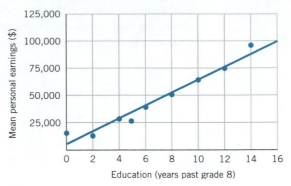

Figure 5 Mean personal earnings vs. education.

The equation of the regression line is

$$\text{mean personal earnings} = 5562 + 5890 \cdot \text{yrs. of educ. past grade 8}$$

$$cc = 0.98.$$

a. Interpret the slope of this new regression line.

b. Compare this regression line with that for median personal earnings vs. education previously cited in the text:

$$\text{median personal earnings} = -2237 + 6592 \cdot \text{yrs. of educ. past grade 8}$$

$$cc = 0.94$$

What do these equations predict for median and mean personal earnings for 12 years of education (high school)? For 18 years of education (master's degree)?

SOLUTION

a. The slope of 5890 suggests that each additional year of education corresponds to a $5890 yearly increase in mean personal earnings.

b. Both equations have cc's close to 1 (one is 0.98, the other 0.94), so both lines are good fits. Using the regression line for means, 12 years of education (or 4 years past grade 8), we get:

$$\text{mean personal earnings} = \$5562 + \$5890\,(4)$$
$$= \$29{,}122$$

Rounding to nearest ten $= \$29{,}120$

Using the regression line for means for 18 years of education (or 10 years past grade 8), we get:

$$\text{mean personal earnings} = \$5562 + \$5890\,(10)$$
$$= \$64{,}462$$

Rounding to nearest ten $= \$64{,}460$

Using the regression lines for medians and rounding to the nearest ten, 12 years of education corresponds to $24,130 and 18 years to $63,680.

Comparing predictions from the mean and median regression lines, we see that the regression line for means predicts higher earnings than the median regression line. This is reasonable since in general, when using income measures, the mean will often exceed the median since medians are not susceptible to outliners.

EXAMPLE 2 Examine the four scatter plots in Figure 6.

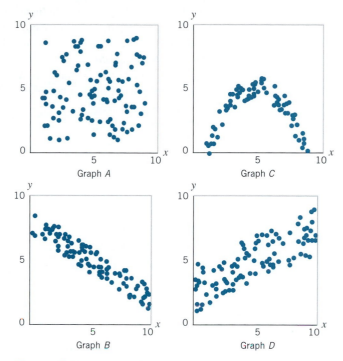

Figure 6 Four scatter plots.

a. Which graph shows a positive linear correlation between x and y? A negative linear correlation? Zero correlation?

b. Which graph shows the closest linear correlation between x and y?

SOLUTION **a.** Graph *D* shows a positive linear correlation between x and y (when one variable increases, the other increases). Graph *B* shows a negative linear correlation (when one variable increases, the other decreases). Both graphs *A* and *C* show zero correlation between x and y. Even though graph *C* shows a pattern in the relationship between x and y, the pattern is not linear.

b. Graph *B*. The correlation coefficient of its regression line would be close to -1, almost a perfect (negative) correlation.

Interpreting Regression Lines: Correlation versus Causation

One is tempted to conclude that increased education *causes* increased earnings. This may be true, but the model we have used does not offer conclusive proof. This model can show how strong or weak a relationship exists between variables but does not answer the question "Why are the variables related?" We need to be cautious in how we interpret our findings.

Regression lines show *correlation,* not *causation.* We say that two events are correlated when there is a statistical link. If we find a regression line with a correlation coefficient that is close to 1 in absolute value, a strong relationship is suggested. In our previous example, education is positively correlated with personal earnings. If education increases, personal earnings increase. Yet this does not prove that education causes an increase in personal earnings. The reverse might be true; that is, an increase in personal earnings might cause an increase in education. The correlation may be due to other factors altogether. It might occur purely by chance or be jointly caused by yet another variable. Perhaps both educational opportunities and earning levels are strongly affected by parental education or a history of family wealth. Thus a third variable, such as parental socioeconomic status, may better account for both more education and higher earnings. We call such a variable that may be affecting the results a *hidden variable*.

Figure 7 shows a clear correlation between the number of radios and the proportion of insane people in England between 1924 and 1937. (People were required to have a license to own a radio.)

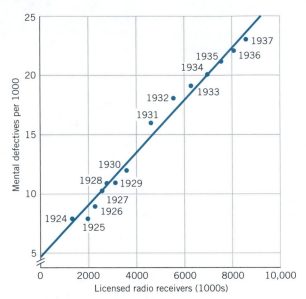

Figure 7 A curious correlation?
Source: E A. Tufte, *Data Analysis for Politics and Policy,* p. 90. Copyright © 1974 by Prentice Hall, Inc., Upper Saddle River, NJ. Reprinted with permission.

Are you convinced that radios cause insanity? Or are both variables just increasing with the years? We tend to accept as reasonable the argument that an increase in education causes an increase in personal earnings, because the results seem intuitively possible and they match our preconceptions. But we balk when asked to believe that an increase in radios causes an increase in insanity. Yet the arguments are based on the same sort of statistical reasoning. The flaw in the reasoning is that statistics can show only that events occur together or are correlated, but *statistics can never prove that one event causes another.* Any time you are tempted to jump to the conclusion that one event causes another because they are correlated, think about the radios in England!

Raising More Questions

When a strong link is found between variables, often the next step is to raise questions whose answers may provide more insight into the nature of the relationship. How can the evidence be strengthened? What if we used the mean instead of the median, restricted ourselves to year-round full-time workers, or used other income measures, such as total personal income or total family income? Will the relationship between education and income still hold? Are there other variables that affect the relationship?

Do Earnings Depend on Age?

We started our exploration by looking at how earnings depend on education, because it seems natural that more education might lead to more earnings. But it is equally plausible that a person's income might depend on his or her age. People may generally earn more as they advance through their working careers, but their earnings usually drop when they eventually retire. We can examine the FAM1000 data to look for evidence to support this hypothesis.

It's hard to see much when we plot all the data points. This time we use *mean* personal earnings and plot it versus age in Figure 8. The graph seems to suggest that up until about age 50, as age increases, mean personal earnings increase. After age 50, as people move into middle age and retirement, mean personal earnings tend to decrease. So age does seem to affect personal earnings, in a way that is roughly consistent with our intuition. But the relationship appears to be nonlinear, and so linear regression may not be a very effective tool to explore this dependence.

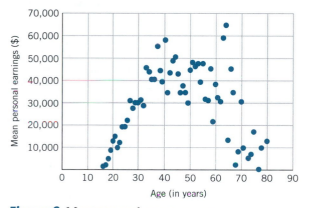

Figure 8 Mean personal wages vs. age.

The FAM1000 data set contains internal relationships that are not obvious on a first analysis. We might, for example, also investigate the relationship between education and age. Age may be acting as a hidden variable influencing the relationship between education and income.

There are a few simple ways to attempt to minimize the effect of age. For example, we can restrict our analysis to individuals who are all roughly the same age. This sample still would include a very diverse collection of people. More sophisticated strategies involve statistical techniques such as *multivariable analysis,* a topic beyond the scope of this course.

Do Earnings Depend upon Gender?

We can continue to look for relationships in the FAM1000 data set by using some of the other variables to sort the data in different ways. For example, we can look at whether the relationship between earnings and education is different for men and women. One way to do this is to compute the mean personal earnings for each year of education for men and women separately and restrict ourselves to those who work full-time year-round.

If we put the data for men and women on the same graph (Figure 9), it is easier to make comparisons. We can see in Figure 9 that for those working full-time year-round, the mean personal earnings of men are consistently higher than the mean personal earnings of women. We can also examine the best-fit lines for mean personal earnings versus education for men and for women shown in Figure 10.

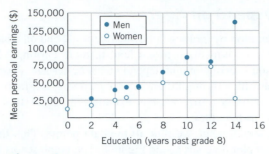

Figure 9 Mean personal earnings vs. education for women and for men (working full-time year-round).

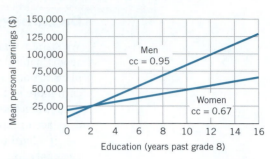

Figure 10 Regression lines for mean personal earnings vs. education for women and for men (working full-time year-round).

The linear model for mean personal earnings for men working full-time is given by

$$P_{\text{men}} = 9965 + 7423E$$

where E = years of education past grade 8 and P_{men} = mean personal earnings for men. The correlation coefficient is 0.95. The rate of change of mean personal earnings with respect to education is approximately \$7423/year. For males in this set, the mean personal earnings increase by roughly \$7423 for each additional year of education.

For women working full-time the comparable linear model is

$$P_{\text{women}} = 19{,}190 + 3000E$$

where E = years of education past grade 8 and P_{women} = mean personal earnings for women. The correlation coefficient is 0.67. As you might predict from the relative status of men and women in the U.S. workforce, the rate of change for women is much lower. For women in this sample, the model predicts that the mean personal earnings increase by only \$3,000 for each additional year of education. In Figure 10, we can see that the regression line for men is steeper than the one for women when plotted on the same grid. In Figure 9, the mean personal earnings for any particular number of years of education are consistently lower for women than for men. The disparity in mean personal earnings between men and women is most dramatic for those with 14 years of education beyond grade 8 or 22 years of education. Although the vertical intercept of the regression line for men's personal earnings is below that for women, after 2 years of education beyond grade 8 the regression line for men lies above that for women.

Going deeper

What other variables have we ignored that we may want to consider in a more refined analysis of the impact of gender on personal earnings? We could, for example, consider type of job or amount of work.

- *Type of job.* Do women and men make the same salaries when they hold the same types of jobs? We could compare only people within the same profession and ask whether the same level of education corresponds to the same level of personal earnings for women as for men. There are many more questions, such as: Are there more men than women in higher-paying professions? Do men and women have the same access to higher-paying jobs, given the same level of education?

- *Amount of work.* In our analysis, we used women and men who were working full-time to explore the impact of gender on earnings. Typically, part-time jobs pay less than full-time jobs, and more women hold part-time jobs than men. In addition, there are usually more women than men who are unemployed. What prediction would you make if all people in FAM1000 were used to study the impact of gender on earnings?

How Good Are the Data?

For this exploration earnings were defined and measured in a number of ways. What issues are raised by the way income in general is defined and measured? What groups of people may be undercounted in the U.S. Current Population Survey? What are some of the current controversies about how the U.S. Bureau of the Census collects the census data? You may want to search online and in the library for articles on the controversies surrounding the census. Who else collects data, and how can you determine if the data are reliable? As access to data is made easier and easier in our Internet society, the ability to assess the reliability and validity of the data becomes more and more important.

How Good Is the Analysis?

What other factors might affect earnings that are not covered by our analysis? What are some limitations to using regression lines to summarize data? Are there hidden variables (such as a history of family wealth or parents' socioeconomic status) that may affect both level of educational attainment and higher earnings? You may want to explore other methods of analysis that address some of these questions and could potentially reveal different patterns.

The following readings at *www.wiley.com/college/kimeclark* are additional resources for examining the relationship between income and educational attainment in the United States.

- The Bureau of Labor Statistics created a website where you can access all the data on earnings and education from the most recent Current Population Survey at *http://data.bls.gov/PDQ/outside.jsp?survey=le*. This site allows you to search according to categories, access historical data, and create graphs.

- U.S. Census Bureau News, *Census Bureau Data Underscore Value of College Degree*. October 2006.

- Bureau of Labor Statistics, *Education Pays*. U.S. Department of Labor. January 2007.

- M. Maier, "Wealth, Income, and Poverty," from *The Data Game: Controversies in Social Science Statistics* (New York: M. E. Sharpe, 1999). Reprinted with permission.

- *Income in the United States: 2002*. Bureau of the Census, Current Population Reports, P60–221. September 2003.

- *The Big Payoff: Educational Attainment and Synthetic Estimates of Work-Life Earnings*. Bureau of the Census, Current Population Reports, P23–210. July 2002.

Exploring on Your Own

Your journey into exploratory data analysis is just beginning. You now have some tools to examine further the complex relationship between education and income. You may want to explore answers to the questions raised above or to your own questions and conjectures. For example, what other variables besides age do you think affect earnings? What other variables besides gender may affect the relationship between education and earnings? How would our analysis change if we used other income measures, such as personal total income or family income?

Working with Partners

You may want to work with a partner so that you can discuss questions that may be worth pursuing and help each other interpret and analyze the findings. In addition, you can compare two regression lines more easily by using two computers or two graphing calculators.

Generating Conjectures

One way to start is to generate conjectures about the effects of other variables or different income measures on the relationship between education and earnings. (See the Data Dictionary in Table 1 for variables included in the FAM1000 Census data.) You can then generate and compare regression lines using the procedures described next.

Procedures for Finding Regression Lines

1. Finding regression lines:
   a. *Using a computer:*
      Open "F3: Regression with Multiple Subsets" in *FAM1000 Census Graphs*. This program allows you to find regression lines for education vs. earnings for different income variables and for different groups of people. Select (by clicking on the appropriate box) one of the four income variables: personal earnings, personal hourly wage, personal total income, or family income. Then select at least two regression lines that it would make sense to compare (e.g., men vs. women, white vs. non-white, two or more regions of the country). You should do some browsing through the various regression line options to pick those that are the most interesting. Print out your regression lines (on overhead transparencies if possible) so that you can present your findings to the class.

   b. *Using graphing calculators and graph link files:* FAM1000 graph link files A to H contain data for generating regression lines for several income variables as a function of years of education. The Graphing Calculator Manual contains descriptions and hardcopy of the files, as well as instructions for downloading and transferring them to TI-83 and TI-84 calculators. Decide on at least two regression lines that are interesting to compare.

2. For each of the regression lines you choose, work together with your partner to record the following information in your notebooks:

   The equation of the regression line
   > What the variables represent
   > A reasonable domain
   > The subset of the data the line represents (men? non-whites?)

   The correlation coefficient
   > Whether or not the line is a good fit and why

   The slope
   > Interpretation of the slope (e.g., for each additional year of education, median personal earnings rise by such and such an amount)

Discussion/Analysis

With your partner, explore ways of comparing the two regression lines. What do the correlation coefficients tell you about the strength of these relationships? How do the two slopes compare? Is one group better off? Is that group better off no matter how many years of education they have? What factors might be hidden or not taken into account?

Were your original conjectures supported by your findings? What additional evidence could be used to support your analysis? Are your findings surprising in any way? If so, why?

You may wish to continue researching questions raised by your analysis by returning to the original FAM1000 data or examining additional sources such as the related readings at *www.wiley.com/college/kimeclark* or the Current Population Survey website at *www.census.gov/cps*.

Results

Prepare a 60-second summary of your results. Discuss with your partner how to present your findings. What are the limitations of the data? What are the strengths and weaknesses of your analysis? What factors are hidden or not taken into account? What questions are raised?

EXERCISES

Technology is required for generating scatter plots and regression lines in Exercises 11, 12, and 18.

1. **a.** Evaluate each of the following:

   $$|0.65| \qquad |-0.68| \qquad |-0.07| \qquad |0.70|$$

   b. List the absolute values in part (a) in ascending order from the smallest to the largest.

2. The accompanying figures show regression lines and corresponding correlation coefficients for four different scatter plots. Which of the lines describes the strongest linear relationship between the variables? Which of the lines describes the weakest linear relationship?

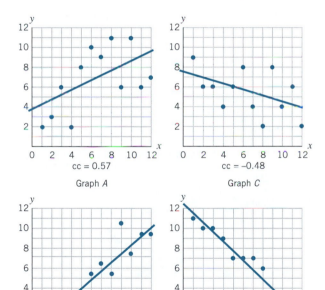

Four regression lines with their correlation coefficients.

3. The following equation represents the best-fit regression line for median personal earnings vs. years of education for the 298 people in the FAM1000 data set who live in the southern region of the United States.

   $$S = -2105 + 6139E$$

   where S = median personal earnings in the South and E = years of education past grade 8, and cc = 0.72.

a. Identify the slope of the regression line, the vertical intercept, and the correlation coefficient.

b. What does the slope mean in this context?

c. By what amount does this regression line predict that median personal earnings for those who live in the South change for 1 additional year of education? For 10 additional years of education?

In Exercises 4 to 6, the data analyzed are from the FAM1000 data files and the equations can be generated using *FAM1000 Census Graphs*. Here, the income measure is personal total income, which includes personal earnings (from work) and other sources of unearned income, such as interest and dividends on investments.

4. The accompanying graph and regression line show median personal total income vs. years of education past grade 8.

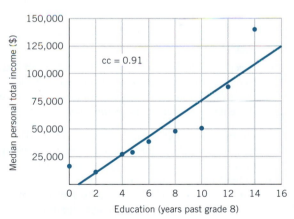

Median personal total income =
$$-3687 + 7994 \cdot \text{yrs. educ. past grade 8}$$

a. What is the slope of the regression line?

b. Interpret the slope in this context.

c. By what amount does this regression line predict that median personal total income changes for 1 additional year of education? For 10 additional years of education?

d. What features of the data are not well described by the regression line?

5. The following equation represents a best-fit regression line for median personal total income of white males vs. years of education past grade eight:

$$\text{median personal total income}_{\text{white males}} =$$
$$-8850 + 10{,}773 \cdot \text{yrs. educ. past grade 8}$$

The correlation coefficient is 0.87 and the sample size is 452 white males.

a. What is the rate of change of median personal total income with respect to years of education?

b. Generate three points that lie on this regression line. Use two of these points to calculate the slope of the regression line.

c. How does this slope relate to your answer to part (a)?

d. Sketch the graph.

6. From the FAM1000 data, the best-fit regression line for median personal total income of white females vs. years of education past grade 8 is:

$$\text{median personal total income}_{\text{white females}} =$$
$$6760 + 4114 \cdot \text{yrs. educ. past grade 8}$$

The correlation coefficient is 0.93 and the sample size is 374 white females.

a. Interpret the number 4114 in this equation.

b. Generate a small table with three points that lie on this regression line. Use two of these points to calculate the slope of the regression line.

c. How does this slope relate to your answer to part (a)?

d. Sketch the graph.

e. Describe some differences between median personal total income vs. education for white females and for white males (see Exercise 5). What are some of the limitations of the model in making this comparison?

7. The term "linear regression" was coined in 1903 by Karl Pearson as part of his efforts to understand the way physical characteristics are passed from generation to generation. He assembled and graphed measurements of the heights of fathers and their fully grown sons from more than a thousand families. The independent variable, F, was the height of the fathers. The dependent variable, S, was the mean height of the sons who all had fathers with the same height. The best-fit line for the data points had a slope of 0.516, which is much less than 1. If, on average, the sons grew to the same height as their fathers, the slope would equal 1. Tall fathers would have tall sons and short fathers would have equally short sons. Instead, the graph shows that whereas the sons of tall fathers are still tall, they are not (on average) as tall as their fathers. Similarly, the sons of short fathers are not as short as their fathers. Pearson termed this *regression;* the heights of sons *regress* back toward the height that is the mean for that population. The equation of this regression line is $S = 33.73 + 0.516F$, where $F =$ height of fathers in inches and $S =$ mean height of sons in inches.

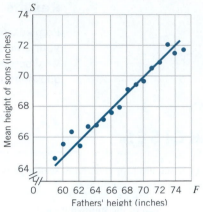

From Snedecor and Cochran, *Statistical Methods,* 8th ed. By permission of the Iowa State University Press. Copyright © 1967.

a. Interpret the number 0.516 in this context.

b. Use the regression line to predict the mean height of sons whose fathers are 64 inches tall and of those whose fathers are 73 inches tall.

c. Predict the height of a son who has the same height as his father.

d. If there were over 1000 families, why are there only 17 data points on this graph?

8. The book *Performing Arts—The Economic Dilemma* studied the economics of concerts, operas, and ballets. It included the following scatter plot and corresponding regression line relating attendance per concert to the number of concerts given, for a major orchestra. What do you think were their conclusions?

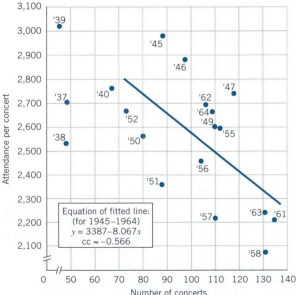

Source: William Baumol and William G. Bowen, *Performing Arts—The Economic Dilemma.* Reprinted with permission from Twentieth Century Fund.

9. (Optional use of technology.) The accompanying graph gives the mean annual cost for tuition and fees at public and private 4-year colleges in the United States since 1985.

Mean Cost of 4-Year Colleges, Public and Private

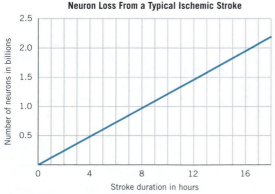

Source: U.S. Bureau of the Census, *Statistical Abstract of the United States: 2006.*

It is clear that the cost of higher education is going up, but public education is still less expensive than private. The graph suggests that costs of both public and private education versus time can be roughly represented as straight lines.

a. By hand, sketch two lines that best represent the data. Calculate the rate of change of education cost per year for public and for private education by estimating the coordinates of two points that lie on the line, and then estimating the slope.

b. Construct an equation for each of your lines in part (a). (Set 1985 as year 0.) If you are using technology, generate two regression lines from the Excel or graph link data file EDUCOSTS and compare these equations with the ones you constructed.

c. If the costs continue to rise at the same rates for both sorts of schools, what would be the respective costs for public and private education in the year 2010? Does this seem plausible to you? Why or why not?

10. Stroke is the third-leading cause of death in the United States, behind heart disease and cancer. The accompanying graph shows the average neuron loss for a typical ischemic stroke.

Neuron Loss From a Typical Ischemic Stroke

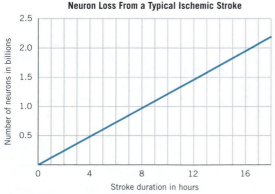

Source: American Heart Association.

a. Find the slope of the line by estimating the coordinates of two points on the line. Interpret the meaning of the slope in this context.

b. Construct an equation for the line, where n = number of neurons in billions and d = duration of stroke in hours.

c. The average human forebrain has about 22 billion neurons and the average stroke lasts about 10 hours. Find the percentage of neurons that are lost from a 10-hour stroke.

11. (Requires technology.) The accompanying table shows (for the years 1965 to 2005 and for people 18 and over) the total percentage of cigarette smokers, the percentage of males who are smokers, and the percentage of females who are smokers.

Percentage of Adult Smokers (18 years and older)

Year	Total Population	Males	Females
1965	42.4	51.9	33.9
1974	37.1	43.1	32.1
1979	33.5	37.5	29.9
1983	32.1	35.1	29.5
1985	30.1	32.6	27.9
1987	28.8	31.2	26.5
1988	28.1	30.8	25.7
1990	25.5	28.4	22.8
1991	25.6	28.1	23.5
1992	26.5	28.6	24.6
1993	25.0	27.7	22.5
1994	25.5	28.2	23.1
1995	24.7	27.0	22.6
2000	23.3	25.7	21.0
2003	21.6	24.1	19.2
2005	20.9	23.9	18.1

Source: U.S. Bureau of the Census, *Statistical Abstract of the United States: 2006.*

a. Construct a scatter plot of the percentage of all smokers 18 and older vs. time.

   i. Calculate the average rate of change from 1965 to 2005.

   ii. Calculate the average rate of change from 1990 to 2005. Be sure to specify the units in each case.

b. On your graph, sketch an approximate regression line. By estimating coordinates of points on your regression line, calculate the average rate of change of the percentage of total smokers with respect to time.

c. Using technology, generate a regression line for the percentage of all smokers 18 and older as a function of time. (Set 1965 as year 0.) Record the equation and the correlation coefficient. How good a fit is this regression line to the data? Compare the rate of change for your hand-generated regression line to the rate of change for the technology-generated regression line.

d. Using technology, generate and record regression lines (and their associated correlation coefficients) for the percentages of both males and females who are smokers vs. time.

e. Write a summary paragraph using the results from your graphs and calculations to describe the trends in smoking from 1965 to 2005.

12. (Requires technology.) The accompanying table shows the calories per minute burned by a 154-pound person moving at speeds from 2.5 to 12 miles/hour (mph). (*Note:* A fast walk is about 5 mph; faster than that is considered jogging or slow running.) Marathons, about 26 miles long, are now run in slightly over 2 hours, so that top distance runners are approaching a speed of 13 mph.

Speed (mph)	Calories per Minute	Speed (mph)	Calories per Minute
2.5	3.0	6.0	12.0
3.0	3.7	7.0	14.0
3.5	4.2	8.0	15.6
4.0	5.5	9.0	17.5
4.5	7.0	10.0	19.6
5.0	8.3	11.0	21.7
5.5	10.1	12.0	24.5

a. Plot the data.

b. Does it look as if the relationship between speed and calories per minute is linear? If so, generate a linear model. Identify the variables and a reasonable domain for the model, and interpret the slope and vertical intercept. How well does your line fit the data?

c. Describe in your own words what the model tells you about the relationship between speed and calories per minute.

13. (Optional use of technology.) The accompanying graph shows the winning running times in minutes for women in the Boston Marathon.

MARATHON

Source: www.bostonmarathon.org/BostonMarathon/PastChampions.asp.

a. Sketch a line that best approximates the data by hand. Set 1972 as year 0 and compute an equation for this line. Interpret the slope of your line in this context. If using technology, generate a regression line from the data in the Excel or graph link file MARATHON and compare its equation with the equation you computed by hand.

b. If the marathon times continue to change at the rate given in your linear model, predict the winning running time for the women's marathon in 2010. Does that seem reasonable? If not, why not?

c. The graph seems to flatten out after about 1986. Based on this trend, what would you predict for the winning running time for the women's marathon in 2010? Does this prediction seem more realistic than your previous prediction?

d. Write a short paragraph summarizing the trends in the Boston Marathon winning times for women.

14. (Optional use of technology.) The accompanying graph shows the world record times for the men's mile. As of January 2006, the 1999 record still stands. Note that several times the standing world record was broken more than once during a year.

MENSMILE

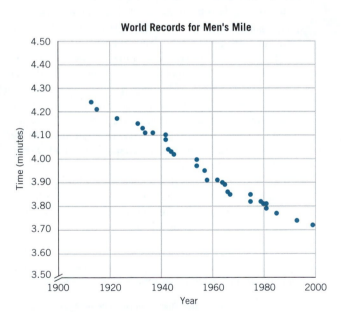

Source: www.runnersworld.com.

a. Generate a line that approximates the data (by hand or, if using technology, with the data in the Excel or graph link file MENSMILE. (Set 1910 or 1913 as year 0). If you are using technology, specify the correlation coefficient. Interpret the slope of your line in this context.

b. If the world record times continue to change at the rate specified in your linear model, predict the record time for the men's mile in 2010. Does your prediction seem reasonable? If not, why not?

c. In what year would your linear model predict the world record to be 0 minutes? Since this is impossible, what do you think is a reasonable domain for your model? Describe what you think would happen in the years after those included in your domain.

d. Write a short paragraph summarizing the trends in the world record times for the men's mile.

15. The temperature at which water boils is affected by the difference in atmospheric pressure at different altitudes above sea level. The classic cookbook *The Joy of Cooking* by Irma S. Rombauer and Marion Rombauer Becker gives the data in the accompanying table (rounded to the nearest degree) on the boiling temperature of water at different altitudes above sea level.

Boiling Temperature of Water

Altitude (ft above sea level)	Temperature Boiling °F
0	212
2,000	208
5,000	203
7,500	198
10,000	194
15,000	185
30,000	158

a. Use the accompanying table for the following:

i. Plot boiling temperature in degrees Fahrenheit, °F, vs. the altitude. Find a formula to describe the boiling temperature of water, in °F, as a function of altitude.

ii. According to your formula, what is the temperature at which water will boil where you live? Can you verify this? What other factors could influence the temperature at which water will boil?

b. The highest point in the United States is Mount McKinley in Alaska, at 20,320 feet above sea level; the lowest point is Death Valley in California, at 285 feet below sea level. You can think of distances below sea level as negative altitudes from sea level. At what temperature in degrees Fahrenheit will water boil in each of these locations according to your formula?

c. Using your formula, find an altitude at which water can be made to boil at 32°F, the freezing point of water at sea level. At what altitude would your formula predict this would happen? (Note that airplane cabins are pressurized to near sea-level atmospheric pressure conditions in order to avoid unhealthy conditions resulting from high altitude.)

16. (Optional use of technology.) The accompanying graph shows the world distance records for the women's long jump. Several times a new long-jump record was set more than once during a given year.

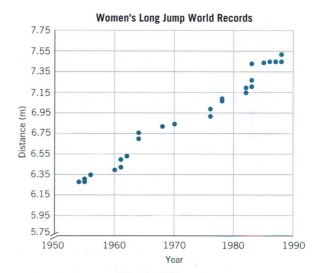

Note: As of January 2006, the 1988 long jump record still stands.
Source: Data extracted from the website at *http://www.uta.fi/~csmipe/sports/eng/mwr.html.*

a. Generate a line that approximates the data (by hand or, if using technology, use the data from the Excel or graph link file LONGJUMP. (Set 1950 or 1954 as year 0). Interpret the slope of your line in this context. If you are using technology, specify the correlation coefficient.

b. If the world record distances continue to change at the rate described in your linear model, predict the world record distance for the women's long jump in the year 2010.

c. What would your model predict for the record in 1943? How does this compare with the actual 1943 record of 6.25 meters? What do you think would be a reasonable domain for your model? What do you think the data would look like for years outside your specified domain?

17. (Optional use of technology.) The accompanying graph shows the increasing number of motor vehicle registrations (cars and trucks) in the United States.

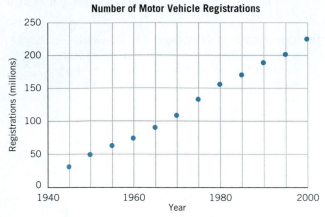

Number of Motor Vehicle Registrations

Source: U.S. Federal Highway Administration, *Highway Statistics,* annual.

a. Using 1945 as the base year, find a linear equation that would be a reasonable model for the data. If using technology, use the data in the Excel or graph link file MOTOR.

b. Interpret the slope of your line in this context.

c. What would your model predict for the number of motor vehicle registrations in 2004? How does this compare with the actual data value of 228.3 million?

d. Using your model, how many motor vehicles will be registered in the United States in 2010? Do you think this is a reasonable prediction? Why or why not?

18. (Requires technology.) Examine the following data on U.S. union membership from 1930 to 2004.

U.S. Union Membership, 1930–2004

Year	Labor Force* (thousands)	Union Members† (thousands)	Percentage of Labor Force
1930	29,424	3,401	11.6
1940	32,376	8,717	26.9
1950	45,222	14,267	31.5
1960	54,234	17,049	31.4
1970	70,920	19,381	27.3
1980	90,564	19,843	21.9
1990	103,905	16,740	16.1
2000	120,786	16,258	13.5
2003	122,481	15,800	12.9
2004	123,564	15,472	12.5

*Does not include agricultural employment; from 1985, does not include self-employed or unemployed persons.
†From 1930 to 1980, includes dues-paying members of traditional trade unions, regardless of employment status; from 1985, includes members of employee associations that engage in collective bargaining with employers.
Source: Bureau of Labor Statistics, U.S. Dept. of Labor.

a. Graph the percentage of labor force in unions vs. time from 1950 to 2004. Measuring time in years since 1950, find a linear regression formula for these data using technology.

b. When does your formula predict that only 10% of the labor force will be unionized?

c. What data would you want to examine to understand why union membership is declining?

19. If a study shows that smoking and lung cancer have a high positive correlation, does this mean that smoking causes lung cancer? Explain your answer.

20. Parental income has been found to have a high positive correlation with their children's academic success. What are two different ways you could interpret this finding?

WHEN LINES MEET: LINEAR SYSTEMS

OVERVIEW

When is solar heating cheaper than conventional heating? Will you pay more tax under a flat tax or a graduated tax plan? We can answer such questions using a system of linear equations or inequalities.

After reading this chapter, you should be able to

- construct, graph, and interpret:
  - systems of linear equations
  - systems of linear inequalities
  - piecewise linear functions

- find a solution for:
  - a system of two linear equations
  - a system of two linear inequalities

3.1   *Systems of Linear Equations*

An Economic Comparison of Solar vs. Conventional Heating Systems

On a planet with limited fuel resources, heating decisions involve both monetary and ecological considerations. Typical costs for three different kinds of heating systems for a small one-bedroom housing unit are given in Table 3.1.

Typical Costs for Three Heating Systems

Type of System	Installation Cost ($)	Operating Cost ($/yr)
Electric	5,000	1,100
Gas	12,000	700
Solar	30,000	150

Table 3.1

Solar heating is clearly the most costly to install and the least expensive to run. Electric heating, conversely, is the cheapest to install and the most expensive to run.

Setting up a system

By converting the information in Table 3.1 into equations, we can find out when the solar heating system begins to pay back its initially higher cost. If no allowance is made for inflation or changes in fuel price,[1] the general equation for the total cost, C, is

$$C = \text{installation cost} + (\text{annual operating cost})(\text{years of operation})$$

If we let n equal the number of years of operation and use the data from Table 3.1, we can construct the following linear equations:

$$C_{\text{electric}} = 5000 + 1100n$$
$$C_{\text{gas}} = 12{,}000 + 700n$$
$$C_{\text{solar}} = 30{,}000 + 150n$$

Together they form a *system of linear equations*. Table 3.2 gives the cost data at 5-year intervals, and Figure 3.1 shows the costs over a 40-year period for the three heating systems.

Heating System Total Costs

Year	Electric ($)	Gas ($)	Solar ($)
0	5,000	12,000	30,000
5	10,500	15,500	30,750
10	16,000	19,000	31,500
15	21,500	22,500	32,250
20	27,000	26,000	33,000
25	32,500	29,500	33,750
30	38,000	33,000	34,500
35	43,500	36,500	35,250
40	49,000	40,000	36,000

Table 3.2

---

[1]A more sophisticated model might include many other factors, such as interest, repair costs, the cost of depleting fuel resources, risks of generating nuclear power, and what economists call opportunity costs.

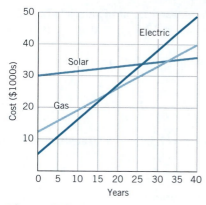

Figure 3.1 Comparison of home heating costs.

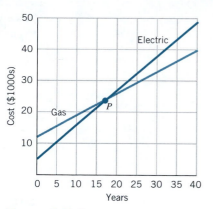

Figure 3.2 Gas versus electric.

Comparing costs

Using Graphs   Let's compare the costs for gas and electric heat. Figure 3.2 shows the graphs of the equations for these two heating systems. The point of intersection, P, shows where the lines predict the *same* total cost for both gas and electricity, given a certain number of years of operation. From the graph, we can estimate the coordinates of the point P:

$$P = (\text{number of years of operation, cost})$$
$$\approx (17, \$24{,}000)$$

The total cost of operation is about \$24,000 for both gas and electric after about 17 years of operation. We can compare the relative costs of each system to the left and right of the point of intersection. Figure 3.2 shows that gas is less expensive than electricity to the right of the intersection point and more expensive than electricity to the left of the intersection point.

Using Equations   Where the gas and electric lines intersect, the coordinates satisfy both equations. At the point of intersection, the total cost of electric heat, C_{electric}, equals the total cost of gas heat, C_{gas}. Thus, the two expressions for the total cost can be set equal to each other to find the exact values for the coordinates of the intersection point:

$$C_{\text{electric}} = 5000 + 1100n \qquad (1)$$
$$C_{\text{gas}} = 12{,}000 + 700n \qquad (2)$$

Set (1) equal to (2)	$C_{\text{electric}} = C_{\text{gas}}$
substitute	$5000 + 1100n = 12{,}000 + 700n$
subtract 5000 from each side	$1100n = 7000 + 700n$
subtract 700n from each side	$400n = 7000$
divide each side by 400	$n = 17.5 \text{ years}$

When $n = 17.5$ years, the total cost for electric or gas heating is the same. The total cost can be found by substituting this value for n in Equation (1) or (2):

Substitute 17.5 for n in Equation (1) $C_{\text{electric}} = 5000 + 1100(17.5)$

$$= 5000 + 19{,}250$$

$$C_{\text{electric}} = \$24{,}250$$

Since we claim that the pair of values (17.5, $24,250) satisfies both equations, we need to check, when $n = 17.5$ years, that C_{gas} is also $24,250:

Substitute 17.5 for n in Equation (2) $C_{gas} = 12,000 + 700(17.5)$

$$= 12,000 + 12,250$$

$$C_{gas} = \$24,250$$

The coordinates (17.5, $24,250) satisfy both equations.

Solutions to a system

When $n = 17.5$ years, then $C_{electric} = C_{gas} = \$24,250$. The point (17.5, $24,250) is called a *solution* to the system of these two equations. After 17.5 years, a total of $24,250 could have been spent on heat for either an electric or a gas heating system.

A Solution to a System of Equations

A pair of real numbers is a *solution* to a system of equations in two variables if and only if the pair of numbers is a solution to each equation in the system.

E X A M P L E   1   **Estimating solutions to systems from graphs**

Using the graphs in Figures 3.3 and 3.4, estimate when the cost will be the same for each of the following systems:

a. Electric vs. solar heating

b. Gas vs. solar heating

You will be asked to find more accurate solutions in the exercises.

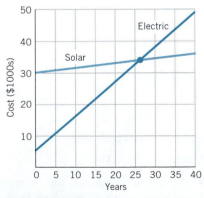

Figure 3.3 Electric versus solar heating.

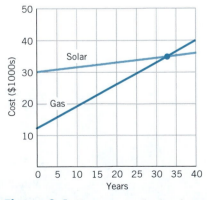

Figure 3.4 Gas versus solar heating.

SOLUTION   **a.** The graphs of electric and solar costs intersect at approximately (26, \$34,000). That means the costs of electric and solar heating are both approximately \$34,000 at about 26 years of operation.

b. Gas and solar heating costs are equivalent at approximately (33, \$35,000), or in 33 years at a cost of \$35,000.

Algebra Aerobics 3.1

1. Our model tells us that electric systems are cheaper than gas for the first 17.5 years of operation. Using Figure 3.1, estimate the interval over which gas is the cheapest of the three heating systems. When does solar heating become the cheapest system compared with the other two heating systems?

2. For the system of equations

$$4x + 3y = 9$$
$$5x + 2y = 13$$

a. Determine whether $(3, -1)$ is a solution.

b. Show why $(1, 4)$ is not a solution for this system.

3. a. Show that the following equations are equivalent:

$$4x = 6 + 3y$$
$$12x - 9y = 18$$

b. How many solutions are there for the system of equations in part (a)?

4. Estimate the solution(s) for each of the systems of equations in Figure 3.5.

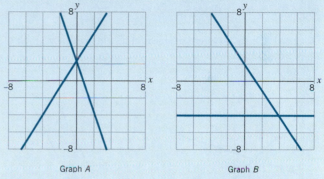

Graph A   Graph B

Figure 3.5 Two graphs of systems of linear equations.

Exercises for Section 3.1

1. a. Determine whether $(5, -10)$ is a solution for the following system of equations:

$$4x - 3y = 50$$
$$2x + 2y = 5$$

b. Explain why $(-10, 5)$ is not a solution for the system in part (a).

2. a. Determine whether $(-2, 3)$ is a solution for the following system of equations:

$$3x + y = -3$$
$$x + 2y = 4$$

b. Explain why $(3, -2)$ is not a solution for the system in part (a).

3. Explain what is meant by "a solution to a system of equations."

4. Estimate the solution to the system of linear equations graphed in the accompanying figure.

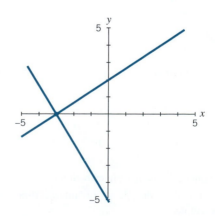

5. a. Estimate the coordinates of the point of intersection on the accompanying graph.

b. Describe what happens to the population of Pittsburgh in relation to the population of Las Vegas to the right and to the left of this intersection point.

Population Trends for Pittsburgh, PA and Las Vegas, NV

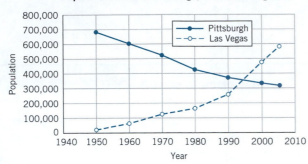

6. a. Match each system of linear equations with the graph of the system.

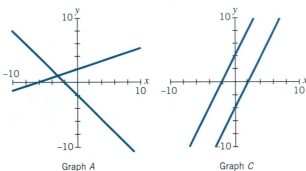

Graph A   Graph C

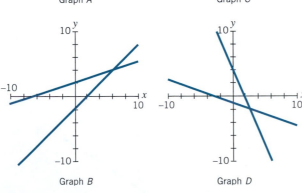

Graph B   Graph D

i. $y = -x - 2$ $y = \frac{1}{3}x + 2$

ii. $y - 2x = 4$ $y - 2x = -4$

iii. $y - x = -2$ $3y - x = 6$

iv. $3y + x = -3$ $y = -2x + 4$

b. Estimate the point of intersection if there is one.

c. Verify that the point of intersection satisfies both equations.

7. Construct a sketch of each system by hand and then estimate the solution(s) to the system (if any).

a. $x + 2y = 1$ **b.** $x + y = 9$
$x + 4y = 3$ $2x - 3y = -2$

8. For the linear system $\begin{cases} x - y = 5 \\ 2x + y = 1 \end{cases}$

a. Graph the system. Estimate the solution for the system and then find the exact solution.

b. Check that your solution satisfies both of the original equations.

9. Create the system of equations that produced the accompanying graph. Estimate the solution for the system from the graph and then confirm using your equations.

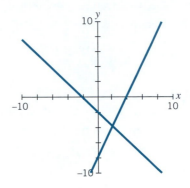

10. The U.S. Bureau of Labor Statistics reports on the percent of all men in the civilian work force, and the corresponding percent of all women. Lines of best-fit on the accompanying graph are used to make predictions about the future of the labor force.

Projected Civilian Labor Participation by Sex

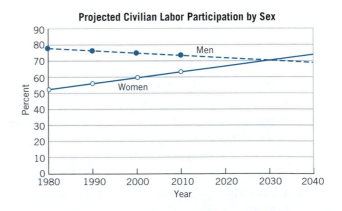

a. Estimate the point of intersection.

b. Describe the meaning of the point of intersection.

3.2 *Finding Solutions to Systems of Linear Equations*

In the heating example, we found a solution for a system of two equations by finding the point where the graphs of the two equations intersect. At the point of intersection, the equations have the same value for the independent variable (input) and the same value for the dependent variable (output).

Visualizing Solutions

For a single linear equation, the graph of its solutions is one line, and every point on that line is a solution. So, one linear equation has an unlimited number of solutions. In a system of two linear equations, a solution must satisfy both equations. We can easily visualize what might happen. If we graph two different straight lines and the lines intersect (Figure 3.6), there is only one solution—at the intersection point. The coordinates of the point of intersection can be estimated by inspecting the graph. For example, in Figure 3.6 the two lines appear to cross near the point $(-1, 1.5)$. If the lines are parallel, they never intersect and there are no solutions (Figure 3.7).

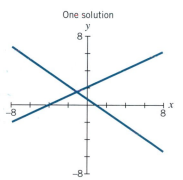

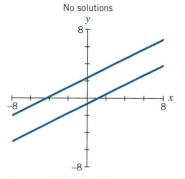

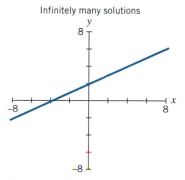

Figure 3.6 Lines intersect at a single point.

Figure 3.7 Parallel lines never intersect.

Figure 3.8 The two equations represent the same line. Every point on the line is a solution to both equations.

How else might the graphs of two lines be related? The two equations could represent the same line. Consider the following two equations:

$$y = 0.5x + 2 \tag{1}$$
$$3y = 1.5x + 6 \tag{2}$$

If we multiply each side of Equation (1) by 3, we obtain Equation (2). The two equations are *equivalent*, since any solution of one equation is also a solution of the other equation. The two equations represent the same line. There are an infinite number of points on that line, and they are all solutions to both equations (Figure 3.8).

The Number of Solutions for a Linear System

On the graph of a system of two linear equations, a *solution* is a point where the two lines intersect. There can be

    One solution, if the lines intersect once

    No solution, if the lines are parallel and distinct

    Infinitely many solutions, if the two lines are the same

Strategies for Finding Solutions

A system of two linear equations can be solved in several ways, each of which will provide the same result. In each case a solution, if it exists, consists of either one or infinitely many pairs of numbers of the form (x, y). The form of the equations can help determine the most efficient strategy. Two of the most common methods are *substitution* and *elimination*.

Substitution method

When at least one of the equations is in (or can easily be converted to) function form, $y = b + mx$, we can use the substitution method. The point of intersection is a solution to both equations, so at that point the two equations have the same value for x and the same value for y. To find values for the intersection point, we can start by substituting the expression for y from one equation into the other equation.

EXAMPLE 1

When both equations are in function form

A long-distance phone plan seen on TV costs $0.03 per minute plus a fixed charge of $0.39 per call. Your current service charges $0.05 per minute plus a fixed charge of $0.20 per call. During one call, after how many minutes would the cost be the same under the two plans?

SOLUTION

If we let n = the number of minutes in one call, then

$$C_1 = 0.39 + 0.03n \qquad \text{models the cost of one call on the TV plan}$$
$$C_2 = 0.20 + 0.05n \qquad \text{models the cost of one call on your current plan}$$

This is the simplest case of substitution, where both equations are already in function form. To find out when the costs of the two plans are equal, that is, when $C_1 = C_2$, we can substitute C_2 for C_1.

Given the equation for C_1	$C_1 = 0.39 + 0.03n$
substitute C_2 for C_1	$0.20 + 0.05n = 0.39 + 0.03n$
solve for n	$0.05n - 0.03n = 0.39 - 0.20$
	$0.02n = 0.19$
	$n = 0.19/0.02 = 9.5 \text{ minutes}$

When $n = 9.5$, the costs of the two plans are equal. To find the cost, we can substitute $n = 9.5$ into the equation for C_1 or C_2. Using C_1, we have

$$C_1 = 0.39 + 0.03(9.5)$$
$$= 0.39 + 0.285$$
$$= \$0.675$$

So for a call lasting 9.5 minutes, the cost on either plan is approximately 68 cents.

Double-Checking the Solution. We can check our answer by substituting $n = 9.5$ into C_2.

$$C_2 = 0.20 + 0.05(9.5)$$
$$= 0.20 + 0.475$$
$$= \$0.675$$

This confirms our original calculation.

EXAMPLE 2

When only one equation is in function form

a. Find the point (if any) where the graphs of the following two linear equations intersect:

$$6x + 7y = 25 \tag{1}$$
$$y = 15 + 2x \tag{2}$$

b. Graph the two equations on the same grid, labeling their intersection point.

SOLUTION   **a.** In Equation (1) substitute the expression for y from Equation (2):

$$6x + 7(15 + 2x) = 25$$

Simplify
$$6x + 105 + 14x = 25$$
$$20x = -80$$
$$x = -4$$

We can use one of the original equations to find the value for y when $x = -4$. Using Equation (2),

$$y = 15 + 2x \qquad\qquad (2)$$

Substitute -4 for x
$$y = 15 + 2(-4)$$

multiply
$$y = 15 - 8$$
$$y = 7$$

Try double-checking your answer in Equation (1).

b. Figure 3.9 shows a graph of the two equations and the intersection point at $(-4, 7)$.

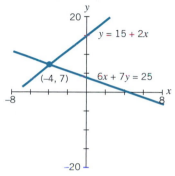

Figure 3.9 Graphs of $6x + 7y = 25$ and $y = 15 + 2x$.

Algebra Aerobics 3.2a

1. Solve for the indicated variable.
   a. $2x + y = 7$ for y
   b. $3x + 5y = 6$ for y
   c. $x - 2y = -1$ for x

2. Determine the number of solutions without solving the system. Justify your answer.
   a. $y = 3x - 5$ **b.** $y = 2x - 4$
       $y = 3x + 8$ $y = 3x - 4$

3. Solve the following systems of equations using the substitution method.
   a. $y = x + 4$ **b.** $y = -1700 + 2100x$
       $y = -2x + 7$ $y = 4700 + 1300x$

   c. $F = C$
   $$F = 32 + \tfrac{9}{5}C$$

   [Part (c) was a question on the TV program *Who Wants to Be a Millionaire?*]

4. Solve the following systems of equations using the substitution method.
   a. $y = x + 3$ **c.** $x = 2y - 5$
       $5y - 2x = 21$ $4y - 3x = 9$
   b. $z = 3w + 1$ **d.** $r - 2s = 5$
       $9w + 4z = 11$ $3r - 10s = 13$

Elimination method

Another method, called *elimination,* can be useful when neither equation is in function form. The strategy is to modify the equations (through multiplication or rearrangement) so that adding (or subtracting) the modified equations eliminates one variable.

EXAMPLE 3 Assume you have $10,000 to invest in an "up market" when the economy is booming. You want to split your investment between conservative bonds and riskier stocks. The bonds will stay fixed in value but return a guaranteed 7% per year in dividends. The stocks pay no dividends, but your return is from the increase in stock value, predicted to be 14% per year. Overall you want a 12% or $1200 return on your $10,000 at the end of one year.

a. How much should you invest in bonds and how much in stocks?

b. What if the economy has a drastic downturn, as it did between 1999 and 2002? Assuming you split your investment as recommended in part (a), what will your return be after one year if your stock value decreased by 10%?

SOLUTION a. If B = $ invested in bonds and S = $ invested in stocks, then

$$B + S = \$10,000 \tag{1}$$

The expected return on your investments after one year is

$$(7\% \text{ of } B) + (14\% \text{ of } S) = \$1200$$

or $\qquad\qquad 0.07B + 0.14S = \$1200 \tag{2}$

We can solve the system by eliminating one variable, in this case B, from both equations. Given the two equations

$$B + S = \$10,000 \tag{1}$$
$$0.07B + 0.14S = \$1200 \tag{2}$$

If we multiply both sides of Equation (1) by 0.07 we get an equivalent Equation (1)*, which has the same coefficient for B as Equation (2). We do this so that we can subtract the equations and eliminate B.

Given	$0.07B + 0.07S = \$700$	(1)*
subtract Equation (2)	$-(0.07B + 0.14S = \$1200)$	(2)
to eliminate B	$-0.07S = -\$500$	

Dividing both sides by -0.07, we have $S \approx \$7143$.

So to achieve your goal of a $1200 return on $10,000 you would need to invest $7143 in stocks and $10,000 - \$7143 = \$2857$ in bonds.

b. If the economy turns sour at the end of the year and the value of your stock drops 10%, then

$$\text{your return} = (7\% \text{ of } \$2857 \text{ from bonds}) - (10\% \text{ of } \$7143 \text{ from stocks})$$
$$\approx \$200 - \$714$$
$$= -\$514$$

So you would lose over $500 on your $10,000 investment that year.

How can you tell if your system has no or infinitely many intersection points?

As we remarked at the beginning of this section, two parallel lines will never intersect and duplicate lines will intersect everywhere, creating infinitely many intersection points. How can you tell whether or not your system of equations falls into either category?

EXAMPLE 4 **A system with no solution: Parallel lines**

a. Solve the following system of two linear equations:

$$y = 20,000 + 1500x \tag{1}$$
$$2y - 3000x = 50,000 \tag{2}$$

b. Graph your results.

SOLUTION   **a.** If we assume the two lines intersect, then the *y* values for Equations (1) and (2) are the same at some point. We can try to find this value for *y* by substituting the expression for *y* from Equation (1) into Equation (2):

$$2(20{,}000 + 1500x) - 3000x = 50{,}000$$

Simplify $\qquad 40{,}000 + 3000x - 3000x = 50{,}000$

false statement $\qquad\qquad\qquad 40{,}000 = 50{,}000 \ (???)$

What could this possibly mean? Where did we go wrong? If we return to the original set of equations and solve Equation (2) for *y* in terms of *x*, we can see why there is no solution for this system of equations.

$$2y - 3000x = 50{,}000 \qquad\qquad (2)$$

Add 3000*x* to both sides $\qquad 2y = 50{,}000 + 3000x$

divide by 2 $\qquad\qquad\qquad y = 25{,}000 + 1500x \qquad (2)^*$

Equations (1) and (2)* (rewritten form of the original Equation (2)) have the same slope of 1500 but different *y*-intercepts.

$$y = 20{,}000 + 1500x \qquad\qquad (1)$$
$$y = 25{,}000 + 1500x \qquad\qquad (2)^*$$

Written in this form, we can see that we have two parallel lines, so the lines never intersect (see Figure 3.10). Our initial premise, that the two lines intersected and thus the two *y*-values were equal at some point, was incorrect.

b. Figure 3.10 shows the graphs of the two lines.

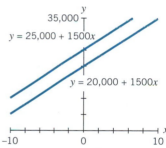

Figure 3.10 Graphs of $y = 25{,}000 + 1500x$ and $y = 20{,}000 + 1500x$.

EXAMPLE 5   **A system with infinitely many solutions: Equivalent equations**
Solve the following system:

$$45x = -y + 33 \qquad\qquad (1)$$
$$2y + 90x = 66 \qquad\qquad (2)$$

SOLUTION   As always, there are multiple ways of solving the system. One strategy is to put both equations in function form:

Solve Equation (1) for *y* $\qquad\qquad 45x = -y + 33 \qquad (1)$

add *y* to both sides $\qquad\qquad\quad y + 45x = 33$

add $-45x$ to both sides $\qquad\qquad\quad y = -45x + 33$

Solve Equation (2) for *y* $\qquad\quad 2y + 90x = 66 \qquad (2)$

add $-90x$ to both sides $\qquad\qquad 2y = -90x + 66$

divide by 2 $\qquad\qquad\qquad\qquad\quad y = -45x + 33$

The two original equations really represent the same line, $y = -45x + 33$, so any of the infinitely many points on the line is a solution to the system (Figure 3.11).

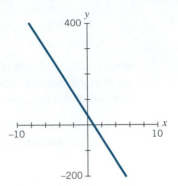

Figure 3.11 Graph of $y = -45x + 33$.

Linear Systems in Economics: Supply and Demand

Economists study the relationship between the price p of an item and the quantity q, the number of items produced. Economists traditionally place quantity q on the horizontal axis and price p on the vertical axis.[2] From the consumer's point of view, an increase in price decreases the quantity demanded. So the consumer's *demand curve* would slope downward (see Figure 3.12).

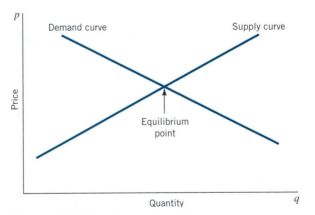

Figure 3.12 Supply and demand curves.

From the manufacturer's (or supplier's) point of view, an increase in price is linked with an increase in the quantity they are willing to supply. So the manufacturer's *supply curve* slopes upward. The intersection point between the demand and supply curves is called the *equilibrium point*. At this point supply equals demand, so both suppliers and consumers are happy with the quantity produced and the price charged.

---

[2]This can be confusing since we usually think of quantity as a function of price.

EXAMPLE 6

Milk supply and demand curves

Loren Tauer[3] studied the U.S. supply and demand curves for milk. If q = billions of pounds of milk and p = dollars per cwt (where 1 cwt = 100 lb), he estimated that the demand function for milk is $p = 55.9867 - 0.2882q$ and the supply function is $p = 0.0865q$.

a. Find the equilibrium point.

b. What will happen if the price of milk is higher than the equilibrium price?

SOLUTION

a. The equilibrium point occurs where

$$\text{supply} = \text{demand}$$

substituting for p $0.0865q = 55.9867 - 0.2882q$

solving for q $0.3747q = 55.9867$

we have $q \approx 149.42$ billions of pounds of milk

If we use the supply function to find p, we have

the supply function $p = 0.0865q$

when $q = 149.42$ $p = 0.0865 \cdot 149.42 \approx \$12.92$ per cwt

If we use the demand function to find p we would also get

the demand function $p = 55.9867 - 0.2882q$

when $q = 149.42$ $p = 55.9867 - (0.2882 \cdot 149.42) \approx \$12.92$

So the equilibrium point is (149.42, $12.92); that is, when the price is $12.92 per cwt, manufacturers are willing to produce and consumers are willing to buy 149.42 billion pounds of milk.

b. If the price of milk rises above $12.92 per cwt to, say, p_1, then, as shown in Figure 3.13, consumers would buy less than 149.42 billion pounds (amount q_1) while manufacturers would be willing to produce more than 149.42 billion pounds (amount q_2).

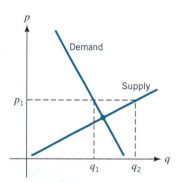

Figure 3.13 At price p_1, consumers would buy q_1 billion pounds of milk, but producers would manufacture q_2 billion pounds.

So there would be a surplus of $q_2 - q_1$ billion pounds of milk, which would drive down the price of milk toward the equilibrium point.

---

[3]Loren W. Tauer, "The value of segmenting the milk market into bST-produced and non-bST produced milk," *Agribusiness 10*(1): 3–12 as quoted in Edmond C. Tomastik, *Calculus: Applications and Technology,* 3rd ed. (Belmont, CA: Thomson Brooks/Cole, 2004).

Algebra Aerobics 3.2b

1. Solve each system of equations using the method you think is most efficient.

   a. $2y - 5x = -1$
   $3y + 5x = 11$

   c. $t = 3r - 4$
   $4t + 6 = 7r$

   b. $3x + 2y = 16$
   $2x - 3y = -11$

   d. $z = 2000 + 0.4(x - 10,000)$
   $z = 800 + 0.2x$

2. Solve each system of equations. If technology is available, check your answers by graphing each system.

   a. $y = 2x + 4$
   $y = -x + 4$

   c. $y = 1500 + 350x$
   $2y = 700x + 3500$

   b. $5y + 30x = 20$
   $y = -6x + 4$

3. Construct a system of two linear equations in two unknowns that has no solution.

4. Determine the number of solutions without solving the system. Explain your reasoning.

   a. $2x + 5y = 7$
   $3x - 8y = -1$

   c. $2x + 3y = 1$
   $4x + 6y = 2$

   b. $3x + y = 6$
   $6x + 2y = 5$

   d. $3x + y = 8$
   $3x + 2y = 8$

5. Solve each of the following systems of equations.

   a. $y = x + 4$
   $\frac{x}{2} + \frac{y}{3} = 3$

   b. $0.5x + 0.7y = 10$
   $30x + 50y = 1000$

6. A small paint dealer has determined that the demand function for interior white paint is $4p + 3q = 240$, where p = dollars/gallon of paint and q = number of gallons.

   a. Find the demand for white paint when the price is $39.00 per gallon.

   b. If consumer demand is for 20 gallons of paint, what price would these consumers be willing to pay?

   c. Sketch the demand function, placing q on the horizontal axis and p on the vertical axis.

   d. The supply function for interior white paint is $p = 0.85q$. Sketch the supply curve on the same graph as the demand curve.

   e. Find the equilibrium point and interpret its meaning.

   f. At a price of $39.00 per gallon of paint, is there a surplus or shortage of supply?

7. Fill in the missing coefficient of x such that there will be an infinite number of solutions to the system of equations:
   $$y = 2x + 4$$
   $$?? \, x = -2y + 8$$

Exercises for Section 3.2

1. Two companies offer starting employees incentives to stay with the company after they are trained for their new jobs. Company A offers an initial hourly wage of $7.00, then increases the hourly wage by $0.15 per month. Company B offers an initial hourly wage of $7.45, then increases the hourly wage by $0.10 per month.

Wage Comparison Company A and Company B

   a. Examine the accompanying graph. After how many months does it appear that the hourly wage will be the same for both companies?

   b. Estimate that hourly wage.

   c. Form two linear functions for the hourly wages in dollars of $W_A(m)$ for company A and $W_B(m)$ for company B after m months of employment.

   d. Does your estimated solution from part (a) satisfy both equations? If not, find the exact solution.

   e. What is the exact hourly wage when the two companies offer the same wage?

   f. Describe the circumstances under which you would rather work for company A. For company B.

2. In the text the following cost equations were given for gas and solar heating:
   $$C_{gas} = 12,000 + 700n$$
   $$C_{solar} = 30,000 + 150n$$

   where n represents the number of years since installation and the cost represents the total accumulated costs up to and including year n.

   a. Sketch the graph of this system of equations.

   b. What do the coefficients 700 and 150 represent on the graph, and what do they represent in terms of heating costs?

   c. What do the constant terms 12,000 and 30,000 represent on the graph? What does the difference between 12,000 and 30,000 say about the costs of gas vs. solar heating?

d. Label the point on the graph where the total accumulated gas and solar heating costs are equal. Make a visual estimate of the coordinates, and interpret what the coordinates mean in terms of heating costs.

e. Use the equations to find a better estimate for the intersection point. To simplify the computations, you may want to round values to two decimal places. Show your work.

f. When is the total cost of solar heating more expensive than gas? When is the total cost of gas heating more expensive than solar?

3. Answer the questions in Exercise 2 (with suitable changes in wording) for the following cost equations for electric and solar heating:

$$C_{electric} = 5000 + 1100n$$
$$C_{solar} = 30,000 + 150n$$

4. Consider the following job offers. At Acme Corporation, you are offered a starting salary of $20,000 per year, with raises of $2500 annually. At Boca Corporation, you are offered $25,000 to start and raises of $2000 annually.

a. Find an equation to represent your salary, $S_A(n)$, after n years of employment with Acme.

b. Find an equation to represent your salary, $S_B(n)$, after n years of employment with Boca.

c. Create a table of values showing your salary at each of these corporations for integer values of n up to 12 years.

d. In what year of employment would the two corporations pay you the same salary?

5. a. Solve the following system algebraically:

$$S = 20,000 + 2500n$$
$$S = 25,000 + 2000n$$

b. Graph the system in part (a) and use the graph to estimate the solution to the system. Check your estimate with your answer in part (a).

6. Predict the number of solutions to each of the following systems. Give reasons for your answer. You don't need to find any actual solutions.

a. $y = 20,000 + 700x$ $y = 15,000 + 800x$
b. $y = 20,000 + 700x$ $y = 15,000 + 700x$
c. $y = 20,000 + 700x$ $y = 20,000 + 800x$

7. For each system:

a. Indicate whether the substitution or elimination method might be easier for finding a solution to the system of equations.

i. $y = \frac{1}{3}x + 6$ **iv.** $y = 2x - 3$
$\quad y = \frac{1}{3}x - 4$ $\quad 4y - 8x = -12$

ii. $2x - y = 5$ **v.** $-3x + y = 4$
$\quad 5x + 2y = 8$ $\quad -3x + y = -2$

iii. $3x + 2y = 2$ **vi.** $3y = 9$
$\quad x = 7y - 30$ $\quad x + 2y = 11$

b. Using your chosen method, find the solution(s), if any, of each system.

8. For each graph, construct the equations for each of the two lines in the system, and then solve the system using your equations.

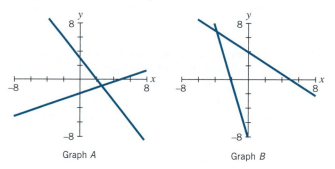

Graph A   Graph B

9. a. Solve the following system algebraically:

$$x + 3y = 6$$
$$5x + 3y = -6$$

b. Graph the system of equations in part (a) and estimate the solution to the system. Check your estimate with your answers in part (a).

10. Calculate the solution(s), if any, to each of the following systems of equations. Use any method you like.

a. $y = -1 - 2x$ **c.** $y = 2200x - 700$
$\quad y = 13 - 2x$ $\quad y = 1300x + 4700$

b. $t = -3 + 4w$ **d.** $3x = 5y$
$\quad -12w + 3t + 9 = 0$ $\quad 4y - 3x = -3$

In some of the following examples you may wish to round off your answers:

e. $y = 2200x - 1800$ **h.** $2x + 3y = 13$
$\quad y = 1300x - 4700$ $\quad 3x + 5y = 21$

f. $y = 4.2 - 1.62x$ **i.** $xy = 1$
$\quad 1.48x - 2y + 4.36 = 0$ $\quad x^2y + 3x = 2$

g. $4r + 5s = 10$ (A nonlinear system!
$\quad 2r - 4s = -3$ *Hint:* Solve $xy = 1$ for y and use substitution.)

11. Assume you have $2000 to invest for 1 year. You can make a safe investment that yields 4% interest a year or a risky investment that yields 8% a year. If you want to combine safe and risky investments to make $100 a year, how much of the $2000 should you invest at the 4% interest? How much at the 8% interest? (*Hint:* Set up a system of two equations in two variables, where one equation represents the total amount of money you have to invest and the other equation represents the total amount of money you want to make on your investments.)

12. Two investments in high-technology companies total $1000. If one investment earns 10% annual interest and the other earns 20%, find the amount of each investment if the total interest earned is $140 for the year (clearly in dot com days).

13. Solve the following systems:

a. $\frac{x}{3} + \frac{y}{2} = 1$ **b.** $\frac{x}{4} + y = 9$
$\quad x - y = \frac{4}{3}$ $\quad y = \frac{x}{2}$

14. For each of the following systems of equations, describe the graph of the system and determine if there is no solution, an infinite number of solutions, or exactly one solution.

    a. $2x + 5y = -10$ **b.** $3x + 4y = 5$ **c.** $2x - y = 5$

       $y = -0.4x - 2$ $3x - 2y = 5$ $6x - 3y = 4$

15. If $y = b + mx$, solve for values for m and b by constructing two linear equations in m and b for the given sets of ordered pairs.

    a. When $x = 2$, $y = -2$ and when $x = -3$, $y = 13$.

    b. When $x = 10$, $y = 38$ and when $x = 1.5$, $y = -4.5$.

16. The following are formulas predicting future raises for four different groups of union employees. N represents the number of years from the start date of all the contracts. Each equation represents the salary that will be earned after N years.

    Group A: Salary $= 30,000 + 1500N$

    Group B: Salary $= 30,000 + 1800N$

    Group C: Salary $= 27,000 + 1500N$

    Group D: Salary $= 21,000 + 2100N$

    a. Will group A ever earn more per year than group B? Explain.

    b. Will group C ever catch up to group A? Explain.

    c. Which group will be making the highest yearly salary in 5 years? How much will that salary be?

    d. Will group D ever catch up to group C? If so, after how many years and at what salary?

    e. How much total salary would an individual in each group have earned 3 years after the contract?

17. A small T-shirt company created the following cost and revenue equations for a line of T-shirts, where cost C is in dollars for producing x units and revenue R is in dollars from selling x units:

$$C = 12.5x + 360 \quad \text{and} \quad R = 15.5x$$

    a. What does 12.5 represent?

    b. What does 15.5 represent?

    c. Find the equilibrium point.

    d. What is the cost of producing x units at the equilibrium point? The revenue at the equilibrium point?

18. A large wholesale nursery sells shrubs to retail stores. The cost $C(x)$ and revenue $R(x)$ equations (in dollars) for x shrubs are

$$C(x) = 15x + 12,000 \quad \text{and} \quad R(x) = 18x$$

    a. Find the equilibrium point.

    b. Explain the meaning of the coordinates for the equilibrium point.

19. The supply and demand equations for a particular bicycle model relate price per bicycle, p (in dollars) and q, the number of units (in thousands). The two equations are

$$p = 250 + 40q \quad \text{Supply}$$
$$p = 510 - 25q \quad \text{Demand}$$

    a. Sketch both equations on the same graph. On your graph identify the supply equation and the demand equation.

    b. Find the equilibrium point and interpret its meaning.

20. For a certain model of DVD-VCR combo player, the following supply and demand equations relate price per player, p (in dollars) and number of players, q (in thousands).

$$p = 80 + 2q \quad \text{Supply}$$
$$p = 185 - 5q \quad \text{Demand}$$

    a. Find the point of equilibrium.

    b. Interpret this result.

21. Explain what is meant by "two equivalent equations." Give an example of two equivalent equations.

22. Construct a problem not found in this text that involves supply and demand where the situation can be modeled with a system of linear equations. Solve your system and verify your results by graphing the system.

23. A restaurant is located on ground that slopes up 1 foot for every 20 horizontal feet. The restaurant is required to build a wheelchair ramp starting from an entry platform that is 3 feet above ground. Current regulations require a wheelchair ramp to rise up 1 foot for every 12 horizontal feet, (See accompanying figure where H = height in feet, d = distance from entry in feet, and the origin is where the H-axis meets the ground.) Where will the new ramp intersect the ground?

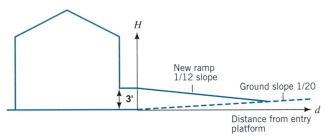

24. A house attic as shown has a roofline with a slope of 5″ up for every 12″ of horizontal run; this is a slope of 5/12. Since the roofline starts at 4′ above the floor, it is not possible to stand in much of the attic space. The owner wants to add a dormer with a 2/12 slope, starting at a 7′ height, to increase the usable space.

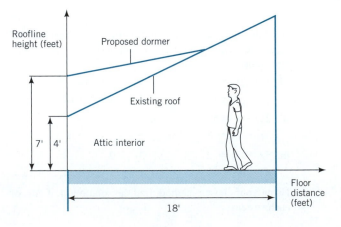

a. With height and floor distance coordinates as shown on the house sketch, find a formula for the original roofline, using R for roof height and F for floor distance from the wall. Also find a formula for the dormer roofline, using D for dormer roof height and F for floor distance from the wall.

b. Make a graph showing the R and D roof height lines for floor distances F from 0′ to 18′.

c. At what height and floor distance will the dormer roofline intersect the existing roofline?

d. How far do you need to measure along the horizontal floor distance to give 6′6″ of head room in the original roofline? What percent of the horizontal floor distance of 18 ft allows less than 6′6″ of head room?

25. Solve the following system of three equations in three variables, using the steps outlined below:

$$2x + 3y - z = 11 \qquad (1)$$
$$5x - 2y + 3z = 35 \qquad (2)$$
$$x - 5y + 4z = 18 \qquad (3)$$

a. Use Equations (1) and (2) to eliminate one variable, creating a new Equation (4) in two variables.

b. Use Equations (1) and (3) to eliminate the same variable as in part (a). You should end up with a new Equation (5) that has the same variables as Equation (4).

c. Equations (4) and (5) represent a system of two equations in two variables. Solve the system.

d. Find the corresponding value for the variable eliminated in part (a).

e. Check your work by making sure your solution works in all three original equations.

26. Using the strategy described in Exercise 25, solve the following system:

$$2a - 3b + c = 4.5 \qquad (1)$$
$$a - 2b + 2c = 0 \qquad (2)$$
$$3a - b + 2c = 0.5 \qquad (3)$$

27. a. Construct a system of linear equations in two variables that has no solution.

b. Construct a system of linear equations in two variables that has exactly one solution.

c. Solve the system of equations you constructed in part (b) by using two different algebraic strategies and by graphing the system of equations. Do your answers all agree?

28. Nenuphar wants to invest a total of $30,000 into two savings accounts, one paying 6% per year in interest and the other paying 9% per year in interest (a more risky investment). If after 1 year she wants the total interest from both accounts to be $2100, how much should she invest in each account?

29. When will the following system of equations have no solution? Justify your answer.

$$y = m_1 x + b_1$$
$$y = m_2 x + b_2$$

30. While totally solar energy–powered home energy systems are quite expensive to install, passive solar systems are much more economical. Many passive solar features can be incorporated at the time of construction with a small additional initial cost to a conventional system. These features enable energy costs to be one-half to one-third of the costs in conventional homes.

Below is the cost analysis from the case study Esperanza del Sol.[4]

Cost of installation of a conventional system:	$10,000
Additional cost to install passive solar features:	$150
Annual energy costs for conventional system:	$740
Annual energy costs of hybrid system with additional passive solar features:	$540

a. Write the cost equation for the conventional system.

b. Write the cost equation for the passive hybrid solar system.

c. When would the total cost of the passive hybrid solar system be the same as the conventional system?

d. After 5 years, what would be the total energy cost of the passive hybrid solar system? What would be the total cost of the conventional system?

31. a. Construct a system of linear equations where both of the following conditions are met:

The coordinates of the point of intersection are (2, 5).

One of the lines has a slope of -4 and the other line has a slope of 3.5.

b. Graph the system of equations you found in part (a). Verify that the coordinates of the point of intersection are the same as the coordinates specified in part (a).

32. A husband drives a heavily loaded truck that can go only 55 mph on a 650-mile turnpike trip. His wife leaves on the same trip 2 hours later in the family car averaging 70 mph. Recall that distance traveled = speed · time traveled.

a. Derive an expression for the distance, D_h, the husband travels in t hours since he started.

b. How many hours has the wife been traveling if the husband has traveled t hours ($t \geq 2$)?

c. Derive an expression for the distance, D_w, that the wife will have traveled while the husband has been traveling for t hours ($t \geq 2$).

d. Graph distance vs. time for husband and wife on the same axes.

e. Calculate when and where the wife will overtake the husband.

f. Suppose the husband and wife wanted to arrive at a restaurant at the same time, and the restaurant is 325 miles from home. How much later should she leave, assuming he still travels at 55 mph and she at 70 mph?

[4]Adapted from *Buildings for a Sustainable America: Case Studies,* American Solar Energy Society, Boulder, CO.

33. Life-and-death travel problems are dealt with by air traffic computers and controllers who are trying to prevent collisions of planes traveling at various speeds in three-dimensional space. To get a taste of what is involved, consider this situation: Airplanes A and B are traveling at the same altitude on the paths shown on the position plot.

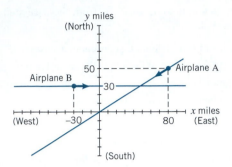

a. Construct two equations that describe the positions of airplanes A and B in x- and y-coordinates. Use y_A and y_B to denote the north/south coordinates of airplanes A and B, respectively.

b. What are the coordinates of the intersection of the airplanes' paths?

c. Airplane A travels at 2 miles/minute and airplane B travels at 6 miles/minute. Clearly, their paths will intersect if they each continue on the same course, but will they arrive at the intersection point at the same time? How far does plane A have to travel to the intersection point? How many minutes will it take to get there? How far does plane B have to travel to the intersection point? How many minutes will it take to get there? (*Hint:* Recall the rule of Pythagoras for finding the hypotenuse of a right triangle: $a^2 + b^2 = c^2$, where c is the hypotenuse and a and b are the other sides.) Is this a safe situation?

34. a. Examine the accompanying figure, where the demand curve has been moved to the right. Does the new demand curve represent an increase or decrease in demand. Why? (*Hint:* Pick an arbitrary price, and see if consumers would want to buy more or less at that price.)

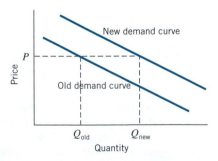

b. Sketch in a possible supply curve identifying the old and new equilibrium points. What does the shift from the old equilibrium point to the new mean for both consumers and suppliers?

35. a. Examine the accompanying figure. Does the new supply curve represent an increase or decrease in supply. Why? (Again, try picking an arbitrary price and see if at that price, the supplier would want to increase or decrease production.)

b. Sketch in a possible demand curve, and label the old and new equilibrium points. What does the shift from the old to the new equilibrium point mean for both consumers and suppliers?

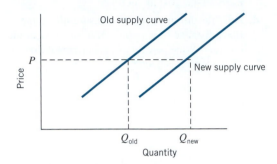

36. In studying populations (human or otherwise), the two primary factors affecting population size are the birth rate and the death rate. There is abundant evidence that, other things being equal, as the population density increases, the birth rate tends to decrease and the death rate tends to increase.[5]

a. Generate a rough sketch showing birth rate as a function of population density. Note that the units for population density on the horizontal axis are the number of individuals for a given area. The units on the vertical axis represent a rate, such as the number of individuals per 1000 people. Now add to your graph a rough sketch of the relationship between death rate and population density. In both cases assume the relationship is linear.

b. At the intersection point of the two lines the growth of the population is zero. Why? (*Note:* We are ignoring all other factors, such as immigration.)

The intersection point is called the *equilibrium point*. At this point the population is said to have stabilized, and the size of the population that corresponds to this point is called the *equilibrium number*.

c. What happens to the equilibrium point if the overall death rate decreases, that is, at each value for population density the death rate is lower? Sketch a graph showing the birth rate and both the original and the changed death rates. Label the graph carefully. Describe the shift in the equilibrium point.

d. What happens to the equilibrium point if the overall death rate increases? Analyze as in part (c).

37. Use the information in Exercise 36 to answer the following questions:

a. What if the overall birth rate increases (that is, if at each population density level the birth rate is higher)? Sketch a

[5]See E. O. Wilson and W. H. Bossert, *A Primer of Population Biology*. Sunderland, MA: Sinauer Associates, 1971, p. 104.

graph showing the death rate and both the original and the changed birth rates. Be sure to label the graph carefully. Describe the shift in the equilibrium point.

b. What happens if the overall birth rate decreases? Analyze as in part (a).

38. In Section 2.7, Exercise 19, you read of a math professor who purchased her condominium in Cambridge, MA, for $70,000 in 1977. Its assessed value has climbed at a steady rate so that it was worth $850,000 as of 2007. Alas, one of her colleagues has not been so fortunate. He bought a house in that same year for $160,000. Not long after his family moved in, rumors began to circulate that the housing complex had been built on the site of a former toxic dump. Although never substantiated, the rumors adversely affected the value of his home, which has steadily decreased in value over the years and in 2007 was worth a meager $40,000. In what year would the two homes have been assessed at the same value?

3.3 Reading between the Lines: Linear Inequalities

Above and Below the Line

Sometimes we are concerned with values that lie above or below a line, or that lie between two lines. To describe these regions we need some mathematical conventions.

Terminology for describing regions

The two linear functions $y_1 = 1 + 3x$ and $y_2 = 5 - x$ are graphed in Figure 3.14. How would you describe the various striped regions?

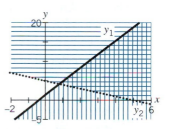

Figure 3.14 Regions bounded by the lines $y_1 = 1 + 3x$ and $y_2 = 5 - x$.

A solid line indicates that the points on the line are included in the area. A dotted line indicates that the points on the line are *not* included in the area. So the vertical-striped region below the solid line y_1 can be described as all points (x, y) that satisfy the inequality

$$y \leq y_1 \qquad \text{Condition (1)}$$

or equivalently $\qquad y \leq 1 + 3x$

The equation $y_1 = 1 + 3x$ is a *boundary line* that is included in the region.

The horizontally striped region above the dotted line y_2 can be described as all points (x, y) that satisfy the inequality

$$y_2 < y \qquad \text{Condition (2)}$$

or $\qquad 5 - x < y$

So $y_2 = 5 - x$ is a boundary line that is not included in the region.

In the cross-hatched region, the y values must satisfy both conditions (1) and (2); that is, we must have

$$y_2 < y \leq y_1$$

or equivalently $\qquad 5 - x < y \leq 1 + 3x$

This is called a *compound inequality*. We could describe this inequality by saying that y is greater than $5 - x$ and less than or equal to $1 + 3x$. So the region can be described as all points (x, y) that satisfy the compound inequality

$$5 - x < y \le 1 + 3x$$

Manipulating Inequalities

Recall that any term may be added to or subtracted from both sides of an inequality without changing the direction of the inequality. The same holds for multiplying or dividing by a positive number. However, multiplying or dividing by a negative number reverses the inequality. For example,

Given the previous inequality	$5 - x < y$
if we wanted to solve for x we could subtract 5 from both sides	$-x < y - 5$
then multiply both sides by -1	$x > -y + 5$

Note that subtracting 5 (or equivalently adding -5) preserved the inequality, but multiplying by -1 reversed the inequality. So "<" in the first two inequalities became ">" in the last inequality.

EXAMPLE 1   The U.S. Army recommends that sleeping bags, which will be used in temperatures between $-40°$ and $+40°$ Fahrenheit, have a thickness of 2.5 inches minus 0.025 times the number of degrees Fahrenheit.

    a. Construct a linear equation that models the recommended sleeping bag thickness as a function of degrees Fahrenheit. Identify the variables and domain of your function.

    b. Graph the model (displaying it over its full domain) and shade in the area where the sleeping bag is not warm enough for the given temperature.

    c. What symbolic expressions would describe the shaded area?

    d. If a manufacturer submitted a sleeping bag to the Army with a thickness of 2.75 inches, would it be suitable for

        i. $-20°$ Fahrenheit?   **ii.** $0°$ Fahrenheit?   **iii.** $+20°$ Fahrenheit?

SOLUTION   **a.** If we let F = number of degrees Fahrenheit and T = thickness of the sleeping bag in inches, then

$$T = 2.5 - 0.025F \qquad \text{where the domain is } -40° \le F \le 40°$$

    b. See Figure 3.15.

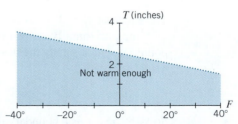

Figure 3.15 T, sleeping bag thickness as a function of F, degrees Fahrenheit.

When the thickness of the sleeping bag, T, is less than the recommended thickness, then the bag will not be warm enough.

c. The shaded region can be described as

$$0 < \text{thickness of sleeping bag} < \text{recommended thickness}$$
$$0 < \qquad\qquad T \qquad\qquad < 2.5 - 0.025F$$

where $-40° \leq F \leq 40°$. We could rephrase this to say that the region is bounded by four lines: $T = 0$, $T = 2.5 - 0.025F$, $F = -40°$, and $F = 40°$. Note that the dotted line indicates that the line itself is not included.

d. If the sleeping bag thickness T is 2.75, we can find the corresponding recommended temperature by solving our equation for F.

Substitute 2.75 for T	$2.75 = 2.5 - 0.025F$
simplify to get	$0.25 = -0.025F$
or	$F = -10°$

So the sleeping bag would not be thick enough for $-20°$F, since the point $(-20°, 2.75'')$ lies in the shaded area below the Army's recommended values. It would be more than thick enough for $0°$F or $20°$F since the points $(0°, 2.75'')$ and $(20°, 2.75'')$ both lie above the shaded area.

Reading between the Lines

EXAMPLE 2   The U.S. Department of Agriculture recommends healthy weight zones for adults based on their height. For men between 60 and 84 inches tall, the recommended lowest weight, W_{lo} (in lb), is

$$W_{lo} = 105 + 4.0H$$

and the recommended highest weight, W_{hi} (in lb), is

$$W_{hi} = 125.4 + 4.6H$$

where H is the number of inches above 60 inches (5 feet).

a. Graph and label the two boundary equations and indicate the underweight, healthy, and overweight zones.

b. Give a mathematical description of the healthy weight zone for men.

c. Two men each weigh 180 lb. One is 5' 10" tall and the other 6' 1" tall. Is either within the healthy weight zone?

d. A 5' 11" man weighs 135 lb. If he gains 2 lb a week, how long will it take him to reach the healthy range?

SOLUTION   **a.** See Figure 3.16.

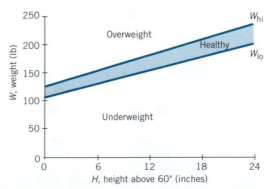

Figure 3.16 Graph of men's healthy weight zone between recommended low (W_{lo}) and high (W_{hi}) weights.

b. Assuming that men's heights generally run between 5 feet (60″) and 7 feet (84″), then H (the height in inches above 60 inches) is bounded by

$$0'' \leq H \leq 24''$$

The recommendations say that a man's weight W (in lb) should be more than or equal to (W_{lo}) and should be less than or equal to W_{hi}. So we have the computed inequality

$$W_{lo} \leq W \leq W_{hi}$$

or $\qquad\qquad\qquad 105 + 4.0H \leq W \leq 125.4 + 4.6H$

c. For a man who is 5′ 10″ (or 70″) tall, $H = 10''$ and his maximum recommended weight, W_{hi}, is $125.4 + (4.6 \cdot 10) = 171.4$ lb. So if he weighs 180 lb, he is overweight.

For a man who is 6′ 1″ (or 73″) tall, $H = 13''$. His minimum recommended weight, W_{lo}, is $105 + (4.0 \cdot 13) = 157$ lb. His maximum recommended weight, W_{hi}, is $125.4 + (4.6 \cdot 13) = 185.2$ lb. So his weight of 180 lb would place him in the healthy zone.

d. If a man is 5′ 11″, his recommended minimum weight, W_{lo} is $105 + (4.0 \cdot 11) = 149$ lb. If he currently weighs 135 lb, he would need to gain at least $149 - 135 = 14$ lb. If he gained 2 lb a week, it would take him 7 weeks to reach the minimum recommended weight of 149 lb.

EXAMPLE 3   Given the inequalities $2x - 3y \leq 12$ and $x + 2y < 4$:

a. Solve each for y.

b. On the same graph, plot the boundary line for each inequality (indicating whether it is solid or dotted) and then shade the region described by each inequality.

c. Write a compound inequality describing the overlapping region.

SOLUTION   **a.** To solve for y in the first inequality $\qquad\qquad 2x - 3y \leq 12$

add $-2x$ to both sides $\qquad\qquad\qquad\qquad\qquad -3y \leq -2x + 12$

divide both sides by -3, reversing the inequality symbol $\qquad\qquad\qquad \dfrac{-3y}{-3} \geq \dfrac{-2x}{-3} + \dfrac{12}{-3}$

then simplify $\qquad\qquad\qquad\qquad\qquad\qquad y \geq \dfrac{2}{3}x - 4$

Solving for y in the second inequality $\qquad\qquad x + 2y < 4$

add $-x$ to both sides $\qquad\qquad\qquad\qquad\qquad 2y < -x + 4$

divide both sides by 2 $\qquad\qquad\qquad\qquad\qquad \dfrac{2y}{2} < \dfrac{-x}{2} + \dfrac{4}{2}$

then simplify $\qquad\qquad\qquad\qquad\qquad\qquad y < -\dfrac{1}{2}x + 2$

b. See Figure 3.17.

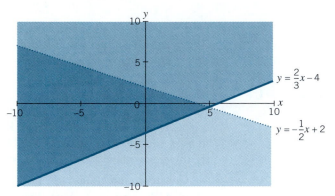

Figure 3.17 The darkest shaded region lies between $y < -\frac{1}{2}x + 2$ and $y \geq \frac{2}{3}x - 4$.

c. The overlapping region consists of all ordered pairs (x, y) such that $\frac{2}{3}x - 4 \leq y < -\frac{1}{2}x + 2$. We could describe the compound inequality by saying "y is greater than or equal to $\frac{2}{3}x - 4$ and less than $-\frac{1}{2}x + 2$."

EXAMPLE 4 Describe the shaded region in Figure 3.18.

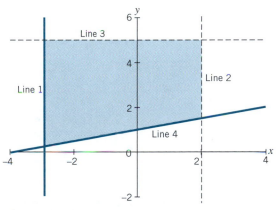

Figure 3.18 A shaded region with four boundary lines.

SOLUTION We need to find the equation for each of the four boundary lines. The two vertical lines are the easiest: Line 1 is the line $x = -3$ (a solid line, so included in the region); Line 2 is the line $x = 2$ (dotted, so excluded from the region). Line 3 is the horizontal line $y = 5$ (dotted, so it is excluded). Line 4 has a vertical intercept of 1. It passes through $(0, 1)$ and $(-4, 0)$ so its slope is $(0 - 1)/(-4 - 0) = 1/4$ or 0.25. So the equation for line 4 is $y = 1 + 0.25x$. The region can be described as all pairs (x, y) that satisfy both compound inequalities:

$$-3 \leq x < 2 \quad \text{and} \quad 1 + 0.25x \leq y < 5$$

Breakeven Points: Regions of Profit or Loss

A simple model for the total cost C to a company producing n units of a product is

$$C = \text{fixed costs} + (\text{cost per unit}) \cdot n$$

A corresponding model for the total revenue R is

$$R = (\text{price per unit}) \cdot n$$

The breakeven point occurs when $C = R$, or the total cost is equal to the total revenue. When $R > C$, or total revenues exceed total cost, the company makes a profit and when $R < C$, the company loses money.

EXAMPLE 5   **Cost versus revenue**

In 1996 two professors from Purdue University reported on their study of fertilizer plants in Indiana.[6] They estimated that for a large-sized fertilizer manufacturing plant the fixed costs were about $450,000 and the additional cost to produce each ton of fertilizer was about $210. The fertilizer was sold at $270 per ton.

a. Construct and graph two equations, one representing the total cost $C(n)$, and the other the revenue $R(n)$, where n is the number of tons of fertilizer produced.

b. What is the breakeven point, where costs equal revenue?

c. Shade between the lines to the right of the breakeven point. What does the region represent?

SOLUTION   **a.** $C(n) = 450{,}000 + 210n$ and $R(n) = 270n$. See Figure 3.19.

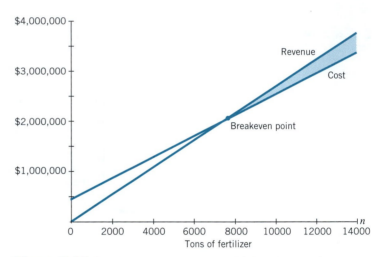

Figure 3.19 Graph of revenue vs. cost with breakeven point.

b. The breakeven point is where   cost = revenue

or $C(n) = R(n)$

substituting for $C(n)$ and $R(n)$ $450{,}000 + 210n = 270n$

solving for n $450{,}000 = 60n$

we get $n = 7500$ tons of fertilizer

Substituting 7500 tons into the revenue equation, we have

$$R(7500) = 270 \cdot 7500 = \$2{,}025{,}000$$

We could have substituted 7500 tons into the cost equation to get the same value.

$$C(7500) = 450{,}000 + 210 \cdot 7500 = \$2{,}025{,}000$$

[6]Duane S. Rogers and Jay T. Aldridge, "Economic impact of storage and handling regulation on retail fertilizer and pesticide plants," *Agribusiness* 12(4): 327–337, as quoted in Edmond C. Tomastik, *Calculus: Applications and Technology*, 3rd ed. (Belmont CA: Thomson Brooks/Cole, 2004).

So the breakeven point is (7500, $2,025,000). At this point the cost of producing 7500 tons of fertilizer and the revenue from selling 7500 tons both equal $2,025,000.

c. In the shaded area between $R(n)$ and $C(n)$ to the right of the breakeven point, $R(n) > C(n)$, so revenue exceeds costs, and the manufacturers are making money. Economists call this the *region of profit*.

Algebra Aerobics 3.3

1. Solve graphically each set of conditions.

   a. $y \geq 2x - 1$
   $y \geq 4 - x$

   b. $y \geq 2x - 1$
   $y \leq 3 - x$

   c. $y \geq 400 + 10x$
   $y \geq 200 + 20x$
   $x \geq 0$
   $y \geq 0$

   d. $y \geq 0$
   $y < -0.5x + 2$
   $y < 0.5x + 2$

   e. $y \leq 0$
   $x \geq 0$
   $y \geq -100 + x$

   f. $0 \leq x \leq 200$
   $y > 2x - 400$
   $y < 100 - 1.5x$

2. Determine which of the following points (if any) satisfy the system of inequalities $y > 2x - 3$ and $y \leq 3x + 8$

   a. (2, 3)   **c.** (0, 8)   **e.** (20, −8)
   b. (−4, 7)   **d.** (−4, −6)   **f.** (1, −1)

3. Determine each inequality that has the given graph in Figure 3.20.

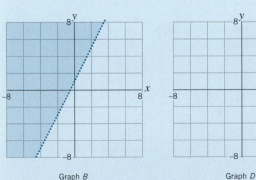

Graph A

Graph C

Graph B

Graph D

Figure 3.20 Graphs of four linear inequalities.

4. A small company sells dulcimer[7] music books on the Internet. Examine the graph in Figure 3.21 of the cost (C) and revenue (R) equations for selling n books.

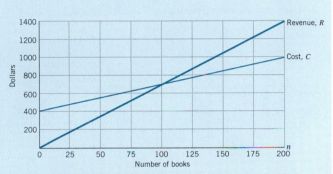

Figure 3.21 Cost and revenue for dulcimer books.

a. Estimate the breakeven point and interpret its meaning.

b. Shade in the region corresponding to losses for the company.

c. What are the fixed costs for selling dulcimer music books?

d. Another company buys dulcimer music books for $3.00 each and sells them for $7.00 each. The fixed cost for this company is $400. Form a system of inequalities that represents when this company would make a profit. Use C_1 for cost and R_1 for revenue for n books.

5. Suppose that the two professors from Example 5 in this section estimated 6 years later that for a large-sized fertilizer manufacturing plant the fixed costs were about $50,000 and the additional cost to produce each ton of fertilizer was about $235. However, market conditions and competition caused the company to continue to sell the fertilizer at $270 per ton.

a. Form the cost function $C(n)$ and revenue function $R(n)$ for n tons of fertilizer.

b. Find the breakeven point and interpret its meaning.

c. Graph the two functions and shade in the region corresponding to profits for the company.

[7]A mountain dulcimer is an Appalachian string instrument, usually with four strings, commonly played on the lap by strumming or plucking.

Exercises for Section 3.3

A graphing program is required for Exercises 19 and 20 and recommended for Exercises 15, 16 & 17. Access to the Internet is required in Exercise 14(c).

1. On different grids, graph and shade in the areas described by the following linear inequalities.

   a. $y < 2x + 2$ **c.** $y < 4$

   b. $y \geq -3x - 3$ **d.** $y \geq \frac{2}{3}x - 4$

2. On different grids, graph and shade in the areas described by the following linear inequalities.

   a. $x - y < 0$ **c.** $3x + 2y > 6$

   b. $x \leq -2$ **d.** $5x - 2y \leq 10$

3. Use inequalities to describe each shaded region.

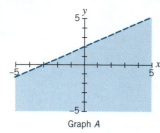
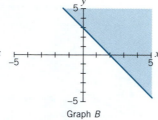

   Graph A   Graph B

4. Use inequalities to describe each shaded region.

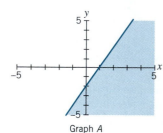
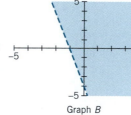

   Graph A   Graph B

5. On different grids, graph each inequality (shading in the appropriate area) and then determine whether or not the origin, the point $(0, 0)$, satisfies the inequality.

   a. $-2x + 6y < 4$ **c.** $y > 3x - 7$

   b. $x \geq 3$ **d.** $y - 3 > x + 2$

6. Determine whether or not the point $(-1, 3)$ satisfies the inequality.

   a. $x - 3y > 6$ **c.** $y \leq 3$

   b. $x < 3$ **d.** $y \leq -\frac{1}{2}x + 3$

7. Explain how you can tell if the region described by the inequality $3x - 5y < 15$ is above or below the boundary line of $3x - 5y = 15$.

8. Shade the region bounded by the inequalities

$$x + 3y \leq 15$$
$$2x + y \leq 15$$
$$x \geq 0$$
$$y \geq 0$$

9. Match each description in parts (a) to (e) with the appropriate compound inequality in parts (f) to (j).

   a. y is greater than 4 and less than $x - 3$.

   b. y is greater than or equal to $x - 3$ and less than -6.

   c. y is less than $2x + 5$ and greater than -6.

   d. y is greater than or equal to $2x + 5$ and less than -6.

   e. y is less than or equal to $x - 3$ and greater than $2x + 5$.

   f. $2x + 5 \leq y < -6$

   g. $4 < y < x - 3$

   h. $2x + 5 < y \leq x - 3$

   i. $x - 3 \leq y < -6$

   j. $-6 < y < 2x + 5$

10. For the inequalities $y > 4x - 3$ and $y \leq -3x + 4$:

    a. Graph the two boundary lines and indicate with different stripes the two regions that satisfy the individual inequalities.

    b. Write the compound inequality for y. Indicate the double-hatched region on the graph that satisfies both inequalities.

    c. What is the point of intersection for the boundary lines?

    d. If $x = 3$, are there any corresponding y values in the region defined in part (b)?

    e. Is the point $(1, 4)$ part of the double-hatched region?

    f. Is the point $(-1, 4)$ part of the double-hatched region?

11. Examine the shaded region in the graph.

    a. Create equations for the boundary lines l_1 and l_2 using y as a function of x.

    b. Determine the compound inequality that created the shaded region.

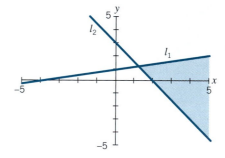

12. Examine the shaded region in the graph. Determine the compound inequality of y in terms of x that created the shaded region.

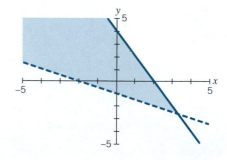

13. The Food and Drug Administration labels suntan products with a sun protection factor (SPF) typically between 2 and 45. Multiplying the SPF by the number of unprotected minutes you can stay in the sun without burning, you are supposed to get the increased number of safe sun minutes. For example, if you can stay unprotected in the sun for 30 minutes without burning and you apply a product with a SPF of 10, then supposedly you can sun safely for $30 \cdot 10 = 300$ minutes or 5 hours.

    Assume that you can stay unprotected in the sun for 20 minutes without burning.

    a. Write an equation that gives the maximum safe sun time T as a function of S, the sun protection factor (SPF).

    b. Graph your equation. What is the suggested domain for S?

    c. Write an inequality that suggests times that would be unsafe to stay out in the sun.

    d. Shade in and label regions on the graph that indicate safe and unsafe regions. (Use two different shadings and remember to include boundaries.)

    e. How would the graph change if you could stay unprotected in the sun for 40 minutes?

    Note that you should be cautious about spending too much time in the sun. Factors such as water, wind, and sun intensity can diminish the effect of SPF products.

14. (Access to Internet required for part (c).) Doctors measure two kinds of cholesterol in the body: low-density lipoproteins, LDL, called "bad cholesterol" and high-density lipoproteins, HDL, called "good cholesterol" because it helps to remove the bad cholesterol from the body. Rather than just measuring the total cholesterol, TC, many doctors use the ratio of TC/HDL to help control heart disease. General guidelines have suggested that men should have TC/HDL of 4.5 or below, and women should have TC/HDL of 4 or below.

    On the following graphs place HDL on the horizontal and TC on the vertical axis.

    a. Construct an equation for men that describes TC as a function of HDL assuming that the ratio of the two numbers is at the recommended maximum for men. Graph the function, using HDL values up to 75. Label on the graph higher-risk and lower-risk areas for heart disease for men.

    b. Construct a similar equation for women that describes TC as a function of HDL assuming that the ratio of the two numbers is at the recommended maximum for women. Graph the function, using HDL values up to 75. Label on the graph higher-risk and lower-risk areas for heart disease for women.

    c. Use the Internet to find the most current cholesterol guidelines.

15. (Graphing program recommended.) The blood alcohol concentration (BAC) limits for drivers vary from state to state, but for drivers under the age of 21 it is commonly set at

0.02. This level (depending upon weight and medication levels) may be exceeded after drinking only one 12-oz can of beer. The formula

$$N = 6.4 + 0.0625(W - 100)$$

gives the number of ounces of beer, N, that will produce a BAC legal limit of 0.02 for an average person of weight W. The formula works best for drivers weighing between 100 and 200 lb.

   a. Write an inequality that describes the condition of too much blood alcohol for drivers under 21 to legally drive.

   b. Graph the corresponding equation and label the areas that represent legally safe to drive, and not legally safe to drive conditions.

   c. How many ounces of beer is it legally safe for a 100-lb person to consume? A 150-lb person? A 200-lb person?

   d. Simplify your formula in part (b) to the standard $y = mx + b$ form.

   e. Would you say that "6 oz of beer + 1 oz for every 20 lb over 100 lb" is a legally safe rule to follow?

16. (Graphing program recommended.) The Ontario Association of Sport and Exercise Sciences recommends the minimum and maximum pulse rates P during aerobic activities, based on age A. The maximum recommended rate, P_{max}, is $0.87(220 - A)$. The minimum recommended pulse rate, P_{min}, is $0.72(220 - A)$.

   a. Convert these formulas to the $y = mx + b$ form.

   b. Graph the formulas for ages 20 to 80 years. Label the regions of the graph that represent too high a pulse rate, the recommended pulse rate, and too low a pulse rate.

   c. What is the maximum pulse rate recommended for a 20-year-old? The minimum for an 80-year-old?

   d. Construct an inequality that describes too low a pulse rate for effective aerobic activity.

   e. Construct an inequality that describes the recommended pulse range.

17. (Graphing program recommended.) We saw in this section the U.S. Department of Agriculture recommendations for healthy weight zones for males based on height. There are comparable recommendations for women between 5' (or 60") and 6'3" (or 75") tall. For women the recommended lowest weight W_{lo} (in lb) is

$$W_{lo} = 100 + 3.5H$$

and the recommended highest weight W_{hi} (in lb) is

$$W_{hi} = 118.2 + 4.2H$$

where H is the number of inches above 60" (or 5 feet).

   a. Graph and label the two equations and indicate the underweight, healthy weight, and overweight zones.

   b. Give a mathematical description of the healthy weight zone for women.

c. Two women have the same weight of 130 lb. One is 5′2″ and the other is 5′5″. Does either one lie within the healthy weight zone? Why?

d. A 5′4″ woman weighs 165 lb. If her doctor puts her on a weight-loss diet of 1.5 lb per week, how many weeks would it take for her to reach the healthy range?

18. Cotton and wool fabrics, unless they have been preshrunk, will shrink when washed and dried at high temperatures. If washed in cold water and drip-dried they will shrink a lot less, and if dry-cleaned, they will not shrink at all. A particular cotton fabric has been found to shrink 8% with a hot wash/dry, and 3% with a cold wash/drip dry.

a. Find a formula to express the hot wash length, H, as a function of the original length, L. Then find a formula for the cold wash length, C, as a function of the original length, L. What formula expresses dry-clean length, D, as a function of the original length?

b. Make a graph with original length, L, on the horizontal axis, up to 60 inches. Plot three lines showing cold wash length, C, hot wash length, H, and dry-clean length, D. Label the lines.

c. If you buy trousers with an original inseam length of 32″, how long will the inseam be after a hot wash? A cold wash?

d. If you need 3 yards of fabric (a yard is 3′) to make a well-fitted garment, how much would you have to buy if you plan to hot-wash the garment?

19. (Graphing program required.) Two professors from Purdue University reported that for a typical small-sized fertilizer plant in Indiana the fixed costs were $235,487 and it cost $206.68 to produce each ton of fertilizer.

a. If the company planned to sell the fertilizer at $266.67 per ton, find the cost, C, and revenue, R, equations for x tons of fertilizer.

b. Graph the cost and revenue equations on the same graph and calculate and interpret the breakeven point.

c. Indicate the region where the company would make a profit and create the inequality to describe the profit region.

20. (Graphing program required.) A company manufactures a particular model of DVD player that sells to retailers for $85. It costs $55 to manufacture each DVD player, and the fixed manufacturing costs are $326,000.

a. Create the revenue function $R(x)$ for selling x number of DVD players.

b. Create the cost function $C(x)$ for manufacturing x DVD players.

c. Plot the cost and revenue functions on the same graph. Estimate and interpret the breakeven point.

d. Shade in the region where the company would make a profit.

e. Shade in the region where the company would experience a loss.

f. What is the inequality that represents the profit region?

21. Describe the shaded region in each graph with the appropriate inequalities.

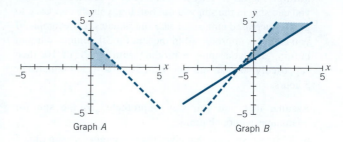

Graph A   Graph B

22. Describe the shaded region. (*Hint:* Create a vertical line through point A, dividing the shaded region into two smaller triangular regions.)

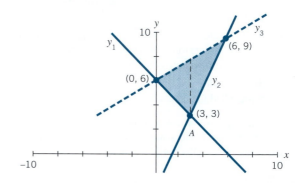

23. A financial advisor has up to $30,000 to invest, with the stipulation that at least $5000 is to be placed in Treasury bonds and at most $15,000 in corporate bonds.

a. Construct a set of inequalities that describes the relationship between buying corporate vs. Treasury bonds where the total amount invested must be less than or equal to $30,000. (Let C be the amount of money invested in corporate bonds, and T the amount invested in Treasury bonds.).

b. Construct a feasible region of investment; that is, shade in the area on a graph that satisfies the spending constraints on both corporate and Treasury bonds. Label the horizontal axis "Amount invested in Treasury bonds" and the vertical axis "Amount invested in corporate bonds."

c. Find all of the intersection points (corner points) of the bounded investment feasibility region and interpret their meanings.

24. A Texas oil supplier sends out at most 10,000 barrels of oil per week. Two distributors need oil. Southern Oil needs at least 2000 barrels of oil per week and Regional Oil needs at least 5000 barrels of oil per week.

a. Let S be the number of barrels of oil sent to Southern Oil and let R be the number of barrels sent to Regional Oil per week. Create a system of inequalities that describes all of the conditions.

b. Graph the feasible region of the system.

c. Choose a point inside the region that would satisfy the conditions and describe its meaning.

3.4 *Systems with Piecewise Linear Functions: Tax Plans*

Graduated vs. Flat Income Tax

The *New York Times* article "How a Flat Tax Would Work for You and for Them" discusses the trade-offs in using a flat tax.

Income taxes may be based on either a flat or a graduated tax rate. With a flat tax rate, no matter what the income level, everyone is taxed at the same percentage. Flat taxes are often said to be unfair to those with lower incomes, because paying a fixed percentage of income is considered more burdensome to someone with a low income than to someone with a high income.

A graduated tax rate means that people with higher incomes are taxed at higher rates. Such a tax is called *progressive* and is generally less popular with those who have high incomes. Whenever the issue appears on the ballot, the pros and cons of the graduated vs. flat tax rate are hotly debated in the news media and paid political broadcasting. Of the forty-one states with a broad-based income tax, thirty-five had a graduated income tax in 2004.

The taxpayer

The *New York Times* article "A Taxation Policy to Make John Stuart Mill Weep" discusses the growing trend by Congress of taxing different sources of income differently.

For the taxpayer there are two primary questions in comparing the effects of flat and graduated tax schemes. For what income level will the taxes be the same under both plans? And, given a certain income level, how will taxes differ under the two plans?

Taxes are influenced by many factors, such as filing status, exemptions, source of income, and deductions. For our comparisons of flat and graduated income tax plans, we examine one filing status and assume that exemptions and deductions have already been subtracted from income.

EXAMPLE 1

Match each graph in Figure 3.22 with the appropriate description.

a. Income taxes are a flat rate of 5% of your income.

b. Income taxes are graduated, with a rate of 5% for first $100,000 of income and a rate of 8% for any additional income > $100,000.

c. Sales taxes are 5% for $0 \le$ purchases $\le$ $100,000 and a flat fee of $5000 for purchases > $100,000.

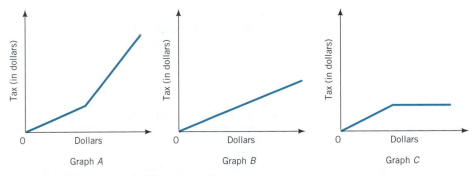

Figure 3.22 Graphs of different tax plans.

SOLUTION

Graph *A* matches description (b).
Graph *B* matches description (a).
Graph *C* matches description (c).

A flat tax model

Flat taxes are a fixed percentage of income. If the flat tax rate is 15% (or 0.15 in decimal form), then flat taxes can be represented as

$$f(i) = 0.15i$$

where i = income. This flat tax plan is represented in Figure 3.23.

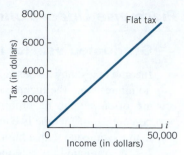

Figure 3.23 Flat tax at a rate of 15%.

A graduated tax model

A Piecewise Linear Function   Under a graduated income tax, the tax rate changes for different income levels. We can use a piecewise linear function to model a graduated tax. Let's consider a graduated tax where the first $10,000 of income is taxed at 10% and any income over $10,000 is taxed at 20%. For example, if your income after deductions is $30,000, then your taxes under this plan are

$$\text{graduated tax} = (10\% \text{ of } \$10,000) + (20\% \text{ of income over } \$10,000)$$
$$= 0.10(\$10,000) + 0.20(\$30,000 - \$10,000)$$
$$= 0.10(\$10,000) + 0.20(\$20,000)$$
$$= \$1000 + \$4000$$
$$= \$5000$$

If an income, i, is over $10,000, then under this plan,

$$\text{graduated tax} = 0.10(\$10,000) + 0.20(i - \$10,000)$$
$$= \$1000 + 0.20(i - \$10,000)$$

The Graph   This graduated tax plan is represented in Table 3.3 and Figure 3.24. The graph of the graduated tax is the result of piecing together two different line segments that represent the two different formulas used to find taxes. The short segment represents taxes for low incomes between $0 and $10,000, and the longer, steeper segment represents taxes for higher incomes that are greater than $10,000.

Taxes under Graduated Tax Plan

Income after Deductions ($)	Taxes ($)
0	0
5,000	500
10,000	1,000
20,000	1,000 + 2,000 = 3,000
30,000	1,000 + 4,000 = 5,000
40,000	1,000 + 6,000 = 7,000
50,000	1,000 + 8,000 = 9,000

Table 3.3

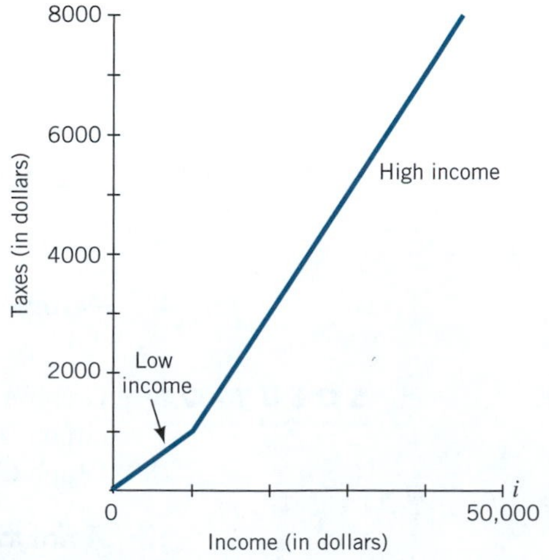

Figure 3.24 Graduated tax.

The Equations   To find an algebraic expression for the graduated tax, we need to use different formulas for different levels of income. Functions that use different formulas for different intervals of the domain are called *piecewise defined*. Since each income determines a unique tax, we can define the graduated tax as a piecewise function, g, of income i as

$$g(i) = \begin{cases} 0.10i & \text{for } i \le \$10{,}000 \\ 1000 + 0.20(i - 10{,}000) & \text{for } i > \$10{,}000 \end{cases}$$

The value of i (the input or independent variable) determines which formula to use to evaluate the function. This function is called a *piecewise linear function,* since each piece consists of a different linear formula. The formula for incomes equal to or below $10,000 is different from the formula for incomes above $10,000.

Evaluating Piecewise Functions   To find $g(\$8000)$, the value of $g(i)$ when $i = \$8000$, use the first formula in the definition since income, i, in this case is less than $10,000:

For $i \le \$10{,}000$ $g(i) = 0.10i$

substituting $8000 for i $g(\$8000) = 0.10(\$8000)$

$= \$800$

To find $g(\$40{,}000)$, we use the second formula in the definition, since in this case income is greater than $10,000:

For $i > \$10{,}000$ $g(i) = \$1000 + 0.20(i - \$10{,}000)$

substituting $40,000 for i $g(\$40{,}000) = \$1000 + 0.20(\$40{,}000 - \$10{,}000)$

$= \$1000 + 0.20(\$30{,}000)$

$= \$1000 + \$6000$

$= \$7000$

Comparing the Two Tax Models

Using graphs

In Figure 3.25 we compare the different tax plans by plotting the flat and graduated tax equations on the same graph.

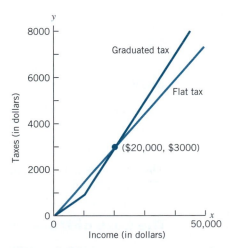

Figure 3.25 Flat tax versus graduated tax.

The intersection points indicate the incomes at which the amount of tax is the same under both plans. From the graph, we can estimate the coordinates of the two points as (0, 0) and ($20,000, $3000). That is, under both plans, with $0 income you pay $0 taxes, and with approximately $20,000 in income you would pay approximately $3000 in taxes. Individual voters want to know what impact these different plans will have on their taxes. From the graph in Figure 3.25, we can see that to the left of the intersection point at ($20,000, $3000), the flat tax is *greater* than the graduated tax for the same income. To the right of this intersection point, the flat tax is *less* than the graduated tax for the same income. So for incomes *less than* $20,000, taxes are *greater* under the flat tax plan, and for incomes *greater than* $20,000, taxes will be *less* under the flat tax plan.

Using equations

To verify the accuracy of our estimates for the coordinates of the point(s) of intersection, we can set $f(i) = g(i)$. We know that $f(i) = 0.15i$. Which of the two expressions do we use for $g(i)$? The answer depends upon what value of income, i, we consider. For $i \le \$10,000$, we have $g(i) = 0.10i$.

If	$f(i) = g(i)$
and $i \le \$10,000$, then	$0.15i = 0.10i$
This can only happen when	$i = 0$

If $i = 0$, both $f(i)$ and $g(i)$ equal 0; therefore, one intersection point is indeed $(0, 0)$.

For $i > \$10,000$, we use $g(i) = \$1000 + 0.20(i - \$10,000)$ and again set $f(i) = g(i)$.

If	$f(i) = g(i)$
and $i > \$10,000$, then	$0.15i = \$1000 + 0.20(i - \$10,000)$
apply the distributive property	$0.15i = \$1000 + 0.20i - (0.20)(\$10,000)$
multiply	$0.15i = \$1000 + 0.20i - \$2000$
combine terms	$0.15i = -\$1000 + 0.20i$
add $-0.20i$ to each side	$-0.05i = -\$1000$
divide by -0.05	$i = -\$1000/(-0.05)$
	$i = \$20,000$

So each plan results in the same tax for an income of $20,000. How much tax is required? We can substitute $20,000 for i into either the flat tax formula or the graduated tax formula for incomes over $10,000 and solve for the tax. Given the flat tax function,

$$f(i) = 0.15i$$

$$\text{if } i = \$20,000, \text{ then} \qquad f(i) = (0.15)(\$20,000)$$
$$= \$3000$$

We can check to make sure that when $i = \$20,000$, the graduated tax, $g(i)$, will also be $3000:

If $i > \$10,000$, then	$g(i) = \$1000 + 0.20(i - \$10,000)$
so, if $i = \$20,000$, then	$g(i) = \$1000 + 0.20(\$20,000 - \$10,000)$
perform operations	$= \$3000$

These calculations confirm that the other intersection point is, as we estimated, ($20,000, $3000).

The Case of Massachusetts

In the state of Massachusetts there has been an ongoing debate about whether or not to change from a flat tax to a graduated income tax[8]. In 1994 Massachusetts voters considered a proposal called Proposition 7. Proposition 7 would have replaced the flat tax rate (at the time of the proposal, 5.95%) with graduated income tax rates (called *marginal* rates) as shown in Table 3.4.

---

[8]For a state-by-state comparison of current income tax rates see *www.taxadmin.org/fta/rate/ind_inc.html*. For a comparison of the recent state tax proposals in Tennessee see *www.yourtax.org*.

Massachusetts Graduated Income Tax Proposal

	Marginal Rate		
Filing Status	5.5%	8.8%	9.8%
Married/joint	<$81,000	$81,000–$150,000	$150,000+
Married/separate	<$40,500	$40,500–$ 75,000	$ 75,000+
Single	<$50,200	$50,200–$ 90,000	$ 90,000+
Head of household	<$60,100	$60,100–$120,000	$120,000+

Table 3.4

Source: Office of the Secretary of State, Michael J. Connolly, Boston, 1994.

In Exploration 3.1 you can analyze the impact of Proposition 7 on people using different filing statuses.

Questions for the taxpayer

For what income and filing status would the taxes be equal under both plans? Who will pay less tax and who will pay more under the graduated income tax plan? We analyze the tax for a single person and leave the analyses of the other filing categories for you to do.

Finding out who pays what

The proposed graduated income tax is designed to tax at higher rates that portion of the individual's income that exceeds a certain threshold. For example, for those who file as single people, the graduated tax rate means that earned income under $50,200 would be taxed at a rate of 5.5%. Any income between $50,200 and $90,000 would be taxed at 8.8%, and any income over $90,000 would be taxed at 9.8%.

Using Equations.   The tax for a single person earning $100,000 would be the sum of three different dollar amounts.

5.5% on the first $50,200 $= (0.055)(\$50,200)$

$= \$2761$

8.8% on the next $39,800 (the portion of income between $50,200 and $90,000)

$= (0.088)(\$39,800)$

$\approx \$3502$

9.8% on the remaining $10,000 $= (0.098)(\$10,000)$

$= \$980$

The total graduated tax would be $2761 + $3502 + $980 = $7243. Under the flat tax rate the same individual would pay 5.95% of $100,000 = 0.0595($100,000), or $5950.

Table 3.5 shows the differences between the flat tax and the proposed graduated tax plan for single people at several different income levels. We can represent flat taxes for single people as a linear function f of income i, where

$$f(i) = 0.0595i$$

Massachusetts Taxes: Flat Rate vs. Graduated Rate for Single People

Income after Exemptions and Deductions ($)	Current Flat Tax at 5.95% ($)	Graduated Tax under Proposition 7 ($)
0	0	0
25,000	1488	1375
50,000	2975	2750
75,000	4463	4943
100,000	5950	7243

Table 3.5

We can write the graduated tax for single people as a piecewise linear function g of income i, where

$$g(i) = \begin{cases} 0.055i & \text{for } 0 \le i < \$50,200 \\ \$2761 + 0.088(i - \$50,200) & \text{for } \$50,200 \le i \le \$90,000 \\ \$6263 + 0.098(i - \$90,000) & \text{for } i > \$90,000 \end{cases}$$

Note that \$2761 in the second formula of the definition is the tax on the first \$50,200 of income (5.5% of \$50,200), and \$6263 in the third formula of the definition is the sum of the taxes on the first \$50,200 and the next \$39,800 of income (5.5% of \$50,200 + 8.8% of \$39,800 ≈ \$2761 + \$3502 = \$6263).

Using Graphs.   For what income would single people pay the same tax under both plans? The flat tax and the graduated income tax for single people are compared in Figure 3.26. An intersection point on the graph indicates where taxes are equal. One intersection point occurs at (0, 0). That makes sense, since under either plan if you have zero income, you pay zero taxes. The second intersection point is at approximately (\$58,000, \$3500). That means for an income of approximately \$58,000, the taxes are the same and are approximately \$3500.

In the Algebra Aerobics you are asked to describe what happens to the right and to the left of the intersection point.

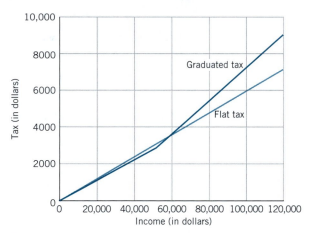

Figure 3.26 Massachusetts graduated tax vs. flat tax.

Algebra Aerobics 3.4

1. Graph the following piecewise functions.

   a. $f(x) = \begin{cases} -x - 1 & \text{for } x \le 0 \\ \frac{1}{2}x - 1 & \text{for } x > 0 \end{cases}$

   b. $g(x) = \begin{cases} 4 & \text{for } 0 \le x \le 5 \\ 2x - 6 & \text{for } x > 5 \end{cases}$

   c. $k(x) = \begin{cases} 0 & \text{for } 0 \le x \le 10 \\ 3x - 30 & \text{for } x > 10 \end{cases}$

2. Use equations to describe the piecewise linear function on each graph in Figure 3.27.

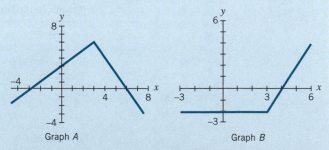

Figure 3.27 Two piecewise linear functions.

3. Evaluate each of the following piecewise defined functions at $x = -5, 0, 2,$ and 10.

a. $P(x) = \begin{cases} 3 & \text{for } x \le 1 \\ 1 - 2x & \text{for } x > 1 \end{cases}$

b. $W(x) = \begin{cases} x - 4 & \text{for } x < 2 \\ x + 4 & \text{for } x \ge 2 \end{cases}$

4. Construct a new graduated tax function, if:

a. The tax is 5% on the first $50,000 of income and 8% on any income in excess of $50,000.

b. The tax is 6% on the first $30,000 of income and 9% on any income in excess of $30,000.

5. Use the graph in Figure 3.26 to predict which tax is larger for each of the following incomes and by approximately how much. Then use the equations defining f and g on page 197 and 198 to check your predictions.

a. $30,000   b. $60,000   c. $120,000

6. Use equations to describe the following graph.

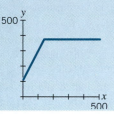

Exercises for Section 3.4

Graphing program recommended for Exercise 12.

1. Consider the following function:

$$f(x) = \begin{cases} 2x + 1 & x \le 0 \\ 3x & x > 0 \end{cases}$$

Evaluate $f(-10), f(-2), f(0), f(2),$ and $f(4)$.

2. Consider the following function:

$$g(x) = \begin{cases} 3 & x \le 1 \\ 4 + 2x & x > 1 \end{cases}$$

Evaluate $g(-5), g(-2), g(0), g(1), g(1.1), g(2),$ and $g(10)$.

3. Match each function with its graph.

a. $f(x) = \begin{cases} x & \text{if } x \le 2 \\ -x + 4 & \text{if } x > 2 \end{cases}$

b. $f(x) = \begin{cases} -x & \text{if } x \le 0 \\ x & \text{if } 0 < x \le 2 \\ -x + 4 & \text{if } x > 2 \end{cases}$

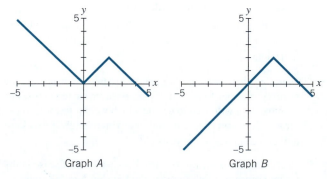

Graph A   Graph B

4. Construct a small table of values and graph the following piecewise linear functions. In each case specify the domain.

a. $f(x) = \begin{cases} 5 & \text{for } x < 10 \\ -15 + 2x & \text{for } x \ge 10 \end{cases}$

b. $g(t) = \begin{cases} 1 - t & \text{for } -10 \le t \le 1 \\ t & \text{for } 1 < t < 10 \end{cases}$

5. Construct a piecewise linear function for each of the accompanying graphs.

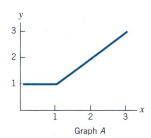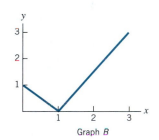

Graph A   Graph B

6. Create a graph for each piecewise function:

a. $f(x) = \begin{cases} 2x + 1 & \text{if } x \le 0 \\ 1 & \text{if } x > 0 \end{cases}$

b. $g(x) = \begin{cases} -x & \text{if } x < 0 \\ x & \text{if } x \ge 0 \end{cases}$

7. Create a graph for each piecewise function:

a. $h(x) = \begin{cases} 20 - 2x & \text{if } 0 \le x < 5 \\ 10 & \text{if } 5 \le x \le 10 \\ 10 + 2(x - 10) & \text{if } x > 10 \end{cases}$

b. $k(x) =$

$\begin{cases} 0.15x & \text{if } 0 \le x \le 20{,}000 \\ 3000 + 0.18(x - 20{,}000) & \text{if } 20{,}000 < x \le 50{,}000 \\ 8400 + 0.23(x - 50{,}000) & \text{if } x > 50{,}000 \end{cases}$

8. Each of the following graphs shows the activity of a hiker relative to her base campsite. Describe her actions first with words and then with a piecewise function.

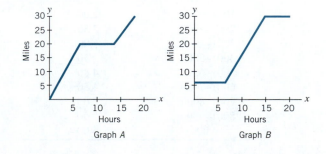

Graph A   Graph B

9. Find out what kind of income tax (if any) your state has. Is it flat or graduated? Construct and graph a function that describes your state's income tax for one filing status. Compute the income tax for various income levels.

10. **a.** Construct a graduated tax function where the tax is 10% on the first $30,000 of income, then 20% on any income in excess of $30,000.

   b. Construct a flat tax function where the tax is 15% of income.

   c. Calculate the tax for both the flat tax function from part (b) and the graduated tax function from part (a) for each of the following incomes: $10,000, $20,000, $30,000, $40,000, and $50,000.

   d. Graph the graduated and flat tax functions on the same grid and estimate the coordinates of the points of intersection. Interpret the points of intersection.

11. Fines for a particular speeding ticket are defined by the following piecewise function, where s is the speed in mph and $F(s)$ is the fine in dollars.

$$F(s) = \begin{cases} 0 & \text{if } s \leq 45 \\ 50 + 5(s - 45) & \text{if } 45 < s \leq 55 \\ 100 + 10(s - 55) & \text{if } 55 < s \leq 65 \\ 200 + 20(s - 65) & \text{if } s > 65 \end{cases}$$

   a. What is the implied posted speed limit for this situation?

   b. Create a table of values for the fine, beginning at 40 mph and incrementing by 5-mph steps up to 80 mph, and then graph $F(s)$.

   c. Describe in words how a speeding fine is calculated.

   d. Explain what 5, 10, and 20 in the formulas for the respective sections of the piecewise function represent.

   e. Find $F(30)$, $F(57)$, and $F(67)$.

   f. Graph $F(s)$.

12. (Graphing program recommended). In 2004, Missouri had a graduated tax plan but it was considering adopting a flat-rate tax of 4% on income after deductions. The tax rate for single people under the graduated plan is shown in the accompanying table. For what income levels would a single person pay less tax under the flat tax plan than under the graduated tax plan?

Missouri State Tax for a Single Person

Income after Deductions ($)	Marginal Tax Rate (%)
≤1000	1.50
1001–2000	2.00
2001–3000	2.50
3001–4000	3.00
4001–5000	3.50
5001–6000	4.00
6001–7000	4.50
7001–8000	5.00
8001–9000	5.50
>9000	6.00

13. You are thinking about replacing your long-distance telephone service. A cell phone company charges a monthly fee of $40 for the first 450 minutes and then charges $0.45 for every minute after 450. Every call is considered a long-distance call. Your local phone company charges you a fee of $60 per month and then $0.05 per minute for every long-distance call.

   a. Assume you will be making only long-distance calls. Create two functions, $C(x)$ for the cell phone plan and $L(x)$ for the local telephone plan, where x is the number of long-distance minutes.

   After how many minutes would the two plans cost the same amount?

   b. Describe when it is more advantageous to use your cell phone for long-distance calls.

   c. Describe when it is more advantageous to use your local phone company to make long-distance calls.

14. The accompanying table, taken from a pediatrics text, provides a set of formulas for the approximate "average" weight and height of normal infants and children.

Age	Weight (lb)
At birth	7
3–12 months	(age in months) + 11
1–5 years	5 · (age in years) + 17
6–12 years	7 · (age in years) + 5

Age	Height (in)
At birth	20
1 year	30
2–12 years	2.5 · (age in years) + 30

Source: R. E. Behrman and V. C. Vaughan (eds.), *Nelson Textbook of Pediatrics,* 12th ed. (Philadelphia: W. B. Saunders, 1983), p. 19.

For children from birth to 12 years of age, construct and graph a piecewise linear function for each of the following (assuming age is the independent variable):

   a. Weight in pounds (How does this model compare to the model for female infants in Chapter 2, Section 5?)

   b. Height in inches

Note that there are certain gaps in the table that need to be resolved in order to construct piecewise linear functions. For example, you will need to decide which weight formula to use for a child who is $5\frac{1}{2}$ years old and which height formula to use for a child who is $1\frac{1}{2}$ years old.

15. Heart health is a prime concern, because heart disease is the leading cause of death in the United States. Aerobic activities such as walking, jogging, and running are recommended for cardiovascular fitness, because they increase the heart's strength and stamina.

   a. A typical training recommendation for a beginner is to walk at a moderate pace of about 3.5 miles/hour (or approximately 0.0583 miles/minute) for 20 minutes. Construct a function that describes the distance traveled

D_{beginner}, in miles, as a function of time, T, in minutes, for someone maintaining this pace. Construct a small table of values and graph the function using a reasonable domain.

b. A more advanced training routine is to walk at a pace of 3.75 miles/hour (or 0.0625 miles/minute) for 10 minutes and then jog at 5.25 miles/hour (or 0.0875 miles/minute) for 10 minutes. Construct a piecewise linear function that gives the total distance, D_{advanced}, as a function of time. T, in minutes. Generate a small table of values and plot the graph of this function on your graph in part (a).

c. Do these two graphs intersect? If so, what do the intersection point(s) represent?

16. A graduated income tax is proposed in Borduria to replace an existing flat rate of 8% on all income. The new proposal states that persons will pay no tax on their first $20,000 of income, 5% on income over $20,000 and less than or equal to $100,000, and 10% on their income over $100,000. (*Note:* Borduria is a fictional totalitarian state in the Balkans that figures in the adventures of Tintin.)

a. Construct a table of values that shows how much tax persons will pay under both the existing 8% flat tax and

the proposed new tax for each of the following incomes: $0, $20,000, $50,000, $100,000, $150,000, $200,000.

b. Construct a graph of tax dollars vs. income for the 8% flat tax.

c. On the same graph plot tax dollars vs. income for the proposed new graduated tax.

d. Construct a function that describes tax dollars under the existing 8% tax as a function of income.

e. Construct a piecewise function that describes tax dollars under the proposed new graduated tax rates as a function of income.

f. Use your graph to estimate the income level for which the taxes are the same under both plans. What plan would benefit people with incomes below your estimate? What plan would benefit people with incomes above your estimate?

g. Use your equations to find the coordinates that represent the point at which the taxes are the same for both plans. Label this point on your graph.

h. If the median income in the state is $27,000 and the mean income is $35,000, do you think the new graduated tax would be voted in by the people? Explain your answer.

CHAPTER SUMMARY

Systems of Linear Equations

A pair of real numbers is a *solution* to a system of linear equations in two variables if and only if the pair of numbers is a solution of each equation. A system of two linear equations in two variables may have one solution, no solutions, or infinitely many solutions.

Infinitely many solutions

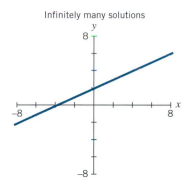

One solution

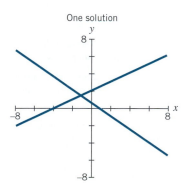

No solutions

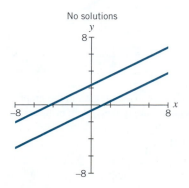

Systems of Linear Inequalities

The solutions to a system of linear inequalities in two variables are pairs of real numbers that satisfy all the inequalities. The solutions typically form a region of the plane and may be represented by a compound inequality such as

$$y_1 < y \le y_2$$
$$5 - x < y \le 1 + 3x$$

The shaded area in the figure represents the solution area for this system.

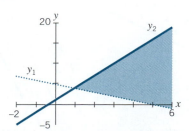

Piecewise Linear Functions

Functions that use different formulas for different intervals of the domain are called *piecewise functions*. Functions constructed out of pieces of several different linear functions are called *piecewise linear functions*.

The accompanying figure shows the graph of a linear piecewise function g of income i, where

$$g(i) = \begin{cases} 0.10i & \text{for } i \leq \$10,000 \\ 1000 + 0.20(i - 10,000) & \text{for } i > \$10,000 \end{cases}$$

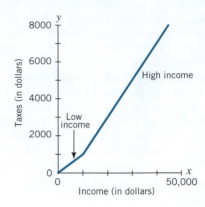

CHECK YOUR UNDERSTANDING

I. Are the statements in Problems 1–29 true or false? Give an explanation for your answer.

  1. Assuming y is a function of x, the number pair $(-5, 3)$ is a solution to the following system of equations:
  $$\begin{cases} 2x - 3y = 21 \\ x = 2y + 3 \end{cases}.$$

  2. Assuming y is a function of x, the number pair $\left(\frac{16}{5}, \frac{2}{5}\right)$ is a solution to the following system of equations:
  $$\begin{cases} y = 10 - 3x \\ 2x - y = 6 \end{cases}$$

  3. The system $\begin{cases} x - \dfrac{5}{y} = 2 \\ \dfrac{x}{3} + 2y = 1 \end{cases}$ is not a linear system of equations.

  4. The system of linear equations $\begin{cases} 2y - \dfrac{x}{3} = -1 \\ \dfrac{6y}{5} - \dfrac{x}{5} = \dfrac{-3}{5} \end{cases}$ has no solution(s).

  5. A system of linear equations in two variables either has a pair of numbers that is a unique solution or has no solution.

  6. A pair of numbers can be a solution to a system of linear equations in two variables if the pair is a solution to at least one of the equations.

Questions 7 and 8 refer to the accompanying graphs of systems of two equations in two variables.

  7. System A has a unique solution at approximately $(50, 25)$.

  8. System B appears to have no solution.

Questions 9 and 10 refer to the supply and demand curves in the accompanying graph.

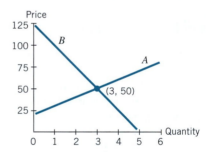

  9. Graph B is the demand curve because as price increases, the quantity demanded decreases.

  10. If three items are produced and sold at a price of $50, the quantity supplied will be equal to the quantity demanded.

  11. The shaded region in the accompanying graph appears to satisfy the linear inequality $12 - x < 3y$.

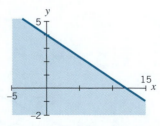

  12. The darkest shaded region in the graph on the next page appears to satisfy the system of linear inequalities
  $$\begin{cases} y \leq 4 - \dfrac{x}{3} \\ y \leq 2 \end{cases}.$$

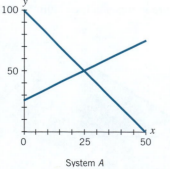

System A

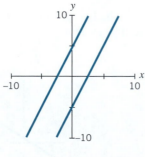

System B

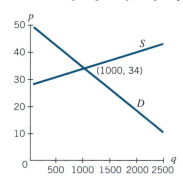

13. The linear inequalities $y \leq 5x - 3$ and $y < 5x - 3$ have exactly the same solutions.

Questions 14 and 15 refer to the accompanying graph of supply and demand, where q = quantity and p = price:

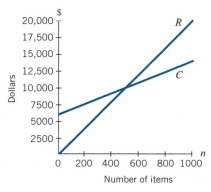

14. The equilibrium point $(1000, 34)$ means that at a price of $1000, suppliers will supply 34 items and consumers will demand 34 items.

15. If the price is less than the equilibrium price, the quantity supplied is less than the quantity demanded and therefore competition will cause the price to increase toward the equilibrium price.

Questions 16–20 refer to the accompanying graph, where R represents total revenue and C represents total costs.

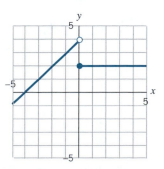

16. The estimated breakeven point is approximately $(500, \$10,000)$.

17. The profit is zero at the breakeven point.

18. If 200 units are produced and sold, the total revenue is larger than the total cost.

19. If the revenue per unit increases, and hence, R, the revenue line, becomes steeper, the breakeven point moves to the right.

20. If 700 units are produced and sold, the company is making about a $5000 profit.

Questions 21–23 refer to the accompanying graph of linear approximations for the body mass index (BMI) of girls age 7–15. B_U gives the minimal recommended BMI and B_N the maximum recommended BMI. B_0 is the dividing line between being somewhat overweight and being obese.

The National Center For Health Statistics calculates the BMI as

$$\text{BMI} = \left[\frac{\text{weight (lb)}}{\text{height (in)}} \right] \cdot \left[\frac{703}{\text{height (in)}} \right]$$

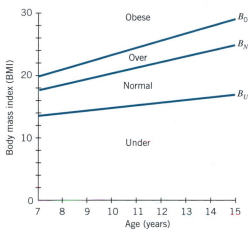

Sources: National Center for Health Statistics, National Center for Chronic Disease Prevention and Health Promotion (2000), *www.cdc.gov/growthcharts.*

21. A 15-year-old girl who weighs 105 pounds and is 55 inches tall has a BMI that is within the normal zone.

22. A 10-year-old girl with BMI = 12 is underweight for her height and age.

23. A mathematical description of the normal BMI zone for girls age 7–15 is $B_U \leq \text{BMI} \leq B_N$.

Questions 24 and 25 refer to the following system of linear inequalities: $\begin{cases} 2x + 3 \leq y \\ y < 5 + x \end{cases}$

24. The boundary line $y = 5 + x$ is not included in the solution region.

25. The pair of numbers $(0, 0)$ is a solution to the system.

Questions 26 and 27 refer to the accompanying graph of the piecewise function $y = f(x)$.

26. $f(-2) = 2$

27. $f(0) = 4$

Questions 28 and 29 refer to the piecewise function

$$f(x) = \begin{cases} 4 - 2x & \text{for } -3 \le x < 2 \\ 0 & \text{for } 2 \le x \le 3 \\ x + 1 & \text{for } 3 < x \le 5 \end{cases}$$

28. $f(-1) = 6$

29. $f(3) = 0$

II. In Problems 30–36, give an example of a function or functions with the specified properties. Express your answer using formulas, and specify the independent and dependent variables.

30. A system of two linear equations in two variables that has no solution.

31. A system of two linear equations in two variables that has an infinite number of solutions, including the pairs (2, 3) and (−1, 9).

32. A system of two linear inequalities in two variables that has no solution.

33. A system of two linear equations in variables c and r with the ordered pair $(-1, 0)$ as a solution, where c is a function of r.

34. A cost and revenue function that has a breakeven point at (100, $5000).

35. A demand equation (downward sloping) that, with the supply function $p = q + 20$, has an equilibrium point at (20 units, $40).

36. A system of linear inequalities that has as its solution all of quadrant IV (the region where $x > 0$ and $y < 0$).

III. Is each of the statements in Problems 37–40 true or false? If a statement is true, explain how you know. If a statement is false, give a counterexample.

37. Linear systems of equations in two variables always have exactly one pair of numbers as a solution.

38. A linear system of equations in two variables can have exactly two pairs of numbers that are solutions to the system.

39. Any solution(s) to a linear system of equations in two variables is either a point or a line.

40. Solutions to a linear system of inequalities in two variables can be a region in the plane.

CHAPTER 3 REVIEW: PUTTING IT ALL TOGETHER

1. The following graph shows worldwide production and consumption of grain in millions of metric tons (MMT). It is based on data from 1998 to May 2006 and includes projections to the end of 2006.

  a. Estimate the maximum production of grain in this time period. In what year did it occur? What was the minimum production and in what year did it occur?

  b. Explain the meaning of the intersection point(s) in this context.

  c. In what year was the difference between consumption and production the largest? Was there a surplus or deficit?

  d. Create a title for the graph.

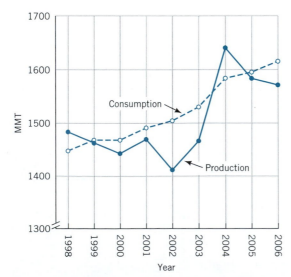

Source: Foreign Agricultural Service Circular Series, FG 05-06, May 2006, *www.fas.usda.gov/grain/circular/2006/05-06/graintoc.htm*.

2. The accompanying graphs show two systems of linear equations.

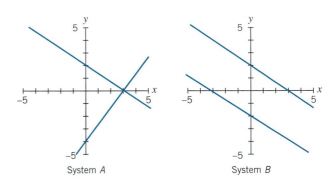

System A   System B

For each system:

  a. Determine the number of solutions using the graph of the system.

  b. Construct the equations of the lines.

  c. Solve the system using your equations from part (b). Explain why your answer makes sense.

3. a. Construct a system of linear equations where both of the following conditions are met:

  i. The coordinates of the point of intersection are (4, 10).

  ii. The two lines are perpendicular to each other and one of the lines has a slope of −4.

  b. Graph the system of equations you found in part (a). Estimate the coordinates of the point of intersection on your graph. Does your estimate confirm your answer for part (a)?

4. New York City taxi fares are as follows: initial fare $2.00, $0.30 per 1/5 mile, and night surcharge (8 p.m.–6 a.m.) of $0.50.[9]

   a. Create and graph the equation for C_d, the cost of a daytime taxi ride for m miles.

   b. Create and graph the equation for C_n, the cost of a night taxi ride for m miles.

   c. How do the two graphs compare?

5. Graph and shade the region bounded by the following inequalities:

$$y < 2$$
$$y \geq -2x + 3$$
$$x \leq 5$$

6. Determine the inequalities that describe the shaded region in the following graph.

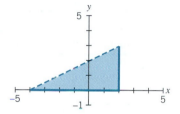

7. A musician produces and sells CDs on her website. She estimates fixed costs of $10,000, with an additional cost of $7 to produce each CD. She currently sells the CDs for $12 each.

   a. Create the cost equation, C, and revenue equation R, in terms of x number of CDs produced and sold.

   b. Find the breakeven point.

   c. By how much should the musician charge for each CD if she wants to break even when producing and selling 1600 CDs, assuming other costs remain the same?

   d. By how much would she need to reduce fixed costs if she wants to break even when producing and selling 1600 CDs, assuming other costs and prices remain the same as originally stated?

8. Data from the U.S. Department of Energy show that gasoline prices vary immensely for different countries. These prices include taxes and are in U.S. dollars/gallon of unleaded regular gasoline, standardized on the U.S. gallon.

Year	Germany	Japan	United States
1990	2.65	3.16	1.16
1992	3.26	3.59	1.13
1994	3.52	4.39	1.11
1996	3.94	3.65	1.23
1998	3.34	2.83	1.06
2000	3.45	3.65	1.51
2002	3.67	3.15	1.36
2004	5.24	3.93	1.88

   a. Compute the average annual rate of change of gas prices for each country using data for 1990 and 2004.

   b. Compute the average (mean) gas price from 1990 to 2004 for each country.

   c. Write a paragraph comparing U.S. gas prices in this 14-year period with prices in Japan and Germany.

9. In 2006 China was the world's second largest consumer of oil, behind the United States.

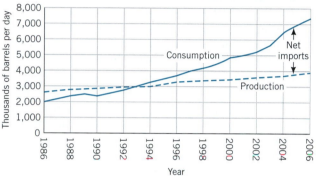

China's Oil Production and Consumption, 1986–2006

Source: EIA International Petroleum Monthly. Data include projections for last four months of 2006.

   a. Estimate the amount of oil consumption and production in China in 1990. What is the net difference between production and consumption? What does this mean?

   b. Estimate the year when oil consumption equaled oil production. Explain the meaning of this condition.

   c. Estimate the amount of oil consumption and oil production in China in 2006. What is the net difference? What does this mean?

10. You keep track of how much gas your car uses and estimate that it gets 32 miles/gallon.

   a. If you start out with a full tank of 14 gallons, write a formula for how many gallons of gas G are left in the tank after you have gone M miles.

   b. You consider borrowing your friend's larger car, which gets 18 miles/gallon and holds 20 gallons. Write a formula for how many gallons of gas G_F are left in the tank after you have gone M miles. Generate a data table and graph the gas remaining (vertical axis) versus miles (horizontal axis) for both cars.

   c. Which car has the longer mileage range?

   d. Is there a distance at which they both have the same amount of gas left? If so, what is it?

11. a. Solve algebraically each of the following systems of linear equations.

   i. $4x + 21y = -5$ **ii.** $2.5a - b = 7$
   $3x + 7y = -10$ $b = \dfrac{a}{6}$

   b. Create a system of linear equations for which there is no solution.

[9]*http://www.ny.com/transportation/taxis/* gives more detailed information on NYC taxi fares.

The accompanying table is for problems 12 and 13.

Activity	Calories Burned per Minute
Running (moderate pace)	10
Swimming laps	8

Let r and s represent the number of minutes spent running and swimming, respectively.

12. Use the table to answer the following questions. Your goal is to run and swim so that you burn at least 800 calories but no more than 1200 calories.

    a. Create a system of inequalities that represents the set of combinations of minutes running and minutes swimming that meet your goal. (*Note:* $r \geq 0$ and $s \geq 0$.)

    b. Graph the solution to the system of inequalities using s on the horizontal axis and r on the vertical axis.

    c. Give an example of one combination of running and swimming times that is in the solution set and one that is not in the solution set.

    d. How would your solution set change if your goal were to burn at least 600 calories but not more than 1000 calories?

13. Use the table to answer the following questions. Your goal is to run and swim so that you spend no more than 60 minutes exercising but burn at least 560 calories.

    a. Create a system of inequalities that represents the set of combinations of minutes running and minutes swimming that meet your goal. (*Note:* $r \geq 0$ and $s \geq 0$.)

    b. Graph the solution to the system of inequalities with s on the horizontal axis and r on the vertical axis.

    c. Give an example of one combination of running and swimming times that is in the solution set and one that is not in the solution set.

    d. How would your solution set change if you spend no more than 70 minutes exercising but still burn at least 560 calories?

14. In a 400-meter relay swim, each team has four swimmers. In sequence, the swimmers each swim 100 meters. The total time (cumulative) and rates for team A are provided in the accompanying table.

Swimmer	Total Time, Cumulative (seconds)	Rate (seconds/meter), Rounded	Total Distance, Cumulative (meters)
1	99	0.99	100
2	201	1.02	200
3	279	0.78	300
4	341	0.62	400

    a. Which swimmer is the fastest? The slowest?

    b. The total swim time function can be written as

$$t(m) = \begin{cases} 0.99m & \text{for } 0 \leq m \leq 100 \\ 99 + 1.02(m - 100) & \text{for } 100 \leq m \leq 200 \\ 201 + 0.78(m - 200) & \text{for } 200 < m \leq 300 \\ 279 + 0.62(m - 300) & \text{for } 300 < m \leq 400 \end{cases}$$

Find and interpret the following: $t(50)$, $t(125)$, $t(250)$, $t(400)$.

15. The accompanying table lists the monthly charge, the number of minutes allowed, and the charge per additional minute for three wireless phone plans.

Cell Phone Plan	Monthly Charge	Number of Minutes	Overtime Charge/Minute
A	$39.99	450	$0.45
B	$59.99	900	$0.40
C	$79.99	1350	$0.35

Let m be the number of minutes per month, $0 \leq m \leq 2500$.

    a. Construct a piecewise function for the cost $A(m)$ for plan A.

    b. Construct a piecewise function for the cost $B(m)$ for plan B.

    c. Construct a piecewise function for the cost $C(m)$ for plan C.

    d. Complete the following table to determine the best plan for each of the estimated number of minutes per month.

Number of Minutes/Month	Cost		
	Plan A	Plan B	Plan C
500			
800			
1000			

16. It is often said that 1 year of a dog's life is equivalent to 7 years of a human's life. A more accurate veterinarian's estimate is that for the first 2 years of a dog's life, each dog year is equivalent to 10.5 years of a human's life, and after 2 years each dog year is equivalent to 4 years of a human's life.

    a. Write formulas to describe human-equivalent years as a function of dog years for the two methods of relating dog years to human years. Use H and D for the popular formula, and H_v and D_v for the veterinary method.

    b. At what dog age do the 2 methods give the same human years, and what human age is that?

17. The time series at the top of the next page shows the price per barrel of crude oil from 1861 to 2006 in both dollars actually spent (*nominal*) and dollars adjusted for inflation (*real* 2006 dollars). Write a 60-second summary about crude oil prices.

18. Suppose a flat tax amounts to 10% of income. Suppose a graduated tax is a fixed $1000 for any income $\leq$ $20,000 plus 20% of any income over $20,000.

    a. Construct functions for the flat tax and the graduated tax.

    b. Construct a small table of values:

Income	Flat Tax	Graduated Tax
$0		
$10,000		
$20,000		
$30,000		
$40,000		

    c. Graph both tax plans and estimate any point(s) at which the two plans would be equal.

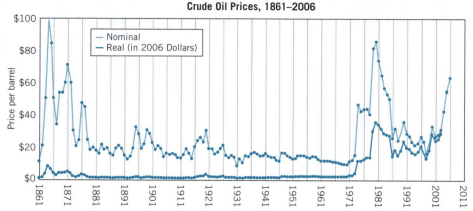

Crude Oil Prices, 1861–2006

Source: http://en.wikipedia.org/wiki/Image:Oil_Prices_1861_2006.jpg Graph created by Michael Ströck, 2006. Released under the GFDL.

d. Use the function definitions to calculate any point(s) at which the two plans would be equal.

e. For what levels of income would the flat tax be more than the graduated tax? For what levels of income would the graduated tax be more than the flat tax.

f. Are there any conditions in which an individual might have negative income under either of the above plans, that is, the amount of taxes would exceed an individual's income? This is not as strange as it sounds. For example, many states impose a minimum corporate tax on a company, even if it is a small, one-person operation with no income in that year.

19. You check around for the best deal on your prescription medicine. At your local pharmacy it costs $4.39 a bottle; by mail-order catalog it costs $3.85 a bottle, but there is a flat shipping charge of $4.00 for any size order; by a source found on the Internet it costs $3.99 a bottle and shipping costs $1.00 for each bottle, but for orders of ten or more bottles it costs $3.79 a bottle, and handling is $2.50 per order plus shipping costs of $1.00 for each bottle.

a. Find a formula to express each of these costs if N is the number of bottles purchased and C_p, C_c, and C_I are the respective costs for ordering from the pharmacy, catalog, and Internet.

b. Graph the costs for purchases up to thirty bottles at a time. Which is the cheapest source if you buy fewer than ten bottles at a time? If you buy more than ten bottles at a time? Explain.

20. Older toilets use about 7 gallons per flush. Since using this much pure water to transport human waste is especially undesirable in areas of the country where water is scarce, new toilets now must meet a water conservation standard and are designed to use 1.6 gallons or less per flush.

a. An old toilet that leaks about 2 cups of water an hour from the tank to the bowl and uses 7 gallons/flush is replaced with a new toilet, which does not leak and uses 1.6 gallons/flush. There are 16 cups in a gallon. For each toilet write an equation for daily water loss, W (gallons), as a function of number of flushes per day, F.

b. Graph the equations in part (a) on the same plot with F on the horizontal axis, where $0 \leq F \leq 25$. From your graph, estimate the amount of water used by each system for 20 flushes a day, and check your estimates with your equations. What is the net difference? What does this mean?

c. If a family flushes the toilet an average of 10 times a day, how much water do they save every day by replacing their leaky old toilet with a new water saver toilet? How much water would be saved over a year?

d. In England during World War II the citizens were asked to flush their toilets only once every five times the toilet was used, in order to save water. A pencil was hung near the toilet, and each user made a vertical mark on the wall, the fifth user made a horizontal line through the last four marks and then flushed the toilet. If this method were used today for the family with the old leaky toilet, would the amount saved be greater than the amount saved with the new toilet if the new toilet is flushed after every use?

21. Regular aerobic exercise at a target heart rate is recommended for maintaining health. Someone starting an exercise program might begin at an intensity level of 60% of the target rate and work up to 70%; athletes need to work at 85% or higher. The American College of Sports Medicine method to compute target heart rate, H (in beats per minute), is based on maximum heart rate, H_{max}, age in years, A, and exercise intensity level, I, where

$$H_{max} = 220 - A$$

and

$$H = I \cdot H_{max}$$

thus

$$H = I \cdot (220 - A)$$

a. Write a formula for beginners' target heart rate H_b if the intensity level is 60%, that is $I = 0.60$.

b. Write a formula for intermediate target heart rate H_i if the intensity level is 70%, that is $I = 0.70$.

c. Write a formula for athletes' target heart rate H_a if the intensity level is 85%, that is $I = 0.85$.

d. Construct a graph showing all three heart rate levels, H_b, H_i, H_a, and the maximum rate, H_{max}, for ages 15 to 75 years. Put age on the horizontal axis. Mark the zone in which athletes should work.

e. Compute the target heart rate for a 20-year-old to work at all three levels. What is H_{max} for a 20-year-old?

f. If a 65-year-old is working at a heart rate of 134 beats/minute, what is her intensity level? Has she exceeded the maximum heart rate for her age? If her 45-year-old son is working at the same heart rate, what is his intensity level? If her 25-year-old granddaughter is also working at the same heart rate, what is her intensity level?

EXPLORATION 3.1

Flat vs. Graduated Income Tax: Who Benefits?

Objectives

- compare the effects of different tax plans on individuals in different income brackets
- interpret intersection points

Material/Equipment

- spreadsheet or graphing calculator (optional). If using a graphing calculator, see examples on graphing piecewise functions in the Graphing Calculator Manual.
- graph paper

Related Readings

"How a Flat Tax Would Work, for You and for Them," *The New York Times*, Jan. 21, 1996.
"Flat Tax Goes from 'Snake Oil' to G.O.P. Tonic," *The New York Times,* Nov. 14, 1999.
"A Taxation Policy to Make John Stuart Mill Weep," *The New York Times,* April 18, 2004.
"Your Real Tax Rate," *www.msnmoney.com*, Feb. 21, 2007.

Procedure

A variety of tax plans were debated in all recent elections for president. (See related readings.) One plan recommended a flat tax of 19% on income after exemptions and deductions. In this exploration we examine who benefits from this flat tax as opposed to the current graduated tax plan. The questions we explore are

For what income will taxes be equal under both plans?

Who will benefit under the graduated income tax plan compared with the flat tax plan?

Who will pay more taxes under the graduated plan compared with the flat tax plan?

In a Small Group or with a Partner

The accompanying table gives the 2006 federal graduated tax rates on income after deductions for single people.

1. Construct a function for flat taxes of 19%, where income, i, is income after deductions.
2. Construct a piecewise linear function for the graduated federal tax for single people in 2006, where income, i, is income after deductions.
3. Graph your two functions on the same grid. Estimate from your graph any intersection points for the two functions.
4. Use your equations to calculate more accurate values for the points of intersection.
5. (Extra credit.) Use your results to make changes in the tax plans. Decide on a different income for which taxes will be equal under both plans. You can use what you know about the distribution of income in the United States from the FAM1000 data to make your decision. Alter one or both of the original functions such that both tax plans will generate the same tax given the income you have chosen.

2006 Tax Rate Schedule for Single Persons

Schedule X—Use if your filing status is **Single**

If the amount on Form 1040, line 40, is: Over—	But not over—	Enter on Form 1040, line 41	of the amount over—
$0	$7,550	10%	$0
$7,550	$30,650	$755+ 15%	$7,550
$30,650	$74,200	$4,220+ 25%	$30,650
$74,200	$154,800	$15,107.50+ 28%	$74,200
$154,800	$336,550	$37,675.50+ 33%	$154,800
$336,550	No limit	$97,653.00+ 35%	$336,550

Source: www.irs.gov.

Analysis

- Interpret your findings. Assume deductions are treated the same under both the flat tax plan and the graduated tax plan. What do the intersection points tell you about the differences between the tax plans?
- What information would be useful in deciding on the merits of each of the plans?
- What if the amount of deductions that most people can take under the flat tax is less than the graduated tax plan? How will your analysis be affected?

Exploration-Linked Homework

Reporting Your Results

Take a stance for or against a flat federal income tax. Using supportive quantitative evidence, write a 60-second summary for a voters' pamphlet advocating your position. Present your arguments to the class.

THE LAWS OF EXPONENTS AND LOGARITHMS: MEASURING THE UNIVERSE

OVERVIEW

Most of the examples we've studied so far have come from the social sciences. In order to delve into the physical and life sciences, we need to compactly describe and compare the extremes in deep time and deep space. In this chapter, we introduce the tools that scientists use to represent very large and very small quantities.

After reading this chapter, you should be able to

- write expressions in scientific notation

- convert between English and metric units

- simplify expressions using the rules of exponents

- compare numbers of widely differing sizes

- calculate logarithms base 10 and plot numbers on a logarithmic scale

4.1 *The Numbers of Science: Measuring Time and Space*

On a daily basis we encounter quantities measured in tenths, tens, hundreds, or perhaps thousands. Finance or politics may bring us news of "1.3 billion people living in China" or "a federal debt of over $7 trillion." In the physical sciences the range of numbers encountered is much larger. *Scientific notation* was developed to provide a way to write numbers compactly and to compare the sizes found in our universe, from the largest object we know—the observable universe—to the tiniest—the minuscule quarks oscillating inside the nucleus of an atom. We use examples from deep space and deep time to demonstrate powers of 10 and the use of scientific notation.

Powers of 10 and the Metric System

The international scientific community and most of the rest of the world use the *metric system,* a system of measurements based on the meter (which is about 39.37 inches, a little over 3 feet). In daily life Americans have resisted converting to the metric system and still use the *English system* of inches, feet, and yards. Table 4.1 shows the conversions for three standard metric units of length: the meter, the kilometer, and the centimeter. For a more complete conversion table, see the inside back cover.

Conversions from Metric to English for Some Standard Units

Metric Unit	Abbreviation	In Meters	Equivalent in English Units	Informal Conversion
meter	m	1 m	3.28 ft	The width of a twin bed, a little more than a yard
kilometer	km	1000 m	0.62 mile	A casual 12-minute walk, a little over half a mile
centimeter	cm	0.01 m	0.39 in	The length of a black ant, a little under half an inch

Table 4.1

Deep space

For an appreciation of the size of things in the universe, we highly recommend the video by Charles and Ray Eames and related book by Philip and Phylis Morrison titled *Powers of Ten: About the Relative Size of Things.*

The Observable Universe. Current measurements with the most advanced scientific instruments generate a best guess for the radius of the observable universe at about 100,000,000,000,000,000,000,000,000 meters, or "one hundred trillion trillion meters." Obviously, we need a more convenient way to read, write, and express this number. To avoid writing a large number of zeros, exponents can be used as a shorthand:

10^{26} can be written as a 1 with twenty-six zeros after it.

10^{26} means: $10 \cdot 10 \cdot 10 \cdot \cdots \cdot 10$, the product of twenty-six 10s.

10^{26} is read as "10 to the twenty-sixth" or "10 to the twenty-sixth power."

So the estimated size of the radius of the observable universe is 10^{26} meters. The sizes of other relatively large objects are listed in Table 4.2.[1]

The Relative Sizes of Large Objects in the Universe

Object	Radius (in meters)
Milky Way	$1,000,000,000,000,000,000,000 = 10^{21}$
Our solar system	$1,000,000,000,000 = 10^{12}$
Our sun	$1,000,000,000 = 10^{9}$
Earth	$10,000,000 = 10^{7}$

Table 4.2

[1]The rough estimates for the sizes of objects in the universe in this section are taken from Timothy Ferris, *Coming of Age in the Milky Way* (New York: Doubleday, 1988).

Us. Human beings are roughly in the middle of the scale of measurable objects in the universe. Human heights, including children's, vary from about one-third of a meter to 2 meters. In the wide scale of objects in the universe, a rough estimate for human height is 1 meter.

To continue the system of writing all sizes using powers of 10, we need a way to express 1 as a power of 10. Since $10^3 = 1000$, $10^2 = 100$, and $10^1 = 10$, a logical way to continue would be to say that $10^0 = 1$. Since reducing a power of 10 by 1 is equivalent to dividing by 10, the following calculations give justification for defining 10^0 as equal to 1.

$$10^2 = \frac{10^3}{10} = \frac{(10)(10)(\cancel{10})}{\cancel{10}} = 100$$

$$10^1 = \frac{10^2}{10} = \frac{(10)(\cancel{10})}{\cancel{10}} = 10$$

$$10^0 = \frac{10^1}{10} = \frac{\cancel{10}}{\cancel{10}} = 1$$

By using negative exponents, we can continue to use powers of 10 to represent numbers less than 1. For consistency, reducing the power by 1 should remain equivalent to dividing by 10. So, continuing the pattern established above, we define $10^{-1} = 1/10$, $10^{-2} = 1/10^2$, and so on. For any positive integer, n, we define

$$10^{-n} = \frac{1}{10^n}$$

DNA Molecules. A DNA strand provides genetic information for a human being. It is made up of a chain of building blocks called nucleotides. The chain is tightly coiled into a double helix, but stretched out it would measure about 0.01 meter in length. How does this DNA length translate to a power of 10? The number 0.01, or one-hundredth, equals $1/10^2$. We can write $1/10^2$ as 10^{-2}. So a DNA strand, uncoiled and measured lengthwise, is approximately 10^{-2} meters, or one centimeter.

Table 4.3 shows the sizes of some objects relative to the size of human beings.

The Relative Sizes of Small Objects in the Universe

Object	Radius (in meters)	
Human beings	$1 = \frac{10}{10} =$	10^0
DNA molecules	$0.01 = \frac{1}{100} = \frac{1}{10^2} =$	10^{-2}
Living cells	$0.000\,01 = \frac{1}{100,000} = \frac{1}{10^5} =$	10^{-5}
Atoms	$0.000\,000\,000\,1 = \frac{1}{10,000,000,000} = \frac{1}{10^{10}} =$	10^{-10}

Table 4.3

The following box gives the definition for various powers of 10.

Powers of 10

When n is a positive integer:
$$10^n = \underbrace{10 \cdot 10 \cdot 10 \cdot \,\cdots\, \cdot 10}_{n \text{ factors}} \quad \text{which can be written as 1 followed by } n \text{ zeros.}$$

$$10^0 = 1$$

$$10^{-n} = \frac{1}{10^n} \quad \text{which can be written as a decimal point followed by } n-1 \text{ zeros and a 1.}$$

Multiplying by 10^n is equivalent to moving the decimal point to the right n places.

Multiplying by 10^{-n} is equivalent to dividing by 10^n, or moving the decimal point to the left n places.

The metric language

By international agreement, standard prefixes specify the power of 10 that is attached to a specific unit of measure. They indicate the number of times the basic unit has been multiplied or divided by 10. Usually these prefixes are attached to metric units of measure, but they are occasionally used with the English system. Table 4.4 gives prefixes and their abbreviations for certain powers of 10. A more complete table is on the inside back cover.

Prefixes for Powers of 10

pico-	p	10^{-12}	(unit)		10^0
nano-	n	10^{-9}	kilo-	k	10^3
micro-	μ	10^{-6}	mega-	M	10^6
milli-	m	10^{-3}	giga-	G	10^9
centi-	c	10^{-2}	tera-	T	10^{12}

Table 4.4

EXAMPLE 1 Indicate the number of meters in each unit of measure: cm, mm, Gm.

SOLUTION

$$1 \text{ cm} = 1 \text{ } centi\text{meter} = 10^{-2} \text{ m} = \frac{1}{10^2} \text{ m} = \frac{1}{100} \text{ m} = 0.01 \text{ meter}$$

$$1 \text{ mm} = 1 \text{ } milli\text{meter} = 10^{-3} \text{ m} = \frac{1}{10^3} \text{ m} = \frac{1}{1000} \text{ m} = 0.001 \text{ meter}$$

$$1 \text{ Gm} = 1 \text{ } giga\text{meter} = 10^9 \text{ m} = 1,000,000,000 \text{ meters}$$

EXAMPLE 2 Translate the following underlined expressions.

A standard CD holds about 700 *mega*bytes of information.

Translation: $700 \cdot 10^6$ bytes or 700,000,000 bytes

A calculator takes about one *milli*second to add or multiply two 10-digit numbers.

Translation: $1 \cdot 10^{-3}$ second or 0.001 second

In Tokyo on January 11, 1999, the NEC company announced that it had developed a *pico*second pulse emission, optical communications laser.

Translation: $1 \cdot 10^{-12}$ second or 0.000 000 000 001 second

It takes a New York City cab driver one *nano*second to beep his horn when the light changes from red to green.

Translation: $1 \cdot 10^{-9}$ second or 0.000 000 001 second

Scientific Notation

In the previous examples we estimated the sizes of objects to the nearest power of 10 without worrying about more precise measurements. For example, we used a gross estimate of 10^7 meters for the measure of the radius of Earth. A more accurate measure is 6,368,000 meters. This number can be written more compactly using *scientific notation* as

$$6.368 \cdot 10^6 \text{ meters}$$

The number 6.368 is called the *coefficient*. The absolute value of the coefficient must always lie between 1 and 10. The power of 10 tells us how many places to shift the decimal point of the coefficient in order to get back to standard decimal form. Here, we would multiply 6.368 times 10^6, which means we would move the decimal place six places to the right, to get 6,368,000 meters.

Any nonzero number, positive or negative, can be written in scientific notation, that is, written as the product of a coefficient N multiplied by 10 to some power, where $1 \le |N| < 10$. Thus 2 million, 2,000,000, and $2 \cdot 10^6$ are all equivalent representations of the same number. The one you choose depends on the context.

In the following examples, you'll learn how to write numbers in scientific notation. Later we'll use scientific notation to simplify operations with very large and very small numbers.

Scientific Notation

A number is in *scientific notation* if it is in the form

$$N \cdot 10^n \qquad \text{where}$$

N is called the *coefficient* and $1 \le |N| < 10$
n is an integer

EXAMPLE 3

The distance to Andromeda, our nearest neighboring galaxy, is 15,000,000,000,000,000,000,000 meters. Express this number in scientific notation.

SOLUTION

The coefficient needs to be a number between 1 and 10. We start by identifying the first nonzero digit and then placing a decimal point right after it to create the coefficient of 1.5. The original number written in scientific notation will be of the form

$$1.5 \cdot 10^?$$

What power of 10 will convert this expression back to the original number? The original number is larger than 1.5, so the exponent will be positive. If we move the decimal place 22 places to the right, we will get back 1.5,000,000,000,000,000,000,000.

This is equivalent to multiplying 1.5 by 10^{22}. So, in scientific notation, 15,000,000,000,000,000,000,000 is written as

$$1.5 \cdot 10^{22}$$

EXAMPLE 4

The radius of a hydrogen atom is 0.000 000 000 052 9 meter across. Express this number in scientific notation.

SOLUTION

The coefficient is 5.29. The original number written in scientific notation will be of the form

$$5.29 \cdot 10^?$$

What power of 10 will convert this expression back to the original number? The original number is smaller than 5.29, so the exponent will be negative. If we move the decimal place 11 places to the left, we will get back 0.000 000 000 05.2 9. This is equivalent to dividing 5.29 by 10^{11} or multiplying it by 10^{-11}:

$$0.000\ 000\ 000\ 052\ 9 = \frac{5.29}{10^{11}}$$

$$= 5.29 \left(\frac{1}{10^{11}} \right)$$

$$= 5.29 \cdot 10^{-11}$$

This number is now in scientific notation.[2]

EXAMPLE 5   Express $-0.000\ 000\ 000\ 052\ 9$ in scientific notation.

SOLUTION   In this case the coefficient, -5.29, is negative. Notice that the absolute value of the coefficient, $|-5.29|$, is equal to 5.29, which is between 1 and 10. In scientific notation, $-0.000\ 000\ 000\ 052\ 9$ is written as

$$-5.29 \cdot 10^{-11}$$

Converting from Standard Decimal Form to Scientific Notation

Place a decimal point to the right of the first nonzero digit, creating the coefficient N, where

$$1 \le |N| < 10.$$

Determine n, the power of 10 needed to convert the coefficient back to the original number.

Write in the form $N \cdot 10^n$, where the exponent n is an integer.

Examples: $346{,}800{,}000 = 3.468 \cdot 10^8$ $0.000\ 008\ 4 = 8.4 \cdot 10^{-6}$

The poem "Imagine" offers a creative look at the Big Bang.

Deep time

The Big Bang.   In 1929 the American astronomer Edwin Hubble published an astounding paper claiming that the universe is expanding. Most astronomers and cosmologists now agree with his once-controversial theory and believe that approximately 13.7 billion years ago the universe began an explosive expansion from an infinitesimally small point. This event is referred to as the "Big Bang," and the universe has been expanding ever since it occurred.[3] Scientific notation can be used to record the progress of the universe since the Big Bang Theory, as shown in Table 4.5.

The Tale of the Universe in Scientific Notation

Object	Age (in years)
Universe	13.7 billion = 13,700,000,000 = $1.37 \cdot 10^{10}$
Earth	4.6 billion = 4,600,000,000 = $4.6 \cdot 10^9$
Human life	100 thousand = 100,000 = $1.0 \cdot 10^5$

Table 4.5

---

[2]Most calculators and computers automatically translate a number into scientific notation when it is too large or small to fit into the display. The notation is often slightly modified by using the letter E (short for "exponent") to replace the expression "times 10 to some power." So $3.0 \cdot 10^{26}$ may appear as 3.0 E+26. The number after the E tells how many places, and the sign (+ or −) indicates in which direction to move the decimal point of the coefficient.

[3]Depending on its total mass and energy, the universe will either expand forever or collapse back upon itself. However, cosmologists are unable to estimate the total mass or total energy of the universe, since they are in the embarrassing position of not being able to find about 90% of either. Scientists call this missing mass *dark matter* and missing energy *dark energy*, which describes not only their invisibility but also the scientists' own mystification.

Carl Sagan's video *Cosmos* and book *Dragons of Eden* condense the life of the universe into one calendar year.

Table 4.5 tells us that humans, *Homo sapiens sapiens,* first walked on Earth about 100,000 or $1.0 \cdot 10^5$ years ago. In the life of the universe, this is almost nothing. If all of time, from the Big Bang to today, were scaled down into a single year, with the Big Bang on January 1, our early human ancestors would not appear until less than 4 minutes before midnight on December 31, New Year's Eve.

Algebra Aerobics 4.1

1. Express as a power of 10:
   a. 10,000,000,000
   b. 0.000 000 000 000 01
   c. 100,000
   d. 0.000 01

2. Express in standard notation (without exponents):
   a. 10^{-8} c. 10^{-4}
   b. 10^{13} d. 10^7

3. Express as a power of 10 and then in standard notation:
   a. A nanosecond in terms of seconds
   b. A kilometer in terms of meters
   c. A gigabyte in terms of bytes

4. Rewrite each measurement in meters, first using a power of 10 and then using standard notation:
   a. 7 cm   b. 9 mm   c. 5 km

5. Avogadro's number is $6.02 \cdot 10^{23}$. A mole of any substance is defined to be Avogadro's number of particles of that substance. Express this number in standard notation.

6. The distance between Earth and its moon is 384,000,000 meters. Express this in scientific notation.

7. An angstrom (denoted by Å), a unit commonly used to measure the size of atoms, is 0.000 000 01 cm. Express its size using scientific notation.

8. The width of a DNA double helix is approximately 2 nanometers, or $2 \cdot 10^{-9}$ meter. Express the width in standard notation.

9. Express in standard notation:
   a. $-7.05 \cdot 10^8$ c. $5.32 \cdot 10^6$
   b. $-4.03 \cdot 10^{-5}$ d. $1.021 \cdot 10^{-7}$

10. Express in scientific notation:
    a. $-43,000,000$ c. $5,830$
    b. $-0.000\,008\,3$ d. $0.000\,000\,024\,1$

11. Express as a power of 10:
    a. $\dfrac{1}{100,000}$
    b. $\dfrac{1}{1,000,000,000}$

Exercises for Section 4.1

1. Write each expression as a power of 10.
   a. $10 \cdot 10 \cdot 10 \cdot 10 \cdot 10 \cdot 10$
   b. $\dfrac{1}{10 \cdot 10 \cdot 10 \cdot 10 \cdot 10}$
   c. one billion
   d. one-thousandth
   e. 10,000,000,000,000
   f. 0.000 000 01

2. Express in standard decimal notation (without exponents):
   a. 10^{-7} c. -10^8 e. 10^{-3}
   b. 10^7 d. -10^{-5} f. 10^5

3. Express each in meters, using powers of 10. (See inside back cover.)
   a. 10 cm   c. 3 terameters
   b. 4 km   d. 6 nanometers

4. Express each unit using a metric prefix. (See inside back cover.)
   a. 10^{-3} seconds
   b. 10^3 grams
   c. 10^2 meters

5. Computer storage is often measured in gigabytes and terabytes. Write these units as powers of 10.

6. Express each of the following using powers of 10.
   a. 10,000,000,000,000   d. $\dfrac{1}{10 \cdot 10 \cdot 10 \cdot 10}$
   b. 0.000 000 000 001   e. one million
   c. $10 \cdot 10 \cdot 10 \cdot 10$ f. one-millionth

7. Write each of the following in scientific notation:
   a. 0.000 29   d. 0.000 000 000 01   g. -0.0049
   b. 654.456   e. 0.000 002 45
   c. 720,000   f. $-1,980,000$

8. Why are the following expressions *not* in scientific notation? Rewrite each in scientific notation.
   a. $25 \cdot 10^4$ c. $0.012 \cdot 10^{-2}$
   b. $0.56 \cdot 10^{-3}$ d. $-425.03 \cdot 10^2$

9. Write each of the following in standard decimal form:
   a. $7.23 \cdot 10^5$ d. $1.5 \cdot 10^6$
   b. $5.26 \cdot 10^{-4}$ e. $1.88 \cdot 10^{-4}$
   c. $1.0 \cdot 10^{-3}$ f. $6.78 \cdot 10^7$

10. Express each in scientific notation. (Refer to the chart in Exploration 4.1.)

    a. The age of the observable universe

    b. The size of the first living organism on Earth

    c. The size of Earth

    d. The age of Pangaea

    e. The size of the first cells with a nucleus

11. Determine if the expressions are true or false. If false, change the right-hand side to make the expression true.

    a. $0.00\,756 = 7.56 \cdot 10^{-2}$

    b. $3.432 \cdot 10^5 = 343{,}200$

    c. 49 megawatts $= 4.9 \cdot 10^6$ watts

    d. $1{,}596{,}000{,}000 = 1.5 \cdot 10^9$

    e. 5 megapixels $= 5.0 \cdot 10^6$ pixels

    f. 6 picoseconds $= 6.0 \cdot 10^{12}$ seconds

12. Express each quantity in scientific notation.

    a. The mass of an electron is about

    0.000 000 000 000 000 000 000 000 001 67 gram.

    b. One cubic inch is approximately 0.000 016 cubic meter.

    c. The radius of a virus is 0.000 000 05 meter.

13. Evaluate:

    a. $|9|$ b. $|-9|$ c. $|-1000|$ d. $-|-1000|$

14. Determine the value of each expression.

    a. $|-5 - 3|$ c. $|2 - 6|$

    b. $|6 - 2|$ d. $-2|-1+3| + |-5|$

15. Determine which statements are true.

    a. $|a - b| = |b - a|$

    b. $|-7a| = 7a$

    c. $2|-1+4| = 2|-1|+2|4|$

    d. $|-2p| = |-2| \cdot |p|$

16. What values for x would make the following true?

    a. $|x| = 7$ c. $|x - 2| = 7$ e. $|2 - x| = 7$

    b. $|x - 1| = 5$ d. $|2x| < 0$ f. $|2x| = 8$

17. Substitute the value $x = 5$ into the statement. Then replace the ? with the sign ($>$, $<$, or $=$) that would make the statement true. Then repeat for $x = -5$.

    a. $|x - 1|$? 5 c. $|x - 1|$? 0 e. $|2x - 1|$? 11

    b. $2|3-x|$? 10 d. $|-x|$? 4 f. $|-x|$? 6

18. a. Generate a small table of values and plot the function $y = |x|$ for $-5 \le x \le 5$.

    b. On the same graph, plot the function $y = |x - 2|$.

19. The accompanying amusing graph shows a roughly linear relationship between the "scientifically" calculated age of Earth and the year the calculation was published. For instance, in about 1935 Ellsworth calculated that Earth was about 2 billion years old. The age is plotted on the horizontal axis and the year the calculation was published on the vertical axis. The triangle on the horizontal coordinate represents the presently accepted age of Earth.

    a. Who calculated that Earth was less than 1 billion years old? Give the coordinates of the points that give this information.

    b. In about what year did scientists start putting the age of Earth at over a billion years? Give the coordinates that represent this point.

    c. On your graph sketch an approximation of a best-fit line for these points. Use two points on the line to calculate the slope of the line.

    d. Interpret the slope of that line in terms of the year of calculation and the estimated age of Earth.

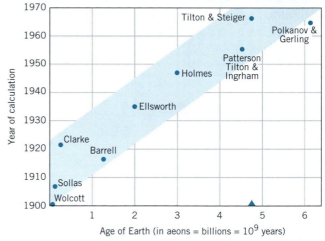

Source: *American Scientist*, Research Triangle Park, NC. Copyright © 1980.

4.2 Positive Integer Exponents

No matter what the base, whether it is 10 or any other number, repeated multiplication leads to *exponentiation*. For example,

$$3 \cdot 3 \cdot 3 \cdot 3 = 3^4$$

Here 4 is the *exponent* of 3, and 3 is called the *base*. In general, if a is a real number and n is a positive integer, then we define a^n as the product of n factors of a.

> **Definition of a^n**
>
> In the expression a^n, the number a is called the *base* and n is called the *exponent* or *power*.
>
> If n is a positive integer, then
>
> $$a^n = \underbrace{a \cdot a \cdot a \cdots a}_{n \text{ factors}} \qquad \text{(the product of } n \text{ factors of } a\text{)}$$

Exponent Rules

In this section we'll see how the rules for manipulating expressions with exponents make sense if we remember what the exponent tells us to do to the base. First we focus on cases where the exponents are positive integers. Later, we extend these rules to cases where the exponents can be any rational numbers, such as negative integers or fractions. In later courses you will extend the rules to all real numbers.

> **Rules for Exponents**
>
> **1.** $a^n \cdot a^m = a^{(n+m)}$
>
> **2.** $\dfrac{a^n}{a^m} = a^{(n-m)} \quad a \neq 0$
>
> **3.** $(a^m)^n = a^{(m \cdot n)}$
>
> **4.** $(ab)^n = a^n b^n$
>
> **5.** $\left(\dfrac{a}{b}\right)^n = \dfrac{a^n}{b^n} \quad b \neq 0$

We show below how Rules 1, 3, and 5 make sense and leave Rules 2 and 4 for you to justify in the exercises.

Rule 1. To justify this rule, think about the total number of times a is a factor when a^n is multiplied by a^m:

$$a^n \cdot a^m = \underbrace{a \cdot a \cdot a \cdots a}_{n \text{ factors}} \cdot \underbrace{a \cdot a \cdots a}_{m \text{ factors}} = \underbrace{a \cdot a \cdot a \cdots a}_{n+m \text{ factors}} = a^{(n+m)}$$

Rule 3. First think about how many times a^m is a factor when we raise it to the nth power:

$$(a^m)^n = \underbrace{a^m \cdot a^m \cdots a^m}_{n \text{ factors of } a^m}$$

Use Rule 1:

$$= a^{\overbrace{(m+m+\cdots+m)}^{n \text{ terms}}}$$

Represent adding m
n times as $m \cdot n$

$$= a^{(m \cdot n)}$$

Rule 5. Remember that the exponent n in the expression $(a/b)^n$ applies to the whole expression within the parentheses:

$$\left(\frac{a}{b}\right)^n = \underbrace{\left(\frac{a}{b}\right) \cdot \left(\frac{a}{b}\right) \cdots \left(\frac{a}{b}\right)}_{n \text{ factors of } a/b}$$

$$= \frac{\overbrace{a \cdot a \cdots a}^{n \text{ factors of } a}}{\underbrace{b \cdot b \cdots b}_{n \text{ factors of } b}}$$

$$= \frac{a^n}{b^n}$$

EXAMPLE 1 Simplify and write as an expression with exponents:

$$7^3 \cdot 7^2 = 7^{3+2} = 7^5 \qquad (x^5)^3 = x^{5\cdot3} = x^{15}$$
$$w^3 \cdot w^5 = w^{3+5} = w^8 \qquad (11^2)^4 = 11^{2\cdot4} = 11^8$$
$$\frac{10^8}{10^3} = 10^{8-3} = 10^5 \qquad \frac{z^8}{z^3} = z^{8-3} = z^5$$

EXAMPLE 2 Simplify:

$$(3a)^4 = 3^4 a^4 = 81a^4$$
$$(-5x)^3 = (-5)^3 x^3 = -125x^3$$
$$\left(\frac{2}{3}\right)^3 = \frac{2^3}{3^3} = \frac{8}{27}$$

EXAMPLE 3 Simplify:

$$\left(\frac{-2a}{3b}\right)^3 = \frac{(-2a)^3}{(3b)^3} = \frac{(-2)^3 a^3}{3^3 b^3} = \frac{-8a^3}{27b^3}$$
$$\frac{-5(x^3)^2}{(2y^2)^3} = \frac{-5x^6}{8y^6}$$

EXAMPLE 4 **Using scientific notation to simplify calculations**
Deneb is 1600 light years from Earth. How far is Earth from Deneb when measured in miles?

SOLUTION The distance that light travels in 1 year, called a *light year,* is approximately 5.88 trillion miles.

Since $\qquad$ 1 light year $= 5,880,000,000,000$ miles

then the distance from Earth to Deneb is

$$1600 \text{ light years} = (1600) \cdot (5,880,000,000,000 \text{ miles})$$
$$= (1.6 \cdot 10^3) \cdot (5.88 \cdot 10^{12} \text{ miles})$$
$$= (1.6 \cdot 5.88) \cdot (10^3 \cdot 10^{12}) \text{ miles}$$
$$\approx 9.4 \cdot 10^{3+12} \text{ miles}$$
$$\approx 9.4 \cdot 10^{15} \text{ miles}$$

Using ratios to compare sizes of objects

In comparing two objects of about the same size, it is common to subtract one size from the other and say, for instance, that one person is 6 inches taller than another. This method of comparison is not effective for objects of vastly different sizes. To say that the difference between the estimated radius of our solar system (1 terameter, or 1,000,000,000,000 meters) and the average size of a human (about 10^0 or 1 meter) is $1,000,000,000,000 - 1 = 999,999,999,999$ meters is not particularly useful. In fact, since our measurement of the solar system certainly isn't accurate to within 1 meter, this difference is meaningless. As shown in the following example, a more useful method for comparing objects of wildly different sizes is to calculate the ratio of the two sizes.

EXAMPLE 5 **The ratio of two quantities**
In April 2007, the U.S. federal government reported that the estimated gross federal debt was \$8.87 trillion and the estimated U.S. population was 301 million. What was the approximate federal debt *per person*?

SOLUTION

$$\frac{\text{federal debt}}{\text{U.S. population}} = \frac{8.87 \cdot 10^{12} \text{ dollars}}{3.01 \cdot 10^{8} \text{ people}}$$

$$= \left(\frac{8.87}{3.01}\right) \cdot \left(\frac{10^{12}}{10^{8}}\right) \frac{\text{dollars}}{\text{people}} \approx 2.95 \cdot 10^{4} \frac{\text{dollars}}{\text{people}}$$

So the federal debt amounted to about $2.95 \cdot 10^{4}$ or $29,500 per person.

EXAMPLE 6 How many times larger is the sun than Earth?

SOLUTION 1 The radius of the sun is approximately 10^{9} meters and the radius of Earth is about 10^{7} meters. One way to answer the question "How many *times* larger is the sun than Earth?" is to form the ratio of the two radii:

$$\frac{\text{radius of the sun}}{\text{radius of Earth}} = \frac{10^{9} \text{ m}}{10^{7} \text{ m}}$$

$$= \frac{10^{9} \text{ m}}{10^{7} \text{ m}} = 10^{9-7} = 10^{2}$$

The units cancel, so 10^{2} is unitless. The radius of the sun is approximately 10^{2}, or 100, times larger than the radius of Earth.

SOLUTION 2 Another way to answer the question is to compare the volumes of the two objects. The sun and Earth are both roughly spherical. The formula for the volume V of a sphere with radius r is $V = \frac{4}{3}\pi r^{3}$.

The radius of the sun is approximately 10^{9} meters and the radius of Earth is about 10^{7} meters. The ratio of the two volumes is

$$\frac{\text{volume of the sun}}{\text{volume of Earth}} = \frac{(4/3)\pi(10^{9})^{3} \text{ m}^{3}}{(4/3)\pi(10^{7})^{3} \text{ m}^{3}}$$

$$= \frac{(10^{9})^{3}}{(10^{7})^{3}} \quad (Note: \frac{4}{3}\pi \text{ and m}^{3} \text{ cancel.})$$

$$= \frac{10^{27}}{10^{21}}$$

$$= 10^{6}$$

So while the radius of the sun is 100 times larger than the radius of Earth, the *volume* of the sun is approximately $10^{6} = 1,000,000$, or 1 million, times larger than the volume of Earth!

Common Errors

The first question to ask in evaluating expressions with exponents is: To what does the exponent apply? Consider the following expressions:

1. $-a^{n} = -(a^{n})$ but $-a^{n} \neq (-a)^{n}$ (unless n is odd)

For example, in the expression -2^{4}, the exponent 4 applies only to 2, not to -2. The order of operations says to compute the power first, before applying the negation sign. So $-2^{4} = -(2^{4}) = -16$. If we want to raise -2 to the fourth power, we write $(-2)^{4} = (-2)(-2)(-2)(-2) = 16$.

In the expression $(-3b)^{2}$, everything inside the parentheses is squared. So $(-3b)^{2} = (-3b)(-3b) = 9b^{2}$. But in the expression $-3b^{2}$, the exponent 2 applies only to the base b.

In the case where n is an *odd integer*, then $(-a)^{n}$ will equal $-(a)^{n}$. For example, $(-2)^{3} = (-2)(-2)(-2) = -8 = -2^{3}$.

2. $ab^n = a(b^n)$ **and** $-ab^n = -a(b^n)$ **but** $ab^n \neq (ab)^n$

Remember, the exponent applies only to the variable to which it is attached. In the expressions ab^n and $-ab^n$, only b is raised to the nth power.

For example,

$$2 \cdot 5^3 = 2 \cdot 125 = 250 \qquad \text{but} \quad (2 \cdot 5)^3 = (10)^3 = 1000$$
$$-2 \cdot 5^3 = -2 \cdot 125 = -250 \qquad \text{but} \quad (-2 \cdot 5)^3 = (-10)^3 = -1000$$

You can use parentheses () to indicate when more than one variable is raised to the nth power.

3. $(ab)^n = a^n b^n$ **but** $(a + b)^n \neq a^n + b^n$ **(if $n \neq 1$)**

For example,

$$(2 \cdot 5)^3 = 2^3 \cdot 5^3 \quad \text{but} \quad (2 + 5)^3 \neq 2^3 + 5^3$$
$$(10)^3 = 8 \cdot 125 \qquad\qquad (7)^3 \neq 8 + 125$$
$$1000 = 1000 \qquad\qquad 343 \neq 133$$

4. $a^n \cdot a^m = a^{n+m}$ **but** $a^n + a^m \neq a^{n+m}$

For example,

$$10^2 \cdot 10^3 = 10^5 \qquad \text{but} \quad 10^2 + 10^3 \neq 10^5$$
$$100 \cdot 1000 = 100{,}000 \qquad 100 + 1000 \neq 100{,}000$$

? **SOMETHING TO THINK ABOUT**

What are some other exceptions to the generalizations made about common errors?

Common Errors Involving Exponents

In general,

$$-a^n \neq (-a)^n \qquad a^n + a^m \neq a^{n+m}$$
$$ab^n \neq (ab)^n \qquad (a + b)^n \neq a^n + b^n$$

Algebra Aerobics 4.2a

1. Simplify where possible, leaving the answer in a form with exponents:

  a. $10^5 \cdot 10^7$ **d.** $5^5 \cdot 6^7$ **g.** $3^4 + 7 \cdot 3^4$

  b. $8^6 \cdot 8^{14}$ **e.** $7^3 + 7^3$ **h.** $2^3 + 2^4$

  c. $z^5 \cdot z^4$ **f.** $5 \cdot 5^6$ **i.** $2^5 + 5^2$

2. Simplify (if possible), leaving the answer in exponent form:

  a. $\dfrac{10^{15}}{10^7}$ **c.** $\dfrac{3^5}{3^4}$ **e.** $\dfrac{5}{5^6}$ **g.** $\dfrac{2^3 \cdot 3^4}{2 \cdot 3^2}$

  b. $\dfrac{8^6}{8^4}$ **d.** $\dfrac{5}{6^7}$ **f.** $\dfrac{3^4}{3}$ **h.** $\dfrac{6}{2^4}$

3. Write each number as a power of 10, then perform the indicated operation. Write your final answer as a power of 10.

  a. $100{,}000 \cdot 1{,}000{,}000$

  b. $1{,}000 \cdot 0.000\,001$

  c. $0.000\,000\,000\,01 \cdot 0.000\,01$

  d. $\dfrac{1{,}000{,}000{,}000}{10{,}000}$

  e. $\dfrac{1{,}000{,}000}{0.001}$

  f. $\dfrac{0.000\,01}{0.0001}$

  g. $\dfrac{0.000\,001}{10{,}000}$

4. Simplify:

  a. $(10^4)^5$ **d.** $(2x)^4$ **g.** $(-3x^2)^3$

  b. $(7^2)^3$ **e.** $(2a^4)^3$ **h.** $((x^3)^2)^4$

  c. $(x^4)^5$ **f.** $(-2a)^3$ **i.** $(-5y^2)^3$

5. Simplify:

  a. $\left(\dfrac{-2x}{4y}\right)^3$ **c.** -5^2 **e.** $(-3yz^2)^4$

  b. $(-5)^2$ **d.** $-3(yz^2)^4$ **f.** $(-3yz^2)^3$

6. A compact disk or CD has a storage capacity of about 737 megabytes ($7.37 \cdot 10^8$ bytes). If a hard drive has a capacity of 40 gigabytes ($4.0 \cdot 10^{10}$ bytes), how many CD's would it take to equal the storage capacity of the hard drive?

7. Write as a single number with no exponents:

  a. $(3 + 5)^3$ **c.** $3 \cdot 5^2$

  b. $3^3 + 5^3$ **d.** $-3 \cdot 5^2$

Estimating Answers

By rounding off numbers and using scientific notation and the rules for exponents, we can often make quick estimates of answers to complicated calculations. In this age of calculators and computers we need to be able to roughly estimate the size of an answer, to make sure our calculations with technology make sense.

EXAMPLE 7

Estimate the value of

$$\frac{(382{,}152) \cdot (490{,}572{,}261)}{(32{,}091) \cdot (1942)}$$

Express your answer in both scientific and standard notation.

SOLUTION

Round each number:

$$\frac{(382{,}152) \cdot (490{,}572{,}261)}{(32{,}091) \cdot (1942)} \approx \frac{(400{,}000) \cdot (500{,}000{,}000)}{(30{,}000) \cdot (2000)}$$

rewrite in scientific notation

$$\approx \frac{(4 \cdot 10^5) \cdot (5 \cdot 10^8)}{(3 \cdot 10^4) \cdot (2 \cdot 10^3)}$$

group the coefficients and the powers of 10

$$\approx \left(\frac{4 \cdot 5}{3 \cdot 2}\right) \cdot \left(\frac{10^5 \cdot 10^8}{10^4 \cdot 10^3}\right)$$

simplify each expression

$$\approx \frac{20}{6} \cdot \frac{10^{13}}{10^7}$$

we get in scientific notation

$$\approx 3.33 \cdot 10^6$$

or in standard notation

$$\approx 3{,}330{,}000$$

Using a calculator on the original problem, we get a more precise answer of 3,008,200.

EXAMPLE 8

As of 2007 the world population was approximately 6.605 billion people. There are roughly 57.9 million square miles of land on Earth, of which about 22% are favorable for agriculture. Estimate how many people per square mile of farmable land there are as of 2007.

SOLUTION

$$\frac{\text{size of world population}}{\text{amount of farmable land}} = \frac{6.605 \text{ billion people}}{22\% \text{ of } 57.9 \text{ million square miles}}$$

rewrite as powers of 10

$$= \frac{6.605 \cdot 10^9 \text{ people}}{(0.22) \cdot (57.9) \cdot 10^6 \text{ mile}^2}$$

round each number

$$\approx \frac{6.6 \cdot 10^9 \text{ people}}{(0.2) \cdot 60 \cdot 10^6 \text{ mile}^2}$$

simplify

$$\approx \frac{6.6 \cdot 10^9 \text{ people}}{12 \cdot 10^6 \text{ mile}^2}$$

we get in scientific notation

$$\approx 0.55 \cdot 10^3 \text{ people/mile}^2$$

or in standard notation

$$\approx 550 \text{ people/mile}^2$$

So there are roughly 550 people/mile^2 of farmable land in the world. Using a calculator and the original numbers, we get a more accurate answer of 519 people/mile^2 of farmable land, which is close to our estimate.

Algebra Aerobics 4.2b

1. Estimate the value of:
   a. $(0.000\ 297\ 6) \cdot (43{,}990{,}000)$
   b. $\dfrac{453{,}897 \cdot 2{,}390{,}702}{0.004\ 38}$
   c. $\dfrac{0.000\ 000\ 319}{162{,}000}$
   d. $28{,}000{,}000 \cdot 7{,}629$
   e. $0.000\ 021 \cdot 391{,}000{,}000$

2. Evaluate the following without the aid of a calculator:
   a. $(3.0 \cdot 10^3)\ (4.0 \cdot 10^2)$
   b. $\dfrac{(5.0 \cdot 10^2)^2}{2.5 \cdot 10^3}$
   c. $\dfrac{2.0 \cdot 10^5}{5.0 \cdot 10^3}$
   d. $(4.0 \cdot 10^2)^3 \cdot (2.0 \cdot 10^3)^2$

3. The radius of Jupiter, the largest of the planets in our solar system, is approximately $7.14 \cdot 10^4$ km. (If r is the radius of a sphere, the sphere's surface area equals $4\pi r^2$ and its volume equals $\frac{4}{3}\pi r^3$.) Assuming Jupiter is roughly spherical,
   a. Estimate the surface area of Jupiter.
   b. Estimate the volume of Jupiter.
   (Express your answers in scientific notation.)

4. Only about three-sevenths of the land favorable for agriculture is actually being farmed. Using the facts in Example 8, estimate the number of people per square mile of farmable land that is being used. Should your estimate be larger or smaller than the ratio of people to farmable land? Explain. (Round your answer to the nearest integer.)

Exercises for Section 4.2

1. Simplify, when possible, writing your answer as an expression with exponents:
   a. $10^4 \cdot 10^3$
   b. $10^4 + 10^3$
   c. $10^3 + 10^3$
   d. $x^5 \cdot x^{10}$
   e. $(x^5)^{10}$
   f. $4^7 + 5^2$
   g. $\dfrac{z^7}{z^2}$
   h. 256^0
   i. $\dfrac{3^5 \cdot 3^2}{3^8}$
   j. $4^5 \cdot (4^2)^3$

2. Simplify:
   a. $(-1)^4$
   b. $-(1)^4$
   c. $(a^4)^3$
   d. $-(2a^2)^3$
   e. $(2a^4)^3$
   f. $(-2a^4)^3$
   g. $(10a^2b^3)^3$
   h. $(2ab)^2 - 3a^2b^2$

3. Simplify:
   a. $(-2a)^4$
   b. $-2(a)^4$
   c. $(-x^5)^3$
   d. $(-2ab^2)^3$
   e. $(2x^4)^5$
   f. $(-4x^3)^2 + x^3(2x^3)$
   g. $(50a^{10})^2$
   h. $(3ab)^3 + ab$

4. Simplify and write each variable as an expression with positive exponents:
   a. $-\left(\dfrac{3}{5}\right)^2$
   b. $\left(\dfrac{-5a^3}{a^2}\right)^4$
   c. $\left(\dfrac{10a^3}{5b}\right)^2$
   d. $\left(\dfrac{-2x^3}{3y^2}\right)^3$

5. Simplify and write each variable as an expression with positive exponents:
   a. $-\left(\dfrac{5}{8}\right)^2$
   b. $\left(\dfrac{3x^3}{5y^2}\right)^3$
   c. $\left(\dfrac{-10x^5}{2b^2}\right)^4$
   d. $\left(\dfrac{-x^5}{x^2}\right)^3$

6. Evaluate and express your answer in standard decimal form:
   a. $-2^4 + 2^2$
   b. $-2^3 + (-4)^2$
   c. $2 \cdot 3^2 - 3(-2)^2$
   d. $-10^4 + 10^5$
   e. $10^3 + 2^3$
   f. $2 \cdot 10^3 + 10^3 + 10^2$
   g. $2 \cdot 10^3 + (-10)^3$
   h. $(1000)^0$

7. Convert each number into scientific notation then perform the indicated operation. Leave your answer in scientific notation.
   a. $2{,}000{,}000 \cdot 4000$
   b. 1.4 million $\div$ 7000
   c. 50 billion $\cdot$ 60 trillion
   d. 2500 billion $\div$ 500 thousand

8. Convert each number into scientific notation and then perform the operation without a calculator.
   a. $60{,}000{,}000{,}000 + 40{,}000{,}000{,}000$
   b. $\dfrac{(20{,}000)^6}{(400)^3}$
   c. $(2{,}000{,}000) \cdot (40{,}000)$

9. Simplify each expression using the properties of exponents.
   a. $(x^5y)(x^6)(x^2y^3)$
   b. $\dfrac{5x^6y^3}{x^2y^2}$
   c. $\left(\dfrac{-2x^5y^5}{x^2y^2}\right)^3$
   d. $(x^2)^5 \cdot (2y^2)^4$
   e. $(3x^2y^5)^4$
   f. $\left(\dfrac{3x^3y}{5xy}\right)^2$

10. Each of the following simplifications contains an error made by students on a test. Find the error and correct the simplification so that the expression becomes true.
   a. $[(x^2)^3]^5 = [x^5]^5 = x^{25}$
   b. $\dfrac{7x^2y^6}{(xy)^2} = \dfrac{7x^2y^6}{x^2y^2} = 7x^4y^8$
   c. $\left(\dfrac{4x^3y^5}{6xy^4}\right)^3 = \left(\dfrac{2x^2y}{3}\right)^3 = \frac{2}{3}x^6y^3$
   d. $(1.1 \cdot 10^6) \cdot (1.1 \cdot 10^4) = 1.1 \cdot 10^6$
   e. $\dfrac{4 \cdot 10^6}{8 \cdot 10^3} = 0.5 \cdot 10^3 = 5.0 \cdot 10^4$
   f. $6 \cdot 10^3 + 7 \cdot 10^5 = 13 \cdot 10^8$

11. Express your answer as a power of 10 and in standard decimal form. (Refer to table on inside back cover.)

    a. How many times larger is a gigabyte of memory than a megabyte?

    b. How many times farther is a kilometer than a dekameter?

    c. How many times heavier is a kilogram than a milligram?

    d. How many times longer is a microsecond than a nanosecond?

12. In 2006 the People's Republic of China was estimated to have about 1,314,000,000 people, and Monaco about 33,000. Monaco has an area of 0.75 miles^2, and China has an area of 3,705,000 miles^2.

    a. Express the populations and geographic areas in scientific notation.

    b. By calculating a ratio, determine how much larger China's population is than Monaco's.

    c. What is the population density (the number of people per square mile) for each country?

    d. Write a paragraph comparing and contrasting the population size and density for these two nations.

13. a. In 2006 Japan had a population of approximately 127.5 million people and a total land area of about 152.5 thousand square miles. What was the population density (the number of people per square mile)?

    b. In 2006 the United States had a population of approximately 300 million people and a total land area of about 3620 thousand square miles. What was the population density of the United States?

    c. Compare the population densities of Japan and the United States.

14. The distance that light travels in 1 year (a light year) is $5.88 \cdot 10^{12}$ miles. If a star is $2.4 \cdot 10^8$ light years from Earth, what is this distance in miles?

15. An average of $1.5 \cdot 10^4$ Coca-Cola beverages were consumed every second worldwide in 2005. There are $8.64 \cdot 10^4$ seconds in a day. What was the daily consumption of Coca-Cola in 2005? (*Source:* World of Coca-Cola® Atlanta)

16. Change each number into scientific notation, then perform the indicated calculation without a calculator.

    a. A $600,000 lottery jackpot is divided among 300 people. What are the winnings per person?

    b. A total of 2500 megawatts are used over 500 hours. What is the rate in watts per hour?

    c. If there were 6 million births in 30 years, what is the birth rate per year?

17. a. For any nonzero real number a, what can we say about the sign of the expression $(-a)^n$ when n is an even integer? What can we say about the sign of $(-a)^n$ when n is an odd integer?

    b. What is the sign of the resulting number if a is a positive number? If a is a negative number? Explain your answer.

18. Round off the numbers and then estimate the value of each of the following expressions without using a calculator. Show

your work, writing your answers in scientific notation. If available, use a calculator to verify your answers.

    a. $(2{,}968{,}001{,}000) \cdot (189{,}000)$

    b. $(0.000\ 079) \cdot (31{,}140{,}284{,}788)$

    c. $\dfrac{4{,}083{,}693 \cdot 49{,}312}{213 \cdot 1945}$

19. Simplify each expression using two different methods, and then compare your answers.

    Method I: Simplify inside the parentheses first, and then apply the exponent rule outside the parentheses.

    Method II: Apply the exponent rule outside the parentheses, and then simplify.

    a. $\left(\dfrac{m^2 n^3}{mn}\right)^2$ 　　　　 b. $\left(\dfrac{2a^2 b^3}{ab^2}\right)^4$

20. Verify that $(a^2)^3 = (a^3)^2$ using the rules of exponents.

21. Verify that $\left(\dfrac{2a^3}{5b^2}\right)^4 = \dfrac{16a^{12}}{625b^8}$ using the rules of exponents.

22. An article in the journal *Nature* (October 2000) analyzed samples of the ballast water from ships arriving in the Chesapeake Bay from foreign ports. It reported that ballast was an important factor in the global distribution of microorganisms. One gallon of ballast water contained on average 3 billion bacteria, including some that cause cholera. The scientists estimated that about 2.5 billion gallons of ballast water are discharged into the Chesapeake Bay each year. Estimate the number of bacteria per year discharged in ballast water into the Chesapeake Bay. Write your answer in scientific notation.

23. Justify the following rule for exponents. If a and b are any nonzero real numbers and n is an integer ≥ 0, then

$$(ab)^n = a^n b^n$$

24. Justify the following rule for exponents. Consider the case of $n \geq m$ and assume m and n are integers > 0. If a is any nonzero real number, then

$$\frac{a^n}{a^m} = a^{(n-m)}$$

25. In 2006 the United Kingdom generated approximately 81 terawatt-hours of nuclear energy for a population of about 60.6 million on 94,525 miles^2. In the same year the United States generated approximately 780 terawatt-hours of nuclear energy for a population of about 300 million on 3,675,031 miles^2. A terawatt is 10^{12} watts.

    a. How many terawatt-hours is the United Kingdom generating per person? How many terawatt-hours is it generating per square mile? Express each in scientific notation.

    b. How many terawatt-hours is the United States generating per person? How many terawatt-hours are we generating per square mile? Express each in scientific notation.

    c. How much nuclear energy is being generated in the United Kingdom per square mile relative to the United States?

    d. Write a brief statement comparing the relative magnitude of generation of nuclear power per person in the United Kingdom and the United States.

26. Hubble's Law states that galaxies are receding from one another at velocities directly proportional to the distances separating them. The accompanying graph illustrates that Hubble's Law holds true across the known universe. The plot includes ten major clusters of galaxies. The boxed area at the lower left represents the galaxies observed by Hubble when he discovered the law. The easiest way to understand this graph is to think of Earth as being at the center of the universe (at 0 distance) and not moving (at 0 velocity). In other words, imagine Earth at the origin of the graph (a favorite fantasy of humans). Think of the horizontal axis as measuring the distance of the galaxy from Earth, and the vertical as measuring the velocity at which a galaxy cluster is moving away from Earth (the recession velocity). Then answer the following questions.

   a. Identify the coordinates of two data points that lie on the regression line drawn on the graph.

   b. Use the coordinates of the points in part (a) to calculate the slope of the line. That slope is called the *Hubble constant*.

   c. What does the slope mean in terms of distance from Earth and recession velocity?

   d. Construct an equation for our line in the form $y = mx + b$. Show your work.

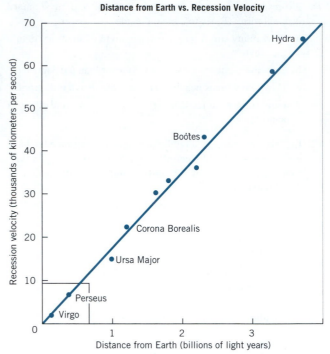

Source: T. Ferris, *Coming of Age in the Milky Way* (New York: William Morrow, 1988). Copyright © by Timothy Ferris. By permission of William Morrow & Company, Inc.

4.3   *Negative Integer Exponents*

The definitions for raising any base to the zero power or to a negative power follow a logic that is similar to the one used to define $10^0 = 1$ and $10^{-n} = \frac{1}{10^n}$.

> **Zero and Negative Exponents**
>
> If a is nonzero and n is a positive integer, then
>
> $$a^0 = 1$$
>
> $$a^{-n} = \frac{1}{a^n}$$

It is important to note that $a^1 = a$, so $a^{-1} = \frac{1}{a^1} = \frac{1}{a}$.

   In the following examples, we show how to apply the five rules for exponents when the exponents are negative integers or zero.

EXAMPLE 1   Simplify $x^2 \cdot x^{-5}$.

SOLUTION   Using Rule 1 for exponents,

$$x^2 \cdot x^{-5} = x^{2+(-5)} = x^{-3}$$

or we can simplify by first writing x^{-5} as $\frac{1}{x^5}$ and then use Rule 2 for exponents:

$$x^2 \cdot x^{-5} = x^2 \cdot \frac{1}{x^5} = \frac{x^2}{x^5} = x^{2-5} = x^{-3}$$

EXAMPLE 2 Simplify. Express your answer with positive exponents.

a. $\dfrac{10^2}{10^6}$ c. $\dfrac{(-5)^2}{(-5)^6}$

b. $\dfrac{6^2}{6^{-7}}$ d. $\dfrac{x^{-2}}{x^4}$

SOLUTION Using Rule 2 for exponents,

a. $\dfrac{10^2}{10^6} = 10^{2-6} = 10^{-4} = \dfrac{1}{10^4}$

b. $\dfrac{6^2}{6^{-7}} = 6^{2-(-7)} = 6^9$

c. $\dfrac{(-5)^2}{(-5)^6} = (-5)^{2-6} = (-5)^{-4} = \dfrac{1}{(-5)^4} = \dfrac{1}{(-1)^4\,(5)^4} = \dfrac{1}{5^4}$

d. $\dfrac{x^{-2}}{x^4} = x^{-2-4} = x^{-6} = \dfrac{1}{x^6}$

EXAMPLE 3 Simplify:

a. $(13^{-8})^3$ b. $(w^2)^{-7}$

SOLUTION Using Rule 3 for exponents,

a. $(13^{-8})^3 = 13^{(-8)3} = 13^{-24}$ b. $(w^2)^{-7} = w^{2(-7)} = w^{-14}$

EXAMPLE 4 Simplify:

$$\frac{v^{-2}(w^5)^2}{(v^{-1})^4\, w^{-3}}$$

SOLUTION Apply Rule 3 twice: $\dfrac{v^{-2}(w^5)^2}{(v^{-1})^4 w^{-3}} = \dfrac{v^{-2}\, w^{10}}{v^{-4}\, w^{-3}}$

Apply Rule 2 twice: $= v^{-2-(-4)}w^{10-(-3)}$

$= v^2 w^{13}$

Evaluating $\left(\dfrac{a}{b}\right)^{-n}$

The rule for applying negative powers is the same whether a is an integer or a fraction:

$$a^{-n} = 1/a^n \quad \text{where } a \neq 0$$

For example,

$$\left(\frac{1}{2}\right)^{-1} = \frac{1}{(1/2)^1} = 1 \div \left(\frac{1}{2}\right) = 1 \cdot \left(\frac{2}{1}\right) = 2$$

In general, if a and b are nonzero, then

$$\left(\frac{a}{b}\right)^{-n} = \frac{1}{(a/b)^n} = 1 \div \left(\frac{a}{b}\right)^n = 1 \cdot \left(\frac{b}{a}\right)^n = \left(\frac{b}{a}\right)^n = \frac{b^n}{a^n}$$

EXAMPLE 5   Simplify:

a. $\left(\dfrac{1}{2}\right)^{-11} \cdot \left(\dfrac{1}{2}\right)^{-2}$ b. $\left(\dfrac{a}{b}\right)^{3} \cdot \left(\dfrac{a}{b}\right)^{-5}$

SOLUTION   a. Using Rule 1 for exponents and the definition of a^{-n},

$$\left(\dfrac{1}{2}\right)^{-11} \cdot \left(\dfrac{1}{2}\right)^{-2} = \left(\dfrac{1}{2}\right)^{-11+(-2)}$$

$$= \left(\dfrac{1}{2}\right)^{-13} = \left(\dfrac{2}{1}\right)^{13} = 2^{13} = 8192$$

b. Using Rules 1 and 5 for exponents and the definition of a^{-n},

$$\left(\dfrac{a}{b}\right)^{3} \cdot \left(\dfrac{a}{b}\right)^{-5} = \left(\dfrac{a}{b}\right)^{3+(-5)}$$

$$= \left(\dfrac{a}{b}\right)^{-2} = \left(\dfrac{b}{a}\right)^{2} = \dfrac{b^{2}}{a^{2}}$$

Algebra Aerobics 4.3

1. Simplify (if possible). Express with a single positive exponent, if possible.

   a. $10^{5} \cdot 10^{-7}$ e. $\dfrac{7^{3}}{7^{3}}$

   b. $\dfrac{11^{6}}{11^{-4}}$ f. $a^{-2} \cdot a^{-3}$

   c. $\dfrac{3^{-5}}{3^{-4}}$ g. $3^{4} \cdot 3^{3}$

   d. $\dfrac{5^{5}}{6^{7}}$ h. $(2^{2} \cdot 3)(2^{6})(2^{4} \cdot 3)$

2. A typical TV signal, traveling at the speed of light, takes $3.3 \cdot 10^{-6}$ seconds to travel 1 kilometer. Estimate how long it would take the signal to travel across the United States (a distance of approximately 4300 kilometers).

3. Distribute and simplify:

   a. $x^{-2}(x^{5} + x^{-6})$

   b. $-a^{2}(b^{2} - 3ab + 5a^{2})$

4. Simplify:

   a. $(10^{4})^{-5}$ d. $\left(\dfrac{8}{x}\right)^{-2}$ g. $\left(\dfrac{3}{2y^{2}}\right)^{-4}$

   b. $(7^{-2})^{-3}$ e. $(2x^{-2})^{-1}$ h. $\dfrac{3}{(2y^{2})^{-4}}$

   c. $(2a^{3})^{-2}$ f. $2(x^{-2})^{-1}$

5. Simplify:

   a. $\dfrac{t^{-3}t^{0}}{(t^{-4})^{3}}$ c. $\dfrac{7^{-8}x^{-1}y^{2}}{7^{-5}xy^{3}}$

   b. $\dfrac{v^{-3}w^{7}}{(v^{-2})^{3}w^{-10}}$ d. $\dfrac{a(5b^{-1}c^{3})^{2}}{5ab^{2}c^{-6}}$

Exercises for Section 4.3

1. Simplify and express your answer using positive exponents. Check your answers by applying the rules for exponents and doing the calculations.

   a. $10^{3} \cdot 10^{-2}$ c. $(10^{-3})^{2}$

   b. $\dfrac{10^{-2}}{10^{3}}$ d. $\dfrac{10^{3}}{10^{-2}}$

2. Simplify and express your answer with positive exponents:

   a. $(x^{-3}) \cdot (x^{4})$ d. $(n^{-2})^{-3}$

   b. $(x^{-3}) \cdot (x^{-2})$ e. $(2n^{-2})^{-3}$

   c. $(x^{2})^{-3}$ f. $n^{-4}(n^{5} - n^{2}) + n^{-3}(n - n^{4})$

3. Simplify where possible. Express your answer with positive exponents.

   a. $\dfrac{2^{3}x^{4}}{2^{5}x^{8}}$ c. $\dfrac{x^{-2}y}{xy^{3}}$ e. $\dfrac{a^{-2}bc^{-5}}{(ab^{2})^{-3}c}$

   b. $\dfrac{x^{4}y^{7}}{x^{3}y^{-5}}$ d. $\dfrac{(x + y)^{4}}{(x + y)^{-7}}$

4. Simplify where possible. Express your answer with positive exponents.

   a. $(3 \cdot 3^{8})^{-2}$ d. $2x^{-3} + 3x(x^{-4})$

   b. $x^{3} \cdot x^{-4} \cdot x^{12}$ e. $10^{-5} + 5^{-2} + 10^{10}$

   c. $2^{6} + 2^{6} + 2^{7} + 2^{-4}$

5. Evaluate and write the result using scientific notation:

   a. $(2.3 \cdot 10^4)(2.0 \cdot 10^6)$

   b. $(3.7 \cdot 10^{-5})(1.1 \cdot 10^8)$

   c. $\dfrac{8.19 \cdot 10^{23}}{5.37 \cdot 10^{12}}$

   d. $\dfrac{3.25 \cdot 10^8}{6.29 \cdot 10^{15}}$

   e. $(6.2 \cdot 10^{52})^3$

   f. $(5.1 \cdot 10^{-11})^2$

6. Write each of the following in scientific notation:

   a. $725 \cdot 10^{23}$

   b. $725 \cdot 10^{-23}$

   c. $\dfrac{1}{725 \cdot 10^{23}}$

   d. $-725 \cdot 10^{23}$

   e. $-725 \cdot 10^{-23}$

7. Change each number to scientific notation, then simplify using rules of exponents. Show your work, recording your final answer in scientific notation.

   a. 10% of 0.000 01

   b. $\dfrac{0.000\ 05}{50{,}000}$

   c. $\dfrac{3}{0.006}$

   d. $\dfrac{8000}{0.000\ 8}$

   e. $\dfrac{0.006\ 4}{8000}$

   f. $5{,}000{,}000 \cdot 40{,}000$

8. Use scientific notation and the rules of exponents to perform the indicated operation without a calculator. Show your work, recording your answer in decimal form.

   a. $\dfrac{20}{200{,}000}$

   b. $\dfrac{0.006}{60{,}000}$

   c. $200 \cdot 0.000\ 007\ 5$

   d. $\dfrac{10{,}000{,}000}{25{,}000}$

   e. $0.06 \cdot 600$

   f. 10% of 0.000 05

9. For each equation determine the value of x that makes it true.

   a. $10^x = 0.000\ 001$

   b. $10^x = \dfrac{1}{1{,}000{,}000}$

   c. $\dfrac{1}{10^x} = 0.000\ 1$

   d. $10^{-x} = 100{,}000$

10. For each equation determine the value of x that makes it true.

   a. $6.3 \cdot 10^x = 0.000\ 63$

   b. $10^{-3} = x$

   c. $5^x = \dfrac{1}{125}$

   d. $x^3 = \dfrac{1}{1000}$

   e. $4^{-3} \cdot 2^{-5} = 2^x$

   f. $9^{-1} \cdot 27^{-2} = 3^x$

11. Simplify the following expressions by using properties of exponents. Write your final answers with only positive exponents.

   a. $\dfrac{(-2x^3y^{-1})^{-3}}{(x^2y^{-2})^0}$

   b. $\dfrac{(-2x^3y^{-1})^{-2}}{(x^2y^{-2})^{-1}}$

   c. $\left(\dfrac{3x^2y^{-5}}{5x^3y^4}\right)^{-1}$

   d. $\left[\left(3x^{-1}z^4\right)^{-2}\right]^{-3}$

12. Each of the following simplifications is false. In each case identify the error and correct it.

   a. $x^{-2}x^{-5} = x^{10}$

   b. $\dfrac{2^{-1}x^2y^{-2}}{x^{-1}y^5} = \dfrac{x^2x^{-1}}{2y^{-2}y^5} = \dfrac{x}{2y^3}$

   c. $(3x^{-1})^2 = \left(\dfrac{1}{3x}\right)^2 = \dfrac{1}{9x^2}$

   d. $(x + y)^{-1} = \dfrac{1}{x} + \dfrac{1}{y}$

13. A TV signal traveling at the speed of light takes about $8 \cdot 10^{-5}$ second to travel 15 miles. How long would it take the signal to travel a distance of 3000 miles?

14. Round off the numbers and then estimate the values of the following expressions without a calculator. Show your work, writing your answers in scientific notation. If available, use a calculator to verify your answers.

   a. $(0.000\ 359) \cdot (0.000\ 002\ 47)$

   b. $\dfrac{0.000\ 007\ 31 \cdot 82{,}560}{1{,}891{,}000}$

15. Simplify and express your answer with positive exponents.

   a. $\dfrac{x^{-2} - y^{-1}}{(xy^2)^{-1}}$

   b. $(5x^{-2}y^{-3})^{-2}$

16. (Requires a calculator that can evaluate powers.) Calculators and spreadsheets use slightly different formats for scientific notation. For example, if you type in Avogadro's number either as 602,000,000,000,000,000,000,000 or as $6.02 \cdot 10^{23}$, the calculator or spreadsheet will display 6.02 E 23, where E stands for "exponent" or power of 10. Perform the following calculations using technology, then write the answer in standard scientific notation rounded to three places.

   a. $\left(\dfrac{9}{5}\right)^{50}$

   b. 2^{35}

   c. $\left(\dfrac{1}{3}\right)^7$

   d. $\dfrac{7}{6^{15}}$

   e. $(5)^{-10}(2)^{10}$

   f. $(-4)^5\left(\dfrac{1}{(16)^{12}}\right)$

17. Describe at least three different methods for entering $5.23 \cdot 10^{-3}$ into a calculator or spreadsheet.

18. Using rules of exponents, show that $\dfrac{9^5}{27^{-7}} = 3^{31}$.

19. Using rules of exponents, show that $\dfrac{1}{x^{-n}} = x^n$.

20. Write an expression that displays the calculation(s) necessary to answer the question. Then use scientific notation and exponent rules to determine the answer.

   a. Find the number of nickels in $500.00.

   b. The circumference of Earth is about 40.2 million meters. Find the radius of Earth, in kilometers, using the formula $C = 2\pi r$.

21. According to the National Confectioners Association, in 2006 there were 35 million pounds (or 9 billion kernels) of candy corn made for Halloween. How many kernels are in a pound?

22. The robot spacecraft NEAR (Near Earth Asteroid Rendezvous) is on a four-year mission through the inner solar system to study asteroids. In February 2001 the spacecraft landed on Eros, a Manhattan-sized asteroid 160 million miles from Earth.

a. If radio messages travel at the speed of light, how long would it take for a message sent back from the NEAR spacecraft to reach the scientists?

b. The near-Earth asteroid Cruithne is now known to be a companion, and an unusual one, of Earth. This asteroid shares Earth's orbit, its motion "choreographed" in such a way as to remain stable and avoid colliding with our planet. At its closest approach Cruithne gets to within 0.1 astronomical unit of Earth (about 15 million kilometers). The asteroid in 2004 was about 0.3 astronomical unit (45 million kilometers) from Earth. If the NEAR spacecraft was in orbit around Cruithne at that time, how long would a radio signal transmitted from Earth take to reach the spacecraft?

4.4 *Converting Units*

Exploration 4.1 will help you understand the relative ages and sizes of objects in our universe and give you practice in scientific notation and unit conversion.

Problems in science constantly require converting back and forth between different units of measure. To do so, we need to be comfortable with the laws of exponents and the basic metric and English units (see Table 4.1 or a more complete table on the inside back cover). The following unit conversion examples describe a strategy based on *conversion factors*.

Converting Units within the Metric System

EXAMPLE 1

Conversion Factors

Light travels at a speed of approximately $3.00 \cdot 10^5$ kilometers per second (km/sec). Describe the speed of light in meters per second (m/sec).

SOLUTION

The prefix *kilo* means thousand. One kilometer (km) is equal to 1000 or 10^3 meters (m):

$$1 \text{ km} = 10^3 \text{ m} \tag{1}$$

Dividing both sides of Equation (1) by 1 km, we can rewrite it as

$$1 = \frac{10^3 \text{ m}}{1 \text{ km}}$$

If instead we divide both sides of Equation (1) by 10^3 m, we get

$$\frac{1 \text{ km}}{10^3 \text{ m}} = 1$$

The ratios $\frac{10^3 \text{ m}}{1 \text{ km}}$ and $\frac{1 \text{ km}}{10^3 \text{ m}}$ are called *conversion factors*, because we can use them to convert between kilometers and meters.

What is the right conversion factor? If units in $\frac{\text{km}}{\text{sec}}$ are multiplied by units in meters per kilometer, we have

$$\frac{\cancel{\text{km}}}{\text{sec}} \cdot \frac{\text{m}}{\cancel{\text{km}}}$$

and the result is in meters per second. So multiplying the speed of light in km/sec by a conversion factor in m/km will give us the correct units of m/sec.

A conversion factor always equals 1. So we will not change the value of the original quantity by multiplying it by a conversion factor. In this case, we use the conversion factor of $\frac{10^3 \text{ m}}{1 \text{ km}}$.

$$3.00 \cdot 10^5 \text{ km/sec} = 3.00 \cdot 10^5 \frac{\cancel{\text{km}}}{\text{sec}} \cdot \frac{10^3 \text{ m}}{1 \cancel{\text{km}}}$$

$$= 3.00 \cdot 10^5 \cdot 10^3 \text{ m/sec}$$

$$= 3.00 \cdot 10^8 \text{ m/sec}$$

Hence light travels at approximately $3.00 \cdot 10^8$ m/sec.

EXAMPLE 2 Check your answer in Example 1 by converting $3.00 \cdot 10^8$ m/sec back to km/sec.

SOLUTION Here we use the same strategy, but now we need to use the other conversion factor. Multiplying $3.00 \cdot 10^8$ m/sec by $(1 \text{ km})/(10^3 \text{ m})$ gives us

$$3.00 \cdot 10^8 \frac{\cancel{m}}{\text{sec}} \cdot \frac{1 \text{ km}}{10^3 \cancel{m}} = 3.00 \cdot \frac{10^8 \text{ km}}{10^3 \text{ sec}}$$

$$= 3.00 \cdot 10^5 \text{ km/sec}$$

which was the original value given for the speed of light.

Converting between the Metric and English Systems

EXAMPLE 3 You're touring Canada, and you see a sign that says it is 130 kilometers to Toronto. How many miles is it to Toronto?

SOLUTION The crucial question is, "What conversion factor should be used?" From Table 4.1 we know that

$$1 \text{ km} \approx 0.62 \text{ mile}$$

This equation can be rewritten in two ways:

$$1 \approx \frac{0.62 \text{ mile}}{1 \text{ km}} \quad \text{or} \quad 1 \approx \frac{1 \text{ km}}{0.62 \text{ mile}}$$

It produces two possible conversion factors:

$$\frac{0.62 \text{ mile}}{1 \text{ km}} \quad \text{and} \quad \frac{1 \text{ km}}{0.62 \text{ mile}}$$

SOMETHING TO THINK ABOUT

Why is the conversion factor 1 km/ 0.62 miles not helpful in solving this problem?

Which one will convert kilometers to miles? We need one with kilometers in the denominator and miles in the numerator, namely $\frac{0.62 \text{ mile}}{1 \text{ km}}$, so that the km will cancel when we multiply by 130 km:

$$130 \cancel{\text{km}} \cdot \frac{0.62 \text{ mile}}{1 \cancel{\text{km}}} = 80.6 \text{ miles}$$

So it is a little over 80 miles to Toronto.

Using Multiple Conversion Factors

EXAMPLE 4 Light travels at $3.00 \cdot 10^5$ km/sec. How many kilometers does light travel in one *year*?

SOLUTION Here our strategy is to use more than one conversion factor to convert from seconds to years. Use your calculator to perform the following calculations:

$$3.00 \cdot 10^5 \frac{\text{km}}{\cancel{\text{sec}}} \cdot \frac{60 \cancel{\text{sec}}}{1 \cancel{\text{min}}} \cdot \frac{60 \cancel{\text{min}}}{1 \cancel{\text{hr}}} \cdot \frac{24 \cancel{\text{hr}}}{1 \cancel{\text{day}}} \cdot \frac{365 \cancel{\text{days}}}{1 \text{ year}} = 94{,}608{,}000 \cdot 10^5 \text{ km/year}$$

$$\approx 9.46 \cdot 10^7 \cdot 10^5 \text{ km/year}$$

$$= 9.46 \cdot 10^{12} \text{ km/year}$$

So a light year, the distance light travels in one year, is approximately equal to $9.46 \cdot 10^{12}$ kilometers.

Algebra Aerobics 4.4

On the back inside cover of the text there are tables containing metric prefixes and conversion facts. Round your answers to two decimal places.

1. Coca-Cola Classic comes in a 2-liter container. How many fluid ounces is that? (*Note:* There are 32 ounces in a quart.)

2. A child's height is 120 cm. How tall is she in inches?

3. Convert to the desired unit:
   a. 12 inches = _____ cm
   b. 100 yards = _____ meters
   c. 20 kilograms = _____ pounds
   d. $40,000 per year = $_____ per hour (assume a 40-hour work week for 52 weeks)
   e. 24 hr/day = _____ sec/day
   f. 1 gallon = _____ ml
   g. 1 mph = _____ ft/sec

4. The distance between the sun and the moon is $3.84 \cdot 10^8$ meters. Express this in kilometers.

5. The mean distance from our sun to Jupiter is $7.8 \cdot 10^8$ kilometers. Express this distance in meters.

6. A light year is about $5.88 \cdot 10^{12}$ miles. Verify that $9.46 \cdot 10^{12}$ kilometers $\approx 5.88 \cdot 10^{12}$ miles.

7. 1 angstrom (Å) $= 10^{-8}$ cm. Express 1 angstrom in meters.

8. If a road sign says the distance to Quebec is 218 km, what is the distance in miles?

9. The distance from Earth to the sun is about 93,000,000 miles. There are 5280 feet in a mile, and a dollar bill is approximately 6 inches long. Estimate how many dollar bills would have to be placed end to end to reach from Earth to the sun.

10. Fill in the missing parts of the following conversion.

$$\frac{2560 \text{ mi}}{4.2 \text{ hrs}} = \frac{2560 \text{ mi}}{4.2 \text{ hrs}} \cdot \frac{1.6 \text{ km}}{?} = \frac{?}{?}$$

$$= \frac{?}{?} \cdot \frac{?}{60 \text{ min}} = \frac{? \text{ km}}{? \text{ min}}$$

11. Anthrax spores, which were inhaled by postal workers, causing severe illness and death, are no larger than 5 microns in diameter. How much larger than a spore is the tip of a pencil that is 1 millimeter in diameter? (*Note:* A micron is the same as a micrometer, μm.)

12. Use the conversion factor of 1 light year $= 9.46 \cdot 10^{12}$ kilometers or $5.88 \cdot 10^{12}$ miles to determine the following.
   a. Alpha Centauri, the nearest star to our sun, is 4.3 light years away. What is the distance in kilometers? How many miles away is it?
   b. The radius of the Milky Way is 10^8 light years. How many meters is that?
   c. Deneb is a star 1600 light years from Earth. How far is that in feet?

13. If 1 angstrom, Å, $= 10^{-10}$ meter, determine the following values.
   a. The radius of a hydrogen atom is 0.5 angstrom. How many meters is the radius?
   b. The radius of a cell is 10^5 angstroms. How many meters is the cell's radius?
   c. A radius of a proton is 0.00001 angstrom. Express the proton's radius in meter.

14. The Harvard Bridge, which connects Cambridge to Boston along Massachusetts Avenue, is literally marked off in units called *Smoots*. A Smoot is equal to about 5.6 feet, the height of an M.I.T. fraternity pledge named Oliver Smoot. The bridge is approximately 364 Smoots long. How long is the bridge in feet? Show all units when doing your conversion.

Exercises for Section 4.4

Use the conversion table on the back cover of the text for problems in this section.

1. Change the following English units to the metric units indicated.
   a. 50 miles to kilometers   d. 12 inches to centimeters
   b. 3 feet to meters   e. 60 feet to meters
   c. 5 pounds to kilograms   f. 4 quarts to liters

2. Change the following metric units to the English units indicated.
   a. 25 kilometers to miles   d. 50 grams to ounces
   b. 700 meters to yards   e. 10 kilograms to pounds
   c. 250 centimeters to inches   f. 10,000 milliliters to quarts

3. For the following questions, make an estimate and then check your estimate using the conversion table on the inside back cover:

   a. One foot is how many centimeters?

   b. One foot is what part of a meter?

4. A football field is 100 yards long. How many meters is this? What part of a kilometer is this?

5. How many droplets of water are in a river that is 100 km long, 250 m wide, and 25 m deep? Assume a droplet is 1 milliliter. (*Note:* one liter = one cubic decimeter and 10 decimeters = 1 meter.)

6. a. A roll of aluminum foil claims to be 50 sq ft or 4.65 m². Show the conversion factors that would verify that these two measurements are equivalent.

   b. One cm³ of aluminum weighs 2.7 grams. If a sheet of aluminum foil is 0.003 8 cm thick, find the weight of the roll of aluminum foil in grams.

7. If a falling object accelerates at the rate of 9.8 meters per second every second, how many feet per second does it accelerate each second?

8. Convert the following to feet and express your answers in scientific notation.

   a. The radius of the solar system is approximately 10^{12} meters.

   b. The radius of a proton is approximately 10^{-15} meter.

9. The speed of light is approximately $1.86 \cdot 10^5$ miles/sec.

   a. Write this number in decimal form and express your answer in words.

   b. Convert the speed of light into meters per year. Show your work.

10. The average distance from Earth to the sun is about 150,000,000 km, and the average distance from the planet Venus to the sun is about 108,000,000 km.

    a. Express these distances in scientific notation.

    b. Divide the distance from Venus to the sun by the distance from Earth to the sun and express your answer in scientific notation.

    c. The distance from Earth to the sun is called 1 astronomical unit (1 A.U.) How many astronomical units is Venus from the sun?

    d. Pluto is 5,900,000,000 km from the sun. How many astronomical units is it from the sun?

11. The distance from Earth to the sun is approximately 150 million kilometers. If the speed of light is $3.00 \cdot 10^5$ km/sec, how long does it take light from the sun to reach Earth? If a solar flare occurs right now, how long would it take for us to see it?

12. Earth travels in an approximately circular orbit around the sun. The average radius of Earth's orbit around the sun is $9.3 \cdot 10^7$ miles. Earth takes one year, or 365 days, to complete one orbit.

    a. Use the formula for the circumference of the circle to determine the distance the Earth travels in one year.

    b. How many hours are in one year?

    c. Speed is distance divided by time. Find the orbital speed of Earth in miles per hour.

13. A barrel of U.S. oil is 42 gallons. A barrel of British oil is 163.655 liters. Which barrel is larger and by how much?

14. A barrel of wheat is 3.2812 bushels (U.S. dry) or 4.0833 cubic feet.

    a. How many cubic feet are in a bushel of wheat?

    b. How many cubic inches are in a barrel?

    c. How many cubic centimeters are in a bushel?

15. In the United States, land is measured in acres and one acre is 43,560 sq ft.

    a. If you buy a one-acre lot that is in the shape of a square, what would be the length of each side in feet?

    b. A newspaper advertisement states that all lots in a new housing development will be a minimum of one and a half acres. Assuming the lot is rectangular and has 150 ft of frontage, how deep will the minimal-size lot be? If the new home owner wants to fence in the lot, how many yards of fencing would be needed?

    c. The metric unit for measuring land is the square hectometer. (A hectometer is a length of 100 meters.) Find the size of a one-acre lot if it were measured in square hectometers.

    d. A hectare is 100 acres. How many one-acre lots can fit in a square mile? How many hectares is that?

16. Estimate the number of heartbeats in a lifetime. Explain your method.

17. A nanosecond is 10^{-9} second. Modern computers can perform on the order of one operation every nanosecond. Approximately how many feet does an electrical signal moving at the speed of light travel in a computer in 1 nanosecond?

18. Since light takes time to travel, everything we see is from the past. When you look in the mirror, you see yourself not as you are, but as you were nanoseconds ago.

    a. Suppose you look up tonight at the bright star Deneb. Deneb is 1600 light years away. When you look at Deneb, how old is the image you are seeing?

    b. Even more disconcerting is the fact that what we see as simultaneous events do not necessarily occur simultaneously. Consider the two stars Betelgeuse and Rigel in the constellation Orion. Betelgeuse is 300 and Rigel 500 light years away. How many years apart were the images generated that we see simultaneously?

19. The world population in 2005 was approximately 6.45 billion people. During that year the Coca-Cola company claimed that 15,000 of their beverages were consumed every second. What was the worldwide consumption of their beverages per year per person in 2005?

20. A homeowner would like to spread shredded bark (mulch) over her flowerbeds. She has three flowerbeds measuring 25 ft by 3 ft, 15 ft by 4 ft, and 30 ft by 1.5 ft. The recommended depth for the mulch is 4 inches, and the shredded bark costs $27.00 per one cubic yard. How much will it cost to cover all of the flowerbeds with shredded bark? (*Note:* You cannot buy a portion of a cubic yard of mulch.)

21. A circular swimming pool is 18 ft in diameter and 4 ft deep.

  a. Determine the volume of the pool in gallons if one gallon is 231 cubic inches.

  b. The pool's filter pump can circulate 2500 gal per hour. How many hours do you need to run the filter in order to filter the number of gallons contained in the pool?

  c. One pound of chlorine shock treatment can treat 10,000 gal. How much of the shock treatment should you use?

22. An angstrom, Å, is a metric unit of length equal to one ten billionth of a meter. It is useful in specifying wavelengths of electromagnetic radiation (e.g., visible light, ultraviolet light, X-rays, and gamma rays).

  a. The visible-light spectrum extends from approximately 3900 angstroms (violet light) to 7700 angstroms (red light) Write this range in centimeters using scientific notation.

  b. Some gamma rays have wavelengths of 0.0001 angstrom. Write this number in centimeters using scientific notation.

  c. The nanometer (nm) is 10 times larger than the angstrom, so 1 nm is equal to how many meters?

23. The National Institutes of Health guidelines suggest that adults over 20 should have a body mass index, or BMI, under 25. This index is created according to the formula

$$\text{BMI} = \frac{\text{weight in kilograms}}{(\text{height in meters})^2}$$

  a. Given that 1 kilogram = 2.2 pounds, and 1 meter = 39.37 inches, calculate the body mass index of President George W. Bush, who is 6 feet tall and in 2003 weighed 194 pounds. According to the guidelines, how would you describe his weight?

  b. Most Americans don't use the metric system. So in order to make the BMI easier to use, convert the formula to an equivalent one using weight in pounds and height in inches. Check your new formula by using Bush's weight and height, and confirm that you get the same BMI.

  c. The following excerpt from the article "America Fattens Up" (*The New York Times*, October 20, 1996) describes a very complicated process for determining your BMI:

  To estimate your body mass index you first need to convert your weight into kilograms by multiplying your weight in pounds by 0.45. Next, find your height in inches. Multiply this number by 0.254 to get meters. Multiply that number by itself and then divide the result into your weight in kilograms. Too complicated? Internet users can get an exact calculation at http://141.106.68.17/bsa.acgl.

  Can you do a better job of describing the process?

  d. A letter to the editor from Brent Kigner, of Oneonta, N.Y., in response to the *New York Times* article says:

  Math intimidates partly because it is often made unnecessarily daunting. Your article "American Fattens Up" convolutes the procedure for calculating the Body Mass Index so much that you suggest readers retreat to the Internet. In fact, the formula is simple: Multiply your weight in pounds by 703, then divide by the square of your height in inches. If the result is above 25, you weigh too much.

  Is Brent Kigner right?

24. Computer technology refers to the storage capacity for information with its own special units. Each minuscule electrical switch is called a "bit" and can be off or on. As the information capacity of computers has increased, the industry has developed some much larger units based on the bit:

1 byte = 8 bits

1 kilobit = 2^{10} bits, or 1024 bits (a kilobit is sometimes abbreviated Kbit)

1 kilobyte = 2^{10} bytes, or 1024 bytes (a kilobyte is sometimes abbreviated Kbyte)

1 megabit = 2^{20} bits, or 1,048,576 bits

1 megabyte = 2^{20} bytes, or 1,048,576 bytes

1 gigabyte = 2^{30} bytes, or 1,073,741,824 bytes

  a. How many kilobytes are there in a megabyte? Express your answer as a power of 2 and in scientific notation.

  b. How many bits are there in a gigabyte? Express your answer as a power of 2 and in scientific notation.

25. The accompanying excerpt is from an article about Planck's length, which at 10^{-35} meter is believed to be the smallest length or size anything can be in the universe (from G. Johnson, "How Is the Universe Built? Grain by Grain," in the science section of the Dec. 7, 1999, *New York Times*, p. D1). Read the accompanying excerpt and then answer the following questions.

  a. How many kilometers is Planck's length?

  b. How many miles is Planck's length?

  c. If light travels at $3 \cdot 10^8$ m/sec, how long will it take light to cross a distance equivalent to Planck's length?

  Slightly smaller than what Americans quaintly insist on calling half an inch, a centimeter (one-hundredth of a meter) is easy enough to see. Divide this small length into 10 equal slices and you are looking, or probably squinting, at a millimeter (one-thousandth, or 10 to the minus 3 meters). By the time you divide one of these tiny units into a thousand minuscule micrometers, you have far exceeded the limits of the finest bifocals. But in the mind's eye, let the cutting continue, chopping the micrometer into a thousand nanometers and the nanometers into a thousand picometers, and those in steps of a thousandfold into femtometers, attometers, zeptameters, and yoctometers. At this point, 10 to the minus 24 meters, about one-billionth the radius of a proton, the roster of Greek names runs out. But go ahead and keep dividing, again and again until you reach a length only one hundred-billionth as large as that tiny amount: 10 to the minus 35 meters. . . . You have finally hit rock bottom: a span called the Planck length, the shortest anything can get. According to recent developments in the quest to devise the "theory of everything," space is not an infinitely divisible continuum. It is not smooth but granular, and the Planck length gives the size of the smallest possible grains.

  The time it takes for a light beam to zip across this ridiculously tiny distance . . . is called Planck time, the shortest possible tick of an imaginary clock.

4.5 *Fractional Exponents*

So far we have derived rules for operating with expressions of the form a^n, where n is any integer. These rules can be extended to expressions of the form $a^{m/n}$, where the exponent is a fraction. We need first to consider what an expression such as $a^{m/n}$ means.

The expression m/n can also be written as $m \cdot (1/n)$ or $(1/n) \cdot m$. If the laws of exponents are consistent, then

$$a^{m/n} = (a^m)^{1/n} = (a^{1/n})^m$$

For example, if we apply Rule 3 for exponents to the expression $(a^{1/3})^2$, then the following is true:

$$(a^{1/3})^2 = a^{(1/3)2} = a^{2/3}$$

Square Roots: Expressions of the Form $a^{1/2}$

The expression $a^{1/2}$ is called the *principal square root* (or just the *square root*) of a and is often written as $\sqrt{a}$. The symbol $\sqrt{}$ is called a *radical*. The principal square root of a is the *nonnegative* number b such that $b^2 = a$. Both the square of -2 and the square of 2 are equal to 4, but the notation $\sqrt{4}$ is defined as *only the positive root*. If both -2 and 2 are to be considered, we write $\pm\sqrt{4}$, which means "plus or minus the square root of 4."

So, $\sqrt{4} = 2$ and $\pm\sqrt{4} = 2$ and -2. When you solve $x^2 = 4$, the solution is 2 and -2.

In the real numbers, $\sqrt{a}$ is not defined when a is negative. For example, $\sqrt{-4}$ is undefined, since there is no real number b such that $b^2 = -4$.

The Square Root

For $a \geq 0$,

$$a^{1/2} = a^{0.5} = \sqrt{a}$$

where $\sqrt{a}$ is the nonnegative number b such that $b^2 = a$.

For example, $25^{1/2} = \sqrt{25} = 5$, since $5^2 = 25$.

Estimating Square Roots.   A number is called *a perfect square* if its square root is an integer. For example, 25 and 36 are both perfect squares since $25 = 5^2$ and $36 = 6^2$, so $\sqrt{25} = 5$ and $\sqrt{36} = 6$. If we don't know the square root of some number x and don't have a calculator handy, we can estimate the square root by bracketing it between two perfect squares, a and b, for which we do know the square roots. If $0 \leq a < x < b$, then $\sqrt{a} < \sqrt{x} < \sqrt{b}$. For example, to estimate $\sqrt{10}$,

we know	$9 < 10 < 16$	where 9 and 16 are perfect squares
so	$\sqrt{9} < \sqrt{10} < \sqrt{16}$	
and	$3 < \sqrt{10} < 4$	

Therefore $\sqrt{10}$ lies somewhere between 3 and 4, probably closer to 3 because 10 is closer to 9 than to 16. According to a calculator, $\sqrt{10} \approx 3.16$.

EXAMPLE 1   Estimate $\sqrt{27}$.

SOLUTION   We know $\qquad\qquad 25 < 27 < 36$

therefore $\qquad\quad \sqrt{25} < \sqrt{27} < \sqrt{36}$

and $\qquad\qquad\qquad 5 < \sqrt{27} < 6$

So $\sqrt{27}$ lies somewhere between 5 and 6. Would you expect $\sqrt{27}$ to be closer to 5 or to 6? Check your answer with a calculator.

Using a calculator

Many calculators and spreadsheet programs have a square root function, often labeled $\sqrt{}$ or perhaps "SQRT." You can also calculate square roots by raising a number to the $\frac{1}{2}$ or 0.5 power using the ^ key, as in 4 ^ 0.5. Try using a calculator to find $\sqrt{4}$ and $\sqrt{9}$.

In any but the simplest cases where the square root is immediately obvious, you will probably use the calculator. For example, use your calculator to find

$$8^{1/2} = \sqrt{8} \approx 2.8284$$

Double-check the answer by verifying that $(2.8284)^2 \approx 8$.

EXAMPLE 2   **Calculating square roots**

The function $S = \sqrt{30d}$ describes the relationship between S, the speed of a car in miles per hour, and d, the distance in feet a car skids after applying the brakes on a dry tar road. Use a calculator to estimate the speed of a car that:

a. Leaves 40-foot-long skid marks on a dry tar road.

b. Leaves 150-foot-long skid marks.

SOLUTION   **a.** If $d = 40$ feet, then $S = \sqrt{30 \cdot 40} = \sqrt{1200} \approx 35$, so the car was traveling at about 35 miles per hour.

b. If $d = 150$ feet, then $S = \sqrt{30 \cdot 150} = \sqrt{4500} \approx 67$, so the car was traveling at almost 70 miles per hour.

EXAMPLE 3   Assuming that the surface area of Earth is approximately 200 million square miles, estimate the radius of Earth.

SOLUTION   **Step 1.** Find the formula for the radius of a sphere.

If we assume that Earth is roughly spherical, we can solve for the radius r in the formula for the surface of a sphere, $S = 4\pi r^2$. We get

$$r = \sqrt{\frac{S}{4\pi}} = \sqrt{\frac{1}{4}} \cdot \sqrt{\frac{S}{\pi}} = \frac{1}{2}\sqrt{\frac{S}{\pi}}$$

Step 2. Estimate the radius of Earth.

Given that Earth's surface area is approximately 200,000,000 square miles, we can use our derived formula to estimate Earth's radius. Substituting for S, we get

$$r = \frac{1}{2}\sqrt{\frac{200,000,000 \text{ miles}^2}{\pi}}$$

$$\approx \frac{1}{2}\sqrt{63,661,977 \text{ miles}^2}$$

$$\approx \frac{1}{2} \cdot 7979 \text{ miles} \approx 3989 \text{ miles}$$

So Earth has a radius of about 4000 miles.

nth Roots: Expressions of the Form $a^{1/n}$

The term $a^{1/n}$ denotes the *n*th root of *a*, often written as $\sqrt[n]{a}$. For $a \geq 0$, the *n*th root of *a* is the nonnegative number whose *n*th power is *a*.

$8^{1/3} = \sqrt[3]{8} = 2$ since $2^3 = 8$ (we call 2 the third or cube root of 8)

$16^{1/4} = \sqrt[4]{16} = 2$ since $2^4 = 16$ (we call 2 the fourth root of 16)

For $a < 0$, if *n* is odd, $\sqrt[n]{a}$ is the negative number whose *n*th power is *a*. Note that if *n* is even, then $\sqrt[n]{a}$ is not a real number when $a < 0$.

$(-8)^{1/3} = \sqrt[3]{-8} = -2$ since $(-2)^3 = -8$.

$(-27)^{1/3} = \sqrt[3]{-27} = -3$ since $(-3)^3 = -27$

$(-16)^{1/4} = \sqrt[4]{-16}$ is not a real number

The *n*th Root

If *a* is a real number and *n* is a positive integer,

$$a^{1/n} = \sqrt[n]{a}, \qquad \text{the } n\text{th root of } a$$

For $a \geq 0$,

$\sqrt[n]{a}$ is the nonnegative number *b* such that $b^n = a$.

For $a < 0$,

If *n* is odd, $\sqrt[n]{a}$ is the negative number *b* such that $b^n = a$.

If *n* is even, $\sqrt[n]{a}$ is not a real number.

If the *n*th root exists, you can find its value on a calculator. For example, to determine a fifth root, raise the number to the $\frac{1}{5}$ or the 0.2 power. So

$$3125^{1/5} = \sqrt[5]{3125} = 5$$

Double-check your answer by verifying that $5^5 = 3125$.

EXAMPLE 4 Simplify:

   a. $625^{1/4}$ **b.** $(-625)^{1/4}$ **c.** $125^{1/3}$ **d.** $(-125)^{1/3}$

SOLUTION   **a.** $625^{1/4} = 5$ since $5^4 = 625$

   b. $(-625)^{1/4}$ does not have a real-number solution

   c. $125^{1/3} = 5$ since $5^3 = 125$

   d. $(-125)^{1/3} = -5$ since $(-5)^3 = -125$

EXAMPLE 5   **a.** The volume of a sphere is given by the equation $V = \frac{4}{3}\pi r^3$. Rewrite the formula, solving for the radius as a function of the volume.

   b. If the volume of a sphere is 370 cubic inches, what is its radius? What common object might have that radius?

   c. What are the dimensions of a cube that contains this volume?

SOLUTION   **a.** Given: $\qquad\qquad\qquad\qquad V = \frac{4}{3}\pi r^3$

       multiply both sides by 3 $\qquad\qquad 3V = 4\pi r^3$

divide by 4π $$\frac{3V}{4\pi} = r^3$$

take the cube root and switch sides $$r = \sqrt[3]{\frac{3V}{4\pi}}$$

b. Substituting 370 for V and 3.14 for π in our derived formula in part (a), we have

$$r \approx \sqrt[3]{\frac{3 \cdot 370}{4 \cdot 3.14}} \approx 4.45 \text{ inches}$$

A regulation-size soccer ball is basically a sphere with a radius of about 4.45 inches.

c. A cube has the same length on all three sides. If s is the side length, then the volume of the desired cube is $s^3 = 370$ cubic inches. So $s = \sqrt[3]{370} \approx 7.18$ inches. A cube of length 7.18 inches on each side would give a volume equivalent to a sphere with a radius of 4.45 inches.

Rules for Radicals

The following rules can help you compute with radicals. They represent extensions of the rules for integer exponents. In the following table we assume that m and n are positive integers and that $\sqrt[n]{a}$ and $\sqrt[n]{b}$ exist.

Rules for Radicals

Example

1. $\sqrt[n]{a} \cdot \sqrt[n]{b} = \sqrt[n]{ab}$ $\qquad\qquad \sqrt{3} \cdot \sqrt{2} = \sqrt{6}$

2. $\dfrac{\sqrt[n]{a}}{\sqrt[n]{b}} = \sqrt[n]{\dfrac{a}{b}} \quad b \neq 0$ $\qquad \dfrac{\sqrt[4]{125}}{\sqrt[4]{25}} = \sqrt[4]{\dfrac{125}{25}} = \sqrt[4]{5}$

3. $(\sqrt[n]{a})^n = \sqrt[n]{a^n} = a \quad a > 0$ $\qquad (\sqrt{7})^2 = \sqrt{7^2} = 7$

EXAMPLE 6 **Simplifying radicals**

Simplify the following radical expressions. Assume all variables are nonnegative real numbers.

a. $\sqrt[3]{625x^4}$ $\qquad$ **b.** $3\sqrt{48} - 5\sqrt{27}$

SOLUTION **a.** Factor 625 $\qquad\qquad\qquad\qquad\qquad \sqrt[3]{625x^4} = \sqrt[3]{5^4 \cdot x^4}$

rewrite using perfect cube factors $\qquad\qquad = \sqrt[3]{5^3 \cdot 5 \cdot x^3 \cdot x}$

use Rule 1 for radicals $\qquad\qquad\qquad\quad = \sqrt[3]{5^3 x^3} \cdot \sqrt[3]{5x}$

extract the perfect cubes (Rule 3), leaving $\qquad = 5x \cdot \sqrt[3]{5x}$
the remaining factors under the radical

b. Find the largest perfect square factors $\quad 3\sqrt{48} - 5\sqrt{27} = 3\sqrt{16 \cdot 3} - 5\sqrt{9 \cdot 3}$

extract the perfect squares (Rule 3) $\qquad\qquad\qquad = 3 \cdot 4\sqrt{3} - 5 \cdot 3\sqrt{3}$

multiply $\qquad\qquad\qquad\qquad\qquad\qquad\qquad = 12\sqrt{3} - 15\sqrt{3}$

use distributive law $\qquad\qquad\qquad\qquad\qquad = (12 - 15)\sqrt{3}$

$\qquad\qquad\qquad\qquad\qquad\qquad\qquad\qquad = -3\sqrt{3}$

Algebra Aerobics 4.5a

1. Evaluate each of the following without a calculator.
   a. $81^{1/2}$ c. $36^{1/2}$ e. $(-36)^{1/2}$
   b. $144^{1/2}$ d. $-49^{1/2}$

2. Assume that all variables represent nonnegative quantities. Then simplify and rewrite the following without radical signs. (Use fractional exponents if necessary.)
   a. $\sqrt{9x}$ c. $\sqrt{36x^2}$ e. $\sqrt{\dfrac{49}{x^2}}$
   b. $\sqrt{\dfrac{x^2}{25}}$ d. $\sqrt{\dfrac{9y^2}{25x^4}}$ f. $\sqrt{\dfrac{4a}{169}}$

3. Use the formula in Example 2 in this section to estimate the following:
   a. The speed of a car that leaves 60-foot-long skid marks on a dry tar road.
   b. The speed of a car that leaves 200-foot-long skid marks on a dry tar road.

4. Without a calculator, find two consecutive integers between which the given square root lies.
   a. $\sqrt{29}$ c. $\sqrt{117}$ e. $\sqrt{39}$
   b. $\sqrt{92}$ d. $\sqrt{79}$

5. Evaluate each of the following without a calculator:
   a. $27^{1/3}$ c. $8^{-1/3}$ e. $27^{-1/3}$ g. $\left(\dfrac{8}{27}\right)^{-1/3}$
   b. $16^{1/4}$ d. $32^{1/5}$ f. $25^{-1/2}$ h. $\left(\dfrac{1}{16}\right)^{1/2}$

6. Evaluate:
   a. $\sqrt[3]{-27}$ c. $(-1000)^{1/3}$ e. $(-8)^{1/3}$
   b. $(-10,000)^{1/4}$ d. $-16^{1/4}$ f. $\sqrt{2500}$

7. Estimate the radius of a spherical balloon with a volume of 2 cubic feet.

8. Simplify if possible.
   a. $\sqrt{9+16}$ c. $\sqrt[3]{-125}$
   b. $-\sqrt{49}$ d. $\sqrt{45}-3\sqrt{125}$

9. Change each radical expression into exponent form, then simplify. Assume all variables are nonnegative.
   a. $\sqrt{36}$ b. $\sqrt[3]{27x^6}$ c. $\sqrt[4]{81a^4b^{12}}$

10. Solve for the indicated variable. Assume all variables represent nonnegative quantities.
    a. $V = \pi r^2 h$ for r d. $c^2 = a^2 + b^2$ for a
    b. $V = \frac{1}{3}\pi r^2 h$ for r e. $S = 6x^2$ for x
    c. $V = s^3$ for s

You may want to do Exploration 4.2 on Kepler's laws of planetary motion after reading this section.

Fractional Powers: Expressions of the Form $a^{m/n}$

In the beginning of this section, we saw that we can write $a^{m/n}$ either as $(a^m)^{1/n}$ or $(a^{1/n})^m$. Writing it as $(a^m)^{1/n}$ means that we would first raise the base, a, to the mth power and then take the nth root of that. Writing it as $(a^{1/n})^m$ implies first finding the nth root of a and then raising that to the mth power. For example,

$$2^{3/2} = (2^3)^{1/2}$$
$$= (8)^{1/2}$$

using a calculator ≈ 2.8284

Equivalently, $2^{3/2} = (2^{1/2})^3$
$$\approx (1.414)^3$$
$$\approx 2.8284$$

We could, of course, use a calculator to compute $2^{3/2}$ (or $2^{1.5}$) directly by raising 2 to the $\frac{3}{2}$ or 1.5 power.

If $a \geq 0$ and m and n are integers ($n \neq 0$), then using radical notation,

$$a^{m/n} = \left(\sqrt[n]{a}\right)^m$$

or equivalently $= \sqrt[n]{a^m}$

Exponents expressed as ratios of the form m/n are called *rational exponents*. The set of laws for simplifying expressions with integer exponents also holds for real exponents, which includes rational exponents.

EXAMPLE 7   Find the product of $(\sqrt{5}) \cdot (\sqrt[3]{5})$, leaving the answer in exponent form.

SOLUTION $$\left(\sqrt{5}\right) \cdot \left(\sqrt[3]{5}\right) = 5^{1/2} \cdot 5^{1/3} = 5^{(1/2)+(1/3)} = 5^{5/6}$$

EXAMPLE 8   According to McMahon and Bonner in *On Size and Life*,[4] common nails range from 1 to 6 inches in length. The weight varies even more, from 11 to 647 nails per pound. Longer nails are relatively thinner than shorter ones. A good approximation of the relationship between length and diameter is given by the equation

$$d = 0.07L^{2/3}$$

where d = diameter and L = length, both in inches. Estimate the diameters of nails that are 1, 3, and 6 inches long.

SOLUTION   When $L = 1$ inch, the diameter $d = 0.07 \cdot (1)^{2/3} = 0.07 \cdot 1 = 0.07$ inches.

When $L = 3$ inches, then $d = 0.07 \cdot (3)^{2/3} \approx 0.07 \cdot 2.08 \approx 0.15$ inches.

When $L = 6$ inches, then $d = 0.07 \cdot (6)^{2/3} \approx 0.07 \cdot 3.30 \approx 0.23$ inches.

Summary of Zero, Negative, and Fractional Exponents

If m and n are integers and $a \neq 0$, then

$$a^0 = 1$$
$$a^{-n} = \frac{1}{a^n}$$
$$a^{1/n} = \sqrt[n]{a}$$
$$a^{m/n} = \sqrt[n]{a^m} = \left(\sqrt[n]{a}\right)^m \qquad a > 0$$

Algebra Aerobics 4.5b

Assume all variables represent positive quantities.

1. Find the product expressed in exponent form:

   a. $\sqrt{2} \cdot \sqrt[3]{2}$ c. $\sqrt{3} \cdot \sqrt[3]{9}$ e. $\sqrt[4]{x^3} \cdot \sqrt{x}$

   b. $\sqrt{5} \cdot \sqrt[4]{5}$ d. $\sqrt[4]{x} \cdot \sqrt[3]{x}$ f. $\sqrt[3]{xy^2} \cdot \sqrt{xy}$

2. Find the quotient by representing the expression in exponent form. Leave the answer in positive exponent form.

   a. $\dfrac{\sqrt{2}}{\sqrt[3]{2}}$ b. $\dfrac{2}{\sqrt[4]{2}}$ c. $\dfrac{\sqrt[4]{5}}{\sqrt[3]{5}}$ d. $\dfrac{\sqrt{x}}{\sqrt[4]{x^3}}$ e. $\dfrac{\sqrt[3]{xy^2}}{\sqrt{xy}}$

3. McMahon and Bonner give the relationship between chest circumference and body weight of adult primates as

   $$c = 17.1w^{3/8}$$

   where w = weight in kilograms and c = chest circumference in centimeters. Estimate the chest circumference of a:

   a. 0.25-kg tamarin   b. 25-kg baboon

4. Simplify each expression by removing all possible factors from the radical.

   a. $\sqrt{20x^2}$ c. $\sqrt[3]{16x^3y^4}$

   b. $\sqrt{75a^3}$ d. $\dfrac{\sqrt[4]{32x^4y^6}}{\sqrt[4]{81x^8y^5}}$

5. Change each radical expression into a form with fractional exponents, then simplify.

   a. $\sqrt{4a^2b^6}$ c. $\sqrt[3]{8.0 \cdot 10^{-9}}$

   b. $\sqrt[4]{16x^4y^6}$ d. $\sqrt{8a^{-4}}$

---

[4]T. A. McMahon and J. Tyler Bonner, *On Size and Life* (New York: Scientific American Library, Scientific American Books, 1983).

Exercises for Section 4.5

1. Evaluate without a calculator:

    a. $100^{1/2}$ **c.** $100^{-1/2}$ **e.** $-1000^{1/3}$

    b. $-100^{1/2}$ **d.** $-100^{-1/2}$ **f.** $(-1000)^{1/3}$

2. Evaluate without a calculator:

    a. $\sqrt{10{,}000}$ **c.** $625^{1/2}$ **e.** $\left(\dfrac{1}{9}\right)^{1/2}$

    b. $\sqrt{-25}$ **d.** $100^{1/2}$ **f.** $\left(\dfrac{625}{100}\right)^{1/2}$

3. Assume that all variables represent positive quantities and simplify.

    a. $\sqrt{\dfrac{a^2b^4}{c^6}}$ **b.** $\sqrt{36x^4y}$ **c.** $\sqrt{\dfrac{49x}{y^6}}$ **d.** $\sqrt{\dfrac{x^4y^2}{100z^6}}$

4. Simplify each expression by removing all possible factors from the radical, then combining any like terms.

    a. $2\sqrt{50} + 12\sqrt{8}$ **c.** $10\sqrt{32} - 6\sqrt{18}$

    b. $3\sqrt{27} - 2\sqrt{75}$ **d.** $2\sqrt[3]{16} + 4\sqrt[3]{54}$

5. Simplify by removing all possible factors for each radical. Assume all variable quantities are positive.

    a. $\sqrt{125a}$ **c.** $\sqrt{8x^3y^2}$

    b. $\sqrt{\dfrac{x^2}{4x^4y^6}}$ **d.** $\sqrt{64x^4y^5}$

6. Estimate the radius, r, of a circular region with an area, A, of 35 ft^2 (where $A = \pi r^2$).

7. Evaluate each expression without using a calculator.

    a. $\sqrt{36 \cdot 10^6}$ **c.** $\sqrt[4]{625 \cdot 10^{20}}$

    b. $\sqrt[3]{8 \cdot 10^9}$ **d.** $\sqrt{1.0 \cdot 10^{-4}}$

8. Calculate the following:

    a. $4^{1/2}$ **c.** $27^{1/3}$ **e.** $8^{2/3}$ **g.** $16^{1/4}$

    b. $-4^{1/2}$ **d.** $-27^{1/3}$ **f.** $-8^{2/3}$ **h.** $16^{3/4}$

9. Calculate:

    a. $\left(\dfrac{1}{100}\right)^{1/2}$ **b.** $25^{-1/2}$ **c.** $\left(\dfrac{9}{16}\right)^{-1/2}$ **d.** $\left(\dfrac{1}{1000}\right)^{1/3}$

10. Estimate the length of a side, s, of a cube with volume, V, of 6 cm^3 (where $V = s^3$).

11. Evaluate when $x = 2$:

    a. $(-x)^2$ **c.** $x^{1/2}$ **e.** $x^{3/2}$

    b. $-x^2$ **d.** $(-x)^{1/2}$ **f.** x^0

12. Determine if the following statements are true or false.

    a. $\sqrt[4]{(3x^2)^4} = 3x^2$

    b. $\sqrt[3]{(x+1)^4} = (x+1)\left(\sqrt[3]{x+1}\right)$

    c. $\sqrt[3]{\dfrac{9}{25}} = \dfrac{\sqrt[3]{45}}{5}$

    d. $\sqrt{15} - \sqrt{3} = \sqrt{12}$

13. Use $<$, $>$, or $=$ to make each statement true.

    a. $\sqrt{3} + \sqrt{7}$ _?_ $\sqrt{3+7}$

    b. $\sqrt{3^2 + 2^2}$ _?_ 5

    c. $\sqrt{5^2 - 4^2}$ _?_ 2

14. Fill in the missing forms in the table.

Radical Form	Rational Exponent Form
a. $\sqrt[3]{64} = 4$	
b.	$-(144)^{1/2} = -12$
c. $\left(\sqrt[4]{81x}\right)^3 = 27 \cdot \sqrt[4]{x^3}$	
d.	$(-243)^{1/5} = -3$
e.	$16^{5/4} = 32$

15. Evaluate:

    a. $27^{2/3}$ **b.** $16^{-3/4}$ **c.** $25^{-3/2}$ **d.** $81^{-3/4}$

16. Without using a calculator, find two *consecutive* integers such that one is smaller and one is larger than each of the following (for example, $3 < \sqrt{11} < 4$). Show your reasoning.

    a. $\sqrt{13}$ **b.** $\sqrt{22}$ **c.** $\sqrt{40}$

17. Estimate the radius of a spherical balloon that has a volume of 4 ft^3.

18. *Constellation.* Reduce each of the following expressions to the form $u^a \cdot m^b$; then plot the exponents as points with coordinates (a, b) on graph paper. Do you recognize the constellation?

    a. $\dfrac{(u^2)^2 \cdot m}{u^2 \cdot m^{-4}}$ **e.** $\dfrac{u^{-3/2} \cdot u^{-7/2} \cdot m^1 \cdot (m^3)^3}{(um)^2}$

    b. $\dfrac{u^{-9/5} \cdot m^3}{(umu^2)^1 \cdot m^{-1}}$ **f.** $\dfrac{1}{u^{12} \cdot m^{-9}}$

    c. $\dfrac{u^2 \cdot u^{-4}}{u^3 \cdot (m^{-2})^3}$ **g.** $\dfrac{(mu)^0 \cdot (u^{10})^{-1} \cdot m^{1/4}}{(m^{-3} \cdot u^{-1/3})^3}$

    d. $\dfrac{(um^2)^3 \cdot u^2}{(um)^4}$

19. An equilateral triangle has sides of length 8 cm.

    a. Find the height of the triangle. (*Hint:* Use the Pythagorean theorem on the inside back cover.)

    b. Find the area A of the triangle if $A = \dfrac{1}{2}bh$.

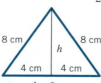

20. An Egyptian pyramid consists of a square base and four triangular sides. A model of a pyramid is constructed using four equilateral triangles each with a side length of 30 inches. Find the surface area of the pyramid model, including the base. (*Note:* Surface area is the sum of the areas of the four triangular sides and the rectangular base. The previous exercise gives the formula for finding the area of a triangle.)

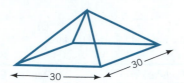

4 ft off the ground. The base of the ramp is 48 ft from the porch. How long is the ramp? (*Hint:* Use the Pythagorean theorem on the inside back cover.)

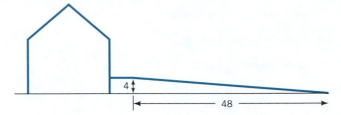

21. The time it takes for one complete swing of a pendulum is called the *period* of its motion. The period T (in seconds) of a swinging pendulum is found using the formula $T = 2\pi\sqrt{\dfrac{L}{32}}$, where L is the length of the pendulum in feet and 32 is the acceleration of gravity in feet per second.^{2}

  a. Find the period of a pendulum whose length is 2 ft 8 in.

  b. How long would a pendulum have to be to have a period of 2 seconds?

22. (Requires the use of a calculator that can evaluate powers.) A wheelchair ramp is constructed at the end of a porch, which is

23. (Requires the use of a calculator that can evaluate powers.) The breaking strength S (in pounds) of a three-strand manila rope is a function of its diameter, D (in inches). The relationship can be described by the equation $S = 1700D^{1.9}$. Calculate the breaking strength when D equals:

  a. 1.5 in   **b.** 2.0 in

24. (Requires the use of a calculator that can evaluate powers.) If a rope is wound around a wooden pole, the number of pounds of frictional force, F, between the pole and the rope is a function of the number of turns, N, according to the equation $F = 14 \cdot 10^{0.70N}$. What is the frictional force when the number of turns is:

  a. 0.5   **b.** 1   **c.** 3

4.6 *Orders of Magnitude*

Comparing Numbers of Widely Differing Sizes

We have seen that a useful method of comparing two objects of widely different sizes is to calculate the ratio rather than the difference of the sizes. The ratio can be estimated by computing *orders of magnitude,* the number of times we would have to multiply or divide by 10 to convert one size to the other. Each factor of 10 represents one order of magnitude.

For example, the radius of the observable universe is approximately 10^{26} meters and the radius of our solar system is approximately 10^{12} meters. To compare the radius of the observable universe to the radius of our solar system, calculate the ratio

$$\frac{\text{radius of the universe}}{\text{radius of our solar system}} \approx \frac{10^{26} \text{ meters}}{10^{12} \text{ meters}}$$

$$\approx 10^{26-12}$$

$$\approx 10^{14}$$

Orders of Magnitude

The radius of the universe is roughly 10^{14} times larger than the radius of the solar system; that is, we would have to multiply the radius of our solar system by 10 fourteen times in order to obtain the radius of the universe. Since each factor of 10 is counted as a single order of magnitude, the radius of the universe is *fourteen orders of magnitude larger* than the radius of our solar system. Equivalently, we could say that the radius of our solar system is *fourteen orders of magnitude smaller* than the radius of the universe.

When something is one order of magnitude larger than a *reference object,* it is 10 times larger. You *multiply* the *reference size* by 10 to get the other size. If the object is two orders of magnitude larger, it is 100 or 10^2 times larger, so you would multiply the reference size by 100. If it is one order of magnitude smaller, it is 10 times smaller, so you would *divide* the reference size by 10. Two orders of magnitude smaller means the reference size is divided by 100 or 10^2.

EXAMPLE 1

The radius of the sun (10^9 meters) is how much larger than the radius of a hydrogen atom (10^{-11} meter)?

SOLUTION

$$\frac{\text{radius of sun}}{\text{radius of the hydrogen atom}} \approx \frac{10^9 \text{ meters}}{10^{-11} \text{ meters}}$$

$$\approx 10^{9-(-11)}$$

$$\approx 10^{20}$$

So the radius of the sun is 10^{20} times, or twenty orders of magnitude, larger than the radius of the hydrogen atom.

EXAMPLE 2

Compare the length of an unwound DNA strand (10^{-2} meter) with the size of a living cell (radius of 10^{-5} meter).

SOLUTION

$$\frac{\text{length of DNA strand}}{\text{radius of the living cell}} \approx \frac{10^{-2} \text{ meter}}{10^{-5} \text{ meter}}$$

$$\approx 10^{-2-(-5)}$$

$$\approx 10^{-2+5}$$

$$\approx 10^3$$

Surprisingly enough, the average width of a living cell is approximately three orders of magnitude *smaller* than one of the single strands of DNA it contains, if the DNA is uncoiled and measured lengthwise.

The Richter Scale

The reading "Earthquake Magnitude Determination" describes how earthquake tremors are measured.

The *Richter scale,* designed by the American Charles Richter in 1935, allows us to compare the magnitudes of earthquakes throughout the world. The Richter scale measures the maximum ground movement (tremors) as recorded on an instrument called a seismograph. Earthquakes vary widely in severity, so Richter designed the scale to measure order-of-magnitude differences. The scale ranges from less than 1 to over 8. Each increase of one unit on the Richter scale represents an increase of ten times, or one order of magnitude, in the maximum tremor size of the earthquake. So an increase from 2.5 to 3.5 indicates a 10-fold increase in maximum tremor size. An increase of two units from 2.5 to 4.5 indicates an increase in maximum tremor size by a factor of 10^2 or 100.

Description of the Richter Scale

Richter Scale Magnitude	Description
2.5	Generally not felt, but recorded on seismographs
3.5	Felt by many people locally
4.5	Felt by all locally; slight local damage may occur
6	Considerable damage in ordinary buildings; a destructive earthquake
7	"Major" earthquake; most masonry and frame structures destroyed; ground badly cracked
8 and above	"Great" earthquake; a few per decade worldwide; total or almost total destruction; bridges collapse, major openings in ground, tremors visible

Table 4.6

Table 4.6 contains some typical values on the Richter scale along with a description of how humans near the center (called the *epicenter*) of an earthquake perceive its effects. There is no theoretical upper limit on the Richter scale. The U.S. Geological Survey reports that the largest measured earthquake in the United States was in Prince William Sound, Alaska, in 1964 (magnitude 9.2), and the largest in the world was in Chile in 1960 (magnitude 9.5).[5]

Graphing Numbers of Widely Differing Sizes: Log Scales

Exploration 4.1 asks you to construct a graph using logarithmic scales on both axes.

If the sizes of various objects in our solar system are plotted on a standard linear axis, we get the uninformative picture shown in Figure 4.1. The largest value stands alone, and all the others are so small when measured in terameters that they all appear to be zero. When objects of widely different orders of magnitude are compared on a linear scale, the effect is similar to pointing out an ant in a picture of a baseball stadium.

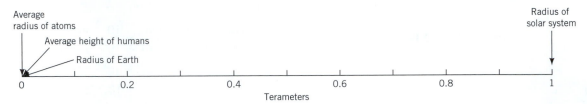

Figure 4.1 Sizes of various objects in the universe on a linear scale.
(*Note:* One terameter $= 10^{12}$ meters.)

A more effective way of plotting sizes with different orders of magnitude is to use an axis that has powers of 10 evenly spaced along it. This is called a *logarithmic* or *log scale*. The plot of the previous data graphed on a logarithmic scale is much more informative (see Figure 4.2).

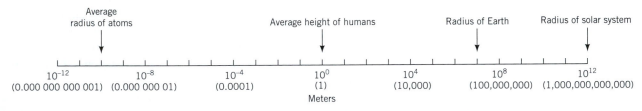

Figure 4.2 Sizes of various objects in the universe on an order-of-magnitude (logarithmic) scale.

Reading Log Scales

Graphing sizes on a log scale can be very useful, but we need to read the scales carefully. When we use a linear scale, each move of one unit to the right is equivalent to *adding* one unit to the number, and each move of k units to the right is equivalent to *adding* k units to the number (Figure 4.3).

---

[5]See the National Earthquake Information Center website at *http://earthquake.usgs.gov/eqcenter/historic_eqs.php.*

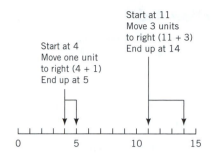

Figure 4.3 Linear scale.

When we use a log scale (see Figure 4.4), we need to remember that one unit of length now represents a change of one order of magnitude. Moving one unit to the right is equivalent to *multiplying by 10*. So moving from 10^4 to 10^5 is equivalent to multiplying 10^4 by 10. Moving three units to the right is equivalent to *multiplying* the starting number by 10^3, or 1000. In effect, a linear scale is an "additive" scale and a logarithmic scale is a "multiplicative" scale.

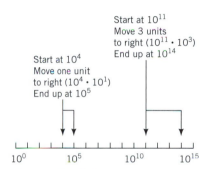

Figure 4.4 Order-of-magnitude (logarithmic) scale.

Algebra Aerobics 4.6

1. In 1987 Los Angeles had an earthquake that measured 5.9 on the Richter scale. In 1988 Armenia had an earthquake that measured 6.9 on the Richter scale. Compare the sizes of the two earthquakes using orders of magnitude.

2. On July 15, 2003, Little Rock, Arkansas, had an earthquake that measured 6.5 on the Richter scale. Compare the size of this earthquake to the largest ever recorded, 9.5 in Chile in 1960.

3. If my salary is $100,000 per year and you make an order of magnitude more, what is your salary? If Henry makes two orders of magnitude less money than I do, what is his salary?

4. For each of the following pairs, determine the order-of-magnitude difference:

   a. The radius of the sun (10^9 meters) and the radius of the Milky Way (10^{21} meters)

   b. The radius of a hydrogen atom (10^{-11} meter) and the radius of a proton (10^{-15} meter)

5. Joe wants to move from Wyoming to California, but he has been advised that houses in California cost an order of magnitude more than houses in Wyoming.

   a. If Joe's house in Wyoming is worth $400,000, how much would a similar house cost in California?

   b. If a house in California sells for $650,000, how much would it cost in Wyoming?

6. How many orders of magnitude greater is a kilometer than a meter? Than a millimeter?

7. By rounding the number to the nearest power of 10, find the approximate location of each of the following on the logarithmic scale in Figure 4.2 on page 244.

   a. The radius of the sun, at approximately 1 billion meters

   b. The radius of a virus, at 0.000 000 7 meter

   c. An object whose radius is two orders of magnitude smaller than that of Earth

Exercises for Section 4.6

1. What is the order-of-magnitude difference between the following units? (Refer to table on inside back cover.)

   a. A millimeter and a gigameter

   b. A second and a day

   c. A square centimeter and an acre (1 acre $= 43,560$ ft^2)

   d. A microfarad and a picofarad

2. Fill in the blanks to make each of the following statements true.

   a. Attaching the prefix "micro" to a unit _____ the size by _____ orders of magnitude.

   b. Attaching the prefix "kilo" to a unit _____ the size by _____ orders of magnitude.

   c. Scientists and engineers have designated prefix multipliers from septillionths (10^{-24}) to septillions (10^{24}), a span of _____ orders of magnitude.

3. Compare the following numbers using orders of magnitude.

   a. 5.261 and 52.61   c. $5.261 \cdot 10^6$ and 526.1

   b. 5261 and 5.261

4. An ant is roughly 10^{-3} meter in length and the average human roughly one meter. How many times longer is a human than an ant?

5. Refer to the chart in Exploration 4.1.

   a. How many orders of magnitude larger is the Milky Way than the first living organism on Earth?

   b. How many orders of magnitude older is the Pleiades (a cluster of stars) than the first *Homo sapiens*?

6. Water boils (changes from a liquid to a gas) at 373 kelvins. The temperature of the core of the sun is 20 million kelvins. By how many orders of magnitude is the sun's core hotter than the boiling temperature of water?

7. An electron weighs about 10^{-27} gram, and a raindrop weighs about 10^{-3} gram. How many times heavier is a raindrop than an electron? How many times lighter is an electron than a raindrop? What is the order-of-magnitude difference?

8. On Nov. 20, 2001, *The New York Times* reported that FBI scientists had found a sealed plastic bag with a letter addressed to Senator Patrick Leahy that was highly contaminated with anthrax. The article said that a sample taken from the bag "showed the presence of 23,000 anthrax spores. This, the scientists said, was roughly three orders of magnitude more spores than found in samples from any of the other 600 bags of mail the bureau examined."

   Estimate the number of spores found in any of the 600 other bags of mail.

9. In the December 1999 issue of the journal *Science,* two Harvard scientists describe a pair of "nanotweezers" they created that are capable of manipulating objects as small as one-50,000th of an inch in width. The scientists used the tweezers to grab and pull clusters of polystyrene molecules, which are of the same size as structures inside cells. A future use of these nanotweezers may be to grab and move components of biological cells.

   a. Express one-50,000th of an inch in scientific notation.

   b. Express the size of objects the tweezers are able to manipulate in meters.

   c. The prefix "nano" refers to nine subdivisions by 10, or a multiple of 10^{-9}. So a nanometer would be 10^{-9} meters. Is the name for the tweezers given by the inventors appropriate?

   d. If not, how many orders of magnitude larger or smaller would the tweezers' ability to manipulate small objects have to be in order to grasp things of nanometer size?

10. Determine the order-of-magnitude difference in the sizes of the radii for:

    a. The solar system (10^{12} meters) compared with Earth (10^7 meters)

    b. Protons (10^{-15} meter) compared with the Milky Way (10^{21} meters)

    c. Atoms (10^{-10} meter) compared with neutrons (10^{-15} meter)

11. To compare the sizes of different objects, we need to use the same unit of measure.

    a. Convert each of these to meters:

       i. The radius of the moon is approximately 1,922,400 yards.

       ii. The radius of Earth is approximately 6400 km.

       iii. The radius of the sun is approximately 432,000 miles.

    b. Determine the order-of-magnitude difference between:

       i. The surface areas of the moon and Earth

       ii. The volumes of the sun and the moon

12. The pH scale measures the hydrogen ion concentration in a liquid, which determines whether the substance is acidic or alkaline. A strong acid solution has a hydrogen ion concentration of 10^{-1} M. One M equals $6.02 \cdot 10^{23}$ particles, such as atoms, ions, molecules, etc., per liter, or 1 mole per liter.[6] A strong alkali solution has a hydrogen ion concentration of 10^{-14} M. Pure water, with a concentration of 10^{-7} M, is neutral. The pH value is the power without the minus sign, so pure water has a pH of 7, acidic substances have a pH less than 7, and alkaline substances have a pH greater than 7.

    a. Tap water has a pH of 5.8. Before the industrial age, rain water commonly had a pH of about 5. With the spread of modern industry, rain in the northeastern United States and parts of Europe now has a pH of about 4, and in extreme cases the pH is about 2. Lemon juice has a pH of 2.1. If acid rain with a pH of 3 is discovered in an area, how much more acidic is it than preindustrial rain?

[6]You may recall from Algebra Aerobics 4.1 that $6.02 \cdot 10^{23}$ is called Avogadro's number. A mole of a substance is defined as Avogadro's number of particles of that substance. M is called a molar unit.

b. Blood has a pH of 7.4; wine has a pH of about 3.4. By how many orders of magnitude is wine more acidic than blood?

13. Which is an additive scale? Explain why. Which is a multiplicative or logarithmic scale? Explain why.

a.
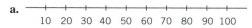
    10 20 30 40 50 60 70 80 90 100

b.
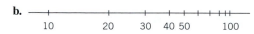
    10   20   30 40 50   100

14. Graph the following on a power-of-10 (logarithmic) scale. (See sample log scale at the end of the exercises.)

   a. 1 meter
   b. 10 meters
   c. 1 hectometer
   d. 1000 kilometers
   e. 10 gigameters

15. (Refer to the chart in Exploration 4.1.) Plot on the logarithmic scale in Figure 4.2 an object whose radius is:

   a. Five orders of magnitude larger than the radius of the first atoms
   b. Twenty orders of magnitude smaller than the radius of the sun

16. **a.** Read the chapter entitled "The Cosmic Calendar" from Carl Sagan's book *The Dragons of Eden*.

   b. Carl Sagan tried to give meaning to the cosmic chronology by imagining the almost 15 billion–year lifetime of the universe compressed into the span of one calendar year. To get a more personal perspective, consider your date of birth as the time at which the Big Bang took place. Map the following five cosmic events onto your own life span:

   i. The Big Bang
   ii. Creation of Earth
   iii. First life on Earth
   iv. First *Homo sapiens*
   v. American Revolution

   Once you have done the necessary mathematical calculations and placed your results on either a chart or a timeline, form a topic sentence and write a playful paragraph about what you were supposedly doing when these cosmic events took place. Hand in your calculations along with your writing.

17. Graph the following on a power-of-10 (logarithmic) scale. (See sample log scale at the end of the exercises.)

   a. 1 watt
   b. 10 kilowatts
   c. 100 billion kilowatts
   d. 1000 terawatts

18. Radio waves, sent from a broadcast station and picked up by the antenna of your radio, are a form of electromagnetic (EM) radiation, as are microwaves, X-rays, and visible, infrared, and ultraviolet light. They all travel at the speed of light. Electromagnetic radiation can be thought of as oscillations like the vibrating strings of a violin or guitar or like ocean swells that have crests and troughs. The distance between the crest or peak of one wave and the next is called the wavelength. The number of times a wave crests per minute, or per second for fast-oscillating waves, is called its frequency. Wavelength and frequency are inversely proportional: the longer the wavelength, the lower the frequency, and vice versa—the faster the oscillation, the shorter the wavelength. For radio waves and other EM, the number of oscillations per second of a wave is measured in hertz, after the German scientist who first demonstrated that electrical waves could transmit information across space. One cycle or oscillation per second equals 1 hertz (Hz).

   For the following exercise you may want to find an old radio or look on a stereo tuner at the AM and FM radio bands. You may see the notation kHz beside the AM band and MHz beside the FM band. AM radio waves oscillate at frequencies measured in the kilohertz range, and FM radio waves oscillate at frequencies measured in the megahertz range.

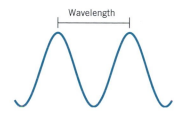

   a. The Boston FM rock station WBCN transmits at 104.1 MHz. Write its frequency in hertz using scientific notation.

   b. The Boston AM radio news station WBZ broadcasts at 1030 kHz. Write its frequency in hertz using scientific notation.

   The wavelength λ (Greek lambda) in meters and frequency μ (Greek mu) in oscillations per second are related by the formula $\lambda = \dfrac{c}{\mu}$ where c is the speed of light in meters per second.

   c. Estimate the wavelength of the WBCN FM radio transmission.

   d. Estimate the wavelength of the WBZ AM radio transmission.

   e. Compare your answers in parts (c) and (d), using orders of magnitude, with the length of a football field (approximately 100 meters).

$10^0 \quad 10^1 \quad 10^2 \quad 10^3 \quad 10^4 \quad 10^5 \quad 10^6 \quad 10^7 \quad 10^8 \quad 10^9 \quad 10^{10} \quad 10^{11} \quad 10^{12} \quad 10^{13} \quad 10^{14} \quad 10^{15}$

Sample log Scale

4.7 Logarithms Base 10

In Section 4.6 we used a logarithmic scale to graph numbers of widely disparate sizes. We labeled the axis with powers of 10, so it was easy to plot numbers such as $1000 = 10^3$ or $100{,}000 = 10^5$ that are integer powers of 10. But how would we plot a number such as $4{,}600{,}000{,}000 = 4.6 \cdot 10^9$, the approximate age of Earth in years? To do that we need to understand logarithms.

Finding the Logarithms of Powers of 10

For handling very large or very small numbers, it is often easier to write the number using powers of 10. For example,

$$100{,}000 = 10^5$$

We say that

$$100{,}000 \text{ equals the base 10 to the fifth power}$$

But we could rephrase this as

$$5 \text{ is the exponent of the base 10 that is needed to produce } 100{,}000$$

The more technical way to say this is

$$5 \text{ is the } logarithm \text{ base 10 of } 100{,}000$$

In symbols we write

$$5 = \log_{10} 100{,}000$$

So the expressions

$$100{,}000 = 10^5 \quad \text{and} \quad 5 = \log_{10} 100{,}000$$

are two ways of saying the same thing. The key point to remember is that a logarithm is an exponent.

Definition of Logarithm

The *logarithm base 10 of x* is the exponent of 10 needed to produce x:

$$\log_{10} x = c \quad \text{means} \quad 10^c = x$$

So to find the logarithm of a number, write it as 10 to some power. The power is the logarithm of the original number.

EXAMPLE 1   Find log (1,000,000,000) without using a calculator.

SOLUTION   Since $\qquad 1{,}000{,}000{,}000 = 10^9$

then $\qquad \log_{10} 1{,}000{,}000{,}000 = 9$

and we say that the logarithm base 10 of 1,000,000,000 is 9. The logarithm of a number tells us the exponent of the number when written as a power of 10. Here the logarithm is 9, so that means that when we write 1,000,000,000 as a power of 10, the exponent is 9.

EXAMPLE 2   Find log 1 without using a calculator.

SOLUTION   Since $\qquad 1 = 10^0$

then $\qquad \log_{10} 1 = 0$

and we say that the logarithm base 10 of 1 is 0. Since logarithms represent exponents, this says that when we write 1 as a power of 10, the exponent is 0.

EXAMPLE 3   How do we calculate the logarithm base 10 of decimals such as 0.000 01?

SOLUTION   Since $\qquad$ $0.000\ 01 = 10^{-5}$

then $\qquad$ $\log_{10} 0.000\ 01 = -5$

and we say that the logarithm base 10 of 0.000 01 is -5.

In the previous example, we found that the log (short for "logarithm") of a number can be negative. This makes sense if we think of logarithms as exponents, since exponents can be any real number. But we cannot take the log of a negative number or zero; that is, $\log_{10} x$ is not defined when $x \leq 0$. Why? If $\log_{10} x = c$, where $x \leq 0$, then $10^c = x$ (a number ≤ 0). But 10 to any power will never produce a number that is negative or zero, so $\log_{10} x$ is not defined if $x \leq 0$.

> $\log_{10} x$ is not defined when $x \leq 0$.

Table 4.7 gives a sample set of values for x and their associated logarithms base 10. To find the logarithm base 10 of x, we write x as a power of 10, and the logarithm is just the exponent.

Most scientific calculators and spreadsheet programs have a LOG function that calculates logarithms base 10. Try using technology to double-check some of the numbers in Table 4.7.

Logarithms of Powers of 10

x	Exponential Notation	$\log_{10} x$
0.0001	10^{-4}	-4
0.001	10^{-3}	-3
0.01	10^{-2}	-2
0.1	10^{-1}	-1
1	10^{0}	0
10	10^{1}	1
100	10^{2}	2
1000	10^{3}	3
10,000	10^{4}	4

Table 4.7

Logarithms base 10 are used frequently in our base 10 number system and are called *common logarithms*. We write $\log_{10} x$ as log x.

Common Logarithms

Logarithms base 10 are called *common logarithms*.

$\qquad$ $\log_{10} x$ is written as log x.

Algebra Aerobics 4.7a

1. Without using a calculator, find the logarithm base 10 of:

   a. 10,000,000   **c.** 10,000   **e.** 1000   **g.** 1
   b. 0.000 000 1   **d.** 0.0001   **f.** 0.001

2. Rewrite the following expressions in an equivalent form using powers of 10:

   a. $\log 100{,}000 = 5$ **c.** $\log 10 = 1$
   b. $\log 0.000\,000\,01 = -8$ **d.** $\log 0.01 = -2$

3. Evaluate without using a calculator. Find a number if its log is:

   a. 3   **c.** 6   **e.** −2
   b. −1   **d.** 0

4. Find c and then rewrite as a logarithm:

   a. $10^c = 1000$ **d.** $10^c = 0.000\,01$
   b. $10^c = 0.001$ **e.** $10^c = 1{,}000{,}000$
   c. $10^c = 100{,}000$ **f.** $10^c = 0.000\,001$

5. Find the value of x that makes the statement true.

   a. $10^{x-3} = 10^2$ **c.** $\log (x - 2) = 1$
   b. $10^{2x-1} = 10^4$ **d.** $\log 5x = -1$

Finding the Logarithm of Any Positive Number

Scientific calculators have a LOG function that will calculate the log of any positive number. However, it's easy to make errors typing in numbers, so it's important not to rely solely on technology-generated answers. To verify that the calculated number is the right order of magnitude, you should estimate the answer without using technology.

E X A M P L E   4   **Estimating, then using technology to calculate logs**

a. Estimate the size of log (2000) and log (0.07).

b. Use a calculator to find the logarithm of 2000 and 0.07.

S O L U T I O N   **a. i.** If we place 2000 between the two closest integer powers of 10, we have

$$1000 < 2000 < 10{,}000$$

Rewriting 1000 and 10,000 as powers of 10 gives

$$10^3 < 2000 < 10^4$$

Taking the log of each term preserves the inequality, so we would expect

$$3 < \log 2000 < 4$$

ii. If we place 0.07 between the two closest integer powers of 10, we have

$$0.01 < 0.07 < 0.10$$

Rewriting 0.01 and 0.10 as powers of 10 gives

$$10^{-2} < 0.07 < 10^{-1}$$

Taking the log of each term preserves the inequality, so we would expect

$$-2 < \log 0.07 < -1$$

b. Using a calculator, we have

i. $\log 2000 \approx 3.301$ **ii.** $\log (0.07) \approx -1.155.$
So our estimates were correct.

E X A M P L E   5   **Calculating logs of very large or small numbers**
Find the logarithm of

a. 3.7 trillion

b. A Planck length of 0.000 000 000 000 000 000 000 000 000 000 000 016 meter

SOLUTION Our strategy in each case is to

- Write the number in scientific notation.
- Then convert the number into a single power of 10.

The resulting exponent is the desired log.

a. In scientific notation 3.7 trillion is $3.7 \cdot 10^{12}$. To convert the entire expression into a single power of 10, we need to first convert the coefficient 3.7 to a power of 10. Using a calculator, we have

$$\log 3.7 \approx 0.568$$

so $$3.7 \approx 10^{0.568}$$

If we substitute for 3.7, $$3.7 \cdot 10^{12} \approx 10^{0.568} \cdot 10^{12}$$
and use rules for exponents, $$= 10^{0.568+12}$$
we have $$= 10^{12.568}$$

So the exponent 12.568 is the desired logarithm.

b. In scientific notation a Planck length is $1.6 \cdot 10^{-35}$ meter. We need to convert 1.6 to a power of 10. Using a calculator, we have $\log(1.6) \approx 0.204$, so $1.6 = 10^{0.204}$. If we

substitute for 1.6 $$1.6 \cdot 10^{-35} \approx 10^{0.204} \cdot 10^{-35}$$
use rules for exponents $$= 10^{0.204-35}$$
and subtract, we get $$= 10^{-34.796}$$

So the exponent -34.796 is the desired log.

So far we have dealt with finding the logarithm of a given number. Logarithms can, of course, occur in expressions involving variables.

EXAMPLE 6 **Finding the number given the log**
Rewrite the following expressions using exponents, and then solve for x without using a calculator.
a. $\log x = 3$ **b.** $\log x = 0$ **c.** $\log x = -2$

SOLUTION **a.** If $\log x = 3$, then $10^3 = x$, so $x = 1000$.
b. If $\log x = 0$, then $10^0 = x$, so $x = 1$.
c. If $\log x = -2$, then $10^{-2} = x$, so $x = 1/10^2 = 1/100 = 0.01$.

EXAMPLE 7 Rewrite the following expressions using logarithms and then solve for x using a calculator. Round off to three decimal places.
a. $10^x = 11$ **b.** $10^x = 0.5$ **c.** $10^x = 0$

SOLUTION **a.** If $10^x = 11$, then $\log 11 = x$. Using a calculator gives $x \approx 1.041$.
b. If $10^x = 0.5$, then $\log 0.5 = x$. Using a calculator gives $x \approx -0.301$.
c. There is no power of 10 that equals 0. Hence there is no solution for x.

Plotting Numbers on a Logarithmic Scale

We are finally prepared to answer the question posed at the very beginning of this section: How can we plot on a logarithmic (or order-of-magnitude) scale a number such as 4.6 million years, the estimated age of Earth?

> **A Strategy for Plotting Numbers on a Logarithmic Scale**
> - Write the number in scientific notation.
> - Then convert the number into a single power of 10.
> - Use the exponent to help plot the number.

In scientific notation the age of Earth equals 4.6 billion = 4,600,000,000 = $4.6 \cdot 10^9$ years. To plot this number on a logarithmic scale, we need to convert 4.6 into a power of 10. Using a calculator, we have log 4.6 ≈ 0.663 , so $4.6 \approx 10^{0.663}$. If we

substitute for 4.6	$4.6 \cdot 10^9 \approx 10^{0.663} \cdot 10^9$
and use rules of exponents	$= 10^{0.663+9}$
we have	$= 10^{9.663}$

The power of 10 seems reasonable since $10^9 < 4.6 \cdot 10^9 < 10^{10}$.

Having converted our original number $4.6 \cdot 10^9$ into $10^{9.663}$, we can plot it on an order-of-magnitude graph between 10^9 and 10^{10} (see Figure 4.5).

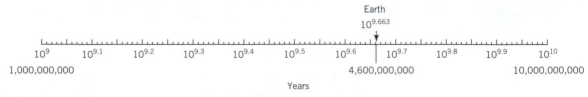

Figure 4.5 Age of Earth plotted on an order-of-magnitude (or logarithmic) scale.

EXAMPLE 8 Plot the numbers 100, 200, 300, 400, 500, 600, 700, 800, 900, and 1000 on a logarithmic scale.

SOLUTION Using our log plotting strategy, we first convert each number to a single power of 10. We have

$$100 = 10^2 \qquad\qquad 600 \approx 10^{2.778}$$
$$200 \approx 10^{2.301} \qquad\quad 700 \approx 10^{2.845}$$
$$300 \approx 10^{2.477} \qquad\quad 800 \approx 10^{2.903}$$
$$400 \approx 10^{2.602} \qquad\quad 900 \approx 10^{2.954}$$
$$500 \approx 10^{2.699} \qquad 1000 = 10^3$$

We can now use the exponents of each power of 10 to plot the numbers directly onto a logarithmic scale (see Figure 4.6).

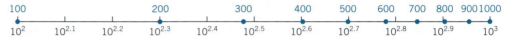

Figure 4.6 A logarithmic plot of the numbers 100, 200, 300, . . . , 1000.

Note that on the logarithmic scale in Figure 4.6 the point halfway between 10^2 and 10^3 is at $10^{2.5} = 316$.

When are numbers evenly spaced on a logarithmic scale?

On a linear (additive) scale the numbers 100, 200, . . . , 1000 would be evenly spaced, since you *add* a constant amount to move from one number to the next. On a logarithmic (multiplicative) scale, numbers that are evenly spaced are generated by *multiplying* by a constant amount to get from one number to the next. For example, the integer powers of 10 are all evenly spaced on a log plot since you multiply each number by 10 to get the next number in the sequence. Similarly, the numbers 100, 200, 400, and 800 are evenly spaced in Figure 4.6 since you multiply by the constant 2 to get from one number in the sequence to the next. The sequence 100, 200, 300, . . . , 1000 is not evenly spaced, since there is not a constant factor that you could multiply one number by to get to the next. In this last sequence, since the multiplication factor needed to move from one number to the next decreases as the numbers approach 1000, the distance between points decreases.

Labeling using only the exponent

Instead of labeling the axis using powers of 10, we can label it using just the exponents of 10 as in Figure 4.7. Remember that exponents are logarithms, which is why we call the scale logarithmic.

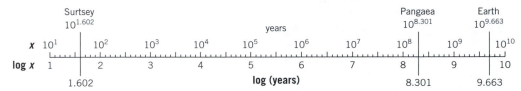

Figure 4.7 The age of Surtsey, Pangaea, and Earth plotted using an order-of-magnitude or logarithmic scale.

EXAMPLE 9

Add on to the logarithmic plot in Figure 4.7 the following numbers:

a. 200 million, the number of years since all of Earth's continents collided to form one giant land mass called Pangaea

b. 40, the number of years since the volcanic island of Surtsey, Earth's newest land mass, emerged near Iceland

SOLUTION

a. In scientific notation 200 million $= 200{,}000{,}000 = 2.0 \cdot 10^8$. Using a calculator, we have $\log 2.0 \approx 0.301$, so $2.0 \approx 10^{0.301}$. Hence $2.0 \cdot 10^8 \approx 10^{0.301} \cdot 10^8 = 10^{8.301}$ is the age of Pangaea in a form easily plotted on a logarithmic scale.

b. In scientific notation $40 = 4.0 \cdot 10^1$. Using a calculator, we have $\log 4.0 \approx 0.602$, so $4.0 \approx 10^{0.602}$. Therefore $4.0 \cdot 10^1 \approx 10^{0.602} \cdot 10^1 = 10^{1.602}$ is the age of Surtsey in a form easily plotted on a log scale.

The two numbers are plotted in Figure 4.7.

Algebra Aerobics 4.7b

Most of these problems require a calculator that can evaluate logs.

1. Use a calculator to estimate each of the following:
   a. $\log 3$ b. $\log 6$ c. $\log 6.37$

2. Use the answers from Problem 1 to estimate values for:
   a. $\log 3{,}000{,}000$ b. $\log 0.006$
   Then use a calculator to check your answers.

3. Write each of the following as a power of 10:
   a. 0.000 000 7 m (the radius of a virus)
   b. 780,000,000 km (the mean distance from our sun to Jupiter)
   c. 0.0042
   d. 5,400,000,000

4. Rewrite the following equations using exponents instead of logarithms. Estimate the solution for x.

Check your estimate with a calculator. Round the value of x to the nearest integer.
   a. $\log x = 4.125$
   b. $\log x = 5.125$
   c. $\log x = 2.125$

5. Rewrite the following equations using logs instead of exponents. Estimate a solution for x and check your estimate with a calculator. Round the value of x to three decimal places.
   a. $10^x = 250$ c. $10^x = 0.075$
   b. $10^x = 250{,}000$ d. $10^x = 0.000\ 075$

6. Write each number as a power of 10 and then plot them all on the logarithmic scale below.
   a. 57   c. 25,000
   b. 182   d. 7,200,000,000

Exercises for Section 4.7

Many of the problems in this section require the use of a calculator that can evaluate logs.

1. Rewrite in an equivalent form using logarithms:
 a. $10^4 = 10,000$ **c.** $10^0 = 1$
 b. $10^{-2} = 0.01$ **d.** $10^{-5} = 0.00001$

2. Use your calculator to evaluate to two decimal places:
 a. $10^{0.4}$ **c.** $10^{0.6}$ **e.** $10^{0.8}$
 b. $10^{0.5}$ **d.** $10^{0.7}$ **f.** $10^{0.9}$

3. Express the number 375 in the form 10^x.

4. Estimate the value of each of the following:
 a. $\log 4000$ **b.** $\log 5,000,000$ **c.** $\log 0.0008$

5. Rewrite the following statements using logs:
 a. $10^2 = 100$ **b.** $10^7 = 10,000,000$ **c.** $10^{-3} = 0.001$
 Rewrite the following statements using exponents:
 d. $\log 10 = 1$ **e.** $\log 10,000 = 4$ **f.** $\log 0.0001 = -4$

6. Evaluate the following without a calculator.
 a. Find the following values:
   i. $\log 100$ **ii.** $\log 1000$ **iii.** $\log 10,000,000$
   What is happening to the values of $\log x$ as x gets larger?
 b. Find the following values:
   i. $\log 0.1$ **ii.** $\log 0.001$ **iii.** $\log 0.00001$
   What is happening to the values of $\log x$ as x gets closer to 0?
 c. What is $\log 0$?
 d. What is $\log(-10)$? What do you know about $\log x$ when x is any negative number?

7. Rewrite the following equations using exponents instead of logs. Estimate a solution for x and then check your estimate with a calculator. Round the value of x to the nearest integer.
 a. $\log x = 1.255$ **c.** $\log x = 4.23$
 b. $\log x = 3.51$ **d.** $\log x = 7.65$

8. Rewrite the following equations using exponents instead of logs. Estimate a solution for x and then check your estimate with a calculator. Round the value of x to the nearest integer.
 a. $\log x = 1.079$ **c.** $\log x = 2.1$
 b. $\log x = 0.699$ **d.** $\log x = 3.1$

9. Rewrite the following equations using logs instead of exponents. Estimate a solution for x and then check your estimate with a calculator. Round the value of x to three decimal places.
 a. $10^x = 12,500$ **c.** $10^x = 597$
 b. $10^x = 3,526,000$ **d.** $10^x = 756,821$

10. Rewrite the following equations using logs instead of exponents. Estimate a solution for x and then check your estimate with a calculator. Round the value of x to three decimal places.
 a. $10^x = 153$ **c.** $10^x = 0.125$
 b. $10^x = 153,000$ **d.** $10^x = 0.00125$

11. Solve for x. (*Hint:* Rewrite each expression so that you can use a calculator to solve for x.)
 a. $\log x = 0.82$ **c.** $\log x = 0.33$
 b. $10^x = 0.012$ **d.** $10^x = 0.25$

12. Without using a calculator, show how you can solve for x.
 a. $10^{x-2} = 100$ **c.** $10^{2x-3} = 1000$
 b. $\log(x - 4) = 1$ **d.** $\log(6 - x) = -2$

13. Without using a calculator show how you can solve for x.
 a. $10^{x-5} = 1000$ **c.** $10^{3x-1} = 0.0001$
 b. $\log(2x + 10) = 2$ **d.** $\log(500 - 25x) = 3$

14. Find the value of x that makes the equation true.
 a. $\log x = -2$ **b.** $\log x = -3$ **c.** $\log x = -4$

15. Without using a calculator, for each number in the form $\log x$, find some integers a and b such that $a < \log x < b$. Justify your answer. Then verify your answers with a calculator.
 a. $\log 11$ **b.** $\log 12,000$ **c.** $\log 0.125$

16. Use a calculator to determine the following logs. Double-check each answer by writing down the equivalent expression using exponents, and then verify this equivalence using a calculator.
 a. $\log 15$ **b.** $\log 15,000$ **c.** $\log 1.5$

17. On a logarithmic scale, what would correspond to moving over to the right:
 a. 0.001 unit   **b.** $\frac{1}{2}$ unit   **c.** 2 units   **d.** 10 units

18. The difference in the noise levels of two sounds is measured in decibels, where decibels $= 10 \log \left(\dfrac{I_2}{I_1} \right)$ and I_1 and I_2 are the intensities of the two sounds. Compare noise levels when $I_1 = 10^{-15}$ watts/cm^2 and $I_2 = 10^{-8}$ watts/cm^2.

19. The concentration of hydrogen ions in a water solution typically ranges from 10 M to 10^{-15} M. (One M equals $6.02 \cdot 10^{23}$ particles, such as atoms, ions, molecules, etc., per liter or 1 mole per liter.) Because of this wide range, chemists use a logarithmic scale, called the pH scale, to measure the concentration (see Exercise 12 of Section 4.6). The formal definition of pH is pH $= -\log[\text{H}^+]$, where $[\text{H}^+]$ denotes the concentration of hydrogen ions. Chemists use the symbol H^+ for hydrogen ions, and the brackets [ ] mean "the concentration of."
 a. Pure water at 25°C has a hydrogen ion concentration of 10^{-7} M. What is the pH?

$$10^0 \quad 10^1 \quad 10^2 \quad 10^3 \quad 10^4 \quad 10^5 \quad 10^6 \quad 10^7 \quad 10^8 \quad 10^9 \quad 10^{10} \quad 10^{11} \quad 10^{12} \quad 10^{13} \quad 10^{14} \quad 10^{15}$$

Sample log Scale

b. In orange juice, $[H^+] \approx 1.4 \cdot 10^{-3}$ M. What is the pH?

c. Household ammonia has a pH of about 11.5. What is its $[H^+]$?

d. Does a higher pH indicate a lower or a higher concentration of hydrogen ions?

e. A solution with a pH > 7 is called basic, one with a pH $= 7$ is called neutral, and one with a pH < 7 is called acidic. Identify pure water, orange juice, and household ammonia as either acidic, neutral, or basic. Then plot their positions on the accompanying scale, which shows both the pH and the hydrogen ion concentration.

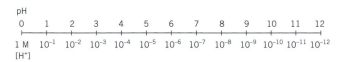

20. a. Place the number 50 on the *additive* scale below.

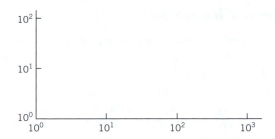

22. Change each number to a power of 10, then plot the numbers on a power-of-10 scale. (See sample log scale on page 254.)

    a. 125   **b.** 372   **c.** 694   **d.** 840

23. Compare the times listed below by plotting them on the same order-of-magnitude scale. (*Hint:* Start by converting all the times to seconds.)

    a. The time of one heartbeat (1 second)

    b. Time to walk from one class to another (10 minutes)

    c. Time to drive across the country (7 days)

    d. One year (365 days)

    e. Time for light to travel to the center of the Milky Way (38,000 years)

    f. Time for light to travel to Andromeda, the nearest large galaxy (2.2 million years)

b. Place the number 50 on the *multiplicative* scale below.

21. The coordinate system below uses multiplicative or log scales on both axes. Position the point whose coordinates are (708, 25).

CHAPTER SUMMARY

Powers of 10

If n is a positive integer, we define

$$10^n = \underbrace{10 \cdot 10 \cdot 10 \cdot \cdots \cdot 10}_{n \text{ factors}}$$

$$10^0 = 1$$

$$10^{-n} = \frac{1}{10^n}$$

Scientific Notation

A number is in scientific notation if it is in the form

$$N \cdot 10^n$$

where N is called the *coefficient*, $1 \le |N| < 10$, and n is an integer.

Example: In scientific notation 67,000,000 is written as $6.7 \cdot 10^7$ and $-0.000\ 000\ 000\ 008\ 1$ is written as $-8.1 \cdot 10^{-12}$.

Powers of a

In the expression a^n, a is called the *base* and n is called the *exponent* or *power*.

If a is nonzero real and n is a positive integer, then

$$a^n = \underbrace{a \cdot a \cdot a \cdots a}_{n \text{ factors}}$$

$$a^0 = 1$$

$$a^{-n} = \frac{1}{a^n}$$

If m and n are positive integers and the base, a, is restricted to values for which the power is defined, then

$$a^{1/2} = \sqrt{a}$$
$$a^{1/n} = \sqrt[n]{a}$$
$$a^{m/n} = (a^m)^{1/n} = (a^{1/n})^m$$
$$= \sqrt[n]{a^m} = (\sqrt[n]{a})^m$$

Rules of Exponents

If a and b are nonzero, then

1. $a^m \cdot a^n = a^{(m+n)}$ **4.** $(ab)^n = a^n b^n$

2. $\dfrac{a^n}{a^m} = a^{(n-m)}$ **5.** $\left(\dfrac{a}{b}\right)^n = \dfrac{a^n}{b^n}$

3. $(a^m)^n = a^{(m \cdot n)}$

Orders of Magnitude

We use *orders of magnitude* when we compare objects of widely different sizes. Each *factor* of 10 is counted as a single order of magnitude.

Example: The radius of the universe is 10^{14} times or fourteen orders of magnitude larger than the radius of the solar system. And vice versa: The radius of the solar system is fourteen orders of magnitude smaller than the radius of the universe.

Logarithms

The *logarithm base 10 of x* is the exponent of 10 needed to produce *x*. So

$$\log_{10} x = c \qquad \text{means} \qquad 10^c = x$$

We say that c is the logarithm base 10 of x.

Example: $\log_{10} 6{,}370{,}000 \approx 6.804$ means that $10^{6.804} \approx 6{,}370{,}000$.

Logarithms base 10 are called *common logarithms*. We usually write $\log_{10} x$ as $\log x$. When $x \leq 0$, $\log x$ is not defined.

Plotting Numbers on a Logarithmic Scale

Logarithmic or powers-of-10 scales are used to graph objects of widely differing sizes. We can plot a number on a log scale by converting the number to a power of 10.

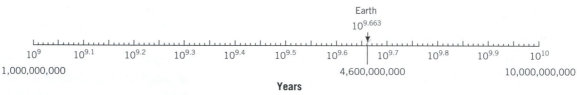

Age of Earth plotted on a logarithmic scale.

CHECK YOUR UNDERSTANDING

I. Are the statements in Problems 1–26 true or false? Give an explanation for your answer.

1. A distance of 10 miles is longer than a distance of 10 kilometers.

2. There are 39 centimeters in 1 inch.

3. 10^{15} is 10 followed by fifteen zeros.

4. $10^0 = 0$.

5. $\dfrac{1}{10^{-m}} = 10^m$.

6. $-0.000\,005\,62 = -5.62 \cdot 10^{-6}$.

7. $15 \cdot 10^4$ is correct scientific notation for the number 150,000.

8. The age of the universe $(1.37 \cdot 10^{10}$ years$)$ is about three times the age of Earth $(4.6 \cdot 10^9$ years$)$.

9. In July 2004, the population of the world (about 6,377,642,000) was approximately three orders of magnitude larger than the population of the United States (about 293,028,000).

10. $-8^2 = (-8)^2$.

11. $\left(\dfrac{5}{3}\right)^{-3} = \dfrac{-15}{-9}$.

12. $10^2 + 10^1 + 10^5 = 10^{2+1+5} = 10^8$.

13. To convert a distance D in kilometers to miles, you could multiply D by $\dfrac{1 \text{ km}}{0.62 \text{ mile}}$.

14. The units of $300 \dfrac{\text{km}}{\text{hr}} \cdot \dfrac{1 \text{ hr}}{60 \text{ min}} \cdot \dfrac{1 \text{ min}}{60 \text{ sec}} \cdot \dfrac{10^3 \text{ m}}{1 \text{ km}}$ are meters per second.

15. $-9 < -\sqrt{75} < -8$.

16. $8^{1/2} = 8^{0.5}$.

17. $\log_{10} 0.0001 = 10^{-4}$.

18. $\log_{10} 1821$ is not defined since 1821 is not a power of 10.

19. $(81)^{1/2} = \pm 9$ because $(9)^2 = 81$ and $(-9)^2 = 81$.

20. $\log 0 = 1$.

21. $\log(-3) = -\log(3)$.

22. $-4 < \log 0.00015 < -3$.

23. $\log 0.143 \approx -0.845$ means that $10^{-0.845} \approx 0.143$.

24. If $P > 0$, $\log P = Q$ means that $10^P = Q$.

25. The following figure illustrates the number 7,500,000 plotted correctly on a logarithmic scale.

26. If $10^x = 36$, then $x \approx 1.556$.

II. In Problems 27–32, give examples with the specified properties.

   27. Populations of two cities A and B, where the population of city A is two orders of magnitude larger than that of city B.

   28. A number x such that $\log x$ lies between 8 and 9.

   29. A number x such that $\log x$ is a negative number.

   30. A positive number b such that $\sqrt{b} > b$.

   31. A non-zero number b such that $b^m = b^n$ for any numbers m and n.

   32. A number b such that $|b| = -b$.

III. Are the statements in Problems 33–42 true or false? If a statement is true, explain how you know. If a statement is false, give a counterexample.

33. If one quantity is four orders of magnitude larger than a second quantity, it is four times as large as the second quantity.

34. $|c| = c$ for any real number c.

35. $\log x$ is defined only for numbers $x > 0$.

36. Raising a number to the $\frac{1}{3}$ power is the same as taking the cube root of that number.

37. $b^m \cdot b^m = b^{m^2}$

38. $(b^p)^q = b^{p+q}$

39. $(b + c)^m = b^m + c^m$

40. $(-b)^q = b^q$

41. $b^m \cdot c^n = (b \cdot c)^{m+n}$

42. If n is odd, $\sqrt[n]{b}$ can be positive, negative, or zero depending on the value of b.

CHAPTER 4 REVIEW: PUTTING IT ALL TOGETHER

1. Evaluate each of the following without a calculator.

   a. $4.2 \cdot 10^3$ **b.** $(-5)2^3$ **c.** -4^2 **d.** $100^{-1/2}$ **e.** $\dfrac{3^5}{3^2}$

2. Use the rules of exponents to simplify the following. Express your answer with positive exponents.

   a. x^4x^3 **c.** $(-2xy^2)^3$ **e.** $(x^{-3}y)(x^2y^{-1/2})$

   b. $\dfrac{10x^2y^4}{5xy^3}$ **d.** $(x^{-1/2})^2$

3. For what integer values of x will the following statement be true?

$(-10)^x = -10^x$

4. a. Show with an example why the following is not a true statement for all values of x: $x^3 + x^5 = x^8$.

   b. For what value of x is the above statement true?

5. Elephant seals can weigh as much as 5000 lb (for males) and 2000 lb (for females). On land, these seals can travel short distances quite quickly, as much as 20 feet in 3 seconds. How many miles per hour is this?

6. An NFL regulation playing field for football is 120 yd (110 m) long including the end zones, and 53 yd 1 ft (48.8 m) wide. An acre is 4840 square yards, and 1 yard = 3 feet.

   a. Which is larger, a football field (including the end zones) or an acre? By how much?

   b. If you bought a house on a square lot that measured half an acre, what would the dimensions of the lot be in feet?

7. In 2006, Tiger Woods was the highest paid athlete in the world (taking into account on and off the field earnings), making $11.9 million in salary and $100 million in endorsements for a total of $111.9 million. By what order of magnitude is his salary greater than that of a minimum-wage worker in the same state making $6.40/hr working 40 hours/week for 52 weeks/year? How many years would the minimum-wage worker have to work to earn what Tiger Woods made in 1 year?

8. The Yangtze River (China) is 6380 km long. The Colorado River is 1400 miles long.

   a. Which river is longer?

   b. Compare the lengths of these rivers using orders of magnitude.

9. Use the accompanying table to answer the following questions.

Country	Area
Russia	17,075,200 km^2
Chile	290,125 mi^2
Canada	3,830,840 mi^2
South Africa	1,184,825 km^2
Norway	323,895 km^2
Monaco	0.5 mi^2

   a. Which country has the largest area? The smallest?

   b. Using scientific notation, arrange the countries from largest area to smallest area.

   c. What is the order-of-magnitude difference between the country with the largest area and the country with the smallest area?

10. The Energy Information Administration of the U.S. Department of Energy estimates that in 2010 the world energy use will be 470.8 quadrillion Btu (British thermal units), where 1 Btu = 0.000 293 1 kWh (kilowatt-hours.)

   a. Express 470.8 quadrillion Btu and 0.000 293 1 kWh in scientific notation.

   b. How many kilowatt-hours are there in 470.8 quadrillion Btu? Give your answer in scientific notation.

11. Is the following statement true or false? "An increase in one order of magnitude is the same as an increase of 100%." If true, explain why. If false, revise the statement to make it true.

12. On October 15, 2006, the *San Francisco Chronicle* published the accompanying graph and table derived from U.S. Bureau of the Census data on the growing size of the U.S. population. Compare the changes from 1915 to 2006 in two different categories, using at least one rate-of-change calculation and one order-of-magnitude comparison. Show your work.

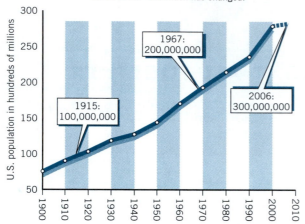

A New Milestone for U.S. Population

The number of residents has tripled since 1915—and much has changed.

	1915 Woodrow Wilson	1967 Lyndon B. Johnson	2006 George W. Bush
President			
Price of new home	$3,200	$24,600	$290,600
Cost of gallon of regular gas	25¢	33¢	$2.25
Cost of a first-class stamp	2¢	5¢	39¢
Average household size	4.5 people	3.3 people	2.6 people
Number of people age 65 and older	4.5 million	19.1 million	36.8 million
Most popular baby names for boys and girls	John and Mary	Michael and Lisa	Jacob and Emily

13. The U.S. Census Bureau estimates that a baby is born somewhere in the country every 7 seconds, a new immigrant arrives every 31 seconds, and someone dies every 13 seconds, for a net average gain of one resident every 10 seconds.

    a. On average, how many babies are born in the country per day?

    b. On average, how many new immigrants arrive per day?

    c. On average, how many people die per day?

14. Temperature can affect the speed of sound. The speed of sound, S (in feet/second), at an air temperature of T (in degrees Celsius) is

$$S = \frac{1087(273 + T)^{0.5}}{16.52}$$

    a. Express T in terms of S.

    b. The speed of sound is often given as 1120 feet/second. At what temperature in degrees Celsius would that be? At what temperature in degrees Fahrenheit would that be? (Recall that degrees Fahrenheit = 1.8(degrees Celsius) + 32.)

15. The radius of Earth is about $6.3 \cdot 10^6$ m and its mass is approximately $5.97 \cdot 10^{24}$ kg. Find its density in kg/m^3 (density = mass/volume).

16. Objects that are less dense than water will float; those that are more dense than water will sink. The density of water is 1.0 g/cm^3. A brick has a mass of 2268 g and a volume of 1230 cm^3. Show that the brick will sink in water (recall density = mass/volume).

17. An adult patient weighs 130 lb. The prescription for a drug is 5 mg per kg of the patient's weight per day. This drug comes in 100-mg tablets. What daily dosage should be prescribed?

18. On March 2, 2007, the *Boston Globe* reported the following:

    An exabyte is 1 quintillion bytes. In 2006 alone, the human race generated 161 exabytes of digital information. So? Well, that's about 3 million times the information in all the books ever written or the equivalent of 12 stacks of books, each extending more than 93 million miles from Earth to the sun.

    a. Use scientific notation to represent the amount of digital information generated in 2006. (One quintillion is 1 followed by eighteen zeros.)

    b. Compare the amount of digital information generated in 2006 with the amount of information in all the books ever written, using orders of magnitude.

    c. Estimate how many miles of books are needed to hold the equivalent information in 161 exabytes. Express your answer in scientific notation.

19. Find the logarithm of each of the following numbers:

    a. 1   b. 1 billion   c. 0.000 001

20. Estimate the following by placing the log between the two closest integer powers of 10.

    a. log 3000   b. log 150,000   c. log 0.05

Item	Value	Value in Scientific Notation
Mass-energy of electron	0.000 000 000 000 051 J	
The kinetic energy of a flying mosquito	0.000 000 160 2 J	
An average person swinging a baseball bat	80 J	
Energy received from the sun at Earth's orbit on one square meter in one second	1,360 J	
Energy released by one gram of TNT	4,184 J	
Energy released by metabolism of one gram of fat	38,000 J	
Approximate annual power usage of a standard clothes dryer	320,000,000 J	

Source: http://en.wikipedia.org.

21. One way of defining the energy unit *joule* (J) is the amount of the energy required to lift a small apple weighing 102 grams one meter above Earth's surface. The accompanying table lists the estimated energy in joules for different situations.

Use the accompanying table to answer the following questions.

a. Write each value in scientific notation.

b. A year's use of a clothes dryer requires how many times the energy of swinging a baseball bat once?

c. Metabolizing one gram of fat releases how many times the kinetic energy of a flying mosquito?

22. Rewrite each number as a power of 10, then create a logarithmic scale and estimate the location of the number on that scale. (*Hint:* log 2 = 0.301.)

   a. 10   **b.** 100   **c.** 200   **d.** 20,000

23. (Requires a scientific calculator.) Some drugs are prescribed in dosages that depend on a patient's BSA, or body surface area, an indicator of metabolic mass. One formula for calculating BSA is BSA = $71.84W^{0.425}H^{0.725}$, where BSA is measured in square centimeters, W is weight in kilograms, and H is height in centimeters.
A patient weighs 180 lb and is 6 feet tall. His dosage of a particular drug is 15 mg/m^2/day (that is, 15 mg per square meter of body surface area per day). What is his daily dosage in mg? (*Source:* DuBois & DuBois, 1916, from *http://en.wikipedia.org/wiki/Body_surface_area# Calculation.*)

EXPLORATION 4.1

The Scale and the Tale of the Universe

Objective

- gain an understanding of the relative sizes and relative ages of objects in the universe using scientific notation and unit conversions

Materials/Equipment

- tape, pins, paper, and string to generate a large wall graph (optional)
- enclosed worksheet and conversion table on inside back cover

Related Readings/Videos

Powers of Ten and "The Cosmic Calendar" from *The Dragons of Eden*
Videos: *Powers of Ten* and *The Cosmic Calendar* in the PBS series *Cosmos*

Related Software

"E1: Tale and Scale of the Universe" in *Exponential & Log Functions*

Procedure

Work in small groups. Each group should work on a separate subset of objects on the accompanying worksheet.

1. Convert the ages and sizes of objects so they can be compared. You can refer to the conversion table that shows equivalences between English and metric units (see inside back cover). In addition, 1 light year $\approx 9.46 \cdot 10^{12}$ km.
2. Generate on the blackboard or on the wall (with string) a blank graph whose axes are marked off in orders of magnitude (integer powers of 10), with the units on the vertical axis representing age of object, ranging from 10^0 to 10^{11} years, and the units on the horizontal axis representing size of object, ranging from 10^{-12} to 10^{27} meters.
3. Each small group should plot the approximate coordinates of their selected objects (size in meters, age in years) on the graph. You might want to draw and label a small picture of your object to plot on your graph.

Discussion/Analysis

- Scan the plotted objects from left to right, looking only at relative sizes. Now scan the plotted objects from top to bottom, considering only relative ages. Does your graph make sense in terms of what you know about the relative sizes and ages of these objects?
- Describe the scale and the tale of the universe.

In Scientific Notation

Object	Age (in years)	Size (of radius)	Age (in years)	Size (in meters)
Observable universe	13.7 billion	10^{26} meters		
Surtsey (Earth's newest land mass)	40 years	0.5 mile		
Pleiades (a galactic cluster)	100 million	32.6 light years		
First living organisms on Earth	4.6 billion	0.000 05 meter		
Pangaea (Earth's prehistoric supercontinent)	200 million	4500 miles		
First *Homo sapiens sapiens*	100 thousand	100 centimeters		
First *Tyrannosaurus rex*	200 million	20 feet		
Eukaryotes (first cells with nuclei)	2 billion	0.000 05 meter		
Earth	4.6 billion	6400 kilometers		
Milky Way galaxy	14 billion	50,000 light years		
First atoms	13.7 billion	0.000 000 0001 meter		
Our sun	5 billion	1 gigameter		
Our solar system	5 billion	1 terameter		

EXPLORATION 4.2

Patterns in the Positions and Motions of the Planets

Objective

• explore patterns in the positions and motions of the planets and discover Kepler's Law

Introduction and Procedure

Four hundred years ago, before Newton's laws of mechanics, Johannes Kepler discovered a law that relates the periods of planets with their average distances from the sun. (A period of a planet is the time it takes the planet to complete one orbit of the sun.) Kepler's strong belief that the solar system was governed by harmonious laws drove him to try to discover hidden patterns and correlations among the positions and motions of the planets. He used the trial-and-error method and continued his search for years.

At the time of his work, Kepler did not know the distance from the sun to each planet in terms of measures of distance such as the kilometer. But he was able to determine the distance from each planet to the sun in terms of the distance from Earth to the sun, now called the astronomical unit, or A.U. for short. One A.U. is the distance from Earth to the sun. The first column in the table below gives the average distance from the sun to each of the planets in astronomical units.

Patterns in the Positions and Motions of the Planets: Kepler's Discovery

Fill in the following table and look for the relationship that Kepler found.

Kepler's Third Law: The First Planet Table (Inner Planetary System)

Planet	Average Distance from Sun (A.U.)*	Cube of the Distance (A.U.^{3})	Orbital Period (years)	Square of the Orbital Period (years^2)
Mercury	0.3870		0.2408	
Venus	0.7232		0.6151	
Earth	1.0000		1.0000	
Mars	1.5233		1.8807	
Jupiter	5.2025		11.8619	
Saturn	9.5387		29.4557	

*1 A.U. $\approx 149.6 \cdot 10^6$ km; 1 year ≈ 365.26 days.
Source: Data from S. Parker and J. Pasachoff, *Encyclopedia of Astronomy,* 2nd ed. (New York: McGraw-Hill, 1993), Table 1, Elements of Planetary Orbits. Copyright © 1993 by McGraw-Hill, Inc. Reprinted with permission.

The planets Uranus and Neptune were discovered after Kepler made his discovery. Check to see whether the relationship you found above holds true for these two planets.

The Second Planet Table (Outer Planetary System)

Planet	Average Distance from Sun (A.U.)	Cube of the Distance (A.U.³)	Orbital Period (years)	Square of the Orbital Period (years²)
Uranus	19.1911		84.0086	
Neptune	30.0601		164.7839	

Source: Data from S. Parker and J. Pasachoff, *Encyclopedia of Astronomy,* 2nd ed. (New York: McGraw-Hill, 1993), Table 1, Elements of Planetary Orbits. Copyright © 1993 by McGraw-Hill, Inc. Reprinted with permission.

Note: Pluto is no longer classified as a planet. It is now called a "dwarf planet."

Summary

- Express your results in words.
- Construct an equation showing the relationship between distance from the sun and orbital period. Solve the equation for distance from the sun. Then solve the equation for orbital period.
- Do your conclusions hold for all of the planets?

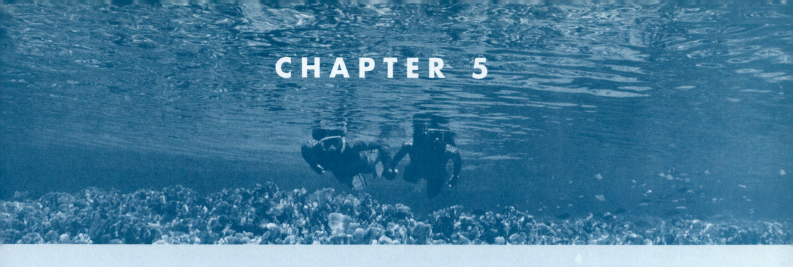

CHAPTER 5

GROWTH AND DECAY: AN INTRODUCTION TO EXPONENTIAL FUNCTIONS

OVERVIEW

Exponential and linear functions are used to describe quantities that change over time. Exponential functions represent quantities that are multiplied by a constant factor during each time period. Linear functions represent quantities to which a fixed amount is added (or subtracted) during each time period. Exponential functions can model such diverse phenomena as bacteria growth, radioactive decay, compound interest rates, inflation, musical pitch, and family trees.

After reading this chapter you should be able to

- recognize the properties of exponential functions and their graphs

- understand the differences between exponential and linear growth

- model growth and decay phenomena with exponential functions

- represent exponential functions using percentages, factors, or rates

- use semi-log plots to determine if data can be modeled by an exponential function

5.1 *Exponential Growth*

The Growth of *E. coli* Bacteria

Measuring and predicting growth is of concern to population biologists, ecologists, demographers, economists, and politicians alike. The growth of bacteria provides a simple model that scientists can generalize to describe the growth of other phenomena such as cells, countries, or capital.

The program "E2: Exponential Growth & Decay" in *Exponential & Log Functions* offers a dramatic visualization of growth and decay.

Bacteria are very tiny, single-celled organisms that are by far the most numerous organisms on Earth. One of the most frequently studied bacteria is *E. coli,* a rod-shaped bacterium approximately 10^{-6} meter (or 1 micrometer) long that inhabits the intestinal tracts of humans and other mammals.[1] The cells of *E. coli* reproduce by a process called fission: The cell splits in half, forming two "daughter cells."

Under ideal conditions *E. coli* divide every 20 minutes. If we start with an initial population of 100 *E. coli* bacteria that doubles every 20-minute time period, we generate the data in Table 5.1. The initial 100 bacteria double to become 200 bacteria at the end of the first time period, double again to become 400 at the end of the second time period, and so on. At the end of the twenty-fourth time period (at $24 \cdot 20$ minutes = 480 minutes, or 8 hours), the initial 100 bacteria in our model have grown to over 1.6 billion bacteria!

Because the numbers become astronomically large so quickly, we run into the problems we saw in Chapter 4 when graphing numbers of widely different sizes. Figure 5.1 shows a graph of the data in Table 5.1 for only the first ten time periods. We can see from the graph that the relationship between number of bacteria and time is not linear. The number of bacteria seems to be increasing more and more rapidly over time.

Growth of *E. coli* Bacteria

Time Periods (of 20 minutes each)	Number of *E. coli* Bacteria
0	100
1	200
2	400
3	800
4	1,600
5	3,200
6	6,400
7	12,800
8	25,600
9	51,200
10	102,400
11	204,800
12	409,600
13	819,200
14	1,638,400
15	3,276,800
16	6,553,600
17	13,107,200
18	26,214,400
19	52,428,800
20	104,857,600
21	209,715,200
22	419,430,400
23	838,860,800
24	1,677,721,600

Table 5.1

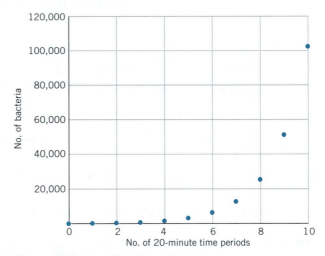

Figure 5.1 Growth of *E. coli* bacteria.

A mathematical model for E. coli growth

Table 5.1 shows us that the initial number of 100 bacteria repeatedly doubles. If we record in a third column (Table 5.2) the number of times we multiply 2 times the original value of 100, we begin to see a pattern emerge.

[1] Most types of *E. coli* are beneficial to humans, aiding in digestion. A few types are lethal. You may have read about deaths resulting from people eating certain deadly strains of *E. coli* bacteria in undercooked hamburgers or tainted spinach. The explosive nature of exponential growth shows how a few dangerous bacteria can multiply rapidly to become a deadly quantity in a very short time.

Pattern in *E. coli* Growth

Number of Time Periods	Number of *E. coli* Bacteria	Generalized Expression
0	100	$100 = 100 \cdot 2^0$
1	200	$100 \cdot 2 = 100 \cdot 2^1$
2	400	$100 \cdot 2 \cdot 2 = 100 \cdot 2^2$
3	800	$100 \cdot 2 \cdot 2 \cdot 2 = 100 \cdot 2^3$
4	1,600	$100 \cdot 2 \cdot 2 \cdot 2 \cdot 2 = 100 \cdot 2^4$
5	3,200	$100 \cdot 2 \cdot 2 \cdot 2 \cdot 2 \cdot 2 = 100 \cdot 2^5$
6	6,400	$100 \cdot 2 \cdot 2 \cdot 2 \cdot 2 \cdot 2 \cdot 2 = 100 \cdot 2^6$
7	12,800	$100 \cdot 2 \cdot 2 \cdot 2 \cdot 2 \cdot 2 \cdot 2 \cdot 2 = 100 \cdot 2^7$
8	25,600	$100 \cdot 2 \cdot 2 \cdot 2 \cdot 2 \cdot 2 \cdot 2 \cdot 2 \cdot 2 = 100 \cdot 2^8$
9	51,200	$100 \cdot 2 \cdot 2 \cdot 2 \cdot 2 \cdot 2 \cdot 2 \cdot 2 \cdot 2 \cdot 2 = 100 \cdot 2^9$
10	102,400	$100 \cdot 2 \cdot 2 \cdot 2 \cdot 2 \cdot 2 \cdot 2 \cdot 2 \cdot 2 \cdot 2 \cdot 2 = 100 \cdot 2^{10}$

Table 5.2

Remembering that 2^0 equals 1 by definition, we can describe the relationship by

$$\text{number of } E.\ coli \text{ bacteria} = 100 \cdot 2^{\text{number of time periods}}$$

If we let N = number of bacteria and t = number of time periods, we can write the equation more compactly as

$$N = 100 \cdot 2^t$$

Since each value of t determines one and only one value for N, the equation represents N as a function of t. The number 100 is the *initial bacteria population*. This function is called *exponential* since the input or independent variable, t, occurs in the exponent of the base 2. The base 2 is the *growth factor*, or the multiple by which the population grows during each time period. The bacteria double every time period, which means that the amount of increase equals the previous amount. Thus it increases at a rate of 100% during each time period. If *E. coli* grew unchecked at this pace, the offspring from one cell would cover Earth with a layer a foot deep in less than 36 hours!

The General Exponential Growth Function

The *E. coli* growth equation

$$N = 100 \cdot 2^t$$

is in the form

$$\text{output} = (\text{initial quantity}) \cdot (\text{growth factor})^{\text{input}}$$

Such an equation, with the input in the exponent and a growth factor >1, describes an *exponential growth function*.

Exponential Growth Function

An exponential growth function can be represented by an equation of the form

$$y = Ca^x \quad (a > 1, C > 0) \qquad \text{where}$$

a, the base, is the *growth factor,* the amount by which y is multiplied when x increases by 1.

C is the *initial value* or y-intercept.

EXAMPLE 1

Constructing an exponential function

If you start with 200 cells that triple during every time period t, an equation to model its growth is

$$N = 200 \cdot 3^t$$

EXAMPLE 2

Interpreting an exponential equation

Interpret the numbers in the following equation:

$$Q = (4 \cdot 10^6) \cdot 2.5^T$$

SOLUTION

$4 \cdot 10^6$, or 4,000,000, represents an initial population that grows by a factor of 2.5 each time period, T. This means the initial population is multiplied by 2.5 each time period, T.

EXAMPLE 3

Take a piece of ordinary paper, about 0.1 mm in thickness. Now start folding it in half, then in half again, and so on. Assume you could continue indefinitely.

a. How would you model the growth in thickness of the folded paper?

b. How many folds would it take to reach:

   i. The height of a human?

   ii. The height of the Matterhorn, a famous mountain in Switzerland?

   iii. The sun?

   iv. The edge of the known universe?

SOLUTION

a. Each time we fold the paper, we double the thickness. If N = number of folds, the initial thickness = 0.1 mm, and T = thickness of the folded paper (in mm), then

$$T = (0.1) \cdot 2^N$$

b. If you start plugging in values for N (this is where technology is useful), it takes 14 folds to reach the height of an average person (well, a short person). At 26 folds, the paper is higher than the Matterhorn. At 51 folds, the paper would reach the sun, and at 54 folds the edge of our solar system. It would take only 84 folds to reach the limits of the Milky Way. A mere 100 folds takes you to the edge of the known universe!

Looking at Real Growth Data for *E. coli* Bacteria

Figure 5.2 shows a plot of real *E. coli* growth over 24 time periods (8 hours). The growth appears to be exponential for the first 12 time periods (4 hours). We can use technology to generate a best-fit exponential function for that section of the curve. The function is

$$N = (1.37 \cdot 10^7) \cdot 1.5^t$$

where N = number of cells per milliliter and t = the number of 20-minute periods. So the initial quantity is $1.37 \cdot 10^7$ (or 13.7 million) cells. More important, the growth factor is 1.5. Every 20 minutes the number of cells is multiplied by 1.5 (a 50% increase), as opposed to the doubling (a 100% increase) in our first, idealized example.

The bacterial growth rate in the laboratory is half that of the idealized data. But even this rate can't be sustained for long. Conditions are rarely ideal; bacteria die, nutrients are used up, space to grow is limited. In the real world, growth that starts out as exponential must eventually slow down (see Figure 5.2). The curve flattens out as the number of bacteria reaches its maximum size, called the *carrying capacity*. The overall shape is called *sigmoid*. The arithmetic of exponentials leads to the inevitable conclusion that in the long term—for bacteria, mosquitoes, or humans—the rate of growth for populations must approach zero.

DATA

ECOLI

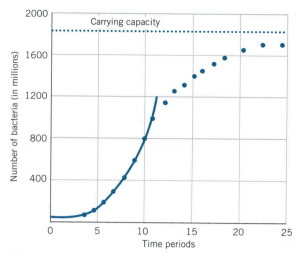

Figure 5.2 Sigmoid curve of real *E. coli* growth over 24 time periods and best-fit exponential for 12 time periods

Algebra Aerobics 5.1

A calculator that can evaluate powers is recommended for Problems 3 and 4.

1. Identify the initial value and the growth factor in each of the following exponential growth functions:

    a. $y = 350 \cdot 5^x$ **c.** $P = (7 \cdot 10^3) \cdot 4^t$

    b. $Q = 25{,}000 \cdot 1.5^t$ **d.** $N = 5000 \cdot 1.025^t$

2. Write an equation for an exponential growth function where:

    a. The initial population is 3000 and the growth factor is 3.

    b. The initial population is $4 \cdot 10^7$ and the growth factor is 1.3.

    c. The initial population is 75 and the population quadruples during each time period.

    d. The initial amount is \$30,000 and the growth factor is 1.12.

3. The population growth of a small country is described by the function $P = 28 \cdot 1.065^t$, where t is in years and P is in millions.

    a. Determine P when $t = 0$. What does that quantity represent?

    b. Determine P when $t = 10, 20$, and 30.

    c. Estimate the value of t that would result in a doubling of the population to 56,000,000. If available, use technology to check your estimate.

4. Fill in a table of values for the amount $A = 80 \cdot 1.06^t$ for values of the time period t of 0, 1, 10, 15, and 20, and then complete the following statements.

    a. The initial amount is ___.

    b. During the first time period, the amount grew from 80 to ___.

    c. Based on this table, the amount doubles between the values $t = $ ___ and $t = $ ___.

    d. Estimate the number of time periods it will take the amount to double.

Exercises for Section 5.1

1. The following exponential functions represent population growth. Identify the initial population and the growth factor.

    a. $Q = 275 \cdot 3^T$ **d.** $A = 25(1.18)^t$

    b. $P = 15{,}000 \cdot 1.04^t$ **e.** $P(t) = 8000(2.718)^t$

    c. $y = (6 \cdot 10^8) \cdot 5^x$ **f.** $f(x) = 4 \cdot 10^5(2.5)^x$

2. Write the equation of each exponential growth function in the form $y = Ca^x$ where:

    a. The initial population is 350 and the growth factor is $\frac{4}{3}$.

    b. The initial population is $5 \cdot 10^9$ and the growth factor is 1.25.

c. The initial population is 150 and the population triples during each time period.

d. The initial population of 2 quadruples every time period.

3. Fill in the missing parts of the table.

Initial Value C	Growth Factor a	Exponential Function $y = Ca^x$
1600	1.05	
		$y = 6.2 \cdot 10^5 (2.07)^x$
	3.25	$y = 1400(\underline{\ \ })^x$

4. The populations of four towns for time t, in years, are given by:

$P_1(t) = 12{,}000(1.05)^t$

$P_2(t) = 6000(1.07)^t$

$P_3(t) = 100{,}000(1.01)^t$

$P_4(t) = 1000(1.9)^t$

a. Which town has the largest initial population?

b. Which town has the largest growth factor?

c. At the end of 10 years, which town would have the largest population?

5. Find C and a such that the function $f(x) = Ca^x$ satisfies the given conditions.

a. $f(0) = 6$ and for each unit increase in x, the output is multiplied by 1.2.

b. $f(0) = 10$ and for each unit increase in x, the output is multiplied by 2.5.

6. In the United States during the decade of the 1990s, live births to unmarried mothers, B, grew according to the exponential model $B = 1.165 \cdot 10^6 (1.013)^t$, where t is the number of years after 1990.

a. What does the model give as the number of live births to unwed mothers in 1990?

b. What was the growth factor?

c. What does the model predict for the number of live births to unwed mothers in 1995? In 2000?

7. A cancer patient's white blood cell count grew exponentially after she had completed chemotherapy treatments. The equation $C = 63(1.17)^d$ describes C, her white blood cell count per milliliter, d days after the treatment was completed.

a. What is the white blood cell count growth factor?

b. What was the initial white blood cell count?

c. Create a table of values that shows the white blood cell counts from day 0 to day 10 after the chemotherapy.

d. From the table of values, approximate when the number of white blood cells doubled.

8. Match each function with its graph.

a. $y = 2(1.5)^x$ c. $y = 2(3)^x$

b. $y = 3(1.5)^x$ d. $y = 4(2)^x$

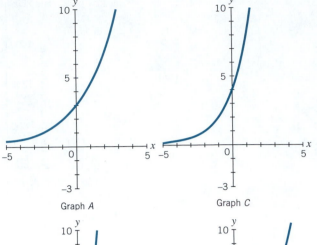

Graph A   Graph C

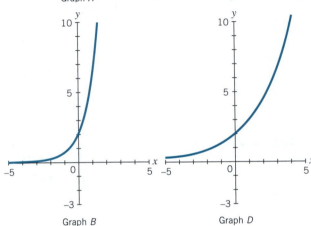

Graph B   Graph D

9. National health care expenditures in 2005 were approximately \$2016 billion and are expected to increase by a factor of 1.076 per year. In 5 years what would be the predicted expenditures?

10. The per-capita consumption of bottled water was 8 gallons in 1990 and has been increasing yearly by a factor of 1.088. What was the per capita consumption of bottled water 12 years later?

11. A tuberculosis culture increases by a factor of 1.185 each hour.

a. If the initial concentration is $5 \cdot 10^3$ cells/ml, construct an exponential function to describe its growth over time.

b. What will the concentration be after 8 hours?

12. An ancient king of Persia was said to have been so grateful to one of his subjects that he allowed the subject to select his own reward. The clever subject asked for a grain of rice on the first square of a chessboard, two grains on the second square, four on the next, and so on.

a. Construct a function that describes the number of grains of rice, G, as a function of the square, n, on the chessboard. (*Note:* There are 64 squares.)

b. Construct a table recording the numbers of grains of rice on the first ten squares.

c. Sketch your function.

d. How many grains of rice would the king have to provide for the 64th (and last) square?

13. A new species of fish is introduced into a pond. The size of the fish population can be modeled by the accompanying graph of the function $P(t)$, where t is time in months.

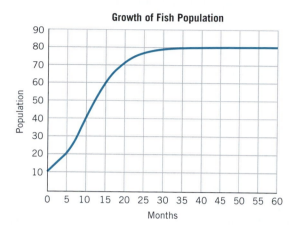

Growth of Fish Population

a. What is the initial population of the fish?

b. How long did it take for the initial fish population to double?

c. What was the sustainable fish population (carrying capacity) of the pond?

d. Estimate the number of months it took for the fish population to reach its sustainable size.

14. The Northern Wildlife Prairie Research Center in Jamestown, North Dakota, measured the weights of three duckling species (wild mallard, gadwall, and blue-winged teal). Each one-day-old duckling weighed about 32 grams. Weights were tracked and are graphed on the following chart. The weight curves appear sigmoidal, each approaching a maximum sustainable (or mature) weight.

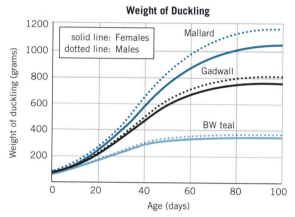

Weight of Duckling

Source: http://www.npwrc.usgs.gov.

a. Estimate the mature weight for the female ducklings in each species.

b. Estimate the mature weight for the male ducklings in each species.

c. Approximately how long did it take for each species of ducklings to attain the mature weight?

5.2 *Linear vs. Exponential Growth Functions*

Linear vs. Exponential Growth

Linear growth is intrinsically *additive*. Linear growth means that for each unit increase in the input, we must *add* a fixed amount (the slope or rate of change) to the value of the output. For example, in the linear function $N = 100 + 2t$, each time we increase t by 1, we add 2 to the value of N. After t time periods, we will have added $2t$ to the initial value of N. Assuming t is a positive integer, we can write this linear function as

$$N = 100 + \underbrace{2 + 2 + \cdots + 2}_{t \text{ times}}$$

Exponential growth is *multiplicative*, which means that for each unit increase in the input, we *multiply* the value of the output by a fixed number (the growth factor). For example, in the exponential function $N = 100 \cdot 2^t$, each time we increase t by 1, we *multiply* the value of N by 2. After t time periods have elapsed, we will have multiplied the value of N by 2^t. If t is a positive integer, we can write this exponential function as

$$N = 100 \cdot \underbrace{2 \cdot 2 \cdot \cdots \cdot 2}_{t \text{ times}}$$

Comparing the general equations

There are both similarities and differences in the general forms of the equations for linear and exponential functions.

If y is a linear function of x, then

$$y = b + mx$$

$$y = y\text{-intercept} + (\text{slope}) \cdot x$$

and if x is a positive integer $\qquad y = \qquad b \qquad + \underbrace{(m + m + m + \cdots + m)}_{x \text{ times}}$

If y is an exponential function of x, then

$$y = C \cdot a^x$$

$$y = y\text{-intercept} \cdot (\text{base})^x$$

and if x is a positive integer $\qquad y = C \cdot \underbrace{(a \cdot a \cdot a \cdot a \cdot \cdots \cdot a)}_{x \text{ times}}$

Linear functions involve repeated additions, whereas exponential functions involve repeated products. In each case, x determines the number of repetitions. For both linear and exponential equations, the vertical intercept gives the initial or starting value.

Linear vs. Exponential Functions

A linear function represents a quantity to which a constant amount is added for each unit increase in the input.

An exponential function represents a quantity that is multiplied by a constant factor for each unit increase in the input.

In the long run, exponential growth will always outpace linear growth

Exponential growth is more rapid as time goes on, as can be seen in Table 5.3. For both the linear function $N = 100 + 2t$ and the exponential function $N = 100 \cdot 2^t$, when $t = 0$, then $N = 100$. After ten time periods, the values for N are strikingly different: 102,400 for the exponential function versus 120 for the linear function. The initial value of 100 has been multiplied by 2 ten times in the exponential function to get $100 \cdot 2^{10}$. In the linear function, 2 has been added to 100 ten times, to get $100 + 2 \cdot 10$.

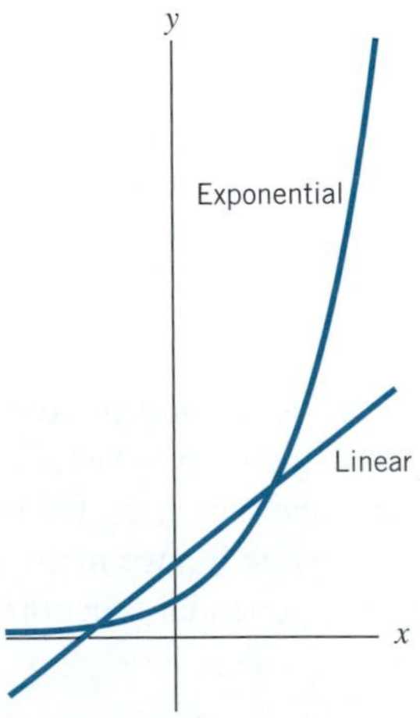

Figure 5.3 Graph comparing linear and exponential growth.

Comparing the Additive Pattern of Linear Growth to the Multiplicative Pattern of Exponential Growth

Time, t	Linear Function, $N = 100 + 2t$ Pattern		Exponential Function, $N = 100 \cdot 2^t$ Pattern	
0	$100 + 2 \cdot 0$	$= 100$	$100 \cdot 2^0$ =	100
1	$100 + 2 \cdot 1$	$= 102$	$100 \cdot 2^1$ =	200
2	$100 + 2 \cdot 2$	$= 104$	$100 \cdot 2^2$ =	400
3	$100 + 2 \cdot 3$	$= 106$	$100 \cdot 2^3$ =	800
4	$100 + 2 \cdot 4$	$= 108$	$100 \cdot 2^4$ =	1,600
5	$100 + 2 \cdot 5$	$= 110$	$100 \cdot 2^5$ =	3,200
6	$100 + 2 \cdot 6$	$= 112$	$100 \cdot 2^6$ =	6,400
7	$100 + 2 \cdot 7$	$= 114$	$100 \cdot 2^7$ =	12,800
8	$100 + 2 \cdot 8$	$= 116$	$100 \cdot 2^8$ =	25,600
9	$100 + 2 \cdot 9$	$= 118$	$100 \cdot 2^9$ =	51,200
10	$100 + 2 \cdot 10$	$= 120$	$100 \cdot 2^{10}$ =	102,400

Table 5.3

If we compare any linear growth function (whose graph will be a straight line with a positive slope) to any exponential growth function (whose graph will curve upward), we see that sooner or later the exponential curve will permanently lie above the linear graph and continue to grow faster and faster (Figure 5.3). The exponential function eventually dominates the linear function.

Comparing the Average Rates of Change

Another way to compare linear and exponential functions is to examine average rates of change. Recall that if N is a function of t, then

$$\text{average rate of change} = \frac{\text{change in } N}{\text{change in } t} = \frac{\Delta N}{\Delta t}$$

Examine Table 5.4, which contains average rates of change for both the linear and exponential functions, where $\Delta t = 1$. For all linear functions, we know that the average rate of change is constant. For the linear function $N = 100 + 2t$, we can tell from the equation, Table 5.4, and Figure 5.4(a) that the (average) rate of change is constant at two units. Average rates of change for the exponential function $N = 100 \cdot 2^t$ are calculated in Table 5.4 and then graphed in Figure 5.4(b). These suggest that the average rates of change of an exponential growth function grow *exponentially*.

Comparing Average-Rate-of-Change Calculations

	Linear Function		Exponential Function	
	$N = 100 + 2t$	Average Rate of Change (between $t - 1$ and t)	$N = 100 \cdot 2^t$	Average Rate of Change (between $t - 1$ and t)
t	N		N	
0	100	n.a.	100	n.a.
1	102	2	200	100
2	104	2	400	200
3	106	2	800	400
4	108	2	1,600	800
5	110	2	3,200	1,600
6	112	2	6,400	3,200
7	114	2	12,800	6,400
8	116	2	25,600	12,800
9	118	2	51,200	25,600
10	120	2	102,400	51,200

Table 5.4

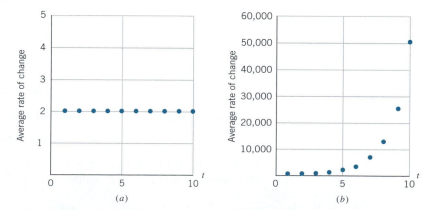

Figure 5.4 (a) Graph of the average rates of change between several pairs of points that satisfy the linear function $N = 100 + 2t$. (b) Graph of the average rates of change between several pairs of points that satisfy the exponential function $N = 100 \cdot 2^t$.

A Linear vs. an Exponential Model through Two Points

A town's population increased from 20,000 to 24,000 over a 5-year period. The town council is concerned about the rapid population growth and wants to predict future population size. The functions most commonly used to predict growth patterns over time are linear and exponential.

If we let t = number of years and P = population size, then when $t = 0$, $P = 20,000$ and when $t = 5$, $P = 24,000$. This gives us two points, (0, 20000) and (5, 24000), which we will use to construct our models.

Constructing a linear model: Adding a constant amount each year

Assuming linear growth, our function will be of the form $P = b + mt$. The initial population, b, is 20,000. The average rate of change or slope, m, through the two points is

$$m = \frac{\text{change in population}}{\text{change in time}} = \frac{24,000 - 20,000}{5 - 0} = \frac{4000}{5} = 800 \text{ people/yr}$$

So the equation is

$$P = 20,000 + 800t$$

This linear model says that the original population of 20,000 is increasing by the constant amount of 800 people each year.

Constructing an exponential model: Multiplying by a constant factor each year

Assuming exponential growth and an initial population of 20,000, our function will be of the form

$$P = 20,000 \cdot a^t$$

where P is the population of the town and t the number of years. To complete our model, we need a value for a, the annual growth factor.

Finding the Growth Factor from Two Data Points.   We have enough information to find the 5-year growth factor. The inputs 0 and 5 years give us two outputs, 20,000 and 24,000 people, respectively. Exponential growth is multiplicative, so for each additional year, the initial population is multiplied by the growth factor, a. After 5 years, the initial population of 20,000 is multiplied by a^5 to get 24,000.

Given $\qquad\qquad 20,000 \cdot a^5 = 24,000$

divide by 20,000 $\qquad\qquad a^5 = \dfrac{24,000}{20,000} = 1.2$

So the 5-year growth factor, a^5, is 1.2. Using a calculator, we can find the annual growth factor, a.

Take the fifth root of both sides $\qquad (a^5)^{1/5} = (1.2)^{1/5}$

$$a^1 \approx 1.0371$$

Now we can represent our exponential function as

$$P = 20,000 \cdot (1.0371)^t$$

where t = number of years and P = population size. Our exponential model tells us the initial population of 20,000 is multiplied by 1.0371 each year.

Making predictions with our models

We can use the models not only to describe past behavior, but also to predict future population sizes (see Table 5.5 and Figure 5.5). The population size is the same in both models for year 0 and year 5. For year 10, the linear and exponential predictions are

Town Population

Year	Linear Model $P = 20{,}000 + 800t$	Exponential Model $P = 20{,}000(1.0371)^t$
0	20,000	20,000
5	24,000	24,000
10	28,000	28,790
15	32,000	34,540
20	36,000	41,440
25	40,000	49,720

Table 5.5

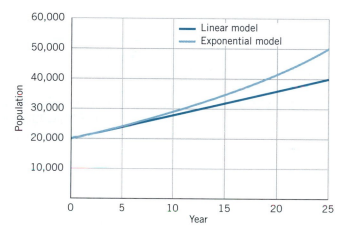

Figure 5.5 Predictions for the town population using linear and exponential models.

fairly close: 28,000 versus approximately 28,790. As we move further beyond year 10, the exponential predictions exceed the linear by an increasingly greater amount. By year 25, the linear model predicts a population of 40,000 whereas the exponential model predicts almost 50,000.

Both models should be considered as generating only crude future estimates, particularly since we constructed the models using only two data points. Clearly, the further out we try to predict, the more unreliable the estimates become.

Identifying Linear vs. Exponential Functions in a Data Table

Table 5.5 shows a linear function side by side with an exponential function. Every 5 years, the linear model *added* 800 people five times, adding a total of 4000 to the population. Every 5 years, the exponential model *multiplied* the population size by the annual growth factor 1.0371 five times; that is, the population size is multiplied by $1.0371^5 \approx 1.2$. The 1-year growth factor $= 1.0371$ and the 5-year growth factor $= 1.0371^5 \approx 1.2$.

If P_1 is the starting population size and P_2 is the population size 5 years later, then

$$\text{In the linear model} \qquad P_1 + 4000 = P_2$$
$$\text{or equivalently} \qquad 4000 = P_2 - P_1$$

$$\text{In the exponential model} \qquad P_1 \cdot 1.2 = P_2$$
$$\text{or equivalently} \qquad 1.2 = P_2/P_1$$

Using these equations we can translate our statements about addition (for linear) and multiplication (for exponential) into statements about differences and ratios.

In general, given any data table of the form (x, y) where Δx is constant, then the data are linear if the *difference* between successive y values is constant. The data are exponential if the *ratio* of successive y values is constant. We can test whether a data table represents a linear or exponential function by using differences or ratios, as long as the Δx values are constant.

EXAMPLE 1   Determine whether each function in the table is linear or exponential.

x	0	10	20	30	40
$f(x)$	5	25	45	65	85
$g(x)$	20	30	45	67.5	101.25

Table 5.6

SOLUTION   The consecutive x values or input in Table 5.6 are ten units apart, so $\Delta x = 10$. Now we compare the output values. The difference between successive values of $f(x)$ is always 20, so $f(x)$ is a linear function. The difference between successive values of $g(x)$ is not constant. For example,

$$g(10) - g(0) = 30 - 20 = 10$$

but $$g(20) - g(10) = 45 - 30 = 15$$

So $g(x)$ cannot be linear. To determine if $g(x)$ is exponential, we calculate the ratios of successive values of $g(x)$. For example,

$$\frac{30}{20} = \frac{45}{30} = \frac{67.5}{45} = \frac{101.25}{67.5} = 1.5$$

The ratios of consecutive y values are constant at 1.5 when $\Delta x = 10$ (this ratio represents the 10-year growth factor). Hence the function is exponential.

For a table describing y as a function of x, where Δx is constant:

If the difference between consecutive y values is constant, the function is linear.

If the ratio of consecutive y values is constant, the function is exponential.

Algebra Aerobics 5.2

1. Fill in the following table and sketch the graph of each function.

t	$N = 10 + 3t$	$N = 10 \cdot 3^t$
0		
1		
2		
3		
4		

2. **a.** Create a table of values (from $t = 0$ to $t = 5$) for the functions $f(t)$, with a vertical intercept at 200 and a constant rate of change of 20, and $g(t)$, with a vertical intercept at 200 and a growth factor of 1.20.

   b. Sketch the graphs of g and f on the same grid.

   c. Compare the graphs.

3. Given the values in Table 5.7, determine which functions (if any) are linear and which are approximately exponential. Justify your answer.

T	f(t)	g(t)	h(t)	p(t)	r(t)
0	10	10	10	10	10
1	16	10	15	100	10.4
2	22	10	22.5	190	10.82
3	28	10	33.75	280	11.25
4	34	10	50.63	370	11.70
5	40	10	75.94	460	12.17

Table 5.7

4. Write an equation for the linear function and the exponential function that pass through the given points.

    a. $(0, 500)$ and $(1, 620)$ **b.** $(0, 3)$ and $(1, 3.2)$

5. In each of the following situations, assume growth is exponential, find the growth factor, and then construct an exponential function that models the situation.

    a. The initial population, P, is 1500 and one time period, t, later, the population grew to 3750.

b. The initial amount, A, is \$80,000 and one time period, t, later, the amount grew by \$2300.

c. A quantity, Q, grew from 30 mg to 32.7 mg after one time period, t.

6. Assume in Table 5.8 that Q is an exponential function of t.

t (years)	Q
0	10
2	20
4	
6	
8	

Table 5.8

a. By what factor is Q multiplied when t increases by 2 years? This is the 2-year growth factor.

b. Fill in the rest of the Q values in the table.

c. What is the annual growth factor?

d. Construct an equation to model the relationship between Q and t.

Exercises for Section 5.2

A graphing program is recommended for Exercises 3 and 11.

1. Determine which of the following functions are linear, which are exponential, and which are neither. In each case identify the vertical intercept.

    a. $C(t) = 3t + 5$ **c.** $y = 5x^2 + 3$ **e.** $P = 7(1.25)^t$

    b. $f(x) = 3(5)^x$ **d.** $Q = 6\left(\frac{3}{2}\right)^t$ **f.** $T = 1.25n$

2. Create a linear or exponential function based on the given conditions.

    a. A function with an average rate of change of 3 and a vertical intercept of 4.

    b. A function with growth factor of 3 and vertical intercept of 4.

    c. A function with slope of 4/3 and initial value of 5.

    d. A function with initial value of 5 and growth factor of 4/3.

3. (Graphing program recommended.) A small village has an initial size of 50 people at time $t = 0$, with t in years.

    a. If the population increases by 5 people per year, find the formula for the population size $P(t)$.

    b. If the population increases by a factor of 1.05 per year, find a new formula $Q(t)$ for the population size.

    c. Plot both functions on the same graph over a 30-year period.

    d. Estimate the coordinates of the point(s) where the graphs intersect. Interpret the meaning of the intersection point(s).

4. A herd of deer has an initial population of 10 at time $t = 0$, with t in years.

    a. If the size of the herd increases by 8 per year, find the formula for the population of deer, $P(t)$, over time.

b. If the size of the herd increases by a factor of 1.8 each year, find the formula for the deer population, $Q(t)$, over time.

c. For each model create a table of values for the deer population for a 10-year period.

d. Using the table, estimate when the two models predict the same population size.

5. Construct both a linear and an exponential function that go through the points $(0, 6)$ and $(1, 9)$.

6. Construct both a linear and an exponential function that go through the points $(0, 200)$ and $(10, 500)$.

7. Find the equation of the linear function and the exponential function that are sketched through two points on each of the following graphs.

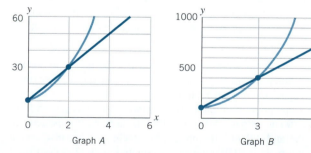

Graph A   Graph B

8. Match the description to any appropriate graph(s) shown below.

    a. When x increases by 10, y increases by 5.

    b. When x increases by 10, y is multiplied by a factor of 5.

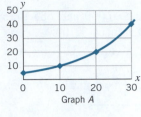

Graph A

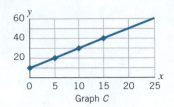

Graph C

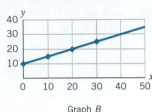

Graph B

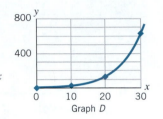

Graph D

9. Each of the following tables contains values representing either linear or exponential functions. Find the equation for each function.

a.

x	-2	-1	0	1	2
$f(x)$	1.12	2.8	7	17.5	43.75

b.

x	-2	-1	0	1	2
$g(x)$	0.1	0.3	0.5	0.7	0.9

10. Each table has values representing either linear or exponential functions. Find the equation for each function.

a.

x	-2	-1	0	1	2
$h(x)$	160	180	200	220	240

b.

x	0	10	20	30	40
$j(x)$	200	230	264.5	304.17	349.8

11. (Graphing program recommended.) Create a table of values for the following functions, then graph the functions.

a. $f(x) = 6 + 1.5x$ c. $h(x) = 1.5(6)^x$

b. $g(x) = 6(1.5)^x$

12. Suppose you are given a table of values of the form (x, y) where Δx, the distance between two consecutive x values, is constant. Why is calculating $y_2 - y_1$, the distance between two consecutive y values, equivalent to calculating the average rate of change between consecutive points?

13. Mute swans were imported from Europe in the nineteenth century to grace ponds. Now there is concern that their population is growing too rapidly, edging out native species. Their population along the Atlantic coast has grown from 5800 in 1986 to 14,313 in 2002. The increase is most acute in the mid-Atlantic region, but Massachusetts has also seen a jump, with 2939 mute swans counted in 2002 as compared with 585 in 1986.

a. Compare the growth factor in the mute swan population for the entire Atlantic coast with that for Massachusetts.

b. Compare the average rate of change in the mute swan population for the entire Atlantic coast with that for Massachusetts.

c. Construct both a linear and an exponential model for the mute swan population in Massachusetts since 1986.

d. Compare the projected populations of mute swans in Massachusetts by the year 2010 as predicted by your linear and exponential models.

14. The price of a home in Medford was $100,000 in 1985 and rose to $200,000 in 2005.

a. Create two models, $f(t)$ assuming linear growth and $g(t)$ assuming exponential growth, where t = number of years after 1985.

b. Fill in the following table representing linear growth and exponential growth for t years after 1985.

t	Linear Growth $f(t) = $ Price (in thousands of dollars)	Exponential Growth $g(t) = $ Price (in thousands of dollars)
0	100	100
10		
20	200	200
30		
40		
50		

c. Which model do you think is more realistic?

15. *The Mass Media*, a student publication at the University of Massachusetts–Boston, reported on a proposed parking fee increase. The university administration recommended gradually increasing the daily parking fee on this campus to $7.00 by the year 2004, followed by an increase of 5% every year after that. Call this plan A. Several other plans were also proposed; one of them, plan B, recommended that every year after 2004 the rate be increased by 50 cents each year.

a. Let $t = 0$ for year 2004 and fill in the chart for parking fees under plans A and B.

Years after 2004	Parking Fee under Plan A	Parking Fee under Plan B
0	$7.00	$7.00
1		
2		
3		
4		

b. Write an equation for parking fees F_A as a function of t (years since 2004) for plan A and an equation F_B for plan B.

c. What will the daily parking fee be by the year 2025 under each plan? (Show your calculations.)

d. Imagine that you are the student representative to the UMass Board of Trustees. Which plan would you recommend for adoption? Explain your reasons and support your position using quantitative arguments.

5.3 *Exponential Decay*

An exponential growth function is of the form $y = Ca^x$, where $a > 1$. Each time x increases by 1, the population is multiplied by a, a number greater than 1, so the population increases. But what if a were positive and less than 1, that is, $0 < a < 1$? Then each time x increases by 1 and the population is multiplied by a, the population size will be reduced. We would have *exponential decay*.

The Decay of Iodine-131

Iodine-131 is one of the radioactive isotopes of iodine that is used in nuclear medicine for diagnosis and treatment. It can be used to test how well the thyroid gland is functioning. An exponential function can be used to describe the decay of iodine-131. After being generated by a fission reaction, iodine-131 decays into a nontoxic, stable isotope. Every 8 days the amount of iodine-131 remaining is cut in half. For example, we show in Table 5.9 and Figure 5.6 how 160 milligrams (mg) of iodine-131 decays over four time periods (or $4 \cdot 8 = 32$ days).

Time Periods (8 days each)	Amount of Iodine-131 (mg)		General Expression
0	160	=	$160 = 160 \left(\frac{1}{2}\right)^0$
1	80	=	$160 \left(\frac{1}{2}\right) = 160 \left(\frac{1}{2}\right)^1$
2	40	=	$160 \left(\frac{1}{2}\right)\left(\frac{1}{2}\right) = 160 \left(\frac{1}{2}\right)^2$
3	20	=	$160 \left(\frac{1}{2}\right)\left(\frac{1}{2}\right)\left(\frac{1}{2}\right) = 160 \left(\frac{1}{2}\right)^3$
4	10	=	$160 \left(\frac{1}{2}\right)\left(\frac{1}{2}\right)\left(\frac{1}{2}\right)\left(\frac{1}{2}\right) = 160 \left(\frac{1}{2}\right)^4$

Table 5.9

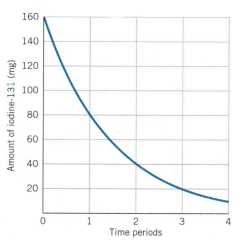

Figure 5.6 Exponential decay of iodine-131.

We can describe the decay of iodine-131 with the equation

$$Q = 160 \cdot \left(\frac{1}{2}\right)^T$$

where Q = quantity of iodine-131 (in mg) and T = number of time periods (of 8 days each). The number 160 is the initial amount of iodine-131 and $\frac{1}{2}$ is the *decay factor*.

The General Exponential Decay Function

The equation for Q is in the form

$$\text{output} = (\text{initial quantity}) \cdot (\text{decay factor})^{\text{input}}$$

This is the standard format for an exponential decay function.

> **Exponential Decay Function**
>
> An *exponential decay function* $y = f(x)$ can be represented by an equation of the form
>
> $$y = Ca^x \qquad (0 < a < 1 \text{ and } C > 0) \qquad \text{where}$$
>
> a is the *decay factor,* the amount by which y is multiplied when x increases by 1.
> C is the *initial value* or y-intercept.

EXAMPLE 1
a. How would you model the decay of 500 mg of iodine-131?
b. How many milligrams would be left after four time periods (or 32 days)?

SOLUTION
a. The initial value is 500 mg, and we know from the previous discussion that the decay factor is $\frac{1}{2}$ for each time period T (of 8 days). So the decay function is $Q = 500 \cdot \left(\frac{1}{2}\right)^T$.
b. When $T = 4$, then $Q = 500 \cdot \left(\frac{1}{2}\right)^4 = 500 \cdot \left(\frac{1}{16}\right) = 31.25$. So after four time periods (32 days) there would be 31.25 mg of iodine-131 remaining.

EXAMPLE 2 Identify the initial value and the decay factor for the function $N = (3 \cdot 10^4) \cdot (0.25)^T$. Interpret the decay factor.

SOLUTION The initial value is $3 \cdot 10^4$ (or 30,000) and the decay factor is 0.25 (or 1/4). So each time T increases by 1, the value for N is one-fourth of its previous value.

EXAMPLE 3 Which of the following exponential functions represent growth and which represent decay? Identify the growth or decay factor.
a. $P(t) = 2000(1.05)^t$ c. $N(t) = 16\left(\frac{2}{3}\right)^t$
b. $Q(t) = 25(0.75)^t$ d. $f(x) = 5(4)^{-x}$

SOLUTION
a. $P(t)$ represents exponential growth because $1.05 > 1$; the term 1.05 is the growth factor.
b. $Q(t)$ represents exponential decay because $0 < 0.75 < 1$; the term 0.75 is the decay factor.
c. $N(t)$ represents exponential decay because $0 < \frac{2}{3} < 1$; the term $\frac{2}{3}$ is the decay factor.
d. We can rewrite the function as $f(x) = 5\left(\frac{1}{4}\right)^x$. Thus $f(x)$ represents exponential decay because $0 < \frac{1}{4} < 1$; the term $\frac{1}{4}$ is the decay factor.

EXAMPLE 4 **Finding an exponential decay function through two points: Measuring caffeine levels**

When you drink an 8-ounce cup of coffee, virtually all the caffeine is absorbed in your gut and passes through the liver and into your bloodstream, acting as a stimulant. The peak blood levels of caffeine are reached in about 30 minutes—and then they begin to fall exponentially. Five hours after the peak, your blood contains 60 milligrams of caffeine. Fifteen hours after the peak your blood contains 15 milligrams of caffeine.
a. Construct a function to model the caffeine decrease in your bloodstream over time.
b. How much caffeine was in the original cup?
c. Generate a sketch of the caffeine levels in your bloodstream over time.

SOLUTION **a.** The maximum caffeine blood levels occur about 30 minutes after drinking the coffee. Since the decrease in caffeine is exponential from that point on, we can model the decline with an exponential decay function. Let t = number of hours after reaching maximum caffeine blood level and C = amount of caffeine in your blood (mg). Five hours after reaching your maximum level, your caffeine level is 60 mg. Fifteen hours after the maximum level, your caffeine level is 15 mg. Assuming C is a function of t, the two points (5, 60) and (15, 15) should satisfy that function.

If we let a = the hourly decay factor, then we can find the 10-hour decay factor, a^{10}, using the ratio of the two caffeine levels.

$$a^{10} = \frac{\text{caffeine at } t = 15 \text{ hr}}{\text{caffeine at } t = 5 \text{ hr}} = \frac{15 \text{ mg}}{60 \text{ mg}} = 0.25$$

So the 10-hour decay factor is $\qquad a^{10} = 0.25$

Taking the 10th root $\qquad (a^{10})^{1/10} = (0.25)^{1/10}$

Using a calculator $\qquad a \approx 0.871,$ the hourly decay factor

So the exponential decay function is of the form

$$C = C_0 \cdot (0.871)^t$$

where C_0 = the peak amount of caffeine and t = number of hours since the peak caffeine level.

b. To find C_0 we can substitute either of the two points (5, 60) and (15, 15) known to satisfy our function.

Given $\qquad C = C_0 \cdot (0.871)^t$

Substitute (5, 60) $\qquad 60 = C_0 \cdot (0.871)^5$

Use a calculator $\qquad 60 \approx C_0 \cdot 0.501$

Solve for C_0 $\qquad C_0 \approx \dfrac{60}{0.501} \approx 120 \text{ mg}$

So the peak amount of caffeine in your blood was approximately 120 mg. Since virtually all the caffeine in the coffee is passed into the bloodstream, your original cup contained about 120 mg of caffeine.[2]

c.

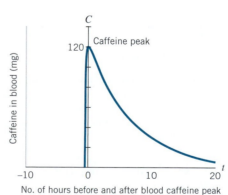

Figure 5.7 Blood caffeine levels after drinking 8 oz of coffee.

[2]Note to coffee drinkers: Consumption of 250 mg of caffeine (2 to 3 cups) is considered a moderate amount; over 800 mg of caffeine is considered excessive. Various factors can influence the rate of caffeine decay. For a pregnant woman or a woman on oral contraceptives, the decrease in caffeine slows down considerably; a little caffeine takes a long time to pass through the body. For smokers, the decrease in caffeine is speeded up, so they can actually drink more coffee without feeling its side effects.

Translating the Growth (or Decay) Factor from n Time Units to One Time Unit

If a is the growth (or decay) factor for n time units, then

$$a^{1/n}$$

is the growth (or decay) factor for one time unit.

Examples: If 2.3 is the *10-year* growth factor, then $2.3^{1/10} \approx 1.0869$ is the *annual* growth factor.

If 0.7 is the *5-month* decay factor, then $0.7^{1/5} \approx 0.9311$ is the *one-month* decay factor.

Algebra Aerobics 5.3

1. Which of the following exponential functions represent growth and which represent decay?

   a. $y = 100 \cdot 3^x$

   b. $f(t) = 75 \cdot \left(\frac{2}{3}\right)^t$

   c. $w = 250 \cdot (0.95)^r$

   d. $g(r) = (2 \cdot 10^6) \cdot (1.15)^r$

   e. $y = (7 \cdot 10^9) \cdot (0.20)^z$

   f. $h(x) = 150 \cdot \left(\frac{5}{2}\right)^x$

2. Write an equation for an exponential decay function where:

   a. The initial population is 2300 and the decay factor is $\frac{1}{3}$.

   b. The initial population is $3 \cdot 10^9$ and the decay factor is 0.35.

   c. The initial population is 375 and the population drops to one-tenth its previous size during each time period.

3. Does the exponential function $y = 12 \cdot (5)^{-x}$ represent growth or decay? (*Hint:* Rewrite the function in the standard form $y = Ca^x$.)

4. Rewrite each of the following using the general form and indicate whether the function represents growth or decay.

   a. $y = 23(2.4)^{-x}$

   b. $f(x) = 8000(0.5)^{-x}$

   c. $P = 52{,}000(1.075)^{-t}$

5. Given $f(x) = 100(0.9)^x$ and $g(x) = 100(0.7)^x$,

   a. Which function decreases more rapidly?

   b. By what percentage does each function decrease each time period?

Exercises for Section 5.3

A graphing program is recommended for Exercise 15.

1. Identify and interpret the decay factor for each of the following functions:

   a. $P = 450(0.43)^t$

   b. $f(t) = 3500(0.95)^t$

   c. $y = 21(3)^{-x}$

2. Which of the following exponential functions represent growth and which decay?

   a. $N = 50 \cdot 2.5^T$

   b. $y = 264(5/2)^x$

   c. $R = 745(1.001)^t$

   d. $g(z) = (3 \cdot 10^5) \cdot (0.8)^z$

   e. $f(T) = (1.5 \cdot 10^{11}) \cdot (0.35)^T$

   f. $h(x) = 2000\left(\frac{2}{3}\right)^x$

3. Write an equation for an exponential decay function where:

   a. The initial population is 10,000 and the decay factor is $\frac{2}{5}$.

   b. The initial population is $2.7 \cdot 10^{13}$ and the decay factor is 0.27.

   c. The initial population is 219 and the population drops to one-tenth its previous size during each time period.

4. The accompanying tables show approximate values for the four exponential functions: $f(x) = 5(2^x)$, $g(x) = 5(0.7^x)$, $h(x) = 6(1.7^x)$, and $j(x) = 6(0.6^x)$. Which table is associated with each function?

Function A		Function B	
x	y	x	y
-2	16.67	-2	10.2
-1	10.00	-1	7.1
0	6.00	0	5.0
1	3.60	1	3.5
2	2.16	2	2.5

Function C	
x	y
−2	1.25
−1	2.50
0	5.00
1	10.00
2	20.00

Function D	
x	y
−2	2.1
−1	3.5
0	6.0
1	10.0
2	17.3

5. Determine which of the following functions are exponential. For each exponential function, identify the growth or decay factor and the vertical intercept.

   a. $y = 5(x^2)$ **c.** $P = 1000(0.999)^t$

   b. $y = 100 \cdot 2^{-x}$

6. Determine which of the following functions are exponential. Identify each exponential function as representing growth or decay and find the vertical intercept.

   a. $A = 100(1.02^t)$ **d.** $y = 100x + 3$

   b. $f(x) = 4(3^x)$ **e.** $M = 2^p$

   c. $g(x) = 0.3(10^x)$ **f.** $y = x^2$

7. Fill in the missing parts of the table.

Initial Value, C	Decay Factor, a	Exponential Function $y = Ca^x$
500	0.95	
		$y = (1.72 \cdot 10^6)(0.75)^x$
	0.25	$y = 1600(\underline{\quad})^x$

8. a. Complete the following table for the exponential function $y = 20(0.75)^x$.

x	y	Difference $y_2 - y_1$	Ratio $\dfrac{y_2}{y_1}$
0	20		
1	15		
2	11.25		
3	8.4375		
4	6.328125		

   b. Choose the correct word in each italicized pair to describe the function:

   For the exponential function $y = 20(0.75)^x$, the differences are *constant/decreasing* in magnitude and the ratios are *constant/decreasing* in magnitude.

9. Match each function with its graph.

   a. $y = 100(0.8)^x$ **b.** $y = 100(0.5)^x$

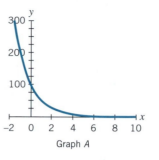

Graph A

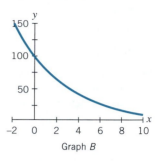

Graph B

10. The per capita (per person) consumption of milk was 27.6 gallons in 1980 and has been steadily decreasing by an annual decay factor of 0.99.

   a. Form an exponential function for per capita milk consumption $M(t)$ for year t after 1980.

   b. According to your function, what was the per capita consumption of milk in 2000? If available, use the Internet to check your predictions.

11. The U.S. Department of Agriculture's data on per capita food commodity consumption for 1980 are listed in the following table.

   a. Using the data in the following table, construct exponential functions for each food category. Then evaluate each function for the year 2000. Assume t is the number of years since 1980.

	Per Capita Consumption (pounds) in 1980	Yearly Growth/Decay Factor	Exponential Function
Beef	72.1	0.994	$B(t) =$
Chicken	32.7	1.024	$C(t) =$
Pork	52.1	0.996	$P(t) =$
Fish	12.4	1.010	$F(t) =$

   b. Which commodities showed exponential growth? Which showed exponential decay?

   c. Write a 60-second summary about the consumption of meat, chicken, and fish from 1980 to 2000.

12. Find the formula for the exponential function that satisfies the given conditions.

   a. $f(0) = 4$ and $f(1) = 2$

   b. $g(0) = 6$ and $g(1) = 3$

   c. $h(0) = 100$ and $h(1) = 75$

   d. $k(0) = 12$ and $k(1) = 7.2$

13. a. A linear function $f(t) = b + mt$ has a slope of -4 and a vertical intercept of 20. Find its equation.

   b. An exponential function $g(t) = Ca^t$ has a decay factor of 1/4 and an initial value of 20. Find its equation.

   c. Plot both functions on the same grid.

14. Which of the following functions (if any) are equivalent? Explain your answer.
    a. $f(x) = 40(0.625)^x$
    b. $g(x) = 40\left(\frac{5}{8}\right)^x$
    c. $h(x) = 40\left(\frac{8}{5}\right)^{-x}$

15. (Graphing program recommended.) Which of the following functions declines more rapidly? Graph the functions on the same grid and check your answer.
    a. $f(x) = 25(5)^{-x}$
    b. $g(x) = 25(0.5)^x$

16. Find the equation of the exponential function through the indicated points in graphs A and B.

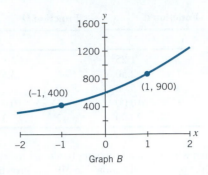

Graph B

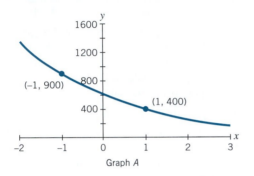

Graph A

17. Plutonium-238 is used in bombs and power plants but is dangerously radioactive. It decays very slowly into nonradio-active materials. If you started with 100 grams today, a year from now you would have 99.2 grams.
    a. Construct an exponential function to describe the decay of plutonium-238 over time.
    b. How much of the original 100 grams of plutonium-238 would be left after 50 years? After 500 years?

18. It takes 1.31 billion years for radioactive potassium-40 to drop to half its original size.
    a. Construct a function to describe the decay of potassium-40.
    b. Approximately what amount of the original potassium-40 would be left after 4 billion years? Justify your answer.

5.4 *Visualizing Exponential Functions*

We can summarize what we've learned so far about exponential functions:

Exponential Functions

Exponential functions can be represented by equations of the form

$$y = Ca^x \qquad (a > 0, a \neq 1) \qquad \text{where}$$

C is the initial value and a is the base

Assuming $C > 0$, then if

$a > 1$, the function represents growth and a is called the *growth factor*.
$0 < a < 1$, the function represents decay and a is called the *decay factor*.

The Effect of the Base *a*

Exponential growth: a > 1

Given an exponential function $y = Ca^x$, when $a > 1$ (and $C > 0$), the function represents growth. The graph of an exponential growth function is concave up and increasing. Figure 5.8 shows how the value of a affects the steepness of the graphs of the following functions:

$$y = 100 \cdot 1.2^x \qquad y = 100 \cdot 1.3^x \qquad y = 100 \cdot 1.4^x$$

Each function has the same initial value of $C = 100$ but different values of a, that is, 1.2, 1.3, and 1.4, respectively. When $a > 1$, the larger the value of a, the more rapid the growth and the more rapidly the graph rises. So as the values of x increase, the values of y not only increase, but increase at an increasing rate.

Exploration 5.1 along with "E3: $y = Ca^x$ Sliders" in *Exponential & Log Functions* allows you to examine the effects of a and C on the graph of an exponential function.

SOMETHING TO THINK ABOUT

What would the graph of the function $y = Ca^x$ look like if $a = 1$? What kind of function would you have?

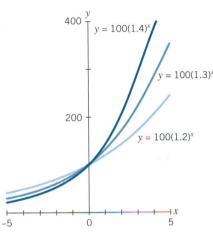

Figure 5.8 Three exponential growth functions.

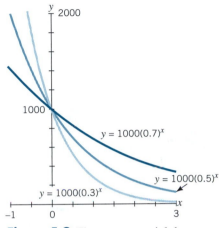

Figure 5.9 Three exponential decay functions.

Exponential decay: $0 < a < 1$

In an exponential function $y = Ca^x$, when a is positive and less than 1 (and $C > 0$), we have decay. The graph of an exponential decay function is concave up but decreasing.

The smaller the value of a, the more rapid the decay. For example, if $a = 0.7$, after each unit increase in x, the value of y would be multiplied by 0.7. So y would drop to 70% of its previous size—a loss of 30%. But if $a = 0.5$, then when x increases by 1, y would be multiplied by 0.5, equivalent to dropping to 50% of its previous size—a loss of 50%.

Figure 5.9 shows how the value of a affects the steepness of the graph. Each function has the same initial population of 1000 but different values of a, that is, 0.3, 0.5, and 0.7, respectively. When $0 < a < 1$, the smaller the value of a, the more rapid the decay and the more rapidly the graph falls.

The Effect of the Initial Value C

In the exponential function $y = Ca^x$, the initial value C is the vertical intercept. When $x = 0$, then

$$\begin{aligned} y &= Ca^0 \\ &= C \cdot 1 \\ &= C \end{aligned}$$

Figure 5.10 compares the graphs of three exponential functions with the same base a of 1.1, but with different C values of 50, 100, and 250, respectively.

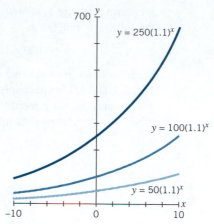

Figure 5.10 Three exponential functions with the same growth factor but different *y*-intercepts, or initial values.

In Exploration 5.1, you can examine what happens when C < 0.

Changing the value of *C* changes where the graph of the function will cross the vertical axis.

E X A M P L E   1   Match each of the graphs (*A* to *D*) in Figure 5.11 to one of the following equations and explain your answers.

a. $f(x) = 1.5(2)^x$ **c.** $j(x) = 5(0.6)^x$

b. $g(x) = 1.5(3)^x$ **d.** $k(x) = 5(0.8)^x$

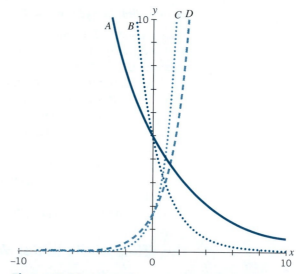

Figure 5.11 Graphs of four exponential functions.

S O L U T I O N   *A* is the graph of $k(x) = 5(0.8)^x$. *B* is the graph of $j(x) = 5(0.6)^x$.

Reasoning: Graphs *A* and *B* both have a vertical intercept of 5 and represent exponential decay. The steeper graph (*B*) must have the smaller decay factor (in this case 0.6).

C is the graph of $g(x) = 1.5(3)^x$. *D* is the graph of $f(x) = 1.5(2)^x$.

Reasoning: Graphs *C* and *D* both have a vertical intercept of 1.5 and both represent exponential growth. The steeper graph (*C*) must have the larger growth factor (in this case 3).

Horizontal Asymptotes

In the exponential decay function $y = 1000 \cdot (\frac{1}{2})^x$, graphed in Figure 5.12, the initial population of 1000 is cut in half each time x increases by 1. So when $x = 0, 1, 2, 3, 4, 5, 6, \ldots$, the corresponding y values are 1000, 500, 250, 125, 62.5, 31.25, 15.625, ... As x gets larger and larger, the y values come closer and closer to, but never reach, zero. Thus as x approaches positive infinity $(x \to +\infty)$, y approaches zero $(y \to 0)$. The graph of the function $y = 1000 (\frac{1}{2})^x$ is said to be *asymptotic* to the x-axis, or we say the x-axis is a *horizontal asymptote* to the graph. In general, a horizontal asymptote is a horizontal line that the graph of a function approaches for extreme values of the input or independent variable.

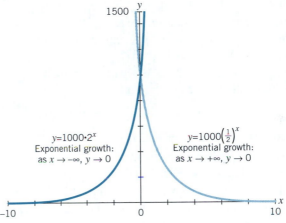

Figure 5.12 The x-axis is a horizontal asymptote for both exponential growth and decay functions of the form $y = Ca^x$.

Exponential decay functions of the form $y = Ca^x$ are asymptotic to the x-axis. Similarly, for exponential growth functions, as x approaches negative infinity $(x \to -\infty)$, y approaches zero $(y \to 0)$. Examine, for instance, the exponential growth function $y = 1000 \cdot 2^x$, also graphed in Figure 5.12. When $x = 0, -1, -2, -3, \ldots$, the corresponding y values are 1000, $1000 \cdot 2^{-1} = 500$, $1000 \cdot 2^{-2} = 250$, $1000 \cdot 2^{-3} = 125, \ldots$. So as $x \to -\infty$, the y values come closer and closer to but never reach zero. So exponential growth functions are also asymptotic to the x-axis.

Graphs of Exponential Functions

For functions in the form $y = Ca^x$,

The value of C tells us where the graph crosses the y-axis.

The value of a affects the steepness of the graph.

Exponential growth $(a > 1, C > 0)$. The larger the value of a, the more rapid the growth and the more rapidly the graph rises.

Exponential decay $(0 < a < 1, C > 0)$. The smaller the value of a, the more rapid the decay and the more rapidly the graph falls.

The graphs are *asymptotic* to the x-axis.

Exponential growth: As $x \to -\infty$, $y \to 0$
Exponential decay: As $x \to +\infty$, $y \to 0$

Algebra Aerobics 5.4

1. a. Draw a rough sketch of each of the following functions all on the same graph:

$$y = 1000(1.5)^x$$
$$y = 1000(1.1)^x$$
$$y = 1000(1.8)^x$$

b. Do the three curves intersect? If so, where?

c. In the first quadrant (where $x > 0$ and $y > 0$), which curve should be on the top? Which in the middle? Which on the bottom?

d. In the second quadrant (where $x < 0$ and $y > 0$), which curve should be on the top? Which in the middle? Which on the bottom?

2. a. Draw a rough sketch of each of the following functions all on the same graph:

$$Q = 250(0.6)^t \quad Q = 250(0.3)^t \quad Q = 250(0.2)^t$$

b. Do the three curves intersect? If so, where?

c. In the first quadrant, which curve should be on the top? Which in the middle? Which on the bottom?

d. In the second quadrant, which curve should be on the top? Which in the middle? Which on the bottom?

3. a. Draw a rough sketch of each of the following functions on the same graph:

$$P = 50 \cdot 3^t \quad \text{and} \quad P = 150 \cdot 3^t$$

b. Do the curves intersect anywhere?

c. Describe when and where one curve lies above the other.

4. Identify any horizontal asymptotes for the following functions:

a. $Q = 100 \cdot 2^t$ **c.** $y = 100 - 15x$

b. $g(r) = (6 \cdot 10^7) \cdot (0.95)^r$

5. Which of these functions has the most rapid growth? Which the most rapid decay?

a. $f(t) = 100 \cdot 1.06^t$

b. $g(t) = 25 \cdot 0.89^t$

c. $h(t) = 4000 \cdot 1.23^t$

d. $r(t) = 45.9 \cdot 0.956^t$

e. $P(t) = 32{,}000 \cdot 1.092^t$

6. Examine the graphs of the exponential growth functions $f(x)$ and $g(x)$ in Figure 5.13.

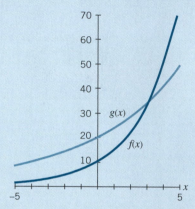

Figure 5.13 Graphs of $f(x)$ and $g(x)$.

a. Which graph has the larger initial value?

b. Which graph has the larger growth factor?

c. As $x \to -\infty$, which graph approaches zero more rapidly?

d. Approximate the point of intersection. After the point of intersection, as $x \to +\infty$, which function has larger values?

Exercises for Section 5.4

A graphing program is required for Exercises 9 and 10, and recommended for Exercises 11, 15, and 16.

1. Each of the following three exponential functions is in the standard form $y = C \cdot a^x$.

$$y = 2^x \qquad y = 5^x \qquad y = 10^x$$

a. In each case identify C and a.

b. Specify whether each function represents growth or decay. In particular, for each unit increase in x, what happens to y?

c. Do all three curves intersect? If so, where?

d. In the first quadrant, which curve should be on top? Which in the middle? Which on the bottom?

e. Describe any horizontal asymptotes.

f. For each function, generate a small table of values.

g. Graph the three functions on the same grid and verify that your predictions in part (d) are correct.

2. Repeat Exercise 1 for the functions

$$y = (0.5)2^x, \ y = 2 \cdot 2^x, \ \text{and} \ y = 5 \cdot 2^x.$$

3. Repeat Exercise 1 for the functions

$$y = 3^x, \quad y = \left(\tfrac{1}{3}\right)^x, \quad \text{and} \quad y = 3 \cdot \left(\tfrac{1}{3}\right)^x.$$

4. Match each equation with its graph (at the top of the next page).

$$f(x) = 30 \cdot 2^x \qquad\qquad h(x) = 100 \cdot 2^x$$
$$g(x) = 30 \cdot (0.5)^x \qquad\quad j(x) = 50 \cdot (0.5)^x$$

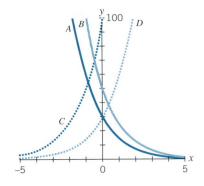

5. Below are graphs of four exponential functions. Match each function with its graph.

$$P = 5 \cdot (0.7)^x \qquad\qquad R = 10 \cdot (1.8)^x$$

$$Q = 5 \cdot (0.4)^x \qquad\qquad S = 5 \cdot (3)^x$$

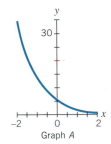

Graph A

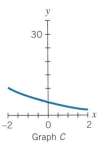

Graph C

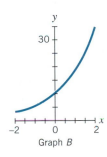

Graph B

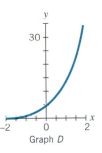

Graph D

6. Examine the accompanying graphs of three exponential growth functions.

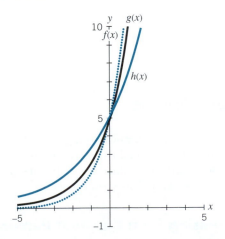

a. Order the functions from smallest growth factor to largest.

b. What point do all of the functions have in common? Will they share any other points?

c. As $x \to -\infty$, will the function with the largest growth factor approach zero more slowly or more quickly than the other functions?

d. For $x > 0$, the graph of which function remains on top?

7. Examine the accompanying graphs of three exponential functions.

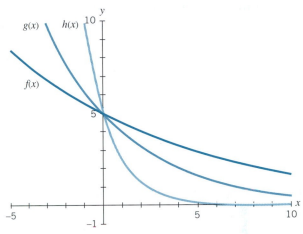

a. Order the functions from smallest decay factor to largest.

b. What point do all of the functions have in common? Will they share any other points?

c. As $x \to +\infty$, will the function with the largest decay factor approach zero more slowly or more quickly than the other functions?

d. When $x < 0$, the graph of which function remains on top?

8. Generate quick sketches of each of the following functions, without the aid of technology.

$$f(x) = 4(3.5)^x \qquad g(x) = 4(0.6)^x$$
$$h(x) = 4 + 3x \qquad k(x) = 4 - 6x$$

a. As $x \to +\infty$, which function(s) approach $+\infty$?

b. As $x \to +\infty$, which function(s) approach 0?

c. As $x \to -\infty$, which function(s) approach $-\infty$?

d. As $x \to -\infty$, which function(s) approach 0?

9. (Graphing program required.) Graph the functions $f(x) = 30 + 5x$ and $g(x) = 3(1.6)^x$ on the same grid. Supply the symbol $<$ or $>$ in the blank that would make the statement true.

a. $f(0)$ _____ $g(0)$ **e.** $f(-6)$ _____ $g(-6)$

b. $f(6)$ _____ $g(6)$ **f.** As $x \to +\infty$, $f(x)$ _____ $g(x)$

c. $f(7)$ _____ $g(7)$ **g.** As $x \to -\infty$, $f(x)$ _____ $g(x)$

d. $f(-5)$ _____ $g(-5)$

10. (Graphing program required.) Graph the functions $f(x) = 6(0.7)^x$ and $g(x) = 6(1.3)^x$ on the same grid. Supply the symbol $<$, $>$, or $=$ in the blank that would make the statement true.

a. $f(0)$ _____ $g(0)$

b. $f(5)$ _____ $g(5)$

c. $f(-5)$ _____ $g(-5)$

d. As $x \to +\infty$, $f(x)$ _____ $g(x)$

e. As $x \to -\infty$, $f(x)$ _____ $g(x)$

11. (Graphing program recommended.)
    a. As $x \to +\infty$, which function will dominate,
       $f(x) = 100 + 500x$ or $g(x) = 2(1.005)^x$?
    b. Determine over which x interval(s) $g(x) > f(x)$.

12. Two cities each have a population of 1.2 million people. City A is growing by a factor of 1.15 every 10 years, while city B is decaying by a factor of 0.85 every 10 years.
    a. Write an exponential function for each city's population $P_A(t)$ and $P_B(t)$ after t years.
    b. For each city's population function generate a table of values for $x = 0$ to $x = 50$, using 10-year intervals, then sketch a graph of each town's population on the same grid.

13. Which function has the steepest graph?

$$F(x) = 100(1.2)^x$$
$$G(x) = 100(0.8)^x$$
$$H(x) = 100(1.2)^{-x}$$

14. Examine the graphs of the following functions.

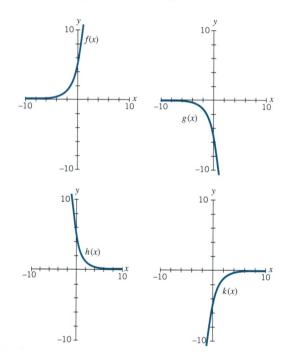

a. Which graphs appear to be mirror images of each other across the y-axis?
b. Which graphs appear to be mirror images of each other across the x-axis?
c. Which graphs appear to be mirror images of each about the origin (i.e., you could translate one into the other by reflecting first about the y-axis, then about the x-axis)?

15. (Graphing program recommended.) On the same graph, sketch $f(x) = 3(1.5)^x$, $g(x) = -3(1.5)^x$, and $h(x) = 3(1.5)^{-x}$.
    a. Which graphs are mirror images of each other across the y-axis?
    b. Which graphs are mirror images of each other across the x-axis?
    c. Which graphs are mirror images of each other about the origin (i.e., you could translate one into the other by reflecting first about the y-axis, then about the x-axis)?
    d. What can you conclude about the graphs of the two functions $f(x) = Ca^x$ and $g(x) = -Ca^x$?
    e. What can you conclude about the graphs of the two functions $f(x) = Ca^x$ and $g(x) = Ca^{-x}$?

16. (Graphing program recommended.) Make a table of values and plot each pair of functions on the same coordinate system.
    a. $y = 2^x$ and $y = 2x$ for $-3 \le x \le 3$
    b. $y = (0.5)^x$ and $y = 0.5x$ for $-3 \le x \le 3$
    c. Which of the four functions that you drew in parts (a) and (b) represent growth?
    d. How many times did the graphs that you drew for part (a) intersect? Find the coordinates of any points of intersection.
    e. How many times did the graphs that you drew for part (b) intersect? Find the coordinates of any points of intersection.

5.5   *Exponential Functions: A Constant Percent Change*

We defined exponential functions in terms of multiplication by a constant (growth or decay) factor. In this section we show how these factors can be translated into a constant percent increase or decrease. While growth and decay factors are needed to construct an equation, journalists are more likely to use percentages.

Exponential Growth: Increasing by a Constant Percent

In Section 5.1 we modeled the growth of *E. coli* bacteria with the exponential function $N = (1.37 \cdot 10^7) \cdot 1.5^t$. The growth factor, 1.5, tells us that for each unit increase in time, the initial number of *E. coli* is multiplied by 1.5. How can we translate this into a statement involving percentages?

If Q is the number of bacteria at any point in time, then one time period later,

$$\text{number of } E.\ coli = Q \cdot (1.5)$$

Rewrite 1.5 as a sum	$= Q(1 + 0.5)$
Use the distributive property	$= 1 \cdot Q + (0.5) \cdot Q$
Rewrite 0.5 as a percentage	$= Q + 50\% \text{ of } Q$

So for each unit increase in time, the number of bacteria increases by 50%. The term 50% is called the *growth rate in percentage form* and 0.5, the equivalent in decimal form, is called the *growth rate in decimal form*.

A quantity that increases by a constant percent over each time interval represents exponential growth. We can represent this growth in terms of growth factors or growth rates. Exponential growth requires a growth factor > 1. If the exponential growth factor $= 1$, the object would never grow, but would stay constant through time. So,

$$\text{growth factor} = 1 + \text{growth rate}$$
$$\text{growth factor} = 1 + r$$

where r is the growth rate in decimal form.

EXAMPLE 1

Converting the growth rate to a growth factor
The U.S. Bureau of the Census has made a number of projections for the first half of this century. For each projection, convert the growth rate into a growth factor.

a. Disposable income is projected to grow at an annual rate of 2.9%.

b. Nonagriculture employment is projected to grow at 1.0% per year.

c. Employment in manufacturing is projected to grow at 0.2% per year.

SOLUTION

a. A growth rate of 2.9% is 0.029 in decimal form. The growth factor is $1 + 0.029 = 1.029$.

b. A growth rate of 1.0% is 0.01 in decimal form. The growth factor is $1 + 0.01 = 1.01$.

c. A growth rate of 0.2% is 0.002 in decimal form. The growth factor is $1 + 0.002 = 1.002$.

EXAMPLE 2

According to UN statistics, in the year 2005 there were an estimated 3.2 billion people living in urban areas. The number of people living in urban areas is projected to grow at a rate of 1.7% per year. If this projection is accurate, how many people will be living in urban areas in 2030?

SOLUTION

If the urban population is increasing by a constant percent each year, the growth is exponential. To construct an exponential function modeling this growth, we need to find the growth factor. We know that the growth rate = 1.7% or 0.017. So,

$$\text{growth factor} = 1 + 0.017 = 1.017$$

Given an initial population of 3.2 billion in 2005, our model is

$$U(n) = 3.2 \cdot (1.017)^n$$

where n = number of years since 2005 and $U(n)$ = urban population (in billions).

Using our model, the urban population in 2030, $U(25) \approx 4.9$ billion people.

Exponential Decay: Decreasing by a Constant Percent

In Section 5.3, we used the exponential function $C = C_0 \cdot (0.871)^t$ (where C_0 = the peak amount of caffeine and t = number of hours since peak caffeine level) to model the amount of caffeine in the body after drinking a cup of coffee. The decay factor, 0.871, tells us that for each unit increase in time, the amount of caffeine is multiplied by 0.871. As with exponential growth, we can translate this statement into one that involves percentages.

If Q is the amount of caffeine at any point in time, then one time period later,

amount of caffeine	$= Q \cdot (0.871)$
Rewrite 0.871 as a difference	$= Q \cdot (1 - 0.129)$
Use the distributive property	$= 1 \cdot Q - (0.129) \cdot Q$
Rewrite 0.129 as a percentage	$= Q - 12.9\% \text{ of } Q$

So for each unit increase in time, the amount of caffeine decreases by 12.9%. The term 12.9% is called the *decay rate in percentage form* and the equivalent decimal form, 0.129, is called the *decay rate in decimal form*.

A quantity that decreases by a constant percent over each time interval represents exponential decay. We can represent this decay in terms of decay factors or decay rates. Exponential decay requires a decay factor between 0 and 1.

$$\text{decay factor} = 1 - \text{decay rate}$$
$$\text{decay factor} = 1 - r$$

where r is the decay rate in decimal form.

Factors, Rates, and Percentages

For an exponential function in the form $y = Ca^x$, where $C > 0$,

The *growth factor* $a > 1$ can be represented as

$$a = 1 + r$$

where r is the *growth rate,* the decimal representation of the percent rate of change.

The *decay factor* a, where $0 < a < 1$, can be represented as

$$a = 1 - r$$

where r is the *decay rate,* the decimal representation of the percent rate of change.

Examples: If a quantity is growing by 3% per month, the growth rate r in decimal form is 0.03, so the growth factor $a = 1 + 0.03 = 1.03$.

If a quantity is decaying by 3% per month, the decay rate r in decimal form is 0.03, so the decay factor $a = 1 - 0.03 = 0.97$.

EXAMPLE 3   **Constructing an exponential decay function**

Sea ice that survives the summer and remains year round, called perennial sea ice, is melting at the alarming rate of 9% per decade, according to a 2006 report by the National Oceanic and Atmospheric Administration. Assuming the percent decrease per decade in sea ice remains constant, construct a function that represents this decline.

SOLUTION   If the percent decrease remains constant, then the melting of sea ice represents an exponential function with a decay rate of 0.09 in decimal form. So,

$$\text{decay factor} = 1 - 0.09 = 0.91$$

If d = number of decades since 2006 and A = amount of sea ice (in millions of acres) in 2006, then we can model the amount of sea ice, $S(d)$, by the exponential function:

$$S(d) = A(0.91)^d$$

EXAMPLE 4   **Growth or decay?**

In the following functions, identify the growth or decay factors and the corresponding growth or decay rates in both percentage and decimal form.

a. $f(x) = 100(0.01)^x$ **c.** $p = 5.34(0.015)^n$

b. $g(x) = 230(3.02)^x$ **d.** $y = 8.75(2.35)^x$

SOLUTION   **a.** The decay factor is 0.01 and if r is the decay rate in decimal form, then

$$\text{decay factor} = 1 - \text{decay rate}$$
$$0.01 = 1 - r$$
$$r = 0.99$$

The decay rate is 0.99, or 99% as a percentage.

b. The growth factor is 3.02 and if r is the growth rate in decimal form, then

$$\text{growth factor} = 1 + \text{growth rate}$$
$$3.02 = 1 + r$$
$$3.02 - 1 = r$$
$$r = 2.02$$

The growth rate in decimal form is 2.02, or 202% as a percentage.

c. The decay factor is 0.015 and the decay rate is 0.985 in decimal form, or 98.5% as a percentage.

d. The growth factor is 2.35 and the growth rate is 1.35 in decimal form, or 135% as a percentage.

Revisiting Linear vs. Exponential Functions

In Section 5.2 we compared linear vs. exponential functions as additive versus multiplicative. We now have another way to compare them.

Linear vs. Exponential Functions

Linear functions represent quantities that increase or decrease by a constant amount. Exponential functions represent quantities that increase or decrease by a constant percent.

EXAMPLE 5   **Identifying linear vs. exponential growth**

According to industry sources, U.S.wireless data service revenues are growing by an amazing 75% annually. India is the biggest growth market, increasing by about 6 million cell phones every month. Which of these two statements represents linear and which represents exponential growth? Justify your answers.

SOLUTION   Quantities that increase by a constant amount represent linear growth. So, an increase of 6 million cell phones every month describes linear growth. Quantities that increase by a constant percent represent exponential growth. So, an increase of 75% annually describes exponential growth.

EXAMPLE 6   Construct an equation for each description of the population of four different cities.

a. In 1950, city A had a population of 123,000 people. Each year since 1950, the population decreased by 0.8%.

b. In 1950, city B had a population of 4,500,000 people. Since 1950, the population has grown by approximately 2000 people per year.

c. In 1950, city C had a population of 625,000. Since 1950, the population has declined by about 5000 people each decade.

d. In 1950, city D had a population of 2.1 million people. Each decade since 1950, the population has increased by 15%.

SOLUTION For each of the following equations, let P = population (in thousands) and t = years since 1950.

a. $P_A = 123(0.992)^t$

b. $P_B = 4500 + 2t$

where the unit for rate of change is people (in thousands) per year.

c. $P_C = 625 - 0.5t$

where the unit for rate of change is people (in thousands) per year. The rate of change is -0.5 people (in thousands) per *year*, since the population declined by 5000 people per *decade*, or 500 people per *year*.

d. growth rate per decade = 0.15
growth factor per decade = $1 + 0.15 = 1.15$
growth factor per year = $(1.15)^{1/10} \approx 1.014$
$P_D = 2100(1.014)^t$

EXAMPLE 7 You are offered a job with a salary of \$30,000 a year and annual raises of 5% for the first 10 years of employment. Show why the average rate of change of your salary in dollars per year is not constant.

SOLUTION Let n = number of years since you signed the contract and $S(n)$ = your salary in dollars after n years. When $n = 0$, your salary is represented by $S(0) = \$30,000$. At the end of your first year of employment, your salary increases by 5%, so

$$S(1) = 30{,}000 + 5\% \text{ of } 30{,}000$$
$$= 30{,}000 + 0.05\,(30{,}000)$$
$$= 30{,}000 + 1500$$
$$= \$31{,}500$$

At the end of your second year, your salary increases again by 5%, so

$$S(2) = 31{,}500 + 5\% \text{ of } 31{,}500$$
$$= 31{,}500 + 0.05(31{,}500)$$
$$= 31{,}500 + 1575$$
$$= \$33{,}075$$

The average rate of change between years 0 and 1 is

$$\frac{S(1) - S(0)}{1} = \$31{,}500 - \$30{,}000 = \$1500$$

The average rate of change between years 1 and 2 is

$$\frac{S(2) - S(1)}{1} = \$33{,}075 - \$31{,}500 = \$1575$$

So, the average rates of change differ. This is to be expected, since in the first year the 5% raise applies only to the initial salary of \$30,000, whereas in the second year the 5% raise applies to both the initial \$30,000 salary and to the \$1500 raise from the first year.

Algebra Aerobics 5.5

A calculator that can evaluate powers is recommended for Problem 4.

1. Complete the following statements.

   a. A growth factor of 1.22 corresponds to a growth rate of ____.

   b. A growth rate of 6.7% corresponds to a growth factor of ____.

   c. A decay factor of 0.972 corresponds to a decay rate of ____.

   d. A decay rate of 12.3% corresponds to a decay factor of ____.

2. Fill in the table below.

3. Determine the growth or decay factor and the growth or decay rate in percentage form.

   a. $f(t) = 100 \cdot 1.06^t$

   b. $g(t) = 25 \cdot 0.89^t$

   c. $h(t) = 4000 \cdot 1.23^t$

   d. $r(t) = 45.9 \cdot 0.956^t$

   e. $P(t) = 32,000 \cdot 1.092^t$

4. Find the percent increase (or decrease) if:

   a. Hourly wages grew from $1.65 per hour to $6.00 per hour in 30 years.

   b. A player's batting average dropped from 0.299 to 0.167 in one season.

   c. The number of new AIDS cases grew from 17 last month to 23 this month.

Exponential Function	Initial Value	Growth or Decay?	Growth or Decay Factor	Growth or Decay Rate (as %)
$A = 4(1.03)^t$				
$A = 10(0.98)^t$				
	1000		1.005	
	30		0.96	
	$50,000	Growth		7.05%
	200 grams	Decay		49%

Exercises for Section 5.5

Access to the internet and a graphing program is required for Exercise 26 part (e). A graphing program is recommended for Exercises 16, 24, and 25.

1. Assume that you start with 1000 units of some quantity Q. Construct an exponential function that will describe the value of Q over time T if, for each unit increase in T, Q increases by:

   a. 300%   b. 30%   c. 3%   d. 0.3%

2. Given each of the following exponential growth functions, identify the growth rate in percentage form:

   a. $Q = 10,000 \cdot 1.5^T$ d. $Q = 10,000 \cdot 2^T$

   b. $Q = 10,000 \cdot 1.05^T$ e. $Q = 10,000 \cdot 2.5^T$

   c. $Q = 10,000 \cdot 1.005^T$ f. $Q = 10,000 \cdot 3^T$

3. Given the following exponential decay functions, identify the decay rate in percentage form.

   a. $Q = 400(0.95)^t$ d. $y = 200(0.655)^x$

   b. $A = 600(0.82)^t$ e. $A = 10(0.996)^T$

   c. $P = 70,000(0.45)^t$ f. $N = 82(0.725)^T$

4. What is the growth or decay factor for each given time period?

   a. Weight increases by 0.2% every 5 days.

   b. Mass decreases by 6.3% every year.

   c. Population increases 23% per decade.

   d. Profit increases 300% per year.

   e. Blood alcohol level decreases 35% per hour.

5. Fill in the following chart and then construct exponential functions for each part (a) to (g).

	Initial Value	Growth or Decay?	Growth or Decay Factor	Growth or Decay Rate (% form)
a.	600		2.06	
b.	1200			200%
c.	6000	Decay		75%
d.	1.5 million	Decay		25%
e.	1.5 million	Growth		25%
f.	7		4.35	
g.	60		0.35	

6. Match the statements (a) through (d) with the correct exponential function in (e) through (h). Assume time t is measured in the unit indicated.

   a. Radon-222 decays by 50% every t days.

   b. Money in a savings account increases by 2.5% per year.

   c. The population increases by 25% per decade.

   d. The pollution in a stream decreases by 25% every year.

   e. $A = 1000(1.025)^t$

   f. $A = 1000(0.75)^t$

   g. $A = 1000(\frac{1}{2})^t$

   h. $A = 1000(1.25)^t$

7. Each of two towns had a population of 12,000 in 1990. By 2000 the population of town A had increased by 12% while the population of town B had decreased by 12%. Assume these growth and decay rates continued.

   a. Write two exponential population models $A(T)$ and $B(T)$ for towns A and B, respectively, where T is the number of decades since 1990.

   b. Write two new exponential models $a(t)$ and $b(t)$ for towns A and B, where t is the number of *years* since 1990.

   c. Now find $A(2)$, $B(2)$, $a(20)$, and $b(20)$ and explain what you have found.

8. On November 25, 2003, National Public Radio did a report on Under Armour, a sports clothing company, stating that their "profits have increased by 1200% in the last 5 years."

   a. Let $P(t)$ represent the profit of the company during every 5-year period, with A_0 the initial amount. Write the exponential model for the company's profit.

   b. Assuming an initial profit of $100,000, what would be the profit in year 5? Year 10?

   c. Determine the *annual* growth rate for Under Armour.

Each of the tables in Exercises 9–12 represents an exponential function. Construct that function and then identify the corresponding growth or decay rate in percentage form.

9.

x	0	1	2	3
y	500.00	425.00	361.25	307.06

10.

x	0	1	2	3
y	500.00	575.00	661.25	760.44

11.

x	0	1	2	3
y	225.00	228.38	231.80	235.28

12.

x	0	1	2	3
y	225.00	221.63	218.30	215.03

13. Generate equations that represent the pollution levels, $P(t)$, as a function of time, t (in years), such that $P(0) = 150$ and:

   a. $P(t)$ triples each year.

   b. $P(t)$ decreases by twelve units each year.

   c. $P(t)$ decreases by 7% each year.

   d. The annual average rate of change of $P(t)$ with respect to t is constant at 1.

14. Given an initial value of 50 units for parts (a)–(d) below, in each case construct a function that represents Q as a function of time t. Assume that when t increases by 1:

   a. $Q(t)$ doubles

   b. $Q(t)$ increases by 5%

   c. $Q(t)$ increases by ten units

   d. $Q(t)$ is multiplied by 2.5

15. Between 1970 and 2000, the United States grew from about 200 million to 280 million, an increase of approximately 40% in this 30-year period. If the population continues to expand by 40% every 30 years, what will the U.S. population be in the year 2030? In 2060? Use technology to estimate when the population of the United States will reach 1 billion.

16. (Graphing program recommended.) According to Mexico's National Institute of Geography, Information and Statistics, in 2005 in Mexico's Quintana Roo state, where the tourist industry in Cancun has created a boom economy, the population was about 1.14 million and growing at a rate of 5.3% per year. In 2005 in Mexico's Baja California state, where many labor-intensive industries are located next to the California border, the population was about 2.8 million and growing at a rate of 3.3% per year. The nation's capital, Mexico City, had 19.2 million inhabitants as of 2005 and the population was declining at a rate of 0.012% per year.

   a. Construct three exponential functions to model the growth or decay of Quintana Roo, Baja California, and Mexico City. Identify your variables and their units.

   b. Use your functions to predict the populations of Quintana Roo, Baja California, and Mexico City in 2010.

17. A pollutant was dumped into a lake, and each year its amount in the lake is reduced by 25%.

   a. Construct a general formula to describe the amount of pollutant after n years if the original amount is A_0.

   b. How long will it take before the pollution is reduced to below 1% of its original level? Justify your answer.

18. A swimming pool is initially shocked with chlorine to bring the chlorine concentration to 3 ppm (parts per million). Chlorine dissipates in reaction to bacteria and sun at a rate of about 15% per day. Above a chlorine concentration of 2 ppm, swimmers experience burning eyes, and below a concentration of 1 ppm, bacteria and algae start to proliferate in the pool environment.

   a. Construct an exponential decay function that describes the chlorine concentration (in parts per million) over time.

   b. Construct a table of values that corresponds to monitoring chlorine concentration for at least a 2-week period.

   c. How many days will it take for the chlorine to reach a level tolerable for swimmers? How many days before bacteria and algae will start to grow and you will need to add more chlorine? Justify your answers.

19. a. If the inflation rate is 0.7% a month, what is it per year?

   b. If the inflation rate is 5% a year, what is it per month?

20. Rewrite each expression so that no fraction appears in the exponent and each expression is in the form a^x.

   a. $3^{x/4}$ b. $2^{x/3}$ c. $\left(\frac{1}{2}\right)^{x/4}$ d. $\left(\frac{1}{4}\right)^{x/2}$

   Hint: Remember that $a^{x/n} = (a^{1/n})^x$

21. a. Show that the following two functions are roughly equivalent.

$$f(x) = 15,000(1.2)^{x/10} \quad \text{and} \quad g(x) = 15,000(1.0184)^x$$

   b. Create a table of values for each function for $0 < x \le 25$.

   c. Would it be correct to say that a 20% increase over 10 years represents a 1.84% annual increase? Explain your answer.

22. The exponential function $Q(T) = 600(1.35)^T$ represents the growth of a species of fish in a lake, where T is measured in 5-year intervals.

   a. Determine $Q(1)$, $Q(2)$, and $Q(3)$.

   b. Find another function $q(t)$, where t is measured in years.

   c. Determine $q(5)$, $q(10)$, and $q(15)$.

   d. Compare your answers in parts (a) and (c) and describe your results.

23. Blood alcohol content (BAC) is the amount of alcohol present in your blood as you drink. It is calculated by determining how many grams of alcohol are present in 100 milliliters (1 deciliter) of blood. So if a person has 0.08 grams of alcohol per 100 milliliters of blood, the BAC is 0.08 g/dl. After a person has stopped drinking, the BAC declines over time as his or her liver metabolizes the alcohol. Metabolism proceeds at a steady rate and is impossible to speed up. For instance, an average (150-lb) male metabolizes about 8 to 12 grams of alcohol an hour (the amount in one bottle of beer[3]). The behavioral effects of alcohol are closely related to the blood alcohol content. For example, if an average 150-lb male drank two bottles of beer within an hour, he would have a BAC level of 0.05 and could suffer from euphoria, inhibition, loss of motor coordination, and overfriendliness. The same male after drinking four bottles of beer in an hour would be legally drunk with a BAC of 0.10. He would likely suffer from impaired motor function and decision making, drowsiness, and slurred speech. After drinking twelve beers in one hour, he will have attained the dosage for stupor (0.30) and possibly death (0.40).

The following table gives the BAC of an initially legally drunk person over time (assuming he doesn't drink any additional alcohol).

Time (hours)	0	1	2	3	4
BAC (g/dl)	0.100	0.067	0.045	0.030	0.020

   a. Graph the data from the table (be sure to carefully label the axes).

   b. Justify the use of an exponential function to model the data. Then construct the function where $B(t)$ is the BAC for time t in hours.

   c. By what percentage does the BAC decrease every hour?

   d. What would be a reasonable domain for your function? What would be a reasonable range?

   e. Assuming the person drinks no more alcohol, when does the BAC reach 0.005 g/dl?

[3]One drink is equal to $1\frac{1}{4}$ oz of 80-proof liquor, 12 oz of beer, or 4 oz of table wine.

24. (Graphing program recommended.) If you have a heart attack and your heart stops beating, the amount of time it takes paramedics to restart your heart with a defibrillator is critical. According to a medical report on the evening news, each minute that passes decreases your chance of survival by 10%. From this wording it is not clear whether the decrease is linear or exponential. Assume that the survival rate is 100% if the defibrillator is used immediately.

   a. Construct and graph a linear function that describes your chances of survival. After how many minutes would your chance of survival be 50% or less?

   b. Construct and graph an exponential function that describes your chances of survival. Now after how many minutes would your chance of survival be 50% or less?

25. (Graphing program recommended.) The infant mortality rate (the number of deaths per 1000 live births) fell in the United States from 7.2 in 1996 to 6.4 in 2006.

   a. Assume that the infant mortality rate is declining linearly over time. Construct an equation modeling the relationship between infant mortality rate and time, where time is measured in years since 1996. Make sure you have clearly identified your variables.

   b. Assuming that the infant mortality rate is declining exponentially over time, construct an equation modeling the relationship, where time is measured in years since 1996.

   c. Graph both of your models on the same grid.

   d. What would each of your models predict for the infant mortality rate in 2010?

26. [Part (e) requires use of the Internet and technology to find a best-fit function.] A "rule of thumb" used by car dealers is that the trade-in value of a car decreases by 30% each year.

   a. Is this decline linear or exponential?

   b. Construct a function that would express the value of the car as a function of years owned.

   c. Suppose you purchase a car for $15,000. What would its value be after 2 years?

   d. Explain how many years it would take for the car in part (c) to be worth less than $1000. Explain how you arrived at your answer.

   e. *Internet search:* Go to the Internet site for the Kelley Blue Book (*www.kbb.com*).

   i. Enter the information about your current car or a car you would like to own. Specify the actual age and mileage of the car. What is the Blue Book value?

   ii. Keeping everything else the same, assume the car is 1 year older and increase the mileage by 10,000. What is the new value?

   iii. Find a best-fit exponential function to model the value of your car as a function of years owned. What is the annual decay rate?

   iv. According to this function, what will the value of your car be 5 years from now?

5.6 *Examples of Exponential Growth and Decay*

Exponential and linear behaviors are ones you will frequently (though perhaps unknowingly) encounter throughout life. Here we'll take a look at some examples showing the wide range of applications of exponential functions.

EXAMPLE 1

Fitting a curve to data: Medicare costs
The costs for almost every aspect of health care in America have risen dramatically over the last 30 years. One of the central issues is the amount of dollars spent on Medicare, a federal program that provides health care for nearly all people age 65 and over. Estimate the average annual percentage increase in Medicare expenses.

SOLUTION

Table 5.10 shows Medicare expenses from 1970 to 2005 as reported by the federal government. In column 2, the years are reinitialized to become the years since 1970.

DATA

See Excel or graph link file
MEDICARE.

Medicare Expenses

Year	Years since 1970	Medicare Expenses (billions of dollars)
1970	0	7.7
1975	5	16.3
1980	10	37.1
1985	15	71.5
1990	20	109.5
1995	25	184.4
2000	30	224.3
2005	35	342.0

Table 5.10
Source: www.census.gov.

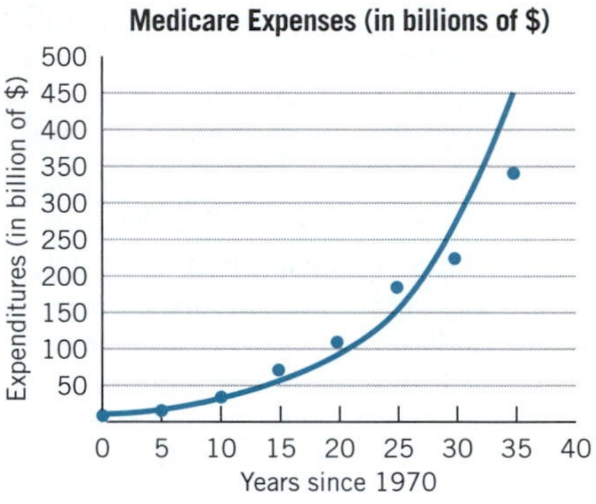

Figure 5.14 Medicare expenses (in billions) with best-fit exponential model.

A curve-fitting program gives a best-fit exponential function as

$$C(n) = 10.6 \cdot (1.11)^n$$

where n = number of years since 1970 and $C(n)$ is the associated cost in billions of dollars. The data and best-fit exponential curve are plotted in Figure 5.14.

In our model the initial value (in year 0 or in 1970) is \$10.6 billion. This differs from the actual value in 1970 of \$7.7 billion, since the model is the best fit to all the data and does not necessarily include any of the original data points. More important, the growth factor is 1.11, so the growth rate is 0.11 in decimal form. So our model estimates that between 1970 and 2005 Medicare expenses were increasing by 11% each year, an amount that far exceeds the general inflation rate.

EXAMPLE 2

Changing the time unit: Credit card debt
In 2007 the average American household owed a balance of almost \$9200 on credit cards.[4] For late payments, credit card companies typically charge a high annual percentage rate (APR) that could be anywhere from 10% to 20%.

a. Assuming you have a \$500 balance on a credit card with a 20% APR, construct a model for your debt over time.

b. What is your *monthly* interest rate?

c. If you made no payments and no charges for 6 months, how much would you owe?

SOLUTION

a. Since the debt is increasing by a constant percent, the growth is exponential. Let T = number of years and D = credit card debt in dollars. An APR of 20% means that

[4]*www.cardweb.com*

the annual growth rate in decimal form is 0.20, and the annual growth factor is 1.20. So given an initial value of $500, an exponential model would be

$$D = 500 \cdot (1.20)^T$$

b. We know that the *yearly* growth factor is 1.20. If a stands for the *monthly* growth factor, then

$$a^{12} = 1.20$$

To find the value for a we can take the twelfth root of both sides.

$$(a^{12})^{1/12} = (1.20)^{1/12}$$

or $\qquad\qquad a \approx 1.015$

Since the monthly growth factor is 1.015, the monthly growth rate is 0.015 in decimal form. So a yearly interest rate of 20% translates into a monthly interest rate of about 1.5%.

c. To find your debt after 6 months, we need the monthly exponential growth function. If we let t = number of *months* and 1.015 is the *monthly* growth factor, then

$$D \approx 500 \cdot (1.015)^t$$

Note that we have transformed debt (D) as a function of number of years (T) into a function of number of months (t). Assuming you made no payments, after 6 months you would owe

$$D \approx 500 \cdot (1.015)^6 = \$547$$

So you would owe $47 in interest (almost 10% of the initial amount) for the "privilege" of borrowing $500 for 6 months.

Half-Life and Doubling Time

Every exponential growth function has a fixed doubling time, and every exponential decay function has a fixed half-life.

> **Doubling Time and Half-Life**
>
> The *doubling time* of an exponentially growing quantity is the time required for the quantity to double in size.
>
> The *half-life* of an exponentially decaying quantity is the time required for one-half of the quantity to decay.

EXAMPLE 3 **Constructing an exponential function given the doubling time or half-life**

In each situation construct an exponential function.

a. The doubling time is 3 years; the initial quantity is 1000.

b. The half-life is 5 days; the initial quantity is 50.

SOLUTION **a.** A doubling time of 3 years describes an exponentially growing quantity that doubles every 3 years. We can construct an exponential function,

$$P(t) = 1000a^t \qquad \text{where } a \text{ is the yearly} \qquad (1)$$
$$\text{growth factor and}$$
$$t = \text{number of years.}$$

Since the doubling time is 3 years, then

$$a^3 = 2$$

Taking the cube root of each side $\quad a = 2^{1/3}$

Substituting in (1) $P(t) = 1000 \cdot (2^{1/3})^t$

Rule 3 for exponents $P(t) = 1000 \cdot 2^{t/3}$

b. A half-life of 5 days means an exponentially decaying quantity that decreases by half every 5 days. We can construct an exponential decay function,

$$Q(t) = 50a^t \qquad \text{where } a \text{ is the daily} \qquad (2)$$
$$\text{decay factor and}$$
$$t = \text{number of days.}$$

Since the half-life is 5 days, then

$$a^5 = \tfrac{1}{2}$$

Taking the fifth root of each side $a = \left(\tfrac{1}{2}\right)^{1/5}$

Substituting in (2) and

Using Rule 3 for exponents $Q(t) = 50 \cdot \left(\tfrac{1}{2}\right)^{t/5}$

Writing an exponential function using the doubling time (or half-life)

If an exponential growth function $G(t) = Ca^t$ has a doubling time of n, then

$$a^n = 2$$
$$a = 2^{1/n}$$

so, $G(t) = C \cdot 2^{t/n}$

If an exponential decay function $D(t) = Ca^t$ has a half-life of n, then

$$a^n = \tfrac{1}{2}$$
$$a = \left(\tfrac{1}{2}\right)^{1/n}$$

so, $D(t) = C \cdot \left(\tfrac{1}{2}\right)^{t/n}$

For example, if $G(t) = 500a^t$ has a doubling time of 6 hours, then $G(t) = 500 \cdot 2^{t/6}$, where $t = $ number of hours.

If $D(t) = 25a^t$ has a half-life of 3 minutes, then $D(t) = 25 \cdot \left(\tfrac{1}{2}\right)^{t/3}$, where $t = $ number of minutes.

EXAMPLE 4   **Half-life: Radioactive decay**

One of the toxic radioactive by-products of nuclear fission is strontium-90. A nuclear accident, like the one in Chernobyl, can release clouds of gas containing strontium-90. The clouds deposit the strontium-90 onto vegetation eaten by cows, and humans ingest strontium-90 from the cows' milk. The strontium-90 then replaces calcium in bones, causing cancer and birth defects. Strontium-90 is particularly insidious because it has a *half-life* of approximately 28 years. That means that every 28 years about half (or 50%) of the existing strontium-90 has decayed into nontoxic, stable zirconium-90, but the other half still remains.

a. Construct a model for the decay of 100 mg of strontium-90 as a function of 28-year time periods.

b. Construct a new model that describes the decay as a function of the number of years. Generate a corresponding table and graph.

SOLUTION   **a.** Strontium-90 decays by 50% every 28 years. If we define our basic time period T as 28 years, then the decay rate is 0.5 and the decay factor is $1 - 0.5 = 0.5$ (or 1/2). Assuming an initial value of 100 mg, the function

$$S = 100 \cdot (0.5)^T$$

gives S, the remaining milligrams of strontium-90 after T time periods (of 28 years).

b. The number 0.5 is the decay factor over a 28-year period. If we let $a = $ *annual* decay factor, then

$$a^{28} = 0.5$$

To find the value for a we can take the 28th root of both sides.

$$(a^{28})^{1/28} = (0.5)^{1/28}$$

or
$$a \approx 0.976$$

So if $t = $ number of *years,* the *yearly* decay factor is 0.976, making our function

$$S \approx 100 \cdot (0.976)^t$$

We now have an equation that describes the amount of strontium-90 as a function of years, t. The decay factor is 0.976, so the decay rate is $1 - 0.976 = 0.024$, or 2.4%. So each year there is 2.4% less strontium-90 than the year before.

Decay of Strontium-90

Time T (28-year time periods)	Time t (years)	Strontium-90 (mg)
0	0	100
1	28	50
2	56	25
3	84	12.5
4	112	6.25
5	140	3.125

Table 5.11

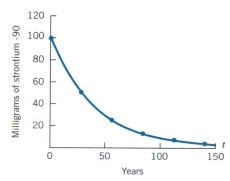

Figure 5.15 Radioactive decay of strontium-90.

Table 5.11 and Figure 5.15 show the decay of 100 mg of strontium-90 over time. After 28 years, half (50 mg) of the original 100 mg still remains. It takes an additional 28 years for the remaining amount to halve again, still leaving 25 grams out of the original 100 after 56 years.

EXAMPLE 5

Identifying the doubling time or half-life
Assuming t is in years, identify the doubling time or half-life for each of the following exponential functions.

a. $Q(t) = 80(2)^{t/12}$ **b.** $P(t) = 5 \cdot 10^6 \left(\frac{1}{2}\right)^{4t}$ **c.** $R(t) = 130(0.5)^{3t}$

SOLUTION

a. When $t = 12$ years, the quantity $Q(12) = 80 \cdot 2^{12/12} = 80 \cdot 2^1 = 160$. So the quantity has doubled from the initial value of 80 and the doubling time is 12 years.

b. When $t = 1/4$, the quantity $P(1/4) = (5 \cdot 10^6)(1/2)^{4(1/4)} = (5 \cdot 10^6)(1/2)^1$. So the quantity is half the initial value of $5 \cdot 10^6$ and the half-life is 1/4 years or 3 months.

c. When $t = 1/3$, the quantity $R(t) = 130(0.5)^{3(1/3)} = 130(0.5)^1 = 65$. So the quantity is half the initial value of 130 and the half-life is 1/3 year or 4 months.

The "rule of 70": A rule of thumb for calculating doubling or halving times

A simple way to understand the significance of constant percent growth rates is to compute the doubling time. A rule of thumb is that a quantity growing at $R\%$ per year has a doubling time of approximately $70/R$ years. If the quantity is growing at $R\%$ per *month*, then $70/R$ gives its doubling time in *months*. The same reasoning holds for any unit of time.

The rule of 70 is easy to apply. For example, if a quantity is growing at a rate of 2% a year, then the doubling time is about 70/2 or approximately 35 years. The rule of 70 also applies when R represents the percentage at which some quantity is decaying. In these cases, $70/R$ equals the half-life.

For now we'll take the rule of 70 on faith. It provides good approximations, especially for smaller values of R (those under 10%). When we return to logarithms in Chapter 6, we'll find out why this rule works.

EXAMPLE 6 Suppose that at age 23 you invest $1000 in a retirement account that grows at 5% per year.

    a. Roughly how long will it take your investment to double?

    b. If you retire at age 65, approximately how much will you have in your account?

SOLUTION **a.** Using the rule of 70, we get 70/5 = 14, so your investment will double approximately every 14 years.

    b. If you retire at 65, then 42 years or three doubling periods will have elapsed ($3 \cdot 14 = 42$). So your $1000 investment (disregarding inflation) will have increased by a factor of 2^3 and be worth about $8000.

EXAMPLE 7 According to Mexico's National Institute of Geography, Information and Statistics (at *www.inegi.gob.mx/difusion/ingles*), in 1997 Mexico's population reached 93.7 million. The annual growth rate is listed as approximately 1.4%. The website states that "if this rate persists, the Mexican population will double in 49.9 years." Does that time period seem about right?

SOLUTION Using the rule of 70, the approximate doubling time for a 1.4% annual growth rate would be 70/1.4 = 50 years. So 49.9 is a reasonable value for the doubling time.

The Rule of 70

The *rule of 70* states that if a quantity is growing (or decaying) at $R\%$ per time period, then the doubling time (or half-life) is approximately

$$\frac{70}{R}$$

provided R is not much bigger than 10.

    Examples: If a quantity is growing at 7% per year, the doubling time is approximately 70/7 = 10 years.

    If a quantity is decaying at 5% per minute, the half-life is approximately 70/5 = 14 minutes.

EXAMPLE 8 Plutonium, the fuel for atomic weapons, has an extraordinarily long half-life, about 24,400 years. Once the radioactive element plutonium is created from uranium, 24,400 years later half the original amount will still remain. You can see why there is concern over stored caches of atomic weapons. Use the rule of 70 to estimate plutonium's annual decay rate.

SOLUTION

The rule of 70 says	$70/R = 24{,}400$
Multiply by R	$70 = 24{,}400R$
Divide and simplify	$R = 70/24{,}400$
	≈ 0.003

Remember that R is already in percentage, not decimal, form. So the annual decay rate is a tiny 0.003%, or three-thousandths of 1%.

? SOMETHING TO THINK ABOUT

An atomic weapon is usually designed with a 1% mass margin. That is, it will remain functional until the original fuel has decayed by more than 1%, leaving less than 99% of the original amount. Estimate how many years a plutonium bomb would remain functional.

Algebra Aerobics 5.6a

A calculator that can evaluate powers is recommended for Problem 2.

1. For each of the following exponential functions (with t in days) determine whether the function represents exponential growth or decay, then estimate the doubling time or half-life. For each function find the initial amount and the amount after one day, one month, and one year. (*Note:* To make calculations easier, use 360 days in one year.)
   a. $f(t) = 300 \cdot 2^{t/30}$
   b. $g(t) = 32 \cdot 0.5^{t/2}$
   c. $P = 32{,}000 \cdot 0.5^{t}$
   d. $h(t) = 40{,}000 \cdot 2^{t/360}$
2. Estimate the doubling time using the rule of 70.
   a. $g(x) = 100(1.02)^x$, where x is in years.
   b. $M = 10{,}000(1.005)^t$, where t is in months.
   c. The annual growth rate is 8.1%.
   d. The annual growth factor is 1.065.
3. Use the rule of 70 to approximate the growth rate when the doubling time is:
   a. 10 years **b.** 5 minutes **c.** 25 seconds
4. Estimate the time it will take an initial quantity to drop to half its value when:
   a. $h(x) = 10(0.95)^x$, where x is in months.
   b. $K = 1000(0.75)^t$, where t is in seconds.
   c. The annual decay rate is 35%.

EXAMPLE 9 **White Blood Cell Counts**

On September 27 a patient was admitted to Brigham and Women's Hospital in Boston for a bone marrow transplant. The transplant was needed to cure myelodysplastic syndrome, in which the patient's own marrow fails to produce enough white blood cells to fight infection.

The patient's bone marrow was intentionally destroyed using chemotherapy and radiation, and on October 3 the donated marrow was injected. Each day, the hospital carefully monitored the patient's white blood cell count to detect when the new marrow became active. The patient's counts are plotted in Figure 5.16. Normal counts for a healthy individual are between 4000 and 10,000 cells per milliliter.

DATA
CELCOUNT.

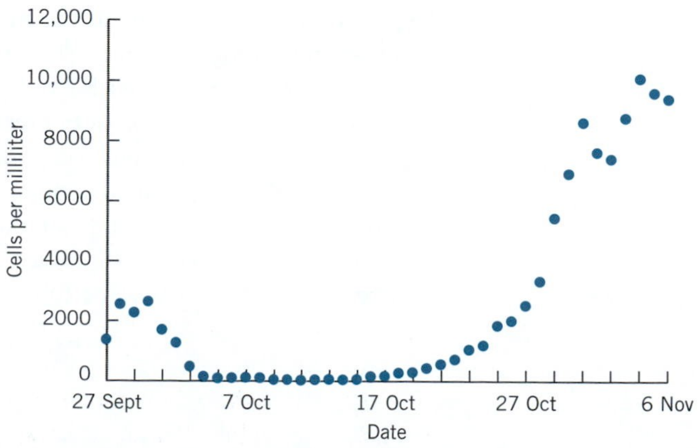

Figure 5.16 White blood cell count.

What was the daily percent increase and estimated doubling time during the period of exponential growth?

SOLUTION   Clearly, the whole data set does not represent exponential growth, but we can reasonably model the data between, say, October 15 and 31 with an exponential function. Figure 5.17 shows a plot of this subset of the original data, along with a computer-generated best-fit exponential function, where t = number of days after October 15.

$$W(t) = 105(1.32)^t$$

The fit looks quite good.

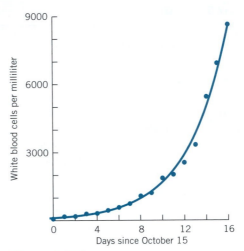

Figure 5.17 White blood cell counts with a best-fit exponential function between October 15 and October 31.

The initial quantity of 105 is the white blood cell count (per milliliter) for the model, not the actual white blood cell count of 70 measured by the hospital. This discrepancy is not unusual; remember that a best-fit function may not necessarily pass through any of the specific data points.

The growth factor is 1.32, so the growth rate is 0.32. So between October 15 and 31, the number of white blood cells was increasing at a rate of 32% a day. According to the rule of 70, the number of white blood cells was doubling roughly every $70/32 = 2.2$ days!

Compound Interest Rates

Exponential functions occur frequently in finance. Probably the most common application is compound interest. For example, suppose you have $100 in a passbook savings account that returns 1% *compounded annually*. That means that at the end of the first year, you earn 1% in interest on the initial $100, called the *principal*. From then on you earn 1% not only on your principal, but also on the interest that you have already earned. This is called compounding and represents exponential growth.

If the growth rate is 1%, or 0.01 in decimal form, then the growth factor is $1 + 0.01 = 1.01$. So each subsequent year, the current value of your savings account will be multiplied by 1.01. The following function models the growth:

$$P_1 = 100 \cdot (1.01)^t$$
$$= 100 \cdot (1 + 0.01)^t$$

$$\text{Value of account} = \left(\begin{array}{c}\text{original}\\\text{investment}\end{array}\right) \cdot (1 + \text{interest rate})^{\text{no. of years}}$$

> **Calculating Compound Interest**
>
> If
>
> $$P_0 = \text{original investment}$$
> $$r = \text{interest rate (in decimal form)}$$
> $$t = \text{time periods at which the interest rate is compounded}$$
>
> the resulting value, P_r, of the investment after t time periods is given by the formula
>
> $$P_r = P_0 \cdot (1 + r)^t$$

EXAMPLE 10

Comparing investments

Suppose you have $10,000 that you could put in either a checking account that earns no interest, a mutual fund that earns 5% a year, or a risky stock investment that you hope will return 15% a year.

a. Model the potential growth of each investment.

b. Compare the investments after 10, 20, 30, and 40 years.

SOLUTION

a. Assuming you make no withdrawals, the money in the checking account will remain constant at $10,000. For both the mutual fund and the stock investment you are expecting a constant annual percent rate, so you can think of them as compounding annually. The mutual fund's predicted annual growth rate is 5%, so its growth factor would be 1.05. The stock's annual growth rate (you hope) will be 15%, so its growth factor would be 1.15. The initial value is $10,000, so after t years

$$\text{value of mutual fund} \quad = 10,000 \cdot (1.05)^t$$
$$\text{value of stock investment} = 10,000 \cdot (1.15)^t$$

SOMETHING TO THINK ABOUT

On March 13, 1997, a resident of Melrose, Massachusetts, "scratched" a $5 state lottery ticket and won $1,000,000. The Massachusetts State Lottery Commission notified her that after deducting taxes, they would send her a check for $33,500 every March 13 for 20 years. Is this fair?

b. Table 5.12 shows the value of each investment over time.

Number of Years	Checking Account (0% per year)	Mutual Fund (5% per year)	Stock (15% per year)
0	$10,000	$10,000	$10,000
10	$10,000	$16,289	$40,456
20	$10,000	$26,533	$163,665
30	$10,000	$43,219	$662,118
40	$10,000	$70,400	$2,678,635

Table 5.12

Forty years from now, your checking account will still have $10,000, but your mutual fund will be worth $70,400. Should you have been lucky enough to have invested in the next Microsoft, 40 years from now, with a 15% interest rate compounded annually, your $10,000 would have become well over $2.5 million.

Inflation and the diminishing dollar

Compound calculations are the same whether you are dealing with inflation or investments. For example, a 5% annual inflation rate would mean that what cost $1 today would cost $1.05 one year from today. If we think of the percentages in Table 5.14 as representing inflation rates, then something that costs $10,000 today in 10 years will cost $10,000, $16,289, or $40,456 if the annual inflation rate is, respectively, 0%, 5%, or 15%.

SOMETHING TO THINK ABOUT

What happens if the inflation rate is higher than the interest rate on your investments?

EXAMPLE 11

How much will your money be worth in 20 years?

Suppose you have a $100,000 nest egg hidden under your bed, awaiting your retirement. In 20 years, how much will your nest egg be worth in today's dollars if the annual inflation rate is:

a. 3%? **b.** 7%?

SOLUTION **a.** Given a 3% inflation rate, what costs $1 today will cost $1.03 next year. Using a ratio to compare costs, we get

$$\frac{\text{cost now}}{\text{cost in 1 year}} = \frac{\$1.00}{\$1.03} \approx 0.971$$

Rewriting $\qquad\qquad$ cost now $\approx 0.971 \cdot (\text{cost in 1 year})$

Our equation shows that next year's dollars need to be multiplied by 0.971 to reduce them to the equivalent value in today's dollars. An annual inflation rate of 3% means that in one year, $1 will be worth only $0.971 \cdot (\$1) = \$0.971$, or about 97 cents in today's dollars. Each additional year reduces the value of your nest egg by a factor of 0.971; in other words, 0.971 is the decay factor. So the value of your nest egg in real dollars is given by

$$V_{3\%}(n) = \$100{,}000 \cdot 0.971^n$$

where $V_{3\%}(n)$ is the real value (measured in today's dollars) of your nest egg after n years with 3% annual inflation. When $n = 20$, we have

$$V_{3\%}(20) = \$100{,}000 \cdot 0.971^{20} \approx \$100{,}000 \cdot 0.555 = \$55{,}500$$

b. Similarly, since $\$1.00/\$1.07 \approx 0.935$, then

$$V_{7\%}(n) = \$100{,}000 \cdot 0.935^n$$

where $V_{7\%}(n)$ is the real value (measured in today's dollars) of your nest egg after n years with 7% annual inflation. When $n = 20$, we have

$$V_{7\%}(20) = \$100{,}000 \cdot 0.935^{20} \approx \$100{,}000 \cdot 0.261 = \$26{,}100$$

So assuming 3% annual inflation, after 20 years your nest egg of $100,000 in real dollars would be almost cut in half. With a 7% inflation your nest egg in real dollars would only be about one-quarter of the original amount.

Real or Constant Dollars To make meaningful comparisons among dollar values in different years, economists use *real* or *constant dollars,* dollars that are adjusted for inflation. For example, the U.S. Census Bureau reported that in 2004 the median income of households was $44,389 and in 1998 only $38,885 in 1998 dollars. This suggests an increase of about $5500. But measured in 2004 dollars, the median income in 1998 was $45,003. So in real (inflation adjusted) dollars, there was a decrease in median household income.

Algebra Aerobics 5.6b

A calculator that can evaluate powers is recommended for Problem 4.

1. Approximately how long would it take for your money to double if the interest rate, compounded annually, were:

  a. 3%? $\qquad$ **b.** 5%? $\qquad$ **c.** 7%?

2. Suppose you are planning to invest a sum of money. Estimate the rate that you need so that your investment doubles in:

  a. 5 years $\qquad$ **b.** 10 years $\qquad$ **c.** 7 years

3. Construct a function that represents the resulting value if you invested $1000 for n years at an annually compounded interest rate of:

  a. 4% $\qquad$ **b.** 11% $\qquad$ **c.** 110%

4. In the early 1980s Brazil's inflation was running rampant at about 10% per *month*.[5] Assuming this inflation rate continued unchecked, construct a function to describe the purchasing power of 100 cruzeiros after n months. (A cruzeiro is a Brazilian monetary unit.) What would 100 cruzeiros be worth after 3 months? After 6 months? After a year?

[5]In such cases sooner or later the government usually intervenes. In 1985 the Brazilian government imposed an anti-inflationary wage and price freeze. When the controls were dropped, inflation soared again, reaching a high in March 1990 of 80% per month! At this level, it begins to matter whether you buy groceries in the morning or wait until that night. On August 2, 1993, the government devalued the currency by defining a new monetary unit, the cruzeiro real, equal to 1000 of the old cruzeiros. Inflation still continued, and on July 1, 1994, yet another unit, the real, was defined equal to 2740 cruzeiros reales. By 1997, inflation had slowed considerably to about 0.1% per month.

5. For each of the following equations, determine the initial investment, the growth factor, and the growth rate, and estimate the time it will take to double the investment.

     a. $A = \$10,000 \cdot 1.065^t$ c. $A = \$300,000 \cdot 1.11^t$

     b. $A = \$25 \cdot 1.08^t$ d. $A = \$200 \cdot 1.092^t$

6. Fill in Table 5.13.

Function	Initial Investment	Growth Factor	Growth Rate	Amount 1 Year Later	Doubling Time
	$50,000		7.2%		
	$100,000			$106,700	
			5.8%	$52,500	
	$3,000	1.13			

Table 5.13

EXAMPLE 12

Musical pitch

There is an exponential relationship between musical octaves and vibration frequency. The vibration frequency of the note A above middle C is 440 cycles per second (or 440 hertz). The vibration frequency doubles at each octave.

   a. Construct a function that gives the vibration frequency as a function of octaves. Construct a corresponding table and graph.

   b. Rewrite the function as a function of individual note frequencies. (*Note:* There are twelve notes to an octave.)

SOLUTION

   a. Let F be the vibration frequency in hertz (Hz) and N the number of octaves above or below the chosen note A. Since the frequency doubles over each octave, the growth factor is 2. If we set the initial frequency at 440 Hz, then we have

$$F = 440 \cdot 2^N$$

Table 5.14 and the graph in Figure 5.18 show a few of the values for the vibration frequency.

"E6: Musical Keyboard Frequencies" in *Exponential & Log Functions* offers a multimedia demonstration of this function.

Musical Octaves

Number of Octaves above or below A (at 440 Hz), N	Vibration Frequency (Hz), F
−3	55
−2	110
−1	220
0	440
1	880
2	1760
3	3520

Table 5.14

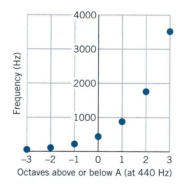

Figure 5.18 Octaves versus frequency.

   b. There are twelve notes in each octave. If we let n = number of notes above or below A, then we have $n = 12N$. Solving for N, we have $N = n/12$. By substituting this expression for N, we can define F as a function of n.

$$F = 440 \cdot 2^{n/12}$$
$$= 440 \cdot 2^{(1/12) \cdot n}$$
$$= 440 \cdot (2^{1/12})^n$$
$$\approx 440 \cdot (1.059)^n$$

So each note on the "even-tempered" scale has a frequency 1.059 times the frequency of the preceding note.

The Malthusian Dilemma

The most famous attempt to predict growth mathematically was made by a British economist and clergyman, Thomas Robert Malthus, in an essay published in 1798. He argued that the growth of the human population would overtake the growth of food supplies, because the population size was *multiplied* by a fixed amount each year, whereas food production only increased by *adding* a fixed amount each year. In other words, he assumed populations grew exponentially and food supplies grew linearly as functions of time. He concluded that humans are condemned always to breed to the point of misery and starvation, unless the population is reduced by other means, such as war or disease.

EXAMPLE 13

Constructing the Malthusian equations

Malthus believed that the population of Great Britain, then about 7,000,000, was growing by 2.8% per year. He counted food supply in units that he defined to be enough food for one person for a year. At that time the food supply was adequate, so he assumed that Britons were producing 7,000,000 food units. He predicted that food production would increase by about 280,000 units a year.

a. Construct functions modeling Britain's population size and the amount of food units over time.

b. Plot the functions from part (a) on the same grid. Estimate when the population would exceed the food supply.

SOLUTION

a. Since the population is assumed to increase by a constant *percent* each year, an exponential model is appropriate. Given an initial size of 7,000,000 and an annual growth rate of 2.8%, the exponential function

$$P(t) = 7,000,000 \cdot (1.028)^t$$

describes the population growth over time, where t is the number of years and $P(t)$ is the corresponding population size.

Since the food units are assumed to grow by a constant *amount* each year, a linear model is appropriate. Given an initial value of 7,000,000 food units and an annual constant rate of change of 280,000, the linear function

$$F(t) = 7,000,000 + 280,000t$$

describes the food unit increase over time, where t = number of years and $F(t)$ = number of food units in year t.

Figure 5.19 reveals that if the formulas were good models, then after about 25 years the population would start to exceed the food supply, and some people would starve.

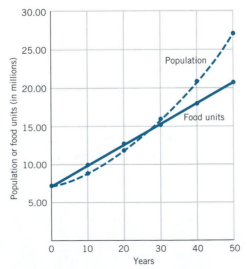

Figure 5.19 Malthus's predictions for population and food.

The two centuries since Malthus published his famous essay have not been kind to his theory. The population of Great Britain in 2002 was about 58 million, whereas Malthus's model predicted over 100 million people before the year 1900. Improved food production techniques and the opening of new lands to agriculture have kept food production in general growing faster than the population. The distribution of food is a problem and famines still occur with unfortunate regularity in parts of the world, but the mass starvation Malthus predicted has not come to pass.

Forming a Fractal Tree

Tree structures offer another useful way of visualizing exponential growth. A computer program can generate a tree by drawing two branches at the end of a trunk, then two smaller branches at the ends of each of those branches, and two smaller branches at the ends of the previous branches, and so on until a branch reaches twig size. This kind of structure produced from self-similar repeating scaled graphic operations is called a *fractal* (Figure 5.20). There are many examples of fractal structures in nature, such as ferns, coastlines, and human lungs.

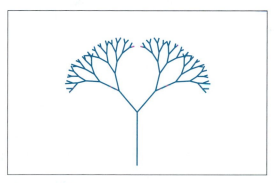

Figure 5.20 A fractal tree.

The fractal tree shown is drawn in successive levels (Figure 5.21):

- At level 0 the program draws the trunk, one line.
- At level 1 it draws two branches on the previous one, for a total of two new lines.
- At level 2 it draws two branches on each of the previous two, for a total of four new lines.
- At level 3 it draws two branches on each of the previous four, for a total of eight new lines.

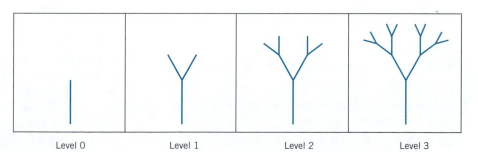

| Level 0 | Level 1 | Level 2 | Level 3 |

Figure 5.21 Forming a fractal tree.

Table 5.15 shows the relationship between the level, L, and the number of new lines, N, at each level. The formula

$$N = 2^L$$

describes the relationship between level, L, and the number of new lines, N. For example, at the fifth level, there would be $2^5 = 32$ new lines.

? SOMETHING TO THINK ABOUT

If you look back at your family tree forty generations ago (roughly 800 years), you had 2^{40} ancestors. This number is larger than all the people that ever lived on the surface of the earth. How can that be?

Level, L	New Lines, N	N as Power of 2
0	1	$1 = 2^0$
1	$2 \cdot 1 = 2$	$2 = 2^1$
2	$2 \cdot 2 = 4$	$4 = 2 \cdot 2 = 2^2$
3	$2 \cdot 4 = 8$	$8 = 2 \cdot 2 \cdot 2 = 2^3$
4	$2 \cdot 8 = 16$	$16 = 2 \cdot 2 \cdot 2 \cdot 2 = 2^4$

Table 5.15

EXAMPLE 14 You can think of Figure 5.20 as depicting your family tree. Each level represents a generation. The trunk is you (at level 0). The first two branches are your parents (at level 1). The next four branches are your grandparents (at level 2), and so on. How many ancestors do you have ten generations back?

SOLUTION The answer is $2^{10} = 1024$; that is, you have 1024 great-great-great-great-great-great-great-great-grandparents.

EXAMPLE 15 An information system that is similar to this process is an emergency phone tree, in which one person calls two others, each of whom calls two others, until everyone in the organization has been called. How many levels of phone calls would be needed to reach an organization with 8000 people?

SOLUTION If we think of Figure 5.20 and Table 5.15 as representing this phone tree, then each new line (or branch) represents a person. We need to count not just the number of new people N at each level L, but also all the previous people called.

At level 0, there is the one person who originates the phone calls. At level 1, there is the original person plus the two he or she called, for a total of $1 + 2 = 3$ people. At level 2, there are $1 + 2 + 4 = 7$ people, etc. At level 11, there are $1 + 2 + 4 + 8 + 16 + 32 + 64 + 128 + 256 + 512 + 1024 + 2048 = 4095$ people who have been called. At level 12, there would be $2^{12} = 4096$ new people called, for a total of $4095 + 4096 = 8191$ people called. So it would take twelve levels of the phone tree to reach 8000 people.

Algebra Aerobics 5.6c

1. Identify the value of x that would make each of the following equations a true statement.

a. $2^x = 32$ **d.** $2^x = 2$ **g.** $2^x = \frac{1}{8}$

b. $2^x = 256$ **e.** $2^x = 1$ **h.** $2^x = \sqrt{2}$

c. $2^x = 1024$ **f.** $2^x = \frac{1}{2}$

2. Assume the tree-drawing process was changed to draw three branches at each level.

a. Draw a trunk and at least two levels of the tree.

b. What would the general formula be for N, the number of new lines, as a function of L, the level?

3. In an emergency phone tree in which one person calls three others, each of whom calls three others, and so on, until everyone in the organization has been called, how many levels of phone calls are required for this phone tree to reach an organization of 8000 people?

Exercises for Section 5.6

Technology for finding a best-fit function is required for Exercises 7, 9, 17, and 18. Internet access is required for Exercises 9 part (c), 29 part (c), and 32 part (e).

1. Which of the following functions have a fixed doubling time? A fixed half-life?

   a. $y = 6(2)^x$ **c.** $Q = 300\left(\frac{1}{2}\right)^T$ **e.** $P = 500 - \frac{1}{2}T$

   b. $y = 5 + 2x$ **d.** $A = 10(2)^{t/5}$ **f.** $N = 50\left(\frac{1}{2}\right)^{t/20}$

2. Identify the doubling time or half-life of each of the following exponential functions. Assume t is in years. [*Hint:* What value of t would give you a growth (or decay) factor of 2 (or 1/2)?]

   a. $Q = 70(2)^t$ **d.** $Q = 100\left(\frac{1}{2}\right)^{t/250}$

   b. $Q = 1000(2)^{t/50}$ **e.** $N = 550(2)^{t/10}$

   c. $Q = 300\left(\frac{1}{2}\right)^t$ **f.** $N = 50\left(\frac{1}{2}\right)^{t/20}$

3. Fill in the following chart. (The first column is done for you.)

	a.	b.	c.	d.
Initial Value	50	1000	4	5000
Doubling time	30 days	7 years	25 minutes	18 months
Exponential function ($f(t) = Ca^{t/n}$)	$f(t) = 50(2)^{t/30}$			
Growth factor per unit of time	$2^{1/30} = 1.0234$ per day			
Growth rate per unit of time (in percentage form)	2.34% per day			

4. Make a table of values for the dotted points on each graph. Using your table, determine if each graph has a fixed doubling time or a fixed half-life. If so, create a function formula for that graph.

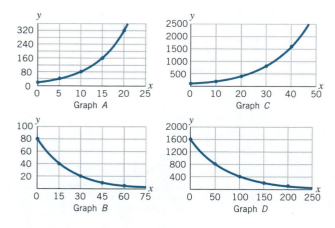

Graph A

Graph C

Graph B

Graph D

5. Insert the symbol $>$, $<$, or $\approx$ (approximately equal) to make the statement true. Assume $x > 0$.

   a. $3(2)^{x/5}$ _____ $3(1.225)^x$

   b. $50\left(\frac{1}{2}\right)^{x/20}$ _____ $50(0.9659)^x$

   c. $200(2)^{x/8}$ _____ $200(1.0905)^x$

   d. $750\left(\frac{1}{2}\right)^{x/165}$ _____ $750(0.911)^x$

6. Lead-206 is not radioactive, so it does not spontaneously decay into lighter elements. Radioactive elements heavier than lead undergo a series of decays, each time changing from a heavier element into a lighter or more stable one. Eventually, the element decays into lead-206 and the process stops. So, over billions of years, the amount of lead in the universe has increased because of the decay of numerous radioactive elements produced by supernova explosions.

Radioactive uranium-238 decays sequentially into thirteen other lighter elements until it stabilizes at lead-206. The half-lives of the fifteen different elements in this decay chain vary from 0.000 164 seconds (from polonium-214 to lead-210) all the way up to 4.47 billion years (from uranium-238 to thorium-234).

   a. Find the decay rate per billion years for uranium-238 to decay into thorium-234.

   b. Find the decay rate per second for polonium-214 to decay into lead-210.

7. (Requires technology to find a best-fit function.) We have seen the accompanying table and graph of the U.S. population at the beginning of Chapter 2.

USPOP

Population of the United States, 1790–2000

Year	Millions	Year	Millions
1790	3.9	1900	76.2
1800	5.3	1910	92.2
1810	7.2	1920	106.0
1820	9.6	1930	123.2
1830	12.9	1940	132.2
1840	17.1	1950	151.3
1850	23.2	1960	179.3
1860	31.4	1970	203.3
1870	39.8	1980	226.5
1880	50.2	1990	248.7
1890	63.0	2000	281.4

Source: U.S. Bureau of the Census, *www.census.gov.*

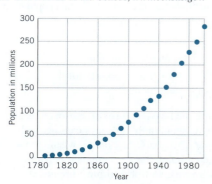

a. Find a best-fit exponential function. (You may want to set 1790 as year 0.) Be sure to clearly identify the variables and their units for time and population. What is the annual growth factor? The growth rate? The estimated initial population?

b. Graph your function and the actual U.S. population data on the same grid. Describe how the estimated population size differs from the actual population size. In what ways is this exponential function a good model for the data? In what ways is it flawed?

c. What would your model predict the population to be in the year 2010? In 2025?

8. (Requires results from Exercise 7.) According to a letter published in the Ann Landers column in the *Boston Globe* on Friday, December 10, 1999, "When Elvis Presley died in 1977, there were 48 professional Elvis impersonators. In 1996, there were 7,328. If this rate of growth continues, by the year 2012, one person in every four will be an Elvis impersonator."

a. What was the growth factor in the number of Elvis impersonators for the 19 years between 1977 and 1996?

b. What would be the *annual* growth factor in number of Elvis impersonators between 1977 and 1996?

c. Construct an exponential function that describes the growth in number of Elvis impersonators since 1977.

d. Use your function to estimate the number of Elvis impersonators in 2012.

e. Use your model for the U.S. population from Exercise 7 to determine if in 2012 one person out of every four will be an Elvis impersonator. Explain your reasoning.

9. (Requires technology to find a best-fit function.) Reliable data on Internet use are hard to find, but *World Telecommunications Indicators* cites estimates of 3 million U.S. users in 1991, 30 million in 1996, 166 million in 2002, 199 million in 2004 and 232 million in 2007.

a. Use technology to plot the data, and generate a best-fit linear and a best-fit exponential function for the data. Which do you think is the better model?

b. What would the linear model predict for Internet usage in 2010? What would the exponential model predict?

c. *Internet use:* Go online and see if you can find the number of current internet users in the U.S. Which of your models turned out to be more accurate?

10. China is the most populous country in the world. In 2000 it had about 1.262 billion people. By 2005 the population had grown to 1.306 billion. Use this information to construct models predicting the size of China's population in the future.

a. Identify your variables and units.

b. Construct a linear model.

c. Construct an exponential model.

d. What will China's population be in 2050 according to each of your models?

11. Tritium, the heaviest form of hydrogen, is a critical element in a hydrogen bomb. It decays exponentially with a half-life of about 12.3 years. Any nation wishing to maintain a viable hydrogen bomb has to replenish its tritium supply roughly every 3 years, so world tritium supplies are closely watched.

Construct an exponential function that shows the remaining amount of tritium as a function of time as 100 grams of tritium decays (about the amount needed for an average size bomb). Be sure to identify the units for your variables.

12. (Graphing program recommended.) Cosmic ray bombardment of the atmosphere produces neutrons, which in turn react with nitrogen to produce radioactive carbon-14. Radioactive carbon-14 enters all living tissue through carbon dioxide (via plants). As long as a plant or animal is alive, carbon-14 is maintained in the organism at a constant level. Once the organism dies, however, carbon-14 decays exponentially into carbon-12. By comparing the amount of carbon-14 to the amount of carbon-12, one can determine approximately how long ago the organism died. Willard Libby won a Nobel Prize for developing this technique for use in dating archaeological specimens. The half-life of carbon-14 is about 5730 years. In answering the following questions, assume that the initial quantity of carbon-14 is 500 milligrams.

a. Construct an exponential function that describes the relationship between A, the amount of carbon-14 in milligrams, and t, the number of 5730-year time periods.

b. Generate a table of values and plot the function. Choose a reasonable set of values for the domain. Remember that the objects we are dating may be up to 50,000 years old.

c. From your graph or table, estimate how many milligrams are left after 15,000 years and after 45,000 years.

d. Now construct an exponential function that describes the relationship between A and T, where T is measured in years. What is the annual decay factor? The annual decay rate?

e. Use your function in part (d) to calculate the number of milligrams that would be left after 15,000 years and after 45,000 years.

13. The body eliminates drugs by metabolism and excretion. To predict how frequently a patient should receive a drug dosage, the physician must determine how long the drug will remain in the body. This is usually done by measuring the half-life of the drug, the time required for the total amount of drug in the body to diminish by one-half.

a. Most drugs are considered eliminated from the body after five half-lives, because the amount remaining is probably too low to cause any beneficial or harmful effects. After five half-lives, what percentage of the original dose is left in the body?

b. The accompanying graph shows a drug's concentration in the body over time, starting with 100 milligrams.

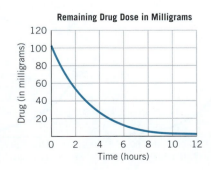

Use the given graph to answer the following questions.

   i. Estimate the half-life of the drug.

  ii. Construct an equation that approximates the curve. Specify the units of your variables.

 iii. How long would it take for five half-lives to occur? Approximately how many milligrams of the original dose would be left then?

 iv. Write a paragraph describing your results to a prospective buyer of the drug.

14. Estimate the doubling time using the rule of 70 when:

  a. $P = 2.1(1.0475)^t$, where t is in years

  b. $Q = 2.1(1.00475)^T$, where T is in years

15. Use the rule of 70 to approximate the growth rate when the doubling time is:

  a. 5730 years   **c.** 5 seconds

  b. 11,460 years   **d.** 10 seconds

16. Estimate the time it will take an initial quantity to drop to half its value when:

  a. $P = 3.02(0.998)^t$, with t in years

  b. $Q = 12(0.75)^T$, with T in decades

17. (Requires technology to find a best-fit function.) Estimates for world population vary, but the data in the accompanying table are reasonable estimates.

WORLDPOP

World Population

Year	Total Population (millions)
1800	980
1850	1260
1900	1650
1950	2520
1970	3700
1980	4440
1990	5270
2000	6080
2005	6480

Source: United Nations Population Division, *www.undp.org/popin.*

a. Enter the data table into a graphing program (you may wish to enter 1800 as 0, 1850 as 50, etc.) or use the data file WORLDPOP in Excel or in graph link form.

b. Generate a best-fit exponential function.

c. Interpret each term in the function, and specify the domain and range of the function.

d. What does your model give for the growth rate?

e. Using the graph of your function, estimate the following:

  i. The world population in 1750, 1920, 2025, and 2050

 ii. The approximate number of years in which world population attained or will attain 1 billion (i.e., 1000 million), 4 billion, and 8 billion

f. Estimate the length of time your model predicts it takes for the population to double from 4 billion to 8 billion people.

18. (Requires technology to find a best-fit function.) In 1911, reindeer were introduced to St. Paul Island, one of the Pribilof Islands, off the coast of Alaska in the Bering Sea. There was plenty of food and no hunting or reindeer predators. The size of the reindeer herd grew rapidly for a number of years, as given in the accompanying table.

REINDEER

Population of Reindeer Herd

Year	Population Size	Year	Population Size
1911	17	1925	246
1912	20	1926	254
1913	42	1927	254
1914	76	1928	314
1915	93	1929	339
1916	110	1930	415
1917	136	1931	466
1918	153	1932	525
1919	170	1933	670
1920	203	1934	831
1921	280	1935	1186
1922	229	1936	1415
1923	161	1937	1737
1924	212	1938	2034

Source: V.B. Scheffer, "The rise and fall of a reindeer herd," *Scientific Monthly, 73:356–362, 1951.*

a. Use the reindeer data file (in Excel or graph link form) to plot the data.

b. Find a best-fit exponential function.

c. How does the predicted population from part (b) differ from the observed ones?

d. Does your answer in part (c) give you any insights into why the model does not fit the observed data perfectly?

e. Estimate the doubling time of this population.

19. In medicine and biological research, radioactive substances are often used for treatment and tests. In the laboratories of a large East Coast university and medical center, any waste containing radioactive material with a half-life under 65 days must be stored for 10 half-lives before it can be disposed of with the non-radioactive trash.

a. By how much does this policy reduce the radioactivity of the waste?

b. Fill out the accompanying chart and develop a general formula for the amount of radioactive pollution at any period, given an initial amount, A_0.

Number of Half-Life Periods	Pollution Amount
0	A_0, original amount
1	$A_1 = 0.5A_0$
2	A_2
3	
4	
Period n	

20. Belgrade, Yugoslavia (from *USA Today*, September 21, 1993):

A 10 billion dinar note hit the streets today. . . . With inflation at 20% per day, the note will soon be as worthless as the 1 billion dinar note issued last month. A year ago the biggest note was 5,000 dinars. . . . In addition to soaring inflation, which doubles prices every 5 days, unemployment is at 50%.

In the excerpt, inflation is described in two very different ways. Identify these two descriptions in the text and determine whether they are equivalent. Justify your answer.

21. It takes 3 months for a malignant lung tumor to double in size. At the time a lung tumor was detected in a patient, its mass was 10 grams.

a. If untreated, determine the size in grams of the tumor at each of the listed times in the table. Find a formula to express the tumor mass M (in grams) at any time t (in months).

t, Time (months)	M, Mass (g)
0	10
3	
6	
9	
12	

b. Lung cancer is fatal when a tumor reaches a mass of 2000 grams. If a patient diagnosed with lung cancer went untreated, estimate how long he or she would survive after the diagnosis.

c. By what percentage of its original size has the 10-gram tumor grown when it reaches 2000 grams?

22. It is now recognized that prolonged exposure to very loud noise can damage hearing. The accompanying table gives the permissible daily exposure hours to very loud noises as recommended by OSHA, the Occupational Safety and Health Administration.

Sound Level, D (decibels)	Maximum Duration, H (hours)
120	0
115	0.25
110	0.5
105	1
100	2
95	4
90	8

a. Examine the data for patterns. How is D progressing? How is H progressing? Do the data represent a growth or a decay phenomenon? Explain your answer.

b. Find a formula for H as a function of D. Fit the data as closely as possible. Graph your formula and the data on the same grid.

23. a. Construct a function that would represent the resulting value if you invested $5000 for n years at an annually compounded interest rate of:

i. 3.5% **ii.** 6.75% **iii.** 12.5%

b. If you make three different $5000 investments today at the three different interest rates listed in part (a), how much will each investment be worth in 40 years?

24. A bank compounds interest annually at 4%.

a. Write an equation for the value V of $100 in t years.

b. Write an equation for the value V of $1000 in t years.

c. After 20 years will the total interest earned on $1000 be ten times the total interest earned on $100? Why or why not?

25. (Graphing program recommended.) You have a chance to invest money in a risky investment at 6% interest compounded annually. Or you can invest your money in a safe investment at 3% interest compounded annually.

a. Write an equation that describes the value of your investment after n years if you invest $100 at 6% compounded annually. Plot the function. Estimate how long it would take to double your money.

b. Write an equation that describes the value of your investment after n years if you invest $200 at 3% compounded annually. Plot the function on the same grid as in part (a). Estimate the time needed to double your investment.

c. Looking at your graph, indicate whether the amount in the first investment in part (a) will ever exceed the amount in the second account in part (b). If so, approximately when?

26. According to the *Arkansas Democrat Gazette* (February 27, 1994):

Jonathan Holdeen thought up a way to end taxes forever. It was disarmingly simple. He would merely set aside some money in trust for the government and leave it there for 500 or 1000 years. Just a penny, Holdeen calculated, could grow to trillions of dollars in that time. But the stash he had in mind would grow much bigger—to quadrillions or quintillions—so big that the government, one day, could pay for all its operations simply from the income. Then taxes could be abolished. And everyone would be better off.

a. Holdeen died in 1967, leaving a trust of $2.8 million that is being managed by his daughter, Janet Adams. In 1994, the trust was worth $21.6 million. The trust was debated in Philadelphia Orphans' Court. Some lawyers who were trying to break the trust said that it is dangerous to let it go on, because "it would sponge up all the money in the world." Is this possible?

b. After 500 years, how much would the trust be worth? Would this be enough to pay off the current national debt (over $7 trillion in 2004)? What about after 1000 years? Describe the model you used to make your predictions.

27. Describe how a 6% inflation rate will erode the value of a dollar over time. Approximately when would a dollar be worth only 50 cents? This is the half-life of the dollar's buying power under 6% inflation.

28. The future value *V* of a savings plan, where regular payments *P* are made *n* times to an account in which the interest rate, *i*, is compounded each payment period, can be calculated using the formula

$$V = P \cdot \frac{(1 + i)^n - 1}{i}$$

The total number of payments, *n*, equals the number of payments per year, *m*, times the number of years, *t*, so

$$n = m \cdot t$$

The interest rate per compounding period, *i*, equals the annual interest rate, *r*, divided by the number of compounding periods a year, *m*, so

$$i = r/m$$

a. Substitute $n = m \cdot t$ and $i = r/m$ in the formula for *V*, getting an expression for *V* in terms of *m*, *t*, and *r*.

b. If a parent plans to build a college fund by putting $50 a month into an account compounded monthly with a 4% annual interest rate, what will be the value of the account in 17 years?

c. Solve the original formula for *P* as a function of *V*, *i*, and *n*.

d. Now you are able to find how much must be paid in every month to meet a particular final goal. If you estimate the child will need $100,000 for college, what monthly payment must the parent make if the interest rate is the same as in part (b)?

29. [Part (c) requires use of the Internet, and technology to find a best-fit function is recommended.] The following data show the total government debt for the United States from 1950 to 2006.

DATA

FEDDEBT

Year	Debt ($ billions)
1950	257
1955	274
1960	291
1965	322
1970	381
1975	542
1980	909
1985	1818
1990	3207
1995	4921
2000	5674
2005	7933
2006	8507

a. By hand or with technology, plot the data in the accompanying table and sketch a curve that approximates the data.

b. Construct an exponential function that models the data. What would your model predict for the current debt?

c. Use the "debt clock" at *www.brillig.com/debt_clock* to find the current debt. How accurate was your prediction in part (b)?

30. The average female adult *Ixodes scapularis* (a deer tick that can carry Lyme disease) lives only a year but can lay up to 10,000 eggs right before she dies. Assume that the ticks all live to adulthood, and half are females that reproduce at the same rate and half the eggs are male.

a. Describe a formula that will tell you how many female ticks there will be in *n* years, if you start with one impregnated tick.

b. How many male ticks will there be in *n* years? How many total ticks in *n* years?

c. If the surface of Earth is approximately $5.089 \cdot 10^{14}$ square meters and an adult tick takes up 0.5 square centimeters of land, approximately how long would it take before the total number of ticks stemming from one generation of ticks would cover the surface of Earth?

31. MCI, a phone company that provides long-distance service, introduced a marketing strategy called "Friends and Family." Each person who signed up received a discounted calling rate to ten specified individuals. The catch was that the ten people also had to join the "Friends and Family" program.

a. Assume that one individual agrees to join the "Friends and Family" program and that this individual recruits ten new members, who in turn each recruit ten new members, and so on. Write a function to describe the number of new people who have signed up for "Friends and Family" at the *n*th round of recruiting.

b. Now write a function that would describe the total number of people (including the originator) signed up after *n* rounds of recruiting.

c. How many "Friends and Family" members, stemming from this one person, will there be after five rounds of recruiting? After ten rounds?

d. Write a 60-second summary of the pros and cons of this recruiting strategy. Why will this strategy eventually collapse?

32. In a chain letter one person writes a letter to a number of other people, *N*, who are each requested to send the letter to *N* other people, and so on. In a simple case with $N = 2$, let's assume person A1 starts the process.

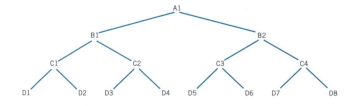

A1 sends to B1 and B2; B1 sends to C1 and C2; B2 sends to C3 and C4; and so on. A typical letter has listed in order the chain of senders who sent the letters. So D7 receives a letter that has A1, B2, and C4 listed.

If these letters request money, they are illegal. A typical request looks like this:

- When you receive this letter, send $10 to the person on the top of the list.

- Copy this letter, but add your name to the bottom of the list and leave off the name at the top of the list.

- Send a copy to two friends within 3 days.

For this problem, assume that all of the above conditions hold.

a. Construct a mathematical model for the number of new people receiving letters at each level L, assuming $N = 2$ as shown in the above tree.

b. If the chain is not broken, how much money should an individual receive?

c. Suppose A1 sent out letters with two additional phony names on the list (say A1a and A1b) with P.O. box addresses she owns. So both B1 and B2 would receive a letter with the list A1, A1a, A1b. If the chain isn't broken, how much money would A1 receive?

d. If the chain continued as described in part (a), how many new people would receive letters at level 25?

e. *Internet search:* Chain letters are an example of a "pyramid growth" scheme. A similar business strategy is multilevel marketing. This marketing method uses the customers to sell the product by giving them a financial incentive to promote the product to potential customers or potential salespeople for the product. (See Exercise 31.) Sometimes the distinction between multilevel marketing and chain letters gets blurred. Search the U.S. Postal Service website (*www.usps.gov*) for "pyramid schemes" to find information about what is legal and what is not. Report what you find.

5.7 *Semi-Log Plots of Exponential Functions*

With exponential growth functions, we often face the same problem that we did in Chapter 4 when we tried to compare the size of an atom with the size of the solar system. The numbers go from very small to very large. In our *E. coli* model, for example, the number of bacteria started at 100 and grew to over 1 billion in twenty-four time periods (see Table 5.1). It is virtually impossible to display the entire data set on a standard graph. Whenever we need to graph numbers of widely varying sizes, we turn to a logarithmic scale.

Previously, using standard linear scales on both axes, we could graph only a subset of the *E. coli* data in order to create a useful graph (see Figure 5.1). However, if we convert the vertical axis to a logarithmic scale, we can plot the entire data set (Figure 5.22). When one axis uses a logarithmic scale and the other a linear scale, the graph is called a *semi-log* (or *log-linear*) plot.

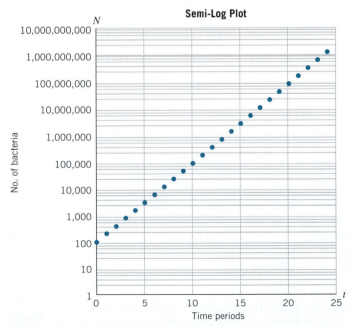

Figure 5.22 Semi-log plot of *E. coli* models data over twenty-four time periods.

Why Does the Graph Appear as a Straight Line?   On a semi-log plot, moving n units horizontally to the right is equivalent to adding n units, but moving up vertically n units is equivalent to multiplying by a factor of n. To stay on a line with slope m, we need to move vertically m units for each unit we move horizontally. So the line tells us that each time we increase the time period by 1, the number of *E. coli* is multiplied by a constant (namely 2). That's precisely the definition of an exponential function.

In general, the graph of any exponential function on a semi-log plot will be a straight line. We'll take a closer look at why this is true in Chapter 6. Since most graphing software easily converts standard linear plots to *log-linear* or *semi-log* plots, this is one of the simplest and most reliable ways to recognize exponential growth in a data set.

> When an exponential function is plotted using a standard linear scale on the horizontal axis and a logarithmic scale on the vertical axis, its graph is a straight line. This type of graph is called a *log-linear* or *semi-log* plot.

E X A M P L E   1

To learn more about Gordon Moore's predictions, read his original paper and a more current interview in *Wired*.

Moore's law

In 1965 Gordon Moore, cofounder of Intel, made his famous prediction that the number of transistors per integrated circuit would increase exponentially over time (doubling every 18 months). This became known as Moore's Law. Figure 5.23 shows the actual increase in the number of transistors over time. Do the data justify his claim of exponential growth?

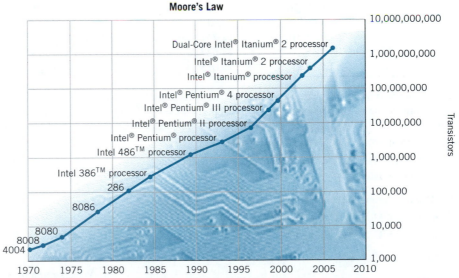

Figure 5.23 The semi-log plot of number of transistors (per circuit board) over time.
Source: Intel's website at *www.intel.com*.

S O L U T I O N   Figure 5.23 shows a semi-log plot of the number of transistors (per integrated circuit) over time. The plot looks basically linear, indicating that the data are exponential in nature.

Algebra Aerobics 5.7

1. **a.** Using the graph in Figure 5.22, estimate the time it takes for the *E. coli* to increase by a factor of 10.

   b. From the original expression for the population, $N = 100 \cdot 2^t$, when does the population increase by a factor of 8?

   c. By a factor of 16?

   d. Are these three answers consistent with each other?

2. What would you expect the graph of $y = 25 \cdot 10^x$ to look like on a semi-log plot? Construct a table of values for x equal to 0, 1, 2, 3, 4 and 5. Plot the graph of this function on the accompanying semi-log grid. Does your graph match your prediction?

3. Use your graph for Problem 2 to estimate the values of y when $x = 3.5$ and when $x = 7$.

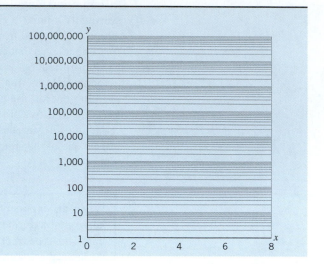

Exercises for Section 5.7

Technology for finding a best-fit function is required for Exercises 5 and 6.

1. (Graphing program recommended.) Below is a table of values for $y = 500(3)^x$ and for log y.

x	y	log y
0	500	2.699
5	121,500	5.085
10	29,524,500	7.470
15	$7.17 \cdot 10^9$	9.856
20	$1.74 \cdot 10^{12}$	12.241
25	$4.24 \cdot 10^{14}$	14.627
30	$1.03 \cdot 10^{17}$	17.013

   a. Plot y vs. x on a linear scale. Remember to identify the largest number you will need to plot before setting up axis scales.

   b. Plot log y vs. x on a semi-log plot with a log scale on the vertical axis and a linear scale on the horizontal axis.

   c. Rewrite the y-values as powers of 10. How do these values relate to log y?

2. Match each function with its semi-log plot.

   a. $y = 200(1.5)^x$ **c.** $y = 200(0.9)^x$

   b. $y = 200(2.5)^x$ **d.** $y = 200(0.5)^x$

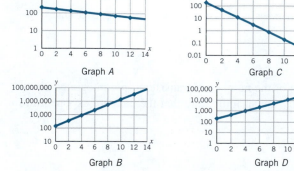

3. The three accompanying graphs are all of the same function, $y = 1000(1.5)^x$.

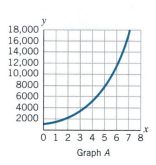

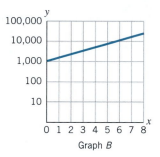

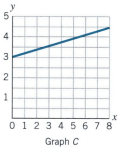

   a. Which graph uses a linear scale for y on the vertical axis? A power-of-10 scale on the vertical axis? Logarithms on the vertical axis?

   b. Why do graphs B and C look the same?

   c. For graphs A and B, estimate the number of units needed on the horizontal scale for the value of $y = 1000$ to increase by a factor of 10.

   d. On which graph is it easier to determine when the function has increased by a factor of 10?

   e. On which graph is it easier to determine when the function has doubled?

4. According to Rubin and Farber's *Pathology*, "Smoking tobacco is the single largest preventable cause of death in the United States, with direct health costs to the economy of tens of billions of dollars a year. Over 400,000 deaths a year—about one sixth of the total mortality in the United States—occur prematurely because of smoking." The accompanying graph compares the risk of dying for smokers, ex-smokers, and nonsmokers. It shows that individuals who have smoked for 2 years are twice as likely to die as a nonsmoker. Someone who has smoked for 14 years is three times more likely to die than a nonsmoker.

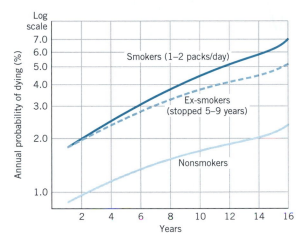

Source: E. Rubin and J. L. Farber, *Pathology*, 3rd ed. (Philadelphia: Lippincott-Raven, 1998), p. 310. Copyright © 1998 by Lippincott-Raven. Reprinted by permission.

a. The graphs for the smokers (one to two packs per day), ex-smokers, and nonsmokers all appear roughly as straight lines on this semi-log plot. What, then, would be appropriate functions to use to model the increased probability of dying over time for all three groups?

b. The plots for smokers and ex-smokers appear roughly as two straight lines that start at the same point, but the graph for smokers has a steeper slope. How would their two function models be the same and how would they be different?

c. The plots of ex-smokers and nonsmokers appear as two straight lines that are roughly equidistant. How would their two function models be the same and how would they be different?

5. (Requires technology to find a best-fit function.)
a. Load the file CELCOUNT, which contains all of the white blood cell counts for the bone marrow transplant patient. Now graph the data on a semi-log plot. Which section(s) of the curve represent exponential growth or decay? Explain how you can tell.

b. Load the file ECOLI, which contains the *E. coli* counts for twenty-four time periods. Graph the *E. coli* data on a semi-log plot. Which section of this curve represents exponential growth or decay? Explain your answer.

6. (Requires technology to find a best-fit function.) The accompanying table shows the U.S. international trade in goods and services. **USTRADE**

U.S. International Trade (Billions of Dollars)

Year	Total Exports	Total Imports
1960	25.9	22.4
1965	35.3	30.6
1970	56.6	54.4
1975	132.6	120.2
1980	271.8	291.2
1985	288.8	410.9
1990	537.2	618.4
1995	793.5	891.0
2000	1070.6	1448.2
2005	1275.2	1992.0

Source: U.S. Department of Commerce, Bureau of Economic Analysis, U.S. Bureau of the Census, *Statistical Abstract of the United States: 2006.*

a. U.S. imports and exports both expanded rapidly between 1960 and 2005. Use technology to plot the total U.S. exports and total U.S. imports over time on the same graph.

b. Now change the vertical axis to a logarithmic scale and generate a semi-log plot of the same data as in part (a). What is the shape of the data now, and what does this suggest would be an appropriate function type to model U.S. exports and imports?

c. Construct appropriate function models for total U.S. imports and for total exports.

d. The difference between the values of exports and imports is called the *trade balance*. If the balance is negative, it is called a *trade deficit*. The balance of trade has been an object of much concern lately. Calculate the trade balance for each year and plot it over time. Describe the overall pattern.

e. We have a trade deficit that has been increasing rapidly in recent years. But for quantities that are growing exponentially, the "relative difference" is much more meaningful than the simple difference. In this case the relative difference is

$$\frac{\text{exports} - \text{imports}}{\text{exports}}$$

This gives the trade balance as a fraction (or if you multiply by 100, as a percentage) of exports.

Calculate the relative difference for each year in the above table and graph it as a function of time. Does this present a more or less worrisome picture? That is, in particular over the last decade, has the relative difference remained stable or is it also rapidly increasing in magnitude?

7. A Fidelity Investments report included the graph, on the following page, illustrating how $10,000 invested in a Fidelity Fund © created on December 1992 would have grown over 10 years. The graph also includes the Standard & Poor's 500 Index[6] (S&P 500) for comparison.

a. What sort of plot is this?

b. The growth from 1993 to 2000 in both the Fidelity Fund and the S&P 500 Index appears roughly linear. What does that tell you?

[6]The Standard & Poor's 500 Index is an index of 500 stocks that is used to measure the performance of the entire U.S. domestic stock market.

c. Between 2000 and 2002, the values of the Fund and 500 Index have a roughly linear decline. What does that tell you?

d. Give at least two reasons why Fidelity would publish this graph.

$10,000 Over 10 Years

Let's say hypothetically that $10,000 was invested in Fidelity Fund on December 31, 1992. The chart shows how the value of your investment would have grown, and also shows how the Standard & Poor's 500 Index did over the same period.

CHAPTER SUMMARY

Exponential Functions

The general form of the equation for an exponential function is

$$y = Ca^x \quad (a > 0 \text{ and } a \neq 1), \quad \text{where}$$

C is the initial value or y-intercept
a is the base and is called the growth (or decay) factor

If $C > 0$ and

$a > 1$, the function represents growth
$0 < a < 1$, the function represents decay

Linear vs. Exponential Growth

Exponential growth is multiplicative, whereas linear growth is additive. Exponential growth involves multiplication by a constant factor for each unit increase in input. Linear growth involves adding a fixed amount for each unit increase in input.

In the long run, any exponential growth function will eventually dominate any linear growth function.

Graphs of Exponential Functions

For functions in the form $y = Ca^x$:
The value of C tells us where the graph crosses the y-axis.
The value of a affects the steepness of the graph.

Exponential growth ($a > 1$, $C > 0$). The larger the value of a, the more rapid the growth and the more rapidly the graph rises.

Exponential decay ($0 < a < 1$, $C > 0$). The smaller the value of a, the more rapid the decay and the more rapidly the graph falls.

The graphs of both exponential growth and decay functions are asymptotic to the x-axis.

Exponential growth: As $x \to -\infty$, $y \to 0$.
Exponential decay: As $x \to +\infty$, $y \to 0$.

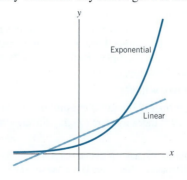

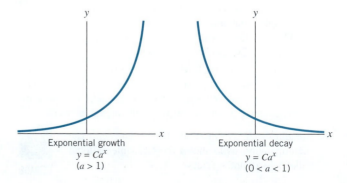

Factors, Rates, and Percents

An exponential function can be represented as a constant percent change. Growth and decay factors can be translated into constant percent increases or decreases, called *growth* or *decay rates*. For an exponential function in the form $y = Ca^x$,

$$\text{growth factor} = 1 + \text{growth rate}$$
$$a = 1 + r$$

where r is the *growth rate* in decimal form.

$$\text{decay factor} = 1 - \text{decay rate}$$
$$a = 1 - r$$

where r is the *decay rate* in decimal form.

Properties of Exponential Functions

The *doubling time* of an exponentially growing quantity is the time required for the quantity to double in size. The *half-life* of an exponentially decaying quantity is the time required for one-half of the quantity to decay.

The *rule of 70* offers a simple way to estimate the doubling time or the half-life. If a quantity is growing at $R\%$ per year, then its doubling time is approximately $70/R$ years. If a quantity is decaying at $R\%$ per month, then $70/R$ gives its half-life in months.

Semi-Log Plots of Exponential Functions

When an exponential function is plotted using a standard scale on the horizontal axis and a logarithmic scale on the vertical axis, its graph is a straight line. This is called a *log-linear* or *semi-log plot*.

CHECK YOUR UNDERSTANDING

I. Is each of the statements in Problems 1–24 true or false? Give an explanation for your answer.

1. If $y = f(x)$ is an exponential function and if increasing x by 1 increases y by a factor of 3, then increasing x by 2 increases y by a factor of 6.

2. If $y = f(x)$ is an exponential function and if increasing x by 1 increases y by 20%, then increasing x by 3 increases y by about 73%.

3. $y = x^3$ is an exponential function.

4. The graph of $y = 10 \cdot 2^x$ is decreasing.

5. The average rate of change between any two points on the graph of $y = 32.5(1.06)^x$ is constant.

6. Of the three exponential functions $y_1 = 5.4(0.8)^x$, $y_2 = 5.4(0.7)^x$, and $y_3 = 5.4(0.3)^x$, y_3 decays the most rapidly.

7. The graph of the exponential function $y = 1.02^x$ lies below the graph of the line $y = x$ for $x > 5$.

8. If $y = 100(0.976)^x$, then as x increases by 1, y decreases by 97.6%.

9. The function in the accompanying figure represents exponential decay with an initial population of 150 and decay factor of 0.8.

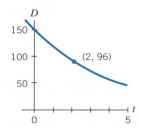

10. If the exponential function that models federal budget expenditures (E in billions of dollars) for a particular department is $E = 134(1.041)^t$, where $t =$ number of years since 1990, then in 1990 expenditures were about $134 billion and were increasing by 4.1% per year.

11. Increasing $1000 by $100 per year for 10 years gives you more than increasing $1000 by 10% per year for 10 years.

12. The value of the dollar with inflation at 2% per month is the same as the value of the dollar with inflation at 24% per year.

13. If M dollars is invested at 6.25% compounded annually, the amount of money, A, in 14 years is $A = M(0.0625)^{14}$.

Problems 14 and 15 refer to the following graph.

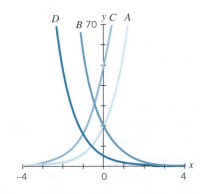

14. Of the functions plotted in the figure, graph C best describes the exponential function $f(x) = 20(3)^x$.

15. Of the functions graphed in the figure, graph B best describes the exponential function $g(x) = 20(3)^{-x}$.

16. If $B = 100(0.4)^x$, then as x increases by 1, B decreases by 60%.

17. If $y = f(x)$ is an exponential function and if increasing x by 1 increases y by 2, then increasing x by 5 increases y by 10.

For Problems 18 and 19, refer to the following table.

Year	1990	1995	2000	2005
Population (000s)	100	127.6	162.9	207.9

18. An exponential function would be a reasonable model for the population data.

19. Assuming this population grows exponentially, the annual growth rate is 6.29% because

$$\frac{\text{population in year 2000}}{\text{population in year 1990}} = \frac{162.9}{100} = 1.629$$

a growth of 62.9% in 10 years, or 62.9%/10 years = 6.29% per year.

20. After 30 years, the amount of interest earned on $5000 invested at 5% interest compounded annually will be half as much as the amount of interest earned on $10,000 invested at the same rate.

21. If a population behaves exponentially over time and if $\frac{\text{population in year 3}}{\text{population in year 0}} = 0.98$, then after 3 years the population will have decreased by 2%.

22. If the half-life of a substance is 10 years, then three half-lives of the substance would be 30 years.

23. If the doubling time of $100 invested at an interest rate r compounded annually is 9 years, then in 27 years the amount of the investment will be $600.

24. The doubling time for the function in the accompanying graph is approximately 10 years.

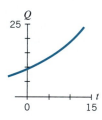

II. In Problems 25–33, give an example of a function with the specified properties. Express your answer using formulas, and specify the independent and dependent variables.

25. An exponential function that has an initial population of 2.2 million people and increases 0.5% per year.

26. An exponential function that has an initial population of 2.2 million people and increases 0.5% per quarter.

27. An exponential function that has an initial value of $1.4 billion and decreases 2.3% per decade.

28. An exponential function that passes through (2, 125) and (4, 5).

29. Two exponential decay functions with the same initial population but with the first decaying at twice the rate of the second.

30. Two functions, one linear and one exponential, that pass through the data points (0, 5) and (3, 0.625).

31. An exponential function that describes the value of a dollar over time with annual inflation of 3%.

32. An exponential function that has a doubling time of 14 years.

33. A function that describes the amount remaining of an initial amount of 200 mg of a substance with a half-life of 20 years.

III. Is each of the statements in Problems 34–43 true or false? If a statement is false, give a counter-example.

34. Exponential functions are multiplicative and linear functions are additive.

35. All exponential functions are increasing.

36. If $P = Ca^t$ describes a population P as a function of t years since 2000, then C is the initial population in the year 2000.

37. Exponential functions $y = Ca^t$ have a fixed or constant percent change per year.

38. Quantities that increase by a constant amount represent linear growth, whereas quantities that increase by a constant percent represent exponential growth.

39. The graph of an exponential growth function plotted on a semi-log plot is concave upward.

40. Eventually, the graph of every exponential function meets the horizontal axis.

41. For exponential functions, the growth rate is the same as the growth factor.

42. Exponential functions of the form $y = C \cdot a^x$ (where $a > 0, a \neq 1$) are always asymptotic to the horizontal axis.

43. Of the two functions in the accompanying figure, only A is decreasing at a constant percent.

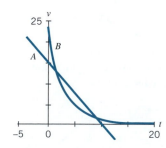

CHAPTER 5 REVIEW: PUTTING IT ALL TOGETHER

Internet access is required for Exercise 9 part (c). A calculator that can evaluate powers is recommended for Exercises 18, 25 part (c), and 27.

1. In each case, generate equations that represent the population, P, as a function of time, t (in years), such that when $t = 0$, $P = 150$ and:

   a. P doubles each year.
   b. P decreases by twelve units each year.
   c. P increases by 5% each year.
   d. The annual average rate of change of P with respect to t is constant at 12.

2. Describe in words the following functions by identifying the type of function, the initial value, and what happens with each unit increase in x.

    a. $f(x) = 100(4)^x$ **b.** $g(x) = 100 - 4x$

3. Match each of the following functions with the appropriate graph.

    a. $y = 2^x$ **b.** $y = 3^x$ **c.** $y = 4^x$

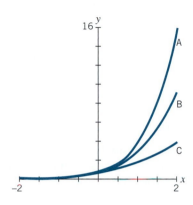

4. Sketch graphs of the following functions on the same grid:

    a. $y = 0.1^x$ **b.** $y = 0.2^x$ **c.** $y = 0.3^x$

5. Studies have shown that lung cancer is directly correlated with smoking. The accompanying graph shows for each year after a smoker has stopped smoking his or her relative risk of lung cancer. The relative risk is the number of times more likely a former smoker is to get lung cancer than someone who has never smoked. For example, if the relative risk is 4.5, then that patient is 4.5 times more likely to get lung cancer than a life-long non-smoker. Time is the measured in number of years since a smoker stopped smoking.

    a. According to the graph, how many times more likely (the relative risk) is a male who just stopped smoking to get lung cancer than a lifelong nonsmoker? How many times more likely is a female who stopped smoking 12 years ago to get lung cancer than someone who never smoked?

    b. Why is it reasonable that the relative risk is declining?

    c. What does it mean if the relative risk is 1? Would you expect the relative risk to go below 1? Explain.

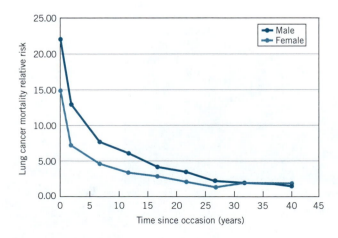

6. A polluter dumped 25,000 grams of a pollutant into a lake. Each year the amount of pollutant in the lake is reduced by 4%. Construct an equation to describe the amount of pollutant after n years.

7. You have a sore throat, so your doctor takes a culture of the bacteria in your throat with a swab and lets the bacteria grow in a Petri dish. If there were originally 500 bacteria in the dish and they grow by 50% per day, describe the relationship between $G(t)$, the number of bacteria in the dish and t, the time since the culture was taken (in days).

8. Which of the following functions are exponential and which are linear? Justify your answers.

x	$f(x)$	$g(x)$	$h(x)$	$j(x)$
0	-500	1	10.25	1
1	-200	2.5	8.25	0.5
2	100	6.25	6.25	0.25
3	400	15.625	4.25	0.125
4	700	39.0625	2.25	0.0625
5	1000	97.65625	0.25	0.03125

9. The United Nations Department of Economic and Social Affairs reported in 1999 that "the world's population stands at 6 billion and is growing at 1.3% per year, for an annual net addition of 78 million people."

    a. This statement actually contains two contradictory descriptions for predicting world population growth. What are they?

    b. In order to decide which of the statements is more accurate, use the information provided to construct a linear model and an exponential model for world population growth.

    c. The U.S. Census Bureau has a website that gives estimates for the current world population at *http://www.census.gov/ ipc/www/popclockworld.html.* Look up the current population. Which of your models is a more accurate predictor of the current world population?

10. Professional photographers consider exposure value (a combination of shutter speed and aperture) in relation to luminance (the amount of light that falls on a certain region) in setting camera controls to produce a desired effect. The data shown in the table give the relationship of the exposure value, E_v, to the luminance, L, for a particular type of film. Find a formula giving L as a function of E_v.

E_v, Exposure Value	L, Luminance (candelas/sq. meter)
0	0.125
1	0.25
2	0.5
3	1
4	2
5	4
6	8
7	16
8	32
9	64
10	128
11	256
12	512

11. According to Optimum Population Trust, the number of motor vehicles in the United Kingdom has nearly doubled every 25 years since 1925. Use the "rule of 70" to estimate the annual percent increase.

12. Mirex, a pesticide used to control fire ants, is rather long-lived in the environment. It takes 12 years for half of the original amount to break down into harmless products. Use the "rule of 70" to estimate the annual percent decrease.

13. The U.S. Department of Human Services reported, "Health care spending in the United States is projected to grow 7.4 percent and surpass $2 trillion in 2005, down from the 7.9 percent growth experienced in 2004." The Department also published the following data about national health care expenditures and projections. Find the annual growth factor and the annual percent growth rate for health care since 2003. Show your calculations, and check to see if they agree with percent growth reported by the U.S. Department of Human Services.

			Projected	
Year	2003	2004	2005	2006
Health care expenditures (billions)	$1,740.6	$1,877.6	$2,016.0	$2,169.5

Source: http://www.cms.hhs.gov/NationalHealthExpendData/.

14. The U.S. Department of Human Services projected that in 2010 national health care expenditures would be approximately $2,887.3 billion. Use your results from Problem 13 to estimate expenditures in 2010, assuming the initial value is $2,169.5 billion (the projected estimate for 2006) and the annual percent growth rate is equal to:

   a. The annual percent growth rate from 2005 to 2006.

   b. The annual percent growth rate from 2004 to 2005.

   c. How close is each projection for 2010 from parts (a) and (b) to the DHS projection? From your calculations, estimate the percent growth rate used by DHS. Assume DHS used an annual percent growth rate from the projected estimate for the initial value in 2006.

15. The Energy Information Administration has made the following projections for energy consumption for the world:

World Energy Consumption

Year	2015	2020	2025	2030
Quadrillion Btu	563	613	665	722

Source: www.eia.doe.gov/iea/.

Would a linear or an exponential function be an appropriate model for these data? Construct a function that models the data.

16. a. Generate a series of numbers N using $N = 2^{0.5n}$, for integer values of n from 0 to 10. Use the value of $2^{0.5} \approx 1.414$ and the rules of exponents to calculate values for N.

   b. Rewrite the N formula using a square root sign.

17. *NUA.com* estimated that there were 605 million Internet users worldwide in 2002. *Useit.com* estimated that the number of Internet users will reach 2 billion in 2015 and 3 billion in 2040.

   a. Given NUA's 2002 estimate of 605 million users, what annual growth rate would give 2 billion users in 2015?

   b. Given Useit's estimate of 2 billion users in 2015, what annual growth rate would give 3 billion users in 2040?

18. (A calculator that can evaluate powers is recommended.) The following graph shows median house prices in the United States from 1968 to 2004.

   a. Estimate from the graph the median price paid for a home in 1968 and in 1993. Use these data points to construct a linear and an exponential model to represent the growth in median price from 1968 to 1993.

   b. Which model is a better predictor of the national median price of a home in 2004? Would you use this model to make predictions for next year? Why or why not?

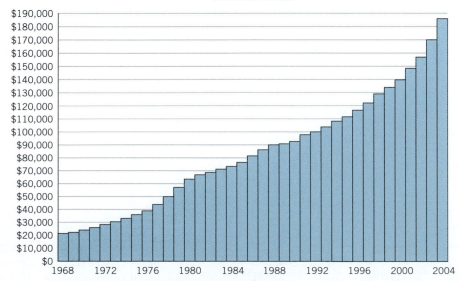

USA Median Prices

Source: National Association of Realtors *www.realtor.org.*

19. Radioactive iodine (I-131), used to test for thyroid problems, has a half-life of 8 days. If you start with 20 grams of radioactive iodine, describe the relationship between $A(t)$, the amount of I-131 (in grams), and t, the time (in days) since you first measured the sample.

20. Xenon gas (Xe-133) is used in medical imaging to study blood flow in the heart and brain. When inhaled, it is quickly absorbed into the bloodstream, and then gradually eliminated from the body (through exhalation). About 5 minutes after inhalation, there is half as much xenon left in the body (i.e., Xe-133 has a *biological* half-life of about 5 minutes). If you originally breathe in and absorb 3 ml. of xenon gas, describe the relationship between X, the amount of xenon gas in your body (in ml), and t, the time (in minutes) since you inhaled it.

21. In the accompanying chart, which graphs approximate exponential growth?

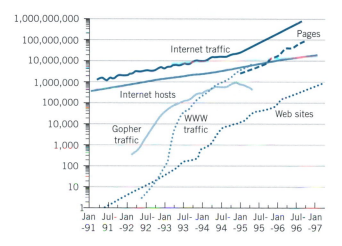

Source: http://www.useit.com/alertbox/9509.html.

22. In 2005 the Government Accountability Office issued a study of textbook prices. The report noted that in the preceding two decades textbook prices had been rising at the rate of 6% per year, roughly double the annual inflation rate of about 3% a year.

a. If a textbook cost $30 in 1985, what would it cost in 2005?

b. Assuming textbook prices continue to rise at 6% per year, if a textbook costs $80 in 2005, what will it cost in 2015?

c. A textbook that was first published in 1990 costs $120 in 2007. Since 1990 its price has increased at the rate of 6% per year. What was its price in 1990?

23. You may have noticed that when you take a branch from certain trees, the branch looks like a miniature version of the tree. When you break off a piece of the branch, that looks like the tree too. Mathematicians call this property self-similarity. The village of Bourton-on-the-Water in southwest England has a wonderful example of self-similarity: it has a 1/10 scale model of itself. Because the

1/10 scale model is a complete model of the town, it must contain a model of itself (that is, a 1/100 scale model of Bourton), and because the 1/100 scale model is also a complete model of Bourton, it also contains a scale model (that is, a 1/1000 scale model of Bourton.)

a. If A_o is the area of the actual village, how does the area of the first 1/10 scale model (where each linear dimension is one-tenth of the actual size) compare with A_o?

b. If the scale models are made of the same materials as the actual village, how does the weight of the building materials in the 1/100 scale model church compare with the weight W_o of the original church?

c. If the actual village is called a level 0 model, the 1/10 model is level 1, the 1/100 model is level 2, and so on, find a formula for the area A_n at any level n of the model as a function of A_o and for the weight W_n at any level n of the model as a function of W_o.

24. Computer worms are annoying and potentially very destructive. They are self-reproducing programs that run independently and travel across network connections. For some worms, such as SoBig, if you are sent an e-mail and you open up an attachment that contains SoBig, two things happen. First, SoBig installs a program that can be remotely activated in the future to send spam messages or shut down your computer. Then SoBig e-mails itself to everyone in your address book.

a. Construct a function that could model the spread of SoBig. Assume that everyone has in his or her address book ten people, none of whom have received SoBig.

b. Assume everyone in America has a computer. How many levels in a fractal tree would it take for SoBig to affect 300,000,000 Americans? (*Hint:* To find the total number of people who receive SoBig you need to add those at level 0, level 1, level, 2, etc.)

c. Why is a worm like SoBig so dangerous?

25. [A calculator that can evaluate powers is recommended for part (c).] Lung cancer is one of the leading causes of death, especially for smokers. There are various forms of cancer that attack the lungs, including cancers that start in some other organ and metastasize to the lungs. Doubling times for lung cancers vary considerably but are likely to fall within the range of 2 to 8 months. Although individual cancers are unpredictable and may speed up or slow down, the following examples give an idea of the range of time possibilities.

a. Two people are found to have lung cancer tumors whose volumes are each estimated at 0.5 cubic centimeters. A former asbestos worker has a tumor with an expected doubling time of 8 months. A heavy smoker has a tumor with an expected doubling time of 2 months. Write two tumor growth exponential functions, $A(t)$ and $S(t)$, for the asbestos worker and smoker, respectively, where t = number of 2-month time periods.

b. Graph $S(t)$ and $A(t)$ over 12 time periods on the same grid and compare the graphs.

c. Assuming both tumors grow in a spherical shape, what would be the diameter of each tumor after one year? (*Note:* The volume of a sphere $V = 4/3\pi r^3$.)

26. Which of the following graphs suggests a linear model for the original data? Which suggests an exponential model for the original data? Justify your answers.

Human Life Expectancy

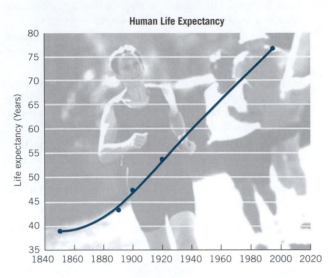

Total U.S. GDP In Constant Dollars

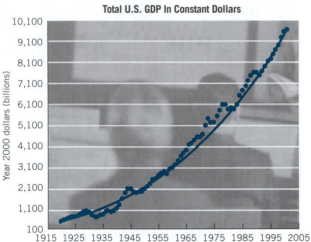

Internet Backbone Bandwidth

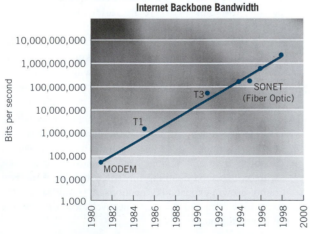

Source: http://www.kurzweilai.net.

27. (A calculator that can evaluate powers is recommended.) In February 2007 the U.S. Intergovernmental Panel on Climate Change reported that human activities have increased greenhouse gases in the atmosphere, resulting in global warming and other climate changes. Methane (a greenhouse gas) in the atmosphere has more than doubled since pre-industrial times from about 750 parts per billion in 1850 to 1,750 in 2005. The following graph shows the increase in methane gas since 1850.

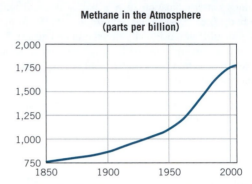

Methane in the Atmosphere (parts per billion)

Source: The New York Times International, February 3, 2007.

a. Construct an exponential equation, $M(t)$ that could be used to model the increase in methane gas in the atmosphere, where t = number of years since 1850.

b. From the graph, estimate the approximate doubling time. Now calculate the approximate doubling time using the "rule of 70." How close are your answers?

c. Describe the exponential growth in methane gas since 1850 in two different ways.

Properties of Exponential Functions

Objectives

- explore the effects of a and C on the graph of the exponential function in the form

$$y = Ca^x \qquad \text{where} \qquad a > 0 \text{ and } a \neq 0$$

Material/Equipment

- computer and software "E3: $y = Ca^x$ Sliders" *in Exponential & Log Functions,* or graphing calculator
- graph paper

Procedure

We start by choosing values for a and C and graphing the resulting equations by hand. From these graphs we make predictions about the effects of a and C on the graphs of other equations. Take notes on your predictions and observations so you can share them with the class. Work in pairs and discuss your predictions with your partner.

Making Predictions

1. Start with the simplest case, where $C = 1$. The equation will now have the form

$$y = a^x$$

Make a data table and by hand sketch on the same grid the graphs for $y = 2^x$ (here $a = 2$) and $y = 3^x$ (here $a = 3$). Use both positive and negative values for x. Predict where the graphs of $y = 2.7^x$ and $y = 5^x$ would be located on your graph. Check your work and predictions with your partner.

x	$y = 2^x$	$y = 3^x$
-2		
-1		
0		
1		
2		
3		

How would you describe your graphs? Do they have a maximum or a minimum value? What happens to y as x *increases?* What happens to y as x *decreases?* Which graph shows y changing faster compared with x?

2. Now create two functions in the form $y = a^x$ where $0 < a < 1$. Create a data table and graph your functions on the same grid. Make predictions for other functions where $C = 1$ and $0 < a < 1$.

3. Now consider the case where C has a value other than 1 for the general exponential function

$$y = Ca^x$$

Create a table of values and sketch the graphs of $y = 0.5(2^x)$ (in this case $C = 0.5$ and $a = 2$) and $y = 3(2^x)$ (in this case $C = 3$ and $a = 2$). What do all these graphs have in common? What do you think will happen when $a = 2$ and $C = 10$? What do you think will happen to the graph if $a = 2$ and $C = -3$? Check your predictions with your partner.

x	$y = 0.5(2^x)$	$y = 3(2^x)$	$y = -3(2^x)$
-2			
-1			
0			
1			
2			
3			

How would you describe your graphs? Do they have a maximum or a minimum value? What happens to y as x *increases?* What happens to y as x *decreases?* What is the y-intercept for each graph?

Testing Your Predictions

Now test your predictions by using a program called "E3: $y = Ca^x$ Sliders" in the *Exponential & Log Functions* software package or by creating graphs using technology.

1. What effect does a have?

Make predictions when $a > 1$ and when $0 < a < 1$, based on the graphs you constructed by hand. Explore what happens when $C = 1$ and you choose different values for a. Check to see whether your observations about the effect of a hold true when $C \neq 1$.

How does changing a change the graph? When does $y = a^x$ describe growth? When does it describe decay? When is it flat? Write a rule that describes what happens when you change the value for a. You only have to deal with cases where $a > 0$.

2. What effect does C have?

Make a prediction based on the graphs you constructed by hand. Now choose a value for a and create a set of functions with different C values. Graph these functions on the same grid.

How does changing C change the graph? What does the value of C tell you about the graphs of functions in the form $y = Ca^x$? Describe your graphs when $C > 0$ and when $C < 0$. Use technology to test your generalizations.

Exploration-Linked Homework

Write a 60-second summary of your results, and present it to the class.

CHAPTER 6

LOGARITHMIC LINKS: LOGARITHMIC AND EXPONENTIAL FUNCTIONS

OVERVIEW

If we know a specific output for an exponential function, how can we find the associated input? To answer this question, we can use logarithmic functions, the close relatives of exponential functions. We return to the *E. coli* model to calculate the doubling time.

After reading this chapter you should be able to

- use logarithms to solve exponential equations

- apply the rules for common and natural (base *e*) logarithms

- create an exponential model for continuous compounding

- understand the properties of logarithmic functions

- describe the relationship between logarithmic and exponential functions

- find the equation of an exponential function on a semi-log plot

6.1 *Using Logarithms to Solve Exponential Equations*

Estimating Solutions to Exponential Equations

So far, we have found a function's output from particular values of the input. For example, in Chapter 5, we modeled the growth of *E. coli* bacteria with the function

$$N = 100 \cdot 2^t$$

where N = number of *E. coli* bacteria and t = time (in 20-minute periods). If the input t is 5, then the corresponding output N is $100 \cdot 2^5 = 100 \cdot 32 = 3200$ bacteria.

Now we do the reverse, that is, find a function's input when we know the output. Starting with a value for N, the output, we find a corresponding value for t, the input. For example, at what time t will the value for N, the number of *E. coli,* be 1000? This turns out to be a harder question to answer than it might first appear. First we will estimate a solution from a table and a graph, and then we will learn how to find an exact solution using logarithms.

From a data table and graph

In Table 6.1, when $t = 3$, $N = 800$. When $t = 4$, $N = 1600$. Since N is steadily increasing, if we know $N = 1000$, then the value of t is somewhere between 3 and 4.

Values for $N = 100 \cdot 2^t$

t	N
0	100
1	200
2	400
3	800
4	1,600
5	3,200
6	6,400
7	12,800
8	25,600
9	51,200
10	102,400

Table 6.1

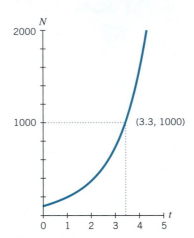

Figure 6.1 Estimating a value for t when $N = 1000$ on a graph of $N = 100 \cdot 2^t$.

We can also estimate the value for t when $N = 1000$ by looking at a graph of the function $N = 100 \cdot 2^t$ (Figure 6.1). By locating the position on the vertical axis where $N = 1000$, we can move over horizontally to find the corresponding point on the function graph. By moving from this point vertically down to the t-axis, we can estimate the t value for this point. The value for t appears to be approximately 3.3, so after about 3.3 time periods (or 66 minutes), the number of bacteria is 1000.

From an equation

An alternative strategy is to set $N = 1000$ and solve the equation for the corresponding value for t.

Start with the equation	$N = 100 \cdot 2^t$
Set $N = 1000$	$1000 = 100 \cdot 2^t$
Divide both sides by 100	$10 = 2^t$

Then we are left with the problem of finding a solution to the equation

$$10 = 2^t$$

We can estimate the value for t that satisfies the equation by bracketing 2^t between consecutive powers of 2. Since $2^3 = 8$ and $2^4 = 16$, then $2^3 < 10 < 2^4$. So $2^3 < 2^t < 2^4$, and therefore t is between 3 and 4, which agrees with our previous estimates. Using this strategy, we can find an approximate solution to the equation $1000 = 100 \cdot 2^t$. Strategies for finding exact solutions to such equations require the use of logarithms.

Algebra Aerobics 6.1a

1. Given the equation $M = 250 \cdot 3^t$, find values for M when:

   a. $t = 0$ **b.** $t = 1$ **c.** $t = 2$ **d.** $t = 3$

2. Use Table 6.1 to determine between which two consecutive integers the value of t lies when N is:

   a. 2000   **b.** 50,000

3. Use the graph of $y = 3^x$ in Figure 6.2 to estimate the value of x in each of the following equations.

   a. $3^x = 7$ **b.** $3^x = 0.5$

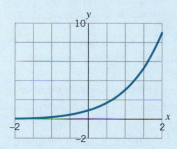

Figure 6.2 Graph of $y = 3^x$.

4. Use Table 6.2 to determine between which two consecutive integers the value of x in each of the following equations lies.

   a. $5^x = 73$ **b.** $5^x = 0.36$

x	5^x
-2	0.04
-1	0.2
0	1
1	5
2	25
3	125
4	625

Table 6.2

5. For each of the following, find two consecutive integers for the exponents a and b that would make the statement true.

   a. $2^a < 13 < 2^b$ **c.** $5^a < 0.24 < 5^b$

   b. $3^a < 99 < 3^b$ **d.** $10^a < 1500 < 10^b$

Rules for Logarithms

In Chapter 4 we defined logarithms. Recall that:

> The *logarithm base 10 of x* is the exponent of 10 needed to produce x.
>
> $$\log_{10} x = c \qquad \text{means that} \quad 10^c = x \qquad \text{where } x > 0$$
>
> Logarithms base 10 are called *common logarithms* and $\log_{10} x$ is written as $\log x$.

So,

$$10^5 = 100,000 \qquad \text{is equivalent to saying that} \qquad \log 100,000 = 5$$

$$10^{-3} = 0.001 \qquad \text{is equivalent to saying that} \qquad \log 0.001 = -3$$

Using a calculator we can solve equations such as

$$10^x = 80$$

Rewrite it in equivalent form $\qquad\qquad \log 80 = x$

use a calculator to compute the log $\qquad 1.903 \approx x$

But to solve equations such as $2^t = 10$ that involve exponential expressions where the base is not 10, we need to know more about logarithms. As the following expressions suggest, the rules of logarithms follow directly from the definition of logarithms and from the rules of exponents.

Rules for Exponents

If a is any positive real number and p and q are any real numbers, then

1. $a^p \cdot a^q = a^{p+q}$ 3. $(a^p)^q = a^{p \cdot q}$

2. $a^p/a^q = a^{p-q}$ 4. $a^0 = 1$

Corresponding Rules for Logarithms

If A and B are positive real numbers and p is any real number, then

1. $\log (A \cdot B) = \log A + \log B$ 3. $\log (A^p) = p \log A$

2. $\log (A/B) = \log A - \log B$ 4. $\log 1 = 0$ (since $10^0 = 1$)

? SOMETHING TO THINK ABOUT

Try expressing in words all the other rules for logarithms in terms of exponents.

Finding the common logarithm of a number means finding its exponent when the number is written as a power of 10. So when you see "logarithm," think "exponent," and the rules of logarithms make sense.

As we learned in Chapter 4, the log of 0 or a negative number is not defined. But when we take the log of a number, we can get 0 or a negative value. For example, $\log 1 = 0$ and $\log 0.1 = -1$.

We will list a rationale for each of the rules of logarithms and prove Rule 1. We leave the other proofs as exercises.

Rule 1 $\qquad\qquad\qquad\qquad \log (A \cdot B) = \log A + \log B$

Rationale

Rule 1 of exponents says that when we multiply two terms with the same base, we keep the base and *add* the exponents; that is, $a^p \cdot a^q = a^{p+q}$.

Rule 1 of logs says that if we rewrite A and B each as a power of 10, then the exponent of $A \cdot B$ is the sum of the exponents of A and B.

Proof

If we let $\log A = x$, then $\qquad\qquad\qquad\qquad 10^x = A$

and if $\log B = y$, then $\qquad\qquad\qquad\qquad 10^y = B$

We have two equal products $\qquad\qquad\quad A \cdot B = 10^x \cdot 10^y$

by laws of exponents $\qquad\qquad\qquad\quad A \cdot B = 10^{x+y}$

Taking the log of each side $\qquad\quad \log(A \cdot B) = \log (10^{x+y})$

by definition of log $\qquad\qquad\qquad \log(A \cdot B) = x + y$

Substituting $\log A$ for x and $\log B$ for y $\quad \log (A \cdot B) = \log A + \log B$

we arrive at our desired result.

Rule 2 $\qquad\qquad\qquad \log (A/B) = \log A - \log B$

Rationale

Rule 2 of exponents says that when we divide terms with the same base, we keep the base and *subtract* the exponents, that is, $a^p/a^q = a^{p-q}$. Rule 2 of logs says that if we write A and B each as a power of 10, then the exponent of A/B equals the exponent of A *minus* the exponent of B.

EXAMPLE 1 **Using the rules for logs**

Indicate whether or not the statement is true using the rules of exponents and the definition of logarithms.

a. $\log (10^2 \cdot 10^3) \overset{?}{=} \log 10^2 + \log 10^3$

b. $\log (10^5 \cdot 10^{-7}) \overset{?}{=} \log 10^5 + \log 10^{-7}$

c. $\log (10^2/10^3) \overset{?}{=} \log 10^2 - \log 10^3$

d. $\log (10^3/10^{-3}) \overset{?}{=} \log 10^3/\log 10^{-3}$

SOLUTION To see if an expression is a true statement, we must verify that the two sides of the expression are equal.

a. $\qquad\qquad\qquad \log (10^2 \cdot 10^3) \overset{?}{=} \log 10^2 + \log 10^3$

Rule 1 of exponents $\qquad \log 10^5 \overset{?}{=} \log 10^2 + \log 10^3$

definition of log $\qquad\qquad\qquad 5 = 2 + 3$

$\qquad\qquad\qquad\qquad\qquad 5 = 5$

Since the values on the two sides of the equation are equal, our original equation is a true statement.

b. $\qquad\qquad\qquad \log (10^5 \cdot 10^{-7}) \overset{?}{=} \log 10^5 + \log 10^{-7}$

Rule 1 of exponents $\qquad \log 10^{-2} \overset{?}{=} \log 10^5 + \log 10^{-7}$

definition of log $\qquad\qquad\qquad -2 = 5 + (-7)$

$\qquad\qquad\qquad\qquad\qquad -2 = -2$

Since the values on the two sides of the equation are equal, our original equation is a true statement.

c. $\qquad\qquad\qquad \log (10^2/10^3) \overset{?}{=} \log 10^2 - \log 10^3$

Rule 2 of exponents $\qquad \log (10^{-1}) \overset{?}{=} \log 10^2 - \log 10^3$

definition of log $\qquad\qquad\qquad -1 = 2 - 3$

$\qquad\qquad\qquad\qquad\qquad -1 = -1$

Since the values on the two sides of the equation are equal, our original equation is a true statement.

d. $\qquad\qquad\qquad \log (10^3/10^{-3}) \overset{?}{=} \log 10^3/\log 10^{-3}$

Rule 2 of exponents $\qquad \log 10^{3-(-3)} \overset{?}{=} \log 10^3/\log 10^{-3}$

combine terms in exponent $\qquad \log 10^6 \overset{?}{=} \log 10^3/\log 10^{-3}$

definition of log $\qquad\qquad\qquad 6 \overset{?}{=} 3/(-3)$

$\qquad\qquad\qquad\qquad\qquad 6 \neq -1$

Since the values on the two sides of the equation are not equal, our original equation is not a true statement.

SOMETHING TO THINK ABOUT

? Why does $\log \sqrt{AB}$ lie halfway between $\log A$ and $\log B$?

Rule 3 $\log A^p = p \log A$

Rationale

Since

$$\log A^2 = \log (A \cdot A) = \log A + \log A = 2 \log A$$

$$\log A^3 = \log (A \cdot A \cdot A) = \log A + \log A + \log A = 3 \log A$$

$$\log A^4 = \log (A \cdot A \cdot A \cdot A) = 4 \log A$$

it seems reasonable to expect that in general

$$\log A^p = p \log A$$

Rule 4 $\log 1 = 0$

Rationale

Since $10^0 = 1$ by definition, the equivalent statement using logarithms is $\log 1 = 0$.

EXAMPLE 2   **Simplifying expressions with logs**

Simplify each expression, and if possible evaluate with a calculator.

$$\log 10^3 = 3 \log 10 = 3 \cdot 1 = 3$$

$$\log 2^3 = 3 \log 2 \approx 3 \cdot 0.301 \approx 0.903$$

$$\log \sqrt{3} = \log 3^{1/2} = \tfrac{1}{2} \log 3 \approx \tfrac{1}{2} \cdot 0.477 \approx 0.239$$

$$\log x^{-1} = (-1) \cdot \log x = -\log x$$

$$\log 0.01^a = a \log 0.01 = a \cdot (-2) = -2a$$

$$(\log 5) \cdot (\log 1) = (\log 5) \cdot 0 = 0$$

EXAMPLE 3   **Expanding expressions with logs**

Use the rules of logarithms to write the following expression as the sum or difference of several logs.

$$\log \left(\frac{x(y-1)^2}{\sqrt{z}} \right)$$

SOLUTION   By Rule 2 $\log \left(\dfrac{x(y-1)^2}{\sqrt{z}} \right) = \log x(y-1)^2 - \log \sqrt{z}$

by Rule 1 and $= \log x + \log (y-1)^2 - \log z^{1/2}$
   exponent notation

by Rule 3 $= \log x + 2 \log (y-1) - \tfrac{1}{2} \log z$

We call this process *expanding the expression*.

EXAMPLE 4   **Contracting expressions with logs**

Use the rules of logarithms to write the following expression as a single logarithm.

$$2 \log x - \log(x-1)$$

SOLUTION   By Rule 3 $2 \log x - \log (x-1) = \log x^2 - \log(x-1)$

by Rule 2 $= \log \left(\dfrac{x^2}{x-1} \right)$

We call this process *contracting the expression*.

Common Error

Probably the most common error in using logarithms stems from confusion over the division property.

$$\log A - \log B = \log\left(\tfrac{A}{B}\right) \qquad \text{but} \quad \log A - \log B \neq \frac{\log A}{\log B}$$

For example,

$$\log 100 - \log 10 = \log\left(\frac{100}{10}\right) = \log 10 = 1$$

$$\text{but} \quad \log 100 - \log 10 \neq \frac{\log 100}{\log 10} = 2$$

EXAMPLE 5 **Solving equations that contain logs**

Solve for x in the equation $\quad \log x + \log x^2 = 15$

SOLUTION

Given	$\log x + \log x^2 = 15$
Rule 1 for logs	$\log x^3 = 15$
Rule 3 for logs	$3 \log x = 15$
divide by 3	$\log x = 5$
rewrite using definition of logs	$x = 10^5$ or $100{,}000$

Algebra Aerobics 6.1b

1. Using only the rules of exponents and the definition of logarithm, verify that:
   a. $\log(10^5/10^7) = \log 10^5 - \log 10^7$
   b. $\log[10^5 \cdot (10^7)^3] = \log 10^5 + 3 \log 10^7$

2. Determine the rule(s) of logarithms that were used in each statement.
   a. $\log 3 = \log 15 - \log 5$
   b. $\log 1024 = 10 \log 2$
   c. $\log \sqrt{31} = \tfrac{1}{2} \log 31$
   d. $\log 30 = \log 2 + \log 3 + \log 5$
   e. $\log 81 - \log 27 = \log 3$

3. Determine if each of the following is true or false. For the true statements tell which rule of logarithms was used.
   a. $\log(x + y) \overset{?}{=} \log x + \log y$
   b. $\log(x - y) \overset{?}{=} \log x - \log y$
   c. $7 \log x \overset{?}{=} \log x^7$
   d. $\log 10^{1.6} \overset{?}{=} 1.6$
   e. $\dfrac{\log 7}{\log 3} \overset{?}{=} \log 7 - \log 3$
   f. $\dfrac{\log 7}{\log 3} \overset{?}{=} \log(7 - 3)$

4. Expand, using the properties of logarithms:
   a. $\log\sqrt{\dfrac{2x - 1}{x + 1}}$
   b. $\log\dfrac{xy}{z}$
   c. $\log\dfrac{x\sqrt{x + 1}}{(x - 1)^2}$
   d. $\log\dfrac{x^2(y - 1)}{y^3 z}$

5. Contract, expressing your answer as a single logarithm: $\tfrac{1}{3}[\log x - \log(x + 1)]$.

6. Use rules of logarithms to combine into a single logarithm (if necessary), then solve for x.
   a. $\log x = 3$
   b. $\log x + \log 5 = 2$
   c. $\log x + \log 5 = \log 2$
   d. $\log x - \log 2 = 1$
   e. $\log x - \log(x - 1) = \log 2$
   f. $\log(2x + 1) - \log(x + 5) = 0$

7. Show that $\log 10^3 - \log 10^2 \neq \dfrac{\log 10^3}{\log 10^2}$

Solving Exponential Equations

Answering our original question: Solving $1000 = 100 \cdot 2^t$

Remember the question that started this chapter? We wanted to find out how many time periods it would take 100 *E. coli* bacteria to become 1000. To find an exact solution, we need to solve the equation $1000 = 100 \cdot 2^t$ or, dividing by 100, the equivalent equation, $10 = 2^t$. We now have the necessary tools.

Given	$10 = 2^t$
take the logarithm of each side	$\log 10 = \log 2^t$
use Rule 3 of logs	$\log 10 = t \log 2$
divide both sides by log 2	$\dfrac{\log 10}{\log 2} = t$
use a calculator	$\dfrac{1}{0.3010} \approx t$
divide	$3.32 \approx t$

which is consistent with our previous estimates of a value for t between 3 and 4, approximately equal to 3.3.

In our original model the time period represents 20 minutes, so 3.32 time periods represents $3.32 \cdot (20 \text{ minutes}) = 66.4$ minutes. So the bacteria would increase from the initial number of 100 to 1000 in a little over 66 minutes.

EXAMPLE 6   **Doubling your money**

As we saw in Chapter 5, the equation $P = 250(1.05)^n$ gives the value of $250 invested at 5% interest (compounded annually) for n years. How many years does it take for the initial $250 investment to double to $500?

SOLUTION   a. *Estimating the answer:* If $R = 5\%$ per year, then the rule of 70 (discussed in Section 5.6) estimates the doubling time as

$$70/R = 70/5 = 14 \text{ years}$$

b. *Calculating a more precise answer:* We can set $P = 500$ and solve the equation.

	$500 = 250(1.05)^n$
Divide both sides by 250	$2 = (1.05)^n$
take the log of both sides	$\log 2 = \log (1.05)^n$
use Rule 3 of logs	$\log 2 = n \log 1.05$
divide by log 1.05	$\dfrac{\log 2}{\log 1.05} = n$
evaluate with a calculator	$\dfrac{0.3010}{0.0212} \approx n$
divide and switch sides	$n \approx 14.2 \text{ years}$

So the estimate of 14 years using the rule of 70 was pretty close.

EXAMPLE 7   Solve the following equation for x in two ways.

$$8^x = 2^{x+1}$$

SOLUTION   *Method 1:* Make the bases the same.

	$8^x = 2^{x+1}$
Make the base the same	$2^{3x} = 2^{x+1}$
set exponents equal	$3x = x + 1$
combine terms	$2x = 1$
solve for x	$x = \dfrac{1}{2}$

Method 2: Use the rules for logarithms.

$$8^x = 2^{x+1}$$

Take logs of both sides	$\log 8^x = \log 2^{x+1}$
Rule 3 of logs	$x \log 8 = (x + 1) \log 2$
distributive law	$x \log 8 = x \log 2 + \log 2$
subtract $x \log 2$ from both sides	$x \log 8 - x \log 2 = \log 2$
factor out x	$x(\log 8 - \log 2) = \log 2$
Rule 2 of logs	$x \log 4 = \log 2$
divide by $\log 4$	$x = \dfrac{\log 2}{\log 4} = \dfrac{\log 2}{\log 2^2} = \dfrac{\log 2}{2 \log 2} = \dfrac{1}{2}$

We can double check by both
evaluating 8^x at $x = \frac{1}{2}$ $8^{1/2} \approx 2.8284$
or evaluating 2^{x+1} at $x = \frac{1}{2}$ $2^{(1/2)+1} = 2^{3/2} = (2^3)^{1/2} = 8^{1/2} \approx 2.8284$

EXAMPLE 8 Time to decay to a specified amount

In Chapter 5 we used the function $f(t) = 100(0.976)^t$ to measure the remaining amount of radioactive material as 100 milligrams (mg) of strontium-90 decayed over time t (in years). How many years would it take for there to be only 10 mg of strontium-90 left?

SOLUTION   Set $f(t) = 10$ and solve the equation.

	$10 = 100(0.976)^t$
Divide both sides by 100	$0.1 = (0.976)^t$
take the log of both sides	$\log 0.1 = \log (0.976)^t$
use Rule 3 of logs	$\log 0.1 = t \log 0.976$
divide by $\log 0.9755$	$\dfrac{\log 0.1}{\log 0.976} = t$
evaluate logs	$\dfrac{-1}{-0.0106} \approx t$
divide and switch sides	$t \approx 94$ years

So it takes almost a century for 100 mg of strontium-90 to decay to 10 mg.

Algebra Aerobics 6.1c

These problems require a calculator that can evaluate logs.

1. Solve the following equations for t.
   a. $60 = 10 \cdot 2^t$ c. $80(0.95)^t = 10$
   b. $500(1.06)^t = 2000$

2. Using the model $N = 100 \cdot 2^t$ for bacteria growth, where t is measured in 20-minute time periods, how long will it take for the bacteria count:
   a. To reach 7000?   b. To reach 12,000?

3. First use the rule of 70 to estimate how long it would take $1000 invested at 6% compounded annually to double to $2000. Then use logs to find a more precise answer.

4. Use the function in Example 8 to determine how long it will take for 100 milligrams of strontium-90 to decay to 1 milligram.

5. Solve each equation for t (in years). Which equation(s) asks you to find the time necessary for the initial amount to double? For the initial amount to drop to half?
   a. $30 = 60(0.95)^t$ c. $500 = 200(1.045)^t$
   b. $16 = 8(1.85)^t$

6. Find the half-life of a substance that decays according to the following models.
   a. $A = 120(0.983)^t$ (t in days)
   b. $A = 0.5(0.92)^t$ (t in hours)
   c. $A = A_0(0.89)^t$ (t in years)

Exercises for Section 6.1

Many of these problems (and those in later sections) require a calculator that can evaluate powers and logs. Some require a graphing program.

1. Use the accompanying table to estimate the number of years it would take $100 to become $300 at the following interest rates compounded annually.

    a. 3%   **b.** 7%

    Compound Interest over 40 Years

Number of Years	Value of $100 at 3% ($)	Value of $100 at 7% ($)
0	100	100
10	134	197
20	181	387
30	243	761
40	326	1497

2. **a.** Generate a table of values to estimate the half-life of a substance that decays according to the function $y = 100(0.8)^x$, where x is the number of time periods, each time period is 12 hours, and y is in grams.

    b. How long will it be before there is less than 1 gram of the substance remaining?

3. The accompanying graph shows the concentration of a drug in the human body as the initial amount of 100 mg dissipates over time. Estimate when the concentration becomes:

    a. 60 mg   **b.** 40 mg   **c.** 20 mg

    Remaining Drug Dosage

    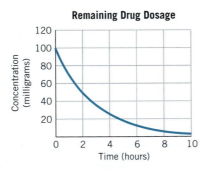

4. (Requires a graphing program.) Assume throughout that x represents time in seconds.

    a. Plot the graph of $y = 6(1.3)^x$ for $0 \le x \le 4$. Estimate the doubling time from the graph.

    b. Now plot $y = 100(1.3)^x$ and estimate the doubling time from the graph.

    c. Compare your answers to parts (a) and (b). What does this tell you?

5. Without a calculator, determine x if we know that log x equals:

    a. -3 **c.** 0   **e.** -1

    b. 6   **d.** 1

6. Given that log $5 \approx 0.699$, without using a calculator determine the value of:

    a. log 25   **b.** $\log \dfrac{1}{25}$ **c.** log 10^{25} **d.** log 0.0025

    Check your answers with a scientific calculator.

7. Expand each logarithm using only the numbers 2, 3, log 2, and log 3.

    a. log 9   **b.** log 18   **c.** log 54

8. Identify the rules of logarithms that were used to expand each expression.

    a. $\log 14 = \log 2 + \log 7$
    b. $\log 14 = \log 28 - \log 2$
    c. $\log 36 = 2 \log 6$
    d. $\log 9z^3 = 2 \log 3 + 3 \log z$
    e. $\log 3x^4 = \log 3 + 4 \log x$
    f. $\log \left(\dfrac{16}{3x} \right) = 4 \log 2 - (\log 3 + \log x)$

9. Use the rules of logarithms to find the value of x. Verify your answer with a calculator.

    a. $\log x = \log 2 + \log 6$
    b. $\log x = \log 24 - \log 2$
    c. $\log x^2 = 2 \log 12$
    d. $\log x = 4 \log 2 - 3 \log 2$

10. Use the rules of logarithms to show that the following are equivalent.

    a. $\log 144 = 2 \log 3 + 4 \log 2$
    b. $7 \log 3 + 5 \log 3 = 12 \log 3$
    c. $2(\log 4 - \log 3) = \log \left(\dfrac{16}{9} \right)$
    d. $-4 \log 3 + \log 3 = \log \left(\dfrac{1}{27} \right)$

11. Prove Rule 2 of logarithms:
    $$\log(A/B) = \log A - \log B \quad (A, B > 0)$$

12. Expand, using the rules of logarithms.

    a. $\log \left(x^2 y^3 \sqrt{z - 1} \right)$ **c.** $\log \left(t^2 \cdot \sqrt[4]{tp^3} \right)$

    b. $\log \dfrac{A}{\sqrt[3]{BC}}$

13. Contract, using the rules of logarithms, and express your answer as a single logarithm.

    a. $3 \log K - 2 \log(K + 3)$
    b. $-\log m + 5 \log(3 + n)$
    c. $4 \log T + \frac{1}{2} \log T$
    d. $\frac{1}{3}(\log x + 2 \log y) - 3(\log x + 2 \log y)$

14. For each of the following equations either prove that it is correct (by using the rules of logarithms and exponents) or else show that it is not correct (by finding numerical values

for the variables that make the values of the two sides of the equation different).

a. $\log\left(\dfrac{x}{y}\right) = \dfrac{\log x}{\log y}$

b. $\log x - \log y = \log\left(\dfrac{x}{y}\right)$

c. $\log(2x) = 2 \log x$

d. $2 \log x = \log(x^2)$

e. $\log\left(\dfrac{x+1}{x+3}\right) = \log(x+1) - \log(x+3)$

f. $\log\left(x\sqrt{x^2+1}\right) = \log x + \dfrac{1}{2}(x^2+1)$

g. $\log(x^2+1) = 2 \log x + \log 1$

15. Prove Rule 3 of common logarithms: $\log A^p = p \log A$ (where $A > 0$).

16. Solve for t.

a. $10^t = 4$

b. $3(2^t) = 21$

c. $1 + 5^t = 3$

d. $10^{-t} = 5$

e. $5^t = 7^{t+1}$

f. $6 \cdot 2^t = 3^{t-1}$

17. Solve for x.

a. $\log x = 3$

b. $\log(x+1) = 3$

c. $3 \log x = 5$

d. $\log(x+1) - \log x = 1$

e. $\log x - \log(x+1) = 1$

18. Solve for x.

a. $2^x = 7$

b. $\left(\sqrt{3}\right)^{x+1} = 9^{2x-1}$

c. $12(1.5)^{x+1} = 13$

d. $\log(x+3) + \log 5 = 2$

e. $\log(x-1) = 2$

19. In Exercise 1, we estimated the number of years it would take $100 to become $300 at each of the interest rates listed below, compounded annually. Now calculate the number of years by constructing and solving the appropriate equations.

a. 3%   **b.** 7%

20. Solve for the indicated variables.

a. $825 = 275 \cdot 3^T$

b. $45{,}000 = 15{,}000 \cdot 1.04^t$

c. $12 \cdot 10^{10} = (6 \cdot 10^8) \cdot 5^x$

d. $100 = 25(1.18)^t$

e. $32{,}000 = 8000(2.718)^t$

f. $8 \cdot 10^5 = 4 \cdot 10^5(2.5)^x$

21. (Requires a graphing program.) Let $f(x) = 500(1.03)^x$ and $g(x) = 4500$.

a. Using technology, graph the two functions on the same screen.

b. Estimate the point of intersection.

c. Solve the equation $4500 = 500(1.03)^x$ using logarithms.

d. Compare your answers.

22. (Requires a graphing program.) Let $f(x) = 100(0.8)^x$ and $g(x) = 10$.

a. Using technology, graph the two functions on the same screen.

b. Estimate the point of intersection.

c. Solve the equation $10 = 100(0.8)^x$ using logarithms.

d. Compare your answers.

23. Find the doubling time or half-life for each of the following functions (where x is in years).

a. $f(x) = 100 \cdot 4^x$

b. $g(x) = 100 \cdot \left(\frac{1}{4}\right)^x$

c. $h(x) = A(4)^x$ (*Hint:* Set $h(x) = 2A$)

d. $j(x) = A \cdot \left(\frac{1}{4}\right)^x$ (*Hint:* Set $j(x) = \frac{1}{2}A$)

24. The yearly per capita consumption of whole milk in the United States reached a peak of 40 gallons in 1945, at the end of World War II. It has been steadily decreasing at a rate of about 2.8% per year.

a. Construct an exponential model $M(t)$ for per capita whole milk consumption (in gallons) where $t =$ years since 1945.

b. Use your model to estimate the year in which per capita whole milk consumption dropped to 7 gallons per person. How does this compare with the actual consumption of 7 gallons per person in 2005?

c. What might have caused this decline?

25. Wikipedia is a popular, free online encyclopedia (at *en.wikipedia.org*) that anyone can edit. (So articles should be taken with "a grain of salt.") One Wikipedia article claims that the number of articles posted on Wikipedia has been growing exponentially since October 23, 2002. At that date there were approximately 90 thousand articles posted, and the growth rate was about 0.2% *per day*.

a. Create an exponential model for the growth in Wikipedia articles.

b. What is the doubling time? Interpret your answer.

26. If the amount of drug remaining in the body after t hours is given by $f(t) = 100\left(\frac{1}{2}\right)^{t/2}$ (graphed in Exercise 3), then calculate:

a. The number of hours it would take for the initial 100 mg to become:

i. 60 mg   **ii.** 40 mg   **iii.** 20 mg

b. The half-life of the drug.

27. In Chapter 5 we saw that the function $N = N_0 \cdot 1.5^t$ described the actual number N of *E. coli* bacteria in an experiment after t time periods (of 20 minutes each) starting with an initial bacteria count of N_0.

a. What is the doubling time?

b. How long would it take for there to be ten times the original number of bacteria?

28. (Requires a graphing program.) A woman starts a training program for a marathon. She starts in the first week by doing 10-mile runs. Each week she increases her run length by 20% of the distance for the previous week.

a. Write a formula for her run distance, D, as a function of week, W.

b. Use technology to graph your function, and then use the graph to estimate the week in which she will reach a marathon length of approximately 26 miles.

c. Now use your formula to calculate the week in which she will start running 26 miles.

29. The half-life of bismuth-214 is about 20 minutes.

    a. Construct a function to model the decay of bismuth-214 over time. Be sure to specify your variables and their units.

    b. For any given sample of bismuth-214, how much is left after 1 hour?

    c. How long will it take to reduce the sample to 25% of its original size?

    d. How long will it take to reduce the sample to 10% of its original size?

30. The atmospheric pressure at sea level is approximately 14.7 lb/in.^2, and the pressure is reduced by half for each 3.6 miles above sea level.

    a. Construct a model that describes the atmospheric pressure as a function of miles above sea level.

    b. At how many miles above sea level will the atmospheric pressure have dropped to half, i.e., to 7.35 lb/in.^2?

31. A department store has a discount basement where the policy is to reduce the selling price, S, of an item by 10% of its current price each week. If the item has not sold after the tenth reduction, the store gives the item to charity.

    a. For a $300 suite, construct a function for the selling price, S, as a function of week, W.

    b. After how many weeks might the suite first be sold for less than $150? What is the selling price at which the suite might be given to charity?

32. If you drop a rubber ball on a hard, level surface, it will usually bounce repeatedly. (See the accompanying graph at the top of the next column.) Each time it bounces, it rebounds to a height that is a percentage of the previous height. This percentage is called the *rebound height*.

    a. Assume you drop the ball from a height of 5 feet and that the rebound height is 60%. Construct a table of values that shows the rebound height for the first four bounces.

    b. Construct a function to model the ball's rebound height, H, on the nth bounce.

    c. How many bounces would it take for the ball's rebound height to be 1 foot or less?

    d. Construct a general function that would model a ball's rebound height H on the nth bounce, where H_0 is the initial height of the ball and r is the ball's rebound height.

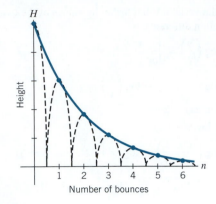

33. The accompanying graph shows fish and shellfish production (in million of pounds) by U.S. companies.

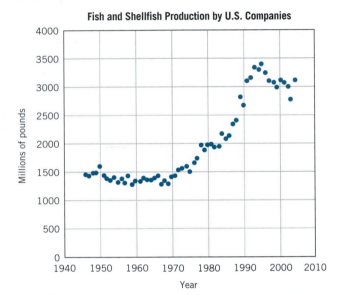

Fish and Shellfish Production by U.S. Companies

Source: U.S. Department of Agriculture, Economic Research Service, Data set on food availability, *www.ers.usda.gov.*

The growth appears to be exponential between 1970 and 1995. The exponential function $F(t)$ models that growth, where $F(t) = 1355(1.036)^t$ where $F(t)$ represents millions of pounds of fish and shellfish produced by U.S. companies since 1970 (where $0 \leq t \leq 25$).

    a. Suppose we used the model to predict when fish production would reach 4000 million pounds. What year would that be? Using the graph, estimate the actual production in that year.

    b. What might have caused the decline in U.S. fish production after 1995? Do you think that America's appetite for fish has waned?

6.2 *Base e and Continuous Compounding*

What is *e*?

Any positive number can be used as the base for an exponential or logarithmic expression. However, there is a "natural" base called *e* that is used in scientific applications. This number was named after Euler, a Swiss mathematician, and its value is approximately 2.71828. (You can use 2.72 as an estimate for *e* in most calculations. A scientific calculator has an e^x key for more accurate computations.) The number *e* is irrational; it cannot be written as the quotient of two integers or as a repeating decimal. Like π, the number *e* is a fundamental mathematical constant.

> The irrational number *e* is a fundamental mathematical constant whose value is approximately 2.71828.

We first learn why *e* is important, and then how to write any exponential function using *e*.

Continuous Compounding

The number *e* arises naturally in cases of continuous growth at a specified rate. For example, suppose we invest $100 (called the principal) in a bank account that pays interest of 6% per year. To compute the amount of money we have at the end of 1 year, we must also know how often the interest is credited (added) to our account, that is, how often it is *compounded*.

What is compound interest?

Compounding Annually.   If we invest $100 in a bank account that pays 6% interest per year, then at the end of one year we would have $100(1.06) = $106. When the interest is applied to your account once a year, we say that the interest is *compounded* annually.

Compounding Twice a Year.   Now suppose that the interest is compounded twice a year. This means that instead of applying the annual rate of 6% once, it is divided by 2 and applied twice, at the end of each 6-month period. At the end of the first 6 months we earn 6%/2 or 3% interest, so our balance is $100(1.03) = $103. At the end of the second 6 months we earn 3% interest on our new balance of $103. So, after 1 year we have

$$\$100 \text{ at } 6\% \text{ interest compounded twice a year} = \$100 \cdot (1.03) \cdot (1.03)$$
$$= \$100 \cdot (1.03)^2$$
$$= \$106.09$$

We earn 9 cents more when interest is credited twice per year than when it is credited once per year. The difference is a result of the interest earned during the second half-year on the $3 in interest credited at the end of the first half-year. In other words, we're starting to earn interest on interest. To earn the same amount with only annual compounding, we would need an interest rate of 6.09%.

When 6% interest is compounded twice a year, then

$$6\% \qquad \text{is the } \textit{nominal interest rate} \text{ (in name only)}$$
$$\text{or the annual percentage rate (APR)}$$

and

$$6.09\% \qquad \text{is the } \textit{effective interest rate}$$
$$\text{or the annual percentage yield (APY)}$$

The effective interest rate is how much you actually earn (or pay) on an account. Banks and credit card companies are required by law to list both the nominal (the APR) and effective (the APY) interest rates.

Compounding Four Times a Year.   Next, suppose that interest is compounded quarterly, or four times per year. In each quarter, we receive one-quarter of 6%, or 1.5% interest. Each quarter, our investment is multiplied by $1 + 0.015 = 1.015$ and, after the first quarter, we earn interest on the interest we have already received. At the end of 1 year we have received interest four times, so our initial $100 investment has become

$$\$100 \cdot (1.015)^4 \approx \$106.14$$

In this case, the effective interest rate (or annual percentage yield) is about 6.14%.

Compounding n Times a Year. We may imagine dividing the year into smaller and smaller time intervals and computing the interest earned at the end of 1 year. The effective interest rate will be slightly more each time (Table 6.3).

Investing $100 for One Year at a Nominal Interest Rate of 6%

Number of Times Interest Computed During the Year	Value of $100 at End of One Year ($)		Effective Annual Interest Rate (%)
1	$100(1 + 0.06) =$	$100(1.06) = 106.00$	6.00
2	$100(1 + 0.06/2)^2 =$	$100(1.03)^2 = 100(1.0609) = 106.09$	6.09
4	$100(1 + 0.06/4)^4 =$	$100(1.015)^4 \approx 100(1.0614) \approx 106.14$	6.14
6	$100(1 + 0.06/6)^6 =$	$100(1.010)^6 \approx 100(1.0615) \approx 106.15$	6.15
12	$100(1 + 0.06/12)^{12} =$	$100(1.005)^{12} \approx 100(1.0617) \approx 106.17$	6.17
24	$100(1 + 0.06/24)^{24} =$	$100(1.0025)^{24} \approx 100(1.0618) \approx 106.18$	6.18
⋮			
n	$100(1 + 0.06/n)^n$		

Table 6.3

In general, if we calculate the interest on $100 n times a year when the nominal interest rate is 6%, we get

$$\$100\left(1 + \frac{0.06}{n}\right)^n$$

At the End of t Years. What if we invest $100 for t years at a nominal interest rate of 6% compounded n times a year? The annual growth factor is $(1 + 0.06/n)^n$; that is, every year the $100 is multiplied by $(1 + 0.06/n)^n$. After t years the $100 is multiplied by $(1 + 0.06/n)^n$ a total of t times, or equivalently, multiplied by $[(1 + 0.06/n)^n]^t = (1 + 0.06/n)^{nt}$. So $100 will be worth

$$\$100(1 + 0.06/n)^{nt}$$

We can generalize our results:

Compounding n Times a Year for t Years

The value P of P_0 dollars (called the principal) invested at a nominal interest rate r (expressed in decimal form) compounded n times a year for t years is

$$P = P_0\left(1 + \frac{r}{n}\right)^{nt}$$

Continuous compounding using e

Imagine increasing the number of periods, n, without limits, so that interest is computed every week, every day, every hour, every second, and so on. The surprising result is that the term by which $100 gets multiplied, namely,

$$\left(1 + \frac{0.06}{n}\right)^n$$

does not get arbitrarily large. Examine Table 6.4.

Value of $(1 + 0.06/n)^n$ as n Increases

Compounding Period	n (Number of Compoundings per Year)	Approximate Value of $(1 + 0.06/n)^n$
Once a day	365	1.0618313
Once an hour	$365 \cdot 24 =$ 8,760	1.0618363
Once a minute	$365 \cdot 24 \cdot 60 =$ 525,600	1.0618365
Once a second	$365 \cdot 24 \cdot 60 \cdot 60 = 31{,}536{,}000$	1.0618365

Table 6.4

Where does the irrational constant e fit in? As n, the number of compounding periods per year, increases, the value of $(1 + 0.06/n)^n$ approaches $1.0618365 \approx e^{0.06}$. You can confirm this on your calculator. As n gets arbitrarily large, we can think of the compounding occurring at each instant. We call this *continuous compounding*.

If we invest \$100 at 6% continuously compounded, at the end of 1 year we will have

$$\$100 \cdot e^{0.06} \approx \$100 \cdot 1.0618365$$

$$= \$106.18365$$

When 6% annual interest is compounded continuously, then

6% is the *nominal interest rate* (in name only)

or the annual percentage rate (APR)

and 6.18365% is the *effective interest rate*

or the annual percentage yield (APY), which tells you how much interest you will earn after one year.

At the End of t Years. If the interest is compounded continuously, the annual growth factor is $e^{0.06}$; that is, every year the \$100 is multiplied by $e^{0.06}$. After t years the \$100 is multiplied by $e^{0.06}$ a total of t times, or equivalently multiplied by $(e^{0.06})^t = e^{0.06t}$. So \$100 will be worth

$$\$100e^{0.06t}$$

Hence, if we invest \$100 over t years at a nominal interest rate of 6%, we will have

$\$100(1 + 0.06/n)^{nt}$ if the interest is compounded n times a year

$\$100e^{0.06t}$ if the interest is compounded continuously

Just as $(1 + 0.06/n)^n$ approaches $e^{0.06}$ as n gets very large, $(1 + r/n)^n$ approaches e^r. So if P_0 dollars are invested at an annual interest rate r (in decimal form) compounded continuously, then after t years we have

$$P_0 \cdot (e^r)^t = P_0 \cdot e^{rt}$$

Compounding Continuously for t Years

The value P of P_0 dollars invested at a nominal interest rate r (expressed in decimal form) *compounded continuously* for t years is

$$P = P_0 \cdot e^{rt}$$

EXAMPLE 1 If you have \$250 to invest, and you are quoted a nominal interest rate of 4%, construct the equations that will tell you how much money you will have if the interest is compounded once a year, quarterly, once a month, or continuously. In each case calculate the value after 10 years.

SOLUTION See Table 6.5.

Investing \$250 at a Nominal Interest Rate of 4% for Different Compounding Intervals

Number of Compoundings per Year	\$ Value after t Years	Approximate \$ Value When $t = 10$ Years
1	$250 \cdot (1 + 0.04)^t = 250 \cdot (1.04)^t$	370.06
4	$250 \cdot (1 + 0.04/4)^{4t} = 250 \cdot (1.01)^{4t}$	372.22
12	$250 \cdot (1 + 0.04/12)^{12t} \approx 250 \cdot (1.00333)^{12t}$	372.71
Continuous	$250 \cdot e^{0.04t}$	372.96

Table 6.5

EXAMPLE 2 **Continuously compounding debt**

Suppose you have a debt on which the nominal annual interest rate (APR) is 7% compounded continuously. What is the effective interest rate (APY)?

SOLUTION The nominal interest rate (APR) of 7% is compounded continuously, so the equation $P = P_0 e^{0.07t}$ describes the amount P that an initial debt P_0 becomes after t years. Using a calculator, we find that $e^{0.07} \approx 1.073$. The equation could be rewritten as $P = P_0(1.073)^t$. So the effective interest rate (APY) on your debt is about 7.3% per year.

EXAMPLE 3 You have a choice between two bank accounts. One is a passbook account in which you receive simple interest of 5% per year, compounded once per year. The other is a 1-year certificate of deposit (CD), which pays interest at the rate of 4.9% per year, compounded continuously. Which account is the better deal?

SOLUTION Since the interest on the passbook account is compounded once a year, the nominal and effective interest rates are both 5%. The equation $P = P_0(1.05)^t$ can be used to describe the amount P that the initial investment P_0 is worth after t years.

The 1-year certificate of deposit has a nominal interest rate of 4.9%. Since this rate is compounded continuously, the equation $P = P_0 e^{0.049t}$ describes the amount P that the initial investment P_0 is worth after t years. Since $e^{0.049} \approx 1.0502$, the equation can also be written as $P = P_0(1.0502)^t$, and the effective interest rate is 5.02%. So the CD is a better deal.

Exponential Functions Base e

The notation of continuous compounding is useful in scientific as well as financial contexts. We can convert any exponential function in the form $f(t) = Ca^t$ into a continuous growth (or decay) function using a power of e as the base. Since $a > 0$, we can always find a value for k such that

$$a = e^k$$

So we can rewrite the function f as

$$f(t) = C(e^k)^t$$
$$= Ce^{kt}$$

In general applications, we call k the *instantaneous* or *continuous growth* (or *decay*) *rate*. The value of k may be given as either a decimal or a percent.

If $k > 0$, then the function represents exponential growth. Why is this true? If an exponential function represents growth, then the growth factor $a > 1$. If we rewrite a as e^k and 1 as e^0, then $e^k > e^0$, so $k > 0$. For example, the equation $P(t) = 100\, e^{0.06t}$ could describe the growth of 100 cells with a continuous growth rate of 0.06 or 6% per time period t.

If $k < 0$, then the function represents exponential decay. For exponential decay, the decay factor a is such that $0 < a < 1$. Rewriting a as e^k and 1 as e^0, we have $0 < e^k < e^0$. We know the value of e^k is always > 0 since e is a positive number. But if $e^k < e^0$, then $k < 0$. For example, the function $Q(t) = 50e^{-0.03t}$ could describe the decay of 50 cells with a continuous decay rate of 0.03 or 3% per time period t.

Continuous Growth and Decay

If $Q(t) = Ce^{kt}$, then k is called the *instantaneous* or *continuous growth* (or *decay*) *rate*.

For exponential growth, k is positive.

For exponential decay, k is negative.

EXAMPLE 4 **Continuous growth or decay rates**

Identify the continuous growth (or decay) rate for each of the following functions and graph each function using technology.

$$f(t) = 100 \cdot e^{0.055t}$$
$$g(t) = 100 \cdot e^{0.02t}$$
$$h(t) = 100 \cdot e^{-0.055t}$$
$$j(t) = 100 \cdot e^{-0.02t}$$

SOLUTION The function f has a continuous growth rate of 0.055 or 5.5% and g has a continuous growth rate of 0.02 or 2%.

The function h has a continuous decay rate of 0.055 or 5.5% and j has a continuous decay rate of 0.02 or 2%.

The graphs of these four functions are shown in Figure 6.3.

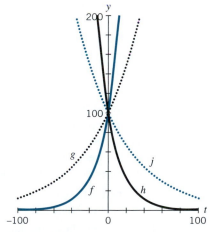

Figure 6.3 Graphs of four exponential functions.

Converting e^k into a

Using the rules of exponents, we can rewrite e^{kt} as $(e^k)^t$. When we know the value of k, we can calculate the value of e^k. For example,

$$P = 100 \cdot e^{0.06t}$$

Rule 3 of exponents $\qquad = 100 \cdot (e^{0.06})^t$

use a calculator to evaluate $e^{0.06}$ $\qquad \approx 100 \cdot 1.0618^t$

The two functions

$$P = 100 \cdot e^{0.06t} \qquad \text{and} \qquad P = 100 \cdot 1.0618^t$$

are equivalent. The first function (base e) suggests growth that occurs *continuously* throughout a time period, so we call 0.06 or 6% the *continuous* growth rate per time period t. The other function suggests growth that happens all at once at the end of each time period, so 0.0618 or 6.18% is just called the growth rate per time period t.

To do the reverse, that is, convert any base a into e^k, we need to know about logarithms base e, which we'll meet in the next section.

EXAMPLE 5   For each of the following functions, identify the *continuous* growth (or decay) rate per year and the growth (or decay) rate based on the growth (or decay) factor. Assume t is measured in years.

 a. $f(t) = 240 \cdot e^{0.127t}$ **b.** $g(t) = 5700 \cdot e^{-0.425t}$

SOLUTION   **a.** The function $f(t)$ has a *continuous* growth rate of 0.127 or 12.7% per year. To find the growth factor, we need to convert $f(t)$ into the form $f(t) = Ca^t$. To do this we need to evaluate $e^{0.127}$. Using a calculator, we find that $e^{0.127} \approx 1.135$. So $f(t) = 240 \cdot e^{0.127t}$ can be rewritten as

$$f(t) = 240 \cdot 1.135^t$$

Since the growth factor $a = 1.135$, the growth rate is 0.135 or 13.5% per year.

 b. The function $g(t)$ has a *continuous* decay rate of 0.425 or 42.5% per year. To find the decay factor, we need to convert $g(t)$ into the form $g(t) = Ca^t$. To do this, we need to evaluate $e^{-0.425}$. Using a calculator, we find that $e^{-0.425} \approx 0.654$. So $g(t) = 5700 \cdot e^{-0.425t}$ can be rewritten as

$$g(t) = 5700 \cdot 0.654^t$$

So the decay factor $a = 0.654$ and the decay rate is $1 - 0.654 = 0.346$ or 34.6% per year.

EXAMPLE 6   **The cost of bottled water**
In 1976 approximately 0.28 billion gallons of bottled water were sold in the United States, according to Beverage Marketing Corp., a New York research and consulting firm. Between 1976 and 2004 the bottled water industry in the United States had a continuous growth rate of about 11.5% a year.

 a. Construct a model that represents the continuous growth of bottled water sales between 1976 and 2004.

 b. Beverage Marketing later reported that in 2005 sales of bottled water were nearly 7.5 billion gallons. If we extrapolate, what would the model predict for sales in 2005? How does this compare with the actual sales?

SOLUTION   **a.** To represent continuous growth, we construct an exponential function using base e and a continuous growth rate of 11.5% a year. If we have an initial value of 0.28 billion gallons and t = number of years since 1976, then

$$f(t) = 0.28 \cdot e^{0.115t} \qquad (0 \leq t \leq 28)$$

models the continuous growth of bottled water sales in the United States between 1976 and 2004.

b. If we extrapolate our model to predict sales for the year 2005, then $t = 2005 - 1976 = 29$ years and

$$f(29) = 0.28 \cdot e^{(0.115)(29)}$$
$$\approx 0.28 \cdot 28.1$$
$$\approx 7.86 \text{ billion gallons}$$

Our model's prediction is somewhat over the actual sales of nearly 7.5 billion gallons in 2005.

You can check the current sales of bottled water at *www.beveragemarketing.com.*

Algebra Aerobics 6.2

Most of these problems require a calculator that can evaluate powers of *e* and, for Problem 4, evaluate logs.

1. Find the amount accumulated after 1 year on an investment of $1000 at 8.5% compounded:

   a. Annually   **b.** Quarterly   **c.** Continuously

2. Find the effective interest rate for each given nominal interest rate that is compounded continuously.

   a. 4%   **b.** 12.5%   **c.** 18%

3. Assume that each of the following describes the value of an investment, *A*, over *t* years. Identify the principal, nominal rate, effective rate, and number of interest periods per year.

   a. $A = 6000 \cdot 1.05^t$ **d.** $A = 50{,}000 \cdot 1.025^{2t}$

   b. $A = 10{,}000 \cdot 1.02^{4t}$ **e.** $A = 125e^{0.076t}$

   c. $A = 500 \cdot 1.01^{12t}$

4. Fill in the missing values, translating from e^k to *a* (the growth or decay factor).

   a. $5e^{0.03t} = 5(\underline{})^t$

   b. $3500e^{0.25t} = 3500(\underline{})^t$

   c. $660e^{1.75t} = 660(\underline{})^t$

   d. $55{,}000e^{-0.07t} = 55{,}000(\underline{})^t$

   e. $125{,}000e^{-0.28t} = 125{,}000(\underline{})^t$

5. The value for *e* is often defined as the number that $(1 + 1/n)^n$ approaches as *n* gets arbitrarily large. Use your calculator to complete Table 6.6 at the bottom of the page. Use your exponent key (x^y or y^x) to evaluate the last column. Is your value consistent with the approximate value for *e* of 2.71828 given in the text?

6. If a principal of $10,000 is invested at the rate of 12% compounded quarterly, the amount accumulated at the end of *t* years is given by the formula

$$A = 10{,}000 \cdot \left(1 + \frac{0.12}{4}\right)^{4t} = 10{,}000(1.03)^{4t}$$

The graph of this function is given in Figure 6.4. Use the graph to estimate for parts (a)–(c) the amount, *A*, accumulated after:

Figure 6.4 Graph of $A = 10{,}000(1.03)^{4t}$.

   a. 1 year   **b.** 5 years   **c.** 10 years

   d. Use the graph to estimate the number of years it will take to double the original investment.

   e. Use the equation to calculate the amount *A* after the years specified in parts (a)–(c) and the doubling time for *A*.

n	1/n	1 + (1/n)	$[1 + (1/n)]^n$
1	1	$1 + 1 = 2$	$2^1 = 2$
100	0.01	$1 + 0.01 = 1.01$	$(1.01)^{100} \approx 2.7048138$
1,000			
1,000,000			
1,000,000,000			

Table 6.6

7. At birth, Maria's parents set aside $8000 in an account designated to help pay for her college education. How much will Maria's account be worth by her 18th birthday if the interest rate was:

   a. 8% compounded quarterly?

   b. 8% compounded continuously?

   c. 8.4% compounded annually?

8. Suppose Maria (from the previous problem) earns a full scholarship and is able to save all the money in her

college account. How much would be in her account on her 30th birthday for each of the rates above?

9. For each of the following functions, identify the continuous growth rate and then determine the effective growth rate. Assume t is measured in years.

   a. $A(t) = Pe^{0.6t}$ **b.** $N(t) = N_0e^{2.3t}$

10. For each of the following functions, identify the continuous decay rate and then determine the effective decay rate. Assume t is measured in years.

   a. $Q(t) = Q_0e^{-0.055t}$ **b.** $P(t) = P_0e^{-0.15t}$

Exercises for Section 6.2

Many of these exercises require a calculator that can evaluate powers (including powers of base e) and logs. One (Exercise 14) requires a graphing program.

1. Use a calculator to find the value of each expression to four decimal places.

   a. $\left(1 + \dfrac{0.035}{2}\right)^2$ **d.** $e^{0.035}$

   b. $\left(1 + \dfrac{0.035}{4}\right)^4$ **e.** How are these values related?

   c. $\left(1 + \dfrac{0.035}{12}\right)^{12}$

2. Assume $y = Pe^{rx}$ represents P dollars invested at an annual interest rate r (in decimal form) compounded continuously for x years. Then for each of the following, calculate the nominal and effective rates in percentage form.

	Nominal Rate (APR)	Effective Rate (APY) (Hint: Evaluate at $e^k = a$)
a.	$y = Pe^{0.025x}$	
b.	$y = Pe^{0.039x}$	
c.	$y = Pe^{0.062x}$	

3. Assume $10,000 is invested at a nominal interest rate of 8.5%. Write the equations that give the value of the money after n years and determine the effective interest rate if the interest is compounded:

   a. Annually   **c.** Quarterly

   b. Semiannually   **d.** Continuously

4. Assume you invest $2000 at 3.5% compounded continuously.

   a. Construct an equation that describes the value of your investment at year t.

   b. How much will $2000 be worth after 1 year? 5 years? 10 years?

5. Construct functions for parts (a) and (b) and compare them in parts (c) and (d).

   a. $25,000 is invested at 5.75% compounded quarterly.

   b. $25,000 is invested at 5.75% compounded continuously.

   c. What is the amount in each account at the end of 5 years?

   d. Explain, using the concept of effective rates, why one amount is larger than the other.

6. Fill in the following chart assuming that the principal is $10,000 in each case.

Nominal Interest Rate (APR)	Compounding Period	Expression for the Value of Your Account after t Years	Effective Interest Rate (APY)
5.25%	Monthly		
		$10,000(1 + 0.045/4)^{4t}$	
	Semiannually		8.16%
3.25%	Daily		
		$10,000(1.02)^t$	

7. Insert the symbol $>$, $<$, or $\approx$ to make the statement true.

   a. $e^{0.045}$ _____ 1.046   **d.** $e^{-0.10}$ _____ 0.90

   b. 1.068 _____ $e^{0.068}$ **e.** 0.8607 _____ $e^{-0.15}$

   c. 1.269 _____ $e^{0.238}$

8. Assume that $5000 was put in each of two accounts. Account A gives 4% interest compounded semiannually. Account B gives 4% compounded continuously.

   a. What are the total amounts in each of the accounts after 10 years?

   b. Show that account B gives 0.04% annually more interest than account A.

9. The half-life of uranium-238 is about 5 billion years. Assume you start with 10 grams of U-238 that decays continuously.

   a. Construct an equation to describe the amount of U-238 remaining after x billion years.

   b. How long would it take for 10 grams of U-238 to become 5 grams?

10. You want to invest money for your newborn child so that she will have $50,000 for college on her 18th birthday. Determine

how much you should invest if the best annual rate that you can get on a secure investment is:

a. 6.5% compounded annually

b. 9% compounded quarterly

c. 7.9% compounded continuously

11. Determine the doubling time for money invested at the rate of 12% compounded:

a. Annually   b. Quarterly

12. a. Phosphorus-32 is used to mark cells in biological experiments. If phosphorus-32 has a continuous daily decay rate of 0.0485 or 4.85%, what is its half-life? (*Hint:* Rewrite the function as $y = Ca^x$ and set $y = 0.5C$.)

b. Phosphorus-32 can be quite dangerous to work with if the experimenter fails to use the proper shields, since its high-energy radiation extends out to 610 cm or about 20 feet. Because disposal of radioactive wastes is increasingly difficult and expensive, laboratories often store the waste until it is within acceptable radioactive levels for disposal with non-radioactive trash. For instance, the rule of thumb for the laboratories of a large East Coast university and medical center is that any waste containing radioactive material with a half-life under 65 days must be stored for 10 half-lives before disposal with the non-radioactive trash.

i. For how many days would phosphorus-32 have to be stored?

ii. What percentage of the original phosphorus-32 would be left at that time?

13. A city of population 1.5 million is expected to experience a 15% decrease in population every 10 years.

a. What is the 10-year decay factor? What is the *yearly* decay factor? The *yearly* decay rate?

b. Use part (a) to create an exponential population model $g(t)$ that gives the population (in millions) after t years.

c. Create an exponential population model $h(t)$ that gives the population (in millions) after t years, assuming a 1.625% continuous yearly decrease.

d. Compare the populations predicted by the two functions after 20 years. What can you conclude?

14. (Requires a graphing program.) Using technology, graph the functions $f(x) = 15,000e^{0.085x}$ and $g(x) = 100,000$ on the same grid.

a. Estimate the point of intersection. (*Hint:* Let x go from 0 to 60.)

b. If $f(x)$ represents the amount of money accumulated by investing at a continuously compounded rate (where x is in years), explain what the point of intersection represents.

15. Rewrite each continuous growth function in its equivalent form $f(t) = Ca^t$. In each case identify the continuous growth rate, and the effective growth rate. (Assume that t is in years.)

a. $P(t) = 500e^{0.02t}$ c. $Q(t) = 45e^{0.06t}$

b. $N(t) = 3000e^{1.5t}$ d. $G(t) = 750e^{0.035t}$

16. Rewrite each continuous decay function in its equivalent form $f(t) = Ca^t$. In each case identify the continuous decay rate and the effective decay rate. (Assume that t is in years.)

a. $P(t) = 600e^{-0.02t}$ c. $Q(t) = 7145e^{-0.06t}$

b. $N(t) = 30,000e^{-0.5t}$ d. $G(t) = 750e^{-0.035t}$

17. Find the nominal interest rate (APR) if a bank advertises that the effective interest rate (APY) on an account compounded continuously is:

a. 3.43% on a checking account

b. 4.6% on a savings account

18. An investment pays 6% compounded four times a year.

a. What is the annual growth factor?

b. What is the annual growth rate?

c. Develop a formula to represent the total value of the investment after each compounding period.

d. If you invest $2000 for a child's college fund, how much will it total after 15 years?

e. For how many years would you have to invest to increase the total to $5000?

6.3 *The Natural Logarithm*

The *common logarithm* uses 10 as a base. The *natural logarithm* uses e as a base and is written $\ln x$ rather than $\log_e x$. Scientific calculators have a key that computes $\ln x$.

> **The Natural Logarithm**
>
> The *logarithm base e* of x is the exponent of e needed to produce x. Logarithms base e are called *natural logarithms* and are written as $\ln x$.
>
> $$\ln x = c \qquad \text{means that} \qquad e^c = x \quad (x > 0)$$

The properties for natural logarithms (base e) are similar to the properties for common logarithms (base 10). Like the common logarithm, $\ln A$ is not defined when $A \le 0$.

If A and B are positive real numbers and p is any real number, then the following rules hold.

Rules of Common Logarithms

1. $\log(A \cdot B) = \log A + \log B$ **3.** $\log A^p = p \log A$

2. $\log(A/B) = \log A - \log B$ **4.** $\log 1 = 0$ (since $10^0 = 1$)

Rules of Natural Logarithms

1. $\ln(A \cdot B) = \ln A + \ln B$ **3.** $\ln A^p = p \ln A$

2. $\ln(A/B) = \ln A - \ln B$ **4.** $\ln 1 = 0$ (since $e^0 = 1$)

In the following examples we show how to use natural logarithms to manipulate expressions and solve exponential equations.

EXAMPLE 1   **Effective vs. nominal rates**

If the effective annual interest rate on an account is 5.21%, estimate the nominal annual interest rate that is compounded continuously.

SOLUTION   A 5.21% effective interest rate is 0.0521 in decimal form. So the equation $P = P_0(1.0521)^x$ represents the value P of P_0 dollars after x years. To find the nominal interest rate that is continuously compounded, we must convert $(1.0521)^x$ into the form e^{rx} or $(e^r)^x$. So e^r must equal 1.0521. We need to solve for r in the equation.

$$1.0521 = e^r$$

Take ln of both sides $\ln 1.0521 = \ln e^r$

use a calculator $0.0508 \approx \ln e^r$

definition of ln $0.0508 \approx r$

So $1.0521 \approx e^{0.0508}$. A nominal interest rate of 0.0508 or 5.08% compounded continuously is equivalent to an effective interest rate of 0.0521 or 5.21%.

EXAMPLE 2   Expand, using the laws of logarithms, the expression: $\ln\sqrt{\dfrac{x + 3}{x - 2}}$.

SOLUTION   Rewrite using exponents $\ln\sqrt{\dfrac{x + 3}{x - 2}} = \ln\left(\dfrac{x + 3}{x - 2}\right)^{1/2}$

Rule 3 of ln $= \dfrac{1}{2}\ln\left(\dfrac{x + 3}{x - 2}\right)$

Rule 2 of ln $= \dfrac{1}{2}[\ln(x + 3) - \ln(x - 2)]$

EXAMPLE 3   Contract, expressing the answer as a single logarithm: $\frac{1}{3}\ln(x - 1) + \frac{1}{3}\ln(x + 1)$

SOLUTION Distributive property $\quad \frac{1}{3}\ln(x-1) + \frac{1}{3}\ln(x+1) = \frac{1}{3}[\ln(x-1) + \ln(x+1)]$

$$= \frac{1}{3}\ln[(x-1)(x+1)]$$

Rule 1 of ln

Rule 3 of ln $\qquad\qquad\qquad\qquad = \ln[(x-1)(x+1)]^{1/3}$

multiply binomials $\qquad\qquad = \ln(x^2-1)^{1/3} \text{ or } \ln\sqrt[3]{x^2-1}$

EXAMPLE 4 Solve the equation $10 = e^t$ for t.

SOLUTION

Given	$10 = e^t$
take ln of both sides	$\ln 10 = \ln e^t$
definition of ln	$\ln 10 = t$
evaluate and switch sides	$t \approx 2.303$

Algebra Aerobics 6.3

Problems 4, 5, 6, and 8 require a calculator that can evaluate natural logs.

1. Evaluate without a calculator:

a. $\ln e^2$
d. $\ln\frac{1}{e^2}$

b. $\ln 1$
e. $\ln\sqrt{e}$

c. $\ln\frac{1}{e}$

2. Expand the following:

a. $\ln\sqrt{xy}$
c. $\ln\left((x+y)^2(x-y)\right)$

b. $\ln\left(\frac{3x^2}{y^3}\right)$
d. $\ln\frac{\sqrt{x+2}}{x(x-1)}$

3. Contract, expressing your answer as a single logarithm:

a. $\ln x + \ln(x-1)$
d. $\frac{1}{2}\ln(x+y)$

b. $\ln(x+1) - \ln x$
e. $\ln x - 2\ln(2x-1)$

c. $2\ln x - 3\ln y$

4. Find the nominal rate on an investment compounded continuously if the effective rate is 6.4%.

5. Determine how long it takes for $10,000 to grow to $50,000 at 7.8% compounded continuously.

6. Solve the following equations for x.

a. $e^{x+1} = 10$
b. $e^{x-2} = 0.5$

7. Determine which of the following are true statements. If the statement is false, rewrite the right-hand side so that the statement becomes true.

a. $\ln 81 \stackrel{?}{=} 4\ln 3$
d. $2\ln 10 \stackrel{?}{=} \ln 20$

b. $\ln 7 \stackrel{?}{=} \frac{\ln 14}{\ln 2}$
e. $\ln\sqrt{e} \stackrel{?}{=} \frac{1}{2}$

c. $\ln 35 \stackrel{?}{=} \ln 5 + \ln 7$
f. $5\ln 2 \stackrel{?}{=} \ln 25$

8. Use the rules of logarithms to contract each expression into a single logarithm (if necessary), then solve for x.

a. $\ln 2 + \ln 6 = x$
c. $\ln(x+1) = 0.9$

b. $\ln 2 + \ln x = 2.48$
d. $\ln 5 - \ln x = -0.06$

Exercises for Section 6.3

Some of these exercises require a calculator that can evaluate powers and logs.

1. Determine the rule(s) of logarithms that were used to expand each expression.

a. $\ln 15 = \ln 3 + \ln 5$
b. $\ln 15 = \ln 30 - \ln 2$
c. $\ln 49 = 2\ln 7$
d. $\ln 25z^3 = 2\ln 5 + 3\ln z$
e. $\ln 5x^4 = \ln 5 + 4\ln x$
f. $\ln\left(\frac{125}{3x}\right) = 3\ln 5 - (\ln 3 + \ln x)$

2. Expand each logarithm using only the numbers 2, 5, ln 2, and ln 5.

a. $\ln 25$
b. $\ln 250$
c. $\ln 625$

3. Write an equivalent expression using exponents.

a. $n = \log 35$
c. $\ln x = \frac{3}{4}$

b. $\ln 75 = x$
d. $\ln\left(\frac{N}{N_0}\right) = -kt$

4. Write an equivalent equation in logarithmic form.

a. $N = 10^{-t/c}$
c. $e^{3x} = 27$

b. $I = I_0 \cdot e^{-k/x}$
d. $\frac{1}{2} = e^{-kt}$

5. Use rules of logarithms to find the value of x. Verify your answer with a calculator.

   a. $\ln x = \ln 2 + \ln 5$

   b. $\ln x = \ln 24 - \ln 2$

   c. $\ln x^2 = 2 \ln 11$

   d. $\ln x = 3 \ln 2 + 2 \ln 6$

   e. $\ln x = 6 \ln 2 - 2 \ln 3$

   f. $\ln x = 4 \ln 2 - 3 \ln 2$

6. Use rules of logarithms to contract to a single logarithm. Use a calculator to verify your answer.

   a. $2 \ln 3 + 4 \ln 2$ **c.** $2(\ln 4 - \ln 3)$

   b. $3 \ln 7 - 5 \ln 3$ **d.** $-4 \ln 3 + \ln 3$

7. Use rules of logarithms to expand.

   a. $\ln\left(\sqrt{4xy}\right)$ **b.** $\ln\left(\dfrac{\sqrt[3]{2x}}{4}\right)$ **c.** $\ln\left(3 \cdot \sqrt[4]{x^3}\right)$

8. Use rules of logarithms to contract to a single logarithm.

   a. $\frac{1}{2} \ln x - 5 \ln y$ **c.** $\ln 2 + \frac{1}{3} \ln x - 4 \ln y$

   b. $\frac{2}{5} \ln x + \frac{4}{5} \ln y$

9. Contract, expressing your answer as a single logarithm.

   a. $\frac{1}{4} \ln (x + 1) + \frac{1}{4} \ln (x - 3)$

   b. $3 \ln R - \frac{1}{2} \ln P$

   c. $\ln N - 2 \ln N_0$

10. Expand:

   a. $\ln \sqrt[3]{\dfrac{x^2 - 1}{x + 2}}$ **b.** $\ln\left(\dfrac{x}{y\sqrt{2}}\right)^2$ **c.** $\ln\left(\dfrac{K^2 L}{M + 1}\right)$

11. Solve for the indicated variable, by changing to logarithmic form. Round your answer to three decimal places.

   a. $e^r = 1.0253$ **b.** $3 = e^{0.5t}$ **c.** $\frac{1}{2} = e^{3x}$

12. Solve for x by changing to exponential form. Round your answer to three decimal places.

   a. $\ln 3x = 1$ **b.** $3 \ln x = 5$ **c.** $\ln 3 + \ln x = 1.5$

13. Solve for x.

   a. $e^x = 10$ **c.** $2 + 4^x = 7$ **e.** $\ln(x + 1) = 3$

   b. $10^x = 3$ **d.** $\ln x = 5$ **f.** $\ln x - \ln(x + 1) = 4$

14. Solve for t.

   a. $5^{t+1} = 6^t$ **d.** $\ln\left(\dfrac{t}{t - 2}\right) = 1$

   b. $e^{t^2} = 4$ **e.** $\ln t - \ln(t - 2) = 1$

   c. $5 \cdot 2^{-t} = 4$

15. Prove that $\ln(A \cdot B) = \ln A + \ln B$, where A and B are positive real numbers.

16. For each of the following equations, either prove that it is correct (by using the rules of logarithms and exponents) or else show that it is not correct (by finding numerical values for x that make the values on the two sides of the equation different).

   a. $e^{x + \ln x} = x \cdot e^x$

   b. $\ln\left(\dfrac{(x + 1)^2}{x}\right) = 2 \ln(x + 1) - \ln x$

   c. $\ln\left(\dfrac{x}{x + 1}\right) = \dfrac{\ln x}{\ln(x + 1)}$

   d. $\ln(x + x^2) = \ln x + \ln x^2$

   e. $\ln(x + x^2) = \ln x + \ln(x + 1)$

17. How long would it take $15,000 to grow to $100,000 if invested at 8.5% compounded continuously?

18. The effective annual interest rate on an account compounded continuously is 4.45%. Estimate the nominal interest rate.

19. The effective annual interest rate on an account compounded continuously is 3.38%. Estimate the nominal interest rate.

20. A town of 10,000 grew to 15,000 in 5 years. Assuming exponential growth:

   a. What is the annual growth rate?

   b. What was its annual continuous growth rate?

21. Solve the following for r.

   a. $1.025 = e^r$ **b.** $\frac{1}{2} = e^r$ **c.** $1.08 = e^r$

22. Solve the following for r.

   a. $0.9 = e^r$ **b.** $2 = e^r$ **c.** $0.75 = e^r$

23. Without using a calculator, identify which functions represent growth and which decay. Then, using a calculator, find the corresponding instantaneous growth or decay rate.

   a. $f(t) = 50e^{(\ln 1.45)t}$ **d.** $j(p) = 2000e^{(\ln 0.3)p}$

   b. $g(x) = 125e^{(\ln 3.15)x}$ **e.** $k(v) = 600e^{(\ln 1.75)v}$

   c. $h(m) = 1500e^{(\ln 0.83)m}$ **f.** $l(x) = e^{(\ln 0.75)x}$

24. Use a calculator to determine the value of each expression. Then rewrite each expression in the form $e^{\ln a}$.

   a. $e^{0.083}$ **b.** $e^{-0.025}$ **c.** $e^{-0.35}$ **d.** $e^{0.83}$

6.4   *Logarithmic Functions*

Exploration 6.1 and course software "E8: Logarithmic Sliders" in *Exponential & Log Functions* will help you understand the properties of logarithmic functions.

Up until now we have been dealing with logarithms of specific numbers, such as log 2 or ln 10. But since for any $x > 0$ there is a unique corresponding value of log x or ln x, we can define two logarithmic functions:

$$y = \log x \qquad \text{and} \qquad y = \ln x \qquad \text{where } x > 0$$

The Graphs of Logarithmic Functions

What will the graphs look like? We know something about the graphs since

Properties of logarithms

If $x > 1$,	$\log x$ and $\ln x$ are both positive
If $x = 1$,	$\log 1 = 0$ and $\ln 1 = 0$
If $0 < x < 1$,	$\log x$ and $\ln x$ are both negative
If $x \leq 0$,	neither logarithm is defined

Table 6.7 and Figure 6.5 show some data points and the graphs of $y = \log x$ and $y = \ln x$.

Evaluating $\log x$ and $\ln x$

x	$y = \log x$	$y = \ln x$
0.001	−3.000	−6.908
0.01	−2.000	−4.605
0.1	−1.000	−2.303
1	0.000	0.000
2	0.301	0.693
3	0.477	1.099
4	0.602	1.386
5	0.699	1.609
6	0.778	1.792
7	0.845	1.946
8	0.903	2.079
9	0.954	2.197
10	1.000	2.303

Table 6.7

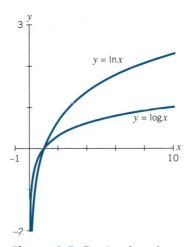

Figure 6.5 Graphs of $y = \log x$ and $y = \ln x$.

The graphs of common and natural logarithms share a distinctive shape. They are both defined only when $x > 0$ and they are both concave down. Both graphs increase throughout with no maximum or minimum value, although they grow more slowly when $x > 1$.

Vertical asymptotes

The graphs of both $y = \log x$ and $y = \ln x$ are vertically asymptotic to the y-axis. This means that as x approaches 0 (through positive values), the graphs come closer and closer to the vertical y-axis but never touch it. For both functions, as $x \to 0$, the values for y get increasingly negative, plunging down near the y-axis toward $-\infty$.

The Graphs of Logarithmic Functions

The graphs of both $y = \log x$ and $y = \ln x$

- lie to the right of the y-axis since they are defined only for $x > 0$
- share a horizontal intercept of $(1, 0)$
- are concave down
- increase throughout with no maximum or minimum
- are asymptotic to the y-axis

EXAMPLE 1 Match each function with the appropriate graph in Figure 6.6.

a. $y = \log(3x)$ **b.** $y = 3 + \log x$ **c.** $y = \log(x^3)$ **d.** $y = 3 \log x$

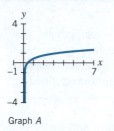
Graph A

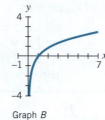
Graph B

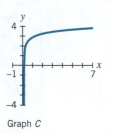
Graph C

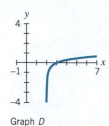
Graph D

Figure 6.6 Four graphs involving logarithms.

SOLUTION **a.** Graph A. When $x = 1/3$, then $y = \log(3 \cdot 1/3) = \log 1 = 0$, so the horizontal intercept is $(1/3, 0)$.

b. Graph C. When $x = 1$, then $y = 3 + \log 1 = 3 + 0 = 3$, so the graph passes through $(1, 3)$.

c. and d. Graph B. Since $\log(x^3) = 3 \log x$, the graphs for functions (c) and (d) are the same. When $x = 1$, then $y = 3 \log 1 = 0$, so the horizontal intercept is $(1, 0)$.

There is no match for Graph D.

EXAMPLE 2 Graph $y = \ln(x - 2)$. Describe its relationship to $y = \ln x$.

SOLUTION Figure 6.7 shows the two graphs.

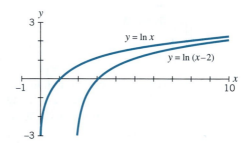

Figure 6.7 Graphs of $y = \ln x$ and $y = \ln(x - 2)$.

The function $y = \ln(x - 2)$ tells us to subtract 2 from x and then apply the function ln. So the graph of $y = \ln(x - 2)$ is the graph of $y = \ln x$ shifted two units to the right. For $y = \ln x$, the horizontal intercept is at 1 since $\ln(1) = 0$. For $y = \ln(x - 2)$, the horizontal intercept is at 3, since $\ln(3 - 2) = \ln 1 = 0$.

The Relationship between Logarithmic and Exponential Functions

Logarithmic vs. exponential growth

Logarithmic and exponential growth are both unbounded. They both increase forever, never reaching a maximum value. But exponential growth is rapid—not only increasing, but doing so at a rate that is speeding up (accelerating). Logarithmic growth is slow—increasing, but at rate that is slowing down (decelerating). For example, if we let $t = 1$, 3, and 5, then 10^t is respectively 10, 1000, and 100,000, but $\log(t)$ is respectively 0, 0.477, and 0.699.

Logarithmic and exponential functions are inverses of each other

What does it mean for two functions to be inverses of each other? It means that what one function does, the other undoes. Log and exponential functions are inverses of each other. For example,

$$\log(10^x) = x \qquad \text{and} \qquad 10^{\log x} = x$$

Why are these true?

Rationale for $\qquad\qquad\qquad\log(10^x) = x$

Finding the logarithm of a number base 10 involves finding the exponent of 10 needed to produce the number. Since 10^x is already written as 10 to the power x, then $\log 10^x = x$.

Rationale for $\qquad\qquad\qquad 10^{\log x} = x$

By definition, $\log x$ is the number such that when 10 is raised to that power the result is x.

Proof   Let $\qquad\qquad\qquad\qquad\qquad\qquad\qquad\qquad y = \log x$

$\qquad$ rewrite using definition of logarithm $\qquad 10^y = x$

$\qquad$ substitute $\log x$ for y $\qquad\qquad\qquad\qquad 10^{\log x} = x$

The two functions

$$y = 10^x \qquad \text{and} \qquad y = \log x$$

are *inverses* of each other. Similarly, for natural logarithms

$$\ln(e^x) = x \qquad \text{and} \qquad e^{\ln x} = x$$

So the functions $\qquad\qquad\qquad y = e^x \qquad \text{and} \qquad y = \ln x$

are also inverses of each other.

> **More Rules for Common and Natural Logarithms**
>
> **5.** $\log(10^x) = x \qquad$ and $\qquad \ln(e^x) = x$
> **6.** $10^{\log x} = x \qquad$ and $\qquad e^{\ln x} = x$

The graphs of two inverse functions are mirror images across the diagonal line y = x

The graphs of two inverse functions such as $y = 10^x$ and $y = \log x$ are mirror images across the line $y = x$. If you imagine folding the graph along the dotted line $y = x$, the two curves would lie right on top of each other. See Figure 6.8 on the following page.

$\qquad$ Why are these graphs mirror images? Choose any point (a, b) on the graph of $y = \log x$. To reach that point you would need to move a units horizontally (on the x-axis) and b units vertically (on the y-axis). Now imagine folding at the dotted line $y = x$. What are the coordinates of the point's mirror image? You would need to move a units vertically (on the y-axis) and b units horizontally (on the x-axis). The mirror-image coordinates would be (b, a). So the points (a, b) and (b, a) are mirror images across the line $y = x$. Table 6.8 lists some pairs of mirror-image points that lie on the graphs of $y = 10^x$ and $y = \log x$.

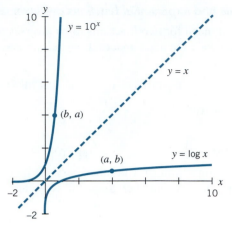

Figure 6.8 The graphs of $y = \log x$ and $y = 10^x$ are mirror images across the dotted line $y = x$.

Mirror-Image Points on $y = 10^x$ and $y = \log x$

$y = 10^x$	$y = \log x$
$(0, 1)$	$(1, 0)$
$(1, 10)$	$(10, 1)$
$(2, 100)$	$(100, 2)$
$(-1, 0.1)$	$(0.1, -1)$
$(-2, 0.01)$	$(0.01, -2)$

Table 6.8

We can use similar arguments to show that the graphs of $y = e^x$ and $y = \ln x$ are mirror images as well.

Inverse Functions and Their Graphs

The functions in each pair

$$y = 10^x \quad \text{and} \quad y = \log x$$
$$y = e^x \quad \text{and} \quad y = \ln x$$

are *inverses* of each other. What one does, the other undoes.

The graphs in each pair are mirror images across the line $y = x$.

EXAMPLE 3   Graph $y = \ln x$ and $y = e^x$. Identify three pairs of points on the function graphs that are mirror images across the line $y = x$.

SOLUTION   Figure 6.9 shows the graphs of $y = \ln x$ and $y = e^x$, and Table 6.9 contains some pairs of points on $y = e^x$ and $y = \ln x$.

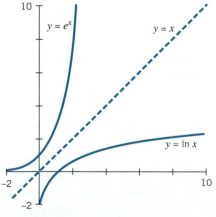

Figure 6.9 Graphs of $y = \ln x$ and $y = e^x$ are mirror images across the dotted line $y = x$.

Mirror-Image Points on $y = e^x$ and $y = \ln x$

$y = e^x$	$y = \ln x$
$(0, 1)$	$(1, 0)$
$(1, e)$	$(e, 1)$
$(-2, e^{-2})$	$(e^{-2}, -2)$

Table 6.9

Applications of Logarithmic Functions

Just as with logarithmic scales, logarithmic functions are used in dealing with quantities that vary widely in size. We'll examine two such functions used to measure acidity and noise levels.

Measuring acidity: The pH scale

Chemists use a logarithmic scale called pH to measure acidity. pH values are defined by the function

$$\text{pH} = -\log[\text{H}^+]$$

where $[\text{H}^+]$ designates the concentration of hydrogen ions. Chemists use the symbol H^+ for hydrogen ions (hydrogen atoms stripped of their one electron), and the brackets [ ] mean "concentration of." Ion concentration $[\text{H}^+]$ is measured in moles per liter, M, where one mole equals $6.022 \cdot 10^{23}$ or Avogadro's number of ions. A pH value is the negative of the logarithm of the number of moles per liter of hydrogen ions.

Table 6.10 and Figure 6.10 show a set of values and a graph for pH. The graph is the standard logarithmic graph flipped over the horizontal axis because of the negative sign in front of the log.

Calculating pH Values

$[\text{H}^+]$ (moles per liter)	pH
10^{-15}	15.000
10^{-10}	10.000
1	0.000
5	−0.699
10	−1.000

Table 6.10

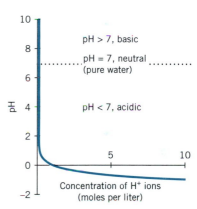

Figure 6.10 Graph of the pH function.

Typically, pH values are between 0 and 14, indicating the level of acidity. Pure water has a pH of 7.0 and is considered *neutral*. A substance with a pH < 7 is called *acidic*. A substance with a pH > 7 is called *basic* or *alkaline*.[1] The lower the pH, the more acidic the substance. The higher the pH, the less acidic and the more alkaline the substance. Table 6.11 shows approximate pH values for some common items. Most foods have a pH between 3 and 7. Substances with a pH below 3 or above 12 can be dangerous to handle with bare hands.

Remember that pH is the *negative* of a logarithmic function. So the larger the hydrogen ion concentration, the smaller the pH. (See Figure 6.10.) Multiplying the hydrogen ion concentration by 10 decreases the pH by 1. So vinegar (with a pH of 3) has a hydrogen ion concentration 10 times greater than that of wine (pH 4) and 10,000 times greater than that of pure water (pH 7).

---

[1] Things that are alkaline tend to be slimy and sticky, like a bar of soap. If you wash your hands with soap and don't rinse, there will be an alkaline residue. Acids can be used to neutralize alkalinity. For example, shampoo often leaves a sticky alkaline residue. So we use acidic conditioners to neutralize the alkalinity.

pH of Various Substances

Acidic (more hydrogen ions than pure water)	pH	Neutral	pH	Basic or Alkaline (fewer hydrogen ions than pure water)	pH
Gastric juice	2	Pure water	7	Egg whites, sea water	8
Coca-Cola, vinegar	3			Soap, baking soda	9
Grapes, wine	4			Detergents, toothpaste, ammonia	10
Coffee, tomatoes	5			Household cleaner	12
Bread	5.5			Caustic oven cleaner	13
Beef, chicken	6				

Table 6.11

EXAMPLE 4 **Comparing hydrogen ion concentrations**

a. Compare the hydrogen ion concentration of Coca-Cola with the hydrogen ion concentrations of coffee and ammonia.

b. Calculate the hydrogen ion concentration of Coca-Cola.

SOLUTION a. Table 6.11 gives the pH of Coca-Cola as 3, coffee as 5 (so both are acidic), and ammonia as 10 (which is alkaline). Increasing the pH by 1 corresponds to decreasing the hydrogen ion concentration by a factor of 10. So Coca-Cola will have more ions than coffee, and even more hydrogen than ammonia. The hydrogen ion concentration of Coca-Cola is $10^{5-3} = 10^2 = 100$ times more than that of coffee, and $10^{10-3} = 10^7 = 10,000,000$ times more than that of ammonia!

b. To calculate the hydrogen ion concentration of Coca-Cola, we need to solve the equation

$$3 = -\log[H^+]$$

Multiply by -1 $\qquad -3 = \log[H^+]$

rewrite as powers of 10 $\qquad 10^{-3} = 10^{\log[H^+]}$

evaluate and use Rule 6 of logs $\qquad 0.001 = [H^+]$

So the hydrogen ion concentration of Coca-Cola is 10^{-3} or 0.001 moles per liter. That means that each liter of Coca-Cola contains $10^{-3} \cdot (6.022 \cdot 10^{23}) = 6.022 \cdot 10^{23-3} = 6.022 \cdot 10^{20}$ hydrogen ions.

EXAMPLE 5 **Calculating the pH level**

Sulfuric acid has a hydrogen ion concentration $[H^+]$ of 0.109 moles per liter. Calculate its pH.

SOLUTION The pH of sulfuric acid equals $-\log 0.109 \approx 0.96$.

The rain in many parts of the world is becoming increasingly acidic. The burning of fossil fuels (such as coal and oil) by power plants and automobile emissions release gaseous impurities into the air. The impurities contain oxides of sulfur and nitrogen that combine with moisture in the air to form droplets of dilute sulfuric and nitric acids. An acid releases hydrogen ions in water. High concentrations of hydrogen ions damage plants and water resources (such as the lakes of New England and Sweden) and erode structures (such as the Parthenon in Athens) by removing oxygen molecules. Some experts feel that the acid rain dilemma may be one of the greatest environmental problems facing the world in the near future.

EXAMPLE 6 **The pH function is not linear**

It is critical that health care professionals understand the nonlinearity of the pH function. An arterial blood pH of 7.35 to 7.45 for a patient is quite normal, whereas a pH of 7.1 means that the patient is severely acidotic and near death.

a. Determine the hydrogen ion concentration, $[H^+]$, first for a patient with an arterial blood pH of 7.4, then for one with a blood pH of 7.1.

b. How many more hydrogen ions are there in blood with a pH of 7.1 than in blood with a pH of 7.4?

SOLUTION To find $[H^+]$ if the pH is 7.4, we must solve the equation

$$7.4 = -\log[H^+]$$

Multiply both sides by -1	$-7.4 = \log[H^+]$
then these powers of 10 are equal	$10^{-7.4} = 10^{\log[H^+]}$
use Rule 6 of logs	$10^{-7.4} = [H^+]$
evaluate and switch sides	$[H^+] \approx 0.000\,000\,040$
	$\approx 4.0 \cdot 10^{-8}\,M$

where M is in moles per liter.

Similarly, if the pH is 7.1, we can solve the equation

$$7.1 = -\log[H^+]$$

to get $\quad [H^+] \approx 7.9 \cdot 10^{-8}\,M$

Comparing the two concentrations gives us

$$\frac{[H^+] \text{ in blood with a pH of } 7.1}{[H^+] \text{ in blood with a pH of } 7.4} \approx \frac{7.9 \cdot 10^{-8}\,M}{4.0 \cdot 10^{-8}\,M} = \frac{7.9}{4.0} \approx 2$$

So there are approximately twice as many hydrogen ions in blood with a pH of 7.1 than in blood with a pH of 7.4.

Measuring noise: The decibel scale

The decibel scale was designed to reflect the human perception of sounds.[2] When it is very quiet, it is easy to notice a small increase in sound intensity. The same increase in intensity in a noisy environment would not be noticed; it would take a much bigger change to be detected by humans. The same is true for light. If a 50-watt light bulb is replaced with a 100-watt light bulb, it is easy to notice the difference in brightness. But if you replaced 500 watts with 550 watts, it would be very hard to distinguish the 50-watt difference. The decibel scale, like the pH scale for acidity or the Richter scale for earthquakes, is logarithmic; that is, it measures order-of-magnitude changes.

Noise levels are measured in units called *decibels,* abbreviated dB. The name is in honor of the inventor of the telephone, Alexander Graham Bell. If we designate I_0 as the intensity of a sound at the threshold of human hearing (10^{-16} watts/cm^2) and we let I represent the intensity of an arbitrary sound (measured in watts/cm^2), then the noise level N of that sound measured in decibels (dB) is defined to be

$$N = 10 \log\left(\frac{I}{I_0}\right)$$

[2]Two scientists, Weber and Fechner, studied the psychological response to intensity changes in stimuli. Their discovery, that the perceived change is proportional to the logarithm of the intensity change of the stimulus, is called the Weber-Fechner stimulus law.

The unitless expression I/I_0 gives the *relative intensity* of a sound compared with the reference value of I_0. For example, if $I/I_0 = 100$, then the noise level, N, is equal to

$$N = 10 \log(100) = 10 \log(10^2) = 10(2) = 20 \text{ dB}$$

Table 6.12 shows relative intensities, the corresponding noise levels (in decibels), and how people perceive these noise levels. Note how much the relative intensity (the ratio I/I_0) of a sound source must increase for people to discern differences. Each time we *add* 10 units on the decibel scale, we *multiply* the relative intensity by 10, increasing it by one order of magnitude.

How Decibel Levels Are Perceived

Relative Intensity I/I_0	Decibels (dB)	Average Perception
1	0	Threshold of hearing
10	10	Soundproof room, very faint
100	20	Whisper, rustle of leaves
1,000	30	Quiet conversation, faint
10,000	40	Quiet home, private office
100,000	50	Average conversation, moderate
1,000,000	60	Noisy home, average office
10,000,000	70	Average radio, average factory, loud
100,000,000	80	Noisy office, average street noise
1,000,000,000	90	Loud truck, police whistle, very loud
10,000,000,000	100	Loud street noise, noisy factory
100,000,000,000	110	Elevated train, deafening
1,000,000,000,000	120	Thunder of artillery, nearby jackhammer
10,000,000,000,000	130	Threshold of pain, ears hurt

Table 6.12

EXAMPLE 7

How loud is a rock band?

What is the decibel level of a typical rock band playing with an intensity of 10^{-5} watts/cm²? How much more intense is the sound of the band than an average conversation?

SOLUTION Given $I_0 = 10^{-16}$ watts/cm² and letting $I = 10^{-5}$ watts/cm² and N represent the decibel level,

by definition	$N = 10 \log\left(\dfrac{I}{I_0}\right)$
substitute for I and I_0	$= 10 \log(10^{-5}/10^{-16})$
Rule 2 of exponents	$= 10 \log(10^{11})$
Rule 5 of logs	$= 10 \cdot 11$
	$= 110 \text{ decibels}$

So the noise level of a typical rock band is about 110 decibels.

According to Table 6.12, an average conversation measures about 50 decibels. So the noise level of the rock band is 60 decibels higher. Each increment of 10 decibels corresponds to a one-order-of-magnitude increase in intensity. So the sound of a rock band is about six orders of magnitude, or 10^6 (a million times), more intense than an average conversation.

EXAMPLE 8   **Perceiving sound**
What's wrong with the following statement? "A jet airplane landing at the local airport makes 120 decibels of noise. If we allow three jets to land at the same time, there will be 360 decibels of noise pollution."

SOLUTION   There will certainly be three times as much sound intensity, but would we perceive it that way? According to Table 6.12, 120 decibels corresponds to a relative intensity of 10^{12}. Three times that relative intensity would equal $3 \cdot 10^{12}$. So the corresponding decibel level would be

$$N = 10 \log(3 \cdot 10^{12})$$

Rule 1 of logs	$= 10(\log 3 + \log(10^{12}))$
use a calculator	$\approx 10(0.477 + 12)$
combine	$\approx 10(12.477)$
multiply and round off	≈ 125 decibels

So three jets landing will produce a decibel level of 125, not 360. We would perceive only a slight increase in the noise level.

Algebra Aerobics 6.4

A graphing program is recommended for Problem 3 and a scientific calculator for Problem 7.

1. How would the graphs of $y = \log x^2$ and $y = 2 \log x$ compare?

2. Draw a rough sketch of the graph $y = -\ln x$. Compare it with the graph of $y = \ln x$.

3. Compare the graphs of $f(x) = \log x$ and $g(x) = \ln x$.
   a. Where do they intersect?
   b. Where does each have an output value of 1? Of 2?
   c. Describe each graph for values of x such that $0 < x < 1$.
   d. Describe each graph for values of x where $x > 1$.

4. Given the accompanying graph of a function f, sketch the graph of its inverse.

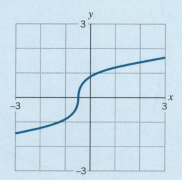

5. Use the rules for logarithms to evaluate the following expressions.
   a. $\log 10^3$ e. $\ln e^5$
   b. $\log 10^{-5}$ f. $\ln e^{0.07}$
   c. $3 \log 10^{0.09}$ g. $\ln e^{3.02} + \ln e^{-0.27}$
   d. $10^{\log 3.4}$ h. $e^{\ln 0.9}$

6. A typical pH value for rain or snow in the northeastern United States is about 4. Is this basic or acidic? What is the corresponding hydrogen ion concentration? How does this compare with the hydrogen ion concentration of pure water?

7. What is the decibel level of a sound whose intensity is $1.5 \cdot 10^{-12}$ watts/cm²?

8. If the intensity of a sound increases by a factor of 100, what is the increase in the decibel level? What if the intensity is increased by a factor of 10,000,000?

Exercises for Section 6.4

Some of these exercises require a calculator that can evaluate powers and logs. Exercise 6 requires a graphing program.

1. Use the rules of logarithms to explain how you can tell which graph is $y = \log x$ and which is $y = \log(5x)$.

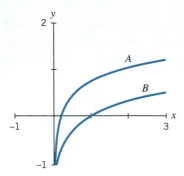

2. Use the rules of logarithms to explain how you can tell which graph is $y = \log x$ and which is $y = \log(x/5)$.

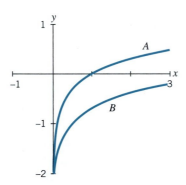

3. The functions $f(x) = \log x$, $g(x) = \log(x - 1)$, and $h(x) = \log(x - 2)$ are graphed below.

   a. Match each function with its graph.

   b. Find the value of x for each function that makes that function equal to zero.

   c. Identify the x-intercept for each function.

   d. Assuming $k > 0$, describe how the graph of $f(x)$ moves if you replace x by $(x - k)$.

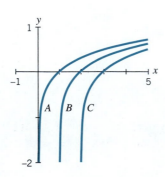

4. The functions $f(x) = \ln x$, $g(x) = \ln(x + 1)$, and $h(x) = \ln(x + 2)$ are graphed below.

   a. Match each function with its graph.

   b. Find the value of x for each function that makes that function equal to zero.

   c. Identify the coordinates of the horizontal intercept for each function.

   d. Assuming $k > 0$, describe how the graph of $f(x)$ moves if you replace x by $(x + k)$.

   e. Determine the y-intercept, if possible, for each function.

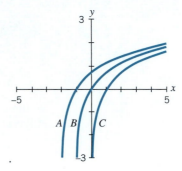

5. Examine the following graphs of four functions.

   a. Which function graphs are mirror images of each other across the y-axis?

   b. Which function graphs are mirror images of each other across the x-axis?

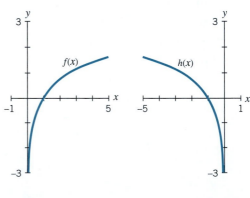

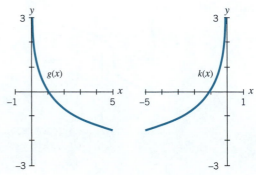

6. (Requires a graphing program.) On the same grid graph $y_1 = \ln(x)$, $y_2 = -\ln(x)$, $y_3 = \ln(-x)$ and $y_4 = -\ln(-x)$.

   a. Which pairs of function graphs are mirror images across the *y*-axis?

   b. Which pairs of function graphs are mirror images across the *x*-axis?

   c. What predictions would you make about the graphs of the functions $f(x) = a\ln x$ and $g(x) = -a\ln x$? Using technology, test your predictions for different values of *a*.

   d. What predictions would you make about the graphs of the functions $f(x) = a\ln x$ and $g(x) = a\ln(-x)$? Using technology, test your predictions for different values of *a*.

7. Logarithms can be constructed using any positive number except 1 as a base:

$$\log_a x = y \text{ means that } a^y = x$$

   a. Complete the accompanying table and sketch the graph of $y = \log_3 x$.

x	$y = \log_3 x$
$\frac{1}{9}$	
$\frac{1}{3}$	
1	
3	
9	
27	

   b. Now make a small table and sketch the graph of $y = \log_4 x$. (*Hint:* To simplify computations, try using powers of 4 for values of *x*.)

8. The stellar magnitude *M* of a star is approximately $-2.5 \log(B/B_0)$, where *B* is the brightness of the star and B_0 is a constant.

   a. If you plotted *B* on the horizontal and *M* on the vertical axis, where would the graph cross the *B* axis?

   b. Without calculating any other coordinates, draw a rough sketch of the graph of *M*. What is the domain?

   c. As the brightness *B* increases, does the magnitude *M* increase or decrease? Is a sixth-magnitude star brighter or dimmer than a first-magnitude star?

   d. If the brightness of a star is increased by a factor of 5, by how much does the magnitude increase or decrease?

9. [*Source:* H. D. Young, *University Physics,* Vol. 1 (Reading, MA: Addison-Wesley, 1992), p. 591] If you listen to a 120-decibel sound for about 10 minutes, your threshold of hearing will typically shift from 0 dB up to 28 dB for a while. If you are exposed to a 92-dB sound for 10 years, your threshold of hearing will be permanently shifted to 28 dB. What intensities correspond to 28 dB and 92 dB?

10. In all of the sound problems so far, we have not taken into account the distance between the sound source and the listener. Sound intensity is inversely proportional to the square of the distance from the sound source; that is, $I = k/r^2$, where *I* is intensity, *r* is the distance from the sound source, and *k* is a constant.

Suppose that you are sitting a distance *R* from the TV, where its sound intensity is I_1. Now you move to a seat twice as far from the TV, a distance $2R$ away, where the sound intensity is I_2.

   a. What is the relationship between I_1 and I_2?

   b. What is the relationship between the decibel levels associated with I_1 and I_2?

11. If there are a number of different sounds being produced simultaneously, the resulting intensity is the sum of the individual intensities. How many decibels louder is the sound of quintuplets crying than the sound of one baby crying?

12. An ulcer patient has been told to avoid acidic foods. If he drinks coffee, with a pH of 5.0, it bothers him, but he can tolerate both tap water, with a pH of 5.8, and milk, with a pH of 6.9.

   a. Will a mixture of half coffee and half milk be at least as tolerable as tap water?

   b. What pH will the half coffee–half milk mixture have?

   c. In order to make 10 oz of a milk-coffee drink with a pH of 5.8, how many ounces of each are required?

13. Lemon juice has a pH of 2.1. If you make diet lemonade by mixing $\frac{1}{4}$ cup of lemon juice with 2 cups of tap water, with a pH of 5.8, will the resulting acidity be more or less than that of orange juice, with a pH of 3?

6.5 Transforming Exponential Functions to Base e

In Section 6.2 we saw that an exponential function can be written in two equivalent forms:

$$f(t) = Ca^t \quad \text{or} \quad f(t) = C(e^k)^t$$

where $e^k = a$. Given a value for *k*, we can evaluate e^k to find *a*. Now we can use logarithms to do the reverse; that is, given a value for *a* we find a value for *k* such that $a = e^k$.

Converting a to e^k

In general, if a is the growth (or decay) factor, then we can always find a value for k such that

$$e^k = a$$

$$\text{by taking ln of both sides} \qquad \ln e^k = \ln a$$

$$\text{using Rule 5 for ln} \qquad k = \ln a$$

So the function $f(t) = Ca^t$ can be rewritten as

$$f(t) = C(e^k)^t$$
$$= C(e^{\ln a})^t$$

where $k = \ln a$.

Recall that k is called the *instantaneous* or *continuous* growth (or decay) rate.

For exponential growth, $a > 1$, so $e^k > 1$ and k is positive.

For exponential decay, $0 < a < 1$, so $0 < e^k < 1$ and k is negative.

Distinguishing Different Rates

For functions in the form $f(t) = C(e^k)^t$, we call the exponent k the *continuous* growth rate (if $k > 0$) or the *continuous* decay rate (if $k < 0$).

For functions in the form $f(t) = Ca^t$, we call the base a the growth (or decay) factor.

$$\text{If } a > 1, \qquad \text{the growth factor} = 1 + \text{growth rate}$$

$$\text{If } 0 < a < 1, \qquad \text{the decay factor} = 1 - \text{decay rate}$$

EXAMPLE 1

a. Rewrite the function $f(t) = 250(1.3)^t$ using base e.

b. Identify the growth rate (based on the growth factor) and the *continuous* growth rate, both per time period t.

SOLUTION

a. The function $f(t) = 250(1.3)^t$ can be rewritten using base e, by

$$\text{substituting } e^{\ln 1.3} \text{ for } 1.3 \qquad f(t) = 250(e^{\ln 1.3})^t$$

$$\text{evaluating } \ln 1.3 \qquad \approx 250(e^{0.262})^t$$

$$\text{using Rule 3 of exponents} \qquad = 250e^{0.262t}$$

b. Since $f(t) = 250(1.3)^t$, the growth rate per time period t is 0.3 or 30%. Since we also have $f(t) = 250e^{0.262t}$, the *continuous* growth rate per time period t is 0.262 or 26.2%.

EXAMPLE 2

a. Rewrite the function $g(t) = 340(0.94)^t$ using base e. Assume t is in years.

b. Identify the decay factor, the decay rate, and the continuous decay rate.

SOLUTION

a. Since the function $g(t) = 340(0.94)^t$ represents exponential decay, the value for k will be negative. The function g can be rewritten using base e, by

$$\text{substituting } e^{\ln 0.94} \text{ for } 0.94 \qquad g(t) = 340(e^{\ln 0.94})^t$$

$$\text{evaluating } \ln 0.94 \qquad \approx 340(e^{-0.062})^t$$

$$\text{using Rule 3 of exponents} \qquad = 340e^{-0.062t}$$

b. The decay factor is 0.94. So the decay rate is $1 - 0.94 = 0.06$ or 6% per year. The *continuous* decay rate is 0.062 or 6.2% per year.

See "E4: $y = Ce^{rx}$ Sliders" and "E5: Comparing $y = Ca^x$ to $y = Ce^{rx}$" in Exponential & Log Functions.

Writing an exponential function in the form $y = Ca^t$ or $y = Ce^{kt}$ (where $k = \ln a$) is a matter of emphasis, since the graphs and functional values are identical. When we use $y = Ca^t$, we may think of the growth taking place at discrete points in time, whereas the form $y = Ce^{kt}$ emphasizes the notion of continuous growth. For example, $y = 100(1.05)^t$ could be interpreted as giving the value of $100 invested for t years at 5% compounded annually, whereas its equivalent form

$$y = 100(e^{\ln 1.05})^t \approx 100e^{0.049t}$$

suggests that the money is invested at 4.9% compounded continuously.

Exponential Functions Using Base e

The exponential function $y = Ca^t$ (where $a > 0$ and $a \neq 1$) can be rewritten as

$$y = C(e^k)^t \quad \text{where } k = \ln a$$

We call k the *instantaneous* or *continuous growth* (or *decay*) *rate.*

Exponential growth:	$a > 1$ and	k is positive
Exponential decay:	$0 < a < 1$ and	k is negative

EXAMPLE 3 **Converting to a continuous growth rate**

Consider the bacterial growth we described with the equation $N = 100 \cdot 2^t$. The bacteria don't all double at the same time, precisely at the beginning of each time period t. A continuous growth pattern is much more likely. Rewrite this equation to reflect a continuous growth rate.

SOLUTION The equation $N = 100 \cdot 2^t$ can be rewritten to reflect a continuous growth rate by converting the base of 2 to its equivalent form using base e. We know the growth factor a can be rewritten as

$$a = e^{\ln a}$$
$$\text{Substitute 2 for } a \qquad 2 = e^{\ln 2}$$
$$\text{evaluate } \ln 2 \qquad 2 \approx e^{0.693}$$

So the equation that reflects a continuous growth rate is
$$N = 100 \cdot e^{0.693t}$$

EXAMPLE 4 **Converting from base a to base e**

In Chapter 5 we saw that the function $f(t) = 100(0.976)^t$ measures the amount of radioactive strontium-90 remaining as 100 milligrams (mg) decay over time t (in years). Rewrite the function using base e.

SOLUTION To rewrite the function $f(t) = 100(0.976)^t$ using base e, we need to convert the base 0.976 to the form e^k, where $k = \ln 0.976$. Using a calculator to evaluate $\ln 0.976$, we get -0.0243. Substituting into our original function, we get

$$f(t) = 100(0.976)^t$$
$$\approx 100(e^{-0.0243})^t$$
$$= 100e^{-0.0243t}$$

Since the original base 0.976 is less than 1, the function represents decay. So, as we would expect, when the function is rewritten using base e, the value of k (in this case -0.0243) is negative.

EXAMPLE 5   Use a continuous compounding model to describe the growth of Medicare expenditures.

SOLUTION   In Chapter 5 we found a best-fit function for Medicare expenditures to be

$$C(n) = 10.6 \cdot (1.11)^n$$

where n = years since 1970 and $C(n)$ = Medicare expenditures in billions of dollars. This implies a growth rate of 11% compounded annually. To describe the same growth in terms of continuous compounding, we need to rewrite the growth factor 1.11 as e^k, where $k = \ln 1.11 \approx 0.104$. Substituting into the original function gives us

$$C(n) = 10.6e^{0.104n}$$

So Medicare expenditures are continuously compounding at about 10.4% per year.

EXAMPLE 6   **Proving the rule of 70**
We now have the tools to prove the rule of 70 introduced in Chapter 5. Recall that the rule said if a quantity is growing (or decaying) at $R\%$ per time period, then the time it takes the quantity to double (or halve) is approximately $70/R$ time periods. For example, if a quantity increases by a rate, R, of 7% each month, the doubling time is about $70/7 = 10$ months.

Proof
Let f be an exponential function of the form

$$f(t) = Ce^{rt}$$

where C is the initial quantity, r is the continuous growth rate, and t represents time. The rate r is in decimal form, and R is the equivalent amount expressed as a percentage. Since $f(t)$ represents exponential growth, $r > 0$. The doubling time for an exponential function is constant, so we need only calculate the time for any given quantity to double. In particular, we can determine the time it takes for the initial amount C (at time $t = 0$) to become twice as large; that is, we can calculate the value for t such that

$$f(t) = 2C$$

Set	$Ce^{rt} = 2C$
divide by C	$e^{rt} = 2$
take ln of both sides	$\ln e^{rt} = \ln 2$
evaluate and use ln Rule 5	$rt \approx 0.693$ (1)

$R = 100r$, so $r = R/100$. If we round 0.693 up to 0.70 and

substitute in Equation (1)	$(R/100) \cdot t \approx 0.70$
multiply both sides by 100	$R \cdot t \approx 70$
divide by R we get	$t \approx 70/R$

So the time t it takes for the initial amount to double is approximately $70/R$, which is what the rule of 70 claims. We leave the similar proof about half-lives, where $r < 0$ represents decay, to the exercises.

Algebra Aerobics 6.5

A calculator that can evaluate logs and powers is required for Problems 2–4.

1. Identify each of the following exponential functions as representing growth or decay. (*Hint:* For parts (d)–(f) use rules for logarithms.)

   a. $M = Ne^{-0.029t}$

   b. $K = 100(0.87)^r$

   c. $Q = 375\,e^{0.055t}$

   d. $P = 250\,e^{(\ln 1.056)t}$

   e. $A = 20\,e^{(\ln 0.834)t}$

   f. $y = (1.2 \cdot 10^5)(e^{\ln 0.752})^t$

2. Rewrite each of the following as a continuous growth or decay model using base e.

   a. $y = 1000(1.062)^t$

   b. $y = 50(0.985)^t$

3. Determine the nominal and effective rates for each of the following. Assume t is measured in years. (*Hint:* To find the effective rate, convert from e^k to a.)

   a. $P = 25{,}000\,e^{(\ln 1.056)t}$

   b. $P = 10e^{(\ln 1.034)t}$

   c. $y = 2000\,e^{(\ln 1.083)t}$

   d. $y = (1.2 \cdot 10^5)(e^{\ln 1.295})^t$

4. Find the growth factor or decay factor for each of the following. (*Hint:* Convert from e^k to a.)

   a. $P = 50{,}000e^{0.08t}$

   b. $y = 30e^{-0.125t}$

Exercises for Section 6.5

Many of these exercises require a calculator that can evaluate powers and logs, some require a graphing program, and Exercise 18 requires technology that can generate a best-fit exponential function.

1. Rewrite each of the following functions using base e.

   a. $N = 10(1.045)^t$

   b. $Q = (5 \cdot 10^{-7}) \cdot (0.072)^A$

   c. $P = 500(2.10)^x$

2. Using technology, graph each of the following. Find, as appropriate, the doubling time or half-life.

   a. $A = 50\,e^{0.025t}$

   b. $A = 100\,e^{-0.046t}$

   c. $P = (3.2 \cdot 10^6)(e^{-0.15})^t$

3. Identify each of the following functions as representing growth or decay:

   a. $Q = Ne^{-0.029t}$

   b. $h(r) = 100(0.87)^r$

   c. $f(t) = 375e^{0.055t}$

4. Identify each function as representing growth or decay. Then determine the annual growth or decay factor, assuming t is in years.

   a. $A = A_0(1.0025)^{20t}$

   b. $A = A_0(1.0006)^{t/360}$

   c. $A = A_0(0.992)^{t/2}$

   d. $A = A_0 e^{-0.063t}$

   e. $A = A_0 e^{0.015t}$

5. For each of the following, find the doubling time, then rewrite each function in the form $P = P_0 e^{rt}$. Assume t is measured in years.

   a. $P = P_0 2^{t/5}$

   b. $P = P_0 2^{t/25}$

   c. $P = P_0 2^{2t}$

6. For each of the following, find the half-life, then rewrite each function in the form $P = P_0 e^{rt}$. Assume t is measured in years.

   a. $P = P_0\left(\frac{1}{2}\right)^{t/10}$

   b. $P = P_0\left(\frac{1}{2}\right)^{t/215}$

   c. $P = P_0\left(\frac{1}{2}\right)^{4t}$

7. The barometric pressure, p, in millimeters of mercury, at height h, in kilometers above sea level, is given by the equation $p = 760e^{-0.128h}$. At what height is the barometric pressure 200 mm?

8. After t days, the amount of thorium-234 in a sample is $A(t) = 35e^{-0.029t}$ micrograms.

   a. How much was there initially?

   b. How much is there after a week?

   c. When is there just 1 microgram left?

   d. What is the half-life of thorium-234?

9. Assume $f(t) = Ce^{rt}$ is an exponential decay function (so $r < 0$). Prove the rule of 70 for halving times; that is, if a quantity is decreasing at $R\%$ per time period t, then the number of time periods it takes for the quantity to halve is approximately $70/R$. (*Hint:* $R = 100r$.)

10. Match the function with its graph.

    a. $f(x) = 10e^{-0.075x}$

    b. $g(x) = 10e^{-0.045x}$

    c. $h(x) = 10e^{-0.025x}$

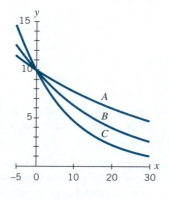

11. Match the function with its graph.

    a. $f(x) = 10e^{0.025x}$ **c.** $h(x) = 10e^{0.075x}$

    b. $g(x) = 10e^{0.045x}$

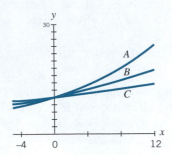

12. The functions $y = 50e^{0.04t}$ and $y = 50(2)^{t/n}$ are two different ways to write the same function.

    a. What does the value 0.04 represent?

    b. Set the functions equal to each other and use rules of natural logarithms to solve for n.

    c. What does the value of n represent?

13. The functions $y = 2500e^{-0.02t}$ and $y = 2500\left(\frac{1}{2}\right)^{t/n}$ are two different ways to write the same function.

    a. What does the value -0.02 represent?

    b. Set the functions equal to each other and use rules of natural logarithms to solve for n.

    c. What does the value of n represent?

14. (Requires a graphing program.) Radioactive lead-210 decays according to the exponential formula $Q = Q_0 e^{-0.0311t}$, where Q_0 is the initial quantity in milligrams and t is in years. What is the half-life of lead-210? Verify your answer by graphing using technology.

15. Radioactive thorium-230 decays according to the formula $Q = Q_0\left(\frac{1}{2}\right)^{t/8000}$, where Q_0 is the initial quantity in milligrams and t is in years.

    a. What is the half-life of thorium-230?

    b. What is the annual decay rate?

    c. Translate the equation into the form $Q = Q_0 e^{rt}$. What does r represent?

16. In 1859, the Victorian landowner Thomas Austin imported 12 wild rabbits into Australia and let them loose to breed. Since they had no natural enemies, the population increased very rapidly. By 1949 there were approximately 600 million rabbits.

    a. Find an exponential function using a continuous growth rate to model this situation.

    b. If the growth had gone unchecked, what would have been the rabbit population in 2000?

    c. *Internet search:* Find out what was done to curb the population of rabbits in Australia, and find the current rabbit population.

17. Biologists believe that, in the deep sea, species density decreases exponentially with the depth. The accompanying graph shows data collected in the North Atlantic. Sketch an exponential decay function through the data. Then identify two points on your curve, and generate two equivalent equations that model the data, one in the form $y = Ca^t$ and the other in the form $y = Ce^{kt}$.

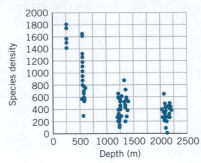

Source: Data collected by Ron Etter, Biology Department. University of Massachusetts–Boston.

18. (Requires technology to find a best-fit exponential.) According to another version of Moore's Law, the computing power built into chips doubles every 18 months (see Section 5.7). The accompanying table shows the computing power of some Intel chips (measured in calculations per second) between 1993 and 2005.

    a. Graph the data points (if possible, on a semi-log plot). Explain why an exponential function would be an appropriate model.

Year	Chip Type	Chip Computing Power (millions of calculations per second)
1993	Pentium ®	66
1997	Pentium II	525
1999	Pentium III	1,700
2000	Pentium 4	3,400
2005	Dual-core Itanium 2	27,079

Source: Intel Corporation.

    b. Construct an exponential function to model the data in the following two different ways. In each case let $t =$ number of years since 1993.

        i. Using the doubling time given by Moore's Law, construct an exponential function, $P_1(t)$, for chip computing power.

        ii. Use technology to find a best-fit exponential function, $P_2(t)$, to the data in the table. Does this model come close to verifying Moore's Law ($P_1(t)$)?

19. The number of neutrons in a nuclear reactor can be predicted from the equation $n = n_0 e^{(\ln 2)t/T}$, where $n =$ number of neutrons at time t (in seconds), $n_0 =$ the number of neutrons at time $t = 0$, and $T =$ the reactor period, the doubling time of the neutrons (in seconds). When $t = 2$ seconds, $n = 11$, and when $t = 22$ seconds, $n = 30$. Find the initial number of neutrons, n_0, and the reactor period, T, both rounded to the nearest whole number.

20. According to Rubin and Farber's *Pathology*, "death from cancer of the lung, more than 85% of which is attributed to cigarette smoking, is today the single most common cancer death in both men and women in the United States." The accompanying graph shows the annual death rate (per thousand) from lung cancer for smokers and nonsmokers.

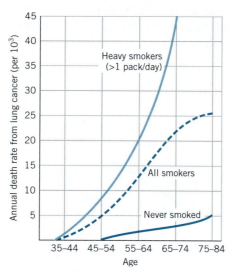

Source: E. Rubin and J. L. Farber, *Pathology,* 3rd ed. (Philadelphia: Lippincott-Raven, 1998), p. 312. Copyright © 1998 by Lippincott-Raven. Reprinted by permission.

a. The death rate for nonsmokers is roughly a linear function of age. After replacing each range of ages with a reasonable middle value (e.g., you could use 60 to approximate 55 to 64), estimate the coordinates of two points on the graph of nonsmokers and construct a linear model. Interpret your results.

b. By contrast, those who smoke more than one pack per day show an exponential rise in the annual death rate from lung cancer. Estimate the coordinates for two points on the graph for heavy smokers, and use the points to construct an exponential model (assume a continuous growth rate). Interpret your results.

21. If an object is put in an environment at a fixed temperature, A (the "ambient temperature"), then its temperature, T, at time t is modeled by Newton's Law of Cooling:

$$T = A + Ce^{-kt}$$

where k is a positive constant. Note that T is a function of t and that as $t \to +\infty$, then $e^{-kt} \to 0$, so the temperature T gets closer and closer to the ambient temperature, A.

a. Assume that a hot cup of tea (at 160°F) is left to cool in a 75°F room. If it takes 10 minutes for it to reach 100°F, determine the constants A, C, and k in the equation for Newton's Law of Cooling. What is Newton's Law of Cooling in this situation?

b. Sketch the graph of your function.

c. What is the temperature of the tea after 20 minutes?

22. Newton's Law of Cooling (see Exercise 21) also works for objects being heated. At time $t = 0$, a potato at 70°F (room temperature) is put in an oven at 375°F. Thirty minutes later, the potato is at 220°.

a. Determine the constants A, C, and k in Newton's Law. Write down Newton's Law for this case.

b. When is the potato at 370°F?

c. When is the potato at 374°F?

d. According to your model, when (if ever) is the potato at 375°F?

e. Sketch a graph of your function in part (a).

6.6   *Using Semi-Log Plots to Construct Exponential Models for Data*

In Chapter 5 we learned that an exponential function appears as a straight line on a semi-log plot (where the logarithmic scale is on the vertical axis). To decide whether an exponential function is an appropriate model for a data set, we can plot the data on a semi-log plot and see if it appears to be linear. This is one of the easiest and most reliable ways to recognize exponential growth in a data set. And, as we learned in Chapters 4 and 5, a logarithmic scale has the added advantage of being able to display clearly a wide range of values.

Why Do Semi-Log Plots of Exponential Functions Produce Straight Lines?

Consider the exponential function

$$y = 3 \cdot 2^x \tag{1}$$

If we take the log of both sides $\log y = \log (3 \cdot 2^x)$

use Rule 1 of logs $= \log 3 + \log (2^x)$

use Rule 3 of logs	$= \log 3 + x \log 2$
evaluate logs and rearrange, we have	$\log y \approx 0.48 + 0.30x$
If we set $Y = \log y$, we have	$Y = 0.48 + 0.30x$ (2)

So Y (or $\log y$) is a linear function of x. Equations (1) and (2) are equivalent to each other. The graph of Equation (2) on a semi-log plot, with Y (or $\log y$) values on the vertical axis, is a straight line (see Figure 6.11). The slope is 0.30 or $\log 2$, the logarithm of the growth factor 2 of Equation (1). The vertical intercept is 0.48 or $\log 3$, the logarithm of the y-intercept 3 of Equation (1).

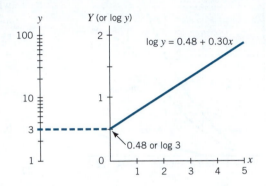

Figure 6.11 The graph of $Y = 0.48 + 0.30x$, showing the relationship between two equivalent logarithmic scales on the vertical axis.

Figure 6.11 shows two equivalent variations of logarithmic scales on the vertical axis. One plots the value of y on a logarithmic scale (using powers of 10), and the other plots the value of $\log y$ (using the exponents of the powers of 10). Spreadsheets and some graphing calculators have the ability to instantly switch axis scales between standard linear and the logarithmic scale using powers of 10. But one can in effect do the same thing by plotting $\log y$ instead of y. Notice that on the vertical $\log y$ scale, the units are now evenly spaced. This allows us to use the standard strategies for finding the slope and vertical intercept of a straight line.

In general, we can translate an exponential function in the form

$$y = Ca^x \quad \text{(where } C \text{ and } a > 0 \text{ and } a \neq 1)$$

into an equivalent linear function

$$\log y = \log C + (\log a)x \qquad \text{or}$$
$$Y = \log C + (\log a)x \qquad \text{where } Y = \log y$$

Finding the Equation of an Exponential Function on a Semi-Log Plot

The graph of an exponential function $y = Ca^x$ (where C and $a > 0$ and $a \neq 1$) appears as a straight line on a semi-log plot. The line's equation is

$$Y = \log C + (\log a)x \quad \text{where } Y = \log y$$

The slope of the line is $\log a$, where a is the growth factor for $y = Ca^x$.
The vertical intercept is $\log C$, where C is the vertical intercept of $y = Ca^x$.

Growth in the Dow Jones

The Dow Jones Industrial Average is based on the stock prices of thirty companies and is commonly used to measure the health of the stock market. Figure 6.12 is a semi-log plot of the Dow between the boom years of 1982 and 2000.

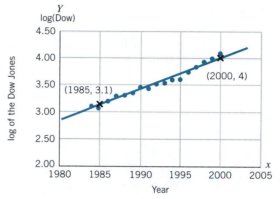

Figure 6.12 Graph of log(Dow) with two points on the best-fit line (not from the original data).
Source: Dow Jones website, *www.djindexes.com.*

The data points lie approximately on a straight line, so an exponential model is appropriate.

EXAMPLE 1

Constructing an exponential function from a line on a semi-log plot
Estimate the annual percent growth rate of the Dow Jones between 1982 and 2000.

SOLUTION

We first construct a linear function for the best-fit line on the semi-log plot, then translate it into an exponential growth function to find the percent growth rate.

Estimating the coordinates of two points (1985, 3.1) and (2000, 4.0) on the best-fit line, *not* from the data (see Figure 6.12), the slope or average rate of change between them is

$$\text{slope} = \frac{\text{change in } Y}{\text{change in } x} = \frac{4.0 - 3.1}{2000 - 1985} = \frac{0.9}{15} = 0.06$$

If we let x = number of years since 1980, the slope remains the same. Reading off the graph, the vertical intercept is approximately 2.8. So the equation for the best-fit line is approximately

$$Y = 2.8 + 0.06x \qquad \text{where } Y = \log(\text{Dow}) \tag{1}$$

We can translate Equation (1) into an exponential function if we

substitute log(Dow) for Y	$\log(\text{Dow}) = 2.8 + 0.06x$
rewrite as a power of 10	$10^{\log(\text{Dow})} = 10^{(2.8+0.06x)}$
use Rule 6 of logs and Rule 1 of exponents	$\text{Dow} = 10^{2.8} \cdot 10^{0.06x}$
Rule 3 of exponents	$\text{Dow} = 10^{2.8} \cdot (10^{0.06})^x$
evaluate using calculator	$\text{Dow} \approx 631 \cdot 1.15^x \tag{2}$

The annual growth factor is approximately 1.15, and the annual growth rate in decimal form is 0.15. So between 1982 and 2000 the Dow Jones grew by about 15% each year.

This strategy is equivalent to using the formula in the previous box, translating $\log(\text{Dow}) = 2.8 + 0.06x$ into the form $\text{Dow} = Ca^x$, by solving $2.8 = \log C$, and $0.06 = \log a$.

Algebra Aerobics 6.6

Many of the problems require a calculator that can evaluate logs and exponents.

1. a. Which of the following three functions would have a straight-line graph on a standard linear plot? On a semi-log plot?

$$y = 3x + 4, \quad y = 4 \cdot 3^x, \quad \log y = (\log 3) \cdot x + \log 4$$

b. For each straight-line graph in part (a), what is the slope of the line. The vertical intercept? (*Hint:* Substitute $Y = \log y$ and $X = \log x$ as needed.)

2. Change each exponential equation to a logarithmic equation.

a. $y = 5(3)^x$ **c.** $y = 10,000(0.9)^x$

b. $y = 1000(5)^x$ **d.** $y = (5 \cdot 10^6)(1.06)^x$

3. Change each logarithmic equation to an exponential equation.

a. $\log y = \log 7 + (\log 2) \cdot x$

b. $\log y = \log 20 + (\log 0.25) \cdot x$

c. $\log y = 6 + (\log 3) \cdot x$

d. $\log y = 6 + \log 5 + (\log 3) \cdot x$

4. a. Below are linear equations of $\log y$ (or Y) in x. Identify the slope and vertical intercept for each.

    i. $\log y = \log 2 + (\log 5) \cdot x$

    ii. $\log y = (\log 0.75) \cdot x + \log 6$

    iii. $\log y = 0.4 + x \log 4$

    iv. $\log y = 3 + \log 2 + (\log 1.05) \cdot x$

b. Use the slope and intercept to create an exponential function for each equation.

5. Solve each equation using the definition of log.

a. $0.301 = \log a$ **c.** $\log a = -0.125$

b. $\log C = 2.72$ **d.** $5 = \log C$

6. Change each function to exponential form, assuming that $Y = \log y$.

a. $Y = 0.301 + 0.477x$

b. $Y = 3 + 0.602x$

c. $Y = 1.398 - 0.046x$

7. For each of the two accompanying graphs examine the scales on the axes. Then decide whether a linear or an exponential function would be the most appropriate model for the data.

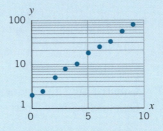 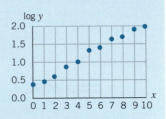

8. Generate an exponential function that could describe the data in the accompanying graph.

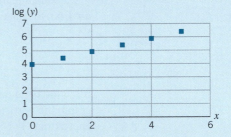

Exercises for Section 6.6

Some exercises require a calculator that can evaluate powers and logs. Exercises 10 and 13 require a graphing program.

1. Match each exponential function in parts (a)–(d) with its logarithmic form in parts (e)–(h).

a. $y = 10,000(2)^x$ **e.** $\log y = 6.477 - 0.097x$

b. $y = 1000(1.4)^x$ **f.** $\log y = 4 + 0.301x$

c. $y = (3 \cdot 10^6)(0.8)^x$ **g.** $\log y = 3 - 0.347x$

d. $y = 1000(0.45)^x$ **h.** $\log y = 3 + 0.146x$

2. Form the exponential function from its logarithmic equivalent for each of the following.

a. $\log y = \log 1400 + (\log 1.06)x$

b. $\log y = \log(25,000) + (\log 0.87)x$

c. $\log y = 2 + (\log 2.5)x$

d. $\log y = 4.25 + (\log 0.63)x$

3. Change each exponential function to its logarithmic equivalent. Round values to the nearest thousandth.

   a. $y = 30,000(2)^x$ **c.** $y = (4.5 \cdot 10^6)(0.7)^x$

   b. $y = 4500(1.4)^x$ **d.** $y = 6000(0.57)^x$

4. Write the exponential equivalent for each function. Assume $Y = \log y$.

   a. $Y = 2.342 + 0.123x$ **c.** $Y = 4.74 - 0.108x$

   b. $Y = 3.322 + 0.544x$ **d.** $Y = 0.7 - 0.004x$

5. Match each exponential function with its semi-log plot.

   a. $y = 30,000(2)^x$ **c.** $y = (4.5 \cdot 10^6)(0.7)^x$

   b. $y = 4500(1.4)^x$ **d.** $y = 6000(0.57)^x$

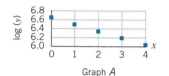

Graph *A*

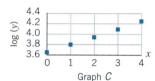

Graph *C*

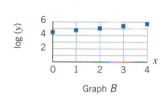

Graph *B*

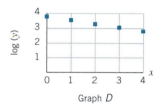

Graph *D*

6. a. From the data in the following table, create a linear equation of Y in terms of x.

x	$\log y$ (or Y)
0	5.00000
1	5.60206
2	6.20412
3	6.80618
4	7.40824

   b. Find the equivalent exponential function of y in terms of x.

7. Determine which data sets (if any) describe y as an exponential function of x, then construct the exponential function. (*Hint:* Find the average rate of change of Y with respect to x.)

a.

x	$\log y$ (or Y)
0	2.30103
10	4.30103
20	4.90309
30	5.25527
40	5.50515

b.

x	$\log y$ (or Y)
0	4.77815
10	3.52876
20	2.27938
30	1.02999
40	-0.21945

8. Construct two functions that are equivalent descriptions of the data in the accompanying graph. Describe the change in $\log y$ (or Y) with respect to x.

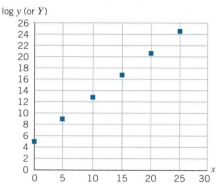

9. The accompanying graph shows the growth of a bacteria population over a 25-day period.

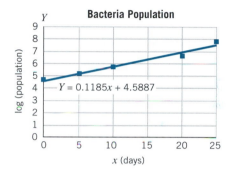

   a. Do the bacteria appear to be growing exponentially? Explain.

   b. Translate the best-fit line shown on the graph into an exponential function and determine the daily growth rate.

10. (Requires a graphing program.) The current population of a city is 1.5 million. Over the next 40 years, the population is expected to decrease by 12% each decade.

   a. Create a function that models the population decline.

   b. Create a table of values at 10-year intervals for the next 40 years.

   c. Graph the population of the city on a semi-log plot.

11. An experiment by a pharmaceutical company tracked the amount of a certain drug (measured in micrograms/liter) in the body over a 30-hour period. The accompanying graph shows the results.

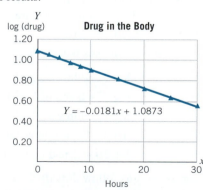

   a. What was the initial amount of the drug?

   b. Is the amount of drug in the body decaying exponentially? If yes, state the decay rate.

   c. If the amount in the body is decaying exponentially, construct an exponential model.

12. The accompanying graph shows the decay of strontium-90.

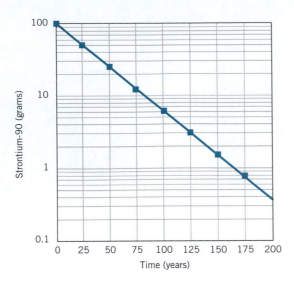

Time (years)

a. Is the graph a semi-log plot?

b. What is the initial amount of strontium-90 for this graph?

c. According to the graph, what is the half-life of strontium-90?

d. What would be the yearly decay rate for strontium-90?

e. Find the exponential model for the decay of strontium-90.

13. (Requires a graphing program.) The data in the following table are from the 2006 *Statistical Abstract* and show the rise of health care costs in the United States since 1960.

Year	U.S. Health Care Expenditures (billions of dollars)
1960	28
1970	75
1980	225
1985	442
1990	717
1995	1020
2000	1359
2005	2016

a. It is clear that the United States is spending more on health care as time goes on. Does this mean that we as individuals are paying more? How would you find out?

b. Is it fair to say that the expenses shown are growing exponentially? Use graphing techniques to find out; measure time in years since 1960.

c. Around 1985 the high cost of health care began to become an increasingly political issue, and insurance companies began to introduce "managed care" in an attempt to cut costs. Find a mathematical model that assumes a continuous growth for health costs from 1960 to 2005. What does your model predict for health costs in 2005? How closely does this reflect the actual cost in 2005?

CHAPTER SUMMARY

Logarithms

The *logarithm base 10 of x* is the *exponent* of 10 needed to produce x.

$$\log x = c \quad \text{means that} \quad 10^c = x$$

Logarithms base 10 are called *common logarithms*.

The *logarithm base e of x* is the *exponent* of e needed to produce x. The number e is an irrational natural constant ≈ 2.71828. Logarithms base e are called *natural logarithms* and are written as $\ln x$:

$$\ln x = c \quad \text{means that} \quad e^c = x$$

If $x > 1$, $\log x$ and $\ln x$ are both positive
If $x = 1$, $\log 1 = 0$ and $\ln 1 = 0$
If $0 < x < 1$, $\log x$ and $\ln x$ are both negative
If $x \leq 0$ neither logarithm is defined
and $\log 10 = 1$ and $\ln e = 1$

The rules for logarithms follow directly from the definition of logarithms and from the rules for exponents.

Rules for Logarithms

If A and B are positive real numbers and p is any real number, then:

For Common Logarithms

1. $\log(A \cdot B) = \log A + \log B$
2. $\log(A/B) = \log A - \log B$
3. $\log A^p = p \log A$
4. $\log 1 = 0$ (since $10^0 = 1$)
5. $\log 10^x = x$
6. $10^{\log x} = x$ $(x > 0)$

For Natural Logarithms

1. $\ln(A \cdot B) = \ln A + \ln B$
2. $\ln(A/B) = \ln A - \ln B$
3. $\ln A^p = p \ln A$
4. $\ln 1 = 0$ (since $e^0 = 1$)
5. $\ln e^x = x$
6. $e^{\ln x} = x$ $(x > 0)$

Solving Exponential Functions Using Logarithms

Logarithms can be used to solve equations such as $10 = 2^t$ by taking the log of each side to get $\log 10 = \log (2^t) \Rightarrow 1 = t \cdot \log 2 \Rightarrow t \approx 3.32$.

Compounding

The value of P_0 dollars invested at a nominal interest rate r (in decimal form) compounded n times a year for t years is

$$P = P_0\left(1 + \frac{r}{n}\right)^{nt}$$

The value for r is called the *annual percentage rate* (APR). The actual interest rate per year is called the *effective interest rate* or the annual percentage yield (APY).

Compounding Continuously

The number e is used to describe *continuous compounding*. For example, the value of P_0 dollars invested at a nominal interest rate r (expressed in decimal form) compounded continuously for t years is

$$P = P_0 e^{rt}$$

Logarithmic Functions

We can define two functions: $y = \log x$ and $y = \ln x$ (where $x > 0$).

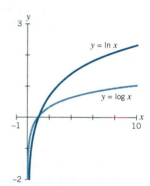

Graphs of $y = \log x$ and $y = \ln x$

The graphs of these functions have similar shapes, and both

- lie to the right of the y-axis
- have a horizontal intercept of $(1, 0)$
- are concave down
- increase throughout with no maximum or minimum
- are asymptotic to the y-axis

Two functions are *inverses* of each other if what one function "does" the other "undoes." The graphs of two inverse functions are mirror images across the line $y = x$. The logarithmic and exponential functions are inverse functions.

Continuous Growth or Decay

Any exponential function $f(t) = Ca^t$ can be rewritten as

$$f(t) = Ce^{kt} \qquad \text{where } k = \ln a$$

We call k the *instantaneous* or *continuous growth* (or *decay*) rate.

Exponential growth: $a > 1$ and k is positive
Exponential decay: $0 < a < 1$ and k is negative

Using Semi-Log Plots

The graph of an exponential function $y = Ca^x$ appears as a straight line on a semi-log plot. The line's equation is

$$Y = \log C + (\log a)x \qquad \text{where } Y = \log y$$

The slope of the line is $\log a$, where a is the growth (or decay) factor for $y = Ca^x$.

The vertical intercept is $\log C$, where C is the vertical intercept of $y = Ca^x$.

CHECK YOUR UNDERSTANDING

I. Is each of the statements in Problems 1–30 true or false? If false, give an explanation for your answer.

1. For the function $y = \log x$, if x is increased by a factor of 10, then y increases by 1.

2. If $\log x = c$, then x is always positive but c can be positive, negative, or zero.

3. $\ln(1.08/2) = \ln(1.08)/\ln(2)$

4. Because the function $y = \log x$ is always increasing, it has no vertical asymptotes.

5. Both $\log 1$ and $\ln 1$ equal 0.

6. $\ln a = b$ means $b^e = a$.

7. $2 < e < 3$

8. $\log(10^2) = (\log 10)^2$

9. $\log(10^3 \cdot 10^5) = 15$

10. $\log\left(\dfrac{10^7}{10^2}\right) = \dfrac{\log(10^7)}{\log(10^2)}$

11. $\log\left(\dfrac{\sqrt{AB}}{(C+2)^3}\right)$
$= \dfrac{1}{2}(\log A + \log B) - 3\log(C+2)$

12. $5\ln A + 2\ln B = \ln(A^5 + B^2)$

13. If $(2.3)^x = 64$, then $x = \dfrac{\log 64}{\log 2.3}$.

14. If $\log 20 = t\log 1.065$, then $t = \log\left(\dfrac{20}{1.065}\right)$.

15. The amount of dollars D compounded continuously for 1 year at a nominal interest rate of 8% yields $D \cdot e^{0.08}$.

16. $\ln t$ exists for any value of t.

17. If $f(t) = 100 \cdot a^t$ and $g(t) = 100 \cdot b^t$ are exponential decay functions and $a < b$, then the half-life of f is longer than the half-life of g.

18. The graphs of the functions $y = \ln x$ and $y = \log x$ increase indefinitely.

19. The graphs of the functions $y = \ln x$ and $y = \log x$ have both vertical and horizontal asymptotes.

20. The graph of the function $y = \ln x$ lies above the graph of the function $y = \log x$ for all $x > 0$.

21. The graph of the function $y = \log x$ is always positive.

22. The graph of the function $y = \ln x$ is always decreasing.

23. The amount of $100 invested at 8% compounded continuously has a doubling time of about 8.7 years. (*Hint:* $e^{0.08} \approx 1.083$)

24. Of the functions of the form $P = P_0 e^{rt}$ graphed in the accompanying figure, graph B has the smallest growth rate, r.

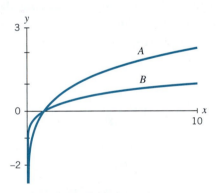

25. Of the two functions $f(x) = \ln x$ and $g(x) = \log x$ graphed in the accompanying figure, graph A is the graph of $f(x) = \ln x$.

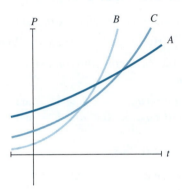

Problems 26–30 refer to the following table and semi-log plot.

U.S. Trade with China (in billions of dollars)

Year	Exports	Imports
2001	19.2	102.3
2002	22.1	125.2
2003	28.4	152.4
2004	34.7	196.7
2005	41.9	243.5
2006	50.0	263.6

Source: U.S. Bureau of the Census, Foreign Trade Statistics, *www.census.gov/foreigntrade/balance.*

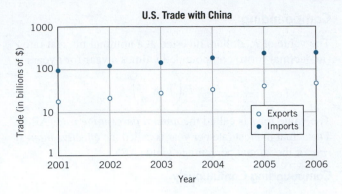

26. Between 2001 and 2006 America's trade deficit ($=$ exports $-$ imports) with China remained roughly constant.

27. We can think of both U.S. exports to China and imports from China (between 2001 and 2006) as functions of the year. The plots of both data sets appear to be approximately linear on a semi-log plot, so both functions are roughly exponential.

28. If $\log(\text{imports}) = 2.015 + 0.086t$ (where $t =$ years since 2001) is the equation of the best-fit line for imports on a semi-log plot, then the annual growth rate for imports is about 8.6%.

29. If we describe the growth in U.S. imports with the equation U.S. imports $= 103.59(1.22)^t$, then the annual growth rate for imports is about 22%.

30. If equations in Problems 28 and 29 are correct, they should be equivalent.

II. In Problems 31–35, give an example of a function or functions with the specified properties. Express your answers using equations, and specify the independent and dependent variables.

31. A function whose graph is identical to the graph of the function $y = \log\sqrt{\dfrac{x}{3}}$.

32. A function (that doesn't use e) whose graph is identical to the graph of the function $y = 50.3e^{0.06t}$. (*Hint:* $e^{0.06} \approx 1.062$)

33. A function (that uses e) whose graph is identical to the graph of the function $y = 100(0.974)^t$. (*Hint:* $0.974 \approx e^{-0.0263}$.)

34. An exponential model describing the number of farms in the United States over time if there were 3.3 million farms in 1966 and 2.1 million farms in 2006. (*Hint:* $\left(\frac{2.1}{3.3}\right)^{(1/40)} \approx 0.989$)

35. An exponential function $y = f(x)$ whose graph on a semi-log plot $(x, \log y)$ is the function $\log y = 2 + 0.08x$. (*Hint:* $10^{0.08} \approx 1.20$)

III. Are the statements in Problems 36–40 true or false? If a statement is true, explain how you know. If a statement is false, give a counterexample.

36. The number e used in the natural logarithm is a variable.

37. Compounding at 3% quarterly is the same as compounding at 12% annually.

38. Assuming your investment is growing, the effective interest rate is always greater than or equal to the nominal interest rate.

39. The graphs of the functions $y = Ca^x$ and $y = Ce^{rx}$, where $r = \ln a$, are identical.

40. The interest rate r for continuous compounding in $y = Ce^{rx}$ is called the nominal interest rate or the instantaneous interest rate, depending on the application.

CHAPTER 6 REVIEW: PUTTING IT ALL TOGETHER

You will need a scientific calculator or its equivalent to evaluate logs and powers of e.

1. Solve each equation for the time t (in months) when the quantity $Q = 100$. Then interpret each result.
   a. $Q = 50 \cdot 1.16^t$ **b.** $Q = 200 \cdot 0.92^t$

2. a. According to the following chart, approximately how many deaths from natural disasters in the 1990s were due to floods? To severe storms? To bushfires?

   b. Why was this bar chart constructed with deaths displayed on a logarithmic scale?

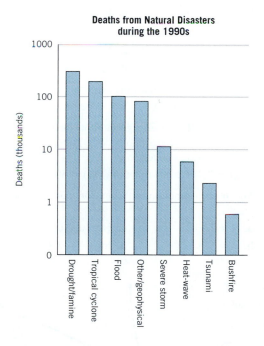

Deaths from Natural Disasters during the 1990s

Source: Australian government, Department of Meteorology.

3. Which of the following expressions are equivalent?
   a. $\log(xy^2)$ **d.** $2 \log x + \log y$
   b. $2 \log(xy)$ **e.** $\log x + 2 \log y$
   c. $\log(xy)^2$ **f.** $2 \log x + 2 \log y$

4. The Lead-based Paint Poisoning Prevention Act (1971) was passed to reduce the toxic blood levels of lead in young children who might eat pieces of peeling lead-based paint. The following chart shows the subsequent drop in the average blood lead level of children, along with additional, related legislative acts.

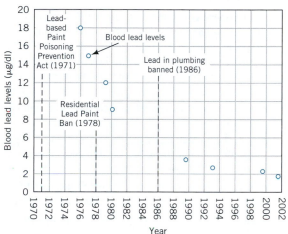

Impact of Lead Poisoning Prevention Policy on Reducing Children's Blood Lead Levels in the United States, 1971–2001

Source: Blood lead levels: National Health and Nutrition Examination Survey. National Center for Health Statistics, Centers for Disease Control and Prevention.

The best-fit exponential decay function for the data points on this graph is $L(t) = 18(0.88)^t$, where $L(t)$ is the average blood lead level (in micrograms per deciliter) of children and $t =$ years since 1976 (the first year data are available).

   a. What is the annual decay rate?

   b. Estimate from the graph when the initial blood lead level (in 1976) would be reduced by 50%.

   c. According to your model, when would the initial blood lead level be reduced by 50%?

5. The equation $A(t) = 325 \cdot (0.5)^t$ and accompanying graph describe the amount of aspirin in your bloodstream at time t (measured in 20-minute periods) after you have ingested a standard aspirin of 325 milligrams (mg).

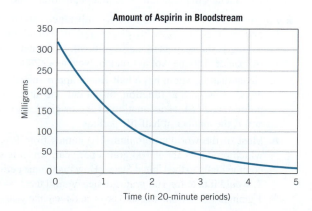

Amount of Aspirin in Bloodstream

a. Using the equation, how much aspirin will remain after one time period? After two time periods?

b. Use the graph to estimate the number of time periods it will take for the aspirin level to reach 100 mg. Now calculate the time using your equation and compare your results.

c. Many doctors suggest that their adult patients take a daily dose of "baby aspirin" (81 mg) for long-term heart protection. Construct a new function $B(t)$ to describe the amount of aspirin in your bloodstream after taking an 81-mg aspirin. (Assume the decay rate is the same as for a 325-mg aspirin.) How do the functions $A(t)$ and $B(t)$ differ, and how are they alike?

6. The Japanese government is concerned about its decreasing population. The Japanese people joke about when Japan will disappear entirely. According to a UN report, the Japanese population is believed to have peaked at 127.5 million in 2005 and, if the current rate continues, will contract to about 105 million people in 2050.

a. What is the 45-year decay factor? The annual decay factor?

b. Use the annual decay factor to construct an exponential decay function to model the Japanese population $J(x)$, where x = years since 2005.

c. Using the rule of 70, estimate the half-life of the population. According to your model, what is the half-life?

7. India and China have the opposite problem from that of Japan (see Problem 6): a huge population that is growing. In 2005 India had an estimated population of 1.08 billion on a land area of 1.2 million square miles, and China had a population of 1.30 billion on 3.7 million square miles.

a. In 2005 did India or China have the larger population density (people per square mile)?

b. China has been trying to slow population growth with its one child-per-family policy. As a result, in 2005 its annual population growth rate (0.6%) was considerably lower than that of India (1.6%). Assuming that each population continues to grow at the 2005 rate, constuct functions $C(x)$ and $I(x)$ for the population growth in China and India, respectively, letting x = years since 2005.

c. Plot each function in part (b) on the same graph, for 30 years after 2005.

d. Looking at the graph, is there a year when India's population is projected to overtake China's? If so, use your models to predict the year and population level.

8. Polonium-210 is a toxic radioactive substance named after Poland by the Curies, who discovered it. Polonium-210 poisoning is the suspected cause of death of former Soviet spy Alexander Litvinenko, in London, November 2006.

a. Given that polonium has a half-life of approximately 138 days, construct a function to model the amount of polonium, $P(T)$, as a function of the original amount A and T, the number of half-life periods.

b. Most of the world's polonium-210 comes from Russia, which produces about 100 grams per year, which is sold commercially to the United States. How many time periods, T, would it take for 100 grams to decay until there is only 1 gram left? Translate this into days, and then into years.

c. The Soviet Union collapsed in 1991. What do you know about any pre-1991 Soviet-era stores of polonium-210?

9. At the birth of their granddaughter, the grandparents create a college fund in her name, investing $10,000 at 7% per year.

a. Construct an equation to model the growth in the account.

b. How much will be in the account in 18 years?

c. If the annual inflation stays at 3%, how much will something that cost $10,000 today cost in 18 years? Calculate the difference between your answer in part (b) and your answer in part (c).

d. Is this equivalent to the return you would have if you invested $10,000 at 4% for 18 years? If not, what does investing $10,000 at 4% represent?

10. Which of the following expressions are equivalent?

a. $\ln(\sqrt{x}/y)$

b. $\left(\frac{1}{2}\right)\ln(x/y)$

c. $\left(\frac{1}{2}\right)\ln x - \ln y$

d. $\left(\frac{1}{2}\right)\ln x/\ln y$

e. $\left(\frac{1}{2}\right)\ln x - \left(\frac{1}{2}\right)\ln y$

f. $\ln(\sqrt{x})/\ln y$

11. Solve each equation for t.

a. $10^t = 2.3$

b. $2\log t + \log 4 = 2$

c. $60 = 30\,e^{0.03t}$

d. $\ln(2t - 5) - \ln(t - 1)) = 0$

12. Simplify these expressions without using a calculator.

a. $\ln(e^2)$

b. $e^{\ln(3)}$

c. $10^{\log(t - 2)}$

d. $\ln(x^2 - x) - \ln x$

13. Convert each of the following expressions to a power of e.

a. 1.5   **b.** 0.7   **c.** 1

14. For each of the following functions identify the growth (or decay) factor, the growth (or decay) rate, and the continuous growth (or decay) rate.

a. $Q = 75(1.02)^t$ **b.** $P(x) = 50e^{-0.3t}$

15. Match each function with one (or more) of the following graphs.

a. $y = 2 + x$ **c.** $y = 100(10)^x$

b. $y = 2 + \log(x)$ **d.** $\log(y) = 2 + x$

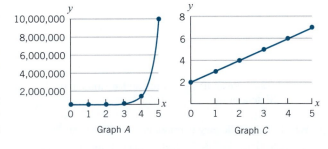

Graph A

Graph C

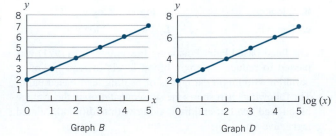

Graph B

Graph D

16. a. In November 2006 the Bank of America offered a 12-month certificate of deposit (CD) at a nominal rate (or annual percentage rate, APR) of 3.6%. What is the effective interest rate (or annual percentage yield, APY) if the bank compounds:

   i. Annually? **ii.** Monthly? **iii.** Continuously?

   b. The Bank lists the APY as 3.66%. Which of the above compounding schedules comes closest to this?

17. For each function, identify the corresponding graph.

   a. $y_1 = 2 \ln x$ **c.** $y_3 = \ln(x + 2)$

   b. $y_2 = 2 + \ln x$ **d.** $y_4 = \ln(x^2)$

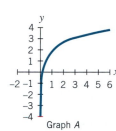

Graph A

Graph C

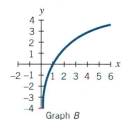

Graph B

Graph D

18. Which pair(s) of graphs show a function and its inverse?

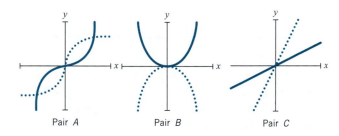

Pair A   Pair B   Pair C

19. If the decibel level moves from 30 (a quiet conversation) to 80 (average street noise), by how many orders of magnitude has the intensity increased? Generate your answer in two ways:

   a. Using the table in the text

   b. Using the definition of decibels

20. Beer has a pH of 4.5 and household lye a pH of 13.5.

   a. Which has the higher hydrogen ion concentration, and by how many orders of magnitude?

   b. What is the difference between the two pH values? How does this relate to your answer in part (a)?

   c. Why is it easier to use the pH number instead of $[H^+]$?

21. In Chapter 4 we encountered the Richter scale, which measures the amplitude of an earthquake. The Richter number R is defined as

$$R = \log(A/A_0)$$

where A is the amplitude of the shockwave caused by the earthquake and A_0 is the reference amplitude, the smallest earthquake amplitude that could be measured by a seismograph at the time this definition was adopted, in 1935.

On October. 17, 1989, a magnitude 6.9 earthquake shook the San Francisco area. Robert Page from the U.S. Geological Survey said, "It was a wakeup call to prepare for the potentially even more devastating shocks that are inevitable in the future." Find the ratio A/A_0 for this earthquake. How many orders of magnitude larger was this earthquake's amplitude compared with the base-level amplitude, A_0?

22. The energy magnitude, M, radiated by an earthquake measures the potential damage to man-made structures. It can be described by the formula

$$M = \left(\tfrac{2}{3}\right) \log E - 2.9$$

where the seismic energy, E, is expressed in joules. Show that for every increase in M of one unit, the associated seismic energy E is increased by about a factor of 32.

23. a. Given the following graph, generate a linear equation for $Y (= \log y)$ in terms of x.

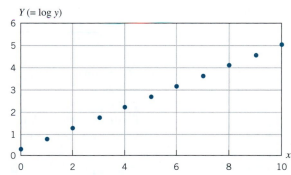

   b. Now substitute $\log y$ for Y and solve your equation for y.

   c. What type of function did you find in part (b)? What does this suggest about functions that appear linear on a semi-log graph (with the log scale on the vertical axis)?

Properties of Logarithmic Functions

Objective

- explore the effects of a and c on the graphs of $y = c \log(ax)$ and $y = c \ln(ax)$

Materials/Equipment

- graphing calculator or computer with "E8: Logarithmic Sliders" in *Exponential & Log Functions* in course software or a function graphing program
- graph paper

Procedure

Making Predictions

1. *The effect of a on the equation $y = \log(ax)$ when $a > 0$ and $x > 0$.*

   a. Why do we need to restrict a and x to positive values?

   b. Using the properties of logarithms, write the expression $\log(ax)$ as the sum of two logs. Discuss with a partner what effect you expect a to have on the graph. Now using $\log 2 \approx 0.301$ and $\log 3 \approx 0.477$, complete the accompanying data table and, by hand, sketch the three graphs on the same grid. Do your results confirm your predictions? What do you expect to happen to the graph of $y = \log(ax)$ if larger and larger positive values are substituted for a?

   Evaluating $y = \log(ax)$ when $a = 1, 2,$ and 3

x	$y = \log x$	$y = \log(2x)$	$y = \log(3x)$
0.001	−3.000		
0.010	−2.000		
0.100	−1.000		
1.000	0.000		
5.000	0.699		
10.000	1.000		

   c. Discuss with your partner how you think the graphs of $y = \log x$ and $y = \log(ax)$ will compare if $0 < a < 1$. Complete the following small data table and sketch the graphs of $y = \log x$ and $y = \log(x/10)$ on the same grid. Do your predictions and your graph agree? Predict what would happen to the graph of $y = \log(ax)$ if smaller and smaller positive values were substituted for a.

   Evaluating $y = \log(ax)$ when $a = 1$ and $\frac{1}{10}$

x	$y = \log x$	$y = \log(x/10)$
0.001	−3.000	
0.010	−2.000	
0.100	−1.000	
1.000	0.000	
5.000	0.699	
10.000	1.000	

   d. How do you think your findings for $y = \log(ax)$ relate to the function $y = \ln(ax)$?

2. *The effect of c on the equation y = c log x*

  a. When $c > 0$: Talk over with your partner your predictions for what will happen as c increases. Fill in the following table and on the same grid draw a quick sketch of the three functions. Were your predictions correct? What do you expect to happen to the graph of $y = c \log x$ as you substitute larger and larger positive values for c?

Evaluating $y = c \log x$ when $c = 1, 2,$ and 3

x	$y = \log x$	$y = 2 \log x$	$y = 3 \log x$
0.001	−3.000		
0.010	−2.000		
0.100	−1.000		
1.000	0.000		
5.000	0.699		
10.000	1.000		

  b. When $c < 0$: Why can c be negative when a has to remain positive? How do you think the graphs of $y = \log x$ and $y = c \log x$ will compare if $c < 0$? What happens if c remains negative but $|c|$ gets larger and larger (for example, $c = -10, -150, -5000$, etc.)?

  c. How do your findings on $y = c \log x$ relate to the function $y = c \ln x$?

3. *Generalizing your results.* Talk over with your partner the effect of varying both a and c on the general functions $y = c \log(ax)$ and $y = c \ln(ax)$. Try predicting the shapes of the graphs of such functions as $y = 3 \log(2x)$ and $y = -2 \log(x/10)$. Have each partner construct a small table, and graph the results of one such function. Compare your findings.

Testing Your Predictions

Test your predictions by either using "E8: Logarithmic Sliders" in *Exponential & Log Functions* or creating your own graphs with a graphing calculator or a function graphing program. Try changing the value for a to create different functions of the form $y = \log(ax)$ or $y = \ln(ax)$. Be sure to try values of $a > 1$ and values such that $0 < a < 1$. Then try changing the value for c in functions of the form $y = c \log x$ and $y = c \ln x$. Let c assume both positive and negative values.

Summarizing Your Results

Write a 60-second summary describing the effect of varying a and c in functions of the form $y = c \log(ax)$ and $y = c \ln(ax)$.

Exploration-Linked Homework

1. Use your knowledge of logarithmic functions to predict the shapes of the graphs of:
  a. *The decibel scale:* given by the function $N = 10 \log \frac{I}{I_0}$, where I is the intensity of a sound and I_0 is the intensity of sound at the threshold of human hearing. (See Section 6.4, pages 359–361.)
  b. *Stellar magnitude:* approximated by the function $M = -2.5 \log \frac{B}{B_0}$, where B is the brightness of a star and B_0 is a constant. (See Section 6.4, Exercise 8, p.363.)

2. Explore the effect of changing the base, that is, the effect of changing b in $y = c \log_b(ax)$. (See Section 6.5, Exercise 7, p. 365.)

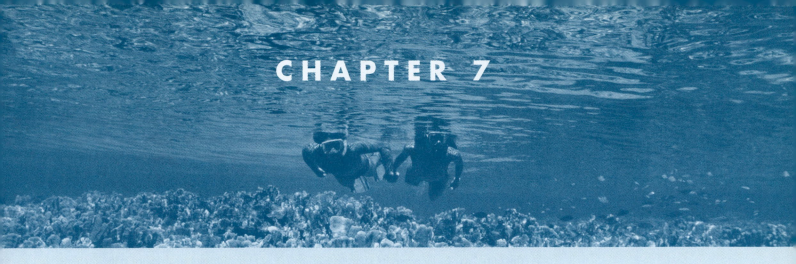

CHAPTER 7

POWER FUNCTIONS

OVERVIEW

Power functions help us answer questions such as "Why is the most important rule of scuba diving 'never hold your breath'?" and "Why do small animals have faster heartbeats and higher metabolic rates than large ones?"

After reading this chapter you should be able to

- recognize the properties of power functions

- construct and interpret graphs of power functions

- understand direct and inverse proportionality

- use logarithmic scales to determine whether a linear, power, or exponential function is the best function model for a set of data

- develop a sense about the relationship between size and shape

7.1   The Tension between Surface Area and Volume

Why do small animals have faster heartbeats and higher metabolic rates than large ones? Why are the shapes of the bodies and organs of large animals often quite different from those of small ones? To find answers to these questions, we examine how the relationship between surface area and volume changes as objects increase in size.

Scaling Up a Cube

Let's look at what happens to the surface area and volume of a simple geometric figure, the cube, as we increase its size. In Figure 7.1 we have drawn a series of cubes where the lengths of the edges are 1, 2, 3, and 4 units.[1]

Figure 7.1 Four cubes for which the lengths of the edges are 1, 2, 3, and 4 units, respectively.

Surface area of a cube

If we were painting a cube, the surface area would tell us how much area we would have to cover. Each cube has six identical faces, so

$$\text{surface area} = 6 \cdot (\text{area of one face})$$

If the edge length of one face of the cube is x, then the surface area of that face is x^2. So the total surface area $S(x)$ of the cube is

$$S(x) = 6x^2$$

$S(x)$ is called a *power function* of degree 2.

The second column of Table 7.1 lists the surface areas of cubes for various edge lengths. Figure 7.2 shows a graph of the function $S(x)$.

Ratio of Surface Area to Volume of Cube with Length of Edge x

Edge Length x	Surface Area $S(x) = 6x^2$	Volume $V(x) = x^3$	$\dfrac{S(x)}{V(x)} = \dfrac{6x^2}{x^3} = \dfrac{6}{x} = R(x)$
1	6	1	6.00
2	24	8	3.00
3	54	27	2.00
4	96	64	1.50
6	216	216	1.00
8	384	512	0.75
10	600	1000	0.60

Table 7.1

[1]For this discussion it doesn't matter which unit we use, but if you prefer, you may think of "unit" as being "centimeter" or "foot."

Since the surface area of a cube with edge length x can be represented as

$$\text{surface area } = \text{ constant} \cdot x^2$$

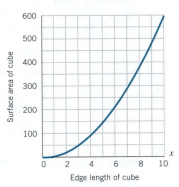

Figure 7.2 Graph of the surface area of a cube, $S(x) = 6x^2$.

we say that the surface area is *directly proportional* to the *square* (or second power) of the length of its edge. In Table 7.1, observe what happens to the surface area when we double the length of an edge. If we double the length from 1 to 2 units, the surface area becomes four times larger, increasing from 6 to 24 square units. If we double the length from 2 to 4 units, the surface area is again four times larger, increasing this time from 24 to 96 square units. In general, if we double the length of the edge from x to $2x$, the surface area will increase by a factor of 2^2, or 4:

Surface area of a cube with edge length x: $S(x) = 6x^2$

Surface area of a cube with edge length $2x$: $S(2x) = 6(2x)^2$

     If we apply rules of exponents $= 6 \cdot 2^2 \cdot x^2$

     simplify and rearrange terms $= 4(6x^2)$

     substitute $S(x)$ for $6x^2$, we get $S(2x) = 4 \cdot S(x)$

Volume of a cube

If the edge length of one face of the cube is x, then the volume $V(x)$ of the cube is given by

$$V(x) = x^3$$

$V(x)$ is an example of a power function of degree 3.

The third column of Table 7.1 lists values for $V(x)$, and Figure 7.3 is the graph of $V(x)$.

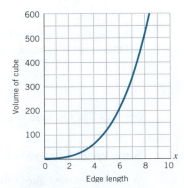

Figure 7.3 Graph of the volume of a cube, $V(x) = x^3$.

Since

$$\text{volume} = \text{constant} \cdot x^3 \qquad \text{(the constant in this case is 1)}$$

we say that the volume is *directly proportional* to the *cube* (or the third power) of the length of its edge.

What happens to the volume when we double the length? In Table 7.1, if we double the length from 1 to 2 units, the volume increases by a factor of 2^3 or 8, increasing from 1 to 8 cubic units. If we double the length from 2 to 4 units, the volume again becomes eight times larger, increasing from 8 to 64 cubic units. In general, if we double the edge length from x to $2x$, the volume will increase by a factor of 8:

Volume of a cube with edge length x:	$V(x) = x^3$
Volume of a cube with edge length $2x$:	$V(2x) = (2x)^3$
If we apply rules of exponents	$= 2^3 \cdot x^3$
simplify	$= 8x^3$
substitute $V(x)$ for x^3, we get	$V(2x) = 8 \cdot V(x)$

Ratio of the surface area to the volume

The ratio of (surface area)/volume generates a new function $R(x)$, where

$$R(x) = \frac{S(x)}{V(x)} = \frac{6x^2}{x^3} = \frac{6}{x} = 6x^{-1}$$

$R(x)$ is called a power function of degree -1. As we increase the value of x, the edge length of the cube, the value of $R(x)$ decreases, as we can see in Table 7.1 and Figure 7.4

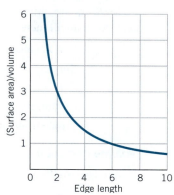

Figure 7.4 Graph of $R(x) =$ (surface area)/volume.

As the size of the cube increases, the volume increases faster than the surface area. For example, if we double the edge length, the surface area increases by a factor of 4, but the volume increases by a factor of 8. So the ratio

$$\frac{\text{surface area}}{\text{volume}}$$

decreases as the edge length increases.

Size and Shape

What we learned about the cube is true for any three-dimensional object, no matter what the shape. In general,

For any shape, as an object becomes larger while keeping the same shape, the ratio of its surface area to its volume decreases.

Thus a larger object has relatively less surface area than a smaller one. This fact allows us to understand some basic principles of biology and to answer the questions we asked at the beginning of this section.

In Exploration 7.1 you can study further the effects of scaling up an object.

Biological functions such as respiration and digestion depend upon surface area but must service the body's entire volume.[2] The biologist J. B. S. Haldane wrote that "comparative anatomy is largely the story of the struggle to increase surface in proportion to volume." This is why the shapes of the bodies and organs of large animals are often quite different from those of small ones. Many large species have adapted by developing complex organs with convoluted exteriors, thus greatly increasing the organs' surface areas. Human lungs, for instance, are heavily convoluted to increase the amount of surface area, thereby increasing the rate of exchange of gases. Stephen Jay Gould wrote that "the villi of our small intestine increase the surface area available for absorption of food."

Body temperature also depends upon the ratio of surface area to volume. Animals generate the heat needed for their volume by metabolic activity and lose heat through their skin surface. Small animals have more surface area in proportion to their volume than do large animals. Since heat is exchanged through the skin, small animals lose heat proportionately faster than large animals and have to work harder to stay warm. Hence their heartbeats and metabolic rates are faster. As a result, smaller animals burn more energy per unit mass than larger animals.

Stephen Jay Gould's essay "Size and Shape" in *Ever Since Darwin: Reflections in Natural History* offers an interesting perspective on the relationship between the size and shape of objects.

Algebra Aerobics 7.1

The inside back cover of the text contains geometric formulas.

1. The function $S(r) = 4\pi r^2$ gives the surface area of a sphere with radius r.

  a. Compare $S(r)$, $S(2r)$, and $S(3r)$.

  b. What happens to the surface area when the radius is doubled? When the radius is tripled?

2. The function $V(r) = \frac{4}{3}\pi r^3$ gives the volume of a sphere with radius r.

  a. Compare $V(r)$, $V(2r)$, and $V(3r)$.

  b. What happens to the volume when the radius is doubled? When the radius is tripled?

3. As the radius of a sphere increases, which grows faster, the surface area or the volume? What happens to the ratio $S(r)/V(r) = $ (surface area)/volume as the radius increases?

4. The two-dimensional analog to the ratio of (surface area)/volume for a sphere is the ratio of circumference/area for a circle.

  a. The circumference $C(r)$ of a circle with radius r is $2\pi r$. Compare $C(r)$, $C(2r)$, and $C(4r)$. What happens to the circumference when the radius is doubled? Quadrupled (multiplied by 4)?

  b. The area $A(r)$ of a circle with radius r is πr^2. Compare $A(r)$, $A(2r)$, and $A(4r)$. What happens to the area when the radius is doubled? Quadrupled (multiplied by 4)?

  c. Which grows faster, the circumference or the area? As a result, what happens to the ratio $C(r)/A(r) = $ circumference/area as the radius increases?

5. Compare a sphere with radius r and a cube with edge length r.

  a. For what value(s) of r, if any, is the volume of the cube equal to the volume of the sphere?

  b. For what value(s) of r, if any, is the volume of the cube greater than the volume of the sphere? Less than?

6. Two different cylinders have the same height of 25 ft, but the base of one has a radius of 5 feet and the base of the other has a radius of 10 feet. Is the volume of one cylinder double the volume of the other? Justify your answer.

7. A cylindrical silo has a radius of 12 feet. If a certain amount of feed fills the silo to a depth of 5 feet, will twice the amount of feed fill the silo to a depth of 10 feet? Justify your answer.

8. The volume V of some regular figures can be defined as the area B of the base times the height h, or $V = Bh$.

  a. Use this definition to find the formulas for the volumes in Figures 7.5 and 7.6.

  b. In each case, if the height is doubled, by what factor is the volume increased?

  c. If all dimensions are doubled, by what factor is the volume increased?

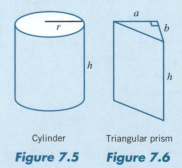

Cylinder   Triangular prism

Figure 7.5   **Figure 7.6**

---

[2] For those who want to investigate how species have adapted and evolved over time, see D. W. Thompson, *On Growth and Form* (New York: Dover, 1992), and T. McMahon and J. Bonner, *On Size and Life* (New York: Scientific American Books, 1983).

Exercises for Section 7.1

A calculator that can evaluate powers is required throughout. Formulas for geometric figures are on the inside back cover of the text.

1. Solve the following formulas for the indicated variable.
   a. $V = lwh$, solve for l
   c. $P = 2l + 2w$, solve for w
   b. $A = \dfrac{hb}{2}$, solve for b
   d. $S = 2x^2 + 4xh$, solve for h

2. In parts (a)–(c), evaluate the functions at R and $3R$.
   a. $C(r) = 2\pi r$, the circumference of a circle with radius r
   b. $A(r) = \pi r^2$, the area of a circle with radius r
   c. $V(r) = \frac{4}{3}\pi r^3$, the volume of a sphere with radius r
   d. Describe what happens to $C(r)$, $A(r)$, and $V(r)$ when the radius triples from R to $3R$.

3. Using the formulas in Exercise 2, simplify the following ratios and identify which ones are increasing as r increases. Describe each result in geometric terms.
   a. $\dfrac{A(r)}{C(r)}$
   b. $\dfrac{V(r)}{A(r)}$
   c. $\dfrac{V(r)}{C(r)}$

4. Consider the formulas $V_1 = \frac{4}{3}\pi r^3$ (volume of a sphere) and $V_2 = \pi r^2 h$ (volume of a cylinder). Simplify the following ratios. Which one(s) increase as r increases (assume h is held constant)? Describe your result in geometric terms.
   a. $\dfrac{V_2}{V_1}$
   b. $\dfrac{V_1}{V_2}$

5. A tiny sphere has a radius of $0.000\ 000\ 000\ 1 = 10^{-10}$ meter, which is roughly equivalent to the radius of a protein molecule. Answer the following questions. Express your answers in scientific notation.
   a. Find its surface area, S, in square meters, where $S = 4\pi r^2$.
   b. Find its volume, V, in cubic meters, where $V = \frac{4}{3}\pi r^3$.
   c. Find the ratio of the surface area to the volume.
   d. As r increases, does the ratio in part (c) increase or decrease?

6. The radius of Earth is about 6400 kilometers. Assume that Earth is spherical. Express your answers to the questions below in scientific notation.
   a. Find the surface area of Earth in square meters.
   b. Find the volume of Earth in cubic meters.
   c. Find the ratio of the surface area to the volume.

7. If the radius of a sphere is x meters, what happens to the surface area and to the volume of a sphere when you:
   a. Quadruple the radius?
   b. Multiply the radius by n?
   c. Divide the radius by 3?
   d. Divide the radius by n?

8. Assume a box has a square base and the length of a side of the base is equal to twice the height of the box.
   a. If the height is 4 inches, what are the dimensions of the base?

   b. Write functions for the surface area and the volume that are dependent on the height, h.
   c. If the volume has increased by a factor of 27, what has happened to the height?
   d. As the height increases, what will happen to the ratio of (surface area)/volume?

9. A box has volume $V = \text{length} \cdot \text{width} \cdot \text{height}$.
   a. Find the volume of a cereal box with dimensions of length $= 19.5$ cm, width $= 5$ cm, and height $= 27$ cm. (Be sure to specify the unit.)
   b. If the length and width are doubled, by what factor is the volume increased?
   c. What are two ways you could increase the volume by a factor of 4 and keep the height the same?

10. A box with a lid has a square base with each base side x inches and a height of h inches.
    a. Write the formula for the volume of the box.
    b. Write the formula for the surface area. (*Hint:* The surface area of the box is the sum of the areas of each side of the box.)
    c. If the side of the base is tripled and the height remains constant, by what factor is the volume increased?
    d. If the side of the base triples (from x to $3x$) with h held constant, what is the change in the surface area?

11. Consider a cylinder with volume $V = \pi r^2 h$. What happens to its volume when you double its height, h? When you double its radius, r?

12. Consider two solid figures, a sphere and cylinder, where each has radius r. The volume of a sphere is $V_s = \frac{4}{3}\pi r^3$ and the volume of a cylinder is $V_c = \pi r^2 h$.

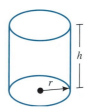

   a. If $h = 1$, when is the volume of the sphere greater than the volume of the cylinder?
   b. What value of h would make the volumes the same?
   c. Complete this statement: "The volume of the cylinder with radius r is greater than the volume of a sphere of radius r, when $h > $ _____."

13. The volume, V, of a cylindrical can is $V = \pi r^2 h$ and the total surface area, S, of the can is

   $$S = \text{area of curved surface} + 2 \cdot (\text{area of base})$$
   $$= 2\pi rh + 2\pi r^2$$

   where r is the radius of the base and h is the height.
   a. Assume the height is three times the radius. Write the volume and the surface area as functions of the radius.

b. As r increases, which grows faster, the volume or the surface area? Explain.

14. (*Hint:* See Exploration 7.1.) To celebrate Groundhog Day, a gourmet candy company makes a solid, quarter-pound chocolate groundhog that is 2 inches tall.

a. If the company wants to introduce a solid chocolate groundhog in the same shape but twice as tall (4 inches high), how much chocolate is required?

b. How tall would a solid chocolate groundhog be if made with twice the original quarter-pound amount (for a total of 8 ounces)?

15. A circle has a radius of x units. Which graph could represent the ratio of area to circumference? Which could represent the ratio of circumference to area? Explain your answers.

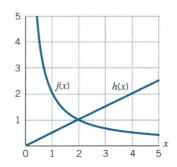

16. A cylinder (with no top or bottom) has a radius of x feet and a height of 3 feet.

a. Construct a function for the volume and another for the surface area of the cylinder.

b. Construct the ratio $\frac{\text{volume}}{\text{surface area}}$. What would happen to the ratio as x increases?

7.2 *Direct Proportionality: Power Functions with Positive Powers*

In the last section we encountered three functions of the edge length x of a cube:

Surface area $\qquad S(x) = 6x^2$

Volume $\qquad V(x) = x^3$

Ratio of $\dfrac{\text{surface area}}{\text{volume}}$ $\qquad R(x) = \dfrac{6}{x}$ or $6x^{-1}$

All three are called *power functions* since they are of the form

$$\text{dependent variable} = \text{constant} \cdot (\text{independent variable})^{\text{power}}$$

or

$$\text{output} = \text{constant} \cdot \text{input}^{\text{power}}$$

> **Power Functions**
>
> A power function $y = f(x)$ can be represented by an equation of the form
>
> $$y = kx^p$$
>
> where k and p are constants.

For the functions: $S(x) = 6x^2$, we have $k = 6$ and $p = 2$.

$V(x) = x^3$, we have $k = 1$ and $p = 3$.

$R(x) = 6x^{-1}$, we have $k = 6$ and $p = -1$.

EXAMPLE 1 Decide whether or not each of the following functions is a power function of the form $y = kx^p$, and if it is, identify the values of k and p.

a. $y = 3x^4$ **b.** $y = 3x^4 + 1$ **c.** $y = 2x^{-4}$ **d.** $y = -6\sqrt{x}$ **e.** $y = 4 \cdot 2^x$

SOLUTION **a.** $y = 3x^4$ is a power function where $k = 3$ and $p = 4$.

b. $y = 3x^4 + 1$ is not a power function since it cannot be written in the form $y = kx^p$.

c. $y = 2x^{-4}$ is a power function where $k = 2$ and $p = -4$.

d. $y = -6\sqrt{x} = -6x^{1/2}$ is a power function where $k = -6$ and $p = \frac{1}{2}$.

e. $y = 4 \cdot 2^x$ is not a power function, since the variable x is in the exponent. It's an exponential function.

Direct Proportionality

In Chapter 2, for linear functions, we said that y is *directly proportional to x* if y equals a constant times x. For example, if $y = 4x$, then y is directly proportional to x. We can extend the same concept to any power function with positive exponents. If $y = kx^p$ and p is positive, we say that y is *directly proportional to x^p*.

In Section 7.1, we saw that the surface area of a cube is directly proportional to the square of its edge length and that the volume is directly proportional to the cube of its edge length. The symbol $\propto$ is used to indicate direct proportionality. For example, if $y = 5x^3$, then y is directly proportional to x^3, which we write as $y \propto x^3$.

Direct Proportionality

If

$$y = kx^p \qquad (k \neq 0 \text{ and } p > 0)$$

we say that y is *directly proportional to x^p*. We write this as

$$y \propto x^p$$

The coefficient k is called the *constant of proportionality*.

EXAMPLE 2   Write formulas to represent the following relationships.

a. The circumference, C, of a circle is directly proportional to its radius, r.

b. The area, A, of a circle is directly proportional to its radius, r, squared.

c. The volume, V, of a liquid flowing through a tube is directly proportional to the fourth power of the radius, r, of the tube.

SOLUTION   **a.** $C = kr$, where $k = 2\pi$.   **c.** $V = kr^4$ for some constant k.

b. $A = kr^2$, where $k = \pi$.

Properties of Direct Proportionality

Consider a general power function $f(x) = kx^p$ where $p > 0$. If we double the input from x to $2x$, we have

evaluating f at $2x$	$f(2x) = k(2x)^p$
using rules of exponents	$= k(2^p x^p)$
and rearranging terms	$= 2^p(kx^p)$
substituting $f(x)$ for kx^p	$= 2^p f(x)$

Doubling the input multiplies the output by 2^p. We saw in Section 7.1 that if we double the input for the function $S(x) = 6x^2$ (the surface area of a cube), the output is multiplied by 2^2 or 4.

If we multiply the input by m, changing the input from x to mx, then

evaluating f at mx	$f(mx) = k(mx)^p$
using rules of exponents	$= k(m^p x^p)$
and rearranging terms	$= m^p(kx^p)$
substituting $f(x)$ for kx^p	$= m^p f(x)$

So multiplying the input by m multiplies the output by m^p. For example, if we triple the input for $S(x) = 6x^2$, the output is multiplied by 3^2 or 9. Note that the value for k (the constant of proportionality) is irrelevant in these calculations.

In general,

> If y is directly proportional to x^p (where $p > 0$), then $y = kx^p$ for some nonzero constant k.
>
> Multiplying the input by m multiplies the output by m^p.
>
> For example, tripling the input multiplies the output by 3^p.

EXAMPLE 3 Given each function, what happens if the input is doubled? Increased by a factor of 10? Cut in half?

a. $y = 2x^4$ **b.** $h(z) = -2x^5$

SOLUTION **a.** If the input is doubled, the output is multiplied by 2^4 or 16. If the input is multiplied by 10, the output is multiplied by 10^4 or 10,000. If the input is cut in half, the output would be multiplied by $\left(\frac{1}{2}\right)^4 = \frac{1}{16}$.

b. If the input is doubled, the output is multiplied by 2^5 or 32. If the input is multiplied by 10, the output is multiplied by 10^5 or 100,000. If the input is cut in half, the output would be multiplied by $\left(\frac{1}{2}\right)^5 = \frac{1}{32}$.

EXAMPLE 4 **a.** What is the difference among $2f(x)$, $f(2x)$, and $f(x) + 2$?

b. If $f(x) = x^2$, evaluate the three expressions in part (a) when $x = 4$.

SOLUTION **a.** $2f(x)$ means to multiply the value of $f(x)$ by 2.
$f(2x)$ means to use $2x$ as the input for the function f.
$f(x) + 2$ means to evaluate $f(x)$ and then add 2.

b. If $f(x) = x^2$ and $x = 4$, then
$$2f(4) = 2 \cdot 4^2 = 2 \cdot 16 = 32$$
$$f(2 \cdot 4) = f(8) = 8^2 = 64, \text{ or equivalently,}$$
$$= (2 \cdot 4)^2 = 2^2 \cdot 4^2 = 4 \cdot 16 = 64$$
$$f(4) + 2 = 4^2 + 2 = 16 + 2 = 18$$

EXAMPLE 5 **Pedaling into the wind**
When you pedal a bicycle, it's much easier to pedal with the wind than into the wind. That's because as the wind blows against an object, it exerts a force upon it. This force is directly proportional to the wind velocity squared.

a. Construct an equation to describe the relationship between the wind force and the wind velocity.

b. If the wind velocity doubles, by how much does the wind force go up?

SOLUTION **a.** $F(v) = kv^2$ for some constant k, where v is the wind velocity and $F(v)$ is the wind force.

b. If the velocity doubles, then the wind force is multiplied by 2^2 or 4. So pedaling into a 20-mph wind requires four times as much effort as pedaling into a 10-mph wind. (See Figure 7.7.)

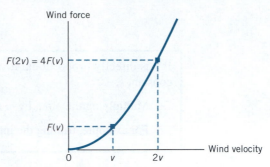

Figure 7.7 Doubling the wind velocity quadruples the wind force.

EXAMPLE 6   **Earth's core**

The radius of the core of Earth is slightly over half the radius of Earth as a whole, yet the core is only about 16% of the total volume of Earth. How is this possible?

SOLUTION   We can think of both Earth and its core as approximately spherical in shape (see Figure 7.8).

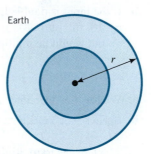

Figure 7.8 A sketch of a spherical Earth with radius r and its core.

If we let $r =$ radius of Earth, then Earth's volume is

$$V_{\text{Earth}} = \tfrac{4}{3}\pi r^3$$

If the radius of the core were exactly half that of Earth $\left(\tfrac{r}{2}\right)$, then the core's volume would be

$$V_{\text{core}} = \tfrac{4}{3}\pi \left(\tfrac{r}{2}\right)^3$$

use rules of exponents
$$= \tfrac{4}{3}\pi \left(\tfrac{r^3}{2^3}\right)$$

multiply and regroup terms
$$= \left(\tfrac{1}{2^3}\right)\left(\tfrac{4}{3}\pi r^3\right)$$

substitute $\tfrac{1}{8}$ for $\left(\tfrac{1}{2^3}\right)$ and V_{Earth} for $\tfrac{4}{3}\pi r^3$
$$= \tfrac{1}{8}V_{\text{Earth}}$$

$$= 12.5\% \text{ of the volume of Earth}$$

Since the radius of the core is slightly more than half the radius of Earth, 16% of Earth's volume is a reasonable estimate for the volume of the core.

EXAMPLE 7   In Chapter 4 we encountered the following formula used by police. It estimates the speed, S, at which a car must have been traveling given the distance, d, the car skidded on a dry tar road after the brakes were applied:

$$S = \sqrt{30d} \approx 5.48d^{1/2}$$

Speed, S, is in miles per hour and distance, d, is in feet.

  a. Use the language of proportionality to describe the relationship between S and d.

  b. If the skid marks were 50 feet long, approximately how fast was the driver going when the brakes were applied?

  c. What happens to S if d doubles? Quadruples?

  d. Is d directly proportional to S?

SOLUTION   **a.** The speed, S, is directly proportional to $d^{1/2}$ and the constant of proportionality is approximately 5.48.

  b. If the length of the skid marks, d, is 50 feet, then the estimated speed of the car $S \approx 5.48 \cdot (50)^{1/2} \approx 5.48(7.07) \approx 39$ mph.

  c. If the skid marks double in length from d to $2d$, then the estimate for the speed of the car increases from $\sqrt{30d}$ to $\sqrt{30 \cdot (2d)} = \sqrt{2} \cdot \sqrt{30d}$. So the estimated speed goes up by a factor of $\sqrt{2} \approx 1.414$ or, equivalently, by about 41.4%.

     If the skid marks quadruple in length from d to $4d$, then the estimated speed of the car goes from $\sqrt{30d}$ to $\sqrt{30 \cdot (4d)} = \sqrt{4} \cdot \sqrt{30d} = 2\sqrt{30d}$. So the speed estimate goes up by a factor of 2, or by 100%.

     For example, if the skid marks doubled from 50 to 100 feet, the speed estimate would go up from 39 mph to $1.414 \cdot 39 \approx 55$ mph. If the skid marks quadrupled from 50 to 200 feet, then the speed estimate would double from about 39 to almost 78 mph.

  d. To determine if d is directly proportional to S, we need to solve our original equation for d.

Given	$S = \sqrt{30d}$
square both sides of the equation	$S^2 = 30d$
divide both sides by 30	$\dfrac{S^2}{30} = d$
or	$d = \dfrac{S^2}{30}$

So d is directly proportional to S^2 but not to S.

Direct Proportionality with More Than One Variable

When a quantity depends directly on more than one other quantity, we no longer have a simple power function. For example, the volume, V, of a cylindrical can depends on both the radius, r, of the base and the height, h. The equation describing this relationship is

$$V = \text{area of base} \cdot \text{height}$$

$$V = \pi r^2 h$$

We say V is directly proportional to both r^2 and h.

EXAMPLE 8 **Deflection**

A person stands at the center of a fir plank (with a cross section of 2″ by 12″) that is anchored at both ends (see Figure 7.9). Architects estimate the downward deflection of the plank using the formula

$$D = (6.6 \cdot 10^{-6}) \cdot P \cdot L^3$$

where D is the deflection in inches, P is the weight of the person in pounds, and L is the length of the plank in feet.[3]

Figure 7.9 A person standing at the center of a plank.

a. Describe the relationship among D, P, and L using the language of proportionality.
b. What is the downward deflection of the plank if the person weighs 200 lb and the plank is 6′ long?
c. What happens to D if we increase P by 50%? If we increase L by 50%?

SOLUTION a. D is directly proportional to P and to L^3.

b. If $P = 200$ lb and $L = 6′$, then

$$D = (6.6 \cdot 10^{-6}) \cdot 200 \cdot 6^3$$

rewrite 200 and rearrange terms $\qquad = (6.6 \cdot 6^3 \cdot 2) \cdot 10^2 \cdot 10^{-6}$

evaluate $\qquad = 2851.2 \cdot 10^{-4}$

$$\approx 0.3 \text{ inches}$$

So the plank is about a third of an inch lower in the middle, under the 200-lb person.

c. Increasing the person's weight by 50% means multiplying the weight by 1.5. Since the deflection is directly proportional to the person's weight, that means that the deflection is multiplied by 1.5 or, equivalently, increased by 50%. So if we replace a 200-lb person (who causes a deflection of 0.3 inches) with someone 300 lb (or 50% heavier), the deflection would become $(1.5) \cdot (0.3) = 0.45$, or almost half an inch.

Increasing the plank's length by 50% means multiplying the length by 1.5. Since the deflection is directly proportional to the cube of the plank length, the deflection is multiplied by $(1.5)^3 \approx 3.4$ or, equivalently, increased by 240%. So for a 200-lb person, if the plank length of 6′ is increased by 50% [to $(1.5) \cdot 6 = 9′$], the projected deflection would be $(3.4) \cdot (0.3 \text{ inches}) = 1.02$ inches, which is at the edge of reliability for this model.

---

[3]The deflection formula for the plank is quite accurate for deflections up to about an inch. After that it starts to produce unrealistic D values very quickly since L is cubed.

Algebra Aerobics 7.2

1. In each case, indicate whether or not the function is a power function. If it is, identify the independent and dependent variables, the constant of proportionality, and the power.
   a. $A = \pi r^2$ c. $z = w^5 + 10$ e. $y = 3x^5$
   b. $y = z^5$ d. $y = 5^x$

2. Identify which (if any) of the following equations represent direct proportionality:
   a. $y = 5.3x^2$ c. $y = 5.3 \cdot 2^x$
   b. $y = 5.3x^2 + 10$

3. Given $g(x) = 5x^3$
   a. Calculate $g(2)$ and compare this value with $g(4)$.
   b. Calculate $g(5)$ and compare this value with $g(10)$.
   c. What happens to the value of $g(x)$ if x doubles in value?
   d. What happens to $g(x)$ if x is divided by 2?

4. Given $h(x) = 0.5x^2$
   a. Calculate $h(2)$ and compare this value with $h(6)$.
   b. Calculate $h(5)$ and compare this value with $h(15)$.
   c. What happens to the value of $h(x)$ if x triples in value?
   d. What happens to $h(x)$ if x is divided by 3?

5. Express each of the following relationships with an equation.
   a. The volume of a sphere is directly proportional to the cube of its radius.
   b. The volume of a prism is directly proportional to its length, width, and height.
   c. The electrostatic force, f, is directly proportional to the particle's charge, c.

6. Express in your own words the relationship between y and x in the following functions.
   a. $y = 3x^5$ b. $y = 2.5x^3$ c. $y = \dfrac{x^5}{4}$

7. Express each of the relationships in Problem 6 in terms of direct proportionality.

8. a. Solve $P = aR^2$ for R.
   b. Solve $V = \left(\frac{1}{3}\right)\pi r^2 h$ for h.

9. If $f(x) = 0.1x^3$, evaluate and describe the difference between the following functions.
   a. $f(2x)$ and $2f(x)$
   b. $f(3x)$ and $3f(x)$
   c. Evaluate each function in part (a) for $x = 5$.

Exercises for Section 7.2

A calculator that can evaluate powers is useful here. A graphing program is recommended for Exercises 15 and 16.

1. Evaluate the following functions when $x = 2$ and $x = -2$:
   a. $f(x) = 5x^2$ c. $h(x) = -5x^2$
   b. $g(x) = 5x^3$ d. $k(x) = -5x^3$

2. Identify which of the following are power functions. For each power function, identify the value of k, the constant of proportionality, and the value of p, the power.
   a. $y = -3x^2$ c. $y = x^2 + 3$
   b. $y = 3x^{10}$ d. $y = 3^x$

3. Each of the following represents direct proportionality with powers of both x and y. Identify the constant of proportionality, k, and then evaluate each function for $x = 4$ and $y = 3$.
   a. $P = \frac{1}{2}x^3y^2$ c. $T = 4\pi x^2 y^3$
   b. $M = 14x^{1/2}y$ d. $N = (18xy)^{1/3}$

4. Find the constant of proportionality, k, for the given conditions.
   a. $y = kx^3$, and $y = 64$ when $x = 2$.
   b. $y = kx^{3/2}$, and $y = 96$ when $x = 16$.
   c. $A = kr^2$, and $A = 4\pi$ when $r = 2$.
   d. $v = kt^2$, and $v = -256$ when $t = 4$.

5. Write the general formula to describe each variation and then solve the variation problem.
   a. y is directly proportional to x, and $y = 2$ when $x = 10$. Find y when x is 20.
   b. p is directly proportional to the square root of s, and $p = 12$ when $s = 4$. Find p when $s = 16$.
   c. A is directly proportional to the cube of r, and $A = 8\pi$ when $r = \sqrt[3]{6}$. Find A when $r = 2 \cdot \sqrt[3]{6}$.
   d. P is directly proportional to the square of m, and $P = 32$ when $m = 8$. If m is halved, what happens to P?

6. The data in the table satisfy the equation $y = kx^n$, where n is a positive integer. Find k and n.

x	2	3	4	5
y	1	2.25	4	6.25

7. Assume Y is directly proportional to X^3.
   a. Express this relationship as a function where Y is the dependent variable.
   b. If $Y = 10$ when $X = 2$, then find the value of the constant of proportionality in part (a).
   c. If X is increased by a factor of 5, what happens to the value of Y?

d. If X is divided by 2, what happens to the value of Y?

e. Rewrite your equation from part (a), solving for X. Is X directly proportional to Y?

8. The distance, d, a ball travels down an inclined plane is directly proportional to the square of the total time, t, of the motion.

a. Express this relationship as a function where d is the dependent variable.

b. If a ball starting at rest travels a total of 4 feet in 0.5 second, find the value of the constant of proportionality in part (a).

c. Complete the equation and solve for t. Is t directly proportional to d?

9. a. Assume L is directly proportional to x^5. What is the effect of doubling x?

b. Assume M is directly proportional to x^p, where p is a positive integer. What is the effect of doubling x?

10. In "Love That Dirty Water" (*Chicago Reader,* April 5, 1996), Scott Berinato interviewed Ernie Vanier, captain of the towboat *Debris Control.* The Captain said, "We've found a lot of bowling balls. You wouldn't think they'd float, but they do." When will a bowling ball float in water? The bowling rule book specifies that a regulation ball must have a circumference of exactly 27 inches. Recall that the circumference of a circle with radius r is $2\pi r$. The volume of a sphere with radius r is $\frac{4}{3}\pi r^3$.

a. What is a regulation bowling ball's radius in inches?

b. What is the volume of a regulation bowling ball in cubic inches? (Retain at least two decimal places in your answer.)

c. What is the weight in pounds of a volume of water equivalent in size to a regulation bowling ball? (Water weighs 0.03612 lb/in³).

d. A bowling ball will float when its weight is less than or equal to the weight of an equivalent volume of water. What is the heaviest weight of a regulation bowling ball that will float in water?

e. Typical men's bowling balls are 15 or 16 pounds. Women commonly use 12-pound bowling balls. What will happen to the men's and to the women's bowling balls when dropped into the water? Will they sink or float?

11. When a variable is directly proportional to the product of two or more variables, we say that the variable is *jointly* proportional to those variables. Express each of the following as an equation.

a. x is jointly proportional to y and the square of z.

b. V is jointly proportional to l, w, and h.

c. w is jointly proportional to the square of x and the cube root of y.

d. The volume of a cylinder is jointly proportional to its height and the square of its radius.

12. Write a general formula to describe each variation. Use the information given to find the constant of proportionality.

a. Q is directly proportional to both the cube root of t and the square of d, and $Q = 18$ when $t = 8$ and $d = 3$.

b. A is directly proportional to both h and the square of the radius, r, and $A = 100\pi$ when $r = 5$ and $h = 2$.

c. V is directly proportional to B and h, and $V = 192$ when $B = 48$ and $h = 4$.

d. T is directly proportional to both the square root of p and the square of u, and $T = 18$ when $p = 4$ and $u = 6$.

13. The cost, C, in dollars, of insulating a wall is directly proportional to the area, A, of the wall (measured in ft²) and the thickness, t, of the insulation (measured in inches).

a. Write a cost equation for insulating a wall.

b. If the insulation costs are $12 when the area is 50 ft² and the thickness is 4 inches, find the constant of proportionality.

c. A storage room is 15 ft by 20 ft with 8-ft-high ceilings. Assuming no windows and one uninsulated door that is 3 feet by 7 feet, what is the total area of the insulated walls for this room?

d. What is the cost of insulating this room with 4-inch insulation? With 6-inch insulation?

14. Suppose you are traveling in your car at speed S and you suddenly brake hard, leaving skid marks on the road. A "rule of thumb" for the distance, D, that the car will skid is given by

$$D = \frac{S^2}{30f}$$

where D = distance the car skids (in feet), S = speed of the car (in miles per hour), and f is a number called the coefficient of friction that depends on the road surface and its condition. For a dry tar road, $f \approx 1.0$. For a wet tar road, $f \approx 0.5$. (We saw a variation of this problem in Example 7, Section 7.2.)

a. What is the equation giving distance skidded as a function of speed for a dry tar road? For a wet tar road?

b. Generate a small table of values for both functions in part (a), including speeds between 0 and 100 miles per hour.

c. Plot both functions on the same grid.

d. Why do you think the coefficient of friction is less for a wet road than for a dry road? What effect does this have on the graph in part (c)?

e. In the accompanying table, estimate the speed given the distances skidded on dry and on wet tar roads. Describe the method you used to find these numbers.

Distance Skidded	Estimated Speed (mph)	
(ft)	Dry Tar	Wet Tar
25		
50		
100		
200		
300		

f. If one car is going twice as fast as another when they both jam on the brakes, how much farther will the faster car skid? Explain. Does your answer depend on whether the road is dry or wet?

15. (Graphing program recommended.) Assume a person weighing P pounds is standing at the center of a 4″ × 12″ fir plank that spans a distance of L feet. The downward deflection $D_{\text{deflection}}$ (in inches) of the plank can be described by

$$D_{\text{deflection}} = (5.25 \cdot 10^{-7}) \cdot P \cdot L^3$$

a. Graph the deflection formula, $D_{\text{deflection}}$, assuming $P = 200$ pounds. Put values of L (from 0 to 25 feet) on the horizontal axis and deflection (in inches) on the vertical axis.

b. A rule used by architects for estimating acceptable deflection, D_{safe}, in inches, of a beam L feet long bent downward as a result of carrying a load is

$$D_{\text{safe}} = 0.05L$$

Add to your graph from part (a) a plot of D_{safe}.

c. Is it safe for a 200-pound person to sit in the middle of a $4'' \times 12''$ fir plank that spans 20 feet? What maximum span is safe for a 200-pound person on a $4'' \times 12''$ plank?

16. (Graphing program recommended.) In Exercise 15 we looked at how much a single load (a person's weight) could bend a plank downward by being placed at its midpoint. Now we look at what a continuous load, such as a solid row of books spread evenly along a shelf, can do. A long row of paperback fiction weighs about 10 pounds for each foot of the row. Typical hardbound books weigh about 20 lb/ft, and oversize hardbounds such as atlases, encyclopedias, and dictionaries weigh around 36 lb/ft. The following function is used to model the deflection D (in inches) of a $1'' \times 12''$ common pine board spanning a length of L feet carrying a continuous row of books:

$$D = (4.87 \cdot 10^{-4}) \cdot W \cdot L^4$$

where W is the weight per foot of the type of books along the shelf. This deflection model is quite good for deflections up to 1 inch; beyond that the fourth power causes the deflection value to increase very rapidly into unrealistic numbers.

a. How much deflection does the formula predict for a shelf span of 30 inches with oversize books? Would you recommend a stronger, thicker shelf?

b. Plot $D_{\text{hardbound}}$, $D_{\text{paperback}}$, and D_{oversize} on the same graph. Put L on the horizontal axis with values up to 4 feet.

c. For each kind of book identify what length, L, will cause a deflection of 0.5 inch in a $1'' \times 12''$ pine shelf.

17. Construct formulas to represent the following relationships.

a. The distance, d, traveled by a falling object is directly proportional to the square of the time, t, traveled.

b. The energy, E, released is directly proportional to the mass, m, of the object and the speed of light, c, squared.

c. The area, A, of a triangle is directly proportional to its base, b, and height, h.

d. The reaction rate, R, is directly proportional to the concentration of oxygen, $[O_2]$, and the square of the concentration of nitric oxide, $[NO]$.

e. When you drop a small sphere into a dense fluid such as oil, it eventually acquires a constant velocity, v, that is directly proportional to the square of its radius, r.

18. a. An insulation blanket for a cylindrical hot water heater is sold in a roll 48 in $\times$ 75 in $\times$ 2 in. Assuming a hot water heater 48 inches high, for what diameter water heater is this insulation blanket made? (Round to the nearest $\frac{1}{2}$ inch.)

b. What is the volume of the hot water heater in part (a)? (*Note:* Ignore the thickness of the heater walls.)

c. If 1 gallon $= 231$ in^3, what is the maximum number of gallons of water that a cylinder the size of this water heater could hold?

19. The volume, V, of a cylinder, with radius r and height h, is given by the formula $V = \pi r^2 h$. Describe what happens to V under the following conditions.

a. The *radius* is doubled; the *radius* is tripled.

b. The *height* is doubled; the *height* is tripled.

c. The *radius* r is multiplied by n, where n is a positive integer.

d. The *height* is multiplied by n, where n is a positive integer.

20. If you were designing cylinders, what are two different ways you could quadruple the volume? (See Exercise 19.)

21. Given $h(x) = 0.5x^2$:

a. Calculate $h(2)$ and compare this value with $h(6)$.

b. Calculate $h(5)$ and compare this value with $h(15)$.

c. What happens to the value of $h(x)$ if x triples in value?

d. What happens to the value of $h(x)$ if x is divided by 3?

22. a. Let $f(x) = 2x^2$. Describe the difference between $f(3x)$ and $3f(x)$.

b. Let $f(x) = 5x^3$. Describe the difference between $f(4x)$ and $4f(x)$.

c. In general, given $f(x) = kx^p$, describe the difference between $f(nx)$ and $nf(x)$, where p and n are positive integers.

7.3 *Visualizing Positive Integer Powers*

What do the graphs of power functions look like when the input values are infinitely large or assume negative values? Let's examine two basic power functions, $f(x) = x^2$ and $g(x) = x^3$.

The Graphs of $f(x) = x^2$ and $g(x) = x^3$

What happens when $x \geq 0$?

For $f(x) = x^2$ and $g(x) = x^3$, as $x \to +\infty$, both x^2 and $x^3 \to +\infty$. If $x = 0$, then x^2 and x^3 are both equal to 0, and if $x = 1$, both x^2 and x^3 equal 1. So $f(x)$ and $g(x)$ intersect at $(0, 0)$ and $(1, 1)$.

We can see from the graph in Figure 7.10 that when $0 < x < 1$, then $x^3 < x^2$, but when $x > 1$, then $x^3 > x^2$. So the graph of $g(x) = x^3$ eventually dominates the graph of $f(x) = x^2$.

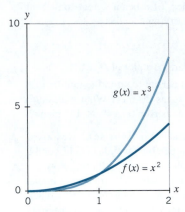

Figure 7.10 Comparing the graphs of $f(x) = x^2$ and $g(x) = x^3$ when $x \geq 0$.

What happens when x < 0?

When x is negative, the functions $f(x) = x^2$ and $g(x) = x^3$ exhibit quite different behaviors (see Table 7.2 and Figure 7.11). When x is negative, x^2 is positive but x^3 is negative. As $x \to -\infty$, $x^2 \to +\infty$ but $x^3 \to -\infty$. So the domain for both functions is $(-\infty, +\infty)$, but the range for $f(x) = x^2$ is $[0, +\infty)$. The range for $g(x) = x^3$ is $(-\infty, +\infty)$.

x	$f(x) = x^2$	$g(x) = x^3$
-4	16	-64
-3	9	-27
-2	4	-8
-1	1	-1
0	0	0
1	1	1
2	4	8
3	9	27
4	16	64

Table 7.2

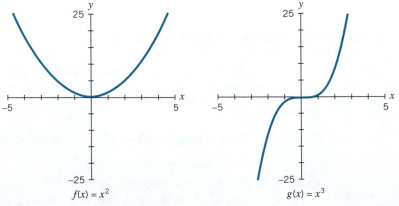

Figure 7.11 The graphs of $f(x) = x^2$ and $g(x) = x^3$.

Exploration 7.2 can help
reinforce your understanding of power functions.

Odd vs. Even Powers

As we saw with the graphs of $f(x) = x^2$ and $g(x) = x^3$, the exponent of the power function affects the shape of the graph. The graphs of power functions with positive even powers are $\cup$-shaped and the graphs of positive odd powers greater than 1 are $\int$-shaped. Why does it matter if the exponent is even or odd?

If n is a positive integer and if x is positive, then x^n is positive. But if x is negative, we have to consider whether n is even or odd. If n is even, then x^n is positive, since

$$(\text{negative number})^{\text{even power}} = \text{positive number}$$

If n is odd, then x^n is negative, since

$$(\text{negative number})^{\text{odd power}} = \text{negative number}$$

So whether the exponent of a power function is odd or even will affect the shape of the graph.

If we graph the simplest power functions, $y = x$, $y = x^2$, $y = x^3$, $y = x^4$, $y = x^5$, $y = x^6$, and so on, we see quickly that the graphs fall into two groups: the odd powers and the even powers (Figure 7.12).

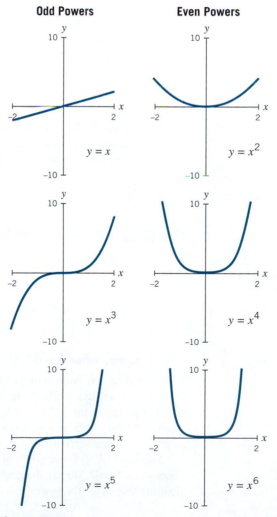

Figure 7.12 Graphs of odd and even power functions.

All the power functions graphed in Figure 7.12 have a positive coefficient k ($=1$). So the graphs of the even powers in Figure 7.12 are all concave up. Graphs of the odd powers (except for the power 1) change concavity at $x = 0$. When $x < 0$, the graphs are concave down. When $x > 0$, the graphs are concave up.

Symmetry of the graphs

For the odd positive powers, the graphs are rotationally symmetric about the origin; that is, if you hold the graph fixed at the origin and then rotate it 180°, you end up with the same graph. As can be seen in Figure 7.13(*a*), the net result is the same as a double reflection, first across the *y*-axis and then across the *x*-axis.

The graphs of even powers are symmetric across the *y*-axis. You can think of the *y*-axis as a dividing line: the "left" side of the graph is a mirror image, or reflection, of the "right" side, as shown in Figure 7.13(*b*).

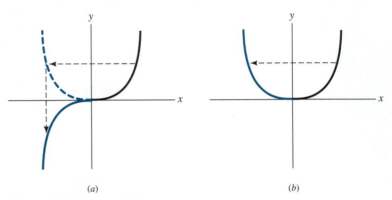

(*a*)   (*b*)

Figure 7.13 (*a*) Power functions with positive odd powers are symmetric about the origin. (*b*) Power functions with positive even powers are symmetric across the *y*-axis.

The program "P1: *k* & *p* Sliders" in *Power Functions* can help you visualize the graphs of $y = kx^p$ for different values of *k* and *p*.

The Effect of the Coefficient *k*

What is the effect of different values for the coefficient *k* on graphs of power functions in the form

$$y = kx^p \quad \text{where } p \text{ is a positive integer and } k \neq 0?$$

What happens when *k* > 0?

We know from our work with power functions of degree 1 (that is, linear functions of the form $y = kx$) that *k* affects the steepness of the line. For all power functions, where $k > 1$, the larger the value for *k*, the more vertical the graph of $y = kx^p$ becomes compared with $y = x^p$. As *k* increases, the steepness of the graph increases. We say the graph is *stretched vertically*.

When $0 < k < 1$, the graph of $y = kx^p$ is flatter than the graph of $y = x^p$ and lies closer to the *x*-axis. We say the graph is *compressed vertically*. The graphs in Figure 7.14 illustrate this effect for power functions of degrees 3 and 4.

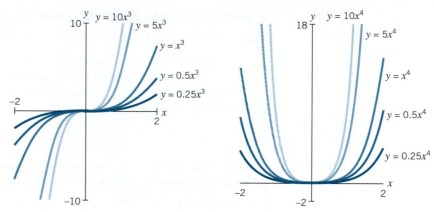

Figure 7.14 When $k > 0$, the larger the value of k, the more closely the power function $y = kx^p$ "hugs" the y-axis for both odd and even powers.

What happens when k < 0?

We know from our work with linear functions that the graphs of $y = kx$ and $y = -kx$ are mirror images across the x-axis. Similarly, the graphs of $y = kx^p$ and $y = -kx^p$ are mirror images of each other across the x-axis. For example, $y = -7x^3$ is the mirror image of $y = 7x^3$. Figure 7.15 shows various pairs of power functions of the type $y = kx^p$ and $y = -kx^p$.

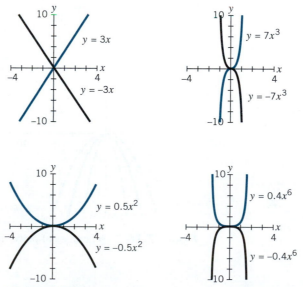

Figure 7.15 In each case, the graphs of $y = kx^p$ and $y = -kx^p$ (shown in blue and black, respectively) are mirror images of each other across the x-axis.

In sum: the value of k can stretch, compress, or reflect the graph

The graph of a power function in the form $y = kx^p$ is the graph of $y = x^p$

- vertically stretched by a factor of k, if $k > 1$
- vertically compressed by a factor of k, if $0 < k < 1$
- vertically stretched or compressed by a factor of $|k|$ and reflected across the x-axis, if $k < 0$

Graphs of Power Functions with Positive Integer Powers

For functions of the form $y = kx^p$ (where $p > 0$ and $k \neq 0$)

The graph goes through the origin.

If the power p is even, the graph is ∪-shaped and symmetric across the y-axis.

If the power p is odd, and > 1, the graph is shaped roughly like ∫ and is symmetric about the origin.

The coefficient k compresses or stretches the graph of $y = x^p$.

The graphs of $y = kx^p$ and $y = -kx^p$ are mirror images of each other across the x-axis.

EXAMPLE 1   Match each set of functions with the appropriate graph in Figure 7.16.

a. $i(x) = -x^2$
$j(x) = -3x^2$
$k(x) = -0.5x^2$

b. $l(x) = x^3$
$m(x) = 3x^3$
$n(x) = 0.5x^3$

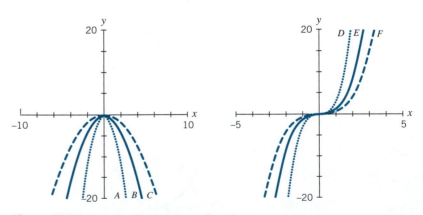

Figure 7.16 Two families of power functions.

SOLUTION

a. $A: j(x) = -3x^2$
$B: i(x) = -x^2$
$C: k(x) = -0.5x^2$

b. $D: m(x) = 3x^3$
$E: l(x) = x^3$
$F: n(x) = 0.5x^3$

Algebra Aerobics 7.3

A graphing program is useful for Problems 5, 6, 9, and 10.

1. If $f(x) = 4x^3$, evaluate the following.

   a. $f(2)$ **b.** $f(-2)$ **c.** $f(s)$ **d.** $f(3s)$

2. If $g(t) = -4t^3$, evaluate the following.

   a. $g(2)$ **b.** $g(-2)$ **c.** $g\left(\frac{1}{2}t\right)$ **d.** $g(5t)$

3. If $f(x) = 3x^2$, evaluate:

   a. $f(4)$ **c.** $f(s)$ **e.** $f(2s)$ **g.** $f\left(\frac{s}{2}\right)$

   b. $f(-4)$ **d.** $2f(s)$ **f.** $f(3s)$ **h.** $f\left(\frac{s}{4}\right)$

4. **a.** For each of the following functions, generate a small table of values, including positive and negative values for x, and sketch the functions on the same grid:

   $$f(x) = 4x^2 \quad \text{and} \quad g(x) = 4x^3$$

   b. What happens to $f(x)$ and to $g(x)$ as $x \to +\infty$?

   c. What happens to $f(x)$ and to $g(x)$ as $x \to -\infty$?

   d. Specify the domain and range of each function.

   e. Where do the graphs of these functions intersect?

   f. For what values of x is $g(x) > f(x)$?

5. Graph both functions $f(x) = 2x^2$ and $g(x) = 2x^3$ over the interval $[-2, 2]$ and then determine values for x where:

   a. $f(x) = g(x)$ **b.** $f(x) > g(x)$ **c.** $f(x) < g(x)$

6. Plot each pair of functions on the same grid.

   a. $y_1 = -2x^2$ and $y_2 = -2x^3$

   b. $y_1 = -0.1x^3$ and $y_2 = -0.1x^4$

7. Without graphing the functions, describe the differences in the graphs of each pair of functions.

   a. $y_1 = 0.2x^3$ and $y_2 = 0.3x^3$

   b. $y_1 = -0.15x^4$ and $y_2 = -0.2x^4$

8. Determine whether p in each power function of the form $y = ax^p$ in Figures 7.17 and 7.18 is even or odd and whether $a > 0$ or $a < 0$. Then find the value of a.

   a.
   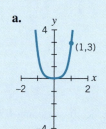

   b.

   Figure 7.17   **Figure 7.18**

9. Draw a quick sketch by hand of the power functions.

   a. $y = x^9$ **b.** $y = x^{10}$

   If possible, check your graphs using technology.

10. Draw a rough sketch of:

    a. $y = x$, $y = 4x$, and $y = -4x$ (all on the same grid)

    b. $y = x^4$, $y = 0.5x^4$, and $y = -0.5x^4$ (all on the same grid)

    If possible, check your graphs using technology.

Exercises for Section 7.3

A graphing program is useful for many of the exercises in this section.

1. Sketch by hand and compare the graphs of the following:

   $$y_1 = x^2 \qquad y_2 = -x^2 \qquad y_3 = 2x^2$$

   $$y_4 = -2x^2 \qquad y_5 = \tfrac{1}{2}x^2 \qquad y_6 = -\tfrac{1}{2}x^2$$

2. (Graphing program option.) For each part sketch by hand the three graphs on the same grid and clearly label each function. Describe how the three graphs are alike and not alike.

   a. $y = x^1$ $y = x^3$ $y = x^5$

   b. $y = x^2$ $y = x^4$ $y = x^6$

   c. $y = x^3$ $y = 2x^3$ $y = -2x^3$

   d. $y = x^2$ $y = 4x^2$ $y = -4x^2$

   If possible, check your results using a graphing program.

3. Match the function with its graph. Explain why you have chosen each graph.

   a. $f(x) = x^2$ **c.** $h(x) = 2x^2$

   b. $g(x) = x^6$ **d.** $j(x) = -2x^2$

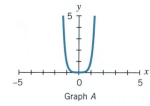

Graph A

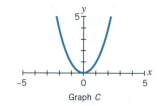

Graph C

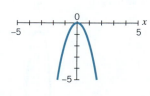

Graph B

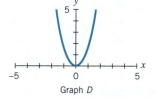

Graph D

4. Match the function with the appropriate graph. Explain why you have chosen each graph.

    a. $f(x) = x^3$ **b.** $g(x) = \frac{1}{3}x^3$ **c.** $h(x) = 3x^3$

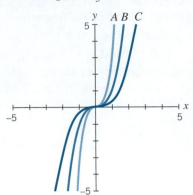

5. Begin with the function $f(x) = x^4$.

    a. Create a new function $g(x)$ by vertically stretching $f(x)$ by a factor of 6.

    b. Create a new function $h(x)$ by vertically compressing $f(x)$ by a factor of $\frac{1}{2}$.

    c. Create a new function $j(x)$ by first vertically stretching $f(x)$ by a factor of 2 and then reflecting it across the x-axis.

6. (Graphing program optional.) Plot the functions $f(x) = x^2$, $g(x) = 3x^2$, and $h(x) = \frac{1}{4}x^2$ on the same grid. Insert the symbol $>$ or $<$ to make the relation true.

    a. For $x > 0$, $g(x)$ ___ $f(x)$ ___ $h(x)$

    b. For $x < 0$, $g(x)$ ___ $f(x)$ ___ $h(x)$

7. (Graphing program recommended.) Plot the functions $f(x) = x^2$ and $g(x) = x^5$ on the same grid. Insert the symbol $>$, $<$, or $=$ to make the relation true.

    a. For $x = 0$, $f(x)$ ___ $g(x)$

    b. For $0 < x < 1$, $f(x)$ ___ $g(x)$

    c. For $x = 1$, $f(x)$ ___ $g(x)$

    d. For $x > 1$, $f(x)$ ___ $g(x)$

    e. For $x < 0$, $f(x)$ ___ $g(x)$

8. Consider the accompanying graph of $f(x) = kx^n$, where n is a positive integer.

    a. Is n even or odd?

    b. Is $k > 0$ or is $k < 0$?

    c. Does $f(-2) = -f(2)$?

    d. Does $f(-x) = -f(x)$?

    e. As $x \to +\infty$, $f(x) \to$ ___

    f. As $x \to -\infty$, $f(x) \to$ ___

9. Consider the accompanying graph of $f(x) = kx^n$, where n is a positive integer.

    a. Is n even or odd?

    b. Is $k > 0$ or is $k < 0$?

    c. Does $f(-2) = f(2)$?

    d. Is $f(-x) = f(x)$?

    e. As $x \to +\infty$, $f(x) \to$ ___

    f. As $x \to -\infty$, $f(x) \to$ ___

10. Consider the accompanying graphs of four functions of the form $y = kx^n$. Which (if any) individual functions:

    a. Are symmetric across the x-axis?

    b. Are symmetric across the y-axis?

    c. Are symmetric about the origin?

    d. Have a positive even power for n?

    e. Have a positive odd power for n?

    f. Have $k < 0$?

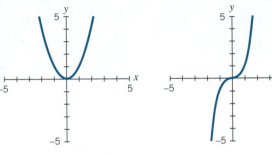

Graph A   Graph C

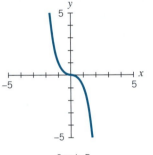

Graph B   Graph D

11. Sketch by hand the graph of each function:

$$f(x) = x^3, \quad g(x) = -x^3, \quad h(x) = \tfrac{1}{2}x^3, \quad j(x) = -2x^3$$

    a. Identify the k value, the constant of proportionality, for each function.

    b. Which graph is a reflection of $f(x)$ across the x-axis?

    c. Which graph is both a stretch and a reflection of $f(x)$ across the x-axis?

    d. Which graph is a compression of $f(x)$?

12. a. Complete the partial graph in the accompanying figure in three different ways by:

    i. Reflecting the graph across the y-axis to create Graph A.

    ii. Reflecting the graph across the x-axis to create Graph B.

    iii. Reflecting the graph first across the y-axis and then across the x-axis to create Graph C.

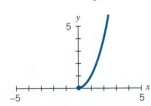

    b. Which of the finished graphs is symmetric across the y-axis?

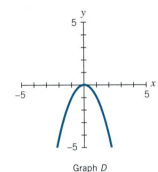

c. Which of the finished graphs is symmetric across the *x*-axis?

d. Which of the finished graphs is symmetric about the origin?

13. (Graphing program optional.) Given $f(x) = 4x^2$, construct a function that is a reflection of $f(x)$ across the horizontal axis. Graph the functions and confirm your answer.

14. (Graphing program optional.) Are $f(x) = x^3$ and $g(x) = -x^3$ reflections of each other across the vertical axis? Graph the functions and confirm your answer.

15. (Graphing program optional.) Evaluate each of the following functions at 0, 0.5, and 1. Then, on the same grid, graph each over the interval [0, 1]. Compare the graphs.

a. $y_1 = x$ $\quad y_2 = x^{1/2}$ $\quad y_3 = x^{1/3}$ $\quad y_4 = x^{1/4}$

b. $y_5 = x^2$ $\quad y_6 = x^3$ $\quad y_7 = x^4$

16. (Graphing program required). Now graph the functions in Exercise 15 over the interval [1, 4] and compare the graphs.

7.4 *Comparing Power and Exponential Functions*

Which Eventually Grows Faster, a Power Function or an Exponential Function?

Although power and exponential functions may appear to be similar in construction, in each function type the independent variable or input assumes a very different role. For power functions the input, *x*, is the *base*, which is raised to a fixed power. Power functions have the form

$$\text{output} = k \cdot (\text{input})^{\text{power}}$$

$$y = kx^p$$

If *k* and *p* are both positive, the power function describes growth.

For exponential functions the input, *x*, is the *exponent* applied to a fixed base. Exponential functions have the form

$$\text{output} = C \cdot a^{\text{input}}$$

$$y = Ca^x$$

If $C > 0$ and $a > 1$, the exponential function describes growth.

Consider the functions $y = x^3$, a power function; $y = 3^x$, an exponential function; and $y = 3x$, a linear function. Table 7.3 compares the role of the independent variable, *x*, in the three functions.

x	Linear Function $y = 3x$ (*x* is multiplied by 3)	Power Function $y = x^3$ (*x* is the *base* raised to the third power)	Exponential Function $y = 3^x$ (*x* is the *exponent* for base 3)
0	$3 \cdot 0 = 0$	$0 \cdot 0 \cdot 0 = 0$	$3^0 = 1$
1	$3 \cdot 1 = 3$	$1 \cdot 1 \cdot 1 = 1$	$3^1 = 3$
2	$3 \cdot 2 = 6$	$2 \cdot 2 \cdot 2 = 8$	$3^2 = 9$
3	$3 \cdot 3 = 9$	$3 \cdot 3 \cdot 3 = 27$	$3^3 = 27$
4	$3 \cdot 4 = 12$	$4 \cdot 4 \cdot 4 = 64$	$3^4 = 81$
5	$3 \cdot 5 = 15$	$5 \cdot 5 \cdot 5 = 125$	$3^5 = 243$

Table 7.3

Visualizing the difference

Table 7.3 and Figure 7.19 show that the power function $y = x^3$ and the exponential function $y = 3^x$ both grow very quickly relative to the linear function $y = 3x$.

Yet there is a vast difference between the growth of an exponential function and the growth of a power function. In Figure 7.20 we zoom out on the graphs. Notice that now the scale on the *x*-axis goes from 0 to 10 (rather than just 0 to 4 as in Figure 7.19) and the *y*-axis extends to 8000, which is still not large enough to show the value of 3^x once *x* is slightly greater than 8. The linear function $y = 3x$ is not shown, since at this scale it would appear to lie flat on the *x*-axis.

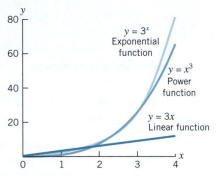

Figure 7.19 A comparison of $y = 3x$, $y = x^3$, and $y = 3^x$.

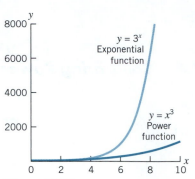

Figure 7.20 "Zooming out" on Figure 7.19.

The exponential function $y = 3^x$ clearly dominates the power function $y = x^3$. The exponential function continues to grow so rapidly that its graph appears almost vertical relative to the graph of the power function.

What if we had picked a larger exponent for the power function? Would the exponential function still overtake the power function? The answer is yes. Let's compare, for instance, the graphs of $y = 3^x$ and $y = x^{10}$. If we could zoom in on the graph in Figure 7.21, we would see that for a while the graph of $y = 3^x$ lies below the graph of $y = x^{10}$. For example, when $x = 2$, then $3^2 < 2^{10}$. But eventually $3^x > x^{10}$. Figure 7.21 shows that somewhere after $x = 30$, the values for 3^x become substantially larger than the values for x^{10}.

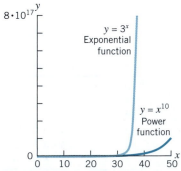

Figure 7.21 Graph of $y = x^{10}$ and $y = 3^x$.

Any exponential growth function will eventually dominate any power function.

EXAMPLE 1   **Comparing power and exponential functions**

a. Construct a table using values of $x \geq 0$ for each of the following functions:

$$y = x^2 \quad \text{and} \quad y = 2^x$$

Then plot the functions on the same grid.

b. If $x > 0$, for what value(s) of x is $x^2 = 2^x$?

c. Does one function eventually dominate? If so, which one and after what value of x?

SOLUTION   **a.** Table 7.4 and Figure 7.22 compare values for $y = x^2$ and $y = 2^x$.

x	$y = x^2$	$y = 2^x$
0	0	1
1	1	2
2	4	4
3	9	8
4	16	16
5	25	32
6	36	64

Table 7.4

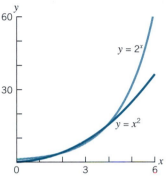

Figure 7.22 Comparison of $y = x^2$ and $y = 2^x$.

b. For positive values of x, if $x = 2$ or $x = 4$, then $x^2 = 2^x$.

c. In Table 7.4 and Figure 7.22, we can see that if $x > 4$, then $2^x > x^2$, so the function $y = 2^x$ will dominate the function $y = x^2$ for $x > 4$.

Algebra Aerobics 7.4

1. a. For each of the following functions construct a table using integer values of x between 0 and 5. Then sketch the functions on the same grid.

$$y = 4^x \quad \text{and} \quad y = x^3$$

b. Does one function eventually dominate? If so, which one and after approximately what value of x?

2. By inspection, determine if each of the graphs in Figure 7.23 is more likely to be the graph of a power function or of an exponential function.

3. Which function eventually dominates?

a. $y = x^{10}$ or $y = 2^x$

b. $y = (1.000\ 005)^x$ or $y = x^{1,000,000}$

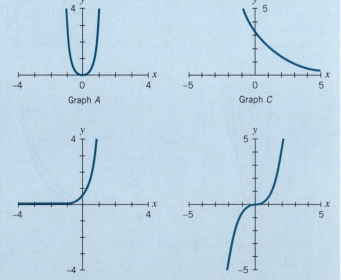

Graph A

Graph C

Graph B

Graph D

Figure 7.23 Graphs of power and exponential functions.

Exercises for Section 7.4

A graphing program is useful for several of the exercises.

1. (Graphing program recommended.) Graph $y = x^4$ and $y = 4^x$ on the same grid.

   a. For positive values of x, where do your graphs intersect? Do they intersect more than once?

   b. For positive values of x, describe what happens to the right and left of any intersection points. You may need to change the scales on the axes or change the windows on a graphing calculator in order to see what is happening.

   c. Which eventually dominates, $y = x^4$ or $y = 4^x$?

2. Examine the following two different versions of the graphs of both $f(x) = 3x^3$ and $g(x) = 3^x$. Notice that each version uses different scales on both axes. Make sure to check each version when answering the following questions.

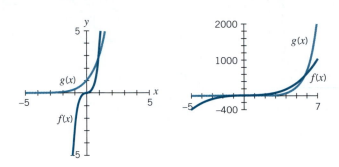

   a. When $x < 0$, how many times do $f(x)$ and $g(x)$ intersect?

   b. When $x > 0$, how many times do $f(x)$ and $g(x)$ intersect?

   c. Estimate the points of intersection.

   d. As $x \to +\infty$, does $f(x) = 3x^3$ or $g(x) = 3^x$ dominate?

   e. Which function, $10 \cdot f(x)$ or $g(x)$, will dominate as $x \to +\infty$?

3. The following figures show two different versions of the graphs of $f(x) = 2x^2$ and $g(x) = 2^x$. Notice that each version uses different scales on the axes. Make sure to check each version when answering the following questions.

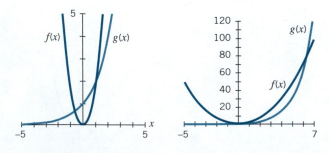

   Supply the appropriate inequality symbol.

   a. As $x \to -\infty$, $f(x)$ ___ $g(x)$.

   b. As $x \to +\infty$, $f(x)$ ___ $g(x)$.

   c. When x lies in the interval $(-\infty, -1)$, then $f(x)$ ___ $g(x)$.

   d. When x lies in the interval $(-0.5, 1)$ then $f(x)$ ___ $g(x)$.

   e. When x lies in the interval $(1, 6)$, then $f(x)$ ___ $g(x)$.

   f. When x lies in the interval $(7, +\infty)$, then $f(x)$ ___ $g(x)$.

4. a. Which eventually dominates, $y = (1.001)^x$ or $y = x^{1000}$?

   b. As the independent variable approaches $+\infty$, which function eventually approaches zero faster, an exponential decay function or a power function with negative integer exponent?

5. (Graphing program optional.) Use a table or graphing utility to determine where $2^x > x^2$ and where $3^x > x^3$ for nonnegative values of x.

6. Match the following data tables with the appropriate function.

   a. $f(x) = 2/x$ b. $f(x) = x^2$ c. $f(x) = 2^x$

x	y	x	y	x	y
1	2	1	2	1	1
2	1	2	4	2	4
3	2/3	3	8	3	9
4	1/2	4	16	4	16
Table 1		**Table 2**		**Table 3**	

7. (Graphing program recommended.) If x is positive, for what values of x is $3 \cdot 2^x < 3 \cdot x^2$? For what values of x is $3 \cdot 2^x > 3 \cdot x^2$?

8. For $x > 0$, match each function with its graph.

   a. $y = x^2$ c. $y = 4x^3$

   b. $y = 2^x$ d. $y = 4(3^x)$

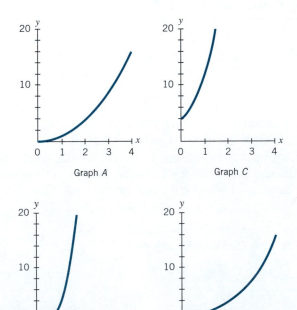

9. Match the function with its graph.

   a. $h(x) = 2x^2$ **b.** $i(x) = 2(3)^x$ **c.** $j(x) = 2x^3$

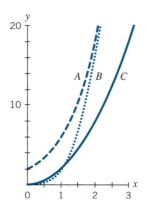

10. Given the functions $f(x) = x^4$ and $g(x) = (4)^x$:

    a. Find $f(3x)$ and $3f(x)$. Summarize the difference between these functions and $f(x)$.

    b. Find $g(3x)$ and $3g(x)$. Summarize the difference between these functions and $g(x)$.

11. Consider a power function $f(x) = x^n$ where n is a positive integer, and an exponential function $g(x) = b^x$ where $b > 1$. Describe how $f(ax)$ and $g(ax)$ differ from $f(x)$ and $g(x)$ (see Exercise 10).

12. Given the points $(1, 2)$ and $(6, 72)$:

    a. Find a linear function of the form $y = ax + b$ that goes through the two points.

    b. Find an exponential function of the form $y = ab^x$ that goes through the two points.

    c. Find a power function of the form $y = ax^p$ that goes through the two points.

13. Find power, exponential, and linear equations that go through the two points $(1, 0.5)$ and $(4, 32)$.

14. Think of the graphs of $f(x) = ax^3$ and $g(x) = a(3)^x$, where $a > 0$, and then decide whether each of the following statements is true or false.

    a. $f(x)$ and $g(x)$ have the same vertical intercept.

    b. $f(x)$ intersects $g(x)$ only once.

    c. As $x \to +\infty$, $g(x) > f(x)$.

    d. As $x \to -\infty$, both $f(x)$ and $g(x)$ approach 0.

15. If you know that one of the graphs in the accompanying figure is a power function of the form $y = ax^p$ and the other is an exponential function of the form $y = Cb^x$, then determine if each of the following is true or false.

    a. $C \geq 1$ **b.** p is an even integer.   **c.** $0 < b < 1$

    d. The graphs of the functions will only intersect twice. (*Hint:* Think of the long-term behavior of these functions.)

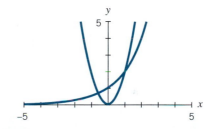

7.5 *Inverse Proportionality: Power Functions with Negative Integer Powers*

Recall that the general form of an equation for a power function is

$$\text{output} = \text{constant} \cdot (\text{input})^{\text{power}}$$

In Sections 7.1 through 7.4 we focused on functions where the power was a positive integer. We now consider power functions where the power is a negative integer.

Using the rules for negative exponents, we can rewrite power functions in the form

$$y = kx^{\text{negative power}}$$

where k is a constant, as

$$y = \frac{k}{x^{\text{positive power}}}$$

For example, $y = 3x^{-2}$ can be rewritten $y = \frac{3}{x^2}$. In this form (with a positive power) it is easier to make calculations and to see what happens to y as x increases or decreases in value.

In Section 7.1 we constructed a function $R(x) = \frac{6}{x} = 6x^{-1}$ from the ratio of surface area to volume of a cube with edge length x. We described $R(x) = 6x^{-1}$ as a power function of degree -1. Repeating the table and graph from Section 7.1 here as Table 7.5 and Figure 7.24 reminds us that as x increases, $R(x)$ decreases. So the shape of this power function with a negative exponent is quite different from the shape of those with a positive exponent.

Edge Length	$\dfrac{\text{Surface Area}}{\text{Volume}}$
x	$R(x) = \frac{6}{x}$
1	6.00
2	3.00
3	2.00
4	1.50
6	1.00
8	0.75
10	0.60

Table 7.5

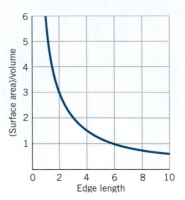

Figure 7.24 The graph of $R(x) = \frac{6}{x}$ shows that as the edge length, x, increases, the ratio of (surface area)/volume decreases.

Inverse Proportionality

The headline for a March 8, 2004, article in *The Economic Times* read "Grace is inversely proportional to the crisis an individual faces." The author was arguing that the greater the crisis, the less gracious an individual is likely to be. Mathematics has a formal definition for the same concept.

For power functions in the form

$$y = \frac{\text{constant}}{x^p}$$

where p is positive, we say that y is *inversely proportional to* x^p. For example, if $y = \frac{8}{x^3}$ we say that y is *inversely proportional to* x^3. If y is inversely proportional to x^p, then as x increases, y decreases.

Direct and Inverse Proportionality

Let p be a positive number and $k \neq 0$.

  If $y = kx^p$, then y is *directly proportional to* x^p.

  If $y = \dfrac{k}{x^p} = kx^{-p}$, then y is *inversely proportional to* x^p.

In both cases k is called the *constant of proportionality*.

EXAMPLE 1   **Examples of direct and inverse proportionality**

Write formulas to represent the following relationships.

a. Boyle's Law says that the volume, V, of a fixed quantity of gas is inversely proportional to the pressure, P, applied to it.

b. The force, F, keeping an electron in orbit is inversely proportional to the square of the distance, d, between the electron and the nucleus.

c. The acceleration, a, of an object is directly proportional to the force, F, applied upon the object and inversely proportional to the object's mass, m.

SOLUTION
a. $V = \frac{k}{P}$ for some constant k c. $a = \frac{kF}{m}$ for some constant k

b. $F = \frac{k}{d^2}$ for some constant k

EXAMPLE 2
In each formula, identify which variables are directly or inversely proportional to each other, and specify the constant of proportionality.

a. $a = \frac{v}{t}$, where a = acceleration, v = velocity, and t = time.

b. $F = \frac{GM_1M_2}{d^2}$, Newton's Law of Universal Gravitation, where F is the force of gravity, G is a gravitational constant, M_1 and M_2 are the masses of two bodies, and d is the distance between the two bodies.

SOLUTION
a. Acceleration, a, is directly proportional to velocity, v, and inversely proportional to time, t. The constant of proportionality is 1.

b. The force of gravity, F, is directly proportional to the masses of the two bodies, M_1 and M_2, and inversely proportional to d^2, the square of the distance between them. The constant of proportionality is G, a gravitational constant.

EXAMPLE 3
You've just found a great house to rent for $2,400 a month. You would need to share the rent with several friends.

a. Construct an equation for the function $R(n)$ that shows the rental cost per person for n people. What kind of function is this?

b. Graph the function for a reasonable domain.

c. Evaluate $R(4)$ and $R(6)$, and describe what they mean in this context.

SOLUTION
a. $R(n) = 2400/n$ is a power function where $R(n)$ is inversely proportional to n.

b.

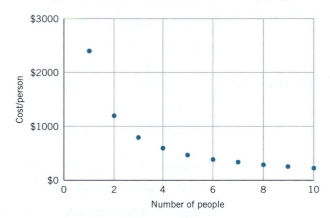

c. $R(4) = 600$, which means that the cost per person would be $600 if there are four renters. $R(6) = 400$, which means that the cost per person would be $400 if there are six renters.

Properties of Inverse Proportionality

If y is *inversely proportional* to x^p, then $y = \frac{k}{x^p}$ (where $p > 0$) for some nonzero constant k.

For an inversely proportional relationship, when we multiply the input by m, the output is multiplied by $\frac{1}{m^p}$ (where $p > 0$). Similar to our argument for direct proportionality, if $f(x) = \frac{k}{x^p}$, then

$$\text{evaluate } f \text{ at } mx \qquad f(mx) = \frac{k}{(mx)^p}$$

$$\text{use rules of exponents} \qquad = \frac{k}{(m^p x^p)}$$

$$\text{factor out } \frac{1}{m^p} \qquad = \left(\frac{1}{m^p}\right) \cdot \left(\frac{k}{x^p}\right)$$

$$\text{substitute } f(x) \text{ for } \left(\frac{k}{x^p}\right) \qquad = \left(\frac{1}{m^p}\right) \cdot f(x)$$

Note that k, the constant of proportionality, does not play a role in these calculations. For example, if $f(x) = \frac{3}{x^2}$, then doubling the input would multiply the output by $\frac{1}{2^2}$ or $\frac{1}{4}$. So if we double the input, the output would be reduced to one-fourth or 25% of the original amount. These doubling calculations do not depend on 3, the constant of proportionality.

If y is inversely proportional to x^p, then $y = \frac{k}{x^p}$, where $p > 0$.

Multiplying the input by m multiplies the output by $\frac{1}{m^p}$.

For example, doubling the input multiplies the output by $\frac{1}{2^p}$.

EXAMPLE 4 For each of the following functions, what happens if the input is doubled? Multiplied by 10? Cut in half?

a. $y = \frac{1}{x^4}$ **b.** $y = 2x^{-3}$

SOLUTION **a.** If the input is doubled, the output is multiplied by $\frac{1}{2^4} = \frac{1}{16}$ or 0.0625. If the input is multiplied by 10, the output is multiplied by $\frac{1}{10^4} = \frac{1}{10,000} = 0.0001$. If the input is cut in half, the output is multiplied by $\frac{1}{(0.5)^4} = \frac{1}{0.0625} = 16$.

b. The function $y = 2x^{-3}$ can be rewritten as $y = \frac{2}{x^3}$. In this form it's easier to see that if the input is doubled, the output is multiplied by $\frac{1}{2^3} = \frac{1}{8} = 0.125$. If the input is multiplied by 10, the output is multiplied by $\frac{1}{10^3} = \frac{1}{1000} = 0.001$. If the input is cut in half, the output is multiplied by $\frac{1}{(0.5)^3} = \frac{1}{0.125} = 8$.

EXAMPLE 5 **The cardinal rule of scuba diving**

The most important rule in scuba diving is "Never, ever, hold your breath." Why?

SOLUTION Think of your lungs as balloons filled with air. As a balloon descends underwater, the surrounding water applies pressure, compressing the balloon. As the balloon ascends, the pressure is lessened and the balloon expands until it attains its original size when it reaches the surface. The volume V of the balloon is inversely proportional to the pressure P; that is, as the pressure increases, the volume decreases. The relationship is given by a special case of Boyle's Law for Gases,

$$V = \frac{1}{P} \quad \text{or equivalently} \quad V = P^{-1}$$

where V is the volume in cubic feet and P the pressure measured in atmospheres (atm). (One atm is 15 lb/in², the atmospheric pressure at Earth's surface.) Each additional 33 feet of water depth increases the pressure by 1 atm. Table 7.6 and Figure 7.25 describe the relationship between pressure and volume.

Pressure vs. Volume

Depth (ft)	Pressure (atm)	Volume (ft³)
0	1	1
33	2	1/2
66	3	1/3
99	4	1/4

Table 7.6

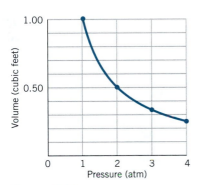

Figure 7.25 Graph of volume vs. pressure for a balloon descending underwater.

We can see in Table 7.6 that if we start with 1 cubic foot of air in the balloon at the surface and descend 33 feet, the surrounding pressure has doubled, from 1 to 2 atm, and the volume of air is cut in half. If the balloon descends to 99 feet, doubling the pressure again from 2 to 4 atm, the volume of air is cut in half again—leaving only one-fourth of the original volume. Why does this matter to divers?

Suppose you are swimming in a pool, take a lung full of air at the surface, and then dive down to the bottom. As you descend, the build-up of pressure will decrease the volume of air in your lungs. When you ascend back to the surface, the volume of air in your lungs will expand back to its original size, and everything is fine.

But when you are scuba diving, you are constantly breathing air that has been pressurized at the surrounding water pressure. If you are scuba diving 33 feet below the surface of the water, the surrounding water pressure is at 2 atm, twice that at the surface. What will happen then if you fill your lungs from your tank, hold your breath, and ascend to the surface? When you reach the surface, the pressure will drop in half, from 2 atm down to 1 atm, so the volume of air in your lungs will double, rupturing your lungs! Hence the first rule of scuba diving is "Never, ever, hold your breath."

EXAMPLE 6 **Designing a soda can**

A designer is asked to redesign a 12-ounce soda can. The volume must remain constant at 22 cubic inches (just enough to hold the 12 ounces and a little air) and the shape must remain cylindrical. What are her options?

SOLUTION If r = radius of the can (in inches) and h = height (in inches), then

$$\text{volume of can} = \text{area of base} \cdot \text{height of can}$$
$$= \pi r^2 h$$

Since the volume of the can must be 22 cubic inches,

$$22 = \pi r^2 h$$

Solving for h gives

$$h = \frac{22}{\pi r^2}$$

So h is inversely proportional to r^2. If we substitute an approximation of 3.14 for π, we get

$$h \approx \frac{22}{3.14 r^2} \approx \frac{7}{r^2}$$

Figure 7.26 shows a graph of the relationship between the height of the can and its radius.

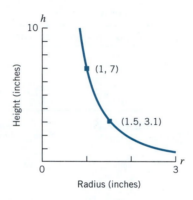

Figure 7.26 The relationship between height, h, and radius, r, of a can holding 22 cubic inches.

The designer can pick any point on the curve to determine the potential dimensions of the can. For example, if $r = 1$, then $h = 7$. So the point $(1, 7)$ on the curve represents a radius of $1''$ (hence a diameter of $2''$) and a height of $7''$. If $r = 1.5$, then $h = \frac{7}{(1.5)^2} = \frac{7}{2.25} \approx 3.1$. The point $(1.5, 3.1)$ on the curve represents a radius of $1.5''$ (diameter of $3''$) and a height of $3.1''$. The two points are labeled on the graph (Figure 7.26), and the corresponding can sizes are drawn in Figure 7.27.

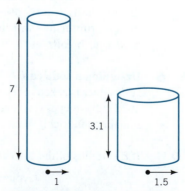

Figure 7.27 Two possible sizes for the soda can.

Inverse Square Laws

When the output is inversely proportional to the square of the input, the functional relationship is called an *inverse square law*. In the preceding example, where we held the volume of the soda can fixed at 22, the resulting function $h = 7/r^2$ is an inverse square law. Inverse square laws are quite common in the sciences.

EXAMPLE 7 **Seeing the light**

The intensity of light is inversely proportional to the square of the distance between the light source and the viewer.

a. Describe light intensity as a function of the distance from the light source.

b. What happens to the intensity of the light if the distance doubles? Triples?

SOLUTION **a.** If d is the distance from the light source, then $I(d)$, the intensity of the light, is inversely proportional to d, so we can write $I(d) = \frac{k}{d^2}$ for some constant k.

b. If the distance doubles, then the light intensity output is multiplied by $\frac{1}{2^2} = \frac{1}{4}$. So the intensity drops to one-fourth of the original intensity.

If the distance triples, then the light intensity is multiplied by $\frac{1}{3^2} = \frac{1}{9}$. So the intensity drops to one-ninth of the original intensity.

For example, if you are reading a book that is 3 feet away from a lamp, and you move the book to 6 feet away (doubling the distance between the book and the light), the light will be one-fourth as intense. If you move the book from 3 to 9 feet away (tripling the distance), the light will be only one-ninth as intense. The reverse is also true; for example, if the book is 6 feet away and the light seems too dim for reading, by cutting the distance in half (to 3 feet), the illumination will be four times as intense.

EXAMPLE 8 **Gravitational force between objects**

The gravitational force between you and Earth is inversely proportional to the square of the distance between you and the center of Earth.

a. Express this relationship as a power function.

b. What happens to the gravitational force as the distance between you and the center of Earth increases by a factor of 10?

c. Why do astronauts appear to be weightless in space?

SOLUTION **a.** The power function

$$F(d) = \frac{k}{d^2} = kd^{-2}$$

describes the gravitational force $F(d)$ between you and Earth in terms of a constant k times d^{-2}, where d is the distance between you and the center of Earth.

b. If the distance between you and Earth's center increases by a factor of 10, then the gravitational force is multiplied by $\frac{1}{10^2} = \frac{1}{100}$; that is, multiplying the distance by 10 decreases the gravitational force to one-hundredth of its original size.

c. Suppose an astronaut starts at the surface of Earth (roughly 4000 miles from Earth's center) and travels to 40,000 miles above Earth's center. She will have increased her distance from Earth's center by a factor or 10. The pull of Earth's gravity there would be one-hundredth that on Earth's surface. So she would appear to be weightless.

Why many inverse square laws work

Inverse square laws in physics often depend on a power source and simple geometry. Imagine a single point as a source of power, emitting perhaps heat, sound, or light. We can think of the power radiating out from the point as passing through an infinite number of concentric spheres. The farther away you are from the point source, the lower the intensity of the power, since it is spread out over the surface area of a sphere that increases in size as you move away from the point source. Therefore, the intensity, I, of the power you receive at any point on the sphere will be a function of the distance you are from the point source. The distance can be thought of as the radius, r, of a sphere with the point source at its center:

$$\text{intensity} = \frac{\text{power from a source}}{\text{surface area of sphere}}$$

$$I = \frac{\text{power}}{4\pi r^2}$$

factor out $\dfrac{1}{r^2}$ $\qquad = \dfrac{\left(\dfrac{\text{power}}{4\pi}\right)}{r^2}$

If the power from the source is constant, we can simplify the expression by substituting a constant $k = \frac{\text{power}}{4\pi}$ and rewrite our equation as

$$I = \frac{k}{r^2}$$

The intensity you receive, I, is inversely proportional to the square of your distance from the source if the power from the point source is constant. If you double the distance, the intensity you receive is one-fourth of the original intensity. If you triple the distance, the intensity you receive is one-ninth of the original intensity. In particular, if $r = 1$, then $I = k$; if r doubles to 2, then $I = \frac{k}{4} = 0.25k$. The graph of $I = \frac{k}{r^2}$ is sketched in Figure 7.28.

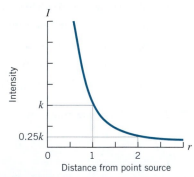

Figure 7.28 Graph of $I = \frac{k}{r^2}$, the relationship between intensity and distance from a point source.

EXAMPLE 9
Direct and inverse proportionality in the same equation: The relative effects of the moon and the sun on tides

The force that creates ocean tides on the surface of Earth varies inversely with the cube of the distance from Earth to any other large body in space and varies directly with the mass of the other body.

a. Construct an equation that describes this relationship.

b. The sun has a mass $2.7 \cdot 10^7$ times larger than the moon's, but the sun is about 390 times farther away from Earth than the moon. Would you expect the sun or the moon to have a greater effect on Earth's tides?

SOLUTION **a.** Using T for the tide-generating force, d for the distance between Earth and another body, and m for the mass of the other body, we have

$$T = \frac{km}{d^3} \quad \text{or} \quad T = kmd^{-3}$$

for some constant of proportionality k.

b. If M_{moon} is the mass of the moon and M_{sun} is the mass of the sun, then

$$M_{sun} = 2.7 \cdot 10^7 \cdot M_{moon}$$

If d_{moon} is the distance from the moon to Earth and d_{sun} is the distance from the sun to Earth, then

$$d_{sun} = 390 d_{moon}$$

If T_{sun} is the tidal force of the sun, then using the formula from part (a), we have

$$T_{sun} = \frac{k \cdot M_{sun}}{(d_{sun})^3}$$

substitute for M_{sun} and d_{sun}

$$= \frac{k \cdot (2.7 \cdot 10^7 \cdot M_{moon})}{(390 d_{moon})^3}$$

regroup terms

$$= \frac{2.7 \cdot 10^7}{390^3} \cdot \frac{k \cdot M_{moon}}{(d_{moon})^3}$$

substitute in T_{moon} and simplify

$$T_{sun} \approx \frac{1}{2} \cdot T_{moon}$$

Thus, despite the fact that the sun is more massive than the moon, because it is much farther away from Earth than the moon, its effect on Earth's tides is about one-half that of the moon.

Algebra Aerobics 7.5

1. Describe each proportionality with a sentence. (Assume k is a constant.)

a. $x = kyz$ **b.** $y = \dfrac{k}{x^2}$ **c.** $D = \dfrac{k\sqrt{y}}{z^3}$

2. Which of the following represent quantities that vary inversely? (Assume k is a constant.)

a. $y = \sqrt{x}$ **c.** $y = kx^{-3}$ **e.** $y = 8 \cdot 2^{-x}$

b. $xy = k$ **d.** $V = \frac{1}{3}\pi r^2 h$

3. You inflate a balloon with 1 ft^3 of compressed air from your scuba tank while diving 99 feet underwater. The pressure at 99 ft^3 is equal to 4 atm. When you have ascended to the surface, by how much will the volume have increased (assuming the balloon doesn't burst)?

4. a. Rewrite each of the following expressions using positive exponents.

i. $15x^{-3}$ **iii.** $3.6x^{-1}$

ii. $-10x^{-4}$ **iv.** $2x^{-2}y^{-3}$

b. Rewrite each of the following expressions using negative exponents.

i. $\dfrac{1.5}{x^2}$ **iii.** $-\dfrac{2}{3x^2}$

ii. $-\dfrac{6}{x^3}$ **iv.** $\dfrac{6}{x^3y^4z}$

5. The time in seconds, t, needed to fill a tank with water is inversely proportional to the square of the diameter, d, of the pipe delivering the water. Write an equation describing this relationship.

6. A light is 4 feet above the book you are reading. The light seems too dim, so you move the light 2 feet closer to the book. What is the change in light intensity?

7. If $g(x) = \frac{3}{x^4}$, what happens to $g(x)$ when:

a. x doubles?   **b.** x is divided by 2?

8. If $h(x) = -\frac{2}{x^3}$, what happens to $h(x)$ if you:

a. Triple x?   **b.** Divide x by 3?

9. Write an equation of variation where:

   a. x is directly proportional to both y and the square of z.

   b. a is directly proportional to both b and the cube of c and inversely proportional to d.

   c. a is directly proportional to both the square of b and the square root of c and inversely proportional to both d and e.

10. Construct an equation of variation and find the constant of proportionality for each of the following situations. (*Hint:* Substitute values for x, y, and z to find the constant of proportionality, k. Then solve for the indicated value.)

   a. x is inversely proportional to the square root of y. When $x = 3$, $y = 4$. Find x when $y = 400$.

   b. x is directly proportional to both y and z. When $x = 3$ and $y = 4$, $z = 0.5$. Find y when $x = 6$ and $z = 2$.

   c. x is directly proportional to the square of y and inversely proportional to the cube root of z. When $x = 6$, $y = 6$ and $z = 8$. Find x when $y = 9$ and $z = 0.027$.

Exercises for Section 7.5

Exercises 21 and 22 require technology that can generate a best-fit power function.

1. For $y = \dfrac{4}{x^3}$ answer the following:

   a. As x changes from 1 to 4, y changes from _____ to _____.

   b. When $x > 0$, as x increases, y _____.

   c. As x changes from 2 to $\frac{1}{2}$, y changes from _____ to _____.

   d. When $x > 0$, as x decreases, y _____.

   e. We say that y is _____ proportional to _____, and _____ is the constant of proportionality.

2. For $f(x) = kx^{-1}$ and $g(x) = kx^{-2}$, find:

   a. $f(2x)$, $f(3x)$, and $f\left(\frac{x}{2}\right)$

   b. $g(2x)$, $g(3x)$, and $g\left(\frac{x}{2}\right)$

3. For $h(x) = kx^{-3}$ and $j(x) = -kx^{-4}$, find:

   a. $h(2x)$, $h(-3x)$, and $h\left(\frac{1}{3}x\right)$

   b. $j(2x)$, $j(-3x)$, and $j\left(\frac{1}{3}x\right)$

4. Find the value for k, the constant of proportionality, if:

   a. $y = \frac{k}{x}$ and $y = 3$ when $x = 2$.

   b. $y = \frac{k}{x^2}$ and $y = \frac{1}{4}$ when $x = 8$.

   c. $y = \frac{k}{x^2}$ and $y = \frac{1}{16}$ when $x = 2$.

   d. $y = \frac{k}{\sqrt{x}}$ and $y = 1$ when $x = 9$.

5. Create an equation that meets the given specifications and then solve for the indicated variable.

   a. If P is inversely proportional to f, and $P = 0.16$ when $f = 0.1$, then what is the value for P when $f = 10$?

   b. If Q is inversely proportional to the square of r, and $Q = 6$ when $r = 3$, then what is the value for Q when $r = 9$?

   c. If S is inversely proportional to w and p, and $S = 8$ when $w = 4$ and $p = \frac{1}{2}$, what is the value for S when $w = 8$ and $p = 1$?

   d. W is inversely proportional to the square root of u and $W = \frac{1}{3}$ when $u = 4$. Find W when $u = 16$.

6. Assume y is inversely proportional to the cube of x.

   a. If x doubles, what happens to y?

   b. If x triples, what happens to y?

   c. If x is halved, what happens to y?

   d. If x is reduced to one-third of its value, what happens to y?

7. a. B is inversely proportional to x^4. What is the effect on B of doubling x?

   b. Z is inversely proportional to x^p, where p is a positive integer. What is the effect on Z of doubling x?

8. The time, t, required to empty a tank is inversely proportional to r, the rate of pumping. If a pump can empty the tank in 30 minutes at a pumping rate of 50 gallons per minute, how long will it take to empty the tank if the pumping rate is doubled?

9. The intensity of light from a point source is inversely proportional to the square of the distance from the light source. If the intensity is 4 watts per square meter at a distance of 6 m from the source, find the intensity at a distance of 8 m from the source. Find the intensity at a distance of 100 m from the source.

10. In Exercise 9, it you wanted to quadruple the intensity, what would need to be done to the distance?

11. A light fixture is mounted flush on a 10-foot-high ceiling over a 3-foot-high counter. How much will the illumination (the light intensity) increase if the light fixture is lowered to 4 feet above the counter?

12. The frequency, F (the number of oscillations per unit of time), of an object of mass m attached to a spring is inversely proportional to the square root of m.

   a. Write an equation describing the relationship.

   b. If a mass of 0.25 kg attached to a spring makes three oscillations per second, find the constant of proportionality.

   c. Find the number of oscillations per second made by a mass of 0.01 kg that is attached to the spring discussed in part (b).

13. Boyle's Law says that if the temperature is held constant, then the volume, V, of a fixed quantity of gas is inversely proportional to the pressure, P. That is, $V = \frac{k}{P}$ for some constant k. What happens to the volume if:

   a. The pressure triples?

   b. The pressure is multiplied by n?

c. The pressure is halved?

d. The pressure is divided by n?

14. The pressure of the atmosphere around us is relatively constant at 15 lb/in² at sea level, or 1 atmosphere of pressure (1 atm). In other words, the column of air above 1 square inch of Earth's surface is exerting 15 pounds of force on that square inch of Earth. Water is considerably more dense. As we saw in Section 7.5, pressure increases at a rate of 1 atm for each additional 33 feet of water. The accompanying table shows a few corresponding values for water depth and pressure.

Water Depth (ft)	Pressure (atm)
0	1
33	2
66	3
99	4

a. What type of relationship does the table describe?

b. Construct an equation that describes pressure, P, as a function of depth, D.

c. In Section 7.5 we looked at a special case of Boyle's Law for the behavior of gases, $P = \frac{1}{V}$ (where V is in cubic feet, P is in atms).

  i. Use Boyle's Law and the equation you found in part (b) to construct an equation for volume, V, as a function of depth, D.

  ii. When $D = 0$ feet, what is V?

  iii. When $D = 66$ feet, what is V?

  iv. If a snorkeler takes a lung full of air at the surface, dives down to 10 feet, and returns to the surface, describe what happens to the volume of air in her lungs.

  v. A large, flexible balloon is filled with a cubic foot of compressed air from a scuba tank at 132 feet below water level, sealed tight, and allowed to ascend to the surface. Use your equation to predict the change in the volume of its air.

15. a. Construct an equation to represent a relationship where x is directly proportional to y and inversely proportional to z.

b. Assume that $x = 4$ when $y = 16$ and $z = 32$. Find k, the constant of proportionality.

c. Using your equation from part (b), find x when $y = 25$ and $z = 5$.

16. a. Construct an equation to represent a relationship where w is directly proportional to both y and z and inversely proportional to the square of x.

b. Assume that $w = 10$ when $y = 12$, $z = 15$, and $x = 6$. Find k, the constant of proportionality.

c. Using your equation from part (b), find x when $w = 2$, $y = 5$, and $z = 6$.

17. Waves on the open ocean travel with a velocity that is directly proportional to the square root of their wavelength,

the distance from one wave crest to the next. [See D. W. Thompson, *On Growth and Form* (New York: Dover Publications, 1992).]

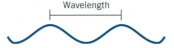
Wavelength

a. If you are in a fixed spot in the ocean, the time interval between successive waves equals $\frac{\text{wave length}}{\text{wave velocity}}$. Show that this time interval is directly proportional to the square root of the wavelength.

b. On one day waves crash on the beach every 3 seconds. On the next day, the waves crash every 6 seconds. On the open ocean, how much farther apart do you expect the wave crests to be on the second day than on the first?

18. When installing Christmas lights on the outside of your house, you read the warning "Do not string more than four sets of lights together." This is because the electrical resistance, R, of wire varies directly with the length of the wire, l, and inversely with the square of the diameter of the wire, d.

a. Construct an equation for electrical wire resistance.

b. If you double the wire diameter, what happens to the resistance?

c. If you increase the length by 25% (say, going from four to five strings of lights), what happens to the resistance?

19. The rate of vibration of a string under constant tension is inversely proportional to the length of the string.

a. Write an equation for the vibration rate of a string, v, as a function of its length, l.

b. If a 48-inch string vibrates 256 times per second, then how long is a string that vibrates 512 times per second?

c. In general, it can be said that if the length of the string increases, the vibration rate will _____.

d. If you want the vibration rate of a string to increase, then you must _____ the length of the string.

e. Playing a stringed instrument, such as a guitar, dulcimer, banjo, or fiddle, requires placing your finger on a fret, effectively shortening the string. Doubling the vibration produces a note pitched one octave higher, and halving the vibration produces a note pitched one octave lower. If the number of vibrations decreased from 440 to 220 vibrations per second, what happened to the length of the string to cause the change in vibration?

20. The weight of a body is inversely proportional to the square of the distance from the body to the center of Earth. Assuming an Earth radius of 4000 miles, a man who weighs 200 pounds on Earth's surface would weigh how much 20 miles above Earth?

21. (Requires technology to generate a best-fit power function.) Oil is forced into a closed tube containing air. The height, H, of the air column is inversely proportional to the pressure, P.

a. Construct a general equation to describe the relationship between H and P.

b. Using technology, construct a best-fit power function for the accompanying data collected on height and pressure.

Height, H (in)	Pressure, P (lb/in²)
13.3	6.7
10.7	8.1
9.2	10.2
7.1	12.9
6.3	14.4
5.1	17.6
4.1	21.4
3.5	25.1
3.0	29.4

22. (Requires technology to generate a best-fit power function.) Recall that Boyle's Law states that for a fixed mass of gas at a constant temperature, the volume, V, of the gas is inversely proportional to the pressure, P, exerted on the gas; that is, $V = \frac{k}{P}$ for some constant k. Use technology to determine a best-fit function model for the accompanying measurements collected on the volume of air as the pressure on it was increased.

Pressure, P (atm)	Volume of Air, V (cm³)
1.0098	20
1.1610	18
1.3776	16
1.6350	14
1.9660	12
2.3828	10
2.9834	8
3.9396	6
5.0428	4
6.2687	2

7.6 Visualizing Negative Integer Power Functions

The Graphs of $f(x) = x^{-1}$ and $g(x) = x^{-2}$

Let's take a close look at two power functions with negative exponents, $f(x) = x^{-1}$ and $g(x) = x^{-2}$. Two important questions are "What happens to these functions when x is close to 0?" and "What happens as x approaches $+\infty$ or $-\infty$?"

What happens when x = 0?

If we rewrite $f(x) = x^{-1}$ as $f(x) = \frac{1}{x}$ and $g(x) = x^{-2}$ as $g(x) = \frac{1}{x^2}$, it is clear that both functions are undefined when $x = 0$. So neither domain includes 0. The graph of each function is split into two pieces: where $x > 0$ and $x < 0$.

What happens when x → +∞ ?

Table 7.7 and Figure 7.29 show that when x is positive and increasing, both $f(x)$ and $g(x)$ are positive and decreasing. As $x \to +\infty$, both $\frac{1}{x}$ and $\frac{1}{x^2}$ grow smaller and smaller, approaching, but never reaching, zero. This means that the range does not include 0. Graphically, as x gets larger and larger, both curves get closer and closer to the x-axis but never touch it. So both graphs are asymptotic to the x-axis.

Values for $f(x)$ and $g(x)$ when $x > 0$

x	$f(x) = \frac{1}{x}$	$g(x) = \frac{1}{x^2}$
0	Undefined	Undefined
1/100	100	10,000
1/4	4	16
1/3	3	9
1/2	2	4
1	1	1
2	1/2	1/4
3	1/3	1/9
4	1/4	1/16
100	1/100	1/10,000

Table 7.7

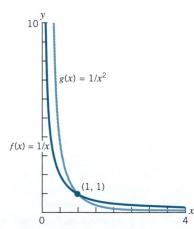

Figure 7.29 Graphs of $f(x) = \frac{1}{x}$ and $g(x) = \frac{1}{x^2}$ when $x > 0$.

What happens when x approaches 0?

We can also see in Table 7.7 and Figure 7.29 that when x is positive and grows smaller and smaller, approaching zero, the values for both $f(x) = \frac{1}{x}$ and $g(x) = \frac{1}{x^2}$ get larger and larger. Let's examine some of the underlying calculations in Table 7.7 for fractional values of x between 0 and 1.

$$f\left(\tfrac{1}{4}\right) = \tfrac{1}{1/4} = 1 \div \tfrac{1}{4} = 1 \cdot \tfrac{4}{1} = 4$$

$$f\left(\tfrac{1}{100}\right) = \tfrac{1}{1/100} = 1 \div \tfrac{1}{100} = 1 \cdot \tfrac{100}{1} = 100$$

$$g\left(\tfrac{1}{4}\right) = \tfrac{1}{(1/4)^2} = 1 \div \tfrac{1}{16} = 1 \cdot \tfrac{16}{1} = 16$$

$$g\left(\tfrac{1}{100}\right) = \tfrac{1}{(1/100)^2} = 1 \div \tfrac{1}{10{,}000} = 1 \cdot \tfrac{10{,}000}{1} = 10{,}000$$

If x is positive and approaches zero, then $f(x) = \frac{1}{x}$ and $g(x) = \frac{1}{x^2}$ approach positive infinity. Graphically, as x gets closer and closer to 0, both curves get closer and closer to the y-axis but never touch it. So both graphs are asymptotic to the y-axis.

What happens when x < 0?

When x is negative, the graphs of $f(x)$ and $g(x)$ have very different shapes. If $x < 0$, then $1/x$ is negative but $1/x^2$ is positive (see Table 7.8 and Figure 7.30). As $x \to 0$ through negative values, $1/x \to -\infty$ and $1/x^2 \to +\infty$. So both $f(x)$ and $g(x)$ are again asymptotic to the y-axis, but in different directions.

Values for $f(x)$ and $g(x)$ when $x < 0$

x	$f(x) = \frac{1}{x}$	$g(x) = \frac{1}{x^2}$
-100	$-1/100$	$1/10{,}000$
-4	$-1/4$	$1/16$
-3	$-1/3$	$1/9$
-2	$-1/2$	$1/4$
-1	-1	1
$-1/2$	-2	4
$-1/3$	-3	9
$-1/4$	-4	16
$-1/100$	-100	$10{,}000$

Table 7.8

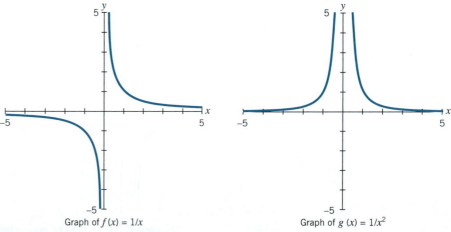

Graph of $f(x) = 1/x$ Graph of $g(x) = 1/x^2$

Figure 7.30 The graphs of $f(x) = \frac{1}{x}$ and $g(x) = \frac{1}{x^2}$.

As $x \to -\infty$, both $1/x$ and $1/x^2 \to 0$. So both functions are again asymptotic to the x-axis, but on different sides of the x-axis.

The graphs of $f(x)$ and $g(x)$, for both positive and negative values of x, are in Figure 7.30. Each graph consists of two pieces, split by the y-axis. When $x > 0$, the graphs are somewhat similar, both positive and concave up. However, when $x < 0$, the graphs lie on opposite sides of the x-axis. One graph is negative and concave down, the other positive and concave up. The graphs of $f(x)$ and $g(x)$ are asymptotic to both the x- and y-axes.

Odd vs. Even Powers

The graphs of power functions with negative powers are all similar in that they are composed of two non-intersecting curves that are each asymptotic to both the x- and y-axes. However, like the positive power functions, the graphs of negative power functions fall into two categories, even and odd powers, that are distinctly different as shown in Figure 7.31.

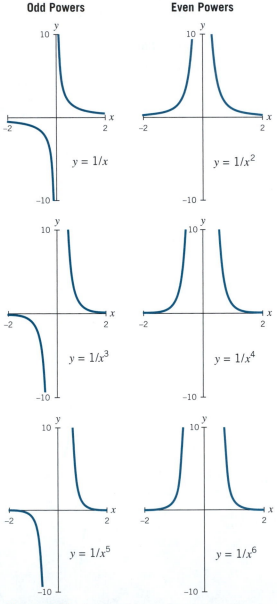

Figure 7.31 Graphs of power functions with even and odd powers.

All of the graphs in Figure 7.31 are split into two pieces and have a positive coefficient $k (= 1)$. So both pieces of the even-power graphs are concave up. However, the two pieces of the odd-power graphs differ in concavity. Here when $x < 0$, the graphs are concave down. When $x > 0$, the graphs are concave up.

Asymptotes

Figure 7.31 confirms that all odd and even negative power functions are asymptotic to both axes. (Note that positive power functions—odd or even—have no asymptotes.) All negative power functions "blow up" at the point 0; that is, the function is not defined at 0, but as x gets closer and closer to 0, the values of $y = k/x^p$ (where $p > 0$) approach either $+\infty$ or $-\infty$. We call 0 a *singularity*. This singularity forces the negative power functions to "explode" at $x = 0$, causing them to be asymptotic to the y-axis.

Similarly, as x approaches either $+\infty$ or $-\infty$ (at either end of the x-axis), the values of y approach, but never reach, 0. This means that the functions are asymptotic to the x-axis.

Symmetry of the graphs

All of the odd negative powers exhibit rotational symmetry about the origin. If you hold the graph fixed at the origin and rotate it 180°, you end up with the same image. The net result is equivalent to a double reflection, first about the y-axis and then about the x-axis.

All of the even negative power functions are symmetric about the y-axis. If you think of the y-axis as the dividing line, the "left" side of the graph is a mirror image, or reflection, of the "right" side.

The Effect of the Coefficient *k*

In Exploration 7.3 you can study more about the effect of *k* on the graphs of negative power functions.

We saw in Section 7.3 that the value of the coefficient k compresses or stretches the graphs of power functions with positive integer powers. We also saw that $y = kx^p$ and $y = -kx^p$ are mirror images of each other across the x-axis. These properties hold true for all power functions.

The graph of power functions in the form $y = \frac{k}{x^p}$ (where $p > 0$ and $k \neq 0$) is the graph of $y = \frac{1}{x^p}$

- vertically stretched by a factor of k, if $k > 1$
- vertically compressed by a factor of k, if $0 < k < 1$
- vertically stretched or compressed by a factor $|k|$, and reflected across the x-axis, if $k < 0$

EXAMPLE 1

a. Describe how the graph of $g(x) = \frac{3}{x^2}$ is related to the graph of $f(x) = \frac{1}{x^2}$. Support your answer by graphing the functions on the same grid using technology.

b. Describe how the graph of $g(x) = -\frac{1}{4x^3}$ is related to the graph of $f(x) = \frac{1}{x^3}$. Support your answer by graphing the functions on the same grid using technology.

SOLUTION

a. $g(x) = \frac{3}{x^2} = 3\left(\frac{1}{x^2}\right)$. Since the coefficient 3 is greater than 1, the graph of $g(x)$ is the graph of $f(x) = \frac{1}{x^2}$ vertically stretched by a factor of 3. See Graph A in Figure 7.32.

b. $g(x) = -\frac{1}{4x^3} = -\frac{1}{4}\left(\frac{1}{x^3}\right)$. Since the coefficient $-\frac{1}{4}$ is negative and $0 < \left|-\frac{1}{4}\right| < 1$, $g(x)$ is the graph of $f(x) = \frac{1}{x^3}$, compressed by a factor of $\frac{1}{4}$ and then reflected across the x-axis. See Graph B in Figure 7.32.

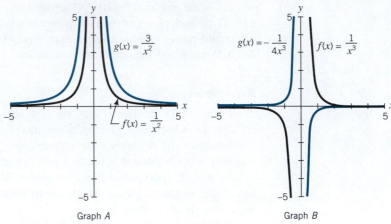

Graph A   Graph B

Figure 7.32 Graph A: The black graph is $f(x) = \frac{1}{x^2}$ and the blue graph is $g(x) = \frac{3}{x^2}$. Graph B: The black graph is $f(x) = \frac{1}{x^3}$ and the blue graph is $g(x) = -\frac{1}{4x^3}$.

Graphs of Power Functions with Negative Integer Powers

The graph of $y = \frac{k}{x^p}$ (where $p > 0$ and $k \neq 0$):

  never goes through the origin, since the function is not defined when $x = 0$,

  is composed of two non-intersecting curves that are each asymptotic to both the x- and y-axes.

If the power p is even, the graph is symmetric across the y-axis.
If the power p is odd, the graph is symmetric about the origin.

The value of the coefficient k compresses or stretches the graph of $y = \frac{1}{x^p}$.

The graphs of $y = \frac{k}{x^p}$ and $y = -\frac{k}{x^p}$ are mirror images of each other across the x-axis.

EXAMPLE 2   For $f(x) = x^{-2}$ and $g(x) = x^{-3}$, construct the following functions and their graphs.

a. $f(-x)$ and $-f(x)$

b. $3f(x)$ and its reflection across the x-axis

c. $g(-x)$ and $-g(x)$

d. $4g(x)$ and its reflection across the x-axis

SOLUTION   **a.** We can rewrite $f(x) = x^{-2}$ as $f(x) = \frac{1}{x^2}$. Then

$$f(-x) = \frac{1}{(-x)^2} = \frac{1}{x^2} \quad \text{and} \quad -f(x) = -\frac{1}{x^2}$$

See Figure 7.33 Graph A.

b. $3f(x) = 3\left(\frac{1}{x^2}\right) = \frac{3}{x^2}$

The function $y = -\frac{3}{x^2}$ is the reflection of $3f(x)$ across the x-axis. See Figure 7.33 Graph B.

c. We can rewrite $g(x) = x^{-3}$ as $g(x) = \frac{1}{x^3}$. Then

$$g(-x) = \frac{1}{(-x)^3} = -\frac{1}{x^3} \quad \text{and} \quad -g(x) = -\frac{1}{x^3}$$

So $g(-x)$ and $-g(x)$ are equivalent functions. See Figure 7.33 Graph C.

d. $4g(x) = 4\left(\frac{1}{x^3}\right) = \frac{4}{x^3}$

The function $y = -\frac{4}{x^3}$ is the reflection of $4g(x)$ across the x-axis. See Figure 7.33 Graph D.

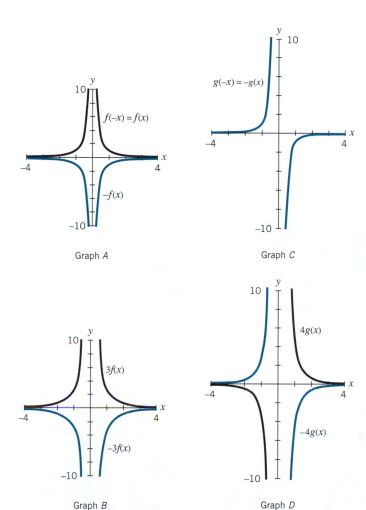

Graph A

Graph C

Graph B

Graph D

Graph of two equivalent functions $g(-x)$ and $-g(x)$.

Figure 7.33 Four graphs of pairs of functions.

EXAMPLE 3   **a.** Graph the power function $f(x) = \frac{2}{x^3}$.

b. Evaluate $f(0.5)$ and $f(-2)$.

c. Use part (b) and the fact that $f(x)$ is a power function to predict the values for $f(10)$ and $f(-0.5)$.

d. Now evaluate $f(10)$ and $f(-0.5)$ to confirm your predictions in part (c).

SOLUTION   **a.** See Figure 7.34 on next page.

b. $f(0.5) = \frac{2}{(0.5)^3} = \frac{2}{(0.125)} = 16$ and $f(-2) = \frac{2}{(-2)^3} = \frac{2}{(-8)} = -0.25$. So the points $(0.5, 16)$ and $(-2, -0.25)$ lie on the graph.

c. Changing the input from 0.5 to 10 is equivalent to multiplying the original input by 20. So the output will be multiplied by $\frac{1}{20^3} = \frac{1}{8000}$. The value of $f(10)$ should be $\left(\frac{1}{8000}\right) \cdot f(0.5) = \left(\frac{1}{8000}\right) \cdot 16 = 0.002$.

Changing the input from 0.5 to -0.5 is equivalent to multiplying the input of 0.5 by -1. The output will be multiplied by $\frac{1}{(-1)^3} = -1$. So $f(-0.5) = -f(0.5) = -16$.

d. Evaluating the functions directly, we have $f(10) = \frac{2}{10^3} = \frac{2}{1000} = 0.002$ and $f(-0.5) = \frac{2}{(-0.5)^3} = \frac{2}{(-0.125)} = -16$, confirming our predictions in part (c).

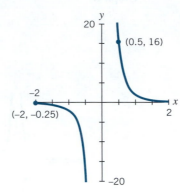

Figure 7.34 The graph of $f(x) = \frac{2}{x^3}$.

Algebra Aerobics 7.6

A graphing program is useful for Problems 3 and 4.

1. Rewrite each of the following expressions with positive exponents and then evaluate each expression for $x = -2$.

   a. x^{-2} **c.** $4x^{-3}$ **e.** $2^{-1}x^{-4}$ **g.** $-2x^{-4}$

   b. x^{-3} **d.** $-4x^{-3}$ **f.** $-2x^{-3}$ **h.** $2x^{-4}$

2. **a.** Select the most likely function type for each graph in Figure 7.35.

   $$y = ax \qquad y = \frac{a}{x} \qquad y = ax^2$$

   $$y = \frac{a}{x^2} \qquad y = ax^3 \qquad y = \frac{a}{x^3}$$

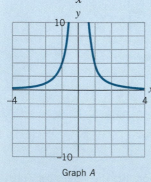
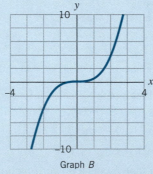

Graph A   Graph B

Figure 7.35 Two unidentified graphs.

   b. After you have decided on the function type, determine the value of a for each graph.

3. Sketch the graph for each equation:

   a. $y = x^{-10}$ **b.** $y = x^{-11}$

4. On the same grid, graph the following equations.

   a. $y = x^{-2}$ and $y = x^{-3}$. Where do these graphs intersect? How are the graphs similar and how are they different?

   b. $y = 4x^{-2}$ and $y = 4x^{-3}$. Where do these graphs intersect? How are the graphs similar and how are they different?

5. Match the appropriate function with each of the graphs in Figure 7.36

   $$f(x) = \frac{5}{x^2} \qquad g(x) = \frac{1}{x^2} \qquad h(x) = \frac{-5}{x^2}$$

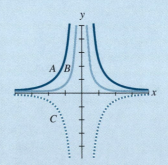

Figure 7.36 Three unidentified graphs.

Exercises for Section 7.6

For many of the exercises in this section the use of a graphing program is required or recommended.

1. A cube of edge length x has a surface area $S(x) = 6x^2$ and a volume $V(x) = x^3$. We constructed the function $R(x) = \frac{S(x)}{V(x)} = \frac{6x^2}{x^3} = \frac{6}{x}$. Consider $R(x)$ as an abstract function. What is the domain? Construct a small table of values, including negative values of x, and plot the graph. Describe what happens to $R(x)$ when x is positive and $x \to 0$. What happens to $R(x)$ when x is negative and $x \to 0$?

2. a. (Graphing program recommended.) Make a table of values and sketch a graph for each of the following functions. Be sure to include negative and positive values for x, as well as values for x that lie close to zero.

$$y_1 = \frac{1}{x} \qquad y_2 = \frac{1}{x^2} \qquad y_3 = \frac{1}{x^3} \qquad y_4 = \frac{1}{x^4}$$

b. Describe the domain and range of each function.

c. Describe the behavior of each function as x approaches positive infinity and as x approaches negative infinity.

d. Describe the behavior of each function when x is near 0.

3. a. (Graphing program recommended.) Generate a table of values and a graph for each of the following functions:

$$g(x) = 5x \qquad h(x) = \frac{x}{5} \qquad t(x) = \frac{1}{x} \qquad f(x) = \frac{5}{x}$$

b. Describe the ways in which the graphs in part (a) are alike and the ways in which they are not alike.

4. (Graphing program optional.) In each part, sketch the three graphs on the same grid and label each function. Describe how the three graphs are similar and how they are different.

a. $y_1 = x^{-1}$ $\qquad y_2 = x^{-3}$ $\qquad y_3 = x^{-5}$

b. $y_1 = x^0$ $\qquad y_2 = x^{-2}$ $\qquad y_3 = x^{-4}$

c. $y_1 = 2x^{-1}$ $\qquad y_2 = 4x^{-1}$ $\qquad y_3 = -2x^{-1}$

5. (Graphing program required.) Using graphing technology, on the same grid graph $y = x^2$ and $y = x^{-2}$.

a. Over what interval does each function increase? Decrease?

b. Where do the graphs intersect?

c. What happens to each function as x approaches positive infinity? Negative infinity?

6. Match each of the following functions with its graph:

a. $y = 3(2^x)$ $\qquad$ **c.** $y = x^{-3}$

b. $y = x^3$ $\qquad$ **d.** $y = x^{-2}$

Graph A

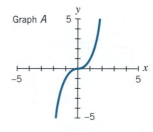

Graph C

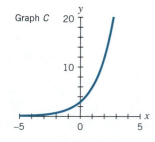

Graph B

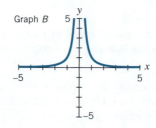

Graph D

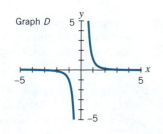

7. Match each function with its graph. Explain your choices.

a. $f(x) = x^{-4}$ $\qquad$ **c.** $h(x) = \dfrac{2}{x^2}$

b. $g(x) = x^{-3}$ $\qquad$ **d.** $j(x) = -\dfrac{2}{x^2}$

Graph A

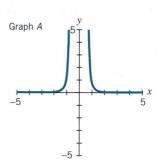

Graph C

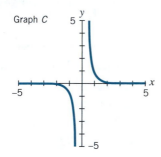

Graph B

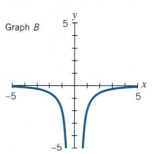

Graph D

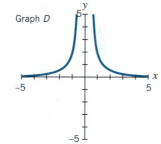

8. Match each function (where $x > 0$) with the appropriate graph. Explain your choices.

a. $f(x) = x^{-3}$ $\qquad$ **b.** $g(x) = \dfrac{1}{x^5}$ $\qquad$ **c.** $h(x) = \dfrac{1}{x}$

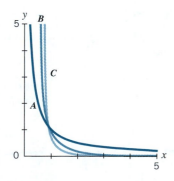

9. Begin with the function $f(x) = x^{-3}$. Then:

a. Create a new function $g(x)$ by vertically stretching $f(x)$ by a factor of 4.

b. Create a new function $h(x)$ by vertically compressing $f(x)$ by a factor of $\frac{1}{2}$.

c. Create a new function $j(x)$ by first vertically stretching $f(x)$ by a factor of 3 and then by reflecting the result across the x-axis.

10. (Graphing program optional.) Graph the functions $f(x) = \frac{1}{x^2}$, $g(x) = \frac{4}{x^2}$, and $h(x) = \frac{1}{4x^2}$.

For parts (a)–(c), insert the symbol $>$ or $<$ to make the relation true.

a. For $x > 0$, $f(x)$ _____ $g(x)$

b. For $x > 0$, $g(x)$ _____ $h(x)$

c. For $x > 0$, $h(x)$ _____ $f(x)$

d. Describe how the graphs are related to each other when $x < 0$. Generalize your findings to any value for the coefficient $k > 0$.

11. (Graphing program required.) Graph the functions $f(x) = \frac{1}{x^5}$ and $g(x) = \frac{1}{x^4}$ and then insert the symbol $>$ or $<$ to make the relation true.

a. For $x > 1$, $f(x)$ _____ $g(x)$

b. For $0 < x < 1$, $f(x)$ _____ $g(x)$

c. For $x < 0$, $f(x)$ _____ $g(x)$

12. Consider the accompanying graph of $f(x) = k \cdot \frac{1}{x^n}$, where n is a positive integer.

a. Is n even or odd?

b. Is $k > 0$ or is $k < 0$?

c. Is $f(-1) > 0$ or is $f(-1) < 0$?

d. Does $f(-x) = -f(x)$?

e. As $x \to +\infty$, $f(x) \to$ ___

f. As $x \to -\infty$, $f(x) \to$ ___

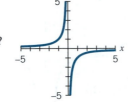

13. Consider the accompanying graph of $f(x) = k \cdot \frac{1}{x^n}$, where n is a positive integer.

a. Is n even or odd?

b. Is $k > 0$ or is $k < 0$?

c. Is $f(-1) > 0$ or is $f(-1) < 0$?

d. Does $f(-x) = f(x)$?

e. As $x \to +\infty$, $f(x) \to$ ___

f. As $x \to -\infty$, $f(x) \to$ ___

14. Consider the following graphs of four functions of the form $y = \frac{k}{x^n}$, where $n > 0$.

Graph A

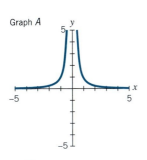

Graph C

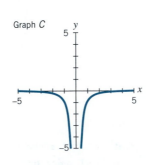

Graph B

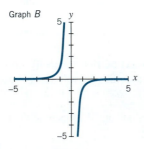

Graph D

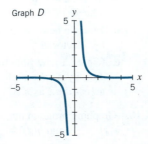

Which functions:

a. Are symmetric across the y-axis?

b. Are symmetric about the origin?

c. Have an even power n?

d. Have an odd power n?

e. Have $k < 0$?

15. (Graphing program optional.) Sketch the graph of the function $f(x) = x^{-3}$. Then consider $f(x)$ in relation to the following functions:

$$g(x) = -x^{-3}, \quad h(x) = \frac{1}{2x^3} \quad \text{and} \quad j(x) = -\frac{2}{x^3}$$

a. Identify k, the constant of proportionality, for each function.

b. Which function has a graph that is a reflection of $f(x)$ across the x-axis?

c. Which function has a graph that is both a stretch and a reflection of $f(x)$ across the x-axis?

d. Which function has a graph that is a compression of $f(x)$?

16. a. Complete the partial graph in the accompanying figure three different ways by adding:

    i. The reflection of the graph across the y-axis to create Graph A.

    ii. The reflection of the graph across the x-axis to create Graph B.

    iii. The result of reflecting the graph first across the y-axis and then across the x-axis to create Graph C.

b. Which of the finished graphs is symmetric across the y-axis?

c. Which of the finished graphs is symmetric about the origin?

d. Which of the finished graphs is symmetric across the x-axis?

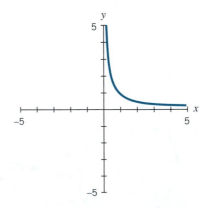

17. (Graphing program required.) Given $f(x) = 4x^{-2}$, construct a function $g(x)$ that is a reflection of $f(x)$ across the horizontal axis. Graph the function and confirm your answer.

18. (Graphing program required.) Are $f(x) = x^{-3}$ and $g(x) = -x^{-3}$ reflections of each other across the vertical axis? Graph the functions and confirm your answer.

19. If $f(x) = \frac{1}{x^4}$ and $g(x) = \frac{1}{x^5}$, construct the following functions.

   a. $f(-x)$ and $-f(x)$ **c.** $g(-x)$ and $-g(x)$

   b. $2f(x)$ and $f(2x)$ **d.** $2g(x)$ and $g(2x)$

   e. The function whose graph is the reflection of $f(x)$ across the x-axis

   f. The function whose graph is the reflection of $g(x)$ across the x-axis

20. (Graphing program required.) Begin with a function $f(x) = \frac{1}{x^3}$. In each part describe how the graph of $g(x)$ is related to the graph of $f(x)$. Support your answer by graphing each function with technology.

   a. $g(x) = \frac{2}{x^3}$ **b.** $g(x) = -\frac{1}{x^3}$ **c.** $g(x) = -\frac{1}{2}x^{-3}$

21. (Graphing program required.) Determine which of the graphs of the following pairs of functions intersect. If the graphs intersect, find the point or points of intersection.

   a. $y = 2x$ $y = 4x^2$ **d.** $y = x^{-1}$ $y = x^{-2}$

   b. $y = 4x^2$ $y = 4x^3$ **e.** $y = 4x^{-2}$ $y = 4x^{-3}$

   c. $y = x^{-2}$ $y = 4x^2$

7.7 Using Logarithmic Scales to Find the Best Functional Model

Looking for Lines

Throughout this course, we have examined several different families of functions: linear, exponential, logarithmic, and power. In this section we address the question "How can we determine a reasonable functional model for a given set of data?" The simple answer is "Look for a way to plot the data as a straight line." We look for straight-line representations not because the world is intrinsically linear, but because straight lines are easy to recognize and manipulate.

In Figure 7.37 we plot the functions (for $x > 0$) of

$$y = 3 + 2x \qquad \text{linear function}$$
$$y = 3 \cdot 2^x \qquad \text{exponential function}$$
$$y = 3x^2 \qquad \text{power function}$$

first with a linear scale on both axes (a standard plot), then with a linear scale on the x-axis and a logarithmic scale on the y-axis (a semi-log plot), and finally with a logarithmic scale on both axes (a *log-log* plot).

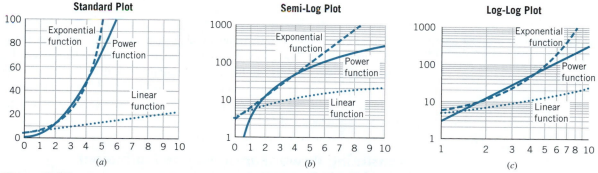

Figure 7.37 (a) *Standard plot.* The graph of a linear function appears as a straight line. (b) *Semi-log plot.* The graph of an exponential function appears as a straight line. (c) *Log-log plot.* The graph of a power function appears as a straight line.

The course software "E11: Semi-Log Plots of $y = Ca^x$" in *Exponential & Log Functions* and "P2: Log-Log Plots of Power Functions" in *Power Functions* can help you visualize the ideas in this section.

On the standard plot in Figure 7.37(a), only the linear function appears as a straight line; the power and exponential functions curve steeply upward. On the semi-log plot in Figure 7.37(b), the exponential function now is a straight line, and the power and linear functions curve downward. As we saw in Chapter 5, exponential growth will always appear as a straight line on a *semi-log* plot, where the independent variable (input) is plotted on a linear scale and the dependent variable (output) on a logarithmic scale. On the log-log plot in Figure 7.37(c), only the power function appears as a straight line.

Why Is a Log-Log Plot of a Power Function a Straight Line?

As we did with exponential functions and semi-log plots in Chapter 6, we can use the properties of logarithms to understand why power functions appear as straight lines on a log-log plot. Let's analyze our power function:

$$y = 3x^2 \qquad (x > 0) \qquad (1)$$

Take the logarithm of both sides $\qquad \log y = \log(3x^2)$

use Rule 1 of logs $\qquad\qquad\qquad = \log 3 + \log x^2$

use Rule 3 of logs $\qquad\qquad\qquad = \log 3 + 2 \log x$

evaluating log 3, we get $\qquad\quad \log y \approx 0.48 + 2 \log x \qquad (2)$

If we let $X = \log x$ and $Y = \log y$, we can rewrite the equation as

$$Y \approx 0.48 + 2X \qquad (3)$$

So Y (or log y) is a linear function of X (or log x). Equations (1), (2), and (3) are all equivalent. The graph of Equation (3) on a log-log plot, with log x values on the horizontal and log y values on the vertical axis, is a straight line (see Figure 7.38). The slope of the line is 2, the power of x in Equation (1). The vertical intercept is 0.48 or log 3, the logarithm of the constant of proportionality in Equation (1).

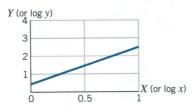

Figure 7.38 The graph of

$$Y \approx 0.48 + 2X.$$

SOMETHING TO THINK ABOUT

On what type of plot would a logarithmic function appear as a straight line?

The slope, 2, described by Equation (3) means that each time X (or log x) increases by one unit, Y (or log y) increases by two units. Remember that a log scale is multiplicative. So, adding 1 to log x is equivalent to multiplying x by 10. Adding 2 to log y is equivalent to multiplying y by $10 \cdot 10$ or 10^2. We can confirm this by examining Equation (1), where we also see that each time x increases by a factor of 10, y increases by a factor of 10^2.

Translating Power Functions into Equivalent Logarithmic Functions

In general, we can translate a power function in the form

$$y = kx^p \qquad \text{where } k \text{ and } x > 0$$

into an equivalent linear function

$$\log y = \log k + p \log x \qquad \text{or}$$
$$Y = \log k + pX \qquad \text{where } X = \log x \text{ and } Y = \log y$$

The graph of the power function will appear as a straight line on a log-log plot with log x on the horizontal and log y on the vertical axis. The slope of the line is the power p of x and the vertical intercept is log k, where k is the constant of proportionality.

Finding the Equation of a Power Function on a Log-Log Plot

The graph of a power function $y = kx^p$, where k and $x > 0$, appears as a straight line on a log-log plot and the equation of the line is

$$Y = \log k + pX \qquad \text{where } X = \log x \text{ and } Y = \log y$$

The slope of the line is p, where p is the power of x for $y = kx^p$.

The vertical intercept is $\log k$, where k is the constant of proportionality for $y = kx^p$.

Analyzing Weight and Height Data

In 1938, Katherine Simmons and T. Wingate Todd measured the average weight and height of children in Ohio between the ages of 3 months and 13 years. Table 7.9 shows their data.

Using a standard plot

Let's examine the relationship between height and weight. Figure 7.39 shows height versus weight for boys and for girls using a standard linear scale on both axes.

DATA See Excel or graph link file CHILDSTA.

Measuring Children

Age	Weight (kg)		Height (cm)	
(yr)	Boys	Girls	Boys	Girls
$\frac{1}{4}$	6.5	5.9	61.3	59.3
1	10.8	9.9	76.1	74.2
2	13.2	12.5	87.4	86.2
3	15.2	14.7	96.2	95.5
4	17.4	16.8	103.9	103.2
5	19.6	19.2	110.9	110.3
6	22.0	22.0	117.2	117.4
7	24.8	24.5	123.9	123.2
8	28.2	27.9	130.1	129.3
9	31.5	32.1	136.0	135.7
10	35.6	35.2	141.4	140.8
11	39.2	39.5	146.5	147.8
12	42.0	46.6	151.1	155.3
13	46.6	52.0	156.7	159.9

Table 7.9
Source: Data adapted from D. W. Thompson, *On Growth and Form* (New York: Dover Paperback, 1992), p. 105.

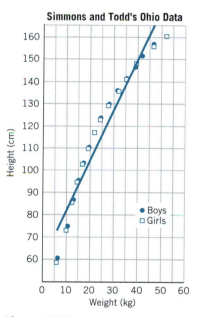

Figure 7.39 Standard linear plot of height vs. weight data for boys and girls.

The filled circles are the data for boys, the open squares the data for girls. Table 7.9 also gives their respective ages, and in general, baby girls are smaller than baby boys and teenage girls are larger than teenage boys. But in the graph in Figure 7.39, we see that at the same weight the heights of boys and girls are roughly the same.

The line in Figure 7.39 shows the linear model that best approximates the combined data. The line does not describe the data very well, and it doesn't seem reasonable that height is a linear function of weight for growing children. Other models may describe the data better.

Using a semi-log plot

SOMETHING TO THINK ABOUT

Why is there no zero on an axis with a logarithmic scale?

Figure 7.40 shows the same data, this time using a linear scale for weight and a logarithmic scale for height. Figure 7.40 also shows a best-fit exponential function, which appears as a straight line on the semi-log plot. This model is not a good fit to the data, and it doesn't seem reasonable to suppose that height is an exponential function of weight.

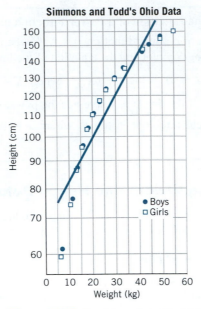

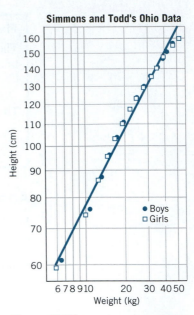

Figure 7.40 Semi-log plot of height vs. weight data for boys and girls.

Figure 7.41 Log-log plot of height vs. weight data.

Using a log-log plot

The same data are plotted again on a log-log plot (a plot with logarithmic scales on both axes) in Figure 7.41. In this case, the line is a reasonably good approximation to the data and certainly the best fit of the three models considered. Since we know that a straight line on a log-log plot represents a power function, we can find the equation for the best-fit straight line on the log-log plot and transform the equation into a power function.

Finding the equation of a power function on a log-log plot

To construct a linear function for the best-fit line on the log-log plot in Figure 7.41, we can regraph the line using $H(= \log h)$ and $W(= \log w)$ scales on the axes, where h = height (cm) and w = weight (kg) (see Figure 7.42).

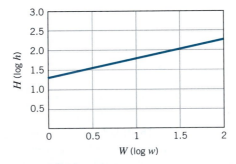

Figure 7.42 Best-fit line for height vs. weight data graphed on a log-log plot.

The equation will then be of the form

$$H = b + mW \tag{1}$$

where $H = \log h$ and $W = \log w$. To find values for the slope m and the vertical intercept b, we can estimate two points on the line, say $(0.5, 1.5)$ and $(1.5, 2.0)$. So

$$\text{slope} = \frac{2.0 - 1.5}{1.5 - 0.5} = 0.5$$

From the graph, the vertical intercept appears to be approximately 1.3. So the linear equation would be

$$H = 1.3 + 0.5W \tag{2}$$

where $H = \log h$ and $W = \log w$. We can transform this equation into an equivalent power function in the form

$$h = kw^p \tag{3}$$

where p is the slope of the best-fit line on the log-log plot and $\log k$ is the vertical intercept (see box on p. 433). The slope of the best-fit line is 0.5, so $p = 0.5$. The vertical intercept is approximately 1.3, so

$$\log k = 1.3$$
$$\text{definition of log} \quad k = 10^{1.3}$$
$$\text{evaluate } 10^{1.3} \quad k \approx 20$$

Substituting $p = 0.5$ and $k = 20$ in Equation (3), we get

$$h = 20w^{0.5} \qquad (\text{or } 20w^{(1/2)}) \tag{4}$$

We now have a power function that models the relationship between height, h, and weight, w.

Interpreting our results

Is it plausible to argue that height is a power law function of weight for growing children? Equation (4) says that

$$h \propto w^{1/2} \qquad \text{or equivalently} \qquad w \propto h^2$$

where the symbol $\propto$ means "is directly proportional to."

Let's think about whether this is a reasonable exponent. Suppose children grew *self-similarly;* that is, they kept the same shape as they grew from the age of 3 months to 13 years. Then, as we have discussed in Section 7.1, their volume, and hence their weight, would be proportional to the cube of their height:

$$w = kh^3 \qquad (\text{self-similar growth})$$

But of course children do not grow self-similarly; they become proportionately more slender as they grow from babies to young adults (see Figure 7.43). Their weight therefore grows less rapidly than would be predicted by self-similar growth, and we expect an exponent less than 3 for weight as a function of height.

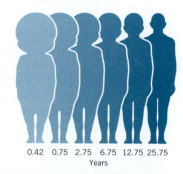

0.42   0.75   2.75   6.75   12.75   25.75
Years

Figure 7.43 The change in human body shape with increasing age.
Source: Medawar, P. B. 1945. "Size and Shape in Age" in *Essays on Growth and Form Presented to D'Arcy Thompson* (W. E. LeGros Clark and P. B. Medawar, eds.), pp. 157–187. Oxford: Clarendon Press.

On the other hand, suppose that children's bodies grew no wider as their height increased. Since weight is proportional to height times cross-sectional area, with constant cross-sectional area we would expect weight to increase linearly with height. So $w \propto h^1$. But certainly, children do become wider as they grow taller, so we expect the exponent of h to be greater than 1. Therefore, the experimentally determined exponent of approximately 2 seems reasonable.

Other considerations

There is a fourth possibility for plotting our height-weight data: We could use a linear scale for the height axis and a logarithmic scale for the weight axis. On such a plot, a straight line will represent a model in which weight is an exponential function of height.

If we generate this plot, the data fall along what is very close to a straight line. Based only on the data, we might argue that an exponential function is a reasonable model. However, it is very difficult to construct a plausible physical explanation of such a model, and it would not help us understand how children grow. It may be, of course, that this relationship reveals some unknown physical or biological law. Perhaps you, the reader, will be the one to find some previously unsuspected explanation.

Allometry: The Effect of Scale

Allometry is the study of how the relationships among different physical attributes of an object change with scale. In biology, allometric studies naturally focus on relationships in living organisms. Biologists look for general "laws" that can describe the relationship, for instance, between height and weight (as in this section) or between surface area and volume (as in Section 7.1) as the organism size increases. The relationships may occur within one species or across species. These laws are characteristically power functions. Sometimes we can predict the exponent by simple reasoning about physical properties and verify our prediction by examining the data; other times we can measure the exponent from data but cannot give a simple explanation for the observed value.

Surface area vs. body mass

In Section 7.1 we made the argument that larger animals have relatively less surface area than smaller ones. Let's see if we can describe this relationship as a power law and then look at some real data to see if our conclusion was reasonable.

Making Predictions.   Since all animals have roughly the same mass per unit volume, their mass should be directly proportional to their volume. We will substitute mass for volume in our discussion since we can determine mass easily by weighing an animal. Since volume is measured in cubic units of some length, if animals of different sizes have roughly the same shape, then we would expect mass, M, to be proportional to the cube of the length, L, of the animal,

$$M \propto L^3 \tag{1}$$

that is, $M = k_1 \cdot L^3$ for some constant k_1. Also, we would expect surface area, S, to be proportional to the square of the length:

$$S \propto L^2 \tag{2}$$

That is, $S = k_2 \cdot L^2$ for some constant k_2.

By taking the cube root and square root, respectively, in relationships (1) and (2), we may rewrite them in the form

$$M^{1/3} \propto L \quad \text{and} \quad S^{1/2} \propto L$$

Since $M^{1/3} \propto L$ and $L \propto S^{1/2}$, we can now eliminate L and combine these two statements into the prediction that

$$S^{1/2} \propto M^{1/3}$$

By squaring both sides, we have the equivalent relationship

$$S \propto M^{2/3} \tag{3}$$

It is useful to eliminate length as a measure of the size of an animal, since it is a little more ambiguous than mass. Should a tail, for instance, be included in the length measurement?

Testing Our Predictions. This prediction, that surface area is directly proportional to the two-thirds power of body mass, can be tested experimentally. Figure 7.44 shows data collected for a wide range of mammals, from mice, with a body mass on the order of 1 gram, to elephants, whose body mass is more than a million grams (1 metric ton). Here surface area, S, measured in square centimeters, is plotted on the vertical axis and body mass, M, in grams, is on the horizontal. The scales on both axes are logarithmic.

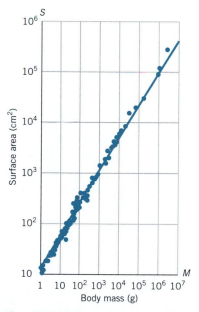

Figure 7.44 A log-log plot of body surface vs. body mass for a wide range of mammals.
Source: McMahon, T. A. 1973. "Size and Shape in Biology," *Science,* 179: 1201–1204.

The data lie pretty much in a straight line. Since this is a log-log plot, that implies that S is directly proportional to M^p for some power p, where p is the slope of the best-fit line. To estimate the slope we can convert the scales on the axes in Figure 7.44 to log(surface area) and log(body mass); that is, we use only the exponents of 10 on both the vertical and the horizontal axes. Then the best-fit line passes through (3, 3) and (6, 5), so the slope is $\frac{5-3}{6-3} = \frac{2}{3}$. This implies that

$$S \propto M^{2/3}$$

confirming our simple prediction of a two-thirds power law. This means that every time the horizontal coordinate increases by a factor of 10^3, the vertical coordinate increases by a factor of 10^2. We have approximate verification of the fact that mammals of different body masses have roughly the same proportion of surface area to the two-thirds power of body mass. This power function is just another way of describing our finding in Section 7.1 on the relationship between surface area and volume.

Metabolic rate vs. body mass

Another example of scaling among animals of widely different sizes is metabolic rate as a function of body mass. This relationship is very important in understanding the mechanisms of energy production in biology. Figure 7.45 shows a log-log plot of metabolic heat production, H, in kilocalories per day, versus body mass, M, in kilograms, for a range of land mammals.

The data fall in a straight line, so we would expect H to be directly proportional to a power of M. The slope of this line, and hence the exponent of M, is $\frac{3}{4} = 0.75$. As body mass increases by four factors of 10, say from 10^{-1} to 10^3 kg, metabolic rate increases by three factors of 10, from 10 to 10^4 kcal/day. Thus $H \propto M^{3/4}$, where M is mass in kilograms and H is metabolic heat production in kilocalories (thousands of calories) per day.

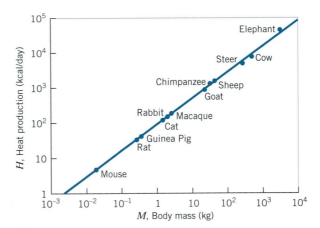

Figure 7.45 A log-log plot of heat production vs. body mass.
Source: Kleiber, M. 1951. "Body Size and Metabolism," *Hilgardia,* 6: 315–353.

Kleiber's Law. This scaling relationship is called "Kleiber's Law," after the American veterinary scientist who first observed it in 1932. It has been verified by many series of subsequent measurements, though its cause is not fully understood.

Animals have evolved biological modifications partially to avoid the consequences of scaling laws. For example, our argument for the two-thirds power law of surface area vs. body mass was based on the assumption that the shapes of animals stay roughly the same as their size increases. But elephants have relatively thicker legs than gazelles in order to provide the extra strength needed to support their much larger weight. Such changes in shape are very important biologically but are too subtle to be seen in the overall trend of the data we have shown. Despite these adaptations, inexorable scaling laws give an upper limit to the size of land animals, since eventually the animal's weight would become too heavy to be supported by its body. That's why large animals such as whales must live in the ocean and why giant creatures in science fiction movies couldn't exist in real life.

Allometric Laws. Allometric laws provide an overview of the effects of scale, valid over several orders of magnitude. They help us compare important traits of elephants and mice, or of children and adults, without getting lost in the details. They help us to understand the limitations imposed by living in three dimensions.

Algebra Aerobics 7.7

A calculator that can evaluate powers and logs is required for Problems 2, 3, and 7.

1. Rewrite each of the following as an equivalent equation by taking the log of both sides.

    a. $y = 3 \cdot 2^x$ **c.** $y = 12 \cdot 10^x$

    b. $y = 4x^3$ **d.** $y = 0.15x^{-2}$

2. Rewrite each of the following so that y is a power or exponential function of x.

    a. $\log y = 0.067 + 1.63 \log x$

    b. $\log y = 2.135 + 1.954x$

    c. $\log y = -1.963 + 0.865x$

    d. $\log y = 0.247 - 0.871 \log x$

3. By inspection, determine whether y is an exponential or a power function of x. Then rewrite the expression so that y is a (power or exponential) function of x.

    a. $\log y = \log 2 + 3 \log x$

    b. $\log y = 2 + x \log 3$

    c. $\log y = 0.031 + 1.25x$

    d. $\log y = 2.457 - 0.732 \log x$

    e. $\log y = -0.289 - 0.983x$

    f. $\log y = -1.47 + 0.654 \log x$

4. **a.** Identify the type of function.

    i. $y = 4x^3$ **ii.** $y = 3x + 4$ **iii.** $y = 4 \cdot 3^x$

    b. Which of the functions in part (a) would have a straight-line graph on a standard linear plot? On a semi-log plot? On a log-log plot? If possible, use technology to check your answers.

    c. For each straight-line graph in part (b), use the original equation in part (a) to predict the slope of the line.

5. Interpret the slopes of 1.2 and 1.0 on the accompanying graph in terms of arm length and body height. [*Note:* Slopes refer to the slopes of the lines, where the units of the axes are translated to log(body height) and log(arm length).]

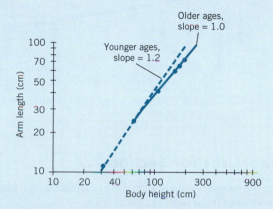

Source: T. A. McMahon and J. T. Bonner, *On Size and Life* (New York: Scientific American Books, 1983), p. 32.

6. For each of the three graphs below, examine the scales on the axes. Then decide whether a linear, exponential, or power function would be the most appropriate model for the data.

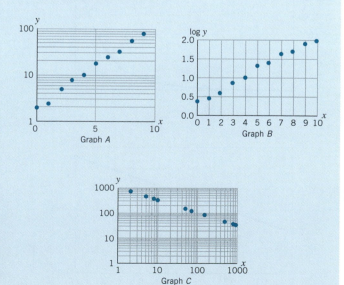

7. a. Using Figure 7.44 on page 437, estimate the surface area for a human being whose mass is 70 kg (or $7 \cdot 10^4$ g).

b. Construct a power function for the plot in Figure 7.44 and use it to calculate the value for part (a).

c. How do your answers in parts (a) and (b) compare?

d. Translate your answers into units of pounds and square inches.

8. The accompanying graph shows the relationship between heart rate (in beats per minute) and body mass (in kilograms) for mammals ranging over many orders of magnitude in size.

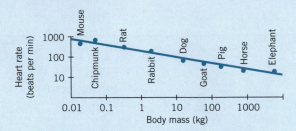

Source: Robert E. Ricklefs, *Ecology* (San Francisco: W. H. Freeman, 1990), p. 66.

a. Would a linear, power, or exponential function be the best model for the relationship between heart rate and body mass?

b. The slope of the best-fit line shown on the graph is -0.23. Construct a function for heart rate in terms of body mass.

c. Interpret the slope of -0.23 in terms of heart rate and body mass.

9. Using the graph in Figure 7.45 on page 438, what rate of heat production does this model predict for a 70-kg human being? Considering that each kilocalorie of heat production requires consumption of one food Calorie, does this seem about right?

Exercises for Section 7.7

Several exercises require a graphing program that can convert axes from a linear to a logarithmic scale.

1. Match each function with its standard linear-scale graph and its logarithmic-scale graph. (*Hint:* Evaluate each function for $x = 4$.)

a. $y = 100(5)^x$ **b.** $y = 100x^5$

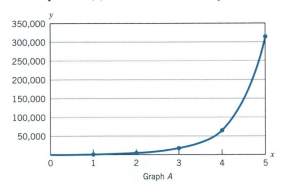

Graph A

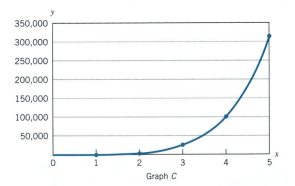

Graph C

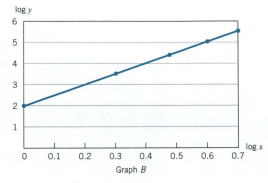

Graph B

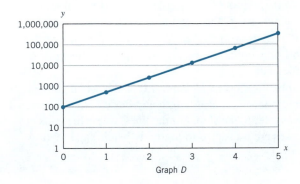

Graph D

2. The accompanying graph shows data on U.S. annual death rate from cancer of the large intestine as a function of age. Would an exponential or power function be a better model for the data? Justify your answer.

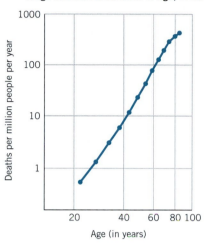

Annual U.S. Death Rate from Cancer of the Large Intestine in Relation to Age, 1968

Source: J. Cairns, *Cancer and Society* (San Francisco: W. H. Freeman and Company, 1978), p. 37.

3. Does the accompanying graph suggest that an exponential or a power function would be a better choice to model infant mortality rates in the United States from 1915 to 1977? Explain your answer.

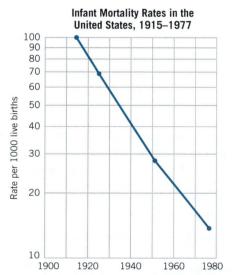

Infant Mortality Rates in the United States, 1915–1977

Source: Healthy People: The Surgeon General's Report on Health Promotion and Disease Prevention, DHEW (PHS) Publication 79-55071, Department of Health, Education, and Welfare, Washington, DC, 1979.

4. Given the following three power functions in the form $y = kx^p$,

$$y_1 = x^3 \qquad y_2 = 5x^3 \qquad y_3 = 2x^4$$

   a. Use the rules of logarithms to change each power function to the form: $\log y = \log k + p \log x$.

   b. Substitute in each equation in part (a), $Y = \log y$ and $X = \log x$ and the value of $\log k$ to obtain a linear function in X and Y.

   c. Compare your functions in parts (a) and (b) to the original functions. What does the value of $\log k$ represent in the linear equation? What does the value of the slope represent in the linear equation?

5. Using rules of logarithms, convert each equation to its power function equivalent in the form $y = kx^p$.

   a. $\log y = \log 4 + 2 \log x$ c. $\log y = \log 1.25 + 4 \log x$

   b. $\log y = \log 2 + 4 \log x$ d. $\log y = \log 0.5 + 3 \log x$

6. Given linear equations in the form $Y = B + mX$, where $Y = \log y$, $X = \log x$, and $B = \log k$, change each linear equation to its power function equivalent, in the form $y = kx^p$. (*Note:* Round values to the nearest hundredth.)

   a. $Y = 0.34 + 4X$ c. $Y = 1.0 + 3X$

   b. $Y = -0.60 + 4X$ d. $Y = 1.0 - 3X$

7. a. Create a table of values for the function $y = x^5$ for $x = 1, 2, 3, 4, 5, 6$. Then plot the values on the given log-log scale.

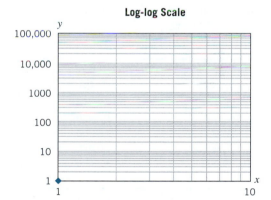

Log-log Scale

   b. Add to the table you created in part (a) the values for $\log x$ and $\log y$. Then graph the results on the $\log y$ vs. $\log x$ scale provided below.

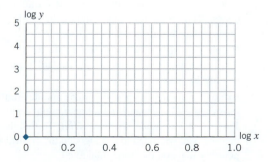

   c. Describe the relationship between the plots in parts (a) and (b).

8. Match each power function with its log y vs. log x plot.

   a. $y = 10x^2$ **c.** $y = \frac{1}{2}x^4$

   b. $y = 4x^3$ **d.** $y = 10x^{-2}$

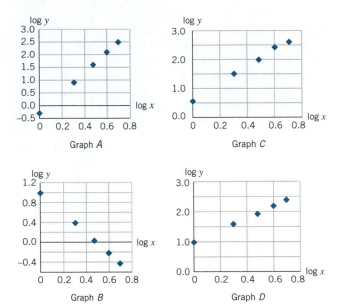

Graph A   Graph C

Graph B   Graph D

9. (Technology recommended for graphing and changing axis scales.) The following data give the typical masses for some birds and their eggs:

Species	Adult Bird Mass (g)	Egg Mass (g)
Ostrich	113,380.0	1,700.0
Goose	4,536.0	165.4
Duck	3,629.0	94.5
Pheasant	1,020.0	34.0
Pigeon	283.0	14.0
Hummingbird	3.6	0.6

Source: W. A. Calder III, *Size, Function and Life History* (Boston: Harvard University Press, 1984).

   a. Create a fourth column in the table with the values for the ratio $\frac{\text{egg mass}}{\text{adult bird mass}}$. Is the ratio the same for all of the birds? (*Note:* The first ratio value should be 0.015.) Write a sentence that describes what you discover.

   b. Graph egg mass (vertical axis) vs. adult bird mass (horizontal axis) three times using standard linear, semi-log, and log-log plots.

   c. Examine the three graphs you made and determine which looks closest to linear. Find a linear equation in the form $Y = b + mX$ to model the line. (*Hint:* To find the slope of the line, you will need to use log y on the vertical axis, and possibly log x on the horizontal axis).

   d. Once you have a linear model, transform it to a form that gives egg mass as a function of adult bird mass.

   e. What egg mass does your formula predict for a 12.7-kilogram turkey?

   f. A giant hummingbird (*Patagona gigas*) lays an egg of mass 2 grams. What size does your formula predict for the mass of the adult bird?

10. Find the equation of the line in each of the accompanying graphs. Rewrite each equation, expressing y in terms of x.

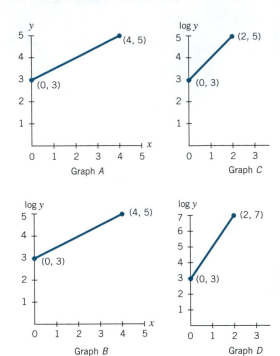

Graph A   Graph C

Graph B   Graph D

11. Use the accompanying graphs to answer the following questions.

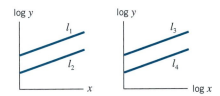

   a. Assume l_1 and l_2 are straight lines that are parallel. In each case, what type of equation would describe y in terms of x? How are the equations corresponding to l_1 and l_2 similar? How are they different?

   b. Assume l_3 and l_4 are also parallel straight lines. For each case, what type of equation would describe y in terms of x? How are the equations corresponding to l_3 and l_4 similar? How are they different?

12. (Technology recommended for graphing and changing axis scales.)

   a. Assume that $f(1) = 5$ and $f(3) = 45$. Then find an equation for f assuming f is:

   i. A linear function

   ii. An exponential function

   iii. A power function

b. Verify that you get a straight line when you plot:

   i. Your linear function on a standard plot

   ii. Your exponential function on a semi-log plot

   iii. Your power function on a log-log plot

13. (Technology optional.) For each table, create a linear, exponential, or power function that best models the data. Support your answer with a graph.

a.

x	1	2	3	4	5
y	3	12	27	48	75

b.

x	1	2	3	4	5
y	8	16	32	64	128

c.

x	1	2	3	4	5
y	9	12	15	18	21

14. (Graphing program required.) Examine the following table.

x	y	log x	log y
1	5	0	0.698 97
2	80	0.301 03	1.903 09
3	405	0.477 12	2.607 46
4	1280	0.602 06	3.107 21
5	3125	0.698 97	3.494 85
6	6480	0.778 15	3.811 58

Treating x as the independent variable:

a. Create a scatter plot of y versus x.

b. Create a scatter plot of log y versus x.

c. Create a scatter plot of log y versus log x.

d. Does the table represent a linear, power, or exponential function? Explain your reasoning.

e. Create the equation that models the data.

15. The accompanying graph shows oxygen consumption versus body mass in mammals.

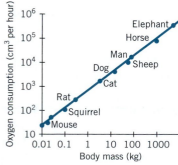

Source: R. E. Ricklefs, *Ecology* (San Francisco: W. H. Freeman, 1990), p. 67.

a. Would a power or an exponential function be the best model of the relationship between oxygen consumption and body mass?

b. The slope of the best-fit line (for log(oxygen) vs. log(mass)) shown on the graph is approximately $\frac{3}{4}$. Construct the basic form of the functional model that you chose in part (a).

c. Interpret the slope in terms of oxygen consumption and body mass. In particular, by how much does oxygen consumption increase when body mass increases by a factor of 10? By a factor of 10^4?

16. a. The left-hand graph below shows the relationship between population density and length of an organism. The slope of the line for log(density) vs. log(length) is -2.25. Express the relationship between population density and length in terms of direct proportionality.

b. The right-hand graph shows the relationship between population density and body mass of mammals. The slope of the line for log(density) vs. log(mass) is -0.75. Express the relationship between population density and body mass in terms of direct proportionality.

c. Are your statements in parts (a) and (b) consistent with the fact that body mass is directly proportional to length³? (*Hint:* Calculate the cube of the length to the -0.75 power.)

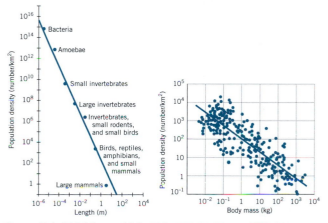

Source: T. A. McMahon and John T. Bonner, *On Size and Life* (New York: Scientific American Books, 1983), p. 228.

17. Public health researchers have developed a new measure of overall population health called "health-adjusted life expectancy" (HALE). HALE represents the number of expected years of life lived in full health. The following chart shows health-adjusted life expectancy versus the health care expenditure per capita for various countries. Note that here the log scale is on the *horizontal* axis.

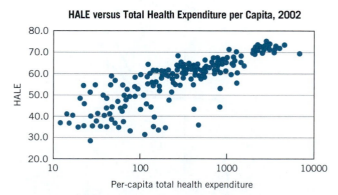

a. Sketch a best-fit line through the data. Replace each label x on the horizontal axis with $\log(x) = X$. (So 10 becomes log 10 = 1, 100 becomes log 100 = 2, etc. You end up with the new labels 1, 2, 3, and 4 on the horizontal axis.)

b. Estimate two points on the line to construct an equation for your line for y in terms of X. (Note that the vertical axis, where $x = 0$, is not on the graph, so you can't use it to estimate the vertical intercept.) Then rewrite your equation as y in terms of $\log x$.

c. Why is it reasonable that HALE is a log function of per capita health expenditure?

d. What does your result suggest about functions that appear to be linear on a semi-log plot, where the logarithmic scale is on the horizontal axis?

CHAPTER SUMMARY

Power Functions

A power function has the form

$$y = kx^p$$

where k is called the *constant of proportionality*.

If $p > 0$ (and $k \neq 0$), then

y is *directly proportional* to x^p if $y = kx^p$

y is *inversely proportional* to x^p if $y = kx^{-p}\left(= \dfrac{k}{x^p}\right)$

Positive Integer Powers

For power functions of the form $y = kx^p$ (where $p > 0$ and $k \neq 0$)

- The graph goes through the origin.
- If the power p is odd (and $p > 1$), the graph is shaped roughly like ∿ and is symmetric about the origin.
- If the power p is even, the graph is ∪-shaped and symmetric across the y-axis.
- The value of the coefficient k compresses or stretches the graph of $y = x^p$.
- The graphs of $y = kx^p$ and $y = -kx^p$ are mirror images of each other across the x-axis.

Even Positive Powers

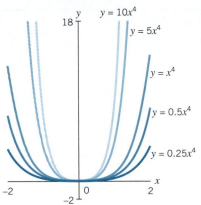

Negative Integer Powers

For power functions of the form $y = \dfrac{k}{x^p}$ or kx^{-p} (where $p > 0$ and $k \neq 0$)

- The graph never goes through the origin, since the function is not defined when $x = 0$.
- The graph is composed of two non-intersecting curves each asymptotic to both the x- and the y-axes.
- If the power p is even, the graph is symmetric across the y-axis.
- If the power p is odd, the graph is symmetric about the origin.
- The value of the coefficient k compresses or stretches the graph of $y = \dfrac{1}{x^p}$.

- The graphs of $y = \dfrac{k}{x^p}$ and $y = -\dfrac{k}{x^p}$ are mirror images of each other across the x-axis.

Odd Positive Powers

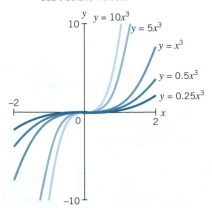

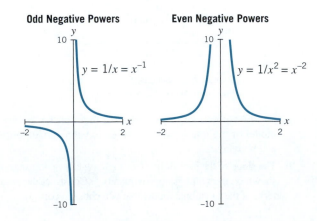

Semi-Log and Log-Log Graphs

To find the best function to model a relationship between two variables, we look for a way to plot the data so they appear as a straight line:

- The graph of a linear function appears as a straight line on a standard plot that has linear scales on both axes.
- The graph of an exponential function appears as a straight line on a semi-log plot that has a linear scale on the horizontal axis and a logarithmic scale on the vertical axis. The slope is the logarithm of the growth (or decay) factor.
- The graph of a power function appears as a straight line on a log-log plot that has logarithmic scales on both axes. The slope is the exponent of the power function.

Finding the Equation of a Power Function on a Log-Log Plot

The graph of a power function $y = kx^p$ (where k and $x > 0$) appears as a straight line on a log-log plot. The equation of the line is

$$Y = \log k + pX \qquad \text{where } X = \log x \text{ and } Y = \log y$$

- The slope of the line is p, where p is the power of x for $y = kx^p$.
- The vertical intercept is $\log k$, where k is the constant of proportionality for $y = kx^p$.

CHECK YOUR UNDERSTANDING

I. Is each of the statements in Problems 1–21 true or false? Give an explanation for your answer.

1. The graph of $y = -3x^5$ is decreasing.
2. The graph of $y = 2x^{-1}$ increases when $x < 0$ and decreases when $x > 0$.
3. The graphs of $y = 3.5x^4$ and $y = 5.1x^2$ intersect only at the origin.
4. If $y = x^3$ and if x increases by 1, then y increases by a factor of 3.
5. If $y = 5x^{-1}$ and if x doubles, then y decreases by a factor of $\frac{1}{2}$.
6. The function $y = 3 \cdot 2^x$ is a power function.
7. The graphs of the functions $y = 3x^2$ and $y = 3x^{-2}$ are reflections of each other across the x-axis.
8. The graphs of the functions $y = 4x^{-1}$ and $y = -4x^{-1}$ are reflections of each other across the x-axis.
9. If $y = x^{-2}$, as x becomes very large, then y approaches zero.
10. If $y = \frac{1}{x^2}$, as x approaches $+\infty$, then y approaches $+\infty$.
11. The function $y = x^{10}$ eventually grows faster than the function $y = 1.5^x$.
12. The functions $y = x^2$ and $y = x^{-1}$ intersect at the point (1, 1).
13. The graph of $y = \frac{2}{x^5}$ on a log-log plot is a straight line with slope of -5.

14. Of the three functions f, g, and h in the accompanying figure, only function h could be a power function.

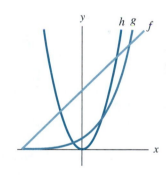

15. If $F = km^{-3}$ and k is a nonzero constant, then F is directly proportional to m.
16. If $K(d) = 3d^{-2}$, then $K(-2) = \frac{-3}{4}$.
17. The graphs of two power functions $f(x) = \frac{1}{x^p}$ and $g(x) = \frac{1}{x^q}$ are shown in the accompanying figure. For these functions, $q < p$.

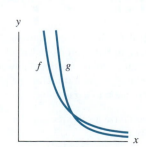

18. If a quantity M is inversely proportional to the cube of a quantity Q, then $M = \frac{k}{Q^3}$ for k, a nonzero constant.

19. In the power function $y = \frac{1}{x^p}$ graphed in the accompanying figure, p is an even positive integer.

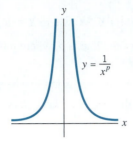

$$y = \frac{1}{x^p}$$

20. If $f(q) = -q^5$, then $f(-2) = -32$.

21. The graphs of the power functions A, B, and C, all of the form $G = k \cdot m^p$, are shown in the accompanying figure. For each of these functions, $k < 0$.

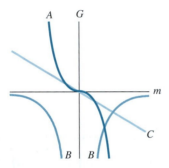

II. In Problems 22–29, construct a function or functions with the specified properties.

22. A power function $y = f(x)$ whose graph is steeper than the graph of $y = 4x^6$ when $x > 1$.

23. A power function $y = g(x)$ whose graph is a reflection of the graph of $y = -3.2x^4$ across the horizontal axis.

24. A power function of even degree whose graph opens downward.

25. A function $w = h(m)$ whose graph will eventually dominate the function $w = 500m^{10}$.

26. A function that is not a power function.

27. A function whose graph on a log-log plot is a straight line.

28. A power function whose graph would be similar to the graph of $y = -\frac{3}{x}$ but would have a different power of x.

29. A power function $T(m)$ where doubling the value of m increases the value of $T(m)$ by a factor of 16.

III. Is each of the statements in Problems 30–40 true or false? If a statement is true, explain why. If a statement is false, give a counterexample.

30. All linear functions are power functions.

31. All power functions $y = kx^p$, where p is a positive integer, pass through the origin.

32. For the two power functions $m(x) = x^p$ and $n(x) = x^q$ in the accompanying figure, both p and q have negative values.

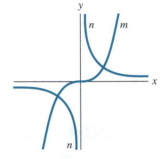

33. The graphs of all power functions $y = kx^p$, where p is a negative integer, lie in quadrants I (where $x \geq 0$ and $y \geq 0$) and III (where $x \leq 0$ and $y \leq 0$).

34. The graphs of all power functions with integer powers are asymptotic to the horizontal axis.

35. The graphs of all power functions $y = kx^p$, where p is a positive even integer, are symmetric across the y-axis.

36. The ratio of (surface area)/volume is less for a larger sphere than for a smaller one.

37. The graphs of the functions $F(m) = a \cdot m^p$ and $G(m) = -a \cdot m^p$, where $a \neq 0$ and p is a positive integer, are reflections of each other across the vertical axis.

38. The graphs of power functions $y = kx^p$, where p is a negative integer, never pass through the origin.

39. The graphs of power functions $f(x) = kx^p$, where p is a positive integer, on a log-log plot are straight lines with a positive slope.

40. The graphs of exponential functions $g(x) = a \cdot b^x$ on a log-log plot are also straight lines.

CHAPTER 7 REVIEW: PUTTING IT ALL TOGETHER

A standard calculator that can do multiplication and division is required throughout. Some problems are identified as needing a calculator that can calculate fractional powers (such as square roots) or a graphing program (that for one problem can change axis scales from linear to logarithmic).

1. Suppose you have a solid rectangular box, with dimensions x by $2x$ by $2x$.

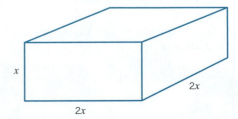

a. What is its volume $V(x)$? If you double the value of x, from X to $2X$, what happens to the volume?

b. What is the surface area $S(x)$? If you double the value of x, from X to $2X$, what happens to the surface area?

c. Construct the ratio $R(x) = S(x)/V(x)$. If you double the value of x, from X to $2X$, what happens to $R(x)$? Describe your result in terms of surface area and volume.

d. As x increases, what happens to
$R(x) = $ (surface area)/volume?

2. Inside a bottle of Bayer aspirin is a small plastic cylinder (about 5/8" in height and 1/2" in diameter) whose tightly packed contents are used to absorb water to keep the aspirin dry.

a. What is the volume (in cubic inches) of the cylinder (and hence a good approximation of the volume of its contents)? [*Note:* The volume of a cylinder with radius r and height h is $\pi r^2 h$.

b. If you examine the contents of the cylinder under a microscope, you would see an incredibly convoluted surface. The total surface area is about the size of a football field (160' wide and 360' long, including the two end zones). What is the area in square feet? In square inches?

c. Interpret your results. Why would the manufacturers want such a large surface area in such a tiny volume?

3. The website for *La Riviera Gourmet* in Lexington, Massachusetts, states, "A 10" cake takes up 3 quarts of volume, whereas a 12" cake takes up 6 quarts of volume. That's why a 12" cake feeds twice as many and costs almost twice as much!" Does this seem reasonable?

4. Which of the following functions represent power functions? Which power functions represent direct and which inverse proportionality?

a. $Q = 4t$ c. $Q = 4 + t^2$ e. $Q = 4t^{-2}$

b. $Q = 4t^2$ d. $Q = 4 \cdot 2^t$

5. A bathtub holds about 15 gallons of water.

a. Construct a function to describe the water volume $V_F(t)$ in gallons at minute t if the tub is being filled at a rate of 3 gallons/minute. What would be a reasonable domain for the function? Graph $V_F(t)$ by hand. Does $V_F(t)$ represent direct proportionality?

b. Assuming the tub is full to the brim, construct a new function $V_D(t)$ to describe the water volume (in gallons) at minute t if the tub has a leak that is draining water at a rate of 0.5 gallons/minute. What would be a reasonable domain for $V_D(t)$? Graph $V_D(t)$ by hand. Does $V_D(t)$ represent inverse proportionality?

6. In some areas wind turbines offer a viable clean source of energy. The amount of power, W, (in watts) delivered by a wind turbine is a function of S, wind speed, and A, the area swept out by the wind turbine rotors. Wind engineers can estimate the wattage using the formula

$$W = 0.0052AS^3$$

where W is in watts, A is in square feet, and S is in miles per hour.

a. Is W directly proportional to A? To S? To S^3?

b. Calculate the circular area A (in square feet) swept out by wind turbine blades with a rotor radius of 10 ft. (Round off to the nearest integer.)

c. Using your answer in part (b), alter the general formula for W to represent the wattage W_{10} generated by a wind turbine with 10-ft blades. Using the new formula, how many watts would be generated by a wind speed of 30 mph? How many kilowatts is that? If a typical suburban house needs on average about 1.8 kilowatts of power at any point in the day, how many houses could this turbine support?

d. If the wind speed doubles, what happens to the wattage output? If the wind speed is cut in half, what happens to the wattage output?

e. Is it better to have ten wind turbines in an area that averages 20-mph winds or two turbines in an area that averages 40-mph winds?

7. Some northern cities celebrate New York's eve with First Night events, which include artists making large, elaborate ice sculptures. (See the accompanying photo from Fairbanks, Alaska.)

An artist planned to build a sculpture 4' high to be carved out of a 4'×2'×2' block of ice. But she decides it would look better proportionally scaled up to 6' high. Her friend says, "You'll have to order 50% more ice if you make it 6 ft instead of 4 ft."

a. Is the friend correct?

b. Give the dimensions of the ice block she should order for the 6' sculpture? What percent increase in the volume of ice would that be?

c. If the temperature gets above freezing and the 6' sculpture loses one-tenth of its volume through melting, how tall will it be then?

8. (Requires a calculator that can compute square roots.) In any stringed instrument, such as a violin, guitar, or ukulele, each note is determined by the frequency F of a vibrating string. F is directly proportional to the square root of the tension T and

inversely proportional to the length L of the string. The higher the frequency, the higher the note.

a. Construct a general formula for F.

b. If the string length is cut in half (by holding down a string in the middle against the neck of the instrument), is the resulting note higher or lower? By how much?

c. If the tension is increased by 10%, will the resulting note be higher or lower? By how much?

9. Construct a formula for each situation.

a. The time, t, taken for a journey is directly proportional to the distance traveled, d, and inversely proportional to the rate r at which you travel.

b. On a map drawn to scale, the distance D_1 between two points is directly proportional to the distance D_2 between the corresponding two points in real life.

c. The resistance R is directly proportional to the voltage, V, squared and inversely proportional to the wattage, W.

10. The following four graphs are power functions with positive exponents (and hence represent direct proportionality). Which graphs depict even and which depict odd power functions?

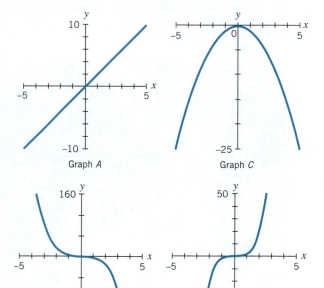

Graph A   Graph C

Graph B   Graph D

11. Describe any symmetries and asymptotes for each function graphed in the preceding problem.

12. Of the following three functions, as $x \to +\infty$, which would eventually dominate (have the largest values for y)?

$$y_1 = x^{1000}, \quad y_2 = 1.000001^x, \quad y_3 = 10^{100,000}x$$

13. (Graphing program optional.)

a. In Nashua, New Hampshire, you can rent an indoor skydiving facility for $450 per half hour. What is the cost per person, $P(n)$, if there are n friends sharing the experience?

b. What is the cost per person for two people? For five people? What is the trade-off between number of people and cost per person?

c. Now consider your function $P(n)$ as an abstract function, where n can be any positive or negative real number. Graph this extended version of $P(n)$ and identify any asymptotes.

14. (Requires a calculator that can compute square roots.) Film studios, photographers, and the lighting industry usually measure light intensity in foot-candles—historically, the amount of light one candle generates on a surface one foot away. Light intensity I (in foot-candles) is directly proportional to candlepower P (which measures how much light is generated by a light source) and inversely proportional to the square of the distance, d, from the light source. (See Section 7.5, Example 7.) The relationship is the inverse square law

$$I = P/d^2$$

where I is measured in foot-candles, P in candlepower, and d in feet.

a. A student's angle-arm lamp can be positioned as close as 1 foot above the work surface and as far as 3 feet, measuring to the center of the bulb. If the 60-watt bulb has 70 candlepower, how many foot-candles illuminate the surface at the lowest and the highest positions?

b. Where would you position the bulb to get 50 foot-candles of illumination, the standard level recommended for reading?

15. (Requires a calculator that can compute square roots.) Surgeons use a lamp that supplies 4000 foot-candles when suspended 3 feet above the operating surface. (See the preceding problem.)

a. What candlepower would the source have to provide?

b. Lighting engineers assume that surgery requires somewhere in the range of 2000 to 5000 foot-candles. How high above the operating surface would you hang the lamp to give 2000 foot-candles?

16. Which of the following graphs could depict power functions with a negative exponent (and hence inverse proportionality)?

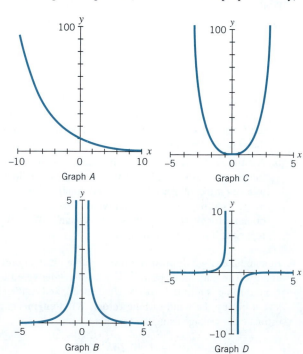

Graph A   Graph C

Graph B   Graph D

17. Describe any apparent symmetries and asymptotes for each of the function graphs in the preceding problem.

18. (Requires a calculator that can compute fractional powers.) Because wind speed increases with height, wind turbines are placed as high as possible, and certainly above the level of any existing structures or trees that may block the wind. In order to estimate typical wind speeds at various heights, wind engineers first measure the average wind speed S_{33} (in mph) at a height of 33 feet. Then they can estimate the wind speed S_h at hight h using the formula

$$S_h = S_{33}(h/33)^{2/7}$$

where h is in feet, and S_h and S_{33} are both in miles per hour.

a. For an area likely to have hurricanes, wind engineers design for a wind speed of 90 mph at a height of 33 feet. Estimate the wind speed under hurricane conditions at the top of a building 120 feet tall.

b. Rearrange the formula to express h in terms of S_h and S_{33}. Using the hurricane conditions in part (a), at what height dose the formula perdict 156-mph winds?

19. a. Given the following graph, what kind of function is Y with respect to X?

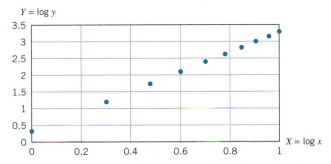

b. Sketch in a best-fit line and then generate its equation using Y in terms of X. [*Hint:* Use the points $(0, 0.3)$ and $(1, 3.3)$.]

c. Transform your equation in part (b) to one using y and x. What sort of function do you have now (for y in terms of x)?

20. (Requires a graphing program that can convert the scale on either axis from linear to logarithmic.) These data give the sound reduction in decibels through walls of different densities, measured in pounds per square foot of the wall. (*Note:* This assumes a fixed depth to the wall material.)

Sound Reduction through Walls

Wall Density (lb/sq. ft)	Sound Loss (Db)
1	22.7
5	32.0
10	37.0
20	41.0
40	45.0
60	48.0
100	51.0
400	60.0

a. Graph these data with wall density on the horizontal axis. Try linear, log-linear (with the log scale on the horizontal and then on the vertical axis), and log-log plots. In which format do the data look most linear?

b. Use the format of the data from part (a) to construct a formula for sound loss as a function of wall density.

c. What does your result suggest about data that appear linear when graphed on a semi-log plot, where the log scale is on the horizontal axis?

21. According to Adrian Bejan's Constructal Theory, the optimal cruising speed v (in meters per second) of flying bodies (whether insects, birds, or airplanes) is directly proportional to the one-sixth power of body mass m (in kilograms). Use the accompanying graph below to confirm or refute this.

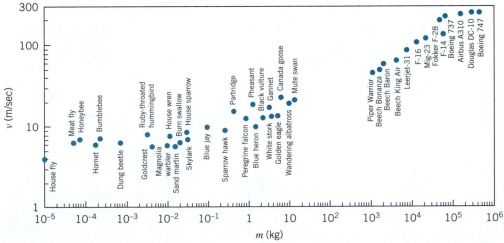

Source: Image from Wikipedia, derived from the work of Adrian Bejan, Professor of Mechanical Engineering at Duke University.

Scaling Objects

Objectives

- find and use general formulas for scaling different types of objects

Procedure

1. *Scaling factors.*

   When a two- or three-dimensional object is enlarged or shrunk, each linear dimension is multiplied by a constant called the *scaling factor*. For the two squares below, the scaling factor, F, is 3. That means that any linear measurement (for example, the length of the side or of the diagonal) is three times larger in the bigger square.

   Scaling up a square by a factor of 3.

   a. What is the relationship between the areas of the two squares above? Show that for all squares

   $$\text{area scaled} = (\text{original area}) \cdot F^2 \qquad \text{where } F \text{ is the scaling factor}$$

   b. Given a scaling factor, F, find an equation to represent the surface area S_1 of a scaled-up cube, in terms of the surface area S_0 of the original cube.

   c. Given a scaling factor, F, find an equation to represent the volume V_1 of a scaled-up cube, in terms of the volume V_0 of the original object.

2. *Representing 3-D objects.*

   We can describe the volume, V, of any three-dimensional object as $V = kL^3$, where V depends on any length, L, that describes the size of the object. The coefficient, k, depends upon the shape of the object and the particular length L and the measurement units we choose. For a statue of a deer, for example, L could represent the deer's width or the length of an antler or the tail. If we let L_0 represent the overall height of the deer in Figure 1, then L_0^3 is the volume of a cube with edge length L_0 that contains the entire deer. The actual volume, V_0, of the deer is some fraction k of L_0^3; that is, $V_0 = kL_0^3$ for some constant k.

   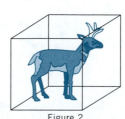

   Figure 1   Figure 2

3. *Finding the scaling factor.*

   If we scale up the deer to a height L_1 (see Figure 2), then the volume of the scaled-up deer equals kL_1^3. The scaled-up object inside the cube will still occupy the same fraction, k, of the volume of the scaled-up cube.

   a. If an original object has volume V_0 and some measure of its length is L_0, then we know $V_0 = kL_0^3$. Rewrite the equation, solving for the coefficient, k, in terms of L_0 and V_0.

   b. Write an equation for the volume, V_1, of a replica scaled to have length L_1. Substitute for k the expression you found in part (a). Simplify the expression such that V_1 is expressed as some term times V_0.

   c. The ratio L_1/L_0 is the *scaling factor*. A scaling factor is the number by which each linear dimension of an original object is *multiplied* when the size is changed. Write an equation for the volume of a scaled object, V_1, in terms of V_0, the volume of the original object, and F, the scaling factor.

4. *Using the scaling factor*

   a. Draw a circle with a 1-inch radius. If you want to draw a circle with an area four times as large, what is the change in the radius?

   b. A photographer wants to blow up a 3-cm by 5-cm photograph. If he wants the final print to be double the area of the original photograph, what scaling factor should he use?

   c. A sculptor is commissioned to make a bronze statue of George Washington sitting on a horse. To fit into its intended location, the final statue must be 15 feet long from the tip of the horse's nose to the end of its tail. In her studio, the sculptor experiments with smaller statues that are only 1 foot long. Suppose the final version of the small statue requires 0.15 cubic feet of molten bronze. When the sculptor is ready to plan construction of the larger statue, how can she figure out how much metal she needs for the full-size one?

   d. One of the famous problems of Greek antiquity was the *duplication of the cube*. Our knowledge of the history of the problem comes down to us from Eratosthenes (circa 284 to 192 B.C.), who is famous for his estimate of the circumference of Earth. According to him, the citizens of Delos were suffering from a plague. They consulted the oracle, who told them that to rid themselves of the plague, they must construct an altar to a particular god. That altar must be the same shape as the existing altar but double the volume. What should the scaling factor be for the new altar? [Adapted from COMAP, *For All Practical Purposes* (New York: W. H. Freeman, 1988), p. 370.]

   e. Pyramids have been built by cultures all over the world, but the largest and most famous ones were built in Egypt and Mexico.

      i. The Great Pyramid of Khufu at Giza in Egypt has a square base 755 feet on a side. It originally rose about 481 feet high (the top 31 feet have been destroyed over time). Find the original volume of the Great Pyramid. [The volume V of a pyramid is given by $V = (B \cdot H)/3$, where B is the area of the base and H is its height.]

      ii. The third largest pyramid at Giza, the Pyramid of Menkaure, occupies approximately one-quarter of the land area covered by the Great Pyramid. Menkaure's pyramid is the same shape as the original shape of the Great Pyramid, but it is scaled down. Find the volume of the Pyramid of Menkaure.

Predicting Properties of Power Functions

Objectives

- construct power functions with positive integer exponents
- find patterns in the graphs of power functions

Materials

- graphing calculator or function graphing program or "P1: k & p Sliders" in *Power Functions*
- graph paper

Procedure

Working in Pairs

1. a. Construct the equation of a power function whose graph is symmetric across the y-axis. Construct a second power function whose graph is symmetric about the origin.

    b. Construct the equations for two power functions whose graphs are mirror images across the x-axis. Construct two more whose graphs are mirror images across the y-axis.

2. Using the same power for each function, construct two different power functions such that:

    a. Both functions have even powers and the graph of one function "hugs" the y-axis more closely than that of the other function.

    b. Both functions have odd powers and the graph of one function "hugs" the y-axis more closely than that of the other function.

3. Using the same value for the coefficient, k, construct two different power functions such that:

    a. Both functions have even powers and the graph of one function "hugs" the y-axis more closely than that of the other function.

    b. Both functions have odd powers and the graph of one function "hugs" the y-axis more closely than that of the other function.

4. a. Choose a value for k that is greater than 1 and construct the following functions:

    i. $y = kx^0 = k$ $y = kx^2$ $y = kx^4$ ii. $y = kx^1$ $y = kx^3$ $y = kx^5$

    b. Graph the functions in part (i) on the same grid. Choose a scale for your axes so you can examine what happens when $0 < x < 1$. Now regraph, choosing a scale for your axes so you can examine what happens when $x > 1$. Then repeat for part (*a*. ii).

    c. Describe your findings.

Class Discussion

Compare your findings. As a class, develop a 60-second summary describing the results.

Exploration-Linked Homework: A Challenge Problem

In Chapter 7 we primarily studied power functions with integer exponents. Try extending your analyses to power functions with positive fractional exponents. Start with fractions that in reduced form have 1 in the numerator. What would the graphs of $y = x^{1/2}$ and $y = x^{1/4}$ look like? What are the domains? The ranges? What would the graphs of $y = x^{1/3}$ or $y = x^{1/5}$ look like? What about their domains and ranges? Generalize your results for fractions of the form $\frac{1}{n}$, where n is a positive integer.

For an even harder challenge, consider fractional exponents of the form m/n.

Visualizing Power Functions with Negative Integer Powers

Objectives

- examine the effect of k on negative integer power functions

Materials

- graphing calculator or function graphing program
- graph paper

Related Software

- "P1: k & p Sliders" in *Power Functions*

Procedure

Class Demonstration: Constructing Negative Integer Power Functions

1. A power function has the form $y = kx^p$, where k and p are constants. Consider the following power functions with negative integer exponents where k is 4 and 6, respectively:

$$y = 4x^{-2} \quad \text{and} \quad y = 6x^{-2}$$

We can also write these as

$$y = \frac{4}{x^2} \quad \text{and} \quad y = \frac{6}{x^2}$$

What are the constraints on the domain and the range for each of these functions?

2. Construct a table of values for both functions using positive and negative values for x. How do different values for k lead to different values for y for these two functions? Sketch a graph for each function. Check your graphs with a function graphing program or graphing calculator.

   a. Describe the overall behavior of these graphs. In each case, when is y increasing? When is y decreasing? How do the graphs behave for values of x near 0?

   b. Describe how these graphs are similar and how they are different.

Working in Small Groups

In the following exploration you will predict the effect of the coefficient k on the graphs of power functions with negative integer exponents. In each part, compare your findings, and then write down your observations.

1. a. Choose a value for p in the function $y = \frac{k}{x^p}$, where p is a positive integer. Construct several functions where p has the same value but k assumes different *positive* values (as in the example above, where $k = 4$ and $k = 6$). What effect do you think k has on the graphs of these equations? Graph the functions. As you choose larger and larger values for k, what happens to the graphs? Try choosing values for k between 0 and 1. What happens to the graphs?

   b. Using the same value of p as in part (a), construct several functions with *negative* values for the constant k. What effect do you think k has on the graphs of these equations? Graph the functions. Describe the effect of k on the graphs of your equations. Do you think your observations about k will hold for any value of p?

c. Choose a new value for p and repeat your experiment. Are your observations still valid? Compare your observations with those of your partners. Have you examined both odd and even negative integer powers?

In your own words, describe the effect of k on the graphs of functions of the form $y = \frac{k}{x^p}$, where p is a positive integer. What is the effect of the sign of the coefficient k?

2. a. Choose a value for k where $k > 1$ and rewrite each function in the form $y = \frac{k}{x^p}$.

   i. $y = kx^0 = k$ $y = kx^{-2}$ $y = kx^{-4}$
   ii. $y = kx^{-1}$ $y = kx^{-3}$ $y = kx^{-5}$

b. Graph the functions in part (i) on the same grid. Choose a scale for your graphs such that you can examine what happens when $0 < x < 1$. Now choose scales for your axes so you can examine what happens when $x > 1$.

c. Graph the functions in part (ii) on the same grid. Choose scales for your graphs such that you can examine what happens when $0 < x < 1$. Now choose scales for your axes so you can examine what happens when $x > 1$.

d. Describe your findings.

Class Discussion

Compare the findings of each of the small groups. As a class, develop a 60-second summary on the effect of k on power functions in the form $y = \frac{k}{x^p}$, where p is a positive integer.

CHAPTER 8

QUADRATICS, POLYNOMIALS, AND BEYOND

OVERVIEW

In this chapter we learn how to transform familiar functions into new ones. We start by adding power functions to create quadratics, polynomials of degree 2, which can be used to describe motion, model traffic flow, or predict the spread of a wildfire. We continue adding power functions to create polynomial functions, and then take the ratio of two polynomial functions to create a rational function. Finally, we look at many ways of transforming or combining functions from all of the families of functions we have studied.

After reading this chapter you should be able to

- understand the behavior and construct graphs of quadratic and other polynomial functions

- determine the vertex and intercepts of a quadratic function

- convert quadratic functions from one form to another

- transform any function using stretches, compressions, shifts, and reflections

- combine any two functions using basic algebraic operations or composition

- identify when a function has an inverse

8.1   *An Introduction to Quadratic Functions*

The Simplest Quadratic

A power function of degree 2 (in the form $y = ax^2$) is also the simplest member of a family of functions called *quadratics.* We've already encountered quadratics such as $S = 6x^2$ (the surface area of a cube) and $S = \pi r^2$ (the area of a circle). Since $y = ax^2$ has an even integer power, its graph has the classic $\cup$ -shaped curve called a *parabola.* (See Figure 8.1.) Certain properties are suggested by the graph:

- The parabola is concave up if $a > 0$ (and concave down if $a < 0$).
- It has a minimum (or maximum) point called the *vertex.*
- It is symmetric across a vertical line called an *axis of symmetry* that runs through the vertex.

A parabola has another unique and useful property. There is an associated point off any parabola called the *focus* or *focal point.* It is located $\left|\frac{1}{4a}\right|$ units above (or below) the vertex and lies within the arms of the parabola.

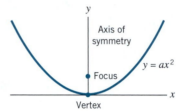

Figure 8.1 The parabolic graph of $y = ax^2$, where $a > 0$.

EXAMPLE 1   Determine whether the graph of $y = 2x^2$ is concave up or down, find its vertex and focal point, and then graph the function.

SOLUTION   Since $2x^2 \geq 0$, the minimum value for $y = 2x^2$ is 0, which will occur when $x = 0$. So the function has a minimum and its vertex is at the origin $(0, 0)$. Since $a > 0$, the graph is concave up. The coefficient, a, of x^2 is 2, so the focal point is $1/(4a) = 1/(4 \cdot 2) = 1/8$ or 0.125 units above the vertex. The coordinates of the focal point (or focus) are $(0, 0.125)$. Figure 8.2 shows the graph of the function.

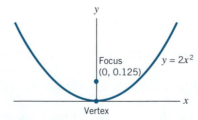

Figure 8.2 Graph of $y = 2x^2$ with vertex at $(0, 0)$ and focal point at $(0, 0.125)$.

The focal point merits its name. A three-dimensional parabolic bowl has vertical cross sections that are all the same parabola. If the bowl is built of a reflective material, the parabolic shape will concentrate parallel rays from the sun or a satellite TV channel at the focal point (see Figure 8.3). This property is used to focus the sun's rays to construct primitive cooking devices or to concentrate electromagnetic waves from satellites or distant stars.

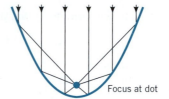

Figure 8.3 A parabola with parallel incoming rays concentrating at its focus.

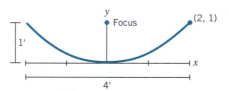

The course software "Q10: Parabolic Reflector" gives an interactive demonstration of the reflective property of parabolas.

This process also works in reverse. For example, if you have a light bulb centered at the focus of a parabola, the light rays will bounce off a reflective parabolic surface as a set of parallel rays. This property is used to construct lamps, car headlights, or radio transmitters that send messages to robot rovers on Mars or broadcast our existence to the universe.

Designing parabolic devices

The easiest way to design a parabolic device is to place the vertex of the parabola at the origin $(0, 0)$—in other words, to construct a function of the form $y = ax^2$.

EXAMPLE 2   **Designing a cooking device**

In developing countries, such as Burkina Faso,[1] solar parabolic cooking devices are cheap ways to cook food and pasteurize water. The cooking devices can be made any size and can even be constructed of hardboard and aluminum foil. If a large parabolic cooking device is built that is 4′ wide and 1′ deep, where should the cooking pot be centered for maximum efficiency?

Figure 8.4 Parabolic cooking device.

SOLUTION   The cooking pot should be placed at the focal point of the parabola, where the sun's rays will be concentrated. If we think of the bottom of the cooker as placed at $(0, 0)$, we can construct a function of the form $y = ax^2$ and use it to find the focal point (see Figure 8.4). To find the value of a, we need the coordinates of a point on the parabolic cross section. Since the cooker is 4′ wide, if we move 2′ to the right of the cooker's base, and 1′ up, we will be at the point $(2, 1)$, which lies on the rim of the cross section. Given the equation

$$y = ax^2$$

let $x = 2$ and $y = 1$ $\qquad 1 = a \cdot 2^2$

evaluate $\qquad\qquad\qquad 1 = a \cdot 4$

solve for a $\qquad\qquad\qquad a = \dfrac{1}{4}$

[1]In 2004 the La Trame documentary film studio released *Bon Appétit, Monsieur Soleil,* a film about solar cooking in Burkina Faso. The studio has produced several versions in different languages, including English. The film covers the essential points of solar cooking, the acceptance of these devices by the population, the advantages achieved by using them (savings in money and time), and their contribution to fighting deforestation. The film is a good tool to inform people living in countries with energy problems about other ways to cook. It also serves to introduce this cooking method to decision makers in government agencies and NGOs.

So the formula for the parabolic oven cross section is $y = \left(\frac{1}{4}\right)x^2$. The focal point is at $\left|\frac{1}{4a}\right| = \left|\frac{1}{4 \cdot \left(\frac{1}{4}\right)}\right| = \frac{1}{1}$ or 1 foot above the vertex, the bottom of the cooker. Since the height of the cooker is also 1 foot, then the pot should be suspended 1 foot above the center, at the level of the rim, for maximum efficiency.

The General Quadratic

A quadratic function may include a linear term added to the ax^2 term. The *general quadratic* in *standard* or *a-b-c form* is

$$y = ax^2 + bx + c \qquad (a \neq 0)$$

where a, b, and c are constants. A quadratic is also called a *polynomial of degree 2*. For example, the quadratic function $y = 3x^2 + 7$ is a polynomial of degree 2 where $a = 3$, $b = 0$, and $c = 7$.

EXAMPLE 3

A quadratic model for tuition revenue

Because of state financial problems, many publicly funded colleges are faced with large budget cuts. To raise more revenue, one state college plans to raise both tuition and student body size over the next 10 years. Its goal is to increase the current tuition of $6000 by $600 a year and increase the student population of 1200 by 40 students a year.

a. Generate two equations, one to describe the projected increases in tuition and the other to describe increases in student body size over time.

b. Generate an equation to describe the projected total tuition revenue over time.

c. Compare the current tuition revenue to the projected revenue in 10 years.

SOLUTION

a. If we let S = student body size and N = number of years after the present, then $S = 1200 + 40N$.

If we let T = cost of tuition in thousands of dollars, then $T = 6 + 0.6N$.

b. The total tuition revenue, R, is the product of the number of students, S, times the cost of tuition, T. So

$$R = S \cdot T$$

Substitute $\qquad = (1200 + 40N)(6 + 0.6N)$

multiply out[2] $\qquad = 7200 + 720N + 240N + 24N^2$

combine terms $\qquad = 7200 + 960N + 24N^2$

So the total revenue R (in thousands of dollars) from tuition is given by the quadratic function

$$R = 7200 + 960N + 24N^2$$

c. At the present time $N = 0$, so the total tuition revenue $R = \$7200$ thousand or $7.2 million. In 10 years, $N = 10$, so

$$R = 7200 + (960 \cdot 10) + (24 \cdot 10^2)$$
$$= 7200 + 9600 + 2400$$
$$= 19{,}200$$

So in 10 years the projected total tuition revenue is $19,200 thousand or $19.2 million.

[2]Recall that the product $(a + b)(c + d)$ is the sum of four terms: the product of the first (F) two terms, the outside (O) terms, the inside (I) terms, and the last (L) two terms of the factors.

$$(a + b)(c + d) = ac + ad + bc + bd$$
$$\qquad\qquad\qquad\quad F \quad O \quad I \quad L$$

Properties of Quadratic Functions

The vertex and the axis of symmetry

The graph of the general quadratic, $y = ax^2 + bx + c$, is a parabola that is symmetric about its axis of symmetry. For any quadratic other than the simplest (of the form $y = ax^2$), the axis of symmetry is not the vertical axis. The vertex lies on the axis of symmetry and is a maximum (if the parabola is concave down) or a minimum (if the parabola is concave up) (see Figure 8.5). In Section 8.2 we'll learn how to find the coordinates of the vertex.

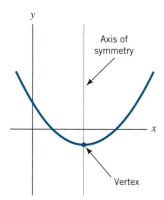

Figure 8.5 Each parabola has a vertex that lies on an axis of symmetry.

Vertical and horizontal intercepts

The general properties of vertical and horizontal intercepts are the same for all functions.

Intercepts of a Function

For any function $f(x)$:

   the vertical intercept is at $f(0)$

   the horizontal intercepts occur at values of x, called *zeros*, where $f(x) = 0$

For a quadratic function $f(x) = ax^2 + bx + c$, we have $f(0) = c$. So $f(x)$ has a vertical intercept at $(0, c)$. We usually shorten this to say that the vertical intercept is at c, since we know the other coordinate is 0.

Because of the $\cup$ shape of its graph, a quadratic function may have no, one, or two horizontal intercepts (see Figure 8.6). If $(r, 0)$ are the coordinates of a horizontal intercept, we abbreviate this to say that the horizontal intercept is at r. These values of r are also called *zeros* of the function since $f(r) = 0$. In Section 8.3 we'll learn how to find the horizontal intercepts.

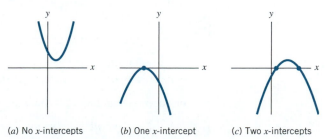

(a) No x-intercepts   (b) One x-intercept   (c) Two x-intercepts

Figure 8.6 Graphs of quadratic functions showing the three possible cases for the number of horizontal or x-intercepts.

Since a parabola is symmetric about its axis of symmetry, each point on one arm of the parabola has a mirror image on the other arm. In particular, if there are two horizontal intercepts at r_1 and r_2, then $(r_1, 0)$ and $(r_2, 0)$ are mirror images across the parabola's axis of symmetry. The horizontal intercepts will lie at an equal distance, d, to the left and right of the axis of symmetry (see Figure 8.7). Equivalently, the axis of symmetry crosses the horizontal axis halfway between r_1 and r_2, at the one-dimensional point where x is $\dfrac{(r_1 + r_2)}{2}$. So the equation of the axis of symmetry is the two-dimensional line $x = \dfrac{(r_1 + r_2)}{2}$.

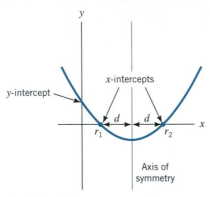

Figure 8.7 Each parabola has a vertical intercept and zero, one, or two horizontal intercepts.

The focal point

The focal point for the general parabola $y = ax^2 + bx + c$ is still $\left|\dfrac{1}{4a}\right|$ units above (or below) the vertex. The distance $\left|\dfrac{1}{4a}\right|$ from the vertex to the focal point is called the *focal length.*

The Quadratic Function and Its Graph

A *quadratic function* can be written in standard form as

$$f(x) = ax^2 + bx + c \qquad \text{(where } a \neq 0)$$

Its graph:

is called a *parabola*

is symmetric about its *axis of symmetry*

has a minimum (if the parabola is concave up) or a maximum (if the parabola is concave down) point called its *vertex*

has one vertical intercept, but may have zero, one, or two horizontal intercepts

has a *focal point* $\left|\dfrac{1}{4a}\right|$ units above the minimum (or below the maximum); the value $\left|\dfrac{1}{4a}\right|$ is called the *focal length.*

Estimating the Vertex and Horizontal Intercepts

Quadratics arise naturally in area and motion problems. The earliest problems we know of that led to quadratic equations are on Babylonian tablets dating from 1700 B.C. The writings suggest a problem similar to the following example.

EXAMPLE 4   **Maximizing area**

What is the maximum rectangular area you can enclose within a fixed perimeter of 24 meters? What are the dimensions of the rectangle with the maximum area?

SOLUTION   If the rectangular region has length L and width W, then

$$2L + 2W = \text{perimeter}$$

Substitute $2L + 2W = 24$

divide by 2 $L + W = 12$

subtract L from both sides $W = 12 - L$

Since area, A, is given by $A = L \cdot W$

substitute for W $= L \cdot (12 - L)$

multiply through $= 12L - L^2$

So area, A, is a quadratic function of L. Table 8.1 shows values of L from 0 to 12 meters and corresponding values for W and A.

Figure 8.8 shows a graph of area, A, versus length, L. In the equation $A = 12L - L^2$ the coefficient of L^2 is negative (-1), so the parabola is concave down. From the table and the graph, it appears that the vertex of the parabola is at $(6, 36)$; that is, at a length of 6 meters the area reaches a maximum of 36 square meters. Since $W = 12 - L$, when $L = 6$, $W = 6$. Hence the rectangle has maximum area when the length equals the width, or in other words when the rectangle is a square.

Length, L (m)	Width, W (m)	Area, A (m^2)
0	12	0
1	11	11
2	10	20
3	9	27
4	8	32
5	7	35
6	6	36
7	5	35
8	4	32
9	3	27
10	2	20
11	1	11
12	0	0

Table 8.1

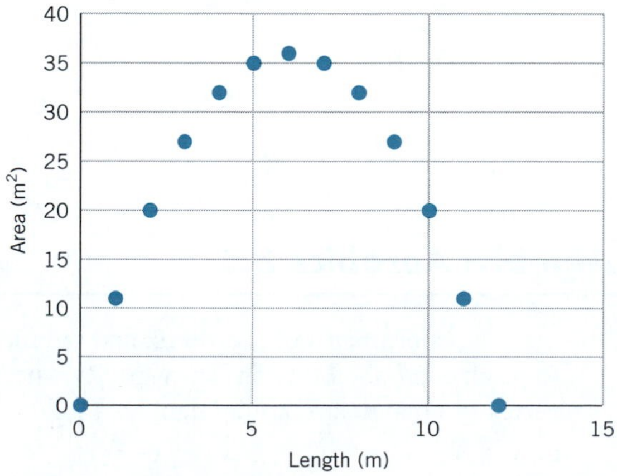

Figure 8.8 Graph of area vs. length for a rectangle.

EXAMPLE 5   **The trajectory of a projectile**

Figure 8.9 shows a plot of the height above the ground (in feet) of a projectile for the first 5 seconds of its trajectory. How could we:

a. Estimate the height from which the projectile was launched?

b. Estimate when the projectile will hit the ground?

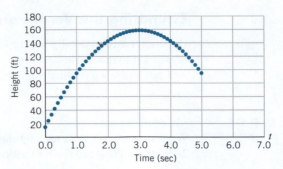

Figure 8.9 The height of a projectile.

SOLUTION

a. The initial height of the projectile corresponds to the vertical intercept of the parabola (where time $t = 0$). A quick look at Figure 8.9 gives us an estimate somewhere between 10 and 20 feet.

If we use technology to generate a best-fit quadratic, we get $H(t) = -16t^2 + 96t + 15$, where $t =$ time (in seconds) and $H(t) =$ height (in feet). Since $H(0) = 15$, then 15 is the vertical intercept. In other words, when $t = 0$ seconds, the initial height is $H(0) = 15$ feet.

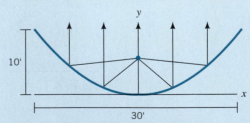

"The Mathematics of Motion" following this chapter offers an extended exploration into constructing mathematical models to describe the motion of freely falling bodies.

b. Figure 8.10 overlays on Figure 8.9 the graph of $H(t) = -16t^2 + 96t + 15$, our function model for the projectile's path. The projectile will hit the ground when the height above the ground $H(t) = 0$. This occurs at a horizontal intercept of the parabola. One horizontal intercept (not shown) would occur at a negative value of time, t, which would be meaningless here. From the graph we can estimate that the other intercept occurs at $t \approx 6.2$ seconds.

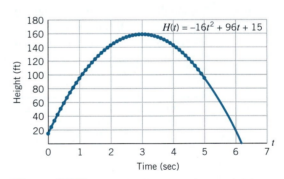

Figure 8.10 A graph of the function model for the projectile's path.

Algebra Aerobics 8.1

1. Find the coordinates of the vertex and the focal point for each quadratic function. Then specify whether each vertex is a maximum or minimum.

   a. $y = 3x^2$ **c.** $y = \frac{1}{24}x^2$

   b. $y = -6x^2$ **d.** $y = -\frac{1}{12}x^2$

2. A designer is planning an outdoor concert place for solo performers. The stage area is to have a parabolic wall 30′ wide by 10′ deep. If the performer stands at the focal point, the sound will be reflected out in parallel sound waves directly to the audience in front. See Figure 8.11.

Figure 8.11 A parabolic design for an outdoor concert stage back wall.

   a. Where would the focal point be?

   b. What is the equation for the reflecting wall?

c. Do you see any problem with this idea for concert acoustics? (*Hint:* What if the audience is noisy?)

3. Evaluate each of the following quadratic functions at 2, −2, 0, and z.

   a. $g(x) = x^2$ d. $m(s) = 5 + 2s - 3s^2$

   b. $h(w) = -w^2$ e. $D(r) = -(r - 3)^2 + 4$

   c. $Q(t) = -t^2 - 3t + 1$ f. $k(x) = 5 - x^2$

4. For each of the graphs of the quadratic functions $f(x)$, $g(x)$, $h(x)$, and $j(x)$ given below, determine:

   i. If the parabola is concave up or concave down

   ii. If the parabola has a maximum or a minimum

   iii. The equation of the axis of symmetry

   iv. The coordinates of the vertex

   v. Estimated coordinates of the horizontal and vertical intercepts

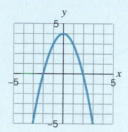

a. Graph of $f(x)$

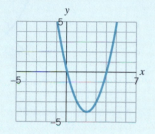

c. Graph of $h(x)$

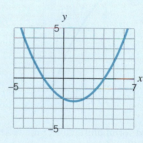

b. Graph of $g(x)$

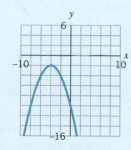

d. Graph of $k(x)$

5. A manufacturer wants to make a camp stove using a parabolic reflector to concentrate sun rays at the focal point of the parabola, where the food to be cooked would be placed. The reflector must be no wider than 24″, and the depth is to be the same as the focal length, f, the distance from the focal point to the bottom. See Figure 8.12.

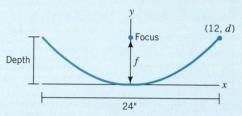

Figure 8.12 A parabolic design for a camp stove.

a. What focal length would the camp stove need to have? (*Hint:* The focal length is $\left|\frac{1}{4a}\right|$).

b. What formula is needed for manufacturing the stove reflector?

6. Put the following quadratic functions into standard a-b-c form:

$$f(x) = (x - 3)(x + 5) \qquad g(x) = (2x + 5)(x + 1)$$

$$h(x) = 10(x - 3)(x - 5) \qquad j(x) = 2(x - 3)(x + 3)$$

Exercises for Section 8.1

The "Extended Exploration: The Mathematics of Motion," which follows this chapter, contains many additional exercises using quadratics to describe freely falling and thrown bodies.

A graphing program is recommended or required for several exercises.

1. From the graph of each quadratic function, identify whether the parabola is concave up or down and hence whether the function has a maximum or minimum. Then estimate the vertex, the axis of symmetry, and any horizontal and vertical intercepts.

a.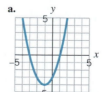

b.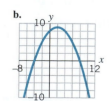

c.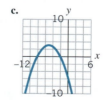

2. Using the quadratic function $y = \frac{1}{8}x^2$:

a. Fill in the values in the table.

x	−12	−8	−4	0	4	8	12
y							

b. Plot the points by hand and sketch the graph of the function.

c. Determine the coordinates of the focal point and place the focal point on the graph.

d. What is the equation of the axis of symmetry?

e. What are the coordinates of the vertex? Is the vertex a maximum or minimum?

3. (Graphing program recommended.) On the same graph, plot the three functions.

$$y_1 = x^2 \qquad y_2 = \tfrac{1}{4}x^2 \qquad y_3 = \tfrac{1}{12}x^2$$

a. Calculate the focal point for each parabola.

b. Compare y_2 and y_3 with y_1.

c. Complete this sentence: As the value of $|a|$ gets smaller, the focal point gets _____ the vertex and the graph gets _____.

4. (Graphing program recommended.) On the same graph, plot the three functions.

$$y_1 = x^2 \qquad y_2 = 4x^2 \qquad y_3 = 8x^2$$

a. Calculate the focal point for each parabola.

b. Compare y_3 and y_2 with y_1.

c. Complete this sentence: As the value of $|a|$ gets larger, the focal point gets _____ to the vertex and the graph of the parabola gets _____.

5. (Graphing program optional.) Given a point P on a parabola of the form $y = ax^2$, find the equation of the parabola. If available, use technology to verify your equations by plotting each parabola.

a. $P = (12, 6)$ **c.** $P = (4, 12)$

b. $P = (-12, -6)$ **d.** $P = (3, 144)$

6. Find the coordinates of the vertex for each quadratic function listed. Then specify whether each vertex is a maximum or minimum.

a. $y = 4x^2$ **c.** $P(n) = \left(\tfrac{1}{12}\right)n^2$

b. $f(x) = -8x^2$ **d.** $Q(t) = -\left(\tfrac{1}{24}\right)t^2$

7. Given the following focal points, write the equation of a parabola in the form $y = ax^2$ by finding a.

a. $(0, 4)$ **c.** $\left(0, \tfrac{1}{16}\right)$

b. $(0, -8)$ **d.** $\left(0, -\tfrac{1}{24}\right)$

8. Find the focal length (the distance from the focal point to the vertex) for each of the following.

a. $f(x) = x^2$ **b.** $f(x) = 3x^2$ **c.** $f(x) = \tfrac{1}{3}x^2$

9. For each of the following functions, evaluate $f(2)$ and $f(-2)$.

a. $f(x) = x^2 - 5x - 2$

b. $f(x) = 3x^2 - x$

c. $f(x) = -x^2 + 4x - 2$

10. A designer proposes a parabolic satellite dish 5 feet in diameter and 15 inches deep.

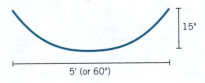

5' (or 60")

a. What is the equation for the cross section of the parabolic dish?

b. What is the focal length (the distance between the vertex and the focal point)?

c. What is the diameter of the dish at the focal point?

11. An electric heater is designed as a parabolic reflector that is 5″ deep. To prevent accidental burns, the centerline of the heating element, placed at the focus, must be set in 1.5″ from the base of the reflector.

a. What equation could you use to design the reflector? (*Hint:* Tip the parabolic reflector so it opens upward.)

b. How wide would the reflector be at its rim?

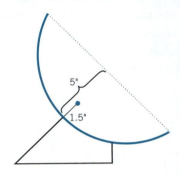

12. A parabolic reflector 3″ in diameter and 2″ deep is proposed for a spotlight.

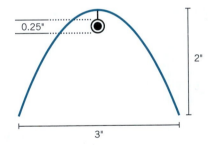

a. What formula is needed for manufacturing the reflector?

b. Where would the focus be?

c. Will a $\tfrac{1}{4}$″-diameter light source fit if it is centered at the focus?

13. A slimline fluorescent bulb $\tfrac{1}{2}$″ in diameter needs 1″ clearance top and bottom in a parabolic reflecting shade.

a. What are the coordinates of the focus for this parabola?

b. What is the equation for the parabolic curve of the reflector?

c. What is the diameter of the opening of the shade?

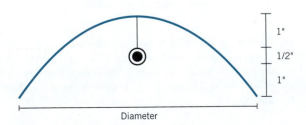

Diameter

14. If we know the radius and depth of a parabolic reflector, we also know where the focus is.

    a. Find a generic formula for the focal length f of a parabolic reflector expressed in terms of its radius R and depth D. The focal length $\left|\frac{1}{4a}\right|$ is the distance between the vertex and the focal point. Assume $a > 0$.

    b. Under what conditions does $f = D$?

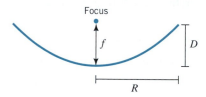

15. Construct several of your own equations of the form $y = ax^2$ and then describe in words how the focal length varies depending on how open or closed the parabolic curve is.

16. For each of the following quadratics with their respective vertices, calculate the distance from the vertex to the focal point. Then determine the coordinates of the focal point.

    a. $f(x) = x^2 - 2x - 3$ with vertex at $(1, -4)$

    b. $g(t) = 2t^2 - 16t + 24$ with vertex at $(4, -8)$

17. Put each of the following quadratics into standard form.

    a. $f(x) = (x + 3)(x - 1)$

    b. $P(t) = (t - 5)(t + 2)$

    c. $H(z) = (2 + z)(1 - z)$

18. Put each of the following quadratics into standard form.

    a. $g(x) = (2x - 1)(x + 3)$

    b. $h(r) = (5r + 2)(2r + 5)$

    c. $R(t) = (5 - 2t)(3 - 4t)$

19. Determine the dimensions for enclosing the maximum area of a rectangle if:

    a. The perimeter is held constant at 200 meters.

    b. The perimeter is held constant at P meters.

20. A gardener wants to grow carrots along the side of her house. To protect the carrots from wild rabbits, the plot must be enclosed by a wire fence. The gardener wants to use 16 feet of fence material left over from a previous project. Assuming that she constructs a rectangular plot, using the side of her house as one edge, estimate the area of the largest plot she can construct.

21. Which of the following are true statements for quadratic functions?

    a. The vertex and focal point always lie on the axis of symmetry.

    b. The graph of a parabola could have three horizontal intercepts.

    c. The graph of a parabola does not necessarily have a vertical intercept.

    d. If $f(2) = 0$, then f has a horizontal intercept at 2.

    e. The focal point always lies above the vertex.

22. The management of a company is negotiating with a union over salary increases for the company's employees for the next 5 years. One plan under consideration gives each worker a bonus of $1500 per year. The company currently employs 1025 workers and pays them an average salary of $30,000 a year. It also plans to increase its workforce by 20 workers a year.

    a. Construct a function $C(t)$ that models the projected cost of this plan (in dollars) as a function of time t (in years).

    b. What will the annual cost be in 5 years?

23. (Graphing program required for part (c).) A landlady currently rents each of her 50 apartments for $1250 per month. She estimates that for each $100 increase in rent, two additional apartments will remain vacant.

    a. Construct a function that represents the revenue $R(n)$ as a function of the number of rent increases, n. (*Hint:* Find the rent per unit after n increases and the number of units rented after n increases.)

    b. After how many rent increases will all the apartments be empty? What is a reasonable domain for this function?

    c. Using technology, plot the function. From the graph, estimate the maximum revenue. Then estimate the number of rent increases that would give you the maximum revenue.

24. (Graphing program required for part (c).)

    a. In economics, revenue R is defined as the amount of money derived from the sale of a product and is equal to the number x of units sold times the selling price p of each unit. What is the equation for revenue?

    b. If the selling price is given by the equation $p = -\frac{1}{10}x + 20$, express revenue R as a function of the number x of units sold.

    c. Using technology, plot the function and estimate the number of units that need to be sold to achieve maximum revenue. Then estimate the maximum revenue.

8.2 *Finding the Vertex: Transformations of y = x²*

In this section we show that every quadratic can be generated through transformations of the basic quadratic $y = x^2$. We start by investigating what happens to the equation of a function when we change its graph.

Stretching and Compressing Vertically

When we stretch or compress a graph we change its shape, creating a graph of a new function. To find the equation of this new function, we need to ask, "What is the relationship between the graphs of the new function and the old function?"

We can build on what we know from Chapter 7, where we learned that multiplying a power function by a constant stretches or compresses its graph. Specifically, for power functions in the form $y = kx^p$, the value of the coefficient k compresses or stretches the graph of $y = x^p$. A quadratic in the form $y = ax^2$ is a power function, so the graph of $y = ax^2$ is the graph of $y = x^2$

> vertically stretched by a factor of a, if $a > 1$
>
> vertically compressed by a factor of a, if $0 < a < 1$

The magnitude (or absolute value) of a tells us how much the graph of $y = x^2$ is stretched or compressed. As $|a|$ increases, a acts as a vertical stretch factor, which pulls harder and harder on the arms of the parabola anchored at the vertex, narrowing the graph. As $|a|$ decreases, a acts as a compression factor, which weighs down the graph, flattening the parabola's arms. So the value of a determines the shape of the parabola. See Figure 8.13.

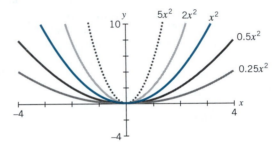

Figure 8.13 The graph of $y = x^2$ vertically compressed and stretched.

Reflections across the Horizontal Axis

If $a < 0$, the graph of $y = ax^2$ is the graph of $y = x^2$ stretched or compressed by a factor of $|a|$ and then reflected across the x-axis. In general, the graphs of $y = ax^2$ and $y = -ax^2$ are reflections of each other across the x-axis. See Figure 8.14.

The sign of a tells us whether the parabola opens up or down and if the vertex represents a maximum or a minimum value of the function. If a is positive, the parabola is concave up (opens upward) and the vertex is a minimum. If a is negative, the parabola is concave down (opens downward) and the vertex is a maximum.

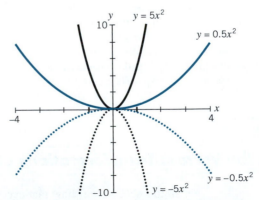

Figure 8.14 Graphs that are reflected across the x-axis.

EXAMPLE 1 Describe how the following pairs of functions are related to each other and to $y = x^2$. Sketch each pair and $y = x^2$ on the same grid.

  a. $y = 3x^2$ and $y = -3x^2$ **b.** $y = 0.3x^2$ and $y = -0.3x^2$

SOLUTION **a.** The functions $y = 3x^2$ and $y = -3x^2$ are reflections of each other across the x-axis. For both functions, the absolute value of the coefficient a of x^2 is 3, so in each case $|a| > 1$. Hence both graphs are the graph of $y = x^2$ stretched by a factor of 3. See Graph A in Figure 8.15.

  b. The functions $y = 0.3x^2$ and $y = -0.3x^2$ are also reflections of each other across the x-axis. For both functions $|a| = 0.3$, so in each case $0 < |a| < 1$. Hence both of their graphs are the graph of $y = x^2$ compressed by a factor of 0.3. See Graph B in Figure 8.15.

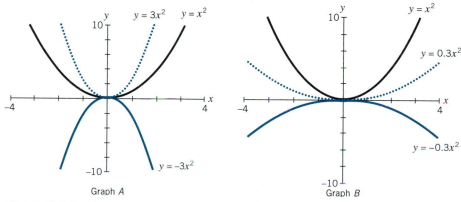

Graph A   Graph B

Figure 8.15 Stretching and compressing $y = x^2$.

Stretching, Compressing, and Reflecting the Graph of Any Function

If f is a function and a is a constant, then the graph of $a \cdot f(x)$ is the graph of $f(x)$

  vertically stretched by a factor of a, if $a > 1$

  vertically compressed by a factor of a, if $0 < a < 1$

  vertically stretched or compressed by a factor of $|a|$ and reflected across x-axis, if $a < 0$

Shifting Vertically and Horizontally

What happens when we keep the shape of the graph of $y = ax^2$ but change its position on the grid, shifting the graph vertically or horizontally? Clearly we will get a graph of a new function where the coordinates of the vertex are no longer (0, 0). Let's see how to construct the equation of the new function from the original function.

Shifting a graph vertically

How does a function change when its graph is shifted up or down? Examine the graphs in Figure 8.16, Graphs A and B.

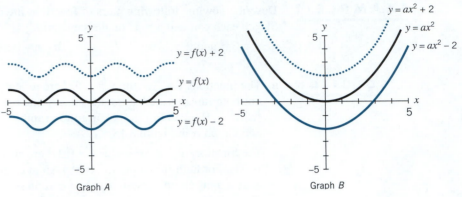

Figure 8.16 Graphs that are shifted vertically.

What happens to the output value, y, when the graph of $y = f(x)$ is shifted up two units? Every y value increases by two units. Translating this shift into equation form, our new function is $y = f(x) + 2$. (See Figure 8.16, Graph A.)

What happens when the graph of $y = f(x)$ is shifted down two units? Our new function is of the form $y = f(x) - 2$. In general, if the vertical shift is k units, then

$$y = f(x) + k$$

is the graph of $y = f(x)$ shifted up by k units if k is positive and shifted down by k units if k is negative.

Figure 8.16, Graph B shows the graph of $f(x) = ax^2$ raised or lowered by two units. If the vertical shift is k units, then the graph of

$$y = ax^2 + k$$

is the graph of $y = ax^2$ shifted up by k units if k is positive and shifted down by k units if k is negative. The vertex is at $(0, k)$.

Algebra Aerobics 8.2a

In Problems 1–5, without drawing the graphs, compare the graph of part (b) to the graph of part (a).

1. **a.** $r(x) = x^2 + 2$ **b.** $s(x) = 2x^2 + 2$
2. **a.** $h(t) = t^2 + 5$ **b.** $k(t) = -t^2 + 5$
3. **a.** $f(z) = -5z^2$ **b.** $g(z) = -0.5z^2$
4. **a.** $f(x) = x^2 + 3x + 2$ **b.** $g(x) = x^2 + 3x + 8$
5. **a.** $f(t) = -3t^2 + t - 5$ **b.** $g(t) = -3t^2 + t - 2$

6. Create a quadratic equation of the form $y = ax^2 + k$ with the given values for a and the vertex. Sketch by hand the graph of each equation.

   a. $a = 3$ and the vertex is at $(0, 5)$.
   b. $a = \frac{1}{3}$ and the vertex is at $(0, -2)$.
   c. $a = -2$ and the vertex is at $(0, 4)$

7. Create new functions by performing the following transformations on $f(x) = x^2$.

   a. $g(x)$ is $f(x)$ stretched by a factor of 3.
   b. $h(x)$ is $f(x)$ stretched by a factor of 5 and reflected across the x-axis.
   c. $j(x)$ is $f(x)$ compressed by a factor of 1/2.
   d. $k(x)$ is $f(x)$ reflected across the x-axis.

8. Create a function in the form $y = ax^2 + k$ for each of the following transformations of $f(x) = x^2$:

   a. Stretched by a factor of 5 and shifted down 2 units
   b. Concave up and shifted 3 units up
   c. Multiplied by a factor of 0.5, concave down, and shifted 4.7 units down
   d. Opens up and is $f(x)$ shifted 71 units down

Shifting a graph horizontally

How does a function change when its graph is shifted to the right or left? Examine the graphs in Figure 8.17, Graphs A and B.

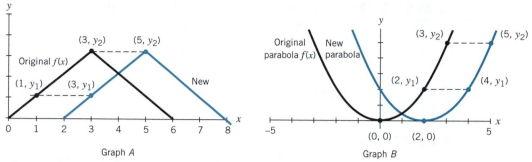

Graph A   Graph B

Figure 8.17 Graphs that are shifted horizontally to the right.

When we shift the graph of $y = f(x)$ two units to the right, how is the new function related to our original function? Recall that adding a constant to the output value of a function shifts its graph vertically. When we shift the graph of $f(x)$ horizontally, the input value, x, is shifted. The dotted lines in Figure 8.17 show that for any particular output value, y, the corresponding input value for the new function is $(x - 2)$. So the graph of $f(x)$ has been shifted to the *right* two units. Translating this shift into equation form, we get

$$y = f(x - 2)$$

At first this may seem counterintuitive. But remember, we are expressing the new function in terms of the original function. If you evaluate the new function at x, the same y-value of the original function $f(x)$ is now at $f(x - 2)$. For example, in Figure 8.17 if you evaluate each of the new functions in Graph A and Graph B at $x = 5$, you get an output of y_2. The same output for the original function occurs at $5 - 2 = 3$.

What if we shift $y = f(x)$ to the *left* two units? We need to think about the effect of the shift and the direction of the shift. If you evaluate the new function at x, the comparable point on the graph of $f(x)$ is $f(x + 2)$. We can think of this shift as replacing x with $x - (-2)$ or $x + 2$, so our new function is

$$y = f(x + 2)$$

Figure 8.18 shows the graph of $y = 3x^2$ after being shifted to the right and then to the left 2 units. Note that the value of the x-coordinate of the vertex changed by ± 2 units. If the horizontal shift is h units, then

$$y = a(x - h)^2$$

is the graph of $y = ax^2$ shifted right by h if h is positive, and shifted left by h if h is negative. The vertex is now at $(h, 0)$. Note that if $h = 2$, the expression $(x - h)$ becomes $x - 2$, and if $h = -2$, the expression $(x - h)$ becomes $x - (-2) = x + 2$.

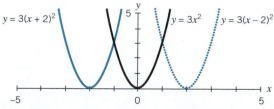

Figure 8.18 Graph of $y = 3x^2$ shifted horizontally to the left and to the right two units.

EXAMPLE 2

a. Identify the vertex for each of the following functions and indicate whether it represents a maximum or minimum value:

$$g(x) = 5(x - 3)^2 \qquad j(x) = 5x^2 + 3$$
$$h(x) = -5(x + 3)^2 \qquad k(x) = -5x^2 - 3$$

b. Describe how the graphs in part (a) are related to the graph of $f(x) = 5x^2$. Specify the order of any transformations.

SOLUTION

a. Vertex for: g is at $(3, 0)$; j is at $(0, 3)$; h is at $(-3, 0)$; k is at $(0, -3)$.

Vertices for g and j represent minimum values since in both cases the coefficient for x^2 is positive ($a = 5$).

Vertices for h and k represent maximum values since in both cases the coefficient for x^2 is negative ($a = -5$).

b. The graph of $f(x) = 5x^2$

shifted horizontally to the right three units is the graph of $g(x)$

shifted vertically up three units is the graph of $j(x)$

shifted horizontally to the left three units and then reflected across the x-axis is the graph of $h(x)$

reflected across the x-axis and then shifted vertically down three units is the graph of $k(x)$

We can generalize to any function $f(x)$.

Horizontal and Vertical Shifts

The graph of $f(x) + k$ is the graph of $f(x)$ shifted vertically $|k|$ units.

If k is positive, the shift is up; if k is negative, the shift is down.

The graph of $f(x - h)$ is the graph of $f(x)$ shifted horizontally $|h|$ units.

If h is positive, the shift is to the right; if h is negative, the shift is to the left.

Using Transformations to Get the Vertex Form

We can use the previous transformations to generate a quadratic function in what is called the *vertex form*. We start by stretching or compressing the graph of $y = x^2$ by a factor of a. So our new function will be

$$y = ax^2$$

Next we shift the graph of the function $y = ax^2$ horizontally h units and vertically k units, to get

$$y = a(x - h)^2 + k$$

The vertex is now at (h, k). This quadratic is in the *vertex* or *a-h-k form*. Its axis of symmetry is the vertical line at $x = h$. See Figure 8.19.

"Q3: a, h, k Sliders," "Q8: $y = ax^2$ vs. $y = a(x - h)^2 + k$," and "Q9: Finding 3 Points: a-h-k Form" in *Quadratic Functions* can help you visualize quadratic functions in the a-h-k form.

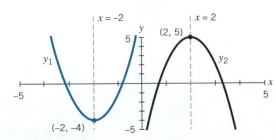

Figure 8.19 Graphs of $y_1 = 3(x + 2)^2 - 4$ and $y_2 = -3(x - 2)^2 + 5$.

> **The Vertex Form of a Quadratic Function**
>
> The *vertex* or *a-h-k form* of the quadratic function is
>
> $$f(x) = a(x - h)^2 + k$$
>
> where the vertex is at (h, k).

EXAMPLE 3 **Transforming graphs**

Given $f(x) = x^2$ and $g(x) = -2(x + 3)^2 + 5$, show how to transform the graph of $f(x)$ into the graph of $g(x)$.

SOLUTION One way to transform the graph of $f(x)$ into the graph of $g(x)$ is shown in Figure 8.20.

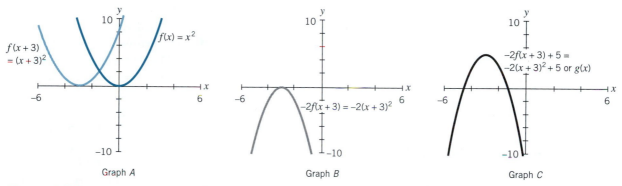

Graph A   Graph B   Graph C

Figure 8.20 The graph of $f(x) = x^2$ in A is shifted horizontally to the left three units, then in B stretched vertically by a factor of 2 and reflected across the x-axis, and finally in C shifted up vertically five units to generate $g(x) = -2(x + 3)^2 + 5$.

EXAMPLE 4 **Finding the function from its graph**

Figure 8.21 shows the graph of $f(x) = 2x^2$ transformed into three new parabolas. Assume that each of the three new graphs retains the shape of $f(x)$.

a. Estimate the coordinates of the vertex for each new parabola.

b. Use your estimates from part (a) to write equations for each parabola in Figure 8.21.

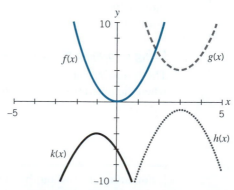

Figure 8.21 Three transformations of $f(x) = 2x^2$

SOLUTION **a.** Vertex for: $g(x)$ is at $(3, 4)$; $h(x)$ is at $(3, -1)$; $k(x)$ is at $(-1, -4)$

b. $g(x) = 2(x - 3)^2 + 4$; $h(x) = -2(x - 3)^2 - 1$; $k(x) = -2(x + 1)^2 - 4$

EXAMPLE 5   **Identifying the vertex**

For the following functions, identify the coordinates of the vertex and specify whether the vertex represents a maximum or minimum.

   a. $y = -5x^2$ **b.** $y = 2(x - 3)^2 - 15$ **c.** $y = -3(x + 5)^2 + 10$

SOLUTION   **a.** The vertex is at $(0, 0)$ and represents a maximum.

   b. The vertex is at $(3, -15)$ and represents a minimum.

   c. The vertex is at $(-5, 10)$ and represents a maximum.

Algebra Aerobics 8.2b

For Problems 1 to 3, without graphing, compare the positions of the vertices of parts (b) and (c) to that of part (a).

1. a. $y = x^2$ **c.** $y = (x - 2)^2$
   b. $y = (x + 3)^2$

2. a. $f(x) = 0.5x^2$ **c.** $f(x) = 0.5(x + 4)^2$
   b. $f(x) = 0.5(x - 1)^2$

3. a. $r = -2t^2$ **c.** $r = -2(t - 0.9)^2$
   b. $r = -2(t + 1.2)^2$

For Problems 4 and 5, without graphing, compare the graphs of parts (b), (c), and (d) to that of part (a).

4. a. $y = x^2$ **c.** $y = (x - 2)^2 + 4$
   b. $y = (x - 2)^2$ **d.** $y = (x - 2)^2 - 3$

5. a. $y = -x^2$ **c.** $y = -(x + 3)^2 - 1$
   b. $y = -(x + 3)^2$ **d.** $y = -(x + 3)^2 + 4$

6. Create new functions by performing the following transformations on $f(x) = x^2$. Give the coordinates of the vertex for each new parabola.

   a. $g(x)$ is $f(x)$ shifted right 2 units, stretched by a factor of 3, and then shifted down by 1 unit.

   b. $h(x)$ is $f(x)$ shifted left 3 units, stretched by a factor of 2, then reflected across the x-axis, and finally shifted up by 5 units.

   c. $j(x)$ is $f(x)$ shifted left 4 units, compressed by a factor of 5, and then shifted down by 3.5 units.

   d. $k(x)$ is $f(x)$ shifted right 1 unit, reflected across the x-axis, and then shifted up by 4 units.

7. Give the coordinates of the vertex for each of the following functions and indicate whether the vertex represents a maximum or minimum.

   a. $y = 2(x - 3)^2 - 4$ **c.** $y = -0.5(x - 4)^2$
   b. $y = -3(x + 1)^2 + 5$ **d.** $y = \frac{2}{3}x^2 - 7$

Finding the Vertex from the Standard Form

What if our quadratic function is in the standard a-b-c form and we want to find the vertex of its graph? We can use a formula to find the coordinates of the vertex or, as we show in the next section, we can convert the quadratic from the standard to vertex form, in which the vertex is easy to identify.

Using a formula to find the vertex

The following formula can be used to find the coordinates of the vertex of a parabola when the function is in standard form.

See the reading "Why the Formula for the Vertex and the Quadratic Formula Work" for a derivation of this vertex formula.

> **Formula for Finding the Vertex from the Standard Form**
>
> The vertex of a quadratic function in the form
>
> $$f(x) = ax^2 + bx + c$$
>
> has coordinates $\left(-\frac{b}{2a}, f\left(-\frac{b}{2a}\right)\right)$

EXAMPLE 6 **Finding the vertex**

Find the vertex and sketch the graph of $f(x) = x^2 - 10x + 100$.

SOLUTION The function f is in standard form, $f(x) = ax^2 + bx + c$. So $a = 1$, $b = -10$, and $c = 100$. Since a is positive, the graph is concave up, so the vertex represents a minimum. Using the formula for the horizontal coordinate of the vertex, we have

$$x = -\frac{b}{2a} = -\frac{(-10)}{2(1)} = \frac{10}{2} = 5$$

To find the vertical coordinate of the vertex, we need to find the value of $f(x)$ when $x = 5$.

Given	$f(x) = x^2 - 10x + 100$
let $x = 5$	$f(5) = 5^2 - 10(5) + 100$
simplify	$= 75$

The coordinates of the vertex are $(5, 75)$ and the y-intercept is at $f(0) = 100$. Figure 8.22 shows a sketch of the graph.

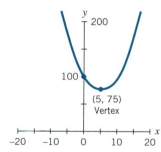

Figure 8.22 Graph of $f(x) = x^2 - 10x + 100$.

EXAMPLE 7 **Measuring traffic flow**

Urban planners and highway designers are interested in maximizing the number of cars that pass along a section of roadway in a certain amount of time. Observations indicate that the primary variable controlling traffic flow (and hence a good choice for the independent variable) is the density of cars on the roadway: the closer each driver is to the car ahead, the more slowly he or she drives. The following quadratic relationship between traffic flow rate and density of cars was derived from observing traffic patterns in the Lincoln Tunnel, which connects New York and New Jersey:

$$t = -0.21d^2 + 34.66d$$

where t = traffic flow rate (cars/hour) and d = density of cars (cars/mile).

a. Find the coordinates of the vertex of the parabola.

b. What does the vertex represent in terms of the traffic flow rate?

c. Graph the function and estimate the value of the horizontal intercepts.

d. What do the horizontal intercepts mean in terms of the traffic flow and density of cars?

SOLUTION **a.** Our equation $t = -0.21d^2 + 34.66d$ is in standard form ($t = ad^2 + bd + c$), so we can use the formula for finding the coordinates of the vertex. Since $a = -0.21$, $b = 34.66$, and $c = 0$, the horizontal coordinate of the vertex is

$$-\frac{b}{2a} = \frac{-34.66}{2(-0.21)}$$

simplify	≈ 82.52
round to the nearest integer	≈ 83

We can find the corresponding value of t by substituting $d = 83$ into our equation:

Given	$t = -0.21d^2 + 34.66d$
substitute for d	$t \approx -0.21(83)^2 + 34.66(83)$
simplify	$= 1430.09$
round to the nearest integer	≈ 1430

So the coordinates of the vertex are approximately (83, 1430).

b. A vertex at (83, 1430) means that when the density is 83 cars/mile, the traffic flow rate reaches a maximum of 1430 cars/hour. When the density is above or below 83 cars/mile, the traffic flow rate is less than 1430 cars/hour.

c. Figure 8.23 shows the graph of $t = -0.21d^2 + 34.66d$. The horizontal intercepts are the values of d when $t = 0$. From the graph we estimate that when $t = 0$, then d is either 0 or about 165.

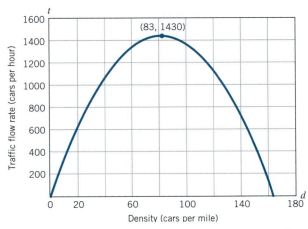

Figure 8.23 The quadratic relationship between traffic flow rate and density.
Source: Adapted from G. B. Whitman, *Linear and Nonlinear Waves* (New York: John Wiley, 1974), p. 68.

d. When the traffic flow rate is equal to zero, traffic is at a standstill. This happens when the density is either zero (when there are no cars) or approximately 165 cars per mile.

Converting between Standard and Vertex Forms

Every quadratic can be written in either standard or vertex form. In this section we examine how to convert from one form to the other.

Converting from a-h-k to a-b-c form

Every function written in the vertex or *a-h-k* form can be rewritten as a function in the standard or *a-b-c* form if we multiply out and group terms with the same power of x.

EXAMPLE 8   **Converting from vertex to standard form**
Rewrite the quadratic $f(x) = 3(x + 7)^2 - 9$ in the *a-b-c* form.

SOLUTION   The function $f(x)$ is in the *a-h-k* form where $a = 3$, $h = -7$, and $k = -9$.

Given	$f(x) = 3(x + 7)^2 - 9$
write out the factors	$= 3(x + 7)(x + 7) - 9$
multiply the factors	$= 3(x^2 + 14x + 49) - 9$
distribute the 3	$= 3x^2 + 42x + 147 - 9$
group the constant terms	$= 3x^2 + 42x + 138$

This function is now in the *a-b-c* format with $a = 3$, $b = 42$, and $c = 138$.

For a graphic illustration of the shift from *a-b-c* to *a-h-k* form, see "Q7: From *a-b-c* to *a-h-k* Form" in *Quadratic Functions*.

Converting from *a-b-c* to *a-h-k* form

Here we examine two strategies that can be used to convert from the standard or *a-b-c* form to the vertex or *a-h-k* form.

Strategy 1: "Completing the Square."

We can convert the function $f(x) = x^2 + 14x + 9$ into *a-h-k* form using a method called *completing the square*.

When a function is in *a-h-k* form, the term $(x - h)^2$ is a perfect square; that is, $(x - h)^2$ is the product of the expression $x - h$ times itself. We can examine separately the expression $x^2 + 14x$ and ask what constant term we would need to add to it in order to make it a perfect square.

A perfect square is in the form

$$(x + m)^2 = x^2 + 2mx + m^2$$

for some number m. Notice that the coefficient of x is two times the number m in our expression for a perfect square. So to turn $x^2 + 14x$ into a perfect square $(x + m)^2$, we need to find m. Since the coefficient of x corresponds to $2m$, then $2m = 14 \Rightarrow m = 7 \Rightarrow m^2 = 49$. So if we add 49 to $x^2 + 14x$ we have a perfect square.

$$(x + 7)^2 = x^2 + 14x + 49$$

To translate $f(x) = x^2 + 14x + 9$ into the *a-k-h* form, we add 49 to make a perfect square, and then subtract 49 to preserve equality. So we have

	$f(x) = x^2 + 14x + 9$
add and subtract 49	$= x^2 + 14x + (49 - 49) + 9$
regroup terms	$= (x^2 + 14x + 49) - 49 + 9$
factor and simplify	$= (x + 7)^2 - 40$

Now $f(x)$ is in *a-h-k* form. The vertex is at $(-7, -40)$.

EXAMPLE 9 **Converting from standard to vertex form**

Convert the function $g(t) = -2t^2 + 12t - 23$ to *a-h-k* form.

SOLUTION This function is more difficult to convert by completing the square, since a is not 1. We first need to factor out -2 *from the t terms only*, getting

$$g(t) = -2(t^2 - 6t) - 23 \tag{1}$$

It is the expression $t^2 - 6t$ for which we must complete the square. Since $\left(\frac{1}{2} \cdot 6\right)^2 = 9$, then adding 9 to $t^2 - 6t$ gives us

$$t^2 - 6t + 9 = (t - 3)^2$$

We must add the constant term 9 *inside the parentheses* in Equation (1) in order to make $t^2 - 6t$ a perfect square. Since everything inside the parentheses is multiplied by -2, we have essentially subtracted 18 from our original function, so we need to add 18 *outside the parentheses* to preserve equality. We have

Given	$g(t) = -2t^2 + 12t - 23$
factor out -2 from t terms	$= -2(t^2 - 6t) - 23$
add 9 inside parentheses and 18 outside parentheses	$= -2(t^2 - 6t + 9) + 18 - 23$
factor and simplify	$= -2(t - 3)^2 - 5$

We now have $g(t)$ in *a-h-k* form. The vertex is at $(3, -5)$.

Strategy 2: Using the Formula for the Vertex.

We could also convert $g(x) = 3x^2 - 12x + 5$ to the *a-h-k* form by using the formula for the coordinates of the vertex. Since the coefficient a is the same in both the *a-b-c* and the *a-h-k* forms, we have $a = 3$. The coordinates of the vertex of a quadratic $f(x)$ in the *a-b-c* form are given by $\left(-\frac{b}{2a}, f\left(\frac{-b}{2a}\right)\right)$.

For $g(x)$ we have $a = 3$, $b = -12$, and $c = 5$. So

$$-\frac{b}{2a} = -\frac{(-12)}{2 \cdot 3} = \frac{12}{6} = 2$$

and

$$g(2) = (3 \cdot 2^2) - (12 \cdot 2) + 5$$
$$= 12 - 24 + 5$$
$$= -7$$

The vertex is at $(2, -7)$. In the *a-h-k* form the vertex is at (h, k), so $h = 2$ and $k = -7$. Substituting for a, h, and k in $g(x) = a(x - h)^2 + k$, we get

$$g(x) = 3(x - 2)^2 - 7$$

and the transformation is complete.

EXAMPLE 10   **Mystery parabola**

Find the equation of the parabola in Figure 8.24. Write it in *a-h-k* and in *a-b-c* form.

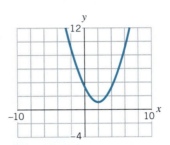

Figure 8.24 A mystery parabola.

SOLUTION   The vertex of the graph appears to be at $(2, 1)$ and the graph is concave up. We can substitute the coordinates of the vertex in the *a-h-k* form of the quadratic equation to get

$$y = a(x - 2)^2 + 1 \tag{1}$$

How can we find the value for a? If we can identify values for any other point (x, y) that lies on the parabola, we can substitute these values into Equation (1) and solve it for a. The y-intercept, estimated at $(0, 3)$, is a convenient point to pick. Setting $x = 0$ and $y = 3$, we get

$$3 = a(0 - 2)^2 + 1$$
$$3 = 4a + 1$$
$$2 = 4a$$
so $$a = 0.5$$

The equation in the *a-h-k* form is

$$y = 0.5(x - 2)^2 + 1$$

If we wanted it in the equivalent *a-b-c* form, we could square, multiply, and collect like terms to get

$$y = 0.5x^2 - 2x + 3$$

> ### The Vertex of a Quadratic Function
>
> A quadratic function in vertex form
>
> $$f(x) = a(x - h)^2 + k \qquad \text{has a vertex at } (h, k)$$
>
> A quadratic function in standard form
>
> $$f(x) = ax^2 + bx + c \qquad \text{has a vertex at } \left(-\frac{b}{2a}, f\left(-\frac{b}{2a}\right)\right)$$
>
> Whether in standard or vertex form, the value of the coefficient a determines if the parabola is relatively narrow or flat.
>
> If $a > 0$, the parabola is concave up and has a minimum at the vertex.
>
> If $a < 0$, the parabola is concave down and has a maximum at the vertex.

SOMETHING TO THINK ABOUT

Using "Q1: a, b, c Sliders" in Quadratic Functions *in the course software, can you describe the effect on the parabola of changing the value for b while you hold a and c fixed?*

EXAMPLE 11

Match the following functions to the graphs in Figure 8.25. What allows you to match them easily?

a. $f(x) = x^2 - 8x + 18$ c. $h(x) = -0.5x^2 + 4x - 6$

b. $g(x) = 2x^2 - 16x + 34$ d. $j(x) = -x^2 + 8x - 14$

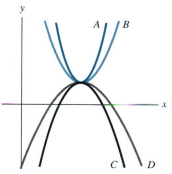

Figure 8.25 Graphs of four quadratic functions.

SOLUTION

All the graphs share the same vertex. Graphs A and B are concave up, so they could match with $f(x)$ or $g(x)$, which both have a positive value of a (the coefficient of x^2). Graphs C and D are concave down, so they could match with $h(x)$ or $j(x)$, which both have a negative value of a.

Graph A is narrower than B, so $g(x)$ matches with graph A and $f(x)$ matches with graph B. Graph C is narrower than D, so $j(x)$ matches with graph C and $h(x)$ matches with graph D. Reasoning: The larger the absolute value of a, the narrower the graph.

Algebra Aerobics 8.2c

Use of a graphing program is optional for Problems 3 and 13.

1. Find the vertex of the graph of each of the following quadratic functions:

   a. $f(x) = 2x^2 - 4$ c. $w = 4t^2 + 1$

   b. $g(z) = -z^2 + 6$

2. Find the vertex of the graph of each of the following functions and then sketch the graphs on the same grid:

   a. $y = x^2 + 3$ b. $y = -x^2 + 3$

3. In parts (a) to (d) determine the vertex and whether the graph is concave up or down. Then predict the number of x-intercepts. Graph the function to confirm your answer. Use technology if available.

   a. $f(x) = -3x^2$ c. $f(x) = x^2 + 4x - 7$

   b. $f(x) = -2x^2 - 5$ d. $f(x) = 4 - x - 2x^2$

4. Without drawing the graph, describe whether the graph of each of the following functions has a maximum or minimum at the vertex and is narrower or broader than $y = x^2$.

   a. $y = 2x^2 - 5$

   b. $y = 0.5x^2 + 2x - 10$

   c. $y = 3 + x - 4x^2$

   d. $y = -0.2x^2 + 11x + 8$

5. For each of the following functions, use the formula $h = -\frac{b}{2a}$ to find the horizontal coordinate of the vertex, and then find its vertical coordinate. Draw a rough sketch of each function.

   a. $y = x^2 + 3x + 2$

   b. $f(x) = 2x^2 - 4x + 5$

   c. $g(t) = -t^2 - 4t - 7$

6. Find the coordinates of the vertex and the vertical intercept, and then sketch the function by hand.

   a. $y = 0.1(x + 5)^2 - 11$

   b. $y = -2(x - 1)^2 + 4$

7. Rewrite each expression in the form $(x + m)^2 - m^2$.

   a. $x^2 + 6x$ **c.** $x^2 - 30x$

   b. $x^2 - 10x$ **d.** $x^2 + x$

8. Convert the following functions to vertex form by completing the square. Identify the stretch factor and the vertex.

   a. $f(x) = x^2 + 2x - 1$

   b. $j(z) = 4z^2 - 8z - 6$

   c. $h(x) = -3x^2 - 12x$

   d. $h(t) = -16t^2 + 96t$

   e. $h(t) = -4.9t^2 - 98t + 200$

9. Express each of the following functions in standard form $y = ax^2 + bx + c$. Identify the stretch or compression factor and the vertex.

   a. $y = 2\left(x - \frac{1}{2}\right)^2 + 5$

   b. $y = -\frac{1}{3}(x + 2)^2 + 4$

   c. $y = 10(x - 5)^2 + 12$

   d. $y = 0.1(x + 0.2)^2 + 3.8$

10. Express each of the following functions in vertex form.

    a. $y = x^2 + 6x + 7$ **b.** $y = 2x^2 + 4x - 11$

11. Convert the following quadratic functions to vertex form. Identify the coordinates of the vertex.

    a. $y = x^2 + 8x + 11$ **b.** $y = 3x^2 + 4x - 2$

12. The daily profit, f (in dollars), of a hot pretzel stand is a function of the price per pretzel, p (in dollars), given by $f(p) = -1875p^2 + 4500p - 2400$.

    a. Find the coordinates of the vertex of the parabola.

    b. Give the maximum profit and the price per pretzel that gives that profit.

13. Find the equations in vertex form of the parabolas that satisfy the following conditions. Then check your solutions by using a graphing program, if available.

    a. The vertex is at $(-1, 4)$ and the parabola passes through the point $(0, 2)$.

    b. The vertex is at $(1, -3)$ and one of the parabola's two horizontal intercepts is at $(-2, 0)$.

Exercises for Section 8.2

A graphing program is required or recommended for several exercises. Technology is needed in Exercises 30, 36, 38, and 40 to generate a best-fit quadratic.

1. Which of the following quadratics have parabolic graphs that are concave up? Concave down? Explain your reasoning.

   a. $y = 20 + x - x^2$ **c.** $y = -3x + 2 + x^2$

   b. $y = 0.5t^2 - 2$ **d.** $y = 3(5 - x)(x + 2)$

2. Match each function with its graph. Explain your reasoning for each choice.

   $$f(x) = x^2 + 3 \qquad g(x) = x^2 - 4 \qquad h(x) = 0.25x^2 + 3$$

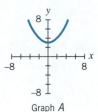
Graph A

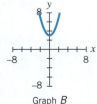
Graph B

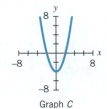
Graph C

3. On the same graph, sketch by hand the plots of the following functions and label each with its equation.

   $$y = 2x^2 \qquad\qquad y = -2x^2 + 3$$
   $$y = -2x^2 \qquad\qquad y = -2x^2 - 3$$

4. In each case sketch by hand a quadratic function $y = f(x)$ with the indicated characteristics.

   a. Does not cross the x-axis and has a negative vertical intercept

   b. Has a vertex at (h, k) where $h < 0$ and $k > 0$ and a positive y-intercept

   c. Crosses the x-axis at $x = 2$ and $x = -4$ and is concave up

   d. Has the same shape as $y = x^2 - 3x$ but is raised up four units

   e. Has the same y-intercept as $y = x^2 - 3x - 2$ but is concave down and narrower

   f. Has the same shape as $y = \frac{x^2}{2}$, but is shifted to the left by three units

5. Identify the stretch/compression factor and the vertex for each of the following.

 a. $y_1 = 0.3(x - 1)^2 + 8$ **c.** $y_3 = 0.01(x + 20)^2$

 b. $y_2 = 30x^2 - 11$ **d.** $y_4 = -6x^2 + 12x$

6. For each of the following functions, identify the vertex and specify whether it represents a maximum or minimum, and then sketch its graph by hand.

 a. $y = (x - 2)^2$ **c.** $y = -2(x + 1)^2 + 5$

 b. $y = \frac{1}{2}(x - 2)^2 + 3$ **d.** $y = -0.4(x - 3)^2 - 1$

7. For each of the following quadratic functions, find the vertex (h, k) and determine if it represents the maximum or minimum of the function.

 a. $f(x) = -2(x - 3)^2 + 5$ **c.** $f(x) = -5(x + 4)^2 - 7$

 b. $f(x) = 1.6(x + 1)^2 + 8$ **d.** $f(x) = 8(x - 2)^2 - 6$

8. For each quadratic function identify the vertex and specify whether it represents a maximum or minimum. Evaluate the function at 0 and then sketch a graph of the function by hand.

 a. $f(t) = 0.25(t - 2)^2 + 1$ **c.** $h(x) = -(x + 3)^2 + 4$

 b. $g(x) = 3 - (x - 5)^2$ **d.** $k(x) = \frac{(x + 5)^2}{3} - \frac{1}{3}$

9. (Graphing program optional.) Create a quadratic function in the vertex form $y = a(x - h)^2 + k$, given the specified values for a and the vertex (h, k). Then rewrite the function in the standard form $y = ax^2 + bx + c$. If available, use technology to check that the graphs of the two forms are the same.

 a. $a = 1, (h, k) = (2, -4)$ **c.** $a = -2, (h, k) = (-3, 1)$

 b. $a = -1, (h, k) = (4, 3)$ **d.** $a = \frac{1}{2}, (h, k) = (-4, 6)$

10. Transform the function $f(x) = x^2$ into a new function $g(x)$ by compressing $f(x)$ by a factor of $\frac{1}{4}$, then shifting the result horizontally left three units, and finally shifting it down by six units. Find the equation of $g(x)$ and sketch it by hand.

11. Transform the function $f(x) = 3x^2$ into a new function $h(x)$ by shifting $f(x)$ horizontally to the right four units, reflecting the result across the x-axis, and then shifting it up by five units.

 a. What is the equation for $h(x)$?

 b. What is the vertex of $h(x)$?

 c. What is the vertical intercept of $h(x)$?

12. For the following quadratic functions in vertex form, $f(x) = a(x - h)^2 + k$, determine the values for a, h, and k. Then compare each to $f(x) = x^2$, and identify which constants represent a stretch/compression factor, or a shift in a particular direction.

 a. $p(x) = 5(x - 4)^2 - 2$

 b. $g(x) = \frac{1}{3}(x + 5)^2 + 4$

 c. $h(x) = -0.25\left(x - \frac{1}{2}\right)^2 + 6$

 d. $k(x) = -3(x + 4)^2 - 3$

13. Using the strategy of "completing the square," fill in the missing numbers that would make the statement true.

 a. $x^2 + 6x + \underline{\hspace{1cm}} = (x + \underline{\hspace{0.7cm}})^2$

 b. $x^2 + 8x + \underline{\hspace{1cm}} = (x + \underline{\hspace{0.7cm}})^2$

 c. $x^2 - 4x + \underline{\hspace{1cm}} = (x - \underline{\hspace{0.7cm}})^2$

 d. $x^2 - 3x + \underline{\hspace{1cm}} = (x - \underline{\hspace{0.7cm}})^2$

 e. $2(x^2 + 2x + \underline{\hspace{1cm}}) = 2(x + \underline{\hspace{0.7cm}})^2$

 f. $-3(x^2 + x + \underline{\hspace{1cm}}) = -3(x + \underline{\hspace{0.7cm}})^2$

14. (Graphing program optional.) For each quadratic function use the method of "completing the square" to convert to the a-h-k form, and then identify the vertex. If available, use technology to confirm that the two forms are the same.

 a. $y = x^2 + 8x + 15$ **d.** $r(s) = -5s^2 + 20s - 10$

 b. $f(x) = x^2 - 4x - 5$ **e.** $z = 2m^2 + 6m - 5$

 c. $p(t) = t^2 - 3t + 2$

15. (Graphing program optional.) For each quadratic function convert to a-h-k form by using $h = -\frac{b}{2a}$ and then find k. If available, use technology to graph the two forms of the function to confirm that they are the same.

 a. $y = x^2 + 6x + 13$ **d.** $p(r) = -3r^2 + 18r - 9$

 b. $f(x) = x^2 - 5x - 5$ **e.** $m(z) = 2z^2 + 8z - 5$

 c. $g(x) = x^2 - 3x + 6$

16. Match each of the following graphs with one of following equations. Explain your reasoning. (*Hint:* Find the vertex.) Note that one function does not have a match.

$$f(x) = 2x^2 - 8x - 2 \qquad h(x) = 0.5x^2 - 2x + 3$$
$$g(x) = 2x^2 - 8x + 3 \qquad i(x) = 0.5x^2 - 2x + 8$$

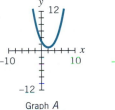

Graph *A*

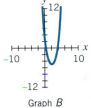

Graph *B*

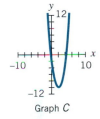

Graph *C*

17. Without drawing the graph, list the following parabolas in order, from the narrowest to the broadest. Verify your results with technology.

 a. $y = x^2 + 20$ **d.** $y = 4x^2$

 b. $y = 0.5x^2 - 1$ **e.** $y = 0.1x^2 + 2$

 c. $y = \frac{1}{3}x^2 + x + 1$ **f.** $y = -2x^2 - 5x + 4$

18. Convert the following functions from the a-b-c or standard form to the a-h-k or vertex form.

 a. $f(x) = x^2 + 6x + 5$ **d.** $y = 2x^2 + 3x - 5$

 b. $g(x) = x^2 - 3x + 7$ **e.** $h(x) = 3x^2 + 6x + 5$

 c. $y = 3x^2 - 12x + 12$ **f.** $y = -x^2 + 5x - 2$

19. (Graphing program required.) Given $f(x) = -x^2 + 8x - 15$:

 a. Estimate by graphing: the x-intercepts, the y-intercept, and the vertex.

 b. Calculate the coordinates of the vertex.

20. Write each of the following quadratic equations in function form (i.e., solve for y in terms of x). Find the vertex and the y-intercept using any method. Finally, using these points, draw a rough sketch of the quadratic function.

 a. $y + 12 = x(x + 1)$ **d.** $y - 8x = x^2 + 15$

 b. $2x^2 + 6x + 14.4 - 2y = 0$ **e.** $y + 1 = (x - 2)(x + 5)$

 c. $y + x^2 - 5x = -6.25$ **f.** $y + 2x(x - 6) = 20$

21. a. Find the equation of the parabola with a vertex of (2, 4) that passes through the point (1, 7).

b. Construct two different quadratic functions both with a vertex at (2, −3) such that the graph of one function is concave up and the graph of the other function is concave down.

c. Find two different equations of a parabola that passes through the points (−2, 5) and (4, 5) and that opens downward. (*Hint:* Find the axis of symmetry.)

22. For each part construct a function that satisfies the given conditions.

a. Has a constant rate of increase of $15,000/year

b. Is a quadratic that opens upward and has a vertex at (1, −4)

c. Is a quadratic that opens downward and the vertex is on the *x*-axis

d. Is a quadratic with a minimum at the point (10, 50) and a stretch factor of 3

e. Is a quadratic with a vertical intercept of (0, 3) that is also the vertex

23. If a parabola is the graph of the equation $y = a(x − 4)^2 − 5$:

a. What are the coordinates of the vertex? Will the vertex change if *a* changes?

b. What is the value of stretch factor *a* if the *y*-intercept is (0, 3)?

c. What is the value of stretch factor *a* if the graph goes through the point (1, −23)?

24. Construct an equation for each of the accompanying parabolas.

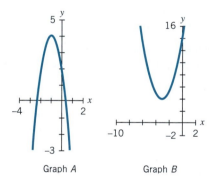

Graph A   Graph B

25. Determine the equation of the parabola whose vertex is at (2, 3) and that passes through the point (4, −1). Show your work, including a sketch of the parabola.

26. Students noticed that the path of water from a water fountain seemed to form a parabolic arc. They set a flat surface at the level of the water spout and measured the maximum height of the water from the flat surface as 8 inches and the distance from the spout to where the water hit the flat surface as 10 inches. Construct a function model for the stream of water.

27. Marketing research by a company has shown that the profit, $P(x)$ (in thousands of dollars), made by the company is related to the amount spent on advertising, *x* (in thousands of dollars), by the equation $P(x) = 230 + 20x − 0.5x^2$. What expenditure (in thousands of dollars) for advertising gives the maximum profit? What is the maximum profit?

28. Tom has a taste for adventure. He decides that he wants to bungee-jump off the Missouri River bridge. At any time *t* (in seconds from the moment he jumps) his height $h(t)$ (in feet above the water level) is given by the function $h(t) = 20.5t^2 − 123t + 190.5$. How close to the water will Tom get?

29. A manager has determined that the revenue $R(x)$ (in millions of dollars) made on the sale of supercomputers is given by $R(x) = 48x − 3x^2$, where *x* represents the number of supercomputers sold. How many supercomputers must be sold to maximize revenue? According to this model, what is the maximum revenue (in millions of dollars)?

30. (Technology required to generate a best-fit quadratic.) The accompanying graph of the data file INCLINE shows the motion of a cart, initially at the bottom of an inclined plane, after it was given a push toward a motion detector positioned at the top of the plane. The distance (in feet) of the cart from the top of the plane is plotted vs. time (in seconds). The motion can be modeled with a quadratic function.

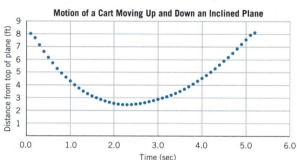

a. Estimate the coordinates of the vertex. Describe what is happening to the cart at the vertex.

b. Use technology to generate a best-fit quadratic to the data.

c. Calculate the coordinates of the vertex.

31. (Graphing program required.) A baseball hit straight up in the air is at a height

$$h = 4 + 50t − 16t^2$$

feet above ground level at time *t* seconds after being hit. (This formula is valid for $t ≥ 0$ until the ball hits the ground.)

a. What is the value of *h* when $t = 0$? What does this value represent in this context?

b. Construct a table of values for $t = 0, 1, 2, 3, 4$. Roughly when does the ball hit the ground? How can you tell?

c. Graph the function. Does the graph confirm your estimate in part (b)?

d. Explain why negative values for *h* make no sense in this situation.

e. Estimate the maximum height that the baseball reaches. When does it reach that height?

32. (Graphing program optional.) The following function represents the relationship between time *t* (in seconds) and height *h* (in feet) for objects thrown upward on *Pluto*. For an initial velocity of 20 ft/sec and an initial height above the ground of 25 feet, we get

$$h = −t^2 + 20t + 25$$

a. Find the coordinates of the point where the graph intersects the *h*-axis.

b. Find the coordinates of the vertex of the parabola.

c. Sketch the graph. Label the axes.

d. Interpret the vertex in terms of time and height.

e. For what values of *t* does the mathematical model make sense?

33. In ancient times, after a bloody defeat that made her flee her city, the queen of Carthage, Dido, found refuge on the shores of Northern Africa. Sympathetic to her plight, the local inhabitants offered to build her a new Carthage along the shores of the Mediterranean Sea. However, her city had to be rectangular in shape, and its perimeter (excluding the coastal side) could be no larger than the length of a ball of string that she could make using fine strips from only one cow hide. Queen Dido made the thinnest string possible, whose length was one mile. Dido used the string to create three non-coastal sides enclosing a rectangular piece of land (assuming the coastline was straight). She made the width exactly half the length. This way, she claimed, she would have the maximum possible area the ball of string would allow her to enclose. Was Dido right?

34. A Norman window has the shape of a rectangle surmounted by a semicircle of diameter equal to the width of the rectangle (see the accompanying figure). If the perimeter of the window is 20 feet (including the semicircle), what dimensions will admit the most light (maximize the area)? (*Hint:* Express *L* in terms of *r*. Recall that the circumference of a circle $= 2\pi r$, and the area of a circle $= \pi r^2$, where *r* is the radius of the circle.)

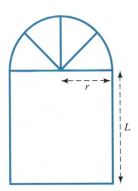

35. A pilot has crashed in the Sahara Desert. She still has her maps and knows her position, but her radio is destroyed. Her only hope for rescue is to hike out to a highway that passes near her position. She needs to determine the closest point on the highway and how far away it is.

a. The highway is a straight line passing through a point 15 miles due north of her and another point 20 miles due east. Draw a sketch of the situation on graph paper, placing the pilot at the origin and labeling the two points on the highway.

b. Construct an equation that represents the highway (using *x* for miles east and *y* for miles north).

c. Now use the Pythagorean Theorem to describe the square of the distance, *d*, of the pilot to any point (x, y) on the highway.

d. Substitute the expression for *y* from part (b) into the equation from part (c) in order to write d^2 as a quadratic in *x*.

e. If we minimize d^2, we minimize the distance *d*. So let $D = d^2$ and write *D* as a quadratic function in *x*. Now find the minimum value for *D*.

f. What are the coordinates of the closest point on the highway, and what is the distance, *d*, to that point?

36. (Requires technology to find a best-fit quadratic.) The accompanying figure is a plot of the data in the file BOUNCE, which shows the height of a bouncing racquetball (in feet) over time (in seconds). The path of the ball between each pair of bounces can be modeled using a quadratic function.

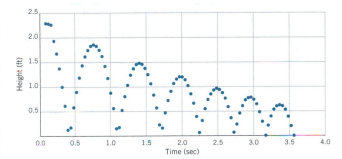

a. Select from the file BOUNCE the subset of the data that represents the motion of the ball between the first and second bounces (that is, between the first and second times the ball hits the floor). Use technology to generate a best-fit quadratic function for this subset.

b. From the graph, estimate the maximum height the ball reaches between the first and second bounces.

c. Use the best-fit function to calculate the maximum height the ball reaches between the first and second bounces.

37. (Graphing program required.) At low speeds an automobile engine is not at its peak efficiency; efficiency initially rises with speed and then declines at higher speeds. When efficiency is at its maximum, the consumption rate of gas (measured in gallons per hour) is at a minimum. The gas consumption rate of a particular car can be modeled by the following equation, where *G* is the gas consumption rate in gallons per hour and *M* is speed in miles per hour:

$$G = 0.0002M^2 - 0.013M + 1.07$$

a. Construct a graph of gas consumption rate versus speed. Estimate the minimum gas consumption rate from your graph and the speed at which it occurs.

b. Using the equation for *G*, calculate the speed at which the gas consumption rate is at its minimum. What is the minimum gas consumption rate?

c. If you travel for 2 hours at peak efficiency, how much gas will you use and how far will you go?

d. If you travel at 60 mph, what is your gas consumption rate? How long does it take to go the same distance that you calculated in part (c)? (Recall that travel distance = speed × time traveled.) How much gas is required for the trip?

e. Compare the answers for parts (c) and (d), which tell you how much gas is used for the same-length trip at two different speeds. Is gas actually saved for the trip by traveling at the speed that gives the minimum gas consumption rate?

f. Using the function G, generate data for gas consumption rate measured in gallons per mile by completing the following table. Plot gallons per mile (on the vertical axis) vs. miles per hour (on the horizontal axis). At what speed is gallons per mile at a minimum?

Speed of Car (mph)	Measures of the Rate of Gas Consumption	
	(gal/hr)	(gal/hr)/mph = gal/mile
0		
10		
20		
30		
40		
50		
60		
70		
80		

g. Add a fourth column to the data table. This time compute miles/gal = mph/(gal/hr). Plot miles per gallon vs. miles per hour. At what speed is miles per gallon at a maximum? This is the inverse of the preceding question; we are normally used to maximizing miles per gallon instead of minimizing gallons per mile. Does your answer make sense in terms of what you found for parts (b) and (f)?

38. (Requires technology to create best-fit functions.) The following data show the average growth of the human embryo prior to birth.

Embryo Age (weeks)	Weight (g)	Length (cm)
8	3	2.5
12	36	9
20	330	25
28	1000	35
36	2400	45
40	3200	50

Source: Reprinted with permission from Kimber et al., *Textbook of Anatomy and Physiology* (Upper Saddle River, NJ: Prentice Hall, 1955), "Embryo Age, Weight and Height," p. 785.

a. Plot weight (on vertical axis) vs. age (on horizontal axis). Then use technology to find the best-fit quadratic model to approximate the data.

b. According to your model, what would an average 32-week-old embryo weigh?

c. Comment on the domain for which your formula is reliable.

d. Plot length versus age; then use technology to construct an appropriate mathematical model for the length as a function of embryo age from 20 to 40 weeks.

e. Using your model, compute the age at which an embryo would be 42.5 centimeters long.

39. A shot-put athlete releases the shot at a speed of 14 meters per second, at an angle of 45 degrees to the horizontal (ground level). The height y (in meters above the ground) of the shot is given by the function

$$y = 2 + x - \tfrac{1}{20}x^2$$

where x is the horizontal distance the shot has traveled (in meters).

a. What was the height of the shot at the moment of release?

b. How high is the shot after it has traveled 4 meters horizontally from the release point? 16 meters?

c. Find the highest point reached by the shot in its flight.

d. Draw a sketch of the height of the shot and indicate how far the shot is from the athlete when it lands.

40. (Technology required to create a best-fit quadratic.) When people meet for the first time, it is customary for all people in the group to shake hands. Below is a table that shows the number of handshakes that occur depending on the size of the group.

Group size	2	3	4	5	6
Total number of handshakes	1	3	6	10	15

a. Draw a scatterplot of the data in the table (with group size on the horizontal axis) and find a quadratic model of best fit.

b. Use the model to find the total number of handshakes for a group of seven people, and for a group of ten people.

c. Factor the model of best fit. Describe in words how you could find the number of handshakes knowing the group size.

8.3 Finding the Horizontal Intercepts

We often want to find the exact values for horizontal intercepts when we use a function to model a real-world situation. We know that a quadratic function may have two, one, or no horizontal intercepts. To find the horizontal intercepts (if any) of a function $f(x)$, we need to find the *zeros* of the function, the value(s) of x such that $f(x) = 0$. If $f(x) = ax^2 + bx + c$, then setting $f(x) = 0$, we have $0 = ax^2 + bx + c$. If, as in the following example, we can factor $ax^2 + bx + c$ as a product of two linear factors (each with an x term), solving for the intercepts is easy. Later on in this chapter we'll use the quadratic formula to find the horizontal intercepts when the function is not easily factored.

Using Factoring to Find the Horizontal Intercepts

E X A M P L E   1

A quadratic model for a battleship gun range

Iowa class battleships have large naval guns that have a muzzle velocity of about 2000 feet per second. If the gun is set to maximize range, then the relationship between the height of the projectile fired and the distance it travels is $h(d) = d - 8 \cdot 10^{-6} d^2$, where d = horizontal distance in feet from the battleship, and $h(d)$ = height in feet of the projectile. What is the maximum range of the battleship gunfire?

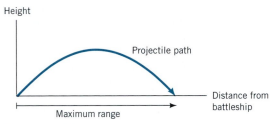

Figure 8.26 Graph of height versus distance for a projectile fired from a battleship.

S O L U T I O N

Assuming that the battleship and the target are at the same level, to find the maximum range we need to find where the projectile will hit the ground. At that point $h(d) = 0$, so the point represents a horizontal intercept on the graph (see Figure 8.26). To find the horizontal intercepts we set $h(d) = 0$ and solve for d.

$$\begin{aligned}
\text{Given the function} &\qquad h(d) = d - 8 \cdot 10^{-6} d^2 \\
\text{if we set } h(d) = 0 &\qquad 0 = d - 8 \cdot 10^{-6} d^2 \\
\text{and factor out } d, \text{ we get} &\qquad 0 = d \cdot (1 - 8 \cdot 10^{-6} d)
\end{aligned}$$

So when $h(d) = 0$, then either

$$d = 0 \text{ (at the gun)} \quad or \quad 1 - 8 \cdot 10^{-6} d = 0 \text{ (when the projectile hits the ground)}$$

$$\begin{aligned}
\text{To solve the second expression for } d &\qquad 1 - 8 \cdot 10^{-6} d = 0 \\
\text{subtract 1 from each side} &\qquad -8 \cdot 10^{-6} d = -1 \\
\text{divide each side by } -8 \cdot 10^{-6} &\qquad d = \frac{-1}{(-8 \cdot 10^{-6})} \\
\text{and simplify} &\qquad = 0.125 \cdot 10^6 \\
\text{to get} &\qquad = 125{,}000 \text{ feet}
\end{aligned}$$

So a projectile fired from this gun is able to hit a target that is 125,000 feet, or almost 24 miles, from the battleship.

Finding the Horizontal Intercepts of a Function

Given a function $f(x)$:

> To find the x-intercepts, set $f(x) = 0$ and solve for x.
>
> Every x-intercept is a *zero* of the function.

One of the properties of real numbers we used in the previous example was the "zero product rule"—the notion that if the product of two terms is 0, then at least one term must be 0.

Zero Product Rule

For any two numbers r and s, if the product $rs = 0$, then r or s or both must equal 0.

We'll use this rule repeatedly in our search for the horizontal intercepts of a function.

Factoring Quadratics

In the battleship example, we wrote $h(d) = d - 8 \cdot 10^{-6}d^2$, which is in standard form, as $h(d) = d(1 - 8 \cdot 10^{-6}d)$, which is in *factored form*. If a quadratic is in factored form, it's easy to find the horizontal intercepts. We set the product equal to 0 and use the zero product rule.

Factoring review

To convert $ax^2 + bx + c$ to factored form requires thinking, practice, and a few hints. It is often a trial-and-error process. We usually restrict ourselves to finding factors with integer coefficients.

First, look for common factors in all of the terms. For example, $10x^2 + 2x$ can be factored as $2x(5x + 1)$.

Second, look for two linear factors. This is easiest to do when the coefficient of x^2 is 1. For example, to factor $x^2 + 7x + 12$, we want to rewrite it as

$$(x + m)(x + n)$$

for some m and n. Note that the coefficients of both x's in the factors equal 1, since x times x is equal to the x^2 in the original expression. If we multiply out the two factors, we get $x^2 + (m + n)x + m \cdot n$. So we need $m + n = 7$ and $m \cdot n = 12$. Since 12 is positive, m and n must have the same sign, either both positive or both negative. But since $m + n$ (=7) is positive, m and n must both be positive. So we consider pairs of positive integers whose product is 12, namely, 1 and 12, 2 and 6, or 3 and 4. We can then narrow our list of factors of 12 to those whose sum equals 7, the coefficient of the x term. Only the factors 3 and 4 fit this criterion. We can factor our polynomial as:

$$x^2 + 7x + 12 = (x + 3)(x + 4)$$

We can check that these factors work by multiplying them out.

Third, look for the special case of the difference of two squares. In this case the middle terms cancel out when multiplying:

$$x^2 - 25 = (x - 5)(x + 5)$$
$$= x^2 - 5x + 5x - 25$$
$$= x^2 - 25$$

In general, $\quad x^2 - n^2 = (x - n)(x + n)$

Guidelines for Factoring $ax^2 + bx + c$

First factor out any common terms.

If $a = 1$,

Find the factors of c that add to give b.

$$x^2 + bx + c = (x + m)(x + n), \quad \text{where } c = mn \text{ and } b = m + n$$

If the quadratic is the difference of two squares (so $b = 0$ and $c = n^2$ for some number n), factor it into the product of a sum and a difference.

$$x^2 - n^2 = (x - n)(x + n)$$

It's a good idea to double-check your answer by multiplying out the factored terms.

EXAMPLE 2 **Finding factors**

Put each of the following functions into factored form and then identify any horizontal intercepts.

a. $f(x) = 300x^2 + 195x$ **c.** $g(z) = -3z^2 + 12z - 12$ **e.** $Q(w) = 2w^2 - 3w - 5$

b. $h(t) = t^2 - 5t + 6$ **d.** $H(v) = -2v^2 + 18$

SOLUTION **a.** In factored form $f(x) = 15x(20x + 13)$. (You can double-check this by multiplying the factors to return to $300x^2 + 195x$.) To find the horizontal intercepts, we need to find values for x such that $f(x) = 0$. If we set $15x(20x + 13) = 0$, then according to the zero product rule, either

$$15x = 0 \quad \text{or} \quad 20x + 13 = 0$$
$$x = 0 \qquad\qquad\qquad 20x = -13$$
$$x = \frac{-13}{20} \quad \text{or} \quad -0.65$$

So there are two horizontal intercepts, at $x = 0$ and $x = -0.65$

b. In factored form $h(t) = (t - 3)(t - 2)$. (Again we can double-check our answer by multiplying the two factors to get $t^2 - 3t - 2t + 6 = t^2 - 5t + 6$.) To find the horizontal intercepts we need to find values for t such that $h(t) = 0$. If we set $(t - 3)(t - 2) = 0$, then either

$$t - 3 = 0 \quad \text{or} \quad t - 2 = 0$$
$$t = 3 \qquad\qquad\qquad t = 2$$

So there are two horizontal intercepts, at $t = 3$ and $t = 2$.

c. In factored form $g(z) = -3(z^2 - 4z + 4)$
$$= -3(z - 2)(z - 2)$$
$$= -3(z - 2)^2$$

To find the horizontal intercepts, we need to find values for z such that $g(z) = 0$. When $-3(z - 2)^2 = 0$, then $z - 2 = 0$, so $z = 2$. Hence $g(z)$ has a single horizontal intercept at $z = 2$.

d. In factored form $H(v) = -2(v^2 - 9)$
$$= -2(v + 3)(v - 3)$$

When $(v + 3)(v - 3) = 0$, then either

$$v + 3 = 0 \quad \text{or} \quad v - 3 = 0$$
$$v = -3 \qquad\qquad\qquad v = 3$$

So $H(v)$ has two horizontal intercepts, at -3 and 3.

e. In factored form $Q(w) = (2w - 5)(w + 1)$. When $(2w - 5)(w + 1) = 0$, then either

$$2w - 5 = 0 \quad \text{or} \quad w + 1 = 0$$
$$2w = 5 \qquad\qquad\qquad w = -1$$
$$w = \tfrac{5}{2}$$

So $Q(w)$ has two horizontal intercepts, at $\tfrac{5}{2}$ (or 2.5) and at -1.

Algebra Aerobics 8.3a

1. Put the function into factored form with integer coefficients and then identify any horizontal intercepts.

   a. $y = -16t^2 + 50t$
   b. $y = t^2 - 25$
   c. $h(z) = z^2 - 3z - 4$
   d. $g(x) = 4x^2 - 9$
   e. $y = 15 - 8x + x^2$
   f. $v(x) = x^2 + 2x + 1$
   g. $p(q) = q^2 - 6q + 9$

2. When possible, put the function into factored form with integer coefficients and then identify any horizontal intercepts.

   a. $f(x) = 5 - x - 4x^2$
   b. $h(t) = 64 - 9t^2$
   c. $y = 10 - 13t - 3t^2$
   d. $z = 4w^2 - 20w + 25$
   e. $y = 2x^2 - 3x - 5$
   f. $Q(t) = 6t^2 + 11t - 10$

3. Identify which of the following quadratic functions can be factored into the product of a sum and a difference, $y = (a + b)(a - b)$, which can be factored into the square of the sum or difference, $y = (a \pm b)^2$, and which can be factored into neither.

   a. $y = x^2 - 9$
   b. $y = x^2 + 4x + 4$
   c. $y = x^2 + 5x + 25$
   d. $y = 9x^2 - 25$
   e. $y = x^2 - 8x + 16$
   f. $y = 16 - 25x^2$
   g. $y = 4 + 16x^2$

4. Find any horizontal and vertical intercepts for the following functions and explain the meaning of each within the context of the problem.

   a. An object is thrown vertically into the air at time $t = 0$ seconds. Its height, $h(t)$, in feet, is given by $h(t) = -16t^2 + 64t$.

   b. The monthly profit $P(q)$ (in dollars) is a function of q, the number of items sold. The relationship is described by $P(q) = -q^2 + 60q - 800$.

5. Match the factored form of the quadratic function with its graph.

   $$y_1 = -x(x - 2)$$
   $$y_2 = (x - 2)(x + 1)$$
   $$y_3 = (x + 4)(x + 1)$$

   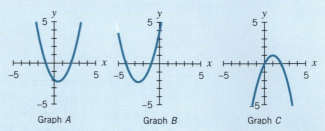

   Graph A   Graph B   Graph C

6. Given $f(x) = x^2 + x - 30$:

   a. Factor $f(x)$.

   b. Find the horizontal intercepts.

   c. Sketch the graph of $f(x)$.

   d. Describe the relationship between the factored form and the horizontal intercepts in this problem.

Using the Quadratic Formula to Find the Horizontal Intercepts

Most quadratic functions are not easily factored. When this is the case, we still set $f(x) = 0$ and solve for x. But now we use the quadratic formula to find the solutions or *roots* to the equation $0 = ax^2 + bx + c$.

> **The Quadratic Formula**
>
> For any quadratic equation of the form $0 = ax^2 + bx + c$ (where $a \neq 0$), the solutions, or *roots,* of the equation are given by
>
> $$x = \frac{-b \pm \sqrt{b^2 - 4ac}}{2a}$$
>
> The term under the radical sign, $b^2 - 4ac$, is called the *discriminant*.

See the reading "Why the Formula for the Vertex and the Quadratic Formula Work" for a derivation of this famous formula.

The symbol $\pm$ lets us use one formula to write the two roots as

$$x = \frac{-b + \sqrt{b^2 - 4ac}}{2a} \quad \text{and} \quad x = \frac{-b - \sqrt{b^2 - 4ac}}{2a}$$

Note on Terminology. The language of roots and zeros can be confusing. The numbers 3 and -3 are called the *roots* or *solutions* of the *equation* $x^2 - 9 = 0$ and the *zeros* of the *function* $f(x) = x^2 - 9$. The zeros of the function $f(x)$ are the roots of the equation $f(x) = 0$.

The discriminant

One shortcut for predicting the number of horizontal intercepts is to use the discriminant, the term $b^2 - 4ac$.

Using the Discriminant

A quadratic function $f(x) = ax^2 + bx + c$ has a discriminant of $b^2 - 4ac$.

If the discriminant > 0, there are two distinct real roots and hence two x-intercepts, at

$$x_1 = \frac{-b + \sqrt{\text{discriminant}}}{2a} \qquad \text{and} \qquad x_2 = \frac{-b - \sqrt{\text{discriminant}}}{2a}$$

If the discriminant $= 0$, there is only one distinct real root and hence only one x-intercept, at

$$x = \frac{-b}{2a}$$

If the discriminant < 0, then $\sqrt{\text{discriminant}}$ is not a real number; so there are no x-intercepts and the zeros of the function are not real numbers.

EXAMPLE 3   Identify whether each of the graphs in Figure 8.27 has a discriminant > 0, < 0, or equal to 0.

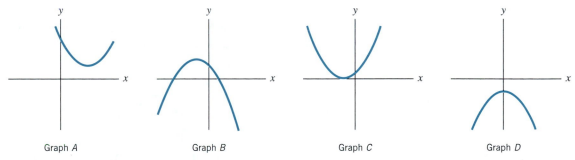

Graph *A*   Graph *B*   Graph *C*   Graph *D*

Figure 8.27 Graphs of four parabolas.

SOLUTION   Graph *B* has two horizontal intercepts, so the discriminant is > 0. Graph *C* has only one horizontal intercept, so the discriminant $= 0$. Graphs *A* and *D* have no horizontal intercepts, so the discriminant is < 0.

EXAMPLE 4   For each of the following functions, use the discriminant to predict the number of horizontal intercepts. If there are any, use the quadratic formula to find them. Then using technology, graph the function to confirm your predictions.

a. $f(z) = z^2 + 3z + 2.25$ **c.** $g(x) = -x^2 - 6x - 10$

b. $h(t) = 34 + 32t - 16t^2$

SOLUTION

a. Setting $f(z) = 0$, we have $0 = z^2 + 3z + 2.25$. Here $a = 1$, $b = 3$, and $c = 2.25$, so the discriminant is $b^2 - 4ac = 3^2 - (4 \cdot 1 \cdot 2.25) = 9 - 9 = 0$. Since $\sqrt{0} = 0$, the quadratic formula says that there is only one root, at $-b/(2a) = -3/(2 \cdot 1) = -1.5$. So $f(z)$ has one real zero, and hence one z-intercept at $(-1.5, 0)$, which must also be the vertex of the parabola. (See Figure 8.28.)

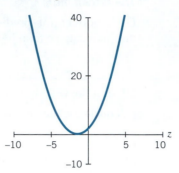

Figure 8.28 Graph of $f(z) = z^2 + 3z + 2.25$, with one horizontal intercept, at the vertex.

b. Setting $h(t) = 0$, we have $0 = 34 + 32t - 16t^2$. If we rearrange the terms as $0 = -16t^2 + 32t + 34$, it's easier to see that we should set $a = -16$, $b = 32$, and $c = 34$. The discriminant is $b^2 - 4ac = (32)^2 - (4 \cdot (-16) \cdot 34) = 1024 + 2176 = 3200$. So there are two real roots:

one at $\dfrac{-32 + \sqrt{3200}}{2(-16)} \approx \dfrac{-32 + 56.6}{-32} = \dfrac{24.6}{-32} \approx -0.77$

the other at $\dfrac{-32 - \sqrt{3200}}{2(-16)} \approx \dfrac{-32 - 56.6}{-32} = \dfrac{-88.6}{-32} \approx 2.77$

Therefore, the parabola for $h(t)$ has two real zeros, and hence two t-intercepts, at approximately $(-0.77, 0)$ and $(2.77, 0)$. (See Figure 8.29.)

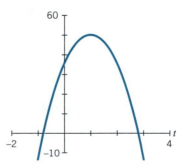

Figure 8.29 Graph of $h(t) = 34 + 32t - 16t^2$, with two horizontal intercepts.

c. Setting $g(x) = 0$, we have $0 = -x^2 - 6x - 10$. Here $a = -1$, $b = -6$, and $c = -10$, so the discriminant is $b^2 - 4ac = (-6)^2 - 4(-1)(-10) = 36 - 40 = -4$. The discriminant is negative, so taking its square root presents a problem. There is no real number r such that $r^2 = -4$. Therefore, the roots at $\dfrac{6 \pm \sqrt{-4}}{-2}$ are not real. Since there are no real zeros for the function, there are no horizontal intercepts. (See Figure 8.30.)

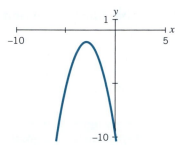

Figure 8.30 Graph of $g(x) = -x^2 - 6x - 10$, with no horizontal intercepts.

Imaginary and complex numbers

Mathematicians were uncomfortable with the notion that certain quadratic equations did not have solutions, so they literally invented a number system in which such equations would be solvable. In the process, they created new numbers, called *imaginary numbers*. The imaginary number i is defined as a number such that

$$i^2 = -1$$

or equivalently

$$i = \sqrt{-1}$$

A number such as $\sqrt{-4}$ is also an imaginary number. We can write $\sqrt{-4}$ as

$$\sqrt{(4)(-1)} = \sqrt{4}\sqrt{-1} = 2\sqrt{-1} = 2i$$

When a number is called imaginary, it sounds as if it does not exist. But imaginary numbers are just as legitimate as real numbers. Imaginary numbers are used to extend the real number system to a larger system called the complex numbers.

Complex Numbers

A complex number is defined as any number that can be written in the form

$$z = a + bi$$

where a and b are real numbers and $i = \sqrt{-1}$.

The real part of z is the number a, and the imaginary part is the number b.

Note that the real numbers are a subset of the complex numbers, since any real number a can be written as $a + 0 \cdot i$.

EXAMPLE 5   Write each expression as a complex number of the form $a + bi$.

a. $-2 + 7i$ **b.** $4 + \sqrt{-9}$ **c.** $13 - \sqrt{36}$ **d.** $\sqrt{-25}$ **e.** $5 + 3i^2$

SOLUTION
a. $-2 + 7i$ is complex and already in $a + bi$ form.
b. $4 + \sqrt{-9} = 4 + \sqrt{9}\sqrt{-1} = 4 + 3i$
c. $13 - \sqrt{36} = 13 - 6 = 7 = 7 + 0 \cdot i$ (a real number)
d. $\sqrt{-25} = \sqrt{25}\sqrt{-1} = 5i = 0 + 5i$ (an imaginary number)
e. Since $i^2 = -1$, then $5 + 3i^2 = 5 + 3(-1) = 5 - 3 = 2 = 2 + 0 \cdot i$ (a real number)

The Factored Form

We started this section by finding the horizontal intercepts from the factored form. We now show that any quadratic function can be put into factored form, whether the zeros are real or complex. The Factor Theorem relates the zeros of a function to the factors of a function.

The Factor Theorem

Given a function $f(x)$, if $f(r) = 0$, then r is a *zero* of the function and $(x - r)$ is a factor of $f(x)$.

Using the Factor Theorem, if r_1 and r_2 are zeros of the function $f(x)$, then both $(x - r_1)$ and $(x - r_2)$ are factors of $f(x)$. So we can write a quadratic function $f(x) = ax^2 + bx + c$ in factored form as $f(x) = a(x - r_1)(x - r_2)$. Note that if you multiply out the factored form, the coefficient of x^2 is a, as it is in the standard form. The factored form is useful when we want to emphasize the zeros of a function.

The Factored Form

The quadratic function $f(x) = ax^2 + bx + c$ can be written in *factored form* as

$$f(x) = a(x - r_1)(x - r_2)$$

where r_1 and r_2 are the *zeros* of $f(x)$.

If r_1 and r_2 are real numbers, then $f(x)$ has x-intercepts at r_1 and r_2.

EXAMPLE 6 **Constructing a quadratic function with complex zeros**

a. Construct a quadratic function that is concave up and has zeros at $5 + i$ and $5 - i$. Put it into standard form.

b. What do we know about the graph of this function? Are there any other functions with the same characteristics?

SOLUTION **a.** The function $f(x) = (x - (5 + i))(x - (5 - i))$ has zeros at $5 + i$ and $5 - i$. If we multiply it out (using the FOIL technique), we get

$$(x - (5 + i))(x - (5 - i)) = x^2 - (5 - i)x - (5 + i)x + (5 + i)(5 - i)$$

use the distributive
law and FOIL again $= x^2 - 5x + ix - 5x - ix + (25 - 5i + 5i - i^2)$

simplify $= x^2 - 10x + (25 - i^2)$

and substitute -1 for i^2 $= x^2 - 10x + 26$

So we can rewrite $f(x)$ in standard form as $f(x) = x^2 - 10x + 26$. (You can double-check this by using the quadratic formula.) Since the coefficient of x^2 is 1, and hence positive, the graph is concave up.

b. Since the zeros are not real, the graph of the function has no x-intercepts.

Any function of the form $af(x) = a(x^2 - 10x + 26)$, where $a > 0$, will be concave up and have zeros at $5 + i$ and $5 - i$. Since there are an infinite number of values of a, there are an infinite number of functions with these characteristics.

EXAMPLE 7 A parabola has horizontal intercepts at $d = -1$ and $d = 2$, and passes through the point $(d, h) = (1.5, 1.25)$.

a. Find the equation for the parabola.

b. Graph the parabola using a dotted line. Now restrict the domain from $d = 0$ to the horizontal intercept that is positive and then color in the corresponding section of the parabola with a solid line. This section of the parabola, drawn with a solid line, describes the path of a water jet located at the center of a circular fountain, where d = distance (in feet) from the base of the fountain, and h = height (in feet) of the water.

c. At what height is the nozzle of the water jet?

d. What is the greatest height the stream of water reaches?

SOLUTION **a.** If the parabola has horizontal intercepts at -1 and 2, then the equation for the parabola is in the form $h = a(d - (-1))(d - 2)$ or $a(d + 1)(d - 2)$. The point $(1.5, 1.25)$ lies on the parabola, so it must satisfy the equation. Hence

given that	$h = a(d + 1)(d - 2)$
if we substitute 1.5 for d and 1.25 for h	$1.25 = a\,(1.5 + 1)(1.5 - 2)$
simplify	$1.25 = a(2.5)(-0.5)$
multiply	$1.25 = a(-1.25)$
and divide by -1.25, we have	$-1 = a$

So the equation of the parabola is $h = -(d + 1)(d - 2)$.

b. See Figure 8.31.

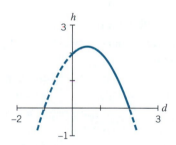

Figure 8.31 Graph of $h = -(d + 1)(d - 2)$.

c. When $d = 0$, $h = -(0 + 1)(0 - 2) = 2$. So the water nozzle is at 2 feet.

d. There are several strategies for finding the vertex. You could calculate it using the formula in Section 8.2. Or recall that the vertex lies on the axis of symmetry, halfway between the two horizontal intercepts (-1 and 2) at $d = \frac{(-1 + 2)}{2} = \frac{1}{2}$ or 0.5. Substituting 0.5 for d in the equation in part (a), we get

$$h = -(0.5 + 1)\,(0.5 - 2)$$
$$= -(1.5)(-1.5)$$
$$= 2.25$$

The coordinates of the vertex are $(0.5, 2.25)$. So the maximum height of the water (at the vertex of the parabola) is 2.25 feet (or 27″), which occurs at 0.5 feet (or 6″) from the base of the fountain.

> **The Horizontal Intercepts of a Quadratic Function**
>
> The graph of a quadratic function may have zero, one, or two horizontal intercepts, which can be found by factoring or using the quadratic formula.
>
> The discriminant in the quadratic formula can be used to predict the number of horizontal intercepts.
>
> Any quadratic function in standard form $f(x) = ax^2 + bx + c$ can be written in *factored form* as
>
> $$f(x) = a(x - r_1)(x - r_2) \quad \text{where } r_1 \text{ and } r_2 \text{ are the zeros of } f(x)$$
>
> If the zeros r_1 and r_2 are real numbers, they are the horizontal intercept(s) of $f(x)$.

Algebra Aerobics 8.3b

1. Find any real numbers that satisfy the following equations.

    a. $4x + 7 = 0$
    b. $4x^2 - 7 = 0$
    c. $4x^2 - 7x = 0$
    d. $2(x + 3) = x^2$
    e. $(2x - 11)^2 = 0$
    f. $(x + 1)^2 = 81$
    g. $x = x^2 - 5$

2. Evaluate the discriminant $b^2 - 4ac$ for each of the following quadratic functions of the form $f(x) = ax^2 + bx + c$. Use the discriminant to determine the nature of the zeros of the function and the number (if any) of horizontal intercepts.

    a. $f(x) = 2x^2 + 3x - 1$
    b. $f(x) = x^2 + 7x + 2$
    c. $f(x) = 4x^2 + 4x + 1$
    d. $f(x) = 2x^2 + x + 5$

3. Find and interpret the horizontal and vertical intercepts for the following height equations.

    a. $h = -4.9t^2 + 50t + 80$ (h is in meters and t is in seconds)
    b. $h = 150 - 80t - 490t^2$ (h is in centimeters and t is in seconds)
    c. $h = -16t^2 + 64t + 3$ (h is in feet and t is in seconds)
    d. $h = 64t - 16t^2$ (h is in feet and t is in seconds)

4. Evaluate the discriminant and then predict the number of x-intercepts for each function. Use the quadratic formula to find all the zeros of each function and then identify the coordinates of any x-intercept(s).

    a. $y = 4 - x - 5x^2$
    b. $y = 4x^2 - 28x + 49$
    c. $y = 2x^2 + 5x + 4$
    d. $y = 2x^2 - 3x - 1$
    e. $y = 2 - 3x^2$

5. From the descriptions given in parts (a) and (b), determine the coordinates of the vertex, find the equation of the parabola, and then sketch the parabola.

    a. A parabola with horizontal intercepts at $x = -2$ and $x = 4$ that passes through the point $(3, 2)$.
    b. A parabola with horizontal intercepts at $x = 2$ and $x = 8$, and a vertical intercept at $y = 10$.

6. Identify the number of x-intercepts for the following functions without converting into standard form.

    a. $y = 3(x - 1)^2 + 5$
    b. $y = -2(x + 4)^2 - 1$
    c. $y = -5(x + 3)^2$
    d. $y = 3(x - 1)^2 - 2$

7. Write an equation for a quadratic function, in factored form, with the specified zeros.

    a. 2 and -3
    b. 0 and -5
    c. 8

8. For each quadratic function in the accompanying graphs, specify the number of real zeros and whether the corresponding discriminant would be positive, negative, or zero.

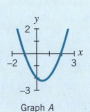
Graph A

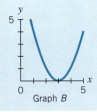
Graph B

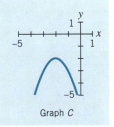
Graph C

Exercises for Section 8.3

Several exercises either require or recommend the use of a graphing program.

1. Solve the following quadratic equations by factoring.

    a. $x^2 - 9 = 0$
    b. $x^2 - 4x = 0$
    c. $3x^2 = 25x$
    d. $x^2 + x = 20$
    e. $4x^2 + 9 = 12x$
    f. $3x^2 = 13x + 10$
    g. $(x + 1)(x + 3) = -1$
    h. $x(x + 2) = 3x(x - 1) - 3$

2. Find the x-intercepts for each of the following functions. Will the vertex lie above, below, or on the x-axis? Find the vertex and sketch the graph, labeling the x-intercepts.

    a. $y = (x + 2)(x + 1)$
    b. $y = 3(1 - 2x)(x + 3)$
    c. $y = -4(x + 3)^2$
    d. $y = \frac{1}{2}(x)(x - 5)$
    e. $q(x) = 2(x - 3)(x + 2)$
    f. $f(x) = -2(5 - x)(3 - 2x)$

3. Factor the quadratic expression and then sketch the graph of the function, labeling the axes and horizontal intercepts.

a. $y = x^2 + 6x + 8$ **d.** $w = t^2 - 25$

b. $z = 3x^2 - 6x - 9$ **e.** $r = 4s^2 - 100$

c. $f(x) = x^2 - 3x - 10$ **f.** $g(x) = 3x^2 - x - 4$

4. a. Construct a quadratic function with zeros at $x = 1$ and $x = 2$.

     b. Is there more than one possible quadratic function for part (a)? Why or why not?

5. (Graphing program required.) Using a graphing program, estimate the real solutions to the following equations. (*Hint:* Think of the equations as resulting from setting $f(x) = 0$.) Verify by factoring, if possible.

a. $x^2 - 5x + 6 = 0$ **d.** $-3x^2 - 12x + 15 = 0$

b. $3x^2 - 2x + 5 = 0$ **e.** $0.05x^2 + 1.1x = 0$

c. $3x^2 - 12x + 12 = 0$ **f.** $-2x^2 - x + 3 = 0$

6. Write each function in factored form, if possible, using integer coefficients.

a. $f(x) = x^2 + 2x - 15$ **d.** $k(p) = p^2 + 5p + 7$

b. $g(x) = x^2 - 6x + 9$ **e.** $l(s) = 5s^2 - 37s - 24$

c. $h(a) = a^2 + 6a - 16$ **f.** $m(t) = 5t^2 + t + 1$

7. Solve the following equations using the quadratic formula. (*Hint:* Rewrite each equation so that one side of the equation is zero.)

a. $6t^2 - 7t = 5$ **e.** $6s^2 - 10 = -17s$

b. $3x(3x - 4) = -4$ **f.** $2t^2 = 3t + 9$

c. $(z + 1)(3z - 2) = 2z + 7$ **g.** $5 = (4x + 1)(x - 3)$

d. $(x + 2)(x + 4) = 1$ **h.** $(2x - 3)^2 = 7$

8. Solve using the quadratic formula.

a. $x^2 - 3x = 12$ **e.** $\dfrac{1}{x - 2} = \dfrac{x + 1}{x - 1}$

b. $3x^2 = 4x + 2$ **f.** $\dfrac{x^2}{3} + \dfrac{x}{2} - \dfrac{1}{6} = 0$

c. $3(x^2 + 1) = x + 2$ **g.** $\dfrac{1}{x^2} - \dfrac{3}{x} = \dfrac{1}{6}$

d. $(3x - 1)(x + 2) = 4$

9. Calculate the coordinates of the x- and y-intercepts for the following quadratics.

a. $y = 3x^2 + 2x - 1$ **c.** $y = (5 - 2x)(3 + 5x)$

b. $y = 3(x - 2)^2 - 1$ **d.** $f(x) = x^2 - 5$

10. Use the discriminant to predict the number of horizontal intercepts for each function. Then use the quadratic formula to find all the zeros. Identify the coordinates of any horizontal or vertical intercepts.

a. $y = 2x^2 + 3x - 5$

b. $f(x) = -16 + 8x - x^2$

c. $f(x) = x^2 + 2x + 2$

d. $y = 2(x - 1)^2 + 1$

e. $g(z) = 5 - 3z - z^2$

f. $f(t) = (t + 2)(t - 4) + 9$

11. In each part (a) to (e), graph a parabola with the given characteristics. Then write an equation of the form $y = ax^2 + bx + c$ for that parabola.

a. $a > 0,\ b^2 - 4ac > 0,\ c > 0$

b. $a > 0,\ b^2 - 4ac > 0,\ c < 0$

c. $a > 0,\ b^2 - 4ac < 0,\ b \neq 0$

d. $a < 0,\ b^2 - 4ac = 0,\ c \neq 0$

e. $b \neq 0,\ c < 0,\ b^2 - 4ac > 0$

12. For each part, draw a rough sketch of a graph of a function of the type $f(x) = ax^2 + bx + c$

a. Where $a > 0, c > 0$, and the function has no real zeros.

b. Where $a < 0, c > 0$, and the function has two real zeros.

c. Where $a > 0$ and the function has one real zero.

13. a. Construct a quadratic function $Q(t)$ with exactly one zero at $t = -1$ and a vertical intercept at -4.

     b. Is there more than one possible quadratic function for part (a)? Why or why not?

     c. Determine the axis of symmetry. Describe the vertex.

14. a. Construct a quadratic function $P(s)$ that goes through the point $(5, -22)$ and has two real zeros, one at $s = -6$ and the other at $s = 4$.

     b. What is the axis of symmetry?

     c. What are the coordinates of the vertex?

     d. What is the vertical intercept?

15. Construct a quadratic function for each of the given graphs. Write the function in both factored form and standard form.

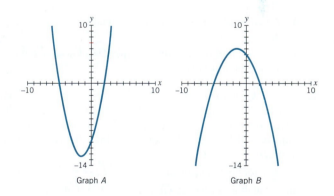

Graph A   Graph B

16. Complete the following table, and then summarize your findings.

$i^1 = \sqrt{-1} = i$	$i^5 = ?$
$i^2 = i \cdot i = -1$	$i^6 = ?$
$i^3 = i \cdot i^2 = ?$	$i^7 = ?$
$i^4 = i^2 \cdot i^2 = ?$	$i^8 = ?$

17. Complex number expressions can be simplified by combining the real parts and then the imaginary parts. Add (or subtract) the following complex numbers and then simplify.

a. $(4 + 3i) + (-5 + 7i)$ **c.** $(7i - 3) + (2 - 4i)$

b. $(-2 + 3i) - (-3 + 3i)$ **d.** $(7i - 3) - (2 - 4i)$

18. Complex number expressions can be multiplied using the distributive property or the FOIL technique. Multiply and simplify the following. (*Note:* $i^2 = -1$.)

 a. $(3 + 2i)(-2 + 3i)$ d. $(5 - 3i)(5 + 3i)$

 b. $(4 - 2i)(3 + i)$ e. $(3 - i)^2$

 c. $(2 + i)(2 - i)$ f. $(4 + 5i)^2$

19. A quadratic function has two complex roots, $r_1 = 1 + i$ and $r_2 = 1 - i$. Use the Factor Theorem to find the equation of this quadratic, assuming $a = 1$, and then put it into standard form.

20. The factored form of a quadratic function is $y = -2(x - (3 + i))(x - (3 - i))$. Answer the following.

 a. Will the graph open up or down? Explain.

 b. What are the zeros of the quadratic function?

 c. Does the graph cross the x-axis? Explain.

 d. Write the quadratic in standard form. (*Hint:* Multiply out; see Exercise 18.)

 e. Verify your answer in part (b) by using the quadratic formula and your answer for part (d).

21. Use the quadratic formula to find the zeros of the function $f(x) = x^2 - 4x + 13$ and then write the function in factored form. Without graphing this function, how can you tell if it intersects the x-axis?

22. Let (h, k) be the coordinates of the vertex of a parabola. Then h is equal to the average of the two real zeros of the function (if they exist). For parts (a) and (b) use this to find h, and then construct an equation in vertex form, $y = a(x - h)^2 + k$.

 a. A parabola with x-intercepts of 4 and 8, and a y-intercept of 32

 b. A parabola with x-intercepts of -3 and 1, and a y-intercept of -1

 c. Can you find the equation of a parabola knowing only its x-intercepts? Explain.

23. Let (h, k) be the coordinates of the vertex of a parabola. Then h equals the average of the two real zeros of the function (if they exist). For each of the following use this to find h, and then put the equations into the vertex form, $y = a(x - h)^2 + k$.

 a. A parabola with equation $y = x^2 + 2x - 8$

 b. A parabola with equation $y = -x^2 - 3x + 4$

24. (Graphing program optional)

 a. Write each of the following functions in both the *a-b-c* and the *a-h-k* forms. Is one form easier than the other for finding the vertex? The x- and y-intercepts?

 $y_1 = 2x^2 - 3x - 20$ $y_3 = 3x^2 + 6x + 3$

 $y_2 = -2(x - 1)^2 - 3$ $y_4 = -(2x + 4)(x - 3)$

 b. Find the vertex and x- and y-intercepts and construct a graph by hand for each function in part (a). If you have access to a graphing program, check your work.

25. Find the equation of the graph of a parabola that has the following properties:

 • The x-intercepts of the graph are at (2, 0) and (3, 0), *and*

 • The parabola is the graph of $y = x^2$ vertically stretched by a factor of 4.

 Explain your reasoning. Sketch the parabola.

26. (Graphing program required for part (b)). We dealt previously with systems of lines and ways to determine the coordinates of points where lines intersect. Once you know the quadratic formula, it's possible to determine where a line and a parabola, or two parabolas, intersect. As with two straight lines, at the point where the graphs of two functions intersect (*if* they intersect), the functions share the same x value and the same y value.

 a. Find the intersection of the parabola $y = 2x^2 - 3x + 5.1$ and the line $y = -4.3x + 10$.

 b. Plot both functions, labeling any intersection point(s).

27. (Graphing program recommended for part (b)).

 a. Find the intersection of the two parabolas $y = 7x^2 - 5x - 9$ and $y = -2x^2 + 4x + 9$.

 b. Plot both functions, labeling any intersection points.

28. Market research suggests that if a particular item is priced at x dollars, then the weekly profit $P(x)$, in thousands of dollars, is given by the function

 $$P(x) = -9 + \tfrac{11}{2}x - \tfrac{1}{2}x^2$$

 a. What price range would yield a profit for this item?

 b. Describe what happens to the profit as the price increases. Why is a quadratic function an appropriate model for profit as a function of price?

 c. What price would yield a maximum profit?

29. A dairy farmer has 1500 feet of fencing. He wants to use all 1500 feet to construct a rectangle and two interior separators that together form three rectangular pens. See the accompanying figure.

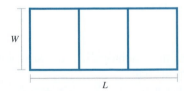

 a. If W is the width of the larger rectangle, express the length, L, of the larger rectangle in terms of W.

 b. Express the total area, $A(W)$, of the three pens as a polynomial in terms of W.

 c. What is the domain of the function $A(W)$?

 d. What are the dimensions of the larger rectangle that give a maximum area? What is the maximum area?

8.4 *The Average Rate of Change of a Quadratic Function*

In previous chapters we argued that the average rate of change of a linear function is constant, and that the average rate of change of an exponential function is exponential. What about the average rate of change of a quadratic function?

EXAMPLE 1

Finding the average rate of change of the simplest quadratic function
Given $y = x^2$, calculate the average rate of change of y with respect to x at unit intervals from -3 to 3. Then calculate the average rate of change of the average rate of change. What do these data points suggest?

SOLUTION

Column 3 in Table 8.2 shows the values for the average rate of change of y with respect to x. For these values the average rate of change is linear, since adding 1 to x increases the average rate of change by 2 over each interval. Column 4 computes the average rate of change of column 3 with respect to x. Since these values all are constant at 2, this confirms that column 3 (the average rate of change) is linear with respect to x. This suggests that when $y = x^2$ the average rate of change of y with respect to x is a linear function.

x	$y = x^2$	Average Rate of Change	Average Rate of Change of the Average Rate of Change
-3	9	n.a.	n.a.
-2	4	$\frac{4-9}{-2-(-3)} = -5$	n.a.
-1	1	$\frac{(1-4)}{1} = -3$	$\frac{-3-(-5)}{-1-(-2)} = 2$
0	0	$\frac{(0-1)}{1} = -1$	$\frac{(-1-(-3))}{1} = 2$
1	1	$\frac{(1-0)}{1} = 1$	$\frac{(1-(-1))}{1} = 2$
2	4	$\frac{(4-1)}{1} = 3$	$\frac{(3-1)}{1} = 2$
3	9	$\frac{(9-4)}{1} = 5$	$\frac{(5-3)}{1} = 2$

Table 8.2

EXAMPLE 2

Finding the average rate of change of a quadratic function
Given $y = 3x^2 - 8x - 23$, calculate the average rate of change of y with respect to x at unit intervals from -3 to 3. Then calculate the average rate of change of the average rate of change. What do the data suggest about the original function?

SOLUTION

Column 3 in Table 8.3 shows the values for the average rate of change. Again the values suggest that the average rate of change is linear, since adding 1 to x increases the average rate of change by 6 over each interval. This is confirmed by column 4, which shows that the average rate of change of the third column with respect to x is constant at 6. This suggests that when $y = 3x^2 - 8x - 23$, the average rate of change of y with respect to x is a linear function.

x	$y = 3x^2 - 8x - 23$	Average Rate of Change	Average Rate of Change of the Average Rate of Change
-3	28	n.a.	n.a.
-2	5	-23	n.a.
-1	-12	-17	6
0	-23	-11	6
1	-28	-5	6
2	-27	1	6
3	-20	7	6

Table 8.3

We have just seen two numerical examples that suggest that the average rate of change of a quadratic function is a linear function. We now show algebraically that this is true for *every* quadratic function.

Suppose we have $y = f(x) = ax^2 + bx + c$. In the previous examples we fixed an interval size of 1 over which to calculate the average rate of change, since it is easy to make comparisons. Now we pick a constant interval size r and for each position x compute the average rate of change of f over the interval from x to $x + r$.

First we must compute $f(x + r)$:

$$f(x + r) = a(x + r)^2 + b(x + r) + c$$

apply exponent
$$= a(x^2 + 2rx + r^2) + b(x + r) + c$$

multiply through
$$= ax^2 + 2arx + ar^2 + bx + br + c$$

regroup terms
$$= ax^2 + bx + c + (2ax + b)r + ar^2$$

Then the average rate of change of $f(x)$ between x and $x + r$ is

$$\frac{\text{change in } f(x)}{\text{change in } x} = \frac{f(x + r) - f(x)}{(x + r) - x}$$

$$= \frac{[ax^2 + bx + c + (2ax + b)r + ar^2] - (ax^2 + bx + c)}{r}$$

$$= \frac{(2ax + b)r + ar^2}{r}$$

$$= 2ax + (b + ar)$$

The average rate of change between $(x, f(x))$ and $(x + r, f(x + r))$ is a linear function of x in its own right, where

$$\text{slope} = 2a \qquad y\text{-intercept} = b + ar$$

Note that the slope depends only on the value of a in the original equation. The y-intercept depends not only on a and b in the original equation, but also on r, the interval size over which we calculate the average rate of change. (See Figure 8.32.)

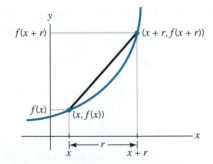

Figure 8.32 The slope of the line segment connecting two points on the parabola separated by a horizontal distance of r is $2ax + (b + ar)$.

If we took smaller and smaller values for r, then the term ar would get closer and closer to zero, and hence the y-intercept $b + ar$ would get closer and closer to b. For very small r's the average rate of change would get closer and closer to the linear expression $2ax + b$.

So given a quadratic function $f(x) = ax^2 + bx + c$, the average rate of change over small intervals around any x value can be approximated by the linear expression $2ax + b$. So we talk about the function $g(x) = 2ax + b$ as representing the average rate of change of $f(x)$ with respect to x. This is a central idea in calculus.

The Average Rate of Change of a Quadratic Function

Given a quadratic function $f(x) = ax^2 + bx + c$, we can think of the linear function $g(x) = 2ax + b$ as representing the average rate of change of $f(x)$ with respect to x.

EXAMPLE 3   Graph each of the following functions. Construct and graph the equation for the average rate of change in each case.

  a. $f(x) = x^2$ **b.** $f(x) = 3x^2 - 8x - 23$

SOLUTION   **a.** The average rate of change of the quadratic function $f(x) = x^2$ is the linear function $g(x) = 2x$ (see Figure 8.33).

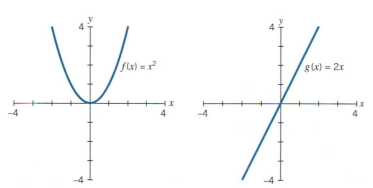

Figure 8.33 Graph of $f(x) = x^2$ and graph of its average rate of change.

b. The average rate of change of the quadratic function $f(x) = 3x^2 - 8x - 23$ is the linear function $g(x) = 6x - 8$ (see Figure 8.34).

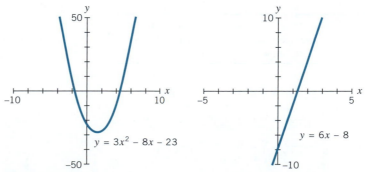

Figure 8.34 Graph of $y = 3x^2 - 8x - 23$ and graph of its average rate of change.

Algebra Aerobics 8.4

1. a. Complete the table below for the function $y = 5 - x^2$.

x	y	Average Rate of Change	Average Rate of Change of Average Rate of Change
-1	4	n.a.	n.a.
0	5	1	n.a.
1	4	-1	-2
2	1		
3			
4			

b. What do the third and fourth columns tell you?

2. Determine from each of the three following tables whether you would expect the original function $y = f(x)$ to be linear, exponential, or quadratic.

a.

x	y	Average Rate of Change	Average Rate of Change of Average Rate of Change
-1	2	n.a.	n.a.
0	1	-1	n.a.
1	6	5	6
2	17	11	6
3	34	17	6
4	57	23	6

b.

x	y	Average Rate of Change	Average Rate of Change of Average Rate of Change
-1	7	n.a.	n.a.
0	4	-3	n.a.
1	1	-3	0
2	-2	-3	0
3	-5	-3	0
4	-8	-3	0

c.

x	y	Average Rate of Change	Average Rate of Change of Average Rate of Change
-1	0.5	n.a.	n.a.
0	1	0.5	n.a.
1	2	1	0.5
2	4	2	1
3	8	4	2
4	16	8	4

3. Construct a function that represents the average rate of change for the following three quadratic functions.

a. $f(t) = t^2 + t$ **c.** $f(x) = 5x^2 + 2x + 7$

b. $f(x) = 3x^2 + 5x$

Exercises for Section 8.4

Exercise 6 requires a graphing program.

1. Complete these sentences.

  a. For a quadratic function, the graph of its average rate of change represents a _____ function.

  b. When a quadratic function is increasing, its average rate of change is _____, and when a quadratic function is decreasing, its average rate of change is _____.

2. Complete the table for the function $y = 3 - x - x^2$.

x	y	Average Rate of Change	Average Rate of Change of Average Rate of Change
-3	-3	n.a.	n.a.
-2	1	$\frac{1 - (-3)}{-2 - (-3)} = \frac{4}{1} = 4$	n.a.
-1	3	$\frac{3 - 1}{-1 - 1(-2)} = \frac{2}{1} = 2$	$\frac{2 - 4}{-1 - (-2)} = \frac{-2}{1} = -2$
0	3	$\frac{3 - 3}{0 - (-1)} = \frac{0}{1} = 0$	$\frac{0 - 2}{0 - (-1)} =$
1	1		
2			
3			

What type of function does the average rate of change represent? What is its slope (the rate of change of the average rate of change)?

3. a. What is the average rate of change of the linear function $y = 3x - 1$? Of $y = -2x + 5$? Of $y = ax + b$? What equation could represent the average rate of change of the general linear function $y = ax + b$? Describe its graph.

  b. We have seen that the average rate of change of a quadratic function $y = ax^2 + bx + c$ can be represented by a linear function of the form $y = 2ax + b$ (where $a \neq 0$). Describe the linear graph. Could such a graph ever represent the average rate of change of a linear function?

  c. What would you guess the equation representing the average rate of change of the cubic function $y = ax^3 + bx^2 + cx + d$ to be?

4. a. Complete the table for the function $f(x) = 3x^2 - 2x - 5$.

x	y	Average Rate of Change	Average Rate of Change of Average Rate of Change
-3	28	n.a.	n.a.
-2	11	$\frac{11-28}{(-2)-(-3)} = -17$	n.a.
-1	0	$\frac{0-11}{(-1)-(-2)} = -11$	$\frac{(-11)-(-17)}{(-1)-(-2)} = 6$
0	-5		
1			
2			
3			

b. What is the slope of the linear function that represents the average rate of change? How is this slope related to the average rate of change of the average rate of change?

c. What type of function represents the average rate of change of $f(x)$?

5. Complete the table for the function $Q = 2t^2 + t + 1$.

a. Plot $Q = 2t^2 + t + 1$. What type of function is this?

b. What does the third column tell you about the function that represents the average rate of change for Q?

t	Q	Average Rate of Change
-3		n.a.
-2		
-1		
0		
1		
2		
3		

6. a. (Graphing program required.) Plot the function $h(t) = 4 + 50t - 16t^2$ for the restricted domain $0 \le t \le 3$.

b. For what interval is this function increasing? Decreasing?

c. Estimate the maximum point.

d. Construct and graph a function $g(t)$ that represents the average rate of change of $h(t)$.

e. What does $g(t)$ tell you about the function $h(t)$?

7. Match the quadratic function with the graph of its average rate of change.

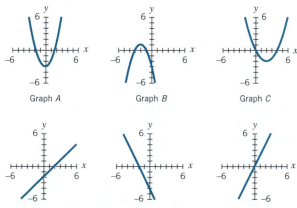

Graph A   Graph B   Graph C

Graph D   Graph E   Graph F

8. Having found the matched pairs of graphs in Exercise 7, explain the relationship between the horizontal intercept of the linear function and the vertex of the quadratic function.

9. Construct a function that represents the average rate of change for each given function.

a. $f(t) = 3t^2 + t$

b. $g(x) = -5x^2 + 0.4x + 3$

c. $h(z) = 2z + 7$

10. Construct a function that represents the average rate of change for each given function.

a. $r(t) = -3t^2 - 7$ **c.** $w(x) = -5x + 0.25$

b. $s(v) = \frac{1}{4}v^2 - \frac{3}{2}v + 5$ **d.** $m(p) = 4p - \frac{1}{2}$

11. Determine from each of the tables whether you would expect the original function $y = f(x)$ to be linear, exponential, or quadratic.

a.

x	y	Average Rate of Change	Average Rate of Change of Average Rate of Change
-1	-2	n.a.	n.a.
0	0	2	n.a.
1	8	8	6
2	22	14	6
3	42	20	6
4	68	26	6

b.

x	y	Average Rate of Change	Average Rate of Change of Average Rate of Change
-1	9	n.a.	n.a.
0	5	-4	n.a.
1	1	-4	0
2	-3	-4	0
3	-7	-4	0
4	-11	-4	0

c.

x	y	Average Rate of Change	Average Rate of Change of Average Rate of Change
-1	0.25	n.a.	n.a.
0	1	0.75	n.a.
1	4	3	2.25
2	16	12	9
3	64	48	36
4	256	192	144

8.5 *An Introduction to Polynomial Functions*

Defining a Polynomial Function

Linear, power, and quadratic functions are all part of a larger family of functions called *polynomials*. We can think of a polynomial as a sum of power functions with positive integer powers. For example, if we sum the power functions

$$y_1 = 2x^4, \quad y_2 = x^3, \quad y_3 = -2x, \quad \text{and} \quad y_4 = 11$$

to get $y = y_1 + y_2 + y_3 + y_4$

we obtain $y = 2x^4 + x^3 - 2x + 11$

This function is called a polynomial of degree 4, since the highest power of the independent variable (x) is 4.

Each of the separate power functions is called a *term* of the polynomial. In standard form we arrange the terms from the highest power down to the lowest.

Definition of a Polynomial Function

A *polynomial function of degree n* is of the form

$$f(x) = a_n x^n + a_{n-1} x^{n-1} + \cdots + a_1 x + a_0$$

where n is a nonnegative integer and $a_n \neq 0$.

> The constants $a_n, a_{n-1}, \ldots, a_0$ are called the *coefficients*.
>
> $a_n x^n$ is called the *leading term*.
>
> a_0 is called the *constant term*.

Polynomials of certain degrees have special names.

Polynomials of Degree	Are Called	Example
0	Constant	$y = 3$
1	Linear	$y = -4x - 8$
2	Quadratic	$y = 3x^2 + 5x - 10$
3	Cubic	$y = 5x^3 - 4x - 14$
4	Quartic	$y = -2x^4 - x^3 + 4x^2 + 4x - 14$
5	Quintic	$y = 8x^5 - 3x^4 + 4x^2 - 4$

Notice that the example for the cubic function, $y = 5x^3 - 4x - 14$, does not include an x^2 term. In this case the coefficient for the x^2 term is zero, since this function could be rewritten as $y = 5x^3 + 0x^2 - 4x - 14$.

Early algebraists believed that higher-degree polynomials were not relevant to the physical world and hence were useless. "Going beyond the cube just as if there were more than three dimensions . . . is against nature."[3] But as we shall see shortly, there are real applications for polynomials with degrees greater than 3.

---

[3]Stifel, as cited by M. Kline, *Mathematical Thought from Ancient to Modern Times* (Oxford: Oxford University Press, 1972).

EXAMPLE 1

Identifying function types

Specify the type(s) of each function (linear, exponential, power, quadratic, and/or polynomial). If it's polynomial, identify the degree and any special name.

a. $y = 3x + 5$ **c.** $Q(t) = 5 \cdot 2^t$ **e.** $g(z) = 2(z^2 - 1)(z^2 + 3)$

b. $f(x) = 4x^2$ **d.** $v = 1 - 13w - 5w^2 + 2w^5$

SOLUTION

All of the functions are polynomials, except for (c).

a. This function is a polynomial of degree 1, so it's a linear function.

b. This function is a polynomial of degree 2, so it's quadratic. It's also a power function.

c. This is an exponential function.

d. This function is a polynomial of degree 5, called a quintic.

e. Multiplying out, we get $g(z) = 2z^4 + 4z^2 - 6$. So this function is a polynomial of degree 4, called a quartic.

EXAMPLE 2

Curing poison ivy

A patient with an acute poison ivy rash is given a 5-day "prednisone taper." The daily dosage of the drug prednisone is respectively 20 mg, 15 mg, 10 mg, 5 mg, and 5 mg. The daily decay rate r (in decimal form) is the rate at which the drug is naturally absorbed and removed from the body. The rate varies from patient to patient. The decay factor, $x = 1 - r$, gives the percentage (in decimal form) of the drug left in the patient's body after each day.

a. If one patient has a prednisone decay rate of 0.35 (or 35%) per day, what is the patient's decay factor? If another has a decay rate of 50%, what would this patient's decay factor be?

b. Construct functions to describe a patient's prednisone level as a function of x, the decay factor, for each of the 5 days of treatment.

c. Use technology to graph the prednisone level on day 5. What type of function is this, and what is its domain in this context?

d. If a patient's daily decay factor is 40%, use the graph to estimate the total dosage of prednisone remaining on day 5. Then use the equation to calculate a more exact answer.

e. Construct a function to describe the patient's prednisone level after 6 days, after 7 days, and after n days. How do the functions for which $n > 5$ differ from those with $n \leq 5$?

SOLUTION

a. A decay rate of 0.35 corresponds to a decay factor of 0.65. A decay rate of 0.50 (or 50%) corresponds to a decay factor of 0.50.

b. On day 1, the patient would receive 20 mg of prednisone. Then on day 2 the patient would have $20x$ remaining of the original 20 mg of prednisone, plus the additional 15 mg he took that day. On day 3 he'd have $20x^2$ left of his 20-mg dosage from day 1, plus $15x$ left from his 15-mg dosage from day 2, plus the 10 mg he took that day. And so on. If we let $\mathrm{Day}_i(x) = $ total amount of prednisone in the body for day i, then

$$\mathrm{Day}_1(x) = 20 \qquad\qquad \mathrm{Day}_4(x) = 20x^3 + 15x^2 + 10x + 5$$
$$\mathrm{Day}_2(x) = 20x + 15 \qquad\qquad \mathrm{Day}_5(x) = 20x^4 + 15x^3 + 10x^2 + 5x + 5$$
$$\mathrm{Day}_3(x) = 20x^2 + 15x + 10$$

c. The graph of $\mathrm{Day}_5(x)$ is given in Figure 8.35. The function is a polynomial of degree 4. Since its domain is restricted to decimal values of decay rates, $0 < x < 1$.

d. Estimating from the graph, a patient with a decay factor of 0.40 has about 10 mg of prednisone left in the body on day 5. Evaluating with a calculator, the same patient on day 5 has

$$\mathrm{Day}_5(0.40) = 20(0.40)^4 + 15(0.40)^3 + 10(0.40)^2 + 5(0.40) + 5 \approx 10.07 \text{ mg}.$$

e. When $n > 5$ there are no more added dosages. Thus $\mathrm{Day}_6(x) = \mathrm{Day}_5(x) \cdot x$, and in general $\mathrm{Day}_n(x) = \mathrm{Day}_{n-1}(x) \cdot x$, if $n > 5$. So for $n \geq 5$,

$$\mathrm{Day}_n(x) = 20x^{n-1} + 15x^{n-2} + 10x^{n-3} + 5x^{n-4} + 5x^{n-5}.$$

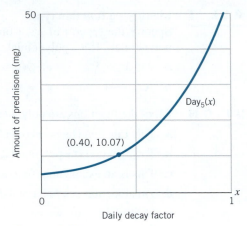

Figure 8.35 Amount of prednisone in the body on day 5 for various decay factors.

Algebra Aerobics 8.5a

1. For each of the following polynomials, specify the degree and evaluate each function at $x = -1$.

   a. $f(x) = 4x^3 + 11x^5 - 11$ **c.** $g(x) = -2x^4 - 20$

   b. $y = 1 + 7x^4 - 5x^3$ **d.** $z = 3x - 4 - 2x^2$

2. Find the degree of the polynomial function without multiplying out.

   a. $f(x) = x^5(3x - 2)^3(5x^2 + 1)$

   b. $g(x) = (x^2 - 1)(x^3 + x - 5)^4$

3. Given the polynomial function

   $f(t) = 0.5 - 2t^5 + 4t^3 - 6t^2 - t$:

   a. What is the degree of the polynomial? What is the name for polynomials of this degree?

   b. What is the leading term?

   c. What is the constant term?

   d. What is $f(0)$? $f(0.5)$? $f(-1)$?

Visualizing Polynomial Functions

What can we predict about the graph of a polynomial function from its equation? Examine the graphs of polynomials of different degrees in Figure 8.36. What can we observe from each of these pairs?

Turning points

The first thing we might notice is the number of turning points on each graph—the number of times the graph bends and changes direction. The quadratics bend once, the cubics seem to bend twice, the quartics three times, one quintic seems to bend four times, and the other quintic appears to bend twice. In general, a polynomial function of degree n will have at most $n - 1$ turning points.

Horizontal intercepts

Second, we might notice the number of times each graph crosses the horizontal axis. Each quadratic crosses at most two times, the cubics each cross at most three times, the quartics cross at most four times, and the quintics cross at most five times. In general, a polynomial function of degree n will cross the horizontal axis at most n times.

Global behavior

Finally, imagine zooming out on a graph to look at it on a global scale. All the polynomials in Figure 8.36 were graphed using small values for x (at most between -10 and 10) in order for us to see all the turning points and horizontal intercepts. What

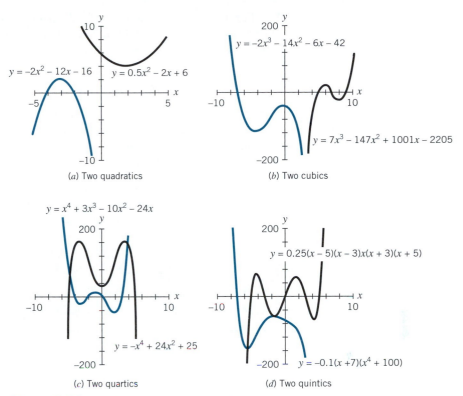

Figure 8.36 Graphs of pairs of polynomial functions (*a*) of degree 2 (quadratics), (*b*) of degree 3 (cubics), (*c*) of degree 4 (quartics), and (*d*) of degree 5 (quintics).

if we graphed the functions using a much wider range of x values, say between -1000 and 1000 or between $-1,000,000$ and 1,000,000?

Figure 8.37 shows two graphs of the same quintic, $y = 0.25(x - 5)(x - 3)x(x + 3)(x + 5)$, that is graphed in Figure 8.36(*d*). The scale for x on the left-hand graph is -10 to 10. The scale for x on the right-hand graph is -1000 to 1000. On the larger-scale graph on the right, the global behavior is clear, but the five horizontal intersection points and four turning points are no longer visible. The dominant features now are the arms of polynomials that extend infinitely upward on the right and downward on the left. Their direction is dictated by the leading term, which eventually dominates all the other terms. So at this large scale, the graph of our quintic function $y = 0.25(x - 5)(x - 3)x(x + 3)(x + 5)$ looks like the graph of $y = 0.25x^5$.

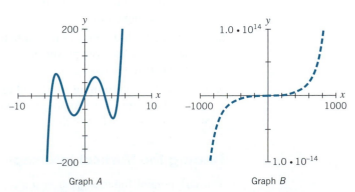

Figure 8.37 Two graphs of the same quintic using different scales.

For polynomial functions of odd degree, the two arms extend in opposite directions, one up and one down. For polynomial functions of even degree, both arms extend in the same direction, either both up or both down (see Figure 8.36). This is because given a polynomial

$$f(x) = a_n x^n + a_{n-1} x^{n-1} + \cdots + a_1 x^1 + a_0$$

of degree n, for large values of x the values of the leading term $a_n x^n$ will dominate the values of the other terms of smaller degree. In other words, for large values of x the function behaves like the power function $g(x) = a_n x^n$. The leading term of a polynomial determines its global shape.

The Graph of a Polynomial Function

The graph of a polynomial function of degree n will

have at most $n - 1$ turning points

cross the horizontal axis at most n times

For large values of x, the graph of $f(x) = a_n x^n + a_{n-1} x^{n-1} + \cdots + a_1 x + a_0$ will resemble the graph of the power function $g(x) = a_n x^n$.

EXAMPLE 3   What are the number of visible turning points and the number of horizontal intercepts for each polynomial graphed in Figure 8.38? What is the minimal degree for each polynomial function?

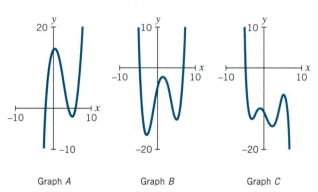

Graph *A*   Graph *B*   Graph *C*

Figure 8.38 Graphs of three polynomial functions.

SOLUTION   **a.** Graph *A* has two turning points, three horizontal intercepts, and the global shape of an increasing odd power function. This is at least a third-degree polynomial with a positive leading term.

b. Graph *B* has three turning points but only two horizontal intercepts and has the global shape of an even power function. This is at least a fourth-degree polynomial with a positive leading term.

c. Graph *C* has four turning points but only one horizontal intercept. It has the global shape of a decreasing odd power function. This is at least a fifth-degree polynomial with a negative leading term.

Finding the Vertical Intercept

The polynomial function $f(x) = a_n x^n + a_{n-1} x^{n-1} + \cdots + a_1 x + a_0$ has a vertical intercept at a_0, since $f(0) = a_0$. For example, the function $f(x) = x^4 - 16$ has a vertical intercept at -16 (see Figure 8.39).

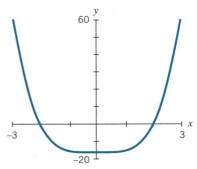

Figure 8.39 The graph of $f(x) = x^4 - 16$ has a vertical intercept at -16 and two horizontal intercepts at 2 and -2.

Finding the Horizontal Intercepts

As with any function, to find the zeros of a polynomial function $f(x)$, we set $f(x) = 0$ and solve for x. If a zero is a real number, it represents a horizontal intercept. For example, revisiting the relatively simple function $f(x) = x^4 - 16$, we can set $f(x) = 0$ and solve the corresponding equation for x to find the zeros.

Setting $f(x) = 0$ we have	$0 = x^4 - 16$
adding 16 to both sides	$16 = x^4$
taking the fourth root	$x = \pm 2$

So -2 and $+2$ are both real zeros, and hence x-intercepts, for the function $f(x) = x^4 - 16$ (see Figure 8.39).

But for a general polynomial function $f(x)$, solving the equation resulting from setting $f(x) = 0$ can be difficult. There is no simple analogue to the quadratic formula for polynomials of degree ≥ 3. The formulas for finding the zeros for third- and fourth-degree polynomials are extremely complicated. For polynomials of degree 5 or higher, there are no general algebraic formulas for finding the zeros. But there are algebraic approximation methods that allow us to compute the values for the real zeros (the horizontal intercepts) accurate to as many decimal places as we wish.

Estimating the horizontal intercepts

The complicated algebraic strategies for finding real zeros for a polynomial function of degree ≥ 3 are beyond the scope of this course. However, we can estimate the horizontal intercepts from a graph.

EXAMPLE 4 **Estimating**
Estimate the horizontal intercepts of the polynomial function in Figure 8.40. What is the minimum degree of the polynomial?

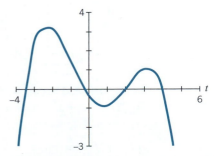

Figure 8.40 Graph of a polynomial function $Q(t)$.

SOLUTION   The t-intercepts are at approximately -3.5, -0.25, 2, and 4. Since the polynomial has four horizontal intercepts and three turning points, it must be at least of degree 4.

EXAMPLE 5   **Zooming in on a graph**

Suppose you make four yearly deposits of $2000 in a risky high-yield account, starting today. What annual interest must you earn if you want to have $10,000 in the account after 4 years?[4]

SOLUTION   Let's track the four separate $2000 deposits made in years 1, 2, 3, and 4. If r is the annual interest rate (in decimal form), then $x = 1 + r$ is the annual multiplier or the growth factor.

The first $2000 invested at the beginning of year 1 will be multiplied by x, the annual growth factor, four times, once for each year. So at the end of 4 years this $2000 investment will be worth $2000x^4$ dollars.

The second $2000 invested at the beginning of year 2 will only have 3 years in which to earn interest, years 2, 3, and 4. At the end of year 4, the second $2000 investment will have been multiplied by the growth factor x only three times, making it worth $2000x^3$ dollars. Similarly, the $2000 investments made at the beginning of years 3 and 4 will be worth $2000x^2$ and $2000x$ dollars, respectively, by the end of year 4. So at the end of the fourth year, assuming you do not make a final deposit, the total dollars in your account are

$$2000x^4 + 2000x^3 + 2000x^2 + 2000x$$

Since you want to have $10,000 in your account after 4 years, our equation is

$$10{,}000 = 2000x^4 + 2000x^3 + 2000x^2 + 2000x \tag{1}$$

We need to solve Equation (1) for x to find the annual interest rate r (where $x = 1 + r$). If we subtract 10,000 from each side and then divide by 2000, we have

$$0 = x^4 + x^3 + x^2 + x - 5 \tag{2}$$

A solution to Equation (2) will be a value x for the function

$$f(x) = x^4 + x^3 + x^2 + x - 5$$

such that $f(x) = 0$. So x-intercepts for $f(x)$ correspond to real solutions to Equation (2).

We can estimate the x-intercepts of $f(x)$ by graphing the function (Figure 8.41) and then zooming in (Figure 8.42).

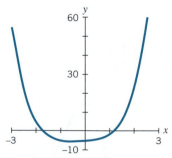

Figure 8.41 Graph of $f(x) = x^4 + x^3 + x^2 + x - 5$.

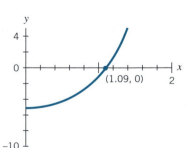

Figure 8.42 Zooming in on the x-intercept between 1 and 2.

[4]Example modeled after a problem in E. Connally, D. Hughes-Hallett, A. M. Gleason, et al., *Functions Modeling Change: A Preparation for Calculus,* 2nd ed. (New York: John Wiley & Sons, 1998), pp. 384–385.

Only positive solutions have meaning in this model, and there is only one positive *x*-intercept, at approximately (1.09, 0). We have $x = 1 + r$, where r is the annual interest rate. Since $x = 1.09$, then $r = x - 1 = 0.09$ in decimal form, or 9% written as a percentage. So you would need an annual interest rate of approximately 9% to end up with $10,000 when investing $2000 each year for 4 years.

Using factoring to find the horizontal intercepts

The Factor Theorem extends to polynomials in general.

The Factored Form of a Polynomial

If $r_1, r_2, \ldots, r_n$ are zeros of a polynomial $f(x) = a_nx^n + a_{n-1}x^{n-1} + \cdots + a_1x + a_0$, then $f(x)$ can be written in *factored form* as

$$f(x) = a_n(x - r_1)(x - r_2) \cdots \cdot (x - r_n)$$

Any real zero is an *x*-intercept for $f(x)$.

So if you are given the zeros, it's easy to construct a function in factored form with those zeros. Conversely, if you have a function in factored form, it is easy to find the zeros.

EXAMPLE 6 **Finding horizontal intercepts**

a. What are the *t*-intercepts of the function $Q(t) = 1.5t(t - 7)(t - 5)(t + 3)$?

b. For large values of *t* (both positive and negative), the graph of $Q(t)$ resembles what power function? Would the left arm of the parabola eventually extend up or down? And the right arm?

SOLUTION **a.** The factors of $Q(t)$ tell us it has real zeros and hence *t*-intercepts at 0, 7, 5, and −3.

b. For large values of *t*, the leading term will dominate. If we put $Q(t)$ into standard form, the leading term will be $1.5t^4$, so in the long run $Q(t)$ behaves like the power function $f(t) = 1.5t^4$. So both arms of the graph of $Q(t)$ will eventually extend upward.

EXAMPLE 7 **Constructing polynomial functions**

Create a polynomial function with *x*-intercepts at 0, −2, and 1.

SOLUTION If $x = 0$ is an *x*-intercept, then *x* is a factor of the polynomial function.
If $x = -2$ is an *x*-intercept, then $(x - (-2))$, or $(x + 2)$, is also a factor.
If $x = 1$ is an *x*-intercept, then $(x - 1)$ is a factor as well.

So the function $f(x) = x(x + 2)(x - 1)$ has *x*-intercepts at 0, −2, and 1.

There are many other polynomial functions with the same horizontal intercepts, as we will see in the next example.

EXAMPLE 8 **Multiple functions with the same horizontal intercepts**

a. Describe two other polynomial functions of degree 3 that have *x*-intercepts at 0, −2, and 1.

b. Construct one that passes through the point (2, −16).

SOLUTION **a.** The functions $g(x) = 3x(x + 2)(x - 1)$ and $h(x) = -5x(x + 2)(x - 1)$ both have x-intercepts at 0, −2, and 1 (see Figure 8.43). Note that both are multiples of the function $f(x) = x(x + 2)(x - 1)$; that is, $g(x) = 3f(x)$ and $h(x) = -5f(x)$. In general, any function of the form $af(x) = ax(x + 2)(x - 1)$ where $a \neq 0$ will have x-intercepts at 0, −2, and −1. So there are infinitely many possibilities. (See Figure 8.43.)

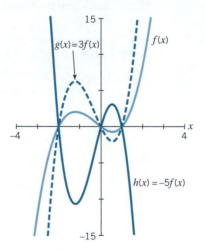

Figure 8.43 The graphs of three polynomial functions all with x-intercepts at 0, −2, and 1.

b. Given the general form of a polynomial with x-intercepts at 0, −2, and 1,

$$f(x) = ax(x + 2)(x - 1)$$

If we substitute in $x = 2$ and $f(x) = -16$ $-16 = a(2)(2 + 2)(2 - 1)$

simplify $-16 = a(8)$

and solve for a, we have $a = -2$

So the polynomial function $f(x) = -2x(x + 2)(x - 1)$ has x-intercepts at 0, −2, and 1 and passes through the point $(2, -16)$ (see Figure 8.44).

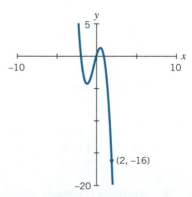

Figure 8.44 A polynomial with x-intercepts at 0, −2, and 1 that passes through $(2, -16)$.

From Examples 7 and 8, we see that we can create a polynomial function with any finite set of real numbers as its zeros. Looking at it another way, we can create new polynomials by multiplying together existing polynomials.

A Summary of Polynomial Behavior

Given a polynomial of the form $f(x) = a_n x^n + a_{n-1} x^{n-1} + \cdots + a_1 x + a_0$,

The graph has at most $n - 1$ turning points and n horizontal intercepts.
For large values of x, the graph will resemble the graph of $g(x) = a_n x^n$.
The *zeros* of $f(x)$ are the values of x such that $f(x) = 0$.
If $r_1, r_2, r_3, \ldots, r_n$ are all the zeros of a function $f(x)$, then $f(x)$ can be written in factored form as $f(x) = a_n(x - r_1)(x - r_2) \cdots \cdots (x - r_n)$.
Each real zero corresponds to an *x-intercept*.
The *vertical intercept* is at $f(0) = a_0$.

Algebra Aerobics 8.5b

A graphing program is required for Problems 1 and 2, and recommended for Problems 4 and 5.

1. Use technology to graph the polynomial function $y = -2(x - 3)^2(x + 4)$.

    a. What is the degree of the function?

    b. How many turning points does it have?

    c. What happens to y as $x \to +\infty$? As $x \to -\infty$?

    d. The graph looks like what power function for very large positive or negative values of x?

    e. What are the horizontal intercepts?

    f. When $x = 0$, what is y? What is this point called?

2. Determine the lowest possible degree and the sign of the leading coefficient for the polynomial functions in the three graphs below. Justify your answers. (*Hint:* Use your knowledge of power functions.)

3. For each of the following functions, identify the vertical intercept. Then use a function graphing program (and its zoom feature) to estimate the number of horizontal intercepts and their approximate values.

    a. $y = 3x^3 - 2x^2 - 3$ b. $f(x) = x^2 + x + 3$

4. Identify the degree and the x-intercepts of each of the following polynomial functions. If technology is available, graph each function to verify your work.

    a. $y = 3x + 6$

    b. $f(x) = (x + 4)(x - 1)$

    c. $g(x) = (x + 5)(x - 3)(2x + 5)$

5. Construct three polynomial functions, all with x-intercepts at -3, 0, 5, and 7. Use technology, if available, to plot them.

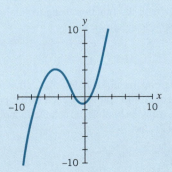

Graph *A*

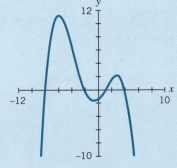

Graph *B*

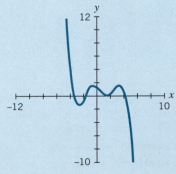

Graph *C*

Exercises for Section 8.5

A calculator that can evaluate powers is required. Several exercises require a graphing program.

1. Identify which of the following are polynomial functions and, for those that are, specify the degree.

a. $y = 3x + 2$ **d.** $y = 3^x - 2$

b. $y = 2 - x^3$ **e.** $y = 3x^5 - 4x^3 - 6x^2 - 12$

c. $y = 5t^2 - 3\sqrt{t^5}$ **f.** $y = 6(x)(x - 5)(2x + 7)$

2. Identify the degree of any of the following functions that are polynomials, and for those that are not polynomials, explain why.

a. $y = 3x^2 - \dfrac{2}{x^3} + 1$ **c.** $y = x^{5/3} + x^2$

b. $y = 2x^5 + \dfrac{x^4}{3} - 3x$ **d.** $y = 2t^3 + 5t^4 + \sqrt{2}$

3. Evaluate the following expressions for $x = 2$ and $x = -2$.

a. x^{-3} **b.** $4x^{-3}$ **c.** $-4x^{-3}$ **d.** $-4x^3$

4. Evaluate the following polynomials for $x = 2$ and $x = -2$, and specify the degree of each polynomial.

a. $y = 3x^2 - 4x + 10$

b. $y = x^3 - 5x^2 + x - 6$

c. $y = -2x^4 - x^2 + 3$

5. Match each of the following functions with its graph.

a. $y = 2x - 3$ **b.** $y = 3(2^x)$ **c.** $y = (x^2 + 1)(x^2 - 4)$

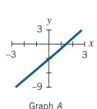

Graph A   Graph B   Graph C

6. Match each of the following functions with its graph.

a. $f(x) = x^2 + 3x + 1$ **c.** $f(x) = -\frac{1}{2}x^3 + x - 3$

b. $f(x) = \frac{1}{3}x^3 + x - 3$

  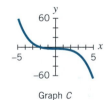

Graph A   Graph B   Graph C

7. For each of the graphs of polynomial functions, at the top of the next column, determine (assuming the arms extend indefinitely in the indicated direction):

   i. The number of turning points

   ii. The number of x-intercepts

   iii. The sign of the leading term

   iv. The minimum degree of the polynomial

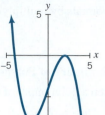 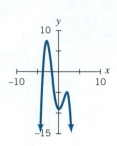 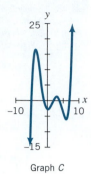

Graph A   Graph B   Graph C

8. Describe how $g(x)$ and $h(x)$ relate to $f(x)$.

$$f(x) = x^5 - 3x^2 + 4$$
$$g(x) = -x^5 + 3x^2 - 4$$
$$h(x) = x^5 - 3x^2 - 2$$

9. Divide the following functions into groups having the same global shape for large values of x. Explain your groupings.

a. $y = (x^2 - 3)(x + 9)$

b. $y = -x^4 + 3$

c. $y = x^5 - 3x^4 - 11x^3 + 3x^2 + 10x$

d. $y = x(3 - x)(x + 1)^2$

e. $y = 7x^3 - 3x^2 - 20x + 5$

f. $y = -(x^2 + 1)(x^2 - 4)$

10. Estimate the maximum number of turning points for each of the polynomial functions. If available, use technology to graph the function to verify the actual number.

a. $y = x^4 - 2x^2 - 5$ **c.** $y = x^3 - 3x^2 + 4$

b. $y = 4t^6 + t^2$ **d.** $y = 5 + x$

11. Describe the behavior of each polynomial function for large values (positive or negative) of the independent variable and estimate the maximum number of turning points. If available, use technology to verify the actual number.

a. $y = -2x^4 + 4x + 3$

b. $y = (t^2 + 1)(t^2 - 1)$

c. $y = x^3 + x + 1$

d. $y = x^5 - 3x^4 - 11x^3 + 3x^2 + 10x$

12. (Graphing program required.) Estimate the maximum number of horizontal intercepts for each of the polynomial functions. Then graph the function using technology to find the actual number. (See Exercise 10.)

a. $y = x^4 - 2x^2 - 5$ **c.** $y = x^3 - 3x^2 + 4$

b. $y = 4t^6 + t^2$ **d.** $y = 5 + x$

13. (Graphing program required.) Estimate the maximum number of horizontal intercepts for each of the polynomial functions. Then, using technology, graph the functions to find their approximate values.

a. $y = -2x^2 + 4x + 3$

b. $y = (t^2 + 1)(t^2 - 1)$

c. $y = x^3 + x + 1$

d. $y = x^5 - 3x^4 - 11x^3 + 3x^2 + 10x$

14. a. (Graphing program required.) Use a function graphing program to estimate the x-intercepts for each of the following. Make a table showing the degree of the polynomial and the number of x-intercepts. What can you conclude?

$$y = 2x + 1 \qquad\qquad y = x^3 - 5x^2 + 3x + 5$$
$$y = x^2 - 3x - 4 \qquad y = 0.5x^4 + x^3 - 6x^2 + x + 3$$

b. Repeat part (a) for the following functions. How do your results compare with those for part (a)? Are there any modifications you need to make to your conclusions in part (a)?

$$y = 3x + 5 \qquad\qquad y = x^3 - 2x^2 - 4x + 8$$
$$y = x^2 + 2x + 3 \qquad y = (x - 2)^2(x + 1)^2$$

15. (Graphing program required.) Use a function graphing program (and its zoom feature) to estimate the number of x-intercepts and their approximate values for:

a. $y = 3x^3 - 2x^2 - 3$ **b.** $f(x) = x^2 + 5x + 3$

16. (Graphing program required.) Identify the x-intercepts of the following functions; then graph the functions to check your work.

a. $y = 3x + 6$

b. $y = (x + 4)(x - 1)$

c. $y = (x + 5)(x - 3)(2x + 5)$

17. a. If the degree of a polynomial is odd, then at least one of its zeros must be real. Explain why this is true.

b. Sketch a polynomial function that has no real zeros and whose degree is:

    i. 2   **ii.** 4

c. Sketch a polynomial function of degree 3 that has exactly:

    i. One real zero   **ii.** Three real zeros

d. Sketch a polynomial function of degree 4 that has exactly two real zeros.

18. In each part, construct a polynomial function with the indicated characteristics.

a. Crosses the x-axis at least three times

b. Crosses the x-axis at -1, 3, and 10

c. Has a y-intercept of 4 and degree of 3

d. Has a y-intercept of -4 and degree of 5

19. Which of the following statements are true about the graph of the polynomial function

$$f(x) = x^3 + bx^2 + cx + d$$

a. It intersects the vertical axis at one and only one point.

b. It intersects the x-axis in at most three points.

c. It intersects the x-axis at least once.

d. The vertical intercept is positive.

e. For large (positive or negative) values of x, the graph looks like $y = x^3$.

f. The origin is a point on the graph.

20. (Graphing program required.) A manufacturer sells children's wooden blocks packed tightly in a cubic tin box with a hinged lid. The blocks cost 3 cents a cubic inch to make. The box and lid material cost 1 cent per square inch. (Assume the sides of the box are so thin that their thickness can be ignored.) It costs 2 cents per linear inch to assemble the box seams. The hinges and clasp on the lid cost $2.50, and the label costs 50 cents. (See the accompanying figure.)

A cubic box with edge length s filled with blocks.

a. If the edge length of the box is s inches, develop a formula for estimating the cost $C(s)$ of making a box that's filled with blocks.

b. Graph the function $C(s)$ for a domain of 0 to 20. What section of the graph corresponds to what the manufacturer actually produces—boxes between 4 and 16 inches in edge length?

c. What is the cost of this product if the cube's edge length is 8 inches?

d. Using the graph of $C(s)$, estimate the edge length of the cube when the total cost is $100.

21. Polynomial expressions of the form $an^3 + bn^2 + cn + d$ can be used to express positive integers, such as 4573, using different powers of 10:

$$4573 = 4 \cdot 1000 + 5 \cdot 100 + 7 \cdot 10 + 3$$

or

$$4573 = 4 \cdot 10^3 + 5 \cdot 10^2 + 7 \cdot 10^1 + 3 \cdot 10^0$$

or if $n = 10$,

$$4573 = 4 \cdot n^3 + 5 \cdot n^2 + 7 \cdot n^1 + 3 \cdot n^0$$

$$= 4n^3 + 5n^2 + 7n + 3$$

Notice that in order to represent any positive number, the coefficient multiplying each power of 10 must be an integer between 0 and 9.

a. Express 8701 as a polynomial in n assuming $n = 10$.

b. Express 239 as a polynomial in n assuming $n = 10$.

Computers use a similar polynomial system, called binary numbers, to represent numbers as sums of powers of 2. The number 2 is used because each minuscule switch in a computer can have one of two states, on or off; the symbol 0 signifies off, and the symbol 1 signifies on. Each binary number is built up from a row of switch positions each set at 0 or 1 as multipliers for different powers of 2. For instance, in the binary number system 13 is represented as 1 1 0 1, which stands for

$$1 \cdot 2^3 + 1 \cdot 2^2 + 0 \cdot 2^1 + 1 \cdot 2^0 =$$
$$1 \cdot 8 + 1 \cdot 4 + 0 \cdot 2 + 1 \cdot 1 =$$
$$13$$

c. What number does the binary notation 1 1 0 0 1 represent?

d. Find a way to write 35 as the sum of powers of 2; then give the binary notation.

22. A typical retirement scheme for state employees is based on three things: age at retirement, highest salary attained, and total years on the job.

Annual retirement allowance =

$$\left(\begin{matrix}\text{total years}\\\text{worked}\end{matrix}\right) \cdot \left(\begin{matrix}\text{retirement}\\\text{age factor}\end{matrix}\right) \cdot \left(\begin{matrix}\text{\% of highest}\\\text{salary}\end{matrix}\right)$$

where the maximum percentage is 80%. The highest salary is typically at retirement.

We define:

total years worked = retirement age − starting age

retirement age factor = 0.001 · (retirement age − 40)

salary at retirement = starting salary + all annual raises

a. For an employee who started at age 30 in 1973 with a salary of $12,000 and who worked steadily, receiving a $2000 raise every year, find a formula to express retirement allowance, R, as a function of employee retirement age, A.

b. Graph R versus A.

c. Construct a function S that shows 80% of the employee's salary at age A and add its graph to the graph of R.

From the graph estimate the age at which the employee annual retirement allowance reaches the limit of 80% of the highest salary.

d. If the rule changes so that instead of highest salary, you use the average of the three highest years of salary, how would your formula for R as a function of A change?

8.6 New Functions from Old

We've seen how the simple quadratic $y = x^2$ can be transformed (through stretching, compressing, and shifting) to generate all quadratic functions. We've also added power functions to generate polynomial functions. In the next three sections we'll look at many ways of transforming any function or combining any two or more functions to create new functions. Our focus, however, will be on combining members from the families of functions we have studied: linear, exponential, logarithmic, power, quadratic, and polynomial.

Transforming a Function

Stretching, compressing, and shifting

Recall from our discussion in Section 8.2 that to stretch or compress the graph of a function, we multiply the function by a constant. To shift the horizontal or vertical position of its graph, we add (or subtract) a constant to either the input or the output of the function.

Stretching, Compressing, or Shifting the Graph of $f(x)$

To stretch or compress: Multiply the output of $f(x)$ *by* a constant to get $af(x)$.
If $|a| > 1$, the graph of $f(x)$ is vertically stretched by a factor of a.
If $0 < |a| < 1$, the graph of $f(x)$ is vertically compressed by a factor of a.

To shift vertically: Add a constant to the output of $f(x)$ to get $f(x) + k$.
If k is positive, the graph of $f(x)$ is shifted up.
If k is negative, the graph of $f(x)$ is shifted down.

To shift horizontally: Subtract a constant from the input of $f(x)$ to get $f(x - h)$.
If h is positive, the graph of $f(x)$ is shifted to the right.
If h is negative, the graph of $f(x)$ is shifted to the left.

Reflections across the horizontal axis

When we studied power functions, we learned that the graphs of $y = kx^p$ and $y = -kx^p$ are reflections of each other across the x-axis. What happens to the equation of any function when we reflect its graph across the horizontal axis?

Figure 8.45 shows two functions, f and g, that are reflections of each other across the x-axis. We can see from the graph that when a point is reflected vertically across the x-axis, the x value stays fixed while the y value changes sign. This means that for each input, the corresponding outputs for f and g are opposites; that is, they have the same magnitude but opposite signs, and are on opposite sides of the x-axis. So $g(x) = -f(x)$.

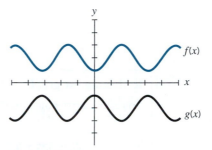

Figure 8.45 The graph of $g(x) = -f(x)$ is a reflection of the graph of $f(x)$ across the horizontal axis.

Reflections across the vertical axis

What if we reflected a function's graph across the vertical axis? What happens to the equation of the original function?

Figure 8.46 shows two functions, f and g, that are reflections of each other across the y-axis. We can see from the graph that when a point is reflected horizontally across the y-axis, the y value stays fixed while the x value changes sign. This means that when f and g have the same output, then the corresponding inputs, or x values, are opposites; that is, they have the same magnitude but opposite signs, and are on opposite sides of the y-axis. So $g(x) = f(-x)$.

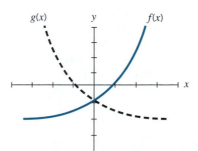

Figure 8.46 The graph of $g(x) = f(-x)$ is a reflection of the graph of $f(x)$ across the vertical axis.

Reflections across both the horizontal and vertical axes

The net result of reflecting the graph of $f(x)$ across both the x- and y-axes is equivalent to rotating the graph $180°$ about the origin, as shown in Figure 8.47. This is a double reflection, so we need to multiply both the input and the output by -1 to get $-f(-x)$.

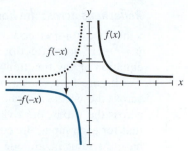

Figure 8.47 The graph of
$-f(-x)$ is the graph of $f(x)$
reflected across the y-axis and then
across the x-axis.

Reflecting the graph of $f(x)$

To reflect across the x-axis:	Multiply the output of $f(x)$ by -1 to get $-f(x)$.
To reflect across the y-axis:	Multiply the input of $f(x)$ by -1 to get $f(-x)$.
To reflect across both the x- and y-axes:	Multiply both the input and output of $f(x)$ by -1 to get $-f(-x)$.

Symmetry

If we reflect the graph of $f(x)$ across the y-axis and get the graph of $f(x)$ again, then $f(x)$ is *symmetric across the y-axis*. Symmetry across the y-axis means that the part of the graph to the left of the y-axis is the mirror image of the part to the right of the y-axis. More formally, it means $f(x) = f(-x)$ for all values of x in the domain of f.

If after we rotate the graph of $f(x)$ $180°$ about the origin we get the graph of $f(x)$, then $f(x)$ is *symmetric about the origin*. We can get the same net result by reflecting the graph of $f(x)$ across both the x- and y-axes. So $f(x) = -f(-x)$ for all values of x in the domain of f.

In Chapter 7 we learned that power functions with even integer powers are symmetric across the y-axis and those with odd integer powers are symmetric about the origin. Figure 8.48 shows the graphs of other functions that have symmetry.

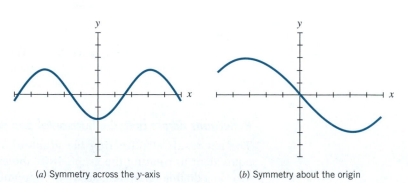

(a) Symmetry across the y-axis   (b) Symmetry about the origin

Figure 8.48 Examples of symmetric graphs.

> **Symmetry across the *y*-Axis and about the Origin**
>
> If for all values of x in the domain of f:
>
> $$f(x) = f(-x) \qquad \text{then the graph of } f \text{ is symmetric across the } y\text{-axis}$$
> $$f(x) = -f(-x) \qquad \text{then the graph of } f \text{ is symmetric about the origin}$$

EXAMPLE 1 **Reflections and symmetry**

a. Describe the functions $g(x)$ and $h(x)$ shown in Figure 8.49 in terms of transformations of $f(x)$.

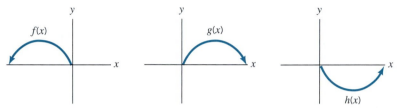

Figure 8.49 Two transformations of $f(x)$.

b. When two objects A and B collide, the force F_1 exerted on object A is equal in magnitude and opposite in direction to the force F_2 exerted on object B, as shown in Figure 8.50(a). The graph in Figure 8.50(b) shows F_1 and F_2 as functions of time. Describe F_1 in terms of F_2.

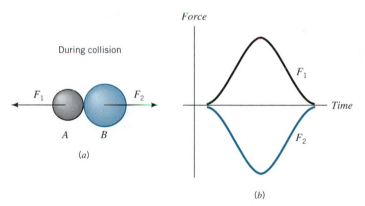

Figure 8.50 Objects colliding.

c. Identify any symmetries in the graphs in Figure 8.51.

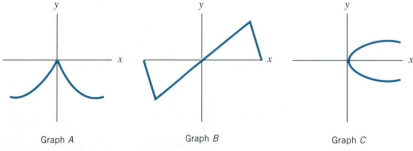

Graph *A* Graph *B* Graph *C*

Figure 8.51 Different symmetries.

SOLUTION **a.** The graph of g is a reflection of the graph of f across the y-axis, so $g(x) = f(-x)$. The graph of h is a refection of the graph of f across both the x- and y-axes, so $h(x) = -f(-x)$.

b. $F_1(t) = -F_2(t)$ since their graphs are reflections of each other across the time axis.

c. A is the graph of a function that is symmetric across the y-axis. B is the graph of a function that is symmetric about the origin. C is a graph that is symmetric across the x-axis, but it is not a function since there are two values of y for each $x > 0$.

EXAMPLE 2 **Shifting and stretching**

a. The time series in Figure 8.52 show two related measures of income inequality in the United States. Zero percent represents perfect equality. Construct an equation that approximates Z_{D+R} in terms of Z_D.

Figure 8.52 Measures of income inequality in the United States.
Source: G. Kluge, *Wealth and People: Inequality Measures* (2002), *www.poorcity.richcity.org/entkiss.htm.*

b. Use the properties of logs and function transformations to predict how the graphs of the following functions will differ from each other. Confirm your predictions by graphing all of the functions on the same grid.

$$y = \log x \qquad y = \log 10x \qquad y = \log 100x \qquad y = 10 \log x$$

SOLUTION **a.** A rough approximation is $Z_{D+R} = Z_D + 1.3$.

b. The graph of $y = \log 10x$ is the graph of $y = \log x$ shifted vertically up one unit, since

Given	$y = \log 10x$
Rule 1 of logs	$= \log x + \log 10$
Evaluate log 10	$= \log x + 1$

Similarly, the graph of $y = \log 100x$ is the graph of $y = \log x$ shifted vertically up two units, since

$$y = \log 100x$$
$$= \log x + \log 100$$
$$= \log x + \log(10^2)$$
$$= \log x + 2$$

The graph of $y = 10 \log x$ is the graph of $y = \log x$ stretched by a factor of 10. Figure 8.53 shows the graphs of the four functions on the same grid.

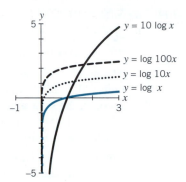

Figure 8.53 Four log functions.

EXAMPLE 3 **Multiple transformations**

a. Without using technology, sketch graphs of each of the following functions:

$$f(x) = x^3 \qquad g(x) = 2f(x + 1) \qquad h(x) = f(-x) - 2$$

b. Construct a new function for each the following transformations of $g(t) = 100(1.03)^t$:

$$3g(t - 2) \quad \text{and} \quad 0.5g(-t) + 3$$

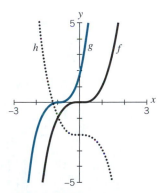

Figure 8.54 Two transformations of $f(x)$.

SOLUTION a. See Figure 8.54.

b. Given $g(t) = 100(1.03)^t$

$$3g(t - 2) = (3)(100)(1.03)^{(t-2)}$$
$$= 300(1.03)^{(t-2)}$$

and

$$0.5g(-t) + 3 = (0.5)(100)(1.03)^{-t} + 3$$
$$= 50(1.03)^{-t} + 3$$
$$= 50\left(\tfrac{1}{1.03}\right)^t + 3$$
$$\approx 50(0.97)^t + 3$$

EXAMPLE 4 **Financial considerations: Horizontal shift of an exponential function**

Twins each receive $10,000 from their grandmother on their shared 21st birthday. One invests her money right away in high-grade corporate bonds that return an annual 6% in interest, which the sister keeps reinvesting. The other twin can't decide what to do, so she puts her money in a non–interest-bearing checking account. At her 33rd birthday, after getting married and having kids, she decides she needs to save for the future. She now invests $10,000 in the equivalent bonds, receiving the same interest rate as her sister.

a. Construct a function $S_1(t)$ that represents how much the first sister's investment is worth t years after her 21st birthday.

b. Use $S_1(t)$ to construct a function $S_2(t)$ for the second sister, who waits until she is 33 to invest her money.

c. Graph $S_1(t)$ and $S_2(t)$ on the same grid. What are the differences and similarities between the graphs in practical terms?

d. Use your graphs to estimate how long it will take for each investment to be worth $40,000.

e. Use your graphs to estimate the difference in the worth of the sisters' investments when they are both 65 years old.

SOLUTION
a. $S_1(t) = 10,000(1.06)^t$ for $t \geq 0$

b. $S_2(t) = S_1(t - 12)$

$\qquad = 10,000(1.06)^{(t-12)}$ for $t \geq 12$

c. The graphs are drawn in Figure 8.55. The graph of S_2 is the graph of S_1 shifted 12 units $(33 - 21 = 12)$ to the right. So at age 33, the second sister has $10,000 in her account, the same amount she had when she was 21 years old. Each successive year the second sister is 12 years behind her twin sister in terms of how much money her investment is worth.

d. It will take about 24 years for each investment to be worth about $40,000. So the first twin would be about 45 and the second twin about 57, or 12 years older than her sister. Remember that $S_2(t) = S_1(t - 12)$, so $S_2(57) = S_1(57 - 12) = S_1(45)$.

Figure 8.55 Twin investments.

e. The difference between the investments when the twins are 65 years old (44 years after their 21st birthday) is about $65,000, since when $t = 44$ years, $S_1(44) \approx \$130,000$ and $S_2(44) \approx \$65,000$. The first twin has almost twice as much as the second.

EXAMPLE 5   **Transformations of a power function**

a. Construct an equation and sketch a graph of the following transformations of $f(x) = 1/x^2$:

$$2f(x - 1) \quad \text{and} \quad 3f(x + 2).$$

b. In what ways will the graph of $g(x) = 0.5f(x - 2)$ differ from the graph of $f(x)$? How will it be similar? Check your predictions by graphing $f(x)$ and $g(x)$ on the same grid.

SOLUTION
a. The graphs are drawn in Figure 8.56, with dotted vertical lines indicating vertical asymptotes.

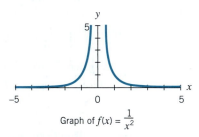

Graph of $f(x) = \dfrac{1}{x^2}$

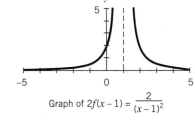

Graph of $2f(x - 1) = \dfrac{2}{(x - 1)^2}$

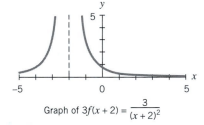

Graph of $3f(x + 2) = \dfrac{3}{(x + 2)^2}$

Figure 8.56 Graph of $f(x) = 1/x^2$ and two transformations of $f(x)$.

b. As shown in Figure 8.57, the graph of $g(x)$ has the same overall shape (containing two parts) as the graph of $f(x)$, but $g(x)$ is $f(x)$ compressed by a factor of 0.5 and shifted horizontally two units to the right. The y-axis is the vertical asymptote for $f(x)$, but the line $x = 2$ is the vertical asymptote (dotted line) for $g(x)$.

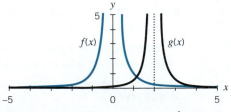

Figure 8.57 Graphs of $f(x) = \dfrac{1}{x^2}$ and $g(x) = 0.5f(x - 2)$.

EXAMPLE 6 **Cooling coffee: Vertical shift of an exponential function**

A cup of hot coffee (at 200° Fahrenheit) is left to stand in a 70° Fahrenheit room. The difference in temperature between the coffee and the room can be modeled by an exponential function where the temperature decay rate is 10% per minute.

a. Construct $D(t)$, the temperature difference, as a function of time t in minutes that the coffee has been left standing.

b. Using $D(t)$, construct a new function $C(t)$ that describes the coffee's temperature as it gradually approaches the temperature of the room.

c. What are the differences and similarities between the graphs of $D(t)$ and $C(t)$? Check your answer by graphing $D(t)$ and $C(t)$ on the same grid.

d. Use the graph of $C(t)$ to estimate how long it will take for the coffee to cool to 150°F, a comfortable temperature for drinking.

SOLUTION **a.** $D(t) = 130(0.9)^t$ since the starting temperature difference is $200° - 70° = 130°$, and the decay factor is $1 - 0.1 = 0.9$.

b. $C(t) = 70 + 130(0.9)^t$

c. The graph of $C(t)$ is the graph of $D(t)$ shifted vertically up 70 units (see Figure 8.58).

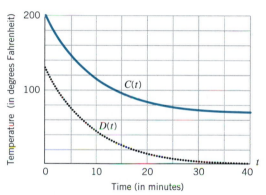

Figure 8.58

d. It will take approximately 5 minutes for the coffee to cool to 150°F.

Algebra Aerobics 8.6

Graphing program optional in Problems 6 and 7.

1. Graphs A, B, and C show $f(x)$ and a transformation of $f(x)$. Identify the transformation as: $-f(x)$, $f(-x)$, or $-f(-x)$.

2. Apply the transformations specified in parts (a)−(e) to $f(x) = 2x - 3$ and $f(x) = 1.5^x$.

a. $f(x + 2)$ **c.** $-f(x)$ **e.** $-f(-x)$
b. $\frac{1}{2}f(x)$ **d.** $f(-x)$

Graph A

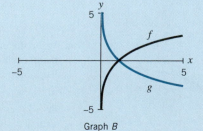

Graph B

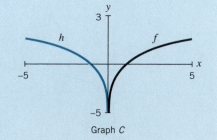

Graph C

3. Identify which of the following graphs are symmetric across the vertical axis. Which are symmetric across the horizontal axis? Which are symmetric about the origin?

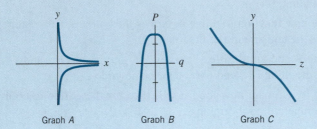

Graph A   Graph B   Graph C

4. Construct a new function from $h(t) = e^t$, where the graph of $h(t)$ is:

  a. First shifted two units right

  b. Then reflected across the t-axis

  c. Then shifted down by one unit

5. Construct a new function from $Q(t) = 2 \cdot 1.06^t$ where the graph of $Q(t)$ is:

  a. First shifted two units to the left

  b. Then shifted down by one unit

  c. Then reflected across the t-axis

(Graphing program optional.) Describe in each part of Problems 6 and 7 how $g(x)$ is related to $f(x)$. If you have a graphing tool available, check your answer by looking at the graphs.

6. a. $f(x) = 5 - x^2$ and $g(x) = x^2 - 5$

  b. $f(x) = 3 \cdot 2^x$ and $g(x) = 3 \cdot 2^{-x}$

7. a. $f(x) = \frac{1}{x-2}$ and $g(x) = \frac{1}{3x-6}$

  b. $f(x) = \ln x$ and $g(x) = \ln 3 + \ln x = \ln 3x$

Exercises for Section 8.6

Graphing program optional for Exercises 7, 16, and 18, and required for Exercises 21 and 22.

1. In each part of the problem the graph of $f(x)$ to the left has been transformed into the graph of $g(x)$ to the right. First describe whether the graph of $f(x)$ was stretched/compressed, reflected, and/or shifted vertically/horizontally to form $g(x)$. Then write the equation for $g(x)$ in terms of $f(x)$.

a.

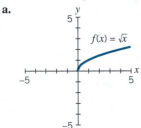
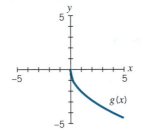

b.

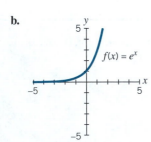
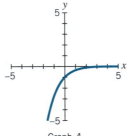

c.

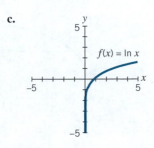
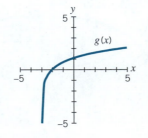

2. Match each of the following functions with its graph. Identify the parent (original) function $p(x)$ and the transformation(s) that took place.

  a. $f(x) = (x - 2)^3 + 4$ **c.** $h(x) = -\ln(x + 1)$

  b. $g(x) = -x^3 - 2$ **d.** $k(x) = -e^{-x}$

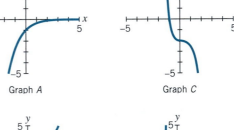

Graph A   Graph C

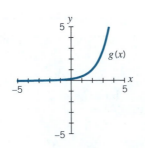
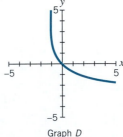

Graph B   Graph D

3. Let $f(x) = x^3$.

  a. Write the equation for the new function $g(x)$ that results from each of the following transformations of $f(x)$. Explain in words the effect of the transformations.

  i. $f(-x)$ **iii.** $f(x + 2)$

  ii. $-2f(x) - 1$ **iv.** $-f(-x)$

  b. Sketch by hand the graph of $f(x)$ and each function in part (a).

4. Explain in words the effect of the following transformations on the graph of $g(t)$.

a. $-5g(t)$ **c.** $-g(-t) - 4$

b. $\frac{1}{5}g(t - 3) + 1$ **d.** $4g(-t) - 2$

5. Decide if each graph (although not necessarily a function) is symmetric across the x-axis, across the y-axis, and/or about the origin.

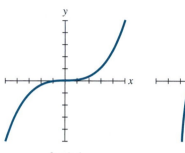

Graph A   Graph C

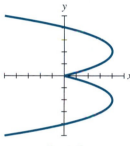

Graph B

6. Complete the partial graph shown in three different ways to create a graph that is:

a. Symmetric across the x-axis

b. Symmetric across the y-axis

c. Symmetric about the origin

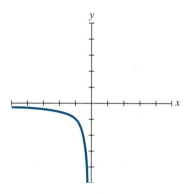

7. (Graphing program optional.) A function is said to be *even* if $f(-x) = f(x)$ and *odd* if $f(-x) = -f(x)$ for all x in f's domain. Use these definitions to:

a. Show that the even power functions are even.

b. Show that the odd power functions are odd.

c. Show whether each of the following functions is even, odd, or neither.

   i. $f(x) = x^4 + x^2$ **iii.** $h(x) = x^4 + x^3$

   ii. $u(x) = x^5 + x^3$ **iv.** $g(x) = 10.3^x$

d. For each function that you have identified as even or odd, what do you predict about the symmetry of its graph? If possible, check your predictions with a function graphing program.

8. Use the function $f(x)$ to create a new function $g(x)$ where the graph of $g(x)$ is:

a. The graph of $f(x)$ shifted three units to the left, then multiplied by 5, and finally shifted down by four units.

b. The graph of $f(x)$ shifted three units to the right, then shifted up by four units, and finally multiplied by five.

9. For each function, construct a new function whose graph is the graph of the original function shifted left by two units, then multiplied by $\frac{1}{3}$, and then shifted down by five units.

a. $f(x) = 60\left(\frac{1}{2}\right)^x$ **b.** $g(x) = 12x^3$ **c.** $y = \log x$

10. If $f(x) = \dfrac{1}{x}$, evaluate:

a. $f(x - 3) + 5$ **c.** $f(t^2 + 2)$

b. $-3f(x + 2) + \frac{1}{2}f(x - 1)$ **d.** $f(x + a) - f(a)$

11. If $f(x) = \dfrac{1}{x^2}$, determine:

a. $f(2 - x) - 1$ **c.** $f(\sqrt{s} - 3)$

b. $\frac{1}{2}f(-t)$ **d.** $f(x + h) - f(x)$

12. If $f(x) = \frac{1}{x}$,

a. Describe the transformations of $f(x)$ used to create the new functions $g(x)$, $h(x)$, and $k(x)$.

$$g(x) = \frac{1}{3x} \qquad h(x) = -\frac{2}{x - 3} \qquad k(x) = \frac{1}{2 - x} + 4$$

b. Determine the domain of each function in part (a).

c. Determine the equation of the vertical asymptote for each function in part (a).

13. Let $k(s) = \frac{1}{s}$. Construct a new function $j(s)$ that is the end result of the transformations of the graph of $k(s)$ described in the following steps. Show your work for each transformation.

a. First shift $k(s)$ to the right by two units.

b. Then compress your result by a factor of 1/3.

c. Reflect across the s-axis.

d. Finally, shift it up four units.

14. If $p(t) = \dfrac{1}{t^2}$, construct an expression for:

a. $p(t + 1)$ **c.** $-2(p(t - 3) + 5)$

b. $3p(t + 2) - 1$

15. Apply the transformations specified in parts (a)–(e) to $f(x) = \ln x$ and $f(x) = \dfrac{1}{x^3}$.

a. $f(x + 2)$ **c.** $-f(x)$ **e.** $-f(-x)$

b. $\frac{1}{2}f(x)$ **d.** $f(-x)$

16. (Graphing program optional.)

    a. Starting with the function $f(x) = e^x$, create a new function $g(x)$ by performing the following transformations. At each step show the transformation in terms of $f(x)$ and e^x,

        i. First shift the graph of $f(x)$ to the left by three units,

        ii. Then compress your result by a factor of 1/4,

        iii. Next reflect it across the x-axis,

        iv. And finally shift it up by five units to create $g(x)$.

    b. Graph $f(x)$ and $g(x)$ on the same grid.

17. Given the function $g(t)$, identify the simplest function $f(t)$ (linear, power, exponential, or logarithmic) from which $g(t)$ could have been constructed. Describe the transformations that changed $f(t)$ to $g(t)$.

    a. $g(t) = \dfrac{t-1}{2}$ **c.** $g(t) = \dfrac{-7}{t-5} - 2$

    b. $g(t) = 3\left(\dfrac{1}{2}\right)^{t+4}$

18. (Graphing program optional.)

    a. On separate grids <u>sketch</u> the graphs of $f(x) = \sqrt{-x+2}$ and $g(x) = -\sqrt{x+2}$.

    b. Using interval notation, describe the domains of $f(x)$ and $g(x)$.

    c. Using interval notation, describe the ranges of $f(x)$ and $g(x)$.

    d. What is the simplest function $h(x)$ from which both $f(x)$ and $g(x)$ could be created?

    e. Describe the transformations of $h(x)$ to obtain $f(x)$. Of $h(x)$ to obtain $g(x)$.

    f. Does the graph of $f(x)$, $g(x)$, or $h(x)$ have any symmetries (across the x- or y-axis, or about the origin)?

19. **a.** Given $f(x) = \ln x$, describe the transformations that created $g(x) = 3f(x+2) - 4$. Find $g(x)$.

    b. Use your knowledge of properties of logarithms to find any vertical and horizontal intercepts for the function $g(x)$.

20. The following two graphs show the hours of daylight during the year for two different locations. One is for a latitude of 40 degrees above the equator (in the Northern Hemisphere), and the other for a latitude of 40 degrees below the equator (in the Southern Hemisphere).

    a. Which graph is associated with which hemisphere?

    b. Using the language of function transformation, describe Graph B in terms of Graph A.

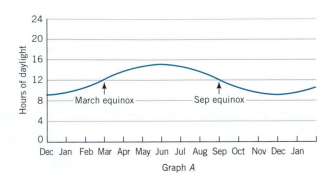

Graph A

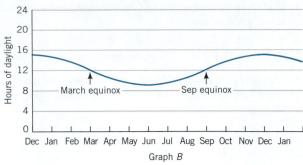

Graph B

Source: www.vcaa.vic.edu.au/prep10/csf/support/sampleunits/ SolarHouseDaylight.xls.

21. (Graphing program required.) If an object is put in an environment with a fixed temperature A (the "ambient temperature"), then the object's temperature, T, at time t is modeled by Newton's Law of Cooling: $T = A + Ce^{-kt}$, where k is a positive constant. (Note that T is a function of t and as $t \to +\infty$, then $e^{-kt} \to 0$, so the temperature T of the object gets closer and closer to the ambient temperature, A.) A corpse is discovered in a motel room at midnight. The corpse's temperature is 80° and the room temperature is 60°. Two hours later the temperature of the corpse had dropped to 75°. (Problem adapted from one in the public domain site S.O.S. Math.)

    a. Using Newton's Law of Cooling, construct an equation to model the temperature T of the corpse over time, t, in hours since the corpse was found.

    b. Then determine the time of death. (Assume the normal body temperature is 98.6°.)

    c. Graph the function from $t = -5$ to $t = 5$, and identify when the person was alive, and the coordinates where the temperature of the corpse was 98.6°, 80°, and 75°.

22. (Graphing program required.) Newton's Law of Cooling (see Exercise 21) also works for objects being heated. Suppose you place a frozen pizza (at 32°) into a preheated oven set at 350°. Thirty minutes later the pizza is at 320°—ready to eat.

    a. Determine the constants A, C, and k in Newton's Law.

    b. Sketch a graph of your function.

    c. From your graph, estimate when the pizza will be at 200° and then calculate the time.

    d. According to your model, if you kept the pizza in the oven indefinitely, would the pizza ever reach 350°? What would be a reasonable domain for the function as a model for cooking pizza?

8.7 Combining Two Functions

We can combine functions the way we combine numbers; that is, we can add, subtract, multiply, or divide two functions.

> **The Algebra of Functions**
>
> If $f(x)$ and $g(x)$ are two functions with the same domain, we can define new functions by:
>
> $$\text{Adding to get } f(x) + g(x) = (f + g)(x)$$
> $$\text{Subtracting to get } f(x) - g(x) = (f - g)(x)$$
> $$\text{Multiplying to get } f(x) \cdot g(x) = (f \cdot g)(x)$$
> $$\text{Dividing to get } \frac{f(x)}{g(x)} = \left(\frac{f}{g}\right)(x) \quad (\text{where } g(x) \neq 0)$$

For example, if $f(x) = x^3$ and $g(x) = 5(2)^x$, then

$$(f + g)(x) = f(x) + g(x) = x^3 + 5(2)^x$$
$$(f - g)(x) = f(x) - g(x) = x^3 - 5(2)^x$$
$$(f \cdot g)(x) = f(x) \cdot g(x) = x^3 \cdot 5(2)^x$$
$$\left(\frac{f}{g}\right)(x) = \frac{f(x)}{g(x)} = \frac{x^3}{5(2)^x}$$

We've already seen several examples of combining functions. We initially described a polynomial function as a sum of power functions. Once a polynomial is put into factored form, we can think of it as a product of linear functions.

EXAMPLE 1

Adding two functions: Skid distances

In Example 7 of Section 7.2, we estimated the skid distance (in feet) of a car after the brakes were applied as $S^2/30$, where S (in mph) is the initial speed of the car. But the total braking distance has an additional component, the distance due to the driver's reaction time before hitting the brakes.

a. It takes the average driver about 0.75 second to react before putting on the brakes. If the car is traveling at S mph, how many feet has the car traveled during the driver's reaction time? Construct the equation for the reaction distance $d_r(S)$ in feet.

b. Construct an equation for the total braking distance as a function of car speed S.

c. If you suddenly see that there is a massive traffic accident 100 feet ahead, can you stop in time if you are traveling at 20 mph? At 40 mph? At 60 mph?

SOLUTION

a. If you are traveling at S mph, the distance traveled during 0.75 second is

$$(0.75 \text{ sec}) \cdot \left(S \frac{\text{miles}}{\text{hour}}\right) = (0.75 \text{ sec}) \cdot \left(S \frac{5280 \text{ feet}}{3600 \text{ sec}}\right) = 1.1S \text{ feet}$$

So $d_r(S) = 1.1S$.

b. The total braking distance, $d_t(S)$, is the sum of two functions: the reaction distance $d_r(S)$ and the skid distance $d_s(S)$.

$$d_t(S) = d_r(S) + d_s(S)$$

substituting, we get $\qquad d_t(S) = 1.1S + \dfrac{S^2}{30}$

where S is in mph and $d_t(S)$ is in feet.

c. If the car is traveling at 20 mph, then the total braking distance is

$$d_t(20) = 1.1(20) + \frac{(20)^2}{30} = 22 + \frac{400}{30} \approx 22 + 13 = 35 \text{ feet}$$

So at 20 mph, you could stop in 35 feet, well before the accident.
At 40 mph, your total braking distance is

$$d_t(40) = 1.1(40) + \frac{(40)^2}{30} = 44 + \frac{1600}{30} \approx 44 + 53 = 97 \text{ feet}$$

So at 40 mph, you would be cutting it very close, stopping only about $100 - 97 = 3$ feet from the accident (too close really, since all the functions are estimates).
At 60 mph, your total braking distance is

$$d_t(60) = 1.1(60) + \frac{(60)^2}{30} = 66 + \frac{3600}{30} = 66 + 120 = 186 \text{ feet}$$

So you will not be able to stop in time.

EXAMPLE 2

Subtracting one function from another: Making a profit
You are the Chair of the Board of the "Friends" of a historical house in Philadelphia. You are trying to increase membership by sending out a mass mailing. The up-front fixed costs are $1000 for designing a logo and printing 5000 brochures (the minimum for a discount rate). The mailing costs are $0.51 per person, which includes stuffing the envelope, adhering address labels, and paying postage. Each person who joins would pay a $35 annual membership fee; however, the usual response rate for such mass mailings is about 3%. How large should your mailing be for you to recover the costs of the mailing through the membership fees?

SOLUTION

If $n =$ the number of brochures mailed, then $C(n)$, the cost of the mailing, is

$$C(n) = \text{fixed costs} + (\text{cost per mailing}) \cdot (\text{number of brochures})$$
$$= \$1000 + \$0.51n$$

The total revenue $R(n)$ is

$$R(n) = (\text{number of responses}) \cdot (\text{membership fee})$$
$$= (3\% \text{ of } n) \cdot (\$35)$$
$$= (0.03n) \cdot (\$35)$$
$$= \$1.05n$$

Your profit $P(n)$ is the difference between the revenue and cost functions.

$$\text{Profit} = \text{Revenue} - \text{Costs}$$
$$P(n) = R(n) - C(n)$$
$$= \$1.05n - (1000 + \$0.51n)$$
$$= \$0.54n - 1000$$

$P(n)$ is a linear function of n, the number of mailings, with a vertical intercept at $-\$1000$ and a positive slope of $\$0.54$ per mailing (see Figure 8.59).

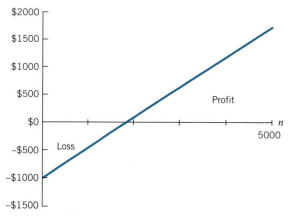

Figure 8.59 Graph of Profit = Revenue − Costs with breakeven point at $n = 1850$.

What is the breakeven point, where revenue equals costs? This happens when the profit (or loss) is 0.

Setting	$P(n) = 0$
substitute for $P(n)$	$\$0.54n - 1000 = 0$
add 1000 to each side	$\$0.54n = 1000$
divide and round to the nearest ten	$n \approx 1850$

The model suggests that you should mail out letters to at least 1850 people to cover your costs. If you mail to more than 1850 people, the model predicts you will have a profit. Mailing to fewer people, you would take a loss, so financially the mass mailing would not be worth doing.

EXAMPLE 3   **The quotient of two functions: The federal debt per person**
Since 1940 both the gross federal debt and the U.S. population have been growing exponentially. Best-fit curves describe the debt (in millions of dollars) as $D(t) = 78,800(1.073)^t$ and the U.S. population size (in millions) as $P(t) = 141(1.012)^t$, where for both functions $t =$ years since 1940.

a. Construct a function for the estimated gross federal debt per person over time.

b. What was the estimated federal debt per person in 1990? In 2007? The projected federal debt per person in 2015?

c. Is the model realistic?

SOLUTION   **a.** $F(t) =$ federal debt per person over time, which equals

$$\frac{\text{federal debt each year}}{\text{U.S. population each year}} = \frac{D(t)}{P(t)} = \frac{78,800(1.073)^t}{141(1.012)^t}$$

b. In 1990 we have $t = 50$. So

$$F(50) = \frac{D(50)}{P(50)} = \frac{78,800(1.073)^{50}}{141(1.012)^{50}} \approx \frac{\$2,670,000 \text{ million}}{256 \text{ million people}} \approx \$10,430 \text{ per person}$$

This means that if we had decided to pay off the gross federal debt in 1990, we would have had to spend the equivalent of over $10,000 per person.

In 2007 we have $t = 67$. So

$$F(67) = \frac{D(67)}{P(67)} = \frac{78,800(1.073)^{67}}{141(1.012)^{67}} \approx \frac{\$88,445,200 \text{ million}}{314 \text{ million people}} \approx \$28,210 \text{ per person}$$

So only 17 years later, the federal debt had almost tripled to over $28,000 per person.

If the same rates of growth continue, then in 2015 (when $t = 75$), we will have

$$F(75) = \frac{D(75)}{P(75)} = \frac{78{,}800(1.073)^{75}}{141(1.012)^{75}} \approx \frac{\$15{,}542{,}000 \text{ million}}{345 \text{ million people}} \approx \$45{,}050 \text{ per person}$$

or over $45,000 of gross federal debt for every man, woman, and child in the United States.

c. There are several U.S. "national debt clocks" (such *www.brilligcom/debt_clock*) that continuously track the gross federal debt and the debt per person. In May 2007 one clock estimated the gross federal debt per person at $29,790, which is slightly over our model's value of $28,000. So the model is reasonably accurate.

　　Check at least one such site and compare its figure for the current debt per person with what our model predicts for the current year.

Rational Functions: The Quotient of Two Polynomials

The quotient of two polynomial functions is called a *rational* function (from the word "ratio"). We've seen simple examples of rational functions before in Chapter 7 in the form of power functions with negative integer powers. For example, $R(x) = 6x^{-1} = \frac{6}{x}$, the ratio of surface area to volume of a cube. A more complex example would be the function

$$S(x) = \frac{x^3 - x + 5}{2x^2 - 7}$$

Rational Functions

A *rational function $R(x)$* is of the form

$$R(x) = \frac{p(x)}{q(x)} \quad (q(x) \neq 0)$$

where $p(x)$ and $q(x)$ are both polynomials.

EXAMPLE 4　**Cost per unit**

The start-up costs for a small pizza company are $100,000, and it costs $3 to produce each additional pizza. (Economists call $3 the *marginal cost*.)

a. Construct a function $C(x)$ for the total cost of producing x pizzas.

b. Then create a function $P(x)$ for the total cost per pizza.

c. As more pizzas are produced, what happens to the cost per pizza?

SOLUTION　a. $C(x) = 100{,}000 + 3x$

b. $P(x) = \dfrac{100{,}000 + 3x}{x}$

c. The cost of producing one pizza is $100,003! But as more pizzas are produced, the total production cost per pizza goes down—allowing the producer to sell pizzas at a cheaper price.

EXAMPLE 5　**A rational function in disguise**

a. Show that $T(x) = \dfrac{1}{x - 2} + 3$ is a rational function.

b. Graph the function. What is the domain of $T(x)$?

SOLUTION   **a.** Given

$$T(x) = \frac{1}{x - 2} + 3$$

multiply 3 by $\dfrac{x - 2}{x - 2}$ $(=1)$

$$= \frac{1}{x - 2} + \frac{3(x - 2)}{(x - 2)}$$

use the distributive law

$$= \frac{1}{(x - 2)} + \frac{3x - 6}{(x - 2)}$$

combine terms with common denominators

$$= \frac{3x - 5}{x - 2}$$

$T(x)$ is now rewritten as the quotient of two polynomials.

b. The graph is shown in Figure 8.60. The domain is all real numbers except 2, since the denominator cannot equal zero.

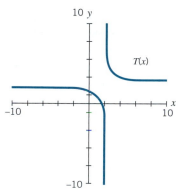

Figure 8.60 The graph of
$$T(x) = \frac{3x - 5}{x - 2}.$$

Visualizing Rational Functions

Horizontal intercepts and vertical asymptotes

The graphs of rational functions can be quite complex. However, if $f(x)$ and $g(x)$ have no common factors, we do know two things about the graph $R(x) = \frac{f(x)}{g(x)}$.

- First, the horizontal intercepts of $f(x)$ are the horizontal intercepts of $R(x)$, since when $f(x) = 0$, $R(x) = 0$.
- Second, when $g(x) = 0$, then $R(x) = \frac{f(x)}{g(x)}$ is not defined since the denominator is 0. The function "blows up" at that value of x, which is called a *singularity*. The vertical line through that value of x is a *vertical asymptote* for the graph; that is, as the values of x approach the vertical asymptote, the values of the function approach $\pm\infty$.

EXAMPLE 6   **Finding horizontal intercepts and vertical asymptotes**

Given the rational function $f(x) = \frac{3x^2 - 13x - 10}{x^2 - 6x - 7}$:

a. Put the numerator and denominator into factored form.

b. What are the horizontal intercepts of $f(x)$?

c. What is the domain of the function? What are its vertical asymptotes?

d. Use technology to graph the function and confirm your answers in parts (b) and (c).

SOLUTION

a. Factoring, we get $f(x) = \dfrac{3x^2 - 13x - 10}{x^2 - 6x - 7} = \dfrac{(3x + 2)(x - 5)}{(x + 1)(x - 7)}$.

b. The horizontal intercepts occur when $f(x) = 0$ or, equivalently, when the numerator $(3x + 2)(x - 5) = 0$. This means either

$$3x + 2 = 0 \qquad \text{or} \qquad x - 5 = 0$$
$$3x = -2 \qquad\qquad\qquad x = 5$$
$$x = -\tfrac{2}{3}$$

So $f(x)$ has two horizontal intercepts, at $(-\tfrac{2}{3}, 0)$ and $(5, 0)$.

c. $f(x)$ is not defined when the denominator $(x + 1)(x - 7)$ is 0. This happens when either

$$x + 1 = 0 \qquad \text{or} \qquad x - 7 = 0$$
$$x = -1 \qquad\qquad\qquad x = 7$$

So the domain is all real numbers except the singularities at $x = -1$ and $x = 7$. The graph of $f(x)$ "blows up" at its singularities, creating two vertical asymptotes at the vertical lines $x = -1$ and $x = 7$.

d. Figure 8.61 shows the graph of this rational function, which confirms our answers in parts (b) and (c).

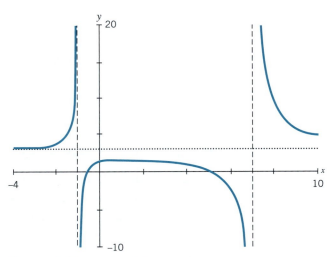

Figure 8.61 Graph of the rational function
$$f(x) = \dfrac{(3x + 2)(x - 5)}{(x + 1)(x - 7)}.$$

The end behavior of rational functions: Additional asymptotes

What happens to the rational function $R(x) = \dfrac{p(x)}{q(x)}$ as $x \to \pm\infty$? This is called its end behavior. As $x \to \pm\infty$, the polynomial expressions $p(x)$ and $q(x)$ become dominated by their leading term—the one with the highest degree. For example, if we look at the rational function in Example 6, where $f(x) = \dfrac{3x^2 - 13x - 10}{x^2 - 6x - 7}$, then as $x \to \pm\infty$, $f(x) \approx \dfrac{3x^2}{x^2} = 3$. So as $x \to \pm\infty$, $f(x) \to 3$ (but never reaches 3). So $f(x)$ is asymptotic to the horizontal line $y = 3$. Examine Figure 8.61 to verify that this conclusion seems reasonable.

EXAMPLE 7 **End behavior**

Describe the behavior of the following rational functions as $x \to \pm\infty$. Then graph each function to confirm your descriptions.

a. $f(x) = \dfrac{1}{2x - 3}$ **b.** $g(x) = \dfrac{6x + 1}{3x + 5}$ **c.** $h(x) = \dfrac{3x^2 + 2x - 4}{x - 1}$

SOLUTION As $x \to \pm\infty$, a rational function's behavior can be approximated by the ratio of the leading terms of the numerator and denominator.

a. As $x \to \pm\infty$, $f(x) = \dfrac{1}{2x - 3} \approx \dfrac{1}{2x}$. So as $x \to \pm\infty$, $f(x) \to 0$ (but never reaches 0). This means that $f(x)$ is asymptotic to the x-axis. (See Figure 8.62(*a*).)

b. As $x \to \pm\infty$, $g(x) = \dfrac{6x + 1}{3x + 5} \approx \dfrac{6x}{3x} = 2$. So as $x \to \pm\infty$, $g(x) \to 2$ (but never reaches 2). This means that $g(x)$ is asymptotic to the horizontal line $y = 2$. (See Figure 8.62(*b*).)

c. As $x \to \pm\infty$, $h(x) = \dfrac{-3x^2 + 2x - 4}{x - 1} \approx \dfrac{-3x^2}{x} = -3x$. So as $x \to \pm\infty$,

$h(x) \to -3x$. This implies that $h(x)$ gets closer to the line $y = -3x$, but never touches it. So $h(x)$ is asymptotic to the line $y = -3x$. (See Figure 8.62(*c*).)

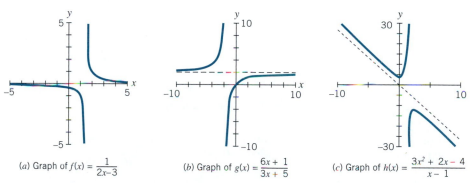

(*a*) Graph of $f(x) = \dfrac{1}{2x-3}$ (*b*) Graph of $g(x) = \dfrac{6x+1}{3x+5}$ (*c*) Graph of $h(x) = \dfrac{3x^2+2x-4}{x-1}$

Figure 8.62 Graphs of three rational functions.

When the end behavior is a horizontal line, we say the rational function has a *horizontal asymptote*. When the end behavior is a polynomial, we say it has an *oblique asymptote*.

Graphs of Rational Functions

For any rational function $R(x) = \dfrac{p(x)}{q(x)}$, where p and q are polynomials with no common factors and $q(x) \neq 0$, then:

The horizontal intercepts of R are the zeros of the numerator p; that is, the values of x where $p(x) = 0$.

The vertical asymptotes of R occur at the zeros of the denominator q; that is, the values of x where $q(x) = 0$.

The end behavior of R can be approximated by the ratio of the leading terms of p and q and is either a *horizontal* or an *oblique asymptote*.

Algebra Aerobics 8.7

Graphing program optional for Problems 5 and 6, and required for Problems 9 and 10.

1. Let $f(x) = x^3$, $g(x) = 2x - 1$, and $h(x) = \frac{1}{x}$. Evaluate each of the following.

   a. $f(2)$ d. $(h \cdot g)(2)$

   b. $g(2)$ e. $(f + g)(2)$

   c. $h(2)$ f. $\left(\frac{h}{g}\right)(2)$

2. Let $Q(t) = 5 + 2t$ and $P(t) = 3t^3$. Evaluate each of the following.

   a. $Q(1) + P(1)$ c. $P(-1) \cdot Q(-1)$

   b. $Q(2) - P(2)$ d. $Q(3)/P(3)$

3. Let $f(t) = 3 - 2t$ and $h(t) = t^2 - 1$. Find general expressions for each of the following.

   a. $f(t) - h(t)$ c. $f(t) \cdot h(t)$

   b. $f(t) + h(t)$ d. $h(t)/f(t)$

4. Given $f(x) = x^2 + 2x - 3$ and $g(x) = \frac{1}{x-1}$, find:

   a. $f(x + 1)$ c. $g(x + 1)$

   b. $f(x) + 1$ d. $g(x) + 1$

5. (Graphing program optional.) Identify any horizontal intercepts and any vertical asymptotes for the following functions. If possible, check your answers by graphing the functions with technology.

   a. $f(x) = \frac{(x - 1)(x + 5)}{(x + 3)}$ b. $g(x) = \frac{(3x + 2)}{(x + 1)(x - 3)}$

6. (Graphing program optional.) Let $h(r) = r^2 - 4r - 12$ and $k(r) = r^2 - 4r + 3$. Construct the functions $f(r)$ and $g(r)$ and identify their horizontal intercepts and vertical asymptotes, if any exist. If possible, check your answers by using a graphing program.

   a. $f(r) = \frac{h(r)}{k(r)}$ b. $g(r) = \frac{k(r)}{h(r)}$

7. Using the function $f(x) = \frac{1}{x^2}$,

   a. Create a new function $g(x) = -f(x + 3) + 1$.

   b. Show that $g(x)$ is a rational function of the form $g(x) = \frac{p(x)}{q(x)}$.

   c. Identify the horizontal intercepts and vertical asymptotes of $g(x)$, if they exist.

8. If $f(x) = \frac{p(x)}{q(x)}$ is a rational function, what would you be looking for if you:

   a. Set $f(x) = 0$?   c. Set $p(x) = 0$?

   b. Evaluated $f(0)$?   d. Set $q(x) = 0$?

(Graphing program required.) For each function in Problems 9 and 10:

   i. Identify any horizontal and vertical intercepts.

   ii. Find any vertical asymptotes.

   iii. Describe the end behavior of the graph (and any other asymptotes).

   iv. Sketch the graph, labeling the intercepts.

9. $f(x) = \dfrac{2x + 6}{x - 3}$

10. $g(x) = \dfrac{x^2 + 2x - 3}{3x - 1}$

Exercises for Section 8.7

Some exercises recommend or require a graphing program.

1. Given $f(t) = 3t^2 + 4t - 5$ and $g(t) = 6t + 1$, find:

   a. $f(t) + g(t)$ b. $g(t) - f(t)$ c. $f(t) \cdot g(t)$ d. $\dfrac{f(t)}{g(t)}$

2. Given $f(m) = \dfrac{3}{m - 4}$ and $g(m) = \dfrac{-3m}{2m - 5}$, find:

   a. $f(m) + g(m)$ c. $(f \cdot g)(m)$ e. $\left(\dfrac{f}{g}\right)(m)$

   b. $(f - g)(m)$ d. $\dfrac{g(m)}{f(m)}$

3. Let $f(x) = 3x^5 + x$ and $g(x) = x^2 - 1$.

   a. Construct the following functions.

   $j(x) = f(x) + g(x)$, $\;k(x) = f(x) - g(x)$, $\;l(x) = f(x) \cdot g(x)$

   b. Evaluate $j(2)$, $k(3)$, and $l(-1)$.

4. If $h(x) = f(x) \cdot g(x) = -x^2 - 3x + 4$, what are possible equations for $f(x)$ and $g(x)$?

5. You own a theater company and you have an upcoming event.

   a. You decide to charge $25 per ticket. Construct a basic ticket revenue function $R(n)$ (in dollars), where n is the number of tickets sold.

   b. You need to pay $500 to keep the box office open for ticket sales. Modify $R(n)$ to reflect this.

   c. You decide to give 30 free tickets to the patrons of your company. Modify your function in part (b) to reflect this.

6. Many colleges around the country are finding they need to buy more computers every year, not only to replace broken or outmoded computers, but also because of the increasing use of computers in classrooms, labs, and studios. A college administrator is preparing a 5-year budget plan. She anticipates that her college, which now has 120 computers, will have to increase that amount by 40 per year for the next 5 years. She currently pays $1000 per computer, but she expects the costs will go up by 3% per year because of inflation.

a. Construct a function $N(t)$ for the number of computers each year as a function of time t (in years since the present).

b. Construct a function $C(t)$ for the individual cost of a computer purchased in year t.

c. Construct a function that will describe the total cost of the computers each year.

d. For year 5, how much money should the budget allow for computers?

7. A worker gets $20/hour for a normal work week of 40 hours and time-and-a-half for overtime. Assuming he works at least 40 hours a week, construct a function describing his weekly paycheck as a function of the number of hours worked.

8. (Graphing program required.) Using the accompanying table, evaluate the following expressions in parts (a)–(d).

x	0	1	2	3	4	5
$f(x)$	-3	-1	5	15	29	47

x	0	1	2	3	4	5
$g(x)$	-3	-5	-11	-21	-35	-53

a. $(f + g)(2)$ **b.** $(g - f)(0)$ **c.** $(f \cdot g)(3)$ **d.** $\left(\dfrac{g}{f}\right)(1)$

9. Use the table in Exercise 8 to create a new table for the functions

a. $h(x) = (f + g)(x)$ **c.** $k(x) = (f \cdot g)(x)$
b. $j(x) = (g - f)(x)$

10. One method of graphing functions is called "addition of ordinates." For example, to graph $y = x + \frac{1}{x}$ using this method, we would first graph $y_1 = x$. On the same coordinate plane, we would then graph $y_2 = \frac{1}{x}$. Then we would estimate the y-coordinates (called ordinates) for several selected x-coordinates by adding geometrically on the graph itself the values of y_1 and y_2 rather than by substituting numerically. This technique is often used in graphing the sum or difference of two different types of functions by hand, without the use of a calculator.

a. Use this technique to sketch the graph of the sum of the two functions graphed in the accompanying figure.

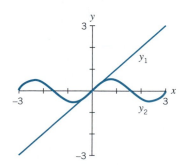

b. Use this technique to sketch the graph of $y = -x^2 + x^3$ for $-2 \le x \le 2$. Then use a graphing tool (if available) and compare.

c. Use this technique to graph $y = 2^x - x^2$ for $-2 \le x \le 5$.

11. Using the accompanying graph of $f(x)$ and $g(x)$, find estimates for the missing values in the following table.

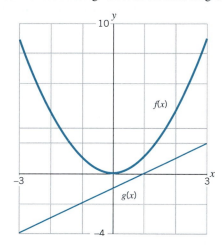

x	-3	-2	-1	0	1	2	3
$f(x)$							
$g(x)$							
$f(x) + g(x)$							
$f(x) - g(x)$							
$f(x) \cdot g(x)$							
$g(x)/f(x)$							

12. From the graph and your results in Exercise 11, find the equations for:

a. $f(x)$ **c.** $(f + g)(x)$ **e.** $(f \cdot g)(x)$

b. $g(x)$ **d.** $(f - g)(x)$ **f.** $\left(\dfrac{g}{f}\right)(x)$

13. The Richland Banquet Hall charges $500 to rent its facility and $40 per person for dinner. The hall holds a minimum of 25 people and a maximum of 100. A sorority decides to hold its formal there, splitting all the costs among the attendees. Let n be the number of people attending the formal.

a. Create a function $C(n)$ for the total cost of renting the hall and serving dinner.

b. Create a function $P(n)$ for the cost per person for the event.

c. What is $P(25)$? $P(100)$? What do these numbers represent?

14. If $f(x) = \frac{1}{x^2}$, find $-f(x) + 2f(x - 3)$; then find a common denominator and combine into one rational expression.

15. For each rational function graphed below, estimate the equation for any vertical or horizontal asymptote(s).

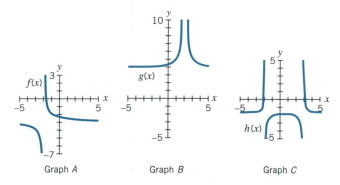

Graph A   Graph B   Graph C

For each of the functions in Exercises 16–18, identify any horizontal intercepts and vertical asymptotes. Then, if possible, use technology to graph each function and verify your results.

16. $f(x) = \dfrac{3x - 13}{x - 4}$

17. $g(x) = -\dfrac{2}{(x + 3)^2}$

18. $h(x) = \dfrac{1}{(x + 3)(x - 1)} - 2$

19. (Graphing program required.)

   a. What is the domain of the rational function $S(x) = \dfrac{x}{x^2 + 1}$?

   b. Does the function have any horizontal intercepts? Any vertical asymptotes?

   c. What is its end behavior?

   d. Graph the function. The graph is one of a set of curves called "serpentine" by Isaac Newton. Why would that name be appropriate?

20. (Graphing program required.) Let $g(x) = \dfrac{18}{(x - 1)^2} - 2$.

   a. We can think of $g(x)$ as being created from transformations of the function $f(x) = \frac{1}{x^2}$. Describe the transformations and then write $g(x)$ as a function of $f(x)$.

   b. Show that $g(x)$ is a rational function of the form $\dfrac{p(x)}{q(x)}$. (*Hint:* Find a common denominator.)

   c. Identify any horizontal intercepts and any vertical asymptotes.

   d. What is its end behavior?

   e. Use graphing technology to confirm your answers and estimate the horizontal asymptote.

21. (Graphing program required.) Construct a rational function $f(x)$ that has horizontal intercepts at $(-3, 0)$ and $(4, 0)$ and vertical asymptotes at the lines $x = 1$ and $x = -5$. Use technology to sketch the graph of $f(x)$.

22. The function $f(x) = 1/x$ was transformed into the function $g(x)$ plotted on the accompanying graph. Construct $g(x)$ in terms of $f(x)$ and then write $g(x)$ in rational function form $g(x) = \dfrac{p(x)}{q(x)}$.

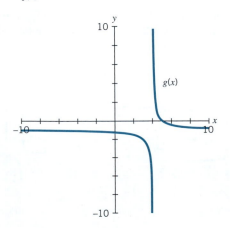

23. Without using technology, match each function with its graph.

   a. $f(x) = -\dfrac{2}{3(x + 2)} + 2$ **c.** $h(x) = \dfrac{x^2 - 9}{5x - 20}$

   b. $g(x) = \dfrac{3x + 5}{x^2 - 4}$

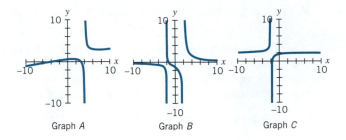

Graph A   Graph B   Graph C

24. (Graphing program required for part (c).) The rational function $g(x) = \frac{4x - 11}{x - 3}$ can be decomposed into a sum by using the following method:

write as sum of a fraction and a constant	$\dfrac{4x - 11}{x - 3} = \dfrac{A}{x - 3} + B$
find the common denominator	$\dfrac{4x - 11}{x - 3} = \dfrac{A}{x - 3} + \dfrac{B(x - 3)}{x - 3}$
multiply and simplify	$\dfrac{4x - 11}{x - 3} = \dfrac{A + Bx - 3B}{x - 3}$
Set numerators equal	$4x - 11 = Bx + A - 3B$
Set x values equal	$4x = Bx$
So	$B = 4$
Set the constants equal substitute 4 for B	$-11 = A - 3B$ $-11 = A - 3(4)$
So	$A = 1$
Therefore	$g(x) = \dfrac{4x - 11}{x - 3} = \dfrac{1}{x - 3} + 4$

which is the graph of $f(x) = \frac{1}{x}$ shifted to the right by three units, then shifted up by four units. See the accompanying graph.

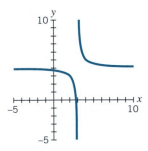

   a. Use the preceding method to decompose $g(x) = \frac{5x + 22}{x - 3}$.

   b. Describe the transformation of the function $f(x) = \frac{1}{x}$ into $g(x) = \frac{5x + 22}{x - 3}$.

   c. Using technology, plot the graphs of $f(x)$ and $g(x)$ to verify that the transformation described in part (b) is correct.

8.8 *Composition and Inverse Functions*

Composing Two Functions

Sometimes it is useful to use the output from one function as the input for another function. As an example, let's look at some simple "parent" functions. Assuming one's mother is a unique person (ignoring complexities such as adoption or cloning), we can define a "mother" function $M(p)$ as the mother of p. Similarly, we can define a "father" function $F(p)$ as the (unique) father of p. Then what would the expression $F(M(p))$ mean? We read the expression from the inside out. Starting with a person p, we apply the mother function to p and then apply the father function to the output. For example, if you are the person p, then

$$F(M(\text{you})) = F(\text{your mother}) = \text{the father of your mother}$$

In other words, this is your grandfather on your mother's side.

When we apply M and then F as above, we call it the *composition* of F and M, a new function denoted by $F \circ M$. We define

$$(F \circ M)(p) = F(M(p))$$

Warning: Be careful not to confuse the product of two functions $f \cdot g$ and the composition $f \circ g$. The product $(f \cdot g)(x) = f(x) \cdot g(x)$ means to evaluate f and g both at x and then multiply the results. The composition $(f \circ g)(x) = f(g(x))$ means to evaluate g at x, and then evaluate f at $g(x)$.

EXAMPLE 1 **Genealogy**
Using the parent functions, who is:

a. $(M \circ F)(\text{you})$?   **b.** $(M \circ M)(\text{you})$?   **c.** Your grandfather on your father's side?

SOLUTION **a.** $(M \circ F)(\text{you}) = M(F(\text{you})) = M(\text{your father}) = $ the mother of your father, or your grandmother on your father's side.

b. $(M \circ M)(\text{you}) = M(M(\text{you})) = M(\text{your mother}) = $ the mother of your mother, or your grandmother on your mother's side.

c. Your grandfather on your father's side is the father of your father, that is, $F(F(\text{you}))$.

The Composition of Two Functions

If $f(x)$ and $g(x)$ are two functions, then the function $f \circ g$, called the *composition of f and g*, is defined by

$$(f \circ g)(x) = f(g(x))$$

EXAMPLE 2 **The order of composition matters**
Let $f(x) = x^2$ and $g(x) = \frac{1}{x+1}$.

a. Evaluate $(f \circ g)(3)$.

b. Determine a general expression for $(f \circ g)(x)$.

c. Does $(f \circ g)(x) = (g \circ f)(x)$? Explain your answer.

SOLUTION **a.** We have:

by definition	$(f \circ g)(3) = f(g(3))$
evaluate $g(x)$ when $x = 3$	$= f\left(\frac{1}{3+1}\right)$
simplify	$= f\left(\frac{1}{4}\right)$
evaluate $f(x)$ when $x = \frac{1}{4}$	$= \left(\frac{1}{4}\right)^2$
or	$= \frac{1}{16}$

b. By definition
$$(f \circ g)(x) = f(g(x))$$

substitute for $g(x)$
$$= f\left(\frac{1}{x+1}\right)$$

evaluate f
$$= \left(\frac{1}{x+1}\right)^2$$

c. By definition
$$(g \circ f)(x) = g(f(x))$$

substitute for $f(x)$
$$= g(x^2)$$

evaluate g
$$= \frac{1}{x^2+1}$$

Since $\left(\frac{1}{x+1}\right)^2 \neq \frac{1}{x^2+1}$ (try evaluating both sides when $x = 1$), then $(f \circ g)(x) \neq (g \circ f)(x)$. So the order of composition does matter.

EXAMPLE 3 **Toxic plumes**

For two weeks during April 2004 the Pentagon, in its efforts to fight terrorism, conducted a study that involved the release of simulated airborne toxins at a fixed starting point. Scientists then measured the dispersion path and spread rate of the pseudo-toxic plume, using sensors placed in concentric circles from the point of release. The circles had radii ranging from a few feet to several thousand feet.

a. Assume that wind speed is not only the major, but the only dispersion factor. If the wind speed is 10 mph, or about 0.17 miles per minute (from any direction), construct an equation that calculates the circular area in square miles (within which the plume would lie) as a function of time in minutes.

b. What circular area would contain the "toxic" plume after 10 minutes? After 30 minutes?

c. If the wind speed doubles, will the area double?

SOLUTION **a.** The area $A(r)$ (in square miles) of the "toxic" circle is a function of its radius r (in miles),

$$A(r) = \pi r^2$$

In this case the radius r is not fixed but is a function $R(t)$ of time t (in minutes). So

$$r = R(t)$$

The wind is traveling at about 0.17 mile per minute. So if we start at the release point (when $t = 0$), then the radius $R(t)$ of the "toxic" circle is

$$R(t) = 0.17t$$

where t is in minutes and $R(t)$ is in miles. Substituting $R(t)$ for r in the equation for $A(r)$ gives us

$$A(R(t)) = A(0.17t) = \pi(0.17t)^2 \approx 0.09t^2$$

which represents the circular area (in square miles) at minute t that encloses the "toxic" plume.

b. Assuming the wind continues to blow at 10 mph (or 0.17 mile per minute), then after 10 minutes the circular area would be $A(R(10)) \approx 0.09(10)^2 = 9$ square miles. After 30 minutes, the circular area would be $A(R(30)) \approx 0.09(30)^2 = 81$ square miles, which is larger than all of Washington, D.C.

c. If the wind doubles from 10 to 20 mph, the radius at any point in time will be twice as large. Our new radius function will now be $R_{new}(t) = 2 \cdot 0.17t = 0.34t$. But the area would be four times as large, since now

$$A(R_{new}(t)) = A(0.34t) = \pi(0.34t)^2 \approx 0.36t^2 = 4 \cdot (0.09t^2) = 4 \cdot A(R(t))$$

or four times the area when the wind was at 10 mph. This makes sense if you think of the composite function $(A \circ R)$ as a direct proportionality with an exponent of 2.

Composing More Than Two Functions

We can think of the transformations of a single function (see Section 8.6) as the result of the composition of two or more functions. For example, the transformation of $f(x) = \frac{1}{x^2}$ to $2f(x - 1) = \frac{2}{(x - 1)^2}$ can be thought of as the composition of three functions:

$$f(x) = \frac{1}{x^2} \quad g(x) = 2x \quad \text{and} \quad h(x) = x - 1$$

Since

$$(g \circ f \circ h)(x) = g(f(h(x))) = g(f(x - 1)) = g\left(\frac{1}{(x - 1)^2}\right) = \frac{2}{(x - 1)^2}$$

It is easy to construct the three functions if you just think of the steps taken to transform $\frac{1}{x^2}$ into $\frac{2}{(x - 1)^2}$. Starting with x, first subtract 1 (applying h), then square and place the result into the denominator (applying f), and finally multiply by 2 (applying g).

EXAMPLE 4 Given $f(x) = \frac{1}{x^3}$, rewrite $\left(\frac{3}{4}\right)f(x + 5) = \frac{3}{4(x + 5)^3}$ as the composition of three functions.

SOLUTION Letting $f(x) = \frac{1}{x^3}$, $g(x) = \left(\frac{3}{4}\right)x$, and $h(x) = x + 5$, the composition

$$(g \circ f \circ h)(x) = g(f(h(x))) = g(f(x + 5)) = g\left(\frac{1}{(x + 5)^3}\right) = \frac{3}{4(x + 5)^3}$$

Algebra Aerobics 8.8a

1. Given $f(x) = 2x + 3$ and $g(x) = x^2 - 4$, find:
   a. $f(g(2))$
   b. $g(f(2))$
   c. $f(g(3))$
   d. $f(f(3))$
   e. $(f \circ g)(x)$
   f. $(g \circ f)(x)$

2. Given $P(t) = \frac{1}{t}$ and $Q(t) = 3t - 5$, find:
   a. $(P \circ Q)(2)$
   b. $(Q \circ P)(2)$
   c. $(Q \circ Q)(3)$
   d. $(P \circ Q)(t)$
   e. $(Q \circ P)(t)$

3. Given $F(x) = \frac{2}{x - 1}$ and $G(x) = 3x - 5$, find:
   a. $(F \circ G)(x)$
   b. $(G \circ F)(x)$
   c. Are the composite functions in parts (a) and (b) equal?

Problems 4, 5, and 6 refer to the accompanying graph of $f(x)$ and $g(x)$.

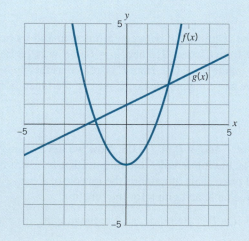

4. Using the graphs of $f(x)$ and $g(x)$, determine the values of:
   a. $f(-2)$
   b. $g(-2)$
   c. $f(0)$
   d. $g(0)$
   e. $(g \circ f)(-2)$
   f. $(f \circ g)(-2)$
   g. $(g \circ f)(0)$
   h. $(f \circ g)(0)$

5. Using your knowledge of quadratic and linear functions, create algebraic expressions for the following:
   a. $f(x)$ b. $g(x)$ c. $(g \circ f)(x)$ d. $(f \circ g)(x)$

6. Using your functions from Problem 5, evaluate $(g \circ f)(-2)$ and $(f \circ g)(-2)$. Do they agree with your answers in Problem 4, parts (e) and (f)?

7. From the table, determine the values of the following compositions in parts (a) and (b).

x	0	1	2	3	4	5
$f(x)$	3	−1	1	3	5	7
$g(x)$	5	3	2	1	0	−1
$h(x)$	2	2	3	4	5	6

   a. $(h \circ f \circ g)(4)$
   b. $(f \circ h \circ g)(1)$

8. Use the given functions f, g, and h to evaluate the compositions.

$$f(x) = \frac{1}{x - 5}, \quad g(x) = \sqrt{x + 1}, \quad h(x) = 3$$

   a. $(h \circ f \circ g)(3)$
   b. $(f \circ g \circ h)(100)$

Inverse Functions: Returning the Original Value

Sometimes when we compose two functions, the input is equal to the output. For example, if $f(x) = \log x$ and $g(x) = 10^x$, then

$$(f \circ g)(x) = f(g(x)) = f(10^x) = \log(10^x) = x$$

and

$$(g \circ f)(x) = g(f(x)) = g(\log x) = 10^{\log x} = x$$

What one function "does," the other "undoes." We say that f and g are *inverse functions* of each other. We denote this as $f^{-1} = g$ and $g^{-1} = f$. The notation f^{-1} is read as "f inverse" or "the inverse of f." A function that has an inverse is called *invertible*.

> **Inverse Functions**
>
> The functions $f(x)$ and $g(x)$ are *inverse functions* of each other if
>
> $$(f \circ g)(x) = x \quad \text{and} \quad (g \circ f)(x) = x$$
>
> We write $f^{-1} = g$ and $g^{-1} = f$.

Warning: The "-1" in the notation f^{-1} can be confusing. It is not an exponent, so f^{-1} does not mean $1/f$. The symbol "-1" simply indicates that f^{-1} is the inverse function of f.

EXAMPLE 5 Show that $f(x) = \ln x$ and $g(x) = e^x$ are inverses of each other.

SOLUTION Since $\quad (f \circ g)(x) = f(g(x)) = f(e^x) = \ln(e^x) = x$

and $\quad (g \circ f)(x) = g(f(x)) = g(\ln x) = e^{\ln x} = x \quad$ (where $x > 0$)

the functions are inverses of each other. So $f^{-1} = g$ (and equivalently $g^{-1} = f$).

EXAMPLE 6 Show that each pair of functions contains a function and its inverse.

a. $f(x) = 2x - 3$ and $g(x) = \frac{x+3}{2}$

b. $h(x) = x^3 + 1$ and $j(x) = (x-1)^{1/3}$

SOLUTION **a.** $(f \circ g)(x) = f(g(x)) = f((x+3)/2) = 2\left(\frac{x+3}{2}\right) - 3 = (x+3) - 3 = x$

$(g \circ f)(x) = g(f(x)) = g(2x-3) = \frac{(2x-3)+3}{2} = x$

b. $(h \circ j)(x) = h(j(x)) = h(x-1)^{1/3} = [(x-1)^{1/3}]^3 + 1 = (x-1) + 1 = x$

$(j \circ h)(x) = j(h(x)) = j(x^3 + 1) = [(x^3 + 1) - 1]^{1/3} = (x^3)^{1/3} = x$

Changing perspectives

To get a better understanding of the relationship between a function and its inverse, let's return to the sales tax example from Chapter 1. There $T = 0.06P$ was the function that calculated the 6% sales tax (T) on the price (P). Here P is treated as the input and T the output; that is, given the price we can compute the tax. If we call the function rule F, we have

$$F(P) = T = 0.06P$$

But we also solved for P in terms of T, to get $P = T/0.06$. Now we have changed our perspective. We are thinking of T as the input and P the output; that is, given the tax paid, we can determine the original price. This function is the inverse of F (i.e., F^{-1}).

Table 8.4 and Figure 8.63 are the table and graph for the function

$$F(P) = T = 0.06P, \text{ where } P \text{ is the input and } T \text{ is the output}$$

Input, P	Output, T
0	0.00
2	0.12
4	0.24
6	0.36
8	0.48
10	0.60

Table 8.4 Table for $T = 0.06P$.

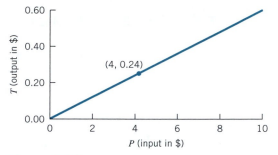

Figure 8.63 Graph of $T = 0.06P$.

Table 8.5 and Figure 8.64 are the table and graph of the inverse function:

$$F^{-1}(T) = P = \frac{T}{0.06}, \qquad \text{where } T \text{ is the input and } P \text{ is the output}$$

Input, T	Output, P
0.00	0
0.12	2
0.24	4
0.36	6
0.48	8
0.60	10

Table 8.5 Table for $P = \frac{T}{0.06}$.

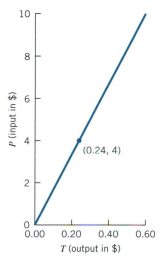

Figure 8.64 Graph of $P = \frac{T}{0.06}$.

The function F and its inverse F^{-1} represent exactly the same information, but from different viewpoints. Comparing Tables 8.4 and 8.5, we see that the columns have been swapped, indicating the reversal of the roles for the input (or independent variable) and the output (the dependent variable). Comparing the graphs (Figures 8.63 and 8.64), we see that the axes have been swapped. So, for example, while the point (4, 0.24) lies on the first graph (Figure 8.63), the point (0.24, 4) lies on the second graph (Figure 8.64). We could write this more formally as

$$F(P) = T \qquad \text{means that} \qquad F^{-1}(T) = P$$

so

$$F(4) = 0.24 \qquad \text{means that} \qquad F^{-1}(0.24) = 4$$

EXAMPLE 7 **Evaluating a function and its inverse**

Given $F(P) = 0.06P$ and $F^{-1}(T) = \frac{T}{0.06}$, evaluate and interpret $F(6)$ and $F^{-1}(6)$.

SOLUTION $F(6) = 0.06 \cdot 6 = \$0.36$, so if the price is \$6, the sales tax is \$0.36.

$F^{-1}(6) = 6/0.06 = \$100$, so if the sales tax is \$6, the price is \$100.

EXAMPLE 8   **Finding the inverse from a table**

If the following table represents a function mapping certain individuals to their cell phone numbers, construct the table of the inverse function.

Individual	Cell Phone Number
Janet Davidson	810 547-1832
John Harbison	919 287-3557
Elaine Woo	202 555-6911
Jon Stewart	212 376-1234

SOLUTION

Cell Phone Number	Individual
810 547-1832	Janet Davidson
919 287-3557	John Harbison
202 555-6911	Elaine Woo
212 376-1234	Jon Stewart

Does every function have an inverse? (Answer: No)

Not every function is invertible.

EXAMPLE 9   Does the function $y = x^2$ have an inverse?

SOLUTION   If we try to solve $y = x^2$ for x in terms of y, we get the equation $x = \pm\sqrt{y}$. But for any positive value of y, there will be two values for x, namely, $+\sqrt{y}$ and $-\sqrt{y}$. For instance, if the input is $y = 100$, then the output could be either $x = -10$ or $x = 10$. So this new equation is not a function, and hence the original function, $y = x^2$, has no inverse.

EXAMPLE 10   Does the mother function, $M(p)$, from the previous section have an inverse?

SOLUTION   Recall that $M(p)$ is the mother of a person p. Each person has only one mother (according to our definition), but each mother could have several children. For example, if both Fred and Jenny have the same mother, Sarah, then $M(\text{Fred}) = \text{Sarah} = M(\text{Jenny})$. (See Figure 8.65.)

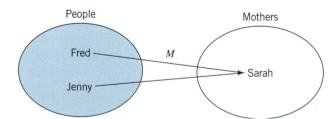

Figure 8.65 The mother function, M, showing that Fred and Jenny have the same mother, Sarah.

There cannot be an inverse function for M, since if the mother, Sarah, were now the input, her output would be both Fred and Jenny.

Functions that have an inverse must be "one-to-one"

In the last two examples (9 and 10), the original function had two different input values associated with the same output value, making it impossible to construct an inverse function. In order to be a function, each input for an inverse function must have a single output. But the inputs for an inverse function are the outputs of the original function. So for a function f to have an inverse function f^{-1}, then f must have only one input for each output value. Technically, that means if $x_1 \neq x_2$, then $f(x_1) \neq f(x_2)$. We call such functions *one-to-one*, which can abbreviated as *1-1*.

A good example of a 1-1 function is Social Security numbers. Each American is assigned a unique Social Security number. (See Figure 8.66.) We can think of this process as a function S that assigns to each person p a unique Social Security number $n = S(p)$. We can define an inverse function $S^{-1}(n) = p$ that maps any existing Social Security number back to its unique owner, p. (See Figure 8.67.)

S: The Social Security function is 1-1

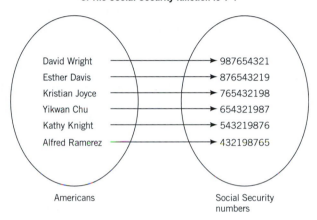

Figure 8.66 Mapping people to their Social Security numbers.

S^{-1}: The inverse of the Social Security function

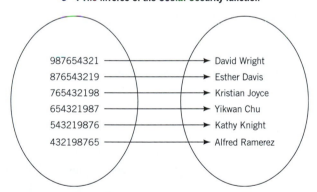

Figure 8.67 Mapping Social Security numbers to people.

The sales tax function is also 1-1 and has an inverse. But we've just seen that the function $y = x^2$ and the mother function, M, are not 1-1 and hence don't have inverses.

A function $f(x)$ is called *one-to-one* (abbreviated as *1-1*) if no two distinct input values are mapped to the same output value; that is, whenever $x_1 \neq x_2$, then $f(x_1) \neq f(x_2)$.

Any one-to-one function f has an inverse function f^{-1} (restricted to an appropriate domain).

Finding the equation for an inverse function

If an invertible function is defined by a simple equation, it can be easy to find its inverse.

EXAMPLE 11

The circumference C of a sphere is a function of the radius r, $C = 2\pi r$. But in real-world situations, it is often easier to measure the circumference than the radius of a sphere.

a. Find the inverse function by solving for r in terms of C.

b. If the circumference of Earth is roughly 25,000 miles, estimate Earth's radius.

SOLUTION

a. C is a 1-1 function of r, so solving for r gives us the inverse function $r = \frac{C}{2\pi}$.

b. Letting $C = 25,000$ miles, we have $r = \frac{25,000}{2\pi} = \frac{12,500}{\pi} \approx 4000$ miles.

It is a little more complicated when we use function notation. To find the inverse of the 1-1 function $f(x) = 1 + \log x$, we need to name the output variable, say y, to get $y = 1 + \log x$. We can then solve this equation for x in terms of y.

Given	$y = 1 + \log x$ (where $x > 0$)
subtract 1 from each side	$y - 1 = \log x$
write as a power of 10	$10^{y-1} = 10^{\log x}$
simplify and switch sides	$x = 10^{y-1}$

So we could write $f^{-1}(y) = 10^{y-1}$ or, more conventionally, since the variables are abstract (and unitless) we could use x as the input to get

$$f^{-1}(x) = 10^{x-1}$$

EXAMPLE 12

a. What is the domain of the function $f(x) = \frac{1}{x+1}$? Is f one-to-one?

b. If so, find its inverse f^{-1} and specify its domain.

SOLUTION

a. The domain of f is all real numbers except -1. The function $f(x)$ is 1-1 since

if $f(x_1) = f(x_2) \Rightarrow \dfrac{1}{x_1 + 1} = \dfrac{1}{x_2 + 1} \Rightarrow x_2 + 1 = x_1 + 1 \Rightarrow x_2 = x_1$. So no two

distinct input values can be mapped to the same output value.

b. Thus f has an inverse function that we can determine by naming the output value $y = f(x)$ and solving the following formula for x in terms of y.

Given	$y = \dfrac{1}{x+1}$ (where $x \neq -1$)
cross-multiply	$x + 1 = \dfrac{1}{y}$
subtract 1 from each side	$x = \dfrac{1}{y} - 1$

So the inverse function is $f^{-1}(y) = \left(\frac{1}{y}\right) - 1$, where the domain is all real numbers except 0. Since this is an abstract function, we can use any variable name for the input. So following the standard convention of using x as the input, we could rewrite f^{-1} as

$$f^{-1}(x) = \frac{1}{x} - 1 \quad \text{(where } x \neq 0)$$

<table>
<tr><td>**?**</td><td>**SOMETHING TO THINK ABOUT**</td></tr>
</table>

Why must the graph of a function pass the horizontal line test in order to have an inverse?

How to tell if graph represents a 1-1 function: The horizontal line test

It is easy to tell from its graph whether a function is 1-1 and hence has an inverse (on an appropriate domain). The graph must pass the *horizontal line test*, which means that no horizontal line can cross the graph twice. (Why?)

Compare Figures 8.68 and 8.69. Figure 8.68 shows the graph (from Section 8.3) of the height $h(d)$ of a projectile fired from a battleship gun as a function of d, its distance from the ship.

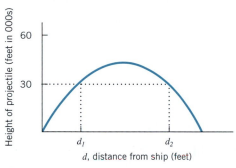

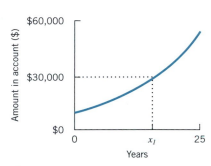

Figure 8.68 Graph of height vs. distance for a projectile fired from a battleship.

Figure 8.69 Amount in an account over time.

Choose any output height value, say 30 (thousand feet), that is below the maximum height. Then there will be two corresponding inputs, that is, two different distances from the battleship, d_1 and d_2, for which the projectile will be at 30 (thousand feet). So $d_1 \neq d_2$, but $h(d_1) = h(d_2) = 30{,}000$. (See Figure 8.68.) So the horizontal line at $h(d) = 30$ (thousand feet) crosses the graph twice. Hence the function h is not 1-1 and cannot have an inverse.

Figure 8.69 shows the amount $A(x)$ at year x, in an account that started with \$10,000 and grew by 7% a year. If we choose any output value (which must be $\geq \$10{,}000$), there will be associated with it a unique input value. For example, if we choose an output of \$30,000, we can see from Figure 8.69 that it will be associated with one and only one input value, labeled x_1. So the function A is 1-1, and there is an inverse function A^{-1} whose domain is restricted to values $\geq \$10{,}000$.

> Recall that a *function* must pass the *vertical line test*; that is, no vertical line crosses its graph more than once.
>
> A *one-to-one function* must pass the *horizontal line test*; that is, no horizontal line crosses its graph more than once.

EXAMPLE 13   Which of the functions in the accompanying graphs have inverses?

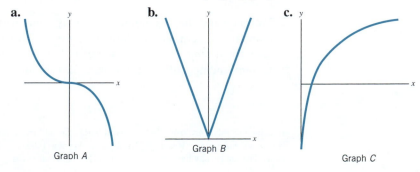

a. Graph A   **b.** Graph B   **c.** Graph C

SOLUTION The functions in Graphs A and C would each have an inverse, since both function graphs pass the horizontal line test. The function in Graph B would not have an inverse, since we could construct a horizontal line (many, in fact) that would cross the graph twice.

EXAMPLE 14 Visualizing their graphs, identify which functions from the following function families have inverses.

    a. Linear functions (of the form $y = b + mx$)

    b. Exponential functions (of the form $y = Ca^x$)

    c. Logarithmic functions (of the form $y = a \log(bx)$ or $y = a \ln(bx)$)

    d. Power functions (of the form $y = kx^a$)

SOLUTION **a.** Every linear function, except those representing horizontal lines (where $m = 0$), has an inverse.

    b, c. Every exponential and logarithmic function has an inverse.

    d. Odd integer power functions have inverses, but even integer power functions (such as $y = ax^2$ or $y = ax^{-2}$) do not.

A Final Example

The following example combines many of the ideas covered in the course—exponential, log, and power functions as well as compositions and inverses. So it's a lengthier example than most, but it involves a reasonable model for an important issue—breast cancer.

EXAMPLE 15 **Modeling breast cancer tumor growth**

    a. Depending on the aggressiveness of a breast tumor, its volume could double in weeks or months. On average, the volume of a breast tumor doubles every 100 days.[5] What is the daily growth factor? The yearly growth factor?

    b. Describe the tumor volume as a function of time. Use an initial volume of 0.06 cubic cm (the minimum tumor volume detectable by a mammogram) and measure time in years after the tumor reached that volume.[6]

    c. Breast tumors are roughly spherical. So their size is usually reported in terms of the diameter, which is easier to measure than the volume. Construct a function that describes the tumor diameter as a function of its volume.

    d. Now compose the functions in parts (b) and (c) to create a new function that gives the tumor diameter as a function of time. What is the minimum tumor diameter a mammogram can detect?

    e. Using technology, graph the function in part (d). What do the points where $t < 0$ represent?

    f. Estimate and then calculate how many years it would take the tumor to reach a diameter of 2 cm, the smallest size detectable by touch. This is also the maximum size for a stage I breast cancer and is often used as the decision point for recommending a lumpectomy vs. a mastectomy.

---

[5]Breast cancer data are from the afterword by Dr. Susan M. Love, MD, of Joyce Wadler's *My Breast: One Woman's Cancer Story* (Reading, MA: Addison-Wesley, 1992).
[6]*Note:* The tumor may have been growing for 10 years before being detectable by a mammogram.

SOLUTION

a. *Growth factors.* Since the tumor volume has a fixed doubling time, its growth is exponential. Since the tumor volume doubles every 100 days, if a is the *daily* growth factor, then

$$2 = a^{100}$$

taking the $\frac{1}{100}$th root $\qquad 2^{1/100} = a$

The *yearly* growth factor is the daily growth factor, $2^{1/100}$, applied 365 times, or

$$(2^{1/100})^{365} = 2^{365/100} = 2^{3.65} \approx 12.6$$

So the yearly growth factor is 12.6. (Recall this means a growth rate of 11.6 (in decimal form) or 1,160% per year!)

b. *Tumor volume as a function of time.* Assuming an initial tumor volume of 0.06 cubic cm and a yearly growth factor of 12.6, the tumor volume $V(t)$ is a function of time t given by

$$V(t) = 0.06(12.6)^t \qquad (1)$$

where $V(t)$ is measured in cubic centimeters and t in number of years since the tumor reached 0.06 cubic cm in volume.

c. *Tumor diameter as a function of volume.* The volume v of a sphere is a function of the radius r, where $v = \left(\frac{4}{3}\right)\pi r^3$. Since we are interested in the diameter d (twice the radius), we can substitute $r = \frac{d}{2}$ into the equation to get

$$v = \frac{4}{3}\pi r^3 = \frac{4}{3}\pi\left(\frac{d}{2}\right)^3 = \frac{4\pi d^3}{3 \cdot 8} = \frac{\pi d^3}{6} \qquad (2)$$

We now have the volume as a function of the diameter.

We need the inverse of that function, one that describes the diameter as a function of the volume. We can solve Equation (2) for d in terms of v:

Given $\qquad\qquad\qquad v = \dfrac{\pi d^3}{6}$

solving for d $\qquad\qquad \dfrac{6v}{\pi} = d^3$

taking the cube root $\qquad d = \sqrt[3]{\dfrac{6v}{\pi}} = \sqrt[3]{\dfrac{6}{\pi}}\sqrt[3]{v} \approx 1.2v^{1/3}$

So using function notation, setting $d = D(v)$, we have a function that describes the tumor diameter $D(v)$ as a function of its volume v:

$$D(v) = 1.2v^{1/3} \qquad (3)$$

where v is in cubic centimeters and $D(v)$ is in centimeters.

d. *Tumor diameter as a function of time.* Since the volume v can be also be written as a function of time as $V(t)$, then the tumor diameter as a function of time is given by the composition of Equations (1) and (3):

$$(D \circ V)(t) = D(V(t)) = D(0.06 \cdot 12.6^t)$$
$$= 1.2(0.06 \cdot 12.6^t)^{1/3}$$
$$= 1.2 \cdot (0.06)^{1/3} \cdot (12.6^{1/3})^t$$
$$\approx 0.5 \cdot 2.3^t$$

where $(D \circ V)(t)$ is the diameter of the tumor at year t. When $t = 0$, then $(D \circ V)(t) \approx 0.5$ cm. So the minimum tumor diameter detectable by mammography is about 0.5 cm, less than a quarter of an inch.

e. *Visualizing tumor diameter over time.* Figure 8.70 shows the growth of a breast tumor over time.

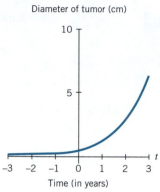

Diameter of tumor (cm)

Time (in years)

Figure 8.70 Graph of breast tumor diameter over time, where $t = 0$ corresponds to the minimum tumor diameter detectable by a mammogram.

When $t < 0$, then $(D \circ V)(t)$ gives the size of the tumor before it was detectable by mammogram. For example, when $t = -5$ (5 years before the tumor had a mammogram-detectable size), the tumor diameter was

$$(D \circ V)(-5) = 0.5 \cdot 2.3^{-5} \approx 0.008 \text{ cm (about three-thousandths of an inch)}$$

f. *Estimating and calculating tumor diameter growth.* Estimating from the graph in Figure 8.70, it would take about 2 years for the diameter to grow from 0.5 cm (when $t = 0$) to 2 cm. To get a more exact number, we can substitute 2 cm for $(D \circ V)(t)$ to get the equation

$$2 = 0.5 \cdot 2.3^t$$

divide by 0.5	$\dfrac{2}{0.5} = 2.3^t$
simplify	$4 = 2.3^t$
take the log of both sides	$\log(4) = \log(2.3^t)$
use rules for logs	$\log(4) = t\log(2.3)$
solve for t	$t = \dfrac{\log(4)}{\log(2.3)}$
use a calculator	$t \approx 1\frac{2}{3}$ years

So it could take almost 2 years for a tumor detectable by a mammogram to grow to a size detectable by touch.

Algebra Aerobics 8.8b

Graphing program required in Problem 10.

1. Let $g(t) = 5 - 2t$ and $h(t) = \frac{5-t}{2}$.

a. Complete the following tables.

t	$g(t)$		t	$h(t)$
0			-1	
1			1	
2			3	
3			5	

b. Find $(g \circ h)(3)$ and $(h \circ g)(3)$.

c. Is $(g \circ h)(t) = (h \circ g)(t)$ for all t?

d. What is the relationship between g and h?

In Problems 2 and 3, verify that f and g are inverse functions by showing that $(f \circ g)(x) = x$ and $(g \circ f)(x) = x$.

2. $f(x) = 2x + 1$ and $g(x) = \frac{x-1}{2}$

3. $f(x) = \sqrt[3]{x} + 1$ and $g(x) = x^3 - 1$

4. Verify that $f(x) = \frac{1}{x-1}$ and $f^{-1}(x) = \frac{1+x}{x}$ are inverse functions of one another.

Find the inverse (if it exists) for each of the functions in Problems 5, 6, and 7.

5. $f(x) = \frac{3}{x} + 5$

6. $g(x) = (x-2)^{2/3}$

7. $h(x) = 5x^3 - 4$

8. State the inverse action for each of the actions described below.

a. Saying "yes"

b. Going to class and then taking the bus home

c. Unlocking the door, opening the door, entering the room, and turning on the light

d. Subtracting 3 from x and multiplying the result by 5

e. Multiplying z by -3 and adding 2

9. Determine from the accompanying function graphs which functions are 1-1.

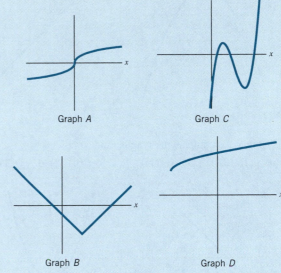

Graph A   Graph C

Graph B   Graph D

10. (Graphing program required.)

a. Sketch a graph of the function $f(x) = (x + 2)^2$. Does f have an inverse?

b. If not, restrict the domain of f so that f^{-1} does exist.

c. Given the restricted domain of f, find the equation for f^{-1} and then graph $f^{-1}(x)$.

11. An oil spill is spreading in a roughly circular shape. The radius, r, is growing by 10 feet per hour. The area $A(r)$ (in square feet) of the spill is a function of the radius r (in feet), given by $A(r) = \pi r^2$.

a. Construct a function $R(t)$ that represents the radius r as a function of time t (in hours since the oil spill).

b. If the oil spill has been spreading for 2 hours, what is the area of the spill?

c. How could you compose the functions A and R to give the area in terms of time t?

Exercises for Section 8.8

1. From the accompanying table, find:

a. $f(g(1))$ **c.** $f(g(0))$ **e.** $f(f(2))$

b. $g(f(1))$ **d.** $g(f(0))$

x	$f(x)$	$g(x)$
0	2	1
1	1	0
2	3	3
3	0	2

2. From the accompanying table, find:

a. $f(g(1))$ **c.** $f(g(0))$ **e.** $f(f(2))$

b. $g(f(1))$ **d.** $g(f(0))$

x	$f(x)$	$g(x)$
-1	-2	1
0	1	2
1	2	-1
2	0	-1

3. Using the accompanying graphs, find:

   a. $g(f(2))$ **b.** $f(g(-1))$ **c.** $g(f(0))$ **d.** $g(f(1))$

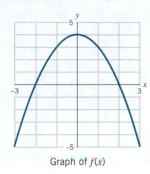 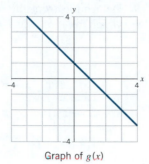

Graph of $f(x)$ Graph of $g(x)$

4. Using the accompanying graphs, find:

   a. $g(f(-2))$ **b.** $f(g(1))$ **c.** $g(f(0))$ **d.** $g(f(1))$

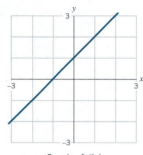 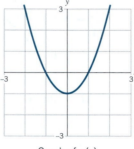

Graph of $f(x)$ Graph of $g(x)$

5. Given $F(x) = 2x + 1$ and $G(x) = \dfrac{x - 1}{x + 2}$, find:

   a. $F(G(1))$ **d.** $F(F(0))$

   b. $G(F(-2))$ **e.** $(F \circ G)(x)$

   c. $F(G(2))$ **f.** $(G \circ F)(x)$

6. Given $f(x) = 3x - 2$ and $g(x) = (x + 1)^2$, find:

   a. $f(g(1))$ **d.** $f(f(2))$

   b. $g(f(1))$ **e.** $(f \circ g)(x)$

   c. $f(g(2))$ **f.** $(g \circ f)(x)$

7. The winds are calm, allowing a forest fire to spread in a circular fashion at 5 feet per minute.

   a. Construct a function $A(r)$ for the circular area burned, where r is the radius. Identify the units for the input and the output of $A(r)$.

   b. Construct a function for the radius $r = R(t)$ for the increase in the fire radius as a function of time t. What are the units now for the input and the output for $R(t)$?

   c. Construct a composite function that gives the burnt area as a function of time. What are the units now for the input and the output?

   d. How much forest area is burned after 10 minutes? One hour?

8. The exchange rate a bank gave for Canadian dollars on March 2, 2007, was 1.18 Canadian dollars for 1 U.S. dollar. The bank also charges a constant fee of 3 U.S. dollars per transaction.

   a. Construct a function F that converts U.S. dollars, d, to Canadian dollars.

   b. Construct a function G that converts Canadian dollars, c, to U.S. dollars.

   c. What would the function $F \circ G$ do? Would its input be U.S. or Canadian dollars (i.e., d or c)? Construct a formula for $F \circ G$.

   d. What would the function $G \circ F$ do? Would its input be U.S. or Canadian dollars (i.e., d or c)? Construct a formula for $G \circ F$.

9. A stone is dropped into a pond, causing a circular ripple that is expanding at a rate of 13 ft/sec. Describe the area of the circle as a function of time.

10. The wind chill temperature is the apparent temperature caused by the extra cooling from the wind. A rule of thumb for estimating the wind chill temperature for an actual temperature t that is above 0°Fahrenheit is $W(t) = t - 1.5S_0$, where S_0 is any given wind speed in miles per hour.

   a. If the wind speed is 25 mph and the actual temperature is 10°F, what is the wind chill temperature?

      We know how to convert Celsius to Fahrenheit; that is, we can write $t = F(x)$, where $F(x) = 32 + \frac{9}{5}x$, with x the number of degrees Celsius and $F(x)$ the equivalent in degrees Fahrenheit.

   b. Construct a function that will give the wind chill temperature as a function of degrees Celsius.

   c. If the wind speed is 40 mph and the actual temperature is $-10°C$, what is the wind chill temperature?

11. Salt is applied to roads to decrease the temperature at which icing occurs. Assume that with no salt, icing occurs at 32°F, and that each unit increase in the density of salt applied decreases the icing temperature by 5°F.

   a. Construct a formula for icing temperature, T, as a function of salt density, s.

   Trucks spread salt on the road, but they do not necessarily spread it uniformly across the road surface. If the edges of the road get half as much salt as the middle, we can describe salt density $S(x)$ as a function of the distance, x, from the center of the road by $S(x) = [1 - \frac{1}{2}(\frac{x}{k})^2]S_d$, where k is the distance from the centerline to the road edges and S_d is the salt density applied in the middle of the road.

   b. What will the expression for $S(x)$ be if the road is 40 feet wide?

   c. What will the value for x be at the middle of the 40-foot-wide road? At the edge of the road? Verify that at the middle of the road the value of the salt density $S(x)$ is S_d and that at the edge the value of $S(x)$ is $\frac{1}{2}S_d$.

   d. Construct a function that describes the icing temperature, T, as a function of x, the distance from the center of the 40-foot-wide road.

   e. What is the icing temperature at the middle of the 40-foot-wide road? At the edge?

12. Using the given functions f, g, and h where

$$f(x) = x + 1 \qquad g(x) = e^x \qquad h(x) = x - 2$$

   a. Create the function $k(x) = (f \circ g \circ h)(x)$.

   b. Describe the transformation from x to $k(x)$.

13. Using the given functions J, K, and L, where

$$J(x) = x^3 \qquad K(x) = \log(x) \qquad L(x) = \tfrac{1}{x}$$

   a. Create the function $M(x) = (L \circ J \circ K)(x)$.

   b. Describe the transformation from x to $M(x)$.

In Exercises 14 and 15, rewrite $j(x)$ as the composition of three functions, f, g, and h.

14. $j(x) = \dfrac{2}{(x-1)^3}$

15. $j(x) = 4e^{x-1}$

In Exercises 16–22, show that the two functions are inverses of each other.

16. $f(x) = 3x + 2$ and $g(x) = \dfrac{x-2}{3}$

17. $f(x) = \sqrt{x-1}$ (where $x > 1$) and $g(x) = x^2 + 1$ (where $x > 0$)

18. $f(x) = 2x - 1$ and $g(x) = \dfrac{x+1}{2}$

19. $f(x) = \sqrt[3]{4x+5}$ and $g(x) = \dfrac{x^3 - 5}{4}$

20. $f(x) = 10^{x/2}$ and $g(x) = \log(x^2)$

21. $F(t) = e^{3t}$ and $G(t) = \ln(t^{1/3})$

22. $H(r) = \tfrac{1}{2}\ln r$ and $J(r) = e^{2r}$

In Exercises 23 and 24, create a table of values for the inverse of the function $f(x)$.

23.

x	$f(x)$
-2	5
-1	1
0	2
1	4

24.

x	$f(x)$
0	5
1	3
-2	2
4	-7

25. Cryptology (the creation and deciphering of codes) is based on 1-1 functions. After you code a message using a 1-1 function, the decoder needs the inverse function in order to retrieve the original message. The following table matches each letter of the alphabet with its coded numerical form.

A	B	C	D	E	F	G	H	I	J	K	L	M
26	25	24	23	22	21	20	19	18	17	16	15	14

N	O	P	Q	R	S	T	U	V	W	X	Y	Z
13	12	11	10	9	8	7	6	5	4	3	2	1

   a. Does this code represent a 1-1 function? Is there an inverse function? If so, what is its domain?

   b. Decode the message "14 26 7 19 9 6 15 22 8."

26. On March 2, 2007, the conversion rate from U.S. dollars to euros was 0.749; that is, on that day you could change \$1 for 0.749 euros, the currency of the European Union.

   a. Was a U.S. dollar worth more or less than 1 euro?

   b. Using the March 2 exchange rate, construct a function $C_1(d)$ that converts d dollars to euros. What is $C_1(1)$? $C_1(25)$?

   c. Now construct a second function $C_2(r)$ that converts r euros back to dollars. What is $C_2(1)$? $C_2(100)$?

   d. Show that C_1 and C_2 are inverses of each other.

   e. Reread the beginning of Exercise 8, which describes a conversion process between Canadian and U.S. dollars. In that process the two formulas are *not* inverses of each other. Why not?

27. Given the accompanying graph of $f(x)$, answer the following.

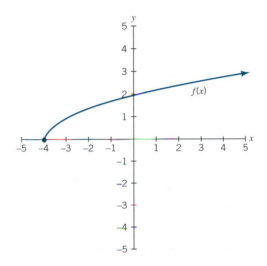

   a. Does $f(x)$ have an inverse? Please explain.

   b. What is the domain of $f(x)$? Estimate the range of $f(x)$?

   c. From the graph, determine $f(-4)$, $f(0)$, and $f(5)$.

   d. Determine $f^{-1}(0)$, $f^{-1}(2)$, and $f^{-1}(3)$.

28. Determine which of the accompanying graphs show functions that are one-to-one.

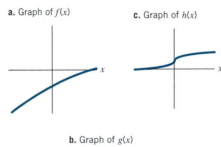

a. Graph of $f(x)$ **c.** Graph of $h(x)$

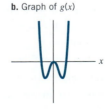

b. Graph of $g(x)$

In Exercises 29–32, for each function Q find Q^{-1}, if it exists. For those functions with inverses, find $Q(3)$ and $Q^{-1}(3)$.

29. $Q(x) = \dfrac{2}{3}x - 5$

30. $Q(x) = 5e^{0.03x}$

31. $Q(x) = \dfrac{x + 3}{x}$

32. Use the graph of $f(x)$ to evaluate each expression.

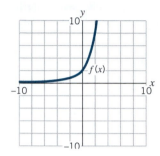

a. $f(2)$ **b.** $f^{-1}(2)$ **c.** $f^{-1}(4)$ **d.** $(f \circ f^{-1})(8)$

33. The following tables represent a function f that converts cups to quarts and a function g that converts quarts to gallons (all measurements are for fluids).

a. Fill in the missing values in the chart. (*Hint:* One quart contains 4 cups, and one gallon contains 4 quarts.)

x (cups)	4	8	16	32
$f(x)$ (quarts)				

x (quarts)	2	4	8	16
$g(x)$ (gallons)				

b. Now evaluate each of the following and identify the units of the results.

   i. $(g \circ f)(8)$ **iii.** $(f^{-1} \circ g^{-1})(1)$

   ii. $g^{-1}(2)$ **iv.** $(f^{-1} \circ g^{-1})(2)$

c. Explain the significance of $(f^{-1} \circ g^{-1})(x)$ in terms of cups, quarts, and gallons.

34. Let $f(x) = mx + b$.

a. Does $f(x)$ always have an inverse? Explain.

b. If $f(x)$ has an inverse, find $f^{-1}(x)$.

c. Using the formula for $f^{-1}(x)$, explain in words how, given any linear equation (under certain constraints), you can find the inverse function knowing the slope m and y-intercept b.

35. If you do an Internet search on formulas for "ideal body weight" (IBW), one that comes up frequently was created by Dr. B. J. Devine. His formula states

IBW for men (in kilograms) =

   50 + (2.3 kg per inch over 5 feet)

IBW for women (in kilograms) =

   45.5 + (2.3 kg per inch over 5 feet)

a. Write the functions for IBW (in kg) for men and women, $W_{\text{men}}(h)$ and $W_{\text{women}}(h)$, where h is a person's height in inches. Give a reasonable domain for each.

b. Evaluate $W_{\text{men}}(70)$ and $W_{\text{women}}(66)$. Describe your results in terms of height and weight.

c. Evaluate $W^{-1}{}_{\text{men}}(77.6)$. What does this tell you?

d. Given that 1 lb = 0.4356 kg, alter the functions to create $W_{\text{newmen}}(h)$ and $W_{\text{newwomen}}(h)$ so that the weight is given in pounds rather than kilograms.

e. Use your functions in part (d) to find $W^{-1}{}_{\text{newwomen}}(125)$. What does this tell you?

[*Note:* More information can be found in the article by M. P. Pari and F. P. Paloucek, "The origin of the 'ideal' body weight equations," Annals of Pharmacology 34 (9), 2000: 10 :1066–69.]

36. The formula for the volume of a cone is $V = \left(\frac{1}{3}\right)\pi r^2 h$. Assume you are holding a 6-inch-high sugar cone for ice cream.

a. Construct a function $V(r)$ for the volume as a function of r. Why don't you need the variable h in this case? Find $V(1.5)$ and explain what have you found (using appropriate units).

b. Evaluate $V^{-1}(25)$. Describe your results. What are the units attached to the number 25?

c. When dealing with abstract functions where $f(x) = y$, we have sometimes used the convention of using x (rather than y) as the input to the inverse function $f^{-1}(x)$. Explain why it does not make sense to interchange V and r here to find the inverse function.

37. In Chapter 6 we learned that a logarithm can be constructed using any positive number (except 1) as a base: $\log_a x = y$ means that $a^y = x$. Show that $F(x) = a^x$ and $G(x) = \log_a x$ are inverse functions. The software "E10: Inverse Functions $y = a^x$ and $y = \log_a x$" in *Exponential and Log Functions* can help you visualize the relationship between the two functions.

38. The Texas Cancer Center website, *www.texascancercenter.com*, notes that the 5-year survival rate for stage I breast cancer (when the tumor diameter is ≤ 2 cm) is about 85%. The 5-year survival rate for stage II breast cancer (when the tumor diameter is ≤ 5 cm and the cancer has not spread to the lymph nodes) is about 65%.

a. Using the equations in Example 15 (the final one in the text) referring to a tumor that doubles in volume every 100 days, how long would it take such a tumor to grow from 0.5 cm in diameter to 5 cm? (Recall that 0.5 cm in diameter corresponds to an initial volume of 0.06 cubic cm, the minimum tumor size detectable by a mammogram.)

b. If the tumor were more aggressive, doubling in volume every 50 days, what would the yearly growth factor be? Use this to construct a new function to reflect the *volume* growth of this more aggressive tumor over time (again using 0.06 cubic cm as the initial volume).

c. Using your function from part (b), construct another function to represent the *diameter* growth over time.

d. How many years would it take the aggressive tumor to grow from 0.5 cm to 5 cm in diameter?

CHAPTER SUMMARY

Quadratic Functions and Their Graphs

A *quadratic function* can be written in *standard form* as

$$f(x) = ax^2 + bx + c \quad \text{(where } a \neq 0\text{)}$$

Its graph

- has a distinctive $\cup$-shape called a *parabola*
- is symmetric across its *axis of symmetry*
- has a minimum or a maximum point called its *vertex*
- is concave up if $a > 0$ and concave down if $a < 0$
- becomes narrower as $|a|$ increases
- has a *focal point* $\left|\frac{1}{4a}\right|$ units above (or below) the vertex on the axis of symmetry

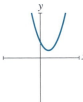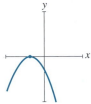

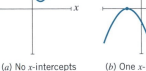

(a) No x-intercepts   (b) One x-intercept   (c) Two x-intercepts
No real zeros   One real zero   Two real zeros

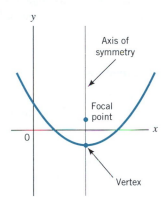

Axis of symmetry

Focal point

Vertex

Finding the Vertex

Any quadratic function $f(x) = ax^2 + bx + c$ can also be written in *vertex* or *a-h-k* form as

$$f(x) = a(x - h)^2 + k$$

where the vertex is at $(h, k) = \left(-\frac{b}{2a}, f\left(-\frac{b}{2a}\right)\right)$. The vertex "anchors" the graph, and the coefficient a determines the shape of the parabola.

Finding the Horizontal Intercepts

Every quadratic function $f(x)$ has two, one, or no horizontal intercepts x. To find the x-intercepts, we set $f(x) = 0$ and solve for x. The solutions are called the *zeros* of the function.

The Factor Theorem says that any quadratic function $f(x) = ax^2 + bx + c$ can be written in *factored form* as

$$f(x) = a(x - r_1)(x - r_2)$$

where r_1 and r_2 are the zeros of $f(x)$.

If the zeros, r_1 and r_2, are real numbers, they are the horizontal intercept(s) of the quadratic function $f(x)$.

The Quadratic Formula

Setting the quadratic function $f(x) = ax^2 + bx + c$ equal to 0 and solving for x using the quadratic formula, gives

$$x = \frac{-b \pm \sqrt{b^2 - 4ac}}{2a}$$

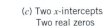

The term $b^2 - 4ac$ is called the *discriminant* and can be used to predict the number of horizontal intercepts (or real zeros) of $f(x)$.

If the discriminant > 0, there are two distinct real roots and hence two x-intercepts.

If the discriminant $= 0$, there is only one distinct real root and hence only one x-intercept.

If the discriminant < 0, then the $\sqrt{b^2 - 4ac}$ is not a real number and hence there are no x-intercepts. These zeros are *complex numbers* of the form $a + bi$, where a and b are real numbers ($b \neq 0$) and $i = \sqrt{-1}$.

The Average Rate of Change of a Quadratic Function

Given a quadratic function $f(x) = ax^2 + bx + c$, the average rate of change between two points on the parabola approaches $2ax + b$ over very small intervals. We can think of the linear function $g(x) = 2ax + b$ as representing the average rate of change of $f(x)$ with respect to x.

Polynomial Functions and Their Graphs

A *polynomial function of degree n* is of the form

$$f(x) = a_n x^n + a_{n-1} x^{n-1} + \cdots + a_1 x + a_0$$

where n is a nonnegative integer and $a_n \neq 0$.

The graph of a polynomial function of degree n will

- Have at most $n - 1$ turning points
- Cross the horizontal axis at most n times

For large values for x, the graph of $f(x) = a_n x^n + a_{n-1} x^{n-1} + \cdots + a_1 x + a_0$ will resemble the graph of the power function $g(x) = a_n x^n$.

If $r_1, r_2, \ldots, r_n$ are zeros of a polynomial $f(x) = a_n x^n + a_{n-1} x^{n-1} + \cdots + a_1 x + a_0$, then $f(x)$ can be written in *factored form* as

$$f(x) = a_n(x - r_1)(x - r_2) \cdot \cdots \cdot (x - r_n)$$

Any real zero is an x-intercept for $f(x)$.

Creating New Functions from Old

Ways to Transform the Graph of f(x)

To stretch or compress:	Multiply the output of $f(x)$ by a constant to get $af(x)$.
To shift vertically:	Add a constant to the output of $f(x)$ to get $f(x) + k$.
To shift horizontally:	Subtract a constant from the input of $f(x)$ to get $f(x - h)$.
To reflect across the x-axis:	Multiply the output of $f(x)$ by -1 to get $-f(x)$.
To reflect across the y-axis:	Multiply the input of $f(x)$ by -1 to get $f(-x)$.

Symmetry

If $f(x) = f(-x)$, then f is symmetric across the y-axis.

If $f(x) = -f(-x)$, then f is symmetric about the origin.

The Algebra of Functions

If $f(x)$ and $g(x)$ are two functions with the same domain, we can define new functions by

Adding to get $f(x) + g(x) = (f + g)(x)$

Subtracting to get $f(x) - g(x) = (f - g)(x)$

Multiplying to get $f(x) \cdot g(x) = (f \cdot g)(x)$

Dividing to get $\dfrac{f(x)}{g(x)} = \left(\dfrac{f}{g}\right)(x)$ (where $g(x) \neq 0$)

A *rational function* $R(x)$ is the quotient of two polynomial functions. If $R(x) = \frac{p(x)}{q(x)}$ (where $p(x)$ and $q(x)$ have no common terms and $q(x) \neq 0$) then

Set $p(x) = 0$ to find horizontal intercepts.

Set $q(x) = 0$ to find vertical asymptotes.

The *end behavior* can be approximated by the ratio of the leading terms of p and q, and is either a *horizontal* or an *oblique* asymptote.

Composition and Inverses

If $f(x)$ and $g(x)$ are two functions, then the function $f \circ g$, called the *composition of f and g*, is defined by

$$(f \circ g)(x) = f(g(x))$$

If both $(f \circ g)(x) = x$ and $(g \circ f)(x) = x$, then the functions f and g are *inverses* of each other, written as $f^{-1} = g$ and $g^{-1} = f$.

A function $f(x)$ is called *one-to-one* (or 1-1) if no two distinct input values are mapped to the same output value. A 1-1 function must pass the horizontal line test.

A function that has an inverse must be one-to-one.

CHECK YOUR UNDERSTANDING

I. Is each of the statements in Problems $1-21$ true or false? Give an explanation for your answer.

1. If $f(t) = 2(t - 1)^2$, then $f(0) = -2$.

2. If the vertical axis is the axis of symmetry for a quadratic function $g(x)$, then $g(-2) = g(2)$.

3. The polynomial function in the accompanying figure has a minimum degree of 5.

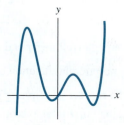

4. The function $y = 3(x - 2)^2 + 5$ has a focal point at $\left(2\frac{1}{12}, 5\right)$.

5. The graph of the quadratic function $y = 2x^2 - 3x + 1$ is steeper than the graph of $y = 3x^2 - 3x + 1$.

6. The graph of $y = 5 - x^2$ is concave down.

7. The graph of $y = x^2 + 2x + 3$ is three units higher than the graph of $y = x^2 + 2x$.

8. The graph of $y = (x + 4)^2$ lies four units to the right of the graph of $y = x^2$.

9. A quadratic function that passes through the points $(1, 5)$ and $(7, 5)$ will have an axis of symmetry at the vertical line $x = 4$.

10. If $f(x) = x^2 - 3x - 4$, then $f(4) = 0$.

11. The function $f(x) = (x + 2)(x + 5)$ has zeros at 2 and 5.

12. In the accompanying figure it appears that $g(x) = -2f(x)$.

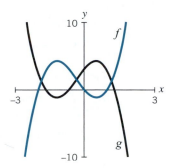

13. The quadratic function $f(x) = 2x^2 - 3x - 1$ has a discriminant with a value of 1.

14. The function $f(x) = 3(x - 1)^2 + 2$ has an axis of symmetry at $x = 1$.

15. The function $f(x) = -2(x + 3)^2 - 1$ has a vertex at $(3, -1)$.

16. There is only one quadratic function $f(x)$ with x-intercepts at 3 and 0.

17. The function $h(t) = t^2 - 3t + 2$ has a zero at $t = 2$ because $h(0) = 2$.

18. In the accompanying figure it appears that $g(x) = -f(x - 1) + 3$.

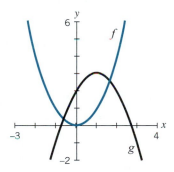

19. Assume the height of a ball thrown vertically upward is modeled by the function $h(t) = -4.9t^2 + 38t + 55$, (where t is time in seconds, and $h(t)$ is the height in meters). Then the ball will hit the ground after approximately 9 seconds.

20. If revenue R (in dollars) from an item sold at price p (in dollars) is modeled by the function $R = p(100 - 5p)$, the revenue will be at a maximum when the price is $10.

21. The functions $m(s) = \frac{1}{2s + 3}$ and $n(s) = \frac{1 - 3s}{2s}$ are inverses of each other.

II. For Problems 22–37 give an example of a function or functions with the specified properties. Express your answer using equations.

22. A polynomial function that does not intersect the horizontal axis.

23. A quadratic function with vertex at the point $(0, 0)$ and with focal point $(0, -1)$.

24. A polynomial function with horizontal intercepts at -1, 3, and 4.

25. Two more polynomial functions with horizontal intercepts at -1, 3, and 4.

26. A quadratic function concave down with a vertex at $(1, 3)$.

27. A quadratic function concave up with its axis of symmetry at the line $x = 3$.

28. A quadratic function concave down with vertical intercept at 2 and zeros at -2 and 2.

29. A quadratic function whose graph will be exactly the same shape as the graph of the function $r = s^2 - s$ but five units higher.

30. A quadratic function $G(x)$ whose graph will be exactly the same shape as the function $F(x) = x^2 + 2x$ but two units to the left.

31. A quadratic function whose graph will be the reflection across the t-axis of the graph of $h(t) = (t - 2)^2$.

32. Two distinct quadratic functions that intersect at the point $(1, 1)$.

33. A quadratic function with one zero at $x = -4$.

34. A polynomial function $h(t)$ that could describe the function in the accompanying figure.

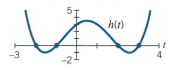

35. Two functions f and g such that $f(x) - g(x) = x^3 + 2x^2 - 5x + 2$.

36. Two functions Q and H such that $(Q \circ H)(t) = \sqrt{3t + 1}$.

37. A rational function with a horizontal intercept at $(2, 0)$ and two vertical asymptotes at $x = 0$ and $x = -3$.

III. Is each of the statements in Problems 38–58 true or false? If a statement is true, explain how you know. If a statement is false, give a counterexample.

38. All polynomial functions are power functions.

39. All linear functions are polynomial functions.

40. All power functions are polynomial functions.

41. All quadratic functions are polynomial functions.

42. The quadratic function with vertex at the origin, $(0, 0)$ and focal point at $(0, 1)$ will be narrower than the quadratic function with the same vertex but with focal point at $(0, 4)$.

43. A polynomial function of degree 4 will always have three turning points.

44. A polynomial function of degree 4 will cross the horizontal axis exactly four times.

45. A polynomial function of odd degree must cross the horizontal axis at least one time.

46. Quadratic functions $f(x) = ax^2 + bx + c$ always have two distinct zeros because the equation $ax^2 + bx + c = 0$ always has two roots, $x = \frac{-b \pm \sqrt{b^2 - 4ac}}{2a}$.

47. If the discriminant is 0, then the associated quadratic function has no horizontal intercept.

48. The leading term determines the global shape of the graph of a polynomial function.

49. The three polynomial functions in the accompanying figure are all of even degree.

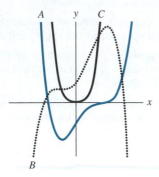

50. Quadratic functions that open upward have a minimum value at the vertex.

51. $(f \circ g)(x) = (g \circ f)(x)$ for any functions f and g.

52. The functions $f(x) = \ln x$ and $g(x) = e^x$ are inverses of each other.

53. $f^{-1}(x) = \frac{1}{f(x)}$ for any function f.

54. If $f(x) = \frac{1}{x}$ and $g(x) = 2x^2 + 1$, then $(f \circ g)(x) = \frac{1}{(2x^2 + 1)}$ and $f(x) \cdot g(x) = \frac{(2x^2 + 1)}{x}$.

55. If $f(x) = f(-x)$, the graph of f is symmetric across the x-axis.

56. If $f(x) = -f(-x)$, the graph of f is symmetric about the origin.

57. A function that passes the vertical line test has an inverse.

58. Every function is 1-1.

CHAPTER 8 REVIEW: PUTTING IT ALL TOGETHER

Some problems are identified as needing a calculator that calculates roots, a graphing program, or technology to generate a best-fit polynomial.

1. For each of the accompanying parabolas, identify the graph as concave up or down, and then estimate the minimum (or maximum) point, the axis of symmetry, and any horizontal intercepts.

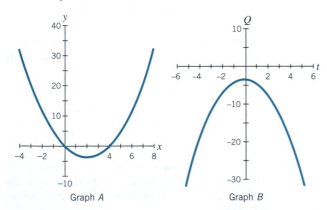

Graph A   Graph B

2. An electric heater is being designed as a parabolic reflector 6″ deep. To prevent accidental burns, the center of the heating element is placed at the focus, which is set 1.5″ from the vertex of the reflector.

a. What equation describes the shape of the reflector?

b. How wide will the reflector be?

c. Sketch an image of the parabolic reflector with the vertex at its origin. Put a circle at the focal point, and label the depth and width of the reflector.

3. (Requires graphing program for parts (e) and (f).) A wood craftsman has created a design for a parquet floor. The pattern for an individual tile is shown in the accompanying image. The square center (x inches wide) of the tile is made from white oak hardwood and is surrounded by 1-inch strips of maple hardwood.

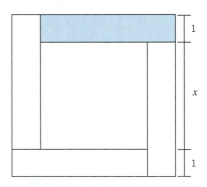

a. What is the area of the interior white oak square (in terms of x)? The area of each of the maple 1-inch strips?

b. White oak costs $2.39 per square foot; maple costs $4.49 per square foot. Calculate the cost per square inch for white oak and for maple.

c. What is the cost for the white oak in one tile of the parquet? What is the cost of the four maple strips in one tile?

d. The approximate labor cost to make each tile is $5.00. Create a cost function $C(x)$ (in dollars) for making one parquet tile. What type of function is this?

e. Use technology to graph the function $C(x)$, where x is the width of the inner white oak square. Use a domain of $0 \le x \le 15$ inches.

f. From your graph, estimate the size of a parquet tile if the total cost (including labor) is to be $7.00 or less per tile.

4. California produces nearly 95% of the processing tomatoes grown in the United States. Therefore, managing irrigation water for tomatoes is a major issue. Agricultural researchers have been able to quantify the relationship between C, the canopy coverage, and K_C, the crop coefficient.[7] Canopy cover is the percentage of the total plot covered by shade produced by the leaves of the plants. The crop coefficient is a measure of the water needed by the plants. (Technically, it is the ratio of the amount of water plants need divided by the amount used for an equivalent area of well-watered grass.) The following graph shows the relationship between the crop canopy and the crop coefficient.

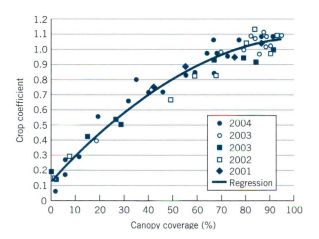

a. As the plants grow, will the canopy cover increase or decrease? Why?

b. The graph suggests that as the plant grows, the crop coefficient increases—but at a decreasing rate. Why would that be true?

c. The best-fit function to the data is a quadratic:

$$K_C = 0.126 + (0.0172)C - (0.0000776)C^2$$
$$\text{where } 0 \le C \le 100\%$$

i. The initial growth stage has 10% canopy coverage. Estimate the corresponding crop coefficient from the graph, then calculate it using the quadratic model. What does this number mean in terms of well-watered lawn grass?

[7]B. R. Hanson and D. M. May, "New crop coefficients developed for high-yield processing tomatoes," *California Agriculture* 60(2), April–June 2006.

ii. In the crop development growth stage, the canopy is between 10% and 75%. Estimate the crop coefficient when the canopy is at 75% and then calculate it. What does this number mean?

5. Construct a function for each parabola $g(x)$ and $h(x)$ in the accompanying graph.

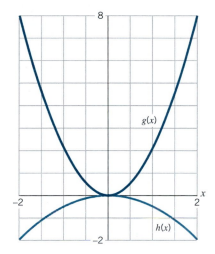

6. (Graphing program optional.) On Earth, the distance d_{Earth} a freely falling object has traveled is a function of time t. It can be modeled by the equation $d_{\text{Earth}} = 16t^2$, where t is in seconds and d_{Earth} is in feet. On Mars, the comparable equation is $d_{\text{Mars}} = 6.1t^2$.

a. Will an object fall faster on Earth or on Mars?

b. Plot the two functions on the same graph for $0 \le t \le 2.5$ seconds.

c. Calculate the number of seconds on Earth it would take a freely falling object to fall 100 feet. Does your graph confirm your answer?

d. How far would a freely falling object fall on Mars during the same time (as in part (c))?

e. On Jupiter, the equation for the distance a freely falling object has traveled (in feet) is given by $d_{\text{Jupiter}} = 40.65t^2$ (where t is in seconds). Will an object fall faster or more slowly on Jupiter than on Earth? How would its graph compare with those of Earth and Mars?

f. If you double the time, what happens to the distance an object has fallen on Earth? On Mars? On Jupiter?

7. a. Identify the coordinates of the vertex for each the following quadratic functions.

$F(x) = x^2$, $\quad G(x) = x^2 + 5$, $\quad H(x) = (x + 2)^2$, and $J(x) = -(x - 1)^2 - 5$

b. Without using technology, draw a rough sketch on the same grid of all the functions for $-4 \le x \le 4$.

c. Describe how the graph of F was transformed into the graphs of G, H, and J, respectively.

8. Find any horizontal intercepts for the following functions.

   a. $y = (x - 3)(2x + 1)$ **c.** $Q(t) = 2t^2 + t - 1$

   b. $G(z) = 2z^2 - z + 3$

9. a. Construct a quadratic function $Q(t)$ that is concave up and has horizontal intercepts at $t = 4$ and $t = -2$. Write it in both factored and standard form. Find its vertex.

   b. Construct a second function $M(t)$ that has the same horizontal intercepts as $Q(t)$ but is steeper. Write $M(t)$ in both factored and standard form. Do the two functions have the same vertex?

   c. Add a term to $Q(t)$ to create a function $P(t)$ that has no horizontal intercepts.

10. Explain why you could (or couldn't) construct a parabola through any three points.

Problems 11, 12, and 13 refer to the accompanying diagram of the cross section of a swimming pool with a reflective parabolic roof.

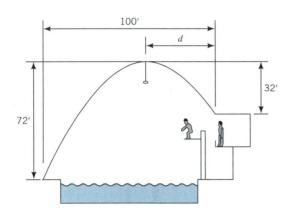

11. Find an equation for the cross section of the parabolic roof of the swimming pool in the diagram. (*Hint:* Place the origin of your coordinate system at the vertex and identify two other points on the parabola in terms of d.)

12. (Requires results of Problem 11.) The pool designer wants to mount a light source at the parabolic focus so that it sheds light evenly on the water surface below. How many feet down from the vertex must that be?

13. A diver jumps up off the high board, which is 25 feet above the surface of the water. Her height, $H(t)$, in feet above the water at t seconds, can be modeled by the function $H(t) = 25 + 12t - 16t^2$.

   a. What will be the highest point above the water of her dive?

   b. When will she hit the water?

14. In the United States a "heat wave" is a period of three or more consecutive days at or above 90°F. A heat wave is often accompanied by high humidity, making the air feel even hotter. The following formula combines an air temperature of 90°F with relative humidity, H, to give the apparent temperature, A, the perceived level of heat:

$$A = 86.61 - 0.132H + 0.0059H^2$$

This formula uses relative humidity as a percentage (e.g., 70% relative humidity appears in the formula as $H = 70$). Remember, this formula applies only for an air temperature of 90°F.

   a. If the relative humidity is 0%, what is the apparent temperature for an air temperature of 90°F? Does the apparent temperature feel lower or higher than the air temperature of 90°F?

   b. On a 90°F day, if the relative humidity is 60%, what is the apparent temperature? How much hotter do you feel?

   c. An apparent temperature of 105°F or above is considered dangerous, especially for children and elders. At what relative humidity on a 90°F day is an apparent temperature of 105°F reached?

15. a. Complete the following table for the function $y = x^2 - 4x$.

x	y	Average Rate of Change	Average Rate of Change of Average Rate of Change
−1	5	n.a.	n.a.
0	0	−5	n.a.
1	−3	−3	$[-3 - (-5)]/(1 - 0) = 2$
2	−4		
3			
4			
5			

   b. If you plotted the points with coordinates of the form (x, average rate of change) and connected adjacent points, what type of function would you get?

   c. Does the fourth column verify your result in part (b)? Why or why not?

16. Using the three accompanying graphs of polynomial functions, determine whether the degree of the polynomial is odd or even, identify its minimum possible degree, and estimate any visible horizontal intercepts of the function.

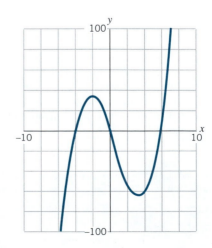

Graph A

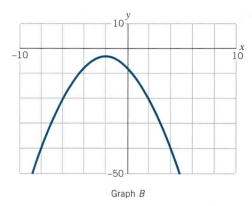

Graph *B*

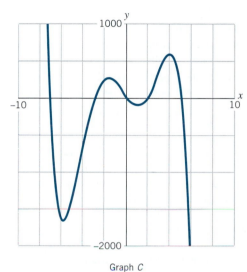

Graph *C*

17. (Graphing technology that can generate a best-fit polynomial is required.) Commercial beekeepers rent out their bees to farmers. The table below gives the average price paid by California almond farmers for each hive of bees (according to Lance Sundberg, who owns and operates the honey business Sunshine Apiary in Columbus, Montana).

	2000	2001	2002	2003	2004	2005	2006
\$/hive	\$51.00	\$51.50	\$52.00	\$52.50	\$53.00	\$82.50	\$128.50

a. Plot the data. What sort of function do you think might best fit the data? Use technology to find that best-fit function $P(t)$, where t = years since 2000.

b. Use your model to predict the average price per hive a beekeeper would get in 2010.

c. On average, there are 2.5 hives per acre of almonds. Predict the amount a beekeeper might get per acre in 2010.

d. Unfortunately, in 2007, for some unknown reason, bees throughout the country have not been returning to their hives after being released. Do you think that the price per hive will go up or down? Why?

18 a. Generate two different polynomials, $M(z)$ and $N(z)$, that have horizontal intercepts at $z = -2, 0$, and 3.

b. Generate a third polynomial, $P(z)$, with the same horizontal intercepts but a higher degree.

19. On the same grid, hand-draw rough sketches of the three functions $f(x) = e^x$, $g(x) = 4e^x$, and $h(x) = -0.5e^x$ for $0 \le x \le 5$. Describe the relationships among the graphs.

20. a. Given the following graph of the function $f(x)$, sketch:

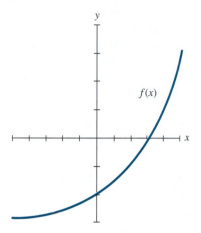

    i. $g(x) = -f(x)$ **ii.** $h(x) = f(-x)$ **iii.** $j(x) = |f(x)|$

b. Describe each function in relation to $f(x)$.

21. (Graphing program required.)

a. If $g(x) = \frac{2}{3(x-4)} - 1$ and $f(x) = \frac{1}{x}$, describe the transformation of the graph of $f(x)$ into the graph of $g(x)$.

b. Rewrite $g(x)$ as a ratio of two polynomials.

c. What is the domain of $g(x)$? Sketch $g(x)$.

d. What are its horizontal and vertical intercepts (if any)?

e. Does $g(x)$ have a vertical asymptote?

f. What is its end behavior?

22. Global warming melts glaciers and polar ice, so scientists predict that the sea level will rise, flooding coastal areas.

a. Assuming Earth is a sphere with radius r and that roughly three-quarters of Earth's surface is ocean (with or without melt water), develop a formula to estimate the volume of melt water necessary to raise the sea level 1 foot. (*Note:* The volume of a sphere is $\frac{4}{3}\pi r^3$.)

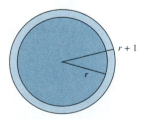

b. Given the Earth's radius is currently about 3959 miles and 1 cubic foot = 7.481 gallons, how many gallons of melt water does your estimate predict? (*Recall:* 1 mile = 5280 feet.)

23. Retirement fund counselors often recommend a mixed portfolio of investments, including some higher-risk investments, which offer higher interest rates, and some more secure investments, with lower interest rates. A woman wants to put half of her \$10,000 savings in a safe 4% fund, and the other half in a riskier

10% fund. She expects to retire in 30 years but would like to know how much she can expect to get if she retires earlier.

a. Create three functions where t is the number of years since the start of the investments and $S(t)$ is the amount of money in the 4% account, $R(t)$ is the amount in the 10% account, and $T(t)$ is the total amount invested in both accounts.

b. On one graph show how the 4% fund, the 10% fund, and the combined fund total accumulate over 30 years.

c. In the worst-case scenario, if she loses all the money in the 10% fund, how much will she be left with in 30 years?

d. You might think that if she is getting 4% on $5000 plus 10% on another $5000, this is the same as getting 14% on $5000. Is it? If not, why not? You can explain your answer using a table and/or a graph.

24. Given the functions $f(x) = x^2$, $g(x) = x + 1$, and $h(x) = -3x$, evaluate each of the following compositions.

a. $(f \circ g)(x)$ and $(g \circ f)(x)$

b. $(h \circ g)(x)$ and $(g \circ h)(x)$

c. $(f \circ g \circ h)(x)$ and $(h \circ g \circ f)(x)$

25. Does the function $f(x) = (x - 2)^3 + 1$ have an inverse? If not, explain why. If so, what is it?

26. Which of these functions has an inverse? If there is one, what is it?

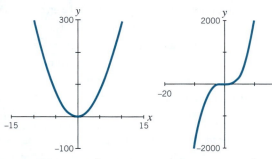

a. Graph of $y = 3x^2 - 1$ b. Graph of $y = 2x^3$

27. When lightning strikes, you seem to see it right away, but the associated thunder often comes a few seconds later. One rule of thumb is that each second of delay represents 1000 feet; that is, if you hear the thunder 3 seconds after the lightning strike, the strike was about 3000 feet away from you.

a. Light travels about 186,000 miles per second, so the light created from a lightning strike a few thousand feet away is seen virtually simultaneously with the strike. However, sound travels much more slowly, at about 761 mph at sea level.

  i. Convert 761 mph into feet/second.

  ii. Now construct a function that gives the distance $D(t)$ (in feet) that the thunder has traveled from the strike site in t seconds.

  iii. Does the rule of thumb seem reasonable?

b. The sound travels in all directions, creating expanding "sound circles" that radiate out from the lightning strike.

  i. Create a function $A(r)$ (in square feet) that gives the area of a sound circle with radius r (in feet).

  ii. We can think of the radius r of the sound circle as $D(t)$, the distance thunder has traveled (in any direction) in t seconds. Substituting r for $D(t)$, construct a composite function $A(D(t))$ to describe the circular area at time t within which the thunder can be detected. What is the circular area (in square feet) within which thunder can be heard 4 seconds after the lightning strike? What is the area in square miles?

  iii. When the time doubles, what happens to the distance the thunder has traveled? What happens to the area within which it can be heard?

How Fast Are You? Using a Ruler to Make a Reaction Timer[1]

Objective

- learn about the properties of freely falling bodies and your own reaction time

Materials/Equipment

- several 12″ rulers
- narrow strips of paper and tape
- calculators

Procedure

General Description

Work in groups of two or three. Each group has a 12″ ruler and will attach a 12″ paper strip to the ruler, adding some marks (specified below). One student drops the ruler between the thumb and forefinger of a second student. The second student tries to catch the ruler as quickly as possible (see image). The reaction time of the second student can be measured by how far the ruler falls before it is caught.

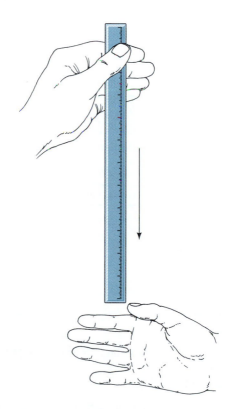

Mathematical Background

Near the surface of Earth, and neglecting air resistance, gravity causes dropped objects to fall approximately according to the formula

$$d = 16t^2$$

[1]This exploration was developed by Karl Schaffer, Mathematics Department, De Anza College, Cupertino, CA.

where d = distance fallen in feet and t = time of fall in seconds.

1. Thus an object in free fall for 1 second will fall 16 feet, and an object in free fall for 0.5 second will fall _____ feet. (Calculate.)

2. Can you give an intuitive explanation for why the object falling for 0.5 second does not fall half as far as the object that fell for 1 second?

3. What assumptions must we make about the shape of the dropped object for it to fall according to this formula? (*Hint:* Will a sheet of paper fall 16 feet in 1 second?)

4. Do you think a heavy object will fall faster or slower or at the same rate as a light object? Explain your reasoning.

5. For each of the following times, use the formula to calculate how far a dropped object will fall.

Time, t	Distance in feet, d	Distance in inches
0.05 second	_____	_____
0.10 second	_____	_____
0.15 second	_____	_____
0.20 second	_____	_____
0.25 second	_____	_____

6. Use tape to attach a strip of paper along the length of the ruler. Think of the ruler as measuring distance fallen. For each distance (in inches) in your previous table, put a mark on the paper that indicates the time corresponding to that distance. So you'll have unevenly spaced marks for the times 0.05 sec, 0.10 sec, up to 0.25 sec. Now the ruler is a reaction timer.

7. One member of your group holds the ruler just above the outstretched thumb and first finger of a second person. The "dropper" suddenly drops the ruler and the "catcher" tries to catch it. Use the time marks on the strip of paper to get an estimate for the reaction time of the catcher. Use the actual number of inches at the point where the catcher caught the ruler to calculate the reaction time. Record the reaction time, and average several tries. If you like, measure the reaction time for someone else in your group.

Person	Distance in inches	Distance in feet, d	Reaction time, t
_____	_____	_____	_____
_____	_____	_____	_____
_____	_____	_____	_____
_____	_____	_____	_____
_____	_____	_____	_____
_____	_____	_____	_____
_____	_____	_____	_____

What was your average reaction time? How did yours compare with that of others in your group?

Further Investigations

1. A popular party trick has one person drop a dollar bill between the fingers of a second person. Usually the bill will fall through the second person's fingers without being caught. How long is a dollar bill, and what must the second person's reaction time be for the bill to be caught? Does the use of money speed up the reaction times you measured with the ruler? Does the bill fall with only negligible air resistance?

2. How do medications or drugs affect our reaction times? Test the reaction times of someone who is taking cold or flu medication or aspirin, or has just drunk a cup of coffee or glass of alcohol (outside of class, of course). Is his or her reaction time impaired? Is there a correlation between the amount of alcohol consumed and reaction time that might enable you to use your ruler

reaction timer as a portable tester to determine whether someone who has consumed alcohol should not drive?

3. Do reaction times measured in this activity improve with practice? Why or why not? (Try it!) Do you think the catcher learns to detect subtle indications of the dropper that she or he is about to drop the ruler? How might these biases be removed from this experiment?

4. Jugglers, athletes, and dancers need to understand, either intuitively or objectively, their own reaction times. In what other occupations is reaction time important? Using the Internet, can you find the reaction times necessary in any of these areas?

5. The *Guinness Book of World Records* lists the fastest times for drawing and firing a gun. Look this up. Are these times consistent with your results?

6. What else might you investigate about reaction times and your ability to measure them?

THE MATHEMATICS OF MOTION

OVERVIEW

In this extended exploration we use the laboratory methods of modern physicists to collect and analyze data about freely falling bodies and then examine the questions asked by Galileo about bodies in motion.

After conducting this exploration, you should be able to

- understand the importance of the scientific method

- describe the relationship between distance and time for freely falling bodies

- derive equations describing the velocity and acceleration of a freely falling body

The Scientific Method

Today we take for granted that scientists study physical phenomena in laboratories using sophisticated equipment. But in the early 1600s, when Galileo did his experiments on motion, the concept of laboratory experiments was unknown. In his attempts to understand nature, Galileo asked questions that could be tested directly in experiments. His use of observation and direct experimentation and his discovery that aspects of nature were subject to quantitative laws were of decisive importance, not only in science but in the broad history of human ideas.

Ancient Greeks and medieval thinkers believed that basic truths existed within the human mind and that these truths could be uncovered through reasoning, not empirical experimentation. Their scientific method has been described as a "qualitative study of nature." Greek and medieval scientists were interested in *why* objects fall. They believed that a heavier object fell faster than a lighter one because "it has weight and it falls to the Earth because it, like every object, seeks its natural place, and the natural place of heavy bodies is the center of the Earth. The natural place of a light body, such as fire, is in the heavens, hence fire rises."[1]

Galileo changed the question from *why* things fall to *how* things fall. This question suggested other questions that could be tested directly by experiment: "By alternating questions and experiments, Galileo was able to identify details in motion no one had previously noticed or tried to observe."[2] His quantitative descriptions of objects in motion led not only to new ways of thinking about motion, but also to new ways of thinking about science. His process of careful observation and testing began the critical transformation of science from a qualitative to a quantitative study of nature.[3] Galileo's decision to search for quantitative descriptions "was the most profound and the most fruitful thought that anyone has had about scientific methodology."[4] This approach became known as the scientific method.

The Free Fall Experiment

Instructions for conducting the free fall experiment are in the last section.

In this extended exploration, you will conduct a modern version of Galileo's free fall experiment. This classic experiment records the distance that a freely falling object falls during each fraction of a second. The experiment can be performed either with a graphing calculator connected to a motion sensor or in a physics laboratory with an apparatus that drops a heavy weight and records its position on a tape.

In this experiment, Galileo sought to answer the following questions:

How can we describe mathematically the distance an object falls over time?
Do freely falling objects fall at a constant speed? If the speed of freely falling objects is not constant, is it increasing at a constant rate?

The software "Q11: Freely Falling Objects" in *Quadratic Functions* provides a simulation of the free fall experiment.

You can try to find answers to these questions by collecting and analyzing your own data or by using the data provided as both Excel and graph link files. Instructions for using technology to collect and analyze data are provided in the last section. The following discussion will help you analyze your results and provide answers to Galileo's questions.

---

[1]M. Kline, *Mathematics for the Nonmathematician* (New York: Dover, 1967), p. 287.
[2]E. Cavicchi, "Watching Galileo's Learning," in the Anthology of Readings on the course website.
[3]Galileo's scientific work was revolutionary in terms not only of science but also of the politics of the time; his work was condemned by the ruling authorities, and he was arrested.
[4]M. Kline, op. cit., p. 288.

Interpreting Data from a Free Fall Experiment

The sketch of a tape given in Figure 1 gives data collected by a group of students from a falling-object experiment. Each dot represents how far the object fell in each succeeding 1/60 of a second.

Since the first few dots are too close together to get accurate measurements, we start measurements at the sixth dot, which we call dot_0. At this point, the object is already in motion. This dot is considered to be the starting point, and the time, t, at dot_0 is set at 0 seconds. The next dot represents the position of the object 1/60 of a second later. Time increases by 1/60 of a second for each successive dot. In addition to assigning a time to each point, we also measure the total distance fallen, d (in cm), from the point designated dot_0. For every dot we have two values: the time, t, and the distance fallen, d. At dot_0, we have $t = 0$ and $d = 0$.

The time and distance measurements from the tape are recorded in Table 1 and plotted on the graph in Figure 2. Time, t, is the independent variable, and distance, d, is the dependent variable. The graph gives a representation of the data collected on distance fallen over time, not a picture of the physical motion of the object. The graph of the data looks more like a curve than a straight line, so we expect the average rates of change between different pairs of points to be different. We know how to calculate the average rate of change between two points and that it represents the slope of a line segment connecting the two points:

$$\text{average rate of change} = \frac{\text{change in distance}}{\text{change in time}} = \text{slope of line segment}$$

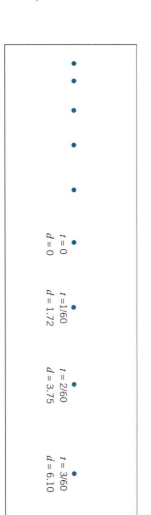

Figure 1 Tape from a free fall experiment.

Time (sec)	Total Distance Fallen (cm)
0.0000	0.00
0.0167	1.72
0.0333	3.75
0.0500	6.10
0.0667	8.67
0.0833	11.58
0.1000	14.71
0.1167	18.10
0.1333	21.77
0.1500	25.71
0.1667	29.90
0.1833	34.45
0.2000	39.22
0.2167	44.22
0.2333	49.58
0.2500	55.15
0.2667	60.99
0.2833	67.11
0.3000	73.48
0.3167	80.10
0.3333	87.05
0.3500	94.23

Table 1

Figure 2 Free fall: distance versus time.

Table 2 and Figure 3 show the increase in the average rate of change over time for three different pairs of points. The time interval nearest the start of the fall shows a relatively small change in the distance per time step and therefore a relatively gentle slope of 188 cm/sec. The time interval farthest from the start of the fall shows a greater change of distance per time step and a much steeper slope of 367 cm/sec.

t	d	Average Rate of Change	
0.0500	6.10	$\dfrac{21.77 - 6.10}{0.1333 - 0.0500}$	≈ 188 cm/sec
0.1333	21.77		
0.0833	11.58	$\dfrac{49.58 - 11.58}{0.2333 - 0.0833}$	≈ 253 cm/sec
0.2333	49.58		
0.2167	44.22	$\dfrac{87.05 - 44.22}{0.3333 - 0.2167}$	≈ 367 cm/sec
0.3333	87.05		

Table 2

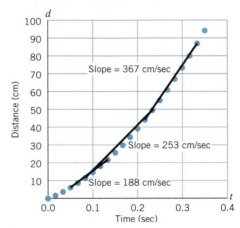

Figure 3 Slopes (or average velocities) between three pairs of end points.

In this experiment the average rate of change has an additional important meaning. For objects in motion, the change in distance divided by the change in time is also called the *average velocity* for that time period. For example, in the calculations in Table 2, the average rate of change of 188 cm/sec represents the average velocity of the falling object between 0.0500 and 0.1333 second.

$$\textit{average velocity} = \frac{\text{change in distance}}{\text{change in time}}$$

Important Questions

Do objects fall at a constant speed?[5] The rate-of-change calculations and the graph in Figure 3 indicate that the average rate of change of position with respect to time—the

---

[5]In everyday usage, "speed" and "velocity" are used interchangeably. In physics, "velocity" gives the direction of motion by the sign of the number—positive for forward, negative for backward. "Speed" means the absolute value, or magnitude, of the velocity. So speed is never negative, whereas velocity can be positive or negative.

velocity—of the falling object is not constant. Moreover, the average velocity appears to be increasing over time. In other words, as the object falls, it is moving faster and faster. Our calculations agree with Galileo's observations. He was the first person to show that the velocity of a freely falling object is not constant.

This finding prompted Galileo to ask more questions. One of these questions was: If the velocity of freely falling bodies is *not constant,* is it increasing at a *constant rate?* Galileo discovered that the velocity of freely falling objects does increase at a constant rate. If the rate of change of velocity with respect to time is constant, then the graph of velocity versus time is a straight line. The slope of that line is constant and equals the rate of change of velocity with respect to time. A theory of gravity has been built around Galileo's discovery of a constant rate of change for the velocity of a freely falling body. This constant of nature, the gravitational constant of Earth, is denoted by g and is approximately 980 cm/sec^2.

Deriving an Equation Relating Distance and Time

Galileo wanted to describe mathematically the distance an object falls over time. Using mathematical and technological tools not available in Galileo's time, we can describe the distance fallen over time in the free fall experiment using a "best-fit" function for our data. Galileo had to describe his finding in words. Galileo described the free fall motion first by direct measurement and then abstractly with a time-squared rule. "This discovery was revolutionary, the first evidence that motion on Earth was subject to mathematical laws."[6]

Using Galileo's finding that distance is related to time by a time-squared rule, we use technology to find the following best-fit quadratic function for the free fall data in Table 1:

$$d = 487.8t^2 + 98.73t - 0.0528$$

Figure 4 shows a plot of the data and the function. If your curve-fitting program does not provide a measure of closeness of fit, such as the correlation coefficient for regression lines, you may have to rely on a visual judgment. Rounding the coefficients to the nearest unit, we obtain the equation

$$d = 488t^2 + 99t - 0$$
$$= 488t^2 + 99t \qquad (1)$$

We now have a mathematical model for our free fall data.

If you are interested in learning more about how Galileo made his discoveries, read Elizabeth Cavicchi's "Watching Galileo's Learning."

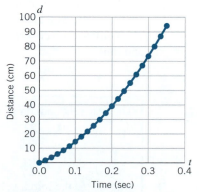

Figure 4 Best-fit function for distance versus time.

[6]E. Cavicchi, "Watching Galileo's Learning," in the Anthology of Readings on the course website.

What are the units for each term of the equation? Since d is in centimeters, each term on the right-hand side of Equation (1) must also be in centimeters. Since t is in seconds, the coefficient, 488, of t^2 must be in centimeters per second squared:

$$\frac{\text{cm}}{\text{sec}^2} \cdot \frac{\text{sec}^2}{1} = \text{cm}$$

The coefficient, 99, of t must be in centimeters per second, and the constant term, 0, in centimeters.

If we ran the experiment again, how would the results compare? In one class, four small groups did the free fall experiment, plotted the data, and found a corresponding best-fit second-degree polynomial. The functions are listed below, along with Equation (1). In each case we have rounded the coefficients to the nearest unit. All of the constant terms rounded to 0.

$$d = 488t^2 + 99t \tag{1}$$
$$d = 486t^2 + 72t \tag{2}$$
$$d = 484t^2 + 173t \tag{3}$$
$$d = 486t^2 + 73t \tag{4}$$
$$d = 495t^2 + 97t \tag{5}$$

Examine the coefficients of each of the terms in these equations. All the functions have similar coefficients for the t^2 term, very different coefficients for the t term, and zero for the constant term. Why is this the case? Using concepts from physics, we can describe what each of the coefficients represents.

The coefficients of the t^2 term found in Equations (1) to (5) are all close to one-half of 980 cm/sec^2, or half of g, Earth's gravitational constant. The data from this simple experiment give very good estimates for $\frac{1}{2}g$.

The coefficient of the t term represents the initial velocity, v_0, of the object when $t = 0$. In Equation (1), $v_0 = 99$ cm/sec. Recall that we didn't start to take measurements until the sixth dot, the dot we called dot_0. So at dot_0, where we set $t = 0$, the object was already in motion with a velocity of approximately 99 cm/sec. The initial velocities, or v_0 values, in Equations (2) to (5) range from 72 to 173 cm/sec. Each v_0 represents approximately how fast the object was moving when $t = 0$, the point chosen to begin recording data in each of the various experiments.

The constant term rounded to zero in each of Equations (1) to (5). On the tape where we set $t = 0$, we set $d = 0$. So we expect that in all our best-fit equations the constant terms, which represent the distance at time zero, are approximately zero. If we substitute zero for t in Equations (1) to (5), the value for d is indeed zero. If we looked at additional experimental results, we might encounter some variation in the constant term, but all should have values of approximately zero.

Galileo's discoveries are the basis for the following equations relating distance and time:

The general equation of motion of freely falling bodies that relates distance fallen, d, to time, t, is

$$d = \frac{1}{2}gt^2 + v_0t$$

where v_0 is the initial velocity and g is the acceleration due to gravity on Earth.

For example, in our original model, $d = 488t^2 + 99t$, the coefficient 488 approximates $\frac{1}{2}g$ (in centimeters per second squared) and 99 approximates the initial velocity (in centimeters per second).

Returning to Galileo's Question

If the velocity for freely falling bodies is not constant, is it increasing at a constant rate? Galileo discovered that the rate of change of the velocity of a freely falling object is constant. In this section we confirm his finding with data from the free fall experiment.

Velocity: Change in Distance over Time

If the rate of change of velocity is constant, then the graph of velocity vs. time should be a straight line. Previously we calculated the average rates of change of distance with respect to time (or average velocities) for three arbitrarily chosen pairs of points. Now, in Table 3 we calculate the average rates of change for all the pairs of adjacent points in our free fall data. The results are in column 4. Since each computed velocity is the average over an interval, for increased precision we associate each velocity with the midpoint time of the interval instead of one of the end points. In Figure 5, we plot velocity from the fourth column against the midpoint times from the third column. The graph is strikingly linear.

See "C3: Average Velocity and Distance" in *Rates of Change.*

Time, t (sec)	Distance Fallen, d (cm)	Midpoint Time, t (sec)	Velocity, v (cm/sec)
0.0000	0.00		
0.0167	1.72	0.0083	103.2
0.0333	3.75	0.0250	121.8
0.0500	6.10	0.0417	141.0
0.0667	8.67	0.0583	154.2
0.0833	11.58	0.0750	174.6
0.1000	14.71	0.0917	187.8
0.1167	18.10	0.1083	203.4
0.1333	21.77	0.1250	220.2
0.1500	25.71	0.1417	236.4
0.1667	29.90	0.1583	251.4
0.1833	34.45	0.1750	273.0
0.2000	39.22	0.1917	286.2
0.2167	44.22	0.2083	300.0
0.2333	49.58	0.2250	321.6
0.2500	55.15	0.2417	334.2
0.2667	60.99	0.2583	350.4
0.2833	67.11	0.2750	367.2
0.3000	73.48	0.2917	382.2
0.3167	80.10	0.3083	397.2
0.3333	87.05	0.3250	417.0
0.3500	94.23	0.3417	430.8

Table 3

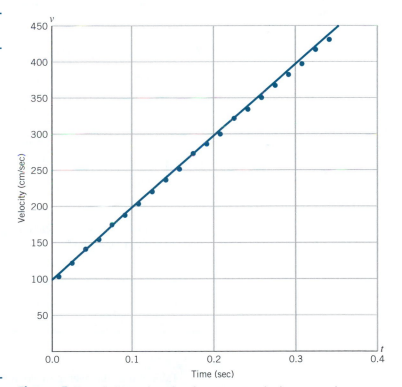

Figure 5 Best-fit linear function for average velocity versus time.

Generating a best-fit linear function and rounding to the nearest unit, we obtain the equation

$$\text{average velocity} = 977t + 98$$

where the average velocity, v, is in centimeters per second, and time t, is in seconds. The graph of this function appears in Figure 5. The slope of the line is constant and equals the rate of change of velocity with respect to time. So although the velocity is not constant, its rate of change with respect to time *is* constant.

The coefficient of t, 977, is the slope of the line and in physical terms represents g, the acceleration due to gravity. The conventional value for g is 980 cm/sec^2. So the velocity of the freely falling object increases by about 980 cm/sec during each second of free fall.

With this equation we can estimate the velocity at any given time t. When $t = 0$, then $v = 98$ cm/sec. This means that the object was already moving at about 98 cm/sec when we set $t = 0$. In our experiment, the velocity when $t = 0$ depends on where we choose to start measuring our dots. If we had chosen a dot closer to the beginning of the free fall, we would have had an initial velocity lower than 98 cm/sec. If we had chosen a dot farther away from the start, we would have had an initial velocity higher than 98 cm/sec. Note that 98 cm/sec closely matches the value of 99 cm/sec in our best-fit quadratic function (Equation 1).

> The general equation that relates, v, the velocity of a freely falling body, to t, time, is
>
> $$v = gt + v_0$$
>
> where v_0 = initial velocity (velocity at time $t = 0$) and g is the acceleration due to gravity.

Acceleration: Change in Velocity over Time

Acceleration means a change in velocity or speed. If you push the accelerator pedal in a car down just a bit, the speed of the car increases slowly. If you floor the pedal, the speed increases rapidly. The rate of change of velocity with respect to time is called *acceleration*. Calculating the average rate of change of velocity with respect to time gives an estimate of acceleration. For example, if a car is traveling at 20 mph and 1 hour later the car has accelerated to 60 mph, then

$$\frac{\text{change in velocity}}{\text{change in time}} = \frac{(60 - 20)\ \text{mph}}{1\ \text{hr}} = (40\ \text{mph})/\text{hr} = 40\ \text{mi/hr}^2$$

In 1 hour, the velocity of the car changed from 20 to 60 mph, so its average acceleration was 40 mph/hr, or 40 mi/hr^2.

> $$\textit{average acceleration} = \frac{\text{change in velocity}}{\text{change in time}}$$

Table 4 uses the average velocity data and midpoint time from Table 3 to calculate average accelerations. Figure 6 shows the plot of average acceleration in centimeters per second squared (the third column) versus time in seconds (the first column).

The data lie along a roughly horizontal line. The average acceleration values vary between a low of 756 cm/sec^2 and a high of 1296 cm/sec^2 with a mean of 982.8. Rounding off, we have

$$\text{acceleration} \approx 980\ \text{cm/sec}^2$$

This expression confirms that for each additional second of free fall, the velocity of the falling object increases by approximately 980 cm/sec. The longer it falls, the faster it goes. We have verified a characteristic feature of gravity near the surface of Earth: It causes objects to fall at a velocity that increases every second by about 980 cm/sec. We say that the acceleration due to gravity near Earth's surface is 980 cm/sec^2.

Midpoint Time, t (sec)	Average Velocity, v (cm/sec)	Average Acceleration (cm/sec^2)
0.0083	103.2	n.a.
0.0250	121.8	1116
0.0417	141.0	1152
0.0583	154.2	792
0.0750	174.6	1224
0.0917	187.8	792
0.1083	203.4	936
0.1250	220.2	1008
0.1417	236.4	972
0.1583	251.4	900
0.1750	273.0	1296
0.1917	286.2	792
0.2083	300.0	828
0.2250	321.6	1296
0.2417	334.2	756
0.2583	350.4	972
0.2750	367.2	1008
0.2917	382.2	900
0.3083	397.2	900
0.3250	417.0	1188
0.3417	430.8	828

Table 4

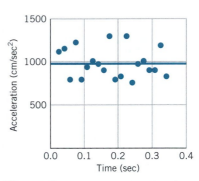

Figure 6 Average acceleration for free fall data.

In order to express g in feet per second squared, we need to convert 980 centimeters to feet. We start with the fact that 1 ft = 30.48 cm. So the conversion factor for centimeters to feet is (1 ft)/(30.48 cm) = 1. If we multiply 980 cm by (1 ft)/(30.48 cm) to convert centimeters to feet, we get

$$980 \text{ cm} = (980 \text{ cm})\left(\frac{1 \text{ ft}}{30.48 \text{ cm}}\right) \approx 32.15 \text{ ft} \approx 32 \text{ ft}$$

So a value of 980 cm/sec^2 for g is equivalent to approximately 32 ft/sec^2.

The numerical value used for the constant g depends on the units being used for the distance, d, and the time, t. The exact value of g also depends on where it is measured.[7]

> The conventional values for g, the acceleration due to gravity near the surface of Earth, are
>
> $$g = 32 \text{ ft/sec}^2$$
>
> or equivalently $\qquad\qquad g = 980 \text{ cm/sec}^2 = 9.8 \text{ m/sec}^2$

---

[7] Because Earth is rotating, is not a perfect sphere, and is not uniformly dense, there are variations in g according to the latitude and elevation. The following are a few examples of local values for g.

Location	North Latitude (deg)	Elevation (m)	g (cm/sec^2)
Panama Canal	9	0	978.243
Jamaica	18	0	978.591
Denver, CO	40	1638	979.609
Pittsburgh, PA	40.5	235	980.118
Cambridge, MA	42	0	980.398
Greenland	70	0	982.534

Source: H. D. Young, *University Physics,* Vol. I, 8th ed. (Reading, MA: Addison-Wesley, 1992), p. 336.

Deriving an Equation for the Height of an Object in Free Fall

Assume we have the following motion equation relating distance fallen, d (in centimeters), and time, t (in seconds):

$$d = 490t^2 + 45t$$

Also assume that when $t = 0$, the height, h, of the object was 110 cm above the ground. Until now, we have considered the distance from the point the object was dropped, a value that *increases* as the object falls. How can we describe a different distance, the *height above ground* of an object, as a function of time, a value that *decreases* as the object falls?

At time zero, the distance fallen is zero and the height above the ground is 110 centimeters. After 0.05 second, the object has fallen about 3.5 centimeters, so its height would be $110 - 3.5 = 106.5$ cm. For an arbitrary distance d, we have $h = 110 - d$. Table 5 gives associated values for time, t, distance fallen, d, and height above ground, h. The graphs in Figure 7 show distance versus time and height versus time.

Time, t (sec)	Distance Fallen, d (cm) $(d = 490t^2 + 45t)$	Height above Ground, h (cm) $(h = 110 - d)$
0.00	0.0	110.0
0.05	3.5	106.5
0.10	9.4	100.6
0.15	17.8	92.2
0.20	28.6	81.4
0.25	41.9	68.1
0.30	57.6	52.4
0.35	75.8	34.2
0.40	96.4	13.6

Table 5

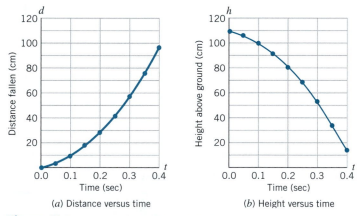

(a) Distance versus time   (b) Height versus time

Figure 7 Representations of free fall data.

How can we convert the equation $d = 490t^2 + 45t$, relating distance fallen and time, to an equation relating height above ground and time? We know that the relationship between height and distance is $h = 110 - d$. We can substitute the expression for d into the height equation:

$$
\begin{aligned}
h &= 110 - d \\
&= 110 - (490t^2 + 45t) \\
&= 110 - 490t^2 - 45t
\end{aligned}
$$

(6)

Switching the order of the terms, we could rewrite this equation as $h = -490t^2 - 45t + 110$. The constant term, here 110 cm, represents the initial height when $t = 0$. By placing the constant term first as in Equation (6), we emphasize 110 cm as the initial or starting value. Height equations often appear in the form $h = c + bt + at^2$ to emphasize the constant term c as the starting height. This is similar to writing linear equations in the form $y = b + mx$ to emphasize the constant term b as the base, or starting, value.

Note that in Equation (6) for height, the coefficients of both t and t^2 are negative. If we consider what happens to the height of an object in free fall, this makes sense. As time increases, the height decreases. (See Table 5 and Figure 7.) When we were measuring the increasing distance an object fell, we did not take into account the direction in which it was going (up or down). We cared only about the magnitudes (the absolute values) of distance and velocity, which were positive. But when we are measuring a decreasing height or distance, we have to worry about direction. In this case we define downward motion to be negative and upward motion to be positive. In the height equation $h = 110 - 45t - 490t^2$, the constant term, the initial height, is 110 cm. The change in height resulting from the initial velocity, $-45t$, is negative because the object was moving down when we started to measure it. The change in height caused by acceleration, $-490t^2$, is also negative because gravity pulls objects downward in what we are now considering as a negative direction, reducing the height of a falling object.

Once we have introduced the notion that downward motion is negative and upward motion is positive, we can also deal with situations in which the initial velocity is upward and the acceleration is downward. The velocity equation for this situation is

$$v = -gt + v_0$$

where v_0 could be either positive or negative, depending on whether the object is thrown upward or downward, and the sign for the g term is negative because gravity accelerates downward in the negative direction.

If we treat upward motion as positive and downward motion as negative, then the acceleration due to gravity is negative. So the general equations of motion of freely falling bodies that relate height, h, and velocity, v, to time, t, are

$$h = h_0 + v_0 t - \tfrac{1}{2}gt^2$$
$$v = -gt + v_0$$

where h_0 = initial height, g = acceleration due to gravity, and v_0 = initial velocity (which can be positive or negative).

Working with an Initial Upward Velocity

If we want to use the general equation to describe the height of a thrown object, we need to understand the meaning of each of the coefficients. Suppose a ball is thrown upward with an initial velocity of 97 cm/sec from a height of 87 cm above the ground. Describe the relationship between the height of the ball and time with an equation.

The initial height of the ball is 87 cm when $t = 0$, so the constant term is 87 cm. The coefficient of t, or the initial-velocity term, is +97 cm/sec since the initial motion is upward. The coefficient of t^2, the gravity term, is -490 cm/sec^2, since gravity causes objects to fall down.

Substituting these values into the equation for height, we get

$$h = 87 + 97t - 490t^2$$

Table 6 gives a series of values for heights corresponding to various times. Figure 8 plots height above ground (cm) vs. time (sec.).

t (sec)	h (cm)
0.00	87.00
0.05	90.63
0.10	91.80
0.15	90.53
0.20	86.80
0.25	80.63
0.30	72.00
0.35	60.93
0.40	47.40
0.45	31.43
0.50	13.00

Table 6

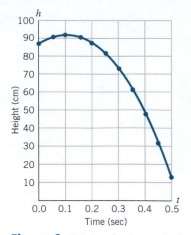

Figure 8 Height of a thrown ball.

The graph of the heights at each time in Figure 8 should not be confused with the trajectory of a thrown object. The actual motion we are talking about is purely vertical—straight up and straight down. The graph shows that the object travels up for a while before it starts to fall. This corresponds with what we all know from practical experience throwing balls. The upward (positive) velocity is decreased by the pull of gravity until the object stops moving upward and begins to fall. The downward (negative) velocity is then increased by the pull of gravity until the object strikes the ground.

Collecting and Analyzing Data from a Free Fall Experiment

Objective

- to describe mathematically how objects fall

Equipment/Materials

- graphing calculator with best-fit function capabilities or computer with spreadsheet and function graphing program
- notebook for recording measurements and results (sample Lab Book on course website)

Equipment needed for collecting data in physics laboratory:

DATA If using precollected data, see the Excel or graph link file FREEFALL.

    a. Free fall apparatus

    b. Meter sticks 2 meters long

    c. Masking tape

Equipment needed for collecting data with CBL® (Calculator-Based Laboratory System®):

    a. CBL® unit with AC-9201 power adapter

    b. Vernier CBL® ultrasonic motion detector

    c. Graphing calculator

    d. Extension cord and some object to drop, such as a pillow or rubber ball

Preparation

If collecting data in a physics laboratory, schedule a time for doing the experiment and have the laboratory assistant available to set up the equipment and assist with the experiment. If collecting data with a CBL® unit with graphing calculator, instructions for using a CBL® unit are in the Instructor's Manual.

Procedure

The following procedures can be used for collecting data in a physics laboratory.[8] If you are collecting data with a CBL®, collect the data and go to the Results section. If you are using the precollected data in the file FREEFALL, go directly to the Results section.

Collecting the data

Since the falling times are too short to record with a stopwatch, we use a free fall apparatus. Every sixtieth of a second a spark jumps between the falling object or "bob" and the vertical metal pole supporting the tape. Each spark burns a small dot on the fixed tape, recording the bob's position. The procedure is to:

1. Position the bob at the top of the column in its holder.
2. Pull the tape down the column so that a fresh tape is ready to receive spark dots.
3. Be sure that the bob is motionless before you turn on the apparatus.
4. Turn on the spark switch and bob release switch as demonstrated by the laboratory assistant.
5. Tear off the length of tape recording the fall of the bob.

Obtaining and recording measurements from the tapes

The tape is a record of the distance fallen by the bob between each sixtieth-of-a-second spark dot. Each pair of students should measure and record the distance between the dots on the tape. Let d = the distance fallen in centimeters and t = time in seconds.

1. Fasten the tape to the table using masking tape.
2. Inspect the tape for missing dots. *Caution:* The sparking apparatus sometimes misses a spark. If this happens, take proper account of it in numbering the dots.
3. Position the 2-meter stick on its edge along the dots on the tape. Use masking tape to fasten the meter stick to the table, making sure that the spots line up in front of the bottom edge of the meter stick so you can read their positions off of the stick.
4. Beginning with the sixth visible dot, mark the time for each spot on the tape; that is, write $t = 0/60$ sec by the sixth dot, $t = 1/60$ by the next dot, $t = 2/60$ by the next dot, and so on, until you reach the end of the tape.

   Note: The first five dots are ignored in order to increase accuracy of measurements. One cannot be sure that the object is released exactly at the time of the spark, instead of between sparks, and the first few dots are too close together to get accurate measurements. When the body passes the sixth dot, it already has some velocity, which we call v_0, and this point is arbitrarily taken as the initial time, $t = 0$.

5. Measure the distances (accurate to a fraction of a millimeter) from the sixth dot to each of the other dots. Record each distance by the appropriate dot on the tape.

---

[8]These procedures are adapted from "Laboratory Notes for Experiment 2: The Kinematics of Free Fall," University of Massachusetts, Boston, Elementary Physics 181.

6. Recheck your measurements.

7. Clean your work area.

Results

Use your notebook to keep a record of your data, observations, graphs, and analysis of the data.

a. Record the data obtained from your measurements on the tape or from using a CBL® unit. If you are entering your data into a function graphing program or a spreadsheet, you can use a printout of the data and staple it into your laboratory notework. Your data should include time, t, and distance fallen, d, as in the following table:

t (sec)	d (cm)
0/60	0
1/60	–
2/60	–
......	–
To last record	–

This table assumes regular time intervals of one-sixtieth of a second. Check your equipment to see whether it uses a different interval size.

b. Note at which dot on the tape you started to make your measurements.

Analysis of data

1. By hand:

a. Graph your data, using the vertical axis for distance fallen, d, in centimeters and the horizontal axis for time, t, in seconds. What does your graph suggest about the average rate of change of distance with respect to time?

b. Calculate the average rate of change for distance, d, with respect to time, t, for three pairs of points from your data table:

$$\text{average rate of change} = \frac{\text{change in distance}}{\text{change in time}} = \frac{\Delta d}{\Delta t}$$

Show your work. This average rate of change is called the *average velocity* of the falling object between these two points. Do your calculations support your answer in part (a)?

c. Jot down your observations from your graph and calculations in your notebook. Staple your graph into your notebook.

2. With graphing calculators or computers:

a. Use technology to graph your data for the free fall experiment. Plot time, t, on the horizontal axis and distance fallen, d, on the vertical axis.

b. Find a best-fit function for distance fallen versus time.

c. Use your spreadsheet or graphing calculator to calculate the average rate of change in distance over each of the small time intervals. This average rate of change is the average velocity over each of these time intervals.

d. Plot average velocity versus time, with time on the horizontal axis and average velocity on the vertical axis.

e. Jot down your observations from your graphs and calculations in your notebook. Be sure to specify the units for any numbers you recorded.

Conclusions

Summarize your conclusions from the experiment:

- Describe what you found out from your graph of distance vs. time and your calculations for the average rate of change of distance with respect to time. Is the average rate of change of distance with respect to time the same for each small time interval?
- What does your graph of the average velocity vs. time tell you about the average velocity of the freely falling body? Is the average rate of change in velocity from one interval to the next roughly constant?
- In light of the readings and class discussion, interpret your graphs for distance and average velocity and interpret the coefficients in the equation you found for distance.

In his own version of this experiment, Galileo sought to answer the following questions:
How can we describe mathematically the distance an object falls over time?
Do freely falling objects fall at a constant speed?
If the velocity of a freely falling object is not constant, is it increasing at a constant rate?
Use your results to answer these questions.

EXERCISES

(A graphing program is optional for many exercises and required for Exercise 30.)

1. Complete the accompanying table. What happens to the average velocity of the object as it falls?

Time (sec)	Distance Fallen (cm)	Average Velocity (average rate of change for the previous 1/30 of a second)
0.0000	0.00	n.a.
0.0333	3.75	$\frac{3.75 - 0.00}{0.0333 - 0.0000} \approx 113$ cm/sec
0.0667	8.67	$\frac{8.67 - 3.75}{0.0667 - 0.0333} \approx 147$ cm/sec
0.1000	14.71	
0.1333	21.77	
0.1667	29.90	

2. The essay "Watching Galileo's Learning" examines the learning process that Galileo went through to come to some of the most remarkable conclusions in the history of science. Write a summary of one of Galileo's conclusions about motion. Include in your summary the process by which Galileo made this discovery and some aspect of your own learning or understanding of Galileo's discovery.

3. (Graphing program optional.) The equation $d = 490t^2 + 50t$ describes the relationship between distance fallen, d, in centimeters, and time, t, in seconds, for a particular freely falling object.

   a. Interpret each of the coefficients and specify its units of measurement.

   b. Generate a table for a few values of t between 0 and 0.3 second.

   c. Graph distance versus time by hand. Check your graph using a computer or graphing calculator if available.

4. A freely falling body has an initial velocity of 125 cm/sec. Assume that $g = 980$ cm/sec^2.

   a. Write an equation that relates d, distance fallen in centimeters, to t, time in seconds.

   b. How far has the body fallen after 1 second? After 3 seconds?

   c. If the initial velocity were 75 cm/sec, how would your equation in part (a) change?

5. (Graphing program optional.) The equation $d = 4.9t^2 + 1.7t$ describes the relationship between distance fallen, d, in meters, and time, t, in seconds, for a particular freely falling object.

   a. Interpret each of the coefficients and specify its units of measurement.

   b. Generate a table for a few values of t between 0 and 0.3 second.

   c. Graph distance versus time by hand. Check your graph using a computer or graphing calculator if available.

   d. Relate your answers to earlier results in this chapter.

6. In the equation of motion $d = \frac{1}{2}gt^2 + v_0t$, we specified that distance was measured in centimeters, velocity in centimeters per second, and time in seconds. Rewrite this as an equation that shows only units of measure. Verify that you get centimeters = centimeters.

7. The equation $d = \frac{1}{2}gt^2 + v_0t$ could also be written using distances measured in meters. Rewrite the equation showing only units of measure and verify that you get meters = meters.

8. The equation $d = \frac{1}{2}gt^2 + v_0t$ could be written using distance measured in feet. Rewrite the equation showing only units of measure and verify that you get feet = feet.

9. A freely falling object has an initial velocity of 50 cm/sec.

   a. Write two motion equations, one relating distance and time and the other relating velocity and time.

   b. How far has the object fallen and what is its velocity after 1 second? After 2.5 seconds? Be sure to identify units in your answers.

10. A freely falling object has an initial velocity of 20 ft/sec.

    a. Write one equation relating distance fallen (in feet) and time (in seconds) and a second equation relating velocity (in feet per second) and time.

    b. How many feet has the object fallen and what is its velocity after 0.5 second? After 2 seconds?

11. (Graphing program optional.) A freely falling object has an initial velocity of 12 ft/sec.

    a. Construct an equation relating distance fallen and time.

    b. Generate a table by hand for a few values of the distance fallen between 0 and 5 seconds.

    c. Graph distance vs. time by hand. Check your graph using a computer or graphing calculator if available.

12. (Graphing program optional.) Use the information in Exercise 11 to do the following:

    a. Construct an equation relating velocity and time.

    b. Generate a table by hand for a few values of velocity between 0 and 5 seconds.

    c. Graph velocity versus time by hand. If possible, check your graph using a computer or graphing calculator.

13. If the equation $d = 4.9t^2 + 11t$ represents the relationship between distance and time for a freely falling body, in what units is distance now being measured? How do you know?

14. The distance that a freely falling object with no initial velocity falls can be modeled by the quadratic function $d = 16t^2$, where t is measured in seconds and d in feet. There is a closely related function $v = 32t$ that gives the velocity, v, in feet per second at time t, for the same freely falling body.

    a. Fill in the missing values in the following table:

Time, t (sec)	Distance, d (ft)	Velocity, v (ft/sec)
1		
1.5		
2		
		80
	144	

    b. When $t = 3$, describe the associated values of d and v and what they tell you about the object at that time.

    c. Sketch both functions, distance versus time and velocity versus time, on two different graphs. Label the points from part (b) on the curves.

    d. You are standing on a bridge looking down at a river. How could you use a pebble to estimate how far you are above the water?

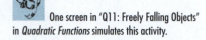

One screen in "Q11: Freely Falling Objects" in *Quadratic Functions* simulates this activity.

15. (This exercise requires a free fall data tape created using a spark timer.)

    a. Make a graph from your tape: Cut the tape with scissors crosswise at each spark dot, so you have a set of strips of paper that are the actual lengths of the distances fallen by the object during each time interval. Arrange them evenly spaced in increasing order, with the bottom of each strip on a horizontal line. The end result should look like a series of steps. You could paste or tape them down on a big piece of paper or newspaper.

    b. Use a straight edge to draw a line that passes through the center of the top of each strip. Is the line a good fit? Each separate strip represents the distance the object fell during a fixed time interval, so we can think of the strips as representing change in distance over time, or average velocity. Interpret the graph of the line you have constructed in terms of the free fall experiment.

16. In the Anthology Reading "Watching Galileo's Learning," Cavicchi notes that Galileo generated a sequence of odd integers from his study of falling bodies. Show that in general the odd integers can be constructed from the difference of the squares of successive integers, that is, that the terms $(n + 1)^2 - n^2$ (where $n = 0, 1, 2, 3, \ldots$) generate a sequence of all the positive odd integers.

17. The data from a free fall tape generate the following equation relating distance fallen in centimeters and time in seconds:

$$d = 485.7t^2 + 7.6t$$

    a. Give a physical interpretation of each of the coefficients along with its appropriate units of measurement.

    b. How far has the object fallen after 0.05 second? 0.10 second? 0.30 second?

18. What would the free fall equation $d = 490t^2 + 90t$ become if d were measured in feet instead of centimeters?

19. In the equation $d = 4.9t^2 + 500t$, time is measured in seconds and distance in meters. What does the number 500 represent?

20. In the height equation $h = 300 + 50t - 4.9t^2$, time is measured in seconds and height in meters.

    a. What does the number 300 represent?

    b. What does the number 50 represent? What does the fact that 50 is positive tell you?

21. (Graphing program optional.) The height of an object that was projected vertically from the ground with initial velocity of 200 m/sec is given by the equation $h = 200t - 4.9t^2$, where t is in seconds.

    a. Find the height of the object after 0.1, 2, and 10 seconds.

    b. Sketch a graph of height vs. time.

    c. Use the graph to determine the maximum height of the projectile and the approximate number of seconds that the object traveled before hitting the ground.

22. (Graphing program optional.) The height of an object that was shot downward from a 200-meter platform with an initial velocity of 50 m/sec is given by the equation $h = -4.9t^2 - 50t + 200$, where t is in seconds and h is in meters. Sketch the graph of height versus time. Use the graph to determine the approximate number of seconds that the object traveled before hitting the ground.

23. (Graphing program optional.) Let $h = 85 - 490t^2$ be a motion equation describing height, h, in centimeters and time, t, in seconds.

   a. Interpret each of the coefficients and specify its units of measurement.

   b. What is the initial velocity?

   c. Generate a table for a few values of t between 0 and 0.3 second.

   d. Graph height versus time by hand. Check your graph using a computer or graphing calculator if possible.

24. (Graphing program optional.) Let $h = 85 + 20t - 490t^2$ be a motion equation describing height, h, in centimeters and time, t, in seconds.

   a. Interpret each of the coefficients and specify its units of measurement.

   b. Generate a table for a few values of t between 0 and 0.3 second.

   c. Graph height versus time by hand. Check your graph using a computer or graphing calculator if possible.

25. At $t = 0$, a ball is thrown upward at a velocity of 10 ft/sec from the top of a building 50 feet high. The ball's height is measured in feet above the ground.

   a. Is the initial velocity positive or negative? Why?

   b. Write the motion equation that describes height, h, at time, t.

26. The concepts of velocity and acceleration are useful in the study of human childhood development. The accompanying figure shows (a) a standard growth curve of weight over time, (b) the rate of change of weight over time (the *growth rate* or *velocity*), and (c) the rate of change of the growth rate over time (or *acceleration*). Describe in your own words what each of the graphs shows about a child's growth.

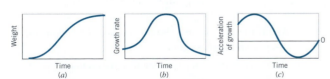

Source: Adapted from B. Bogin, "The Evolution of Human Childhood," *BioScience*, Vol. 40, p. 16.

27. The relationship between the velocity of a freely falling object and time is given by

$$v = -gt - 66$$

where g is the acceleration due to gravity and the units for velocity are centimeters per second.

   a. What value for g should be used in the equation?

   b. Generate a table of values for t and v, letting t range from 0 to 4 seconds.

   c. Graph velocity vs. time by hand and interpret your graph.

   d. What was the initial condition? Was the object dropped or thrown? Explain your reasoning.

28. A certain baseball is at height $h = 4 + 64t - 16t^2$ feet at time t in seconds. Compute the average velocity over each of the following time intervals and indicate for which intervals the baseball is rising and for which it is falling. In which interval was the average velocity the greatest?

   a. $t = 0$ to $t = 0.5$ **e.** $t = 2$ to $t = 3$

   b. $t = 0$ to $t = 0.1$ **f.** $t = 1$ to $t = 3$

   c. $t = 0$ to $t = 1$ **g.** $t = 4$ to $t = 4.01$

   d. $t = 1$ to $t = 2$

29. At $t = 0$, an object is in free fall 150 cm above the ground, falling at a rate of 25 cm/sec. Its height, h, is measured in centimeters above the ground.

   a. Is its velocity positive or negative? Why?

   b. Construct an equation that describes its height, h, at time t.

   c. What is the average velocity from $t = 0$ to $t = \frac{1}{2}$? How does it compare with the initial velocity?

30. (Graphing program required.) *The force of acceleration on other planets.* We have seen that the function $d = \frac{1}{2}gt^2 + v_0t$ (where g is the acceleration due to Earth's gravity and v_0 is the object's initial velocity) is a mathematical model for the relationship between time and distance fallen by freely falling bodies near Earth's surface. This relationship also holds for freely falling bodies near the surfaces of other planets. We just replace g, the acceleration of Earth's gravitational field, with the acceleration for the planet under consideration. The following table gives the acceleration due to gravity for planets in our solar system:

Acceleration Due to Gravity

	m/sec²	ft/sec²
Mercury	3.7	12.1
Venus	8.9	29.1
Earth	9.8	32.1
Mars	3.7	12.1
Jupiter	24.8	81.3
Saturn	10.4	34.1
Uranus	8.5	27.9
Neptune	11.6	38.1

Note: Pluto is no longer considered a planet.
Source: The Astronomical Almanac, U.S. Naval Observatory, 1981.

   a. Choose units of measurement (meters or feet) and three of the planets (other than Earth). For each of these planets, find an equation for the relationship between the distance an object falls and time. Construct a table as shown at the top of the next page. Assume for the moment that the initial velocity of the freely falling object is 0.

Name of Planet	Function Relating Distance and Time (sec)	Units for Distance

b. Using a graphing program, plot the three functions, with time on the horizontal axis and distance on the vertical axis. What domain makes sense for your models? Why?

c. On which of your planets will an object fall the farthest in a given time? On which will it fall the least distance in a given time?

d. Examine the graphs and think about the *similarities* that they share. Describe their general shape. What happens to d as the value for t increases?

e. Think about the *differences* among the three curves. What effect does the coefficient of the t^2 term have on the shape of the graph; that is, when the coefficient gets larger (or smaller), how is the shape of the curve affected? Which graph shows d increasing the fastest compared with t?

31. Suppose an object is moving with constant acceleration, a, and its motion is initially observed at a moment when its velocity is v_0. We set time, t, equal to 0, at this point when velocity equals v_0. Then its velocity t seconds after the initial observation is $V(t) = at + v_0$. (Note that the product of acceleration and time is velocity.) Now suppose we want to find its average velocity between time 0 and time t. The average velocity can be measured in two ways. First, we can find the average of the initial and final velocities by calculating a numerical average or mean; that is, we add the two velocities and divide by 2. So, between time 0 and time t,

$$\text{average velocity} = \frac{v_0 + V(t)}{2} \qquad (1)$$

We can also find the average velocity by dividing the change in distance by the change in time. Thus, between time 0 and time t,

$$\text{average velocity} = \frac{\Delta \text{distance}}{\Delta \text{time}} = \frac{d - 0}{t - 0} = \frac{d}{t} \qquad (2)$$

If we substitute the expression for average velocity (from time 0 to time t) given by Equation (1) into Equation (2), we get

$$\frac{d}{t} = \frac{v_0 + V(t)}{2} \qquad (3)$$

We know that $V(t) = at + v_0$. Substitute this expression for $V(t)$ in Equation (3) and solve for d. Interpret your results.

32. In 1974 in Anaheim, California, Nolan Ryan threw a baseball at just over 100 mph. If he had thrown the ball straight upward at this speed, it would have risen to a height of over 335 feet and taken just over 9 seconds to fall back to

Earth. Choose another planet and see what would have happened if he had been able to throw a baseball straight up at 100 mph on that planet. In your computations, use the table for the acceleration due to gravity on other planets from Exercise 30.

33. An object that is moving horizontally along the ground is observed to have (an initial) velocity of 60 cm/sec and to be accelerating at a constant rate of 10 cm/sec².

a. Determine its velocity after 5 seconds, after 60 seconds, and after t seconds.

b. Find the average velocity for the object between 0 and 5 seconds.

34. (Requires results from Exercise 33.) Find the distance traveled by the object described in the previous exercise after 5 seconds by using two different methods.

a. Use the formula distance = rate · time. For the rate, use the average velocity found in Exercise 33(b). For time, use 5 seconds.

b. Write an equation of motion $d = \frac{1}{2}at^2 + v_0 t$ using $a = 10$ cm/sec² and $v_0 = 60$ cm/sec and evaluate when $t = 5$. Does your answer agree with part (a)?

35. An object is observed to have an initial velocity of 200 m/sec and to be accelerating at 60 m/sec².

a. Write an equation for its velocity after t seconds.

b. Write an equation for the distance traveled after t seconds.

36. You may have noticed that when a basketball player or dancer jumps straight up in the air, in the middle of a blurred impression of vertical movement, the jumper appears to "hang" for an instant at the top of the jump.

a. If a player jumps 3 feet straight up, generate equations that describe his height above ground and velocity during his jump. What initial upward velocity must the player have to achieve a 3-foot-high jump?

b. How long does the total jump take from takeoff to landing? What is the player's downward velocity at landing?

c. How much vertical distance is traveled in the first third of the total time that the jump takes? In the middle third? In the last third?

d. Now explain in words why it is that the jumper appears suspended in space at the top of the jump.

37. Old Faithful, the most famous geyser at Yellowstone National Park, regularly shoots up a jet of water 120 feet high.

a. At what speed must the stream of water be traveling out of the ground to go that high?

b. How long does it take to reach its maximum height?

38. A vehicle trip is composed of the following parts:

i. Accelerate from 0 to 30 mph in 1 minute.

ii. Travel at 30 mph for 12 minutes.

iii. Accelerate from 30 to 50 mph in $\frac{1}{2}$ minute.

iv. Travel at 50 mph for 6 minutes.

v. Decelerate from 50 to 0 mph in $\frac{1}{2}$ minute.

a. Sketch a graph of speed versus time for the trip.

b. What are the average velocities for parts (i), (iii), and (v) of the trip?

c. How much distance is covered in each part of the trip, and what is the total trip distance?

39. In general, for straight motion of a vehicle with constant acceleration, a, the velocity, v, at any time, t, is the original velocity, v_0, plus acceleration multiplied by time: $v = v_0 + at$. The distance traveled in time t is $d = v_0 t + \frac{1}{2}at^2$.

a. A criminal going at speed v_c passes a police car and immediately accelerates with constant acceleration a_c. If the police car has constant acceleration $a_p > a_c$, starting from 0 mph, how long will it take to pass the criminal? Give t in terms of v_c, a_p, and a_c.

b. At what time are the police and the criminal traveling at the same speed? If they are traveling at the same speed, does it mean the police have caught up with the criminal? Explain.

Tables for Exploration 2.1 (p. 139) from:

The University of Massachusetts Boston

STATISTICAL PORTRAIT FALL 2006

Office of Institutional Research and Policy Studies (OIRP)

Undergraduate Admissions Summary

	1996	1997	1998	1999	2000	2001	2002	2003	2004	2005
FRESHMEN										
Applied	2,740	2,668	2,977	3,461	3,478	3,902	3,838	4,187	3,530	3,870
Decision Ready	2,347	2,305	2,466	2,724	2,667	2,652	2,704	2,834	2,903	3,174
Admitted	1,541	1,441	1,460	1,694	1,562	1,539	1,478	1,561	1,553	1,920
Admit Rate	65.7%	62.5%	59.2%	62.2%	58.6%	58.0%	54.7%	55.1%	53.5%	60.5%
Enrolled	743	637	674	789	706	701	576	610	565	781
Admitted but Deferred	57	44	61	30	32	14	23	13	30	15
Yield Rate	48.2%	44.2%	46.2%	46.6%	45.2%	45.5%	39.0%	39.1%	36.4%	40.7%
Not Admitted: Denied	806	864	1,006	1,030	1,105	1,113	1,226	1,273	1,350	1,254
Inc. Applications	393	363	511	737	811	1,250	1,134	1,353	627	696
TRANSFERS										
Applied	2,890	2,994	3,515	3,790	4,008	4,172	4,063	4,076	3,390	3,317
Decision Ready	2,510	2,576	2,872	2,964	3,038	2,916	2,892	2,779	2,697	2,639
Admitted	2,234	2,224	2,479	2,601	2,631	2,564	2,378	2,360	2,125	2,089
Admit Rate	89.0%	86.3%	86.3%	87.8%	86.6%	87.9%	82.2%	84.9%	78.8%	79.2%
Enrolled	1,371	1,347	1,574	1,590	1,556	1,542	1,382	1,339	1,193	1,326
Admitted but Deferred	121	102	161	89	113	45	92	20	67	51
Yield Rate	61.4%	60.6%	63.5%	61.1%	59.1%	60.1%	58.1%	56.7%	56.1%	63.5%
Not Admitted: Denied	276	352	393	363	407	352	514	419	572	550
Inc. Applications	380	418	643	826	970	1,256	1,171	1,297	693	678
TOTAL UNDERGRADUATES										
Applied	5,630	5,662	6,492	7,251	7,486	8,074	7,901	8,263	6,920	7,187
Decision Ready	4,857	4,881	5,338	5,688	5,705	5,568	5,596	5,613	5,600	5,813
Admitted	3,775	3,665	3,939	4,295	4,193	4,103	3,856	3,921	3,678	4,009
Admit Rate	77.7%	75.1%	73.8%	75.5%	73.5%	73.7%	68.9%	69.9%	65.7%	69.0%
Enrolled	2,114	1,984	2,248	2,379	2,262	2,243	1,958	1,949	1,758	2,107
Admitted but Deferred	178	146	222	119	145	59	115	33	97	66
Yield Rate	56.0%	54.1%	57.1%	55.4%	53.9%	54.7%	50.8%	49.7%	47.8%	52.6%
Not Admitted: Denied	1,082	1,216	1,399	1,393	1,512	1,465	1,740	1,692	1,922	1,804
Inc. Applications	773	781	1,154	1,563	1,781	2,506	2,305	2,650	1,320	1,374

Trend in New Student Race/Ethnicity in the College of Liberal Arts

	2001	2002	2003	2004	2005
Native American	0.6%	0.8%	0.4%	0.6%	0.6%
Asian/Pacific Islander	11.9%	11.9%	12.9%	12.8%	12.5%
Black	15.3%	14.0%	13.3%	15.3%	12.7%
Hispanic	7.0%	8.8%	8.0%	7.1%	11.8%
Cape Verdean	0.9%	1.0%	1.8%	1.9%	1.4%
Total minority	**35.7%**	**36.5%**	**36.4%**	**37.7%**	**39.0%**
White	55.7%	58.4%	60.1%	58.6%	59.1%
Non-resident alien	8.7%	5.1%	3.4%	3.7%	2.0%
Known race [N]	[1,509]	[1,244]	[960]	[885]	[1034]

SAT Scores of New Freshmen by College/Program

SAT Scores of New Freshmen by College/Program, 10-Year Trend (Excluding the DSP Program, Learning Disabled and Foreign Students)

		1996	1997	1998	1999	2000	2001	2002	2003	2004	2005
College of Liberal Arts	SATVerbal	518	502	513	521	515	524	520	522	528	540
	SATMath	500	504	511	513	529	536	532	519	517	539
	Combined	1,018	1,006	1,024	1,034	1,044	1,060	1,052	1,041	1,045	1,079
	[N]	[288]	[244]	[328]	[418]	[363]	[369]	[317]	[224]	[201]	[278]
College of Management	SATVerbal	510	450	468	481	516	511	515	492	487	511
	SATMath	541	521	546	537	548	556	546	542	523	554
	Combined	1,051	971	1,014	1,018	1,064	1,067	1,061	1,034	1,010	1,065
	[N]	[34]	[51]	[54]	[43]	[42]	[34]	[37]	[36]	[35]	[59]

SOLUTIONS

CHAPTER 1

Section 1.1

Algebra Aerobics 1.1

1.

Fraction	Decimal	Percent
$\frac{7}{12}$	0.583	58.3%
$\frac{1}{40}$	0.025	2.5%
$\frac{1}{50}$	0.02	2%
$\frac{1}{200}$	0.005	0.5%
$\frac{7}{20}$	0.35	35%
$\frac{1}{125}$	0.008	0.8%

2. a. 500 people

b. 38.94 or 39 students

c. 37.5%

3. mean = $25,040; median = $20,000

4. mean GPA ≈ 2.06; median GPA = 2.0

5. a. The numbers (in millions) for the Hispanic population.

b. 2000: $\frac{35.3}{0.125}$ = 282.4 million; 2005: $\frac{42.7}{0.144}$ ≈ 296.5 million

6. a. Frequency Count (FC) for Age 1–20 interval is 38% of total: 38% of 137 = 0.38(137) ≈ 52.

FC for Age 61–80 interval is total FC minus all the others: 137 − (52 + 35 + 28) = 137 − 115 = 22.

Relative Frequency (RF) for each interval is its FC divided by total:

RF of (21–40) interval is: $\frac{35}{137}$ ≈ 0.255 ≈ 26%

RF of (41–60) interval is: $\frac{28}{137}$ ≈ 0.204 ≈ 20%

RF of (61–80) interval is: $\frac{22}{137}$ ≈ 0.161 ≈ 16%

Age	Frequency Count	Relative Frequency (%)
1–20	52	38
21–40	35	26
41–60	28	20
61–80	22	16
Total	137	100

Table 1.3

b. 20% + 16% = 36%

7.

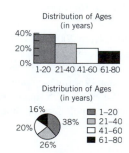

8. Table for the histogram in Figure 1.6

Age	Relative Frequency (%)	Frequency Count
1–20	20	(0.20)(1352) ≈ 270
21–40	35	(0.35)(1352) ≈ 473
41–60	30	(0.30)(1352) ≈ 406
61–80	15	(0.15)(1352) ≈ 203
Total	100	1352

9. a. sum = $8750, so mean = $8750/9 ≈ $972.22; median = $300

b. sum = 4.7, so mean = 4.7 ÷ 8 ≈ 0.59; median = (0.4 + 0.5)/2 = 0.45

10. One of the values ($6,000) is much higher than the others which forces a high value for the mean. In cases like this, the median is generally a better choice for measuring central tendency.

Exercises for Section 1.1

1. a. 34.7% of 13- to 17-year-old females spent at least 3 hours per day watching TV in 2006.

b. 39.9%

c. The total numbers of male and female teenagers in that age range.

3. a. Paying off bills/debts. The other categories add up to 62%, not 66%.

b. They do not add up to 100%. Also, the pie chart is 3D and tipped. Thus the front slices are disproportionately larger than the other slices.

5. a. 45

b. Quantitative data

c. $\frac{3}{45}$ ≈ 6.7%

d.

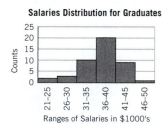

Salaries Distribution for Graduates

7. a. In housing; 36.16%

b. 0.1438 · $35,000 = $5033

c. Answers will vary. Americans spend over 36% of their income for housing.

9. a. Mean = 386/7 = 55.14; median = 46.

b. Changing any entry in the list that is greater than the median to something still higher will not change the median of the list but will increase the mean. The same effect can be had if an entry less than the median is increased to a value that is still less than or equal to the median.

11. The mean annual salary is $24,700 and the median annual salary is $18,000. The mean is heavily weighted by the two high salaries. The mean salary is more attractive but is not likely to be an accurate indicator.

13. No answer is given here. (In general, when answers from students can vary quite a bit, either a typical answer or none is given.)

15. The mean age in the United States is slightly higher than the median age since there are a lot of older Americans (including the baby boomers), which pulls the mean age higher than the median. In developing countries, the mean age will be less than the median age since there are a lot of younger people and this pulls the mean lower than the median. Answers will vary by State and will depend on how the ages are distributed.

17. He is correct, provided the person leaving state A has an IQ that is below the average IQ of people in state A and above the average IQ of the people in state B.

19. The mean net worth of a group, e.g., American families, is heavily biased upward by the very high incomes of a relatively small subset of the group. The median net worth of a group such as this is not as biased. The two measures would be the same if net worths were distributed symmetrically about the mean.

21. **a.** $\left(\sum_{i=1}^{5} x_i\right)/5$ **b.** $\left(\sum_{i=1}^{n} t_i\right)/n$ **c.** $\sum_{k=1}^{5} 2k = 30$

23. Here is a progressive table of the work needed. The mean age of 36.6 years was obtained by dividing total people-years by the total number of persons. Total people-years was obtained by adding up the products of midpoint ages by the population counts in each age group.

Age (years)	Population (thousands)	Midpoint Age	Product of Midpoint Age and Population (people-years)
Under 10	39,677	4.5	178,546
10 to 19	41,875	14.5	607,187
20 to 29	40,532	24.5	993,034
30 to 39	41,523	34.5	1,432,543
40 to 49	45,179	44.5	2,010,465
50 to 59	35,986	54.5	1,961,237
60 to 74	31,052	67	2,080,484
75 to 84	12,971	79.5	1,031,194
85 and over	4,860	92.5	449,550
Total	293,655	n.a.	10,744,242
		Mean age	36.6

25. If we assume an estimated allowance of $5 for all students with an allowance in the $0–9 category, an estimate of $15 for all students with an allowance in the $10–19 category, etc., then the mean $\approx$ ($5 · 8 + $15 · 6 + $25 · 12 + $35 · 14 + $45 · 9)/49 = 1325/49 $\approx$ $27. The median lies somewhere between $20 and $29, probably closer to $29.

27. Answer is omitted. Student answers will vary.

29. Some factors to note are given below. You may find others. It is important in giving your answer to cite some population numbers and specific age brackets when making specific comparisons.

   1) In Ghana there is a steady decline in the number in each age group from 5 to 9 up to over 80 (from approx. 2.7 million to 0.1 million), whereas in the United States the number of people in each age category is generally the same or more than the number of people 0 to 4 years of age (approx. 19 million) up to the ages 45 to 49 (approx. 23 million).

   2) At almost all age levels there seems to be an even distribution of males and females in Ghana; it is somewhat the same in the United States except that from age 60 onward there is a dominance of females in each age category.

   3) There is a small bulge in population in the teens as compared to those younger and those just older in the United States (by about 2 million) as compared to a steady decline in Ghana in the population in all age groups once past the 5 to 9 age bracket.

Section 1.2

Algebra Aerobics 1.2a

1. **a.** The median net worth of households, after decreasing from 1988 to 1993, increased from 1993 to 2000. It reached a low of about $43,600 in 1993 and increased to $55,000 in 2000.

   b. Many factors would be useful, including size of households and number of wage earners in household.

2. **a.** 1975

   b. 2030

   c. 55 years

3. **a.** approximately 11 billion

   b. approximately 1 billion

   c. approximately $11 - 1 = 10$ billion

   d. The total world population increased rapidly from 1950 to 2000, and is projected to continue to increase but at a decreasing rate, reaching approximately 11 billion in the year 2150.

4. **a.** $\frac{\text{pop } 2000}{\text{pop } 1900} \approx \frac{6}{1.5} = 4$. The population of the world in year 2000 was approximately 4 times greater than in 1900. The difference $\approx$ 4.5 billion.

   b. $\frac{\text{pop } 2100}{\text{pop } 2000} \approx \frac{10.5}{6} = 1.75$, so the population in 2100 was approximately 1.75 times greater than in 2000. The difference $\approx$ 4.5 billion.

   c. The population from 2000 to 2100 is expected to grow by 4.5 billion people, which is the same as the 4.5 billion increase from 1900 to 2000. The world population in 2000 was approximately 4 times greater than in 1900 and it is projected to be about 1.75 times greater in 2100 than in 2000. While the population continues to increase, the rate of increase is slowing down.

Algebra Aerobics 1.2b

1. a. Square the value of x, then multiply that result by 3, then subtract the value of x and add $+1$.

b. $(0, 1)$ is the only one of those ordered pairs that is a solution.

c.

x	-3	-2	-1	0	1	2	3
y	31	15	5	1	3	11	25

2. a. Subtract 1 from the value of x, then square the result.

b. $(0, 1)$ and $(1, 0)$ are the only ones of those ordered pairs that are solutions.

c.

x	-3	-2	-1	0	1	2	3
y	16	9	4	1	0	1	4

3.

x	-4	-2	-1	0	1	2	4
y_1	16	10	7	4	1	-2	-8
y_2	-15	3	6	5	0	-9	-39

a. & b.

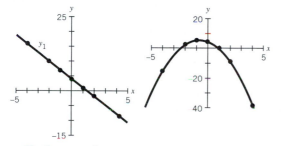

c. Yes for y_1; no for y_2.

d. No for y_1; yes for y_2.

e. No, since $4 - 3(-3) = 13$, $-2(-3)^2 - 3(-3) + 5 = -4$

4.

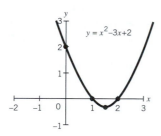

a. $y = -1/4$

b. Many answers: $(0, 0)$ and $(-2, 5)$ are examples

Exercises for Section 1.2

1. a. Answers may vary.

1) AIDS cases in the United States have nearly halved from 1993 to 2004, since they dropped from an all-time high of 79,879 in 1993 to 42,514 in 2004.

2) AIDS cases increased slightly from 2001 to 2004.

3) AIDS cases reached a low of about 39,200 in 2001.

b. Student answers will vary. Possible topic sentences: AIDS cases in the United States went down from 1993 to 2001 but then started to rise very slowly. The three points

mentioned in part (a) could form the rest of the paragraph. A concluding sentence might be: "It is worrisome that the number of cases has been on the rise from 2001 to 2004. The number of cases has increased by a little over 3000 from 2001 to 2004."

3. a.

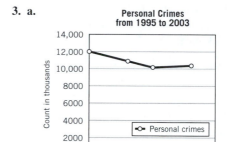

b. In 1995 there were approximately 6.7 times more property crimes than violent crimes. In 2003 there were approximately 7.6 times more property crimes than violent crimes.

c. Answers may vary. Some observations could be the following. Property crime, from 1995 to 2003, was over six times as great as personal crime. Both property crimes and personal crimes decreased from 1995 to 2003, but from 1995 to 2003 the ratio of property to personal crimes actually increased.

5. a. As men grow older, the risk of cancer goes up dramatically—for example, from 1 in 25,000 at age 45, to 1 in 25 at age 65, to 1 in 6 at age 80.

b.

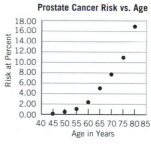

c. The ratio of risks for a 50- vs. a 45-year-old man is $0.21\%/0.004\% = 52.5$. This means that a 50-year-old man is 52.5 times more likely to have had prostate cancer than a 45-year-old man. The ratio of risks for a 55- vs. 50-year-old man is $0.83\%/0.21\% \approx 4$. So a 55-year-old man is only 4 times as likely to have had prostate cancer than a 50-year-old man. At first this may seem contradictory. But the biggest incremental risk occurs between 45 and 50. So while the absolute risk continues to rise (note that the percent risk is cumulative), the ratios of the percent risk decrease over subsequent 5-year age intervals.

d. The medical profession has recommended this test for all men from age 40 up. The insurance companies think that this is too expensive. Your answer may be different.

7. a. Except for a dip in the count for 12- to 19-year-olds in the 1976 to 1980 period there has been a dramatic rise in the percentage of children who are overweight.

b. It increased in going from every time interval to the next, except in going from the first (1963 to 1970) to the second (1971 to 1974), when the percentage stayed the same.

c. From the period of 1963 to 1970 to the period of 1971 to 1974.

d. During the 1976–1980 period.

e. The goal is reasonable, but the data suggest that the trend in obesity will keep rising.

9. a. True; B is the newer car and it costs more than A.

b. True; A is the slower car in cruising speed and it has the larger size.

c. False; A is the larger car but it is older than B.

d. True; A carries more passengers and it is less expensive.

e. Your answers will vary. You may mention that the larger range car also has the larger passenger capacity.

f. Your answers will vary. Much depends on what features you value. There are many trade-offs.

11. a. Only $(1, -3)$ satisfies the formula.

b. There are many answers: e.g., $(0, 2)$, $(-1, 7)$. In general, pick a value of T and plug it into the formula to compute the corresponding value of R.

c. Here is the scatter plot for the three points given:

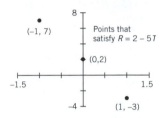

d. The plot suggests that solutions could be found on the straight line through these three points by eyeballing the line. Another source would be using the formula with other values of T to generate the corresponding values of R.

13. a. Add 1 to the value of x; divide the result by what one gets by subtracting 1 from the value of x.

b. $(5, 1.5)$

c. $(2, 3)$

d. No, the formula is not defined if $x = 1$.

15. a. $x = 0$ implies $y = 0$

b. If $x > 0$ then $y < 0$.

c. If $x < 0$ then $y < 0$.

d. No

17. a. Only $(-1, 3)$ satisfies $y = 2x + 5$.

b. $(1, 0)$ and $(2, 3)$ satisfy $y = x^2 - 1$.

c. $(-1, 3)$ and $(2, 3)$ satisfy $y = x^2 - x + 1$

d. Only $(1, 2)$ satisfies $y = 4/(x + 1)$.

19. a. There are far more people now than in 1971, and those who are most susceptible, the elderly, are now a greater proportion of the population than in 1971. Thus, it is not necessarily the case that proportionately twice as many now as then will have cancer. Also, better cancer detection

tools could mean more frequent prevention of death. Lastly, not all cancers are fatal.

b. Most of the observations given by William M. London are cited in part (a) but his arguments make claims that are not buttressed clearly enough by data. We need to have data for cancer incidence, cancer treatment, cancer conquest, and cancer fatality, for all age groups by sex and type of cancer before we can make any solid overall claims either way.

c. The statements given above in parts (a) and (b) can be used to forge a paragraph.

Section 1.3

Algebra Aerobics 1.3

1. Table A represents a function. For each input value there is one and only one output value.

Table B does not represent a function. For the input of 2 there are two outputs, 7 and 8.

2. There are two output values, 5 and 7, for the input of 1. For the table to represent a function, there can be only one output value for each input. If you change 5 to 7 *or* change 7 to 5, then there will be only one output value for the input of 1.

3. Yes, it passes the vertical line test.

4. Graph B represents a function, while graphs A and C do not represent functions since they fail the vertical line test.

5. Neither is a function of the other. The graph fails the vertical line test, so weight is not a function of height. At height 51 inches there are two weights (115 and 120 pounds) and at height 56 inches there are two weights (135 and 140 pounds). If we reverse the axes, so that weight is on the horizontal axis and height is on the vertical axis, that graph will also fail the vertical line test, since the weight of 140 pounds has two corresponding heights of 56 inches and 58 inches.

6. a. D is a function of Y, since each value of Y determines a unique value of D.

b. Y is not a function of D, since one value of D, \$2.70, yields two values for Y, 1993 and 1997.

7. a.

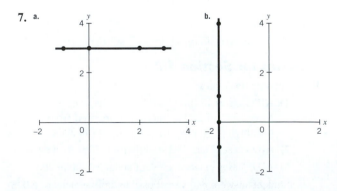

a. The line containing the points is the graph of a function since for each value of x we have a unique value of y. The line is horizontal. Its equation can be written as $y = 3$.

b. The line containing the points is not the graph of a function since for one value of x we have infinitely many values of y. The line is vertical. Its equation can be written as $x = -2$.

8. a. Tip = 15% of the cost of meal, so $T = 0.15M$. Independent variable: M (meal price); Dependent variable: T (tip).

The equation is a function since to each value of M there corresponds a unique value of T.

b. $T = (0.15)(\$8) = \$1.20$

c. $T = (0.15)(\$26.42) = \$3.96$. One would probably round that up to $4.00.

Exercises for Section 1.3

1. a. Yes, each date has one and only one temperature.

b. No, the temperature 27°C goes with two different dates.

3. a. function (all input values are different and thus each input has one and only one output)

b. function [same reason as in part (a)]

c. not a function [same input values have different output values]

d. not a function [same reason as in part (c)]

5. a. Not a function: fails the vertical line test, e.g., look at the y-axis.

b. Function: passes the vertical line test.

c. Not a function: fails the vertical line test, e.g., look at the y-axis

7. a. The formulas are: $y = x + 5$; $y = x^2 + 1$; $y = 3$

b. All three represent y as a function of x; each input of x has only one output y.

9. a. $S_1 = 0.90 \cdot P$; \$90

b. $S_2 = (0.90)^2 \cdot P$; \$81

c. $S_3 = (0.90)^3 \cdot P$; \$72.90

d. $S_5 = (0.90)^5 \cdot P$; \$59.05; 40.95%

11. y is a function of x in parts (a), (b), and (c) but not in (d). In (d), for example, if $x = 1$, then $y = \pm 1$.

13. a. Since the dosage depends on the weight, the logical choice for the independent variable is W (expressed in kilograms) and for the dependent variable is D (expressed in milligrams).

b. In this formula, each value of W determines a unique dosage D, so D is a function of W.

c. The following table and graph are representations of the function. Since the points $(0, 0)$ and $(10, 500)$ are not included in the model, these points are represented with a hollow circle on the graph.

W (kg)	D (mg)
0	0
2	100
4	200
6	300
8	400

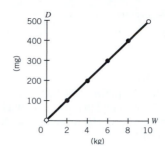

d. The domain and range are now all the real numbers. The table can now include values of W that are less than or equal to zero and values greater than or equal to 10. The graph will continue indefinitely in both directions.

Section 1.4

Algebra Aerobics 1.4a

1. $g(0) = 0$
$g(-1) = -3$
$g(1) = 3$
$g(20) = 60$
$g(100) = 300$

2. $f(0) = (0)^2 - 5(0) + 6 = 6$, so $f(0) = 6$
$f(1) = (1)^2 - 5(1) + 6 = 2$, so $f(1) = 2$
$f(-3) = (-3)^2 - 5(-3) + 6 = 30$, so $f(-3) = 30$

3. $f(0) = \frac{2}{0-1} = -2$, so $f(0) = -2$;
$f(-1) = \frac{2}{(-1)-1} = -1$, so $f(-1) = -1$;
$f(1) = \frac{2}{(1)-1} = \frac{2}{0}$, so $f(1)$ is undefined;
$f(-3) = \frac{2}{(-3)-1} = -\frac{1}{2}$, so $f(-3) = -1/2$

4. $5 - 2t = 3$, so $t = 1$; $3t - 9 = 3$, so $t = 4$; $5t - 12 = 3$, so $t = 3$.

5. $2(x - 1) - 3(y + 5) = 10 \Rightarrow 2x - 2 - 3y - 15 = 10 \Rightarrow$
$2x - 17 - 3y = 10$; $2x - 27 = 3y \Rightarrow y = \frac{2x - 27}{3}$
So y is a function of x. $f(x) = \frac{2x - 27}{3}$.

6. $x^2 + 2x - y + 4 = 0 \Rightarrow x^2 + 2x + 4 = y$.
So y is a function of x. $f(x) = x^2 + 2x + 4$

7. $7x - 2y = 5 \Rightarrow -2y = -7x + 5 \Rightarrow y = (-7x + 5)/(-2) \Rightarrow$
$y = \frac{7x - 5}{2}$
So y is a function of x. $f(x) = \frac{7x - 5}{2}$

8. $f(-4) = 2$; $f(-1) = -1$; $f(0) = -2$; $f(3) = 1$. When $x = -2$ or $x = 2$, then $f(x) = 0$.

9. $f(0) = 20$, $f(20) = 0$, $f(x) = 10$ when $x = 10$ and $x = 30$. It is a function because for every value of x, there is one and only one value of $f(x)$.

Algebra Aerobics 1.4b

1. a. $(2, \infty)$

b. $[4, 20)$

c. $(-\infty, 0] \cup (500, +\infty)$

2. a. $-3 \le x < 10$

 b. $-2.5 < x \le 6.8$

 c. $x \le 5$ or $x \ge 12$

3. a. $[2.5, 3.6]$

 b. $[0.333, 1.000]$ (*Note:* The highest possible batting average is 1.000, meaning that the batter had a hit every time he has been at bat. This is commonly known as "batting a thousand.")

 c. $[35000, 50000]$

4. $2(x + 1) + 3y = 5 \Rightarrow 2x + 2 + 3y = 5 \Rightarrow$

$$3y = -2x + 3 \Rightarrow$$
$$y = \frac{-2x + 3}{3}$$

So y is a function of x. The domain and range are all real numbers. $f(x) = \frac{-2x + 3}{3}$

5. $x + 2y = 3x - 4 \Rightarrow 2y = 2x - 4 \Rightarrow y = x - 2$

So y is a function of x. The domain and range are all real numbers. $f(x) = x - 2$

6. $y = \sqrt{x}$

So y is a function of x. The domain and range are all real numbers ≥ 0. $f(x) = \sqrt{x}$

7. $2xy = 6 \Rightarrow y = \frac{6}{2x} = \frac{3}{x}$. So y is a function of x. The domain is all real numbers except 0. The range is all real numbers except 0. $f(x) = \frac{3}{x}$

8. $6\left(\frac{x}{2} + \frac{y}{3}\right) = 6(1) \Rightarrow 3x + 2y = 6 \Rightarrow$

$$2y = 6 - 3x \Rightarrow y = \frac{6 - 3x}{2}$$

So y is a function of x. The domain and range are all real numbers. $f(x) = \frac{6 - 3x}{2}$

9. $f(x)$ is undefined at $x = -5$; domain: $(-\infty, -5)\cup(-5, +\infty)$; range: all real numbers except 0.

$g(x)$ is undefined at $x = -1$; domain: $(-\infty, -1)\cup(-1, +\infty)$; range: all real numbers except 0.

$h(x)$ is undefined at $x < 10$; domain: $[10, +\infty)$; range: $[0, +\infty)$.

Exercises for Section 1.4

1. a. $T(0) = 2, T(-1) = 6, T(1) = 0, T(-5) = 42$

3. a. Tax $= 0.16 \cdot$ Income

 b. Income is the independent variable and Tax is the dependent variable.

 c. Yes, the formula represents a function: for each input there is only one output.

 d. As the Income gets closer and closer to $20,000, the Tax gets closer and closer to $3200. In fact the Tax, at some point will round off to $3200.00, even though the Income is not quite $20,000. Its domain is [0, 20000] and its range is [0, 3200].

5. The equation is $C = 2.00 + 0.32M$. It represents a function. The independent variable is M measured in miles. The dependent variable is C measured in dollars. Here is a table of values.

Miles	Cost ($)	Miles	Cost ($)
0	2.00	30	11.60
10	5.20	40	14.80
20	8.40	50	18.00

The graph is in the accompanying diagram. Some of the table values are marked.

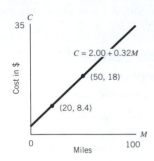

7. a. $f(2) = 4$ **c.** $f(0) = 2$

 b. $f(-1) = 4$ **d.** $f(-5) = 32.$

9. a. $p(-4) = 0.063, p(5) = 32$ and $p(1) = 2.$

 b. $n = 1$ only

11. a. $f(-2) = 5, f(-1) = 0, f(0) = -3,$ and $f(1) = -4.$

 b. $f(x) = -3$ if and only if $x = 0$ or 2.

 c. The range of f is from -4 to $+\infty$ since we may assume that its arms extend out indefinitely.

13. $f(0) = 1, f(1) = 1,$ and $f(-2) = 25$

15. a. $f(0) = 1, g(0) = 1$

 b. $f(-2) = 2, g(-3) = 10$

 c. $f(2) = 0, f(1) = 0.5$

 d. $f(3) = -0.5, g(3) = 10$

17. a. $x = -2$ and 2   **b.** $x = 2$ **c.** $x = 0$ and 2

Section 1.5

Algebra Aerobics 1.5

1. *Possible titles:*

 a. From 2004 to 2006 Prices Decreased 50% for LCD TVs

 b. iTunes Sales Increase Tenfold From December 2003 to January 2005

2. a. maximum value: approx. $46,000 in 1999; (1999, $46000)

 b. minimum value: approx. $40,000 in 1993; (1993, $40000)

 c. Median household income decreased from 1990 to 1993, reaching a low point of about $40,000 in 1993. From 1993 to 1999 it increased steadily, reaching a high point of about $46,000 in 1999, which was followed by a decrease until 2004.

3. Graph B is the best match for the situation. It is the only graph that represents the child stopping at the top of the slide with a speed of zero for a few minutes.

4.

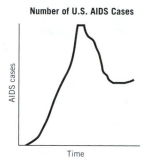

Number of U.S. AIDS Cases

5. a. Graph that is concave up with minimum at $(-2, 1)$

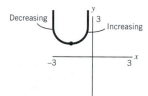

b. Graph that is concave down with maximum at $(3, -2)$

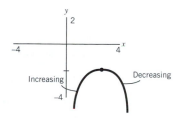

Exercises for Section 1.5

1. a. Graph B   **c.** Graph A
  b. Graph A   **d.** Graph B

3. The price of Rockwell Collins hovered under $30 at the beginning of the week and then rose to $32 on Wednesday and then slowly climbed for the rest of the week. The price of Transkaryotic, by contrast, rose from $16.50 on Friday to $17 on Monday and then dropped to $13 on Tuesday and hovered near $13 for the rest of the week.

5. a. Positive over $(-5, 0)$ and $(5, +\infty)$
  b. Negative over $(0, 5)$ and $(-\infty, -5)$
  c. Decreasing over $(-3, 3)$
  d. Increasing over $(-\infty, -3)$ and $(3, +\infty)$
  e. There is no minimum.
  f. There is no maximum.

7. a. A, domain $= (-\infty, +\infty)$ and range $= (-\infty, 2)$
    B, domain $= [0, +\infty)$ and range $= [0, +\infty)$
    C, domain $= (-\infty, +\infty)$ and range $= (0, +\infty)$
    D, domain $= (-\infty, +\infty)$ and range $= [2, +\infty)$
  b. A, $(-5, -1)$, B, $(0, +\infty)$, C, $(-\infty, +\infty)$, D, $(-\infty, +\infty)$
  c. A, $(-\infty, -5)$ and $(-1, +\infty)$ B, nowhere
    C, nowhere   D, nowhere

9. a. $[-6, -3)$ and $(5, 11)$
  b. $(-3, 5)$ and $(11, 12]$
  c. $(-6, 2)$ and $(8, 12)$
  d. $(2, 8)$

e. Concave down approximately over interval $(0, 5)$; concave up approximately over interval $(5, 8)$.
  f. $f(x) = 4$ when $x = 1$ and 3.
  g. $f(-8)$ is not defined.

11. a. Graph B seems best. It indicates several stops, rises and falls in speed and, most importantly, it is the only one that ends with a stop.
  b. Graph E is suitable. The horizontal parts on the graph indicate a time at which the bus stopped. Graphs D and F seem out of place. They would indicate that the bus went backwards.

13. a. Graph of $h(x) = x^4$

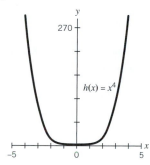

  b. Graph of $k(x) = x^4 - 24x + 50$

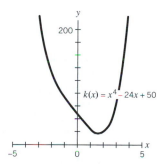

The graphs of $h(x)$ and $k(x)$ are everywhere concave up.

15. a. 1996 to 1997, 1998 to 1999, 2000 to 2001.
  b. 1995 to 1996, 1997 to 1998, 1999 to 2000, 2001 to 2003.
  c. 17 billion dollars in 1999.
  d. 10 billion dollars in 2003.

17. a. Johnsonville's population goes from $2.4 \cdot 100{,}000 = 240{,}000$ to $5.8 \cdot 100{,}000 = 580{,}000$. Palm City's population ranges from $1.8 \cdot 100{,}000 = 180{,}000$ to a high of $3.8 \cdot 100{,}000 = 380{,}000$. (This notation adheres to what is found in the graph.)
  b. The population of Palm City increased from 1900 to 1930.
  c. The population of Palm City decreased from 1930 to 1990.
  d. The two populations were equal sometime around 1940.

19. a. Yes, P is a function of Y since the inputs are all distinct.
  b. The domain is the set of years from 1990 to 1995 inclusive; the range is the set of corresponding values, namely $\{-0.5, 0, 1.2, 1.4, 2.3\}$.
  c. The maximum P value is 2.3; it occurs when $Y = 1991$.
  d. P is increasing from 1990 to 1991 and from 1993 to 1994. It is decreasing from 1991 to 1993 and 1994 to 1995.
  e. Y is not a function of P, since the two inputs of 1.4 have different outputs.

21. a. From 1750 to 2100.

   b. Yes, from 2100 to 2200.

   c. Student answers will vary; it is hoped that something is said about the dramatic increase and the slowing down noted in parts (a) and (b).

23. a. Male and female enlisted reserve personnel reached their respective maxima in 1992.

   b. The maximum for males was approximately 1350 (thousand) and for women it was approximately 215 (thousand).

   c. For men, there was a steady rise from 1990 to 1992 and then a gradual decline, reaching a low of 800,000 in 2004. For women there was a sharp rise from 1990 to 1992; then a steady decline to 175,000 until 1998; and then a slight rise from 1998 to 2001, when it climbed to 175,000 and a gradual decline since then to 2004, when it reached 170,000.

   d. Answers will vary.

25.

Here is one graph. Phrases like "after a while" and "for a while" were loosely interpreted. As for "levels of mobility," the idea was to give a good picture by means of a graph without being held to numerical values. Time intervals, when explicitly given, were respected. There are discontinuities in the going-to-crutches phases.

27. Here are the graphs asked for:

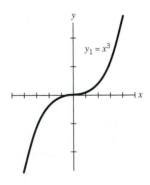

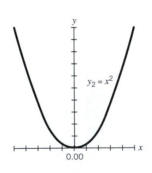

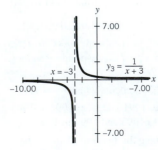

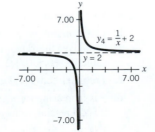

a. As $x \to +\infty$

   y_1 approaches $+\infty$, y_2 approaches $+\infty$,
   y_3 approaches 0; y_4 approaches 2

b. As $x \to -\infty$

   y_1 approaches $-\infty$, y_2 approaches $+\infty$,
   y_3 approaches 0 y_4 approaches 2.

29. Clearly one cannot simulate the session here. But one can give the graphs of f and g (see the accompanying diagram) and note that the graph of g is the mirror image of the graph of f with respect to the x-axis.

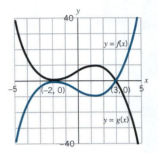

Ch. 1: Check Your Understanding

1. True	**8.** False	**15.** True	**22.** True
2. False	**9.** True	**16.** False	**23.** True
3. True	**10.** False	**17.** True	**24.** False
4. True	**11.** True	**18.** False	**25.** True
5. True	**12.** False	**19.** False	**26.** True
6. True	**13.** False	**20.** True	**27.** True
7. False	**14.** True	**21.** True	

28. Possible answer: $z^4 = 2w + 3$

29. Possible answer:

z	1	2	3	4
w	10	-3	9	-3

30. Possible answer:

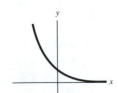

31. Possible answer:

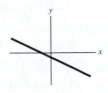

32. Possible answer:

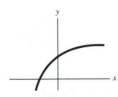

33. Possible answer:

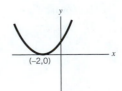

34. Possible answer: The number of wolf pups born and the number that have survived in Yellowstone National Park have both increased significantly since 1995.

35. False   **37.** False   **39.** False   **41.** True

36. False   **38.** True   **40.** False   **42.** True

Ch. 1 Review: Putting It All Together

1. a. 25%

   b. 20%

   c. 40 lb is a smaller percentage of 200 lb (his weight at age 60) than it is of 160 lb (his weight at age 20).

3. This could be explained by changes in the distribution of incomes. For example, some incomes (originally below the median) could have increased to more than the previous median (to raise the median), while other incomes decreased on either side of the median (to lower the mean). To simultaneously decrease the mean while raising the median, the decrease in incomes would need to be greater than the increase in incomes. For example, for a small set of numbers such as {5, 15, 20, 50, 110}, the median is 20 and a mean is 40. If we change the set such that 5 increases to 25 and 110 decreases to 80, then the new set of {15, 20, 25, 50, 80} would have a median of 25 and a mean of 38.

5. a. Heart disease

   b.

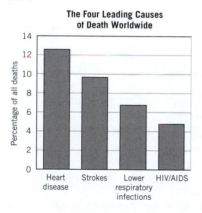

The Four Leading Causes of Death Worldwide

   c. Sixty-six percent of the deaths were not accounted for. For example, tuberculosis, malaria, lung cancer, war, and car accidents are not included. No other cause can account for more than 5% of deaths, or else it would supersede HIV/AIDS and be one of the four leading causes.

7. $E = 0.025P$

9. Answers will vary. Possible title and summary: "The Popularity of the Name Emma over Time." Emma was a very popular name in 1880, when it was given to about 9,000 out of every million babies. Its use fell rapidly until it bottomed out in the 1970s, when it was given to about 200 out of every million babies. It has been rising in popularity since the 1980s, although it was only about half as popular in 2004 as it had been in 1880.

11. a. i. Function

   ii. Not a function

iii. Function

iv. Not a function

   b, c. Answers will vary.

13. a. Approx. 51%.

   b. In November 2004.

   c. There was a major increase between September and November of 2004; other lesser increases occurred between June and August of 2004 and between February 2005 and April 2005.

   d. In June 2005 there was a difference of about 30% (approx. 63% minus 33%).

15. a.

d	1	2	3	4	10	20	100
y	8	16	24	32	80	160	800

   b. $y = 8d$ (d independent, y dependent)

   c.

y	8	12	16	20	50	80	100
d	1	1.5	2	2.5	6.25	10	12.5

   d. $d = (1/8)y$ (y independent, d dependant)

17.

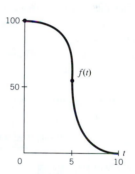

19. a. Yes, it passes the vertical line test. Domain: [1970, 2010]; approx. range: [41, 63].

   b. Approx. 1990; approx. 63 years of age.

   c. Approx. 2003 (projected); approx. 41 years of age. AIDS is a likely candidate.

   d. Increasing from approx. 1970 to 1990 and projected to increase from 2003 to 2010; decreasing from 1990 to 2003.

   e. Concave down over interval (1970, 1997); concave up over interval (1997, 2010).

   f. In 1970 the life expectancy in Botswana was about 52 years, as opposed to about 57 years worldwide. Over the next 20 years this gap gradually closed. In 1990 the life expectancy in Botswana was about 63, and 64 was the worldwide average. However, after 1990 the life expectancy in Botswana began to rapidly decline; this decline was expected to continue until about 2003, at which time it was projected to begin to increase slowly.

21. a. It means that 3% of the worldwide out-of-school population is in Latin America or the Caribbean.

   b. It does not. It tells us only what percentage of the worldwide out-of-school population is in sub-Saharan Africa.

c. This graph does not show the number of children out of school worldwide or within each region. Nor does it show the percentage of children within each region who are out of school. The number of children out of school in a particular region may be relatively small compared to the total worldwide out-of-school population, but quite large compared to the total population of the region. This could be confusing.

d. Sub-Saharan Africa has the largest percentage of the world's out-of-school population. Seventy-three percent of the world's out-of-school population reside in sub-Saharan Africa or South Asia. Industrialized nations represent the smallest percentage.

e. Answers will vary. One idea is that UNICEF should consider focusing on sub-Saharan Africa and South Asia. This is where almost three-quarters of the children who are out of school can be found. Evidence from the graph should be cited.

23. a. $f(0) = -\frac{1}{2}$, $f(-1) = -\frac{1}{3}$, $f(2)$ is undefined

b. Domain: all real numbers, $x \neq 2$
Range: all real numbers, $f(x) \neq 0$.

25. a. February has the maximum cost and July the lowest. This makes sense, as heating costs would be less in the warm months and more in the cold ones.

b. Rounded off: mean = $104.67 and median = $86. Since the mean is higher than the median, together the mean and median suggest that there are a few expensive months, which account for most of the cost of gas for the year.

c.

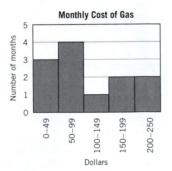

Monthly Cost of Gas

d. Answers will vary. Some patterns: For one-third of the year, the cost of gas is between $150 and $250. Gas costs peak in February and then start to decrease throughout the spring and summer. In August they begin to increase, continuing throughout the fall and early winter.

CHAPTER 2

Section 2.1

Algebra Aerobics 2.1

1. $\frac{(143 - 135)\,\text{lb}}{5\,\text{yr}} = 1.6\frac{\text{lb}}{\text{yr}}$

2. a. i. $\frac{\$(820.2 - 537.2)\text{billion}}{(2006 - 1990)} = \frac{\$283\text{ billion}}{16\text{ yr}} \approx \$17.69$ billion/yr

ii. $\frac{\$(1{,}273.2 - 618.4)\text{billion}}{(2006 - 1990)} = \frac{\$654.8\text{ billion}}{16\text{ yr}} \approx \$40.93$ billion/yr

iii. $\frac{\$(-453.0 - (-81.2))\text{billion}}{(2006 - 1990)} = \frac{-\$371.8\text{ billion}}{16\text{ yr}}$

$\approx -\$23.24$ billion/yr

b. While both exports and imports were increasing, imports were increasing at a faster rate, and hence, the trade balance was decreasing. The trade deficit increased by over $23 billion/year from 1990 to 2006.

3. a. $\frac{(41.8 - 52.1)\text{thousand deaths}}{20\text{ yrs}} \approx \frac{-0.52\text{ thousand deaths}}{\text{yr}}$,
or a decrease of 520 deaths per year.

b. $\frac{(42.6 - 41.8)\text{thousand deaths}}{4\text{ yrs}} = \frac{0.2\text{ thousand deaths}}{\text{yr}}$,
or an increase of 200 deaths per year.

4. 60 mph/5 sec = 12 mph per second

5. $\frac{(\$867 - \$689)}{6\text{ yrs}} \approx \$29.67$/yr

6. $(978 - 1056)/4 = -78/4 = -19.5$. On the average, his performance was declining by 19.5 yards per year.

7. $(30{,}000 - 150{,}000)/10 = -12{,}000$ elephants per year. On the average, the African elephant population is decreasing by 12,000 elephants per year.

Exercises for Section 2.1

1. a. inches/pound   **b.** minutes/inch   **c.** pounds/inch

3. 212 miles/10.8 gal. ≈ 19.6 miles per gallon

5. $\frac{7.5 - 9.4}{2000 - 1960} = \frac{-1.9}{40} = -0.0475$ pounds per year

7. a. Math: $\frac{504 - 498}{2005 - 2000} = \frac{6}{5} = 1.2$ points per year.

b. Verbal: $\frac{505 - 504}{2005 - 2000} = \frac{1}{5} = 0.2$ point per year.

9. a. $\frac{13{,}600{,}000 - 630{,}000}{2005 - 1985} = \frac{12{,}970{,}000}{20} = 648{,}500$ computers per year.

b. $\frac{4.0 - 84.1}{2005 - 1985} = \frac{-80.1}{20} = -4.005 \approx -4$ students per computer per year, or a decrease of 4 students per computer each year.

11. a. $\frac{500{,}000 - 4{,}000{,}000}{2000 - 1930} = \frac{-3{,}500{,}000}{70} = -50{,}000$;
i.e., on average there were 50,000 fewer elephants per year.

b. In 1980 the rate was twice as large as the rate computed in part (a) for 1930 to 2000. That means that either before or after the 1980s, the average rate of decline must have been much smaller.

13. a. For whites: $\frac{85.8 - 26.1}{2004 - 1940} = \frac{59.7}{64} \approx 0.9$ percentage points per year.

For blacks: $\frac{80.6 - 7.3}{2004 - 1940} = \frac{73.3}{64} \approx 1.1$ percentage points per year.

For Asian/Pacific Islanders: $\frac{85.0 - 22.6}{2004 - 1940} = \frac{62.4}{64} \approx 1.0$ percentage point per year.

For all: $\frac{85.2 - 24.5}{2004 - 1940} = \frac{60.7}{64} \approx 0.9$ percentage point per year.

b. In 2007:

For whites: $85.8 + 3 \cdot 0.9 = 88.5\%$ will have completed 4 years of high school.

For blacks: $80.6 + 3 \cdot 1.1 = 83.9\%$.

For Asian/Pacific Islanders: $85.0 + 3 \cdot 1.0 = 88.0\%$.

For all: $85.2 + 3 \cdot 0.9 = 87.9\%$.

c. There was a major increase in the percentage of those who completed 4 years of high school or more in all four categories. Over the 64-year period the increases ranged from 59.7 percentage points for whites to 73.3 percentage points for blacks. Blacks showed the highest average gain: 1.14 percentage points per year. Other comments could be made.

d. For x years since 1940:

Whites: $100 = 26.1 + 0.9x \Rightarrow x \approx 82.1$ years, or in early 2022.

Blacks: $100 = 7.3 + 1.1x \Rightarrow x \approx 84.3$ years, or in early 2024.

Asian Pacific Islanders: $100 = 22.6 + 1.0x \Rightarrow x = 77.4$ years, or in mid-2017.

All: $100 = 24.5 + 0.9x \Rightarrow x = 83.9$, or in late 2023.

These 100% predictions may not make much sense since the time is too far into the future to allow a reliable prediction.

15. a. White females had the highest life expectancy in 1900 and in 2005. Black males had the lowest life expectancy in 1900 and in 2005.

b. For white males: $\frac{75.4 - 46.6}{2005 - 1900} = \frac{28.8}{105} \approx 0.27$ years per year.

For white females: $\frac{81.1 - 48.7}{2005 - 1900} = \frac{32.4}{105} \approx 0.31$ years per year.

For black males: $\frac{69.9 - 32.5}{2005 - 1900} = \frac{37.4}{105} \approx 0.36$ years per year.

For black females: $\frac{76.8 - 33.5}{2005 - 1900} = \frac{43.3}{105} \approx 0.41$ years per year.

Thus black females had the largest average rate of change in life expectancy between 1900 and 2005.

c. In all four groups the average rate of change of life expectancy rose over the 105 years. White males had the smallest increase, 0.27 years per year, and black females had the largest increase, 0.41 years per year.

Section 2.2

Algebra Aerobics 2.2

1. a.

World Population

Year	Total Population (in millions)	Average Rate of Change over Prior 50 yrs.
1800	980	n.a.
1850	1260	$\frac{(1260 - 980)\text{ million}}{(1850 - 1800)} = \frac{280\text{ million}}{50\text{ yrs}}$ $= 5.6$ million/yr
1900	1650	$\frac{(1650 - 1260)\text{ million}}{(1990 - 1850)} = \frac{390\text{ million}}{50\text{ yrs}}$ $= 7.8$ million/yr
1950	2520	$\frac{(2520 - 1650)\text{ million}}{(1950 - 1900)} = \frac{870\text{ million}}{50\text{ yrs}}$ $= 17.4$ million/yr
2000	6090	$\frac{(6090 - 2520)\text{ million}}{(2000 - 1950)} = \frac{3570\text{ million}}{50\text{ yrs}}$ $= 71.4$ million/yr
2050	9076	$\frac{(9076 - 6090)\text{ million}}{(2050 - 2000)} = \frac{2986\text{ million}}{50\text{ yrs}}$ $= 59.72$ million/yr

b. Graph of average rate of change over prior 50 yrs.

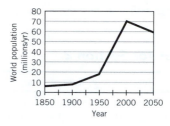

c. During 1950–2000, average annual rate of change = 71.4 million/yr was the greatest.

d. The average rate of change increased in the time interval 1800 to 2000, yet despite a projected increase in world population from 2000 to 2050, the rate of change is projected to decrease to 59.72 million/yr.

2. From 2003 to 2005 the graph would rise, indicating that profits increased. From 2005 to 2006, the graph would be flat, so there would be no change in profits. From 2006 to 2007, the graph would fall, indicating that profits decreased. (Note, however, that a lower profit does not imply a loss.)

3. a. **High School Completers**

Year	Number (thousands)	Average Rate of Change (thousands/yr)
1960	1679	n.a.
1970	2757	107.8
1980	3089	33.2
1990	2355	−73.4
2000	2756	40.1
2004	2752	−1.0

Table 2.7

b. The number of individuals completing high school each year rose between 1960 and 1980, when it peaked at 3,089,000. The 1960s had the greatest rate of change, with an annual increase of 107.8 (thousand) per year. The rate slowed during the 1970s.

In the decade from 1980 to 1990, the number of high school completers each year showed a drastic decline on average of 73,400 per year. The trend reversed in the next decade (1990 to 2000), increasing on average by 40,100 per year. Between 2000 and 2004 the numbers remained almost the same, with only a slight decrease of 4000 over the 4 years.

c. If the average rate of change is positive, there is an increase in the number of high school completers. An example would be from 1970 to 1980, where the number of high school completers increased from 2,757,000 to 3,089,000.

d. If the average rate of change is negative, there is a decrease in the number of high school completers. An example would be from 1980 to 1990, where the number of high school completers decreased from 3,089,000 to 2,355,000.

e. The growth is slowing down.

Exercises for Section 2.2

1.

x	$f(x)$	Avg. Rate of Change
0	0	n.a.
1	1	1
2	8	7
3	27	19
4	64	37
5	125	61

a. The function is increasing.

b. The average rate of change is also increasing.

3. a.

Year	Registered Motor Vehicles (millions)	Average Annual Rate of Change (over prior decade)
1960	74	n.a.
1970	108	3.4
1980	156	4.8
1990	189	3.3
2000	218	2.9

b. From 1990 to 2000.

c. From 1970 to 1980.

d. Student paragraphs might include the observation that from 1960 to 2000 there was a continuous rise in the number of registered motor vehicles in the United States (from 74 million to 218 million). The largest average rate of change (4.8 million per year) was in the decade from 1970 to 1980 and the lowest average rate of change (2.9 million per year) was from 1990 to 2000.

5.

I.

x	$f(x)$	Average Rate of Change
0	5	n.a.
10	25	2
20	45	2
30	65	2
40	85	2
50	105	2

a. $f(x)$ is increasing.

b. The average rate of change is constant.

II.

x	$g(x)$	Average Rate of Change
0	270	n.a.
10	240	−3
20	210	−3
30	180	−3
40	150	−3
50	120	−3

a. $g(x)$ is decreasing.

b. The average rate of change is constant.

7. a. Graph *B*   **b.** Graph *A*   **c.** Graph *C*

9. a.

Table A

x	y	Average Rate of Change
0	2	n.a.
1	5	3
2	8	3
3	11	3
4	14	3
5	17	3
6	20	3

b. The average rate of change is constant.

c. Its graph is a straight line.

Table B

x	y	Average Rate of Change
0	0	n.a.
1	0.5	0.5
2	2	1.5
3	4.5	2.5
4	8	3.5
5	12.5	4.5
6	18	5.5

b. The average rate of change is increasing.

c. Its graph is concave up.

11. a. In 1920 there were $\frac{27,791,000}{106,000,000} = 0.3$ copy of newspapers printed per person (about one-third of a copy per person, or equivalently, roughly one copy for every three people).

In 2000 there were $\frac{55,800,000}{281,400,000} = 0.2$ copy of newspapers printed per person (about one-fifth of a copy per person, or equivalently, about one copy for every five people).

b.

Year	No. of TV Stations	Avg. Annual Rate of Change (over prior decade)	No. of Newspapers Published	Avg. Annual Rate of Change (over prior decade)
1950	98	n.a.	1772	n.a.
1960	515	41.7	1763	−0.9
1970	677	16.2	1748	−1.5
1980	734	5.7	1745	−0.3
1990	1092	35.8	1611	−13.4
2000	1248	15.6	1480	−13.1

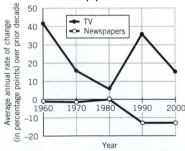

Average Rates of Change for Number of TV Stations and Newspapers Published

c. From the table in part (b), the average rate of change from 1990 to 2000 was 15.6 stations per year. At this rate, in 2010 there will be $1248 + 15.6 \cdot 10 = 1404$ stations. This could be off, given that the avg. rate per year for each decade vacillates a fair amount.

d. It seems that news is being more disseminated through TV stations than through newspapers. But there is not enough information to really give an answer here. A great increase in TVs does not necessarily indicate a great increase in getting the news via TV. Many other programs besides the news are watched on TV.

Section 2.3

Algebra Aerobics 2.3

1. a.   **i.** $m = (11 - 1)/(8 - 4) = 10/4 = 2\frac{1}{2}$

     ii. $m = (6 - 6)/[2 - (-3)] = 0$

     iii. $m = [-3 - (-1)]/[0 - (-5)] = -\frac{2}{5}$

   b.   **i.** $m = (1 - 11)/(4 - 8) = (-10)/(-4) = 2\frac{1}{2}$

     ii. $m = (6 - 6)/(-3 - 2) = 0$

     iii. $m = [-1 - (-3)]/(-5 - 0) = 2/(-5) = -\frac{2}{5}$

2. Between 1999 and 2002 the slope is approximately zero.

3. positive: 2001–2002, 2004–2005

   negative: 1998–2000, 2003–2004

   zero: 2000–2001, 2002–2003

4. $\frac{\Delta y}{\Delta x} = 4 \Rightarrow \frac{y - (-2)}{5 - 3} = 4 \Rightarrow \frac{y + 2}{2} = 4 \Rightarrow y + 2 = 8$

   $\Rightarrow y = 6.$

5. $\frac{\Delta y}{\Delta x} = \frac{(9 + 2h) - 9}{(2 + h) - 2} = \frac{2h}{h} = 2$

6. a. Points lie on the graph $\Rightarrow$ the coordinates (x, y) satisfy the equation $y = x^2$ since

$$0 = 0^2, \quad 1 = 1^2, \quad 4 = 2^2, \quad 9 = 3^2.$$

   b. P_1 and P_2: $m = \frac{1 - 0}{1 - 0} = 1$; P_2 and P_3: $m = \frac{4 - 1}{2 - 1} = 3$; P_3 and P_4: $m = \frac{9 - 4}{3 - 2} = 5$.

   c. The positive slopes suggest that the function increases between $x = 0$ and $x = 3$; because the slopes increase in size as we move further to the right, the graph rises at an increasing rate.

Exercises for Section 2.3

1. a. $m = \frac{3 - (-6)}{2 - (-5)} = \frac{9}{7}$ **b.** $m = \frac{-3 - 6}{2 - (-5)} = \frac{-9}{7}$

3. a. Graph A crosses the y-axis at $(0, -4)$ and goes through $(1, -4)$. So the slope is $\frac{(-4) - (-4)}{1 - 0} = 0$.

   b. Graph B crosses the x-axis at $(2, 0)$ and the y-axis at $(0, -8)$. So the slope is $\frac{-8 - 0}{0 - 2} = 4$.

5. a. slope of segment $A = [2 - (-6)]/[-4 - (-8)] = 8/4 = 2$

   slope of segment $B = [-8 - 2]/[0 - (-4)] = -10/4 = -2.5$

   slope of segment $C = [-8 - (-8)]/[2 - 0] = 0/2 = 0$

   slope of segment $D = [-6 - (-8)]/[6 - 2] = 2/4 = 0.5$

   slope of segment $E = [6 - (-6)]/[10 - 6] = 12/4 = 3$

   slope of segment $F = [-16 - 6]/[12 - 10] = -22/2 = -11$

b. The slope of segment F is steepest.

c. Segment C has a slope equal to 0.

7. a. $m = 75/10 = 7.5$

   b. $70 - y = -4 \Rightarrow y = 74$

   c. $32 = 4(28 - x) \Rightarrow 32 = 112 - 4x \Rightarrow x = 20$

   d. $6 = 0.6(x - 10) \Rightarrow 12 = 0.6x \Rightarrow x = 20$

9. a. $-4 = (1 - t)/(-2 - 3) \Rightarrow 20 = 1 - t \Rightarrow t = -19$

   b. $2/3 = (9 - 6)/(t - 5) \Rightarrow 2(t - 5) = 9 \Rightarrow t - 5 = 4.5 \Rightarrow$
     $t = 9.5$

11. a. $m_1 = (7 - 3)/(4 - 2) = 4/2 = 2$; $m_2 = (15 - 7)/(8 - 4) = 2$; collinear

   b. $m_1 = (4 - 1)/(2 + 3) = 3/5$; $m_2 = (8 - 1)/(7 + 3) = 0.7$; not collinear

13. a. $m = \frac{-2 - 0}{6 - 0} = -\frac{1}{3}$ **b.** $m = \frac{7 - 0}{-4 - 0} = -\frac{7}{4}$

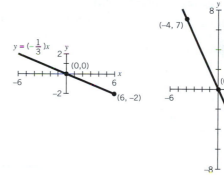

15. a. $\frac{0 - \sqrt{2}}{\sqrt{2} - 0} = -1$

   b. $\frac{0 + 3/2}{-3/2 - 0} = -1$

   c. $\frac{0 - b}{b - 0} = -\frac{b}{b} = -1$

   d. All slopes are -1, since each pair of points is of the form $(0, a)$ and $(a, 0)$.

17. The points in parts (a) and (c).

19. a. 1/10

   b. 1/12

   c. old: $\frac{3}{\text{run}} = \frac{1}{10} \Rightarrow$ run $= 30$ ft.; new: $\frac{3}{\text{run}} = \frac{1}{12} \Rightarrow$ run $= 36$ ft.

21. Student answers will vary.

Section 2.4

Algebra Aerobics 2.4

1. a. (1960, 22.2) and (2000, 11.3)

   b. (1970, 12.6) and (2005, 12.6)

   c. (2000, 11.3) and (2005, 12.6)

2. a. Between 2005 and 2006 the stock price surged from $1.02 to $1.12 per share, a 9.8% increase or, equivalently, an increase of $0.10 per share. See Graph A.

b. Between 2006 and 2007 the stock price dropped drastically from $1.12 to $1.08 per share, about a 3.6% decrease or, equivalently, a decrease of $0.04 per share. See Graph *B*.

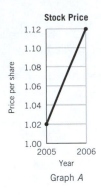

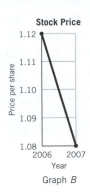

Graph *A*   Graph *B*

3. a. You could draw a graph that is cropped and stretched vertically, as shown in Graph *A*.

b. You could show a graph that is not too steep (stretched horizontally or compressed vertically) and emphasize the decline in the number of casualties in week three, as in Graph *B*.

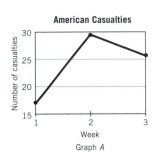

Graph *A*   Graph *B*

4. Some of the strategies are: use of dramatic language (the title "Gold Explodes" in big bold type, followed by "Experts Predict $1,500.00 an Ounce"; cropped vertical axis on graph (it starts at $300, not $0, making the graph look steeper); use of powerful graphics (the arrow that increases in size), suggesting dynamic growth.

Exercises for Section 2.4

1. Graph *A*: "30-Year Mortgage Rates Steadily Climb"

Graph *B*: "30-Year Mortgage Rates Rise Sharply"

Graph *C*: "30-Year Mortgage Rates Show Little Gain"

3. a. Graph *B* appears to have the steeper slope.

b. The slope of Graph *A* is -6 and the slope of Graph *B* is -4. So Graph *A* actually has the steeper slope.

5. a. From 1984 to 2005 the federal appropriations for the most part rose, but the reading scores stayed the same.

b. There is no relationship between the scale on the left (for federal funding) and the scale on the right (for NAEP scores). The NAEP scale was arbitrarily started at 185. Had it started at 0, the NAEP scores would still remain constant, but would lie much higher on the graph.

7. a.

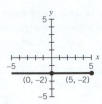

b, c. The task is impossible. The slope of the line is 0, so it is not possible to make a graph through the given points appear to have a large positive slope.

9. "Precipitous," "dire," and "catastrophic" come to mind. Students will probably think of others.

11. a. The number of persons with AIDS who are still alive has been steadily growing since 1995 by approximately $\frac{425{,}000 - 200{,}000}{2004 - 1995} = 25{,}000$ people per year. Also, deaths from AIDS have declined since 1993 from 55,000 to less than 20,000 (using left scale).

b. Despite the progress in treating AIDS and education about it, there has been an increase in cases diagnosed since 2000 of about 1000 per year.

c. AIDS cases increased until 1993, when new treatments, public awareness, education, and testing for the disease may have caused a decrease in the number of new cases.

13. Graph *A* gives the appearance that the percentage has declined quite a bit, while Graph *B* gives the impression that the percentage has not declined much at all. The difference is in the vertical scale: it is greatly magnified in Graph *A*.

Section 2.5

Algebra Aerobics 2.5

1. The weight of a 4.5-month-old baby girl appears to be ≈ 14 lb. From the equation, the exact weight is

$W = 7.0 + 1.5(4.5) = 13.75$ lb. Our estimate is within 0.25 lb of the exact weight.

2. The age appears to be ≈ 2.5 months. Solve $11 = 7 + 1.5A \Rightarrow 11 - 7 = 1.5A \Rightarrow 4 = 1.5A \Rightarrow A = 2.67$ months.

3. a. The units for 15 are dollars/person. The units for 10 are dollars.

b. dollars = (dollars/person)(persons) + dollars

4. a. The units for 1200 are dollars and for 50 are dollars/month.

b. dollars = dollars + $\left(\frac{\text{dollars}}{\text{month}}\right)$ months

5. a. 0.8 million dollars per year is the slope or average rate of change in the sales per year.

b. 19 million dollars represents the sales this year.

c. In 3 years $S = 0.8(3) + 19 = 21.4$ million dollars

d. dollars = $\left(\frac{\text{dollars}}{\text{year}}\right)$ years + dollars, where all values for dollars are in millions.

6. a. The average cost to operate a car is $0.45 per mile, and the units are dollars per mile.

b. $0.45(25{,}000) = \$11{,}250$

c. dollars = $\left(\frac{\text{dollars}}{\text{mile}}\right)$ miles

7. a. $840

  b. The beginning mortgage is $302,400.

  c. After 10 years, $B = \$302,400 - 840(12)(10) = \$201,600$;
  After 20 years, $B = \$302,400 - 840(12)(20) = \$100,800$;
  After 30 years, $B = \$302,400 - 840(12)(30) = \$0$.

8. a. $m = 5, b = 3$

  b. $m = 3, b = 5$

  c. $m = 5, b = 0$

  d. $m = 0, b = 3$

  e. $m = -1, b = 7.0$

  f. $m = -11, b = 10$

  g. $m = -2/3, b = 1$

  h. $m = 5, b = -3$ since $2y = 10x - 6 \Rightarrow y = 5x - 3$

9. a. $f(x)$ is a linear function because it is represented by an equation of the form $y = mx + b$, where here $b = 50$ and $m = -25$.

  b. $f(0) = 50 - 25(0) = 50 - 0 = 50$
  $f(2) = 50 - 25(2) = 50 - 50 = 0$

  c. Since the line passes through $(0, 50)$ and $(2, 0)$, the slope m is: $m = \frac{50 - 0}{0 - 2} = -25$.

10. a. linear: $m = 3, b = 5$

  b. linear: $m = 1, b = 0$

  c. not linear

  d. linear: $m = -2/3, b = 4$

11. a. $y = 3x + 4$

  b. $y = -x$

  c. $y = -3$

12. Graph A:　$y = 4x + 3$

  Graph B:　$y = -2x + 0 = -2x$

  Graph C:　$y = 0x + 3 = 3$

  Graph D:　$y = x + 0 = x$

Exercises for Section 2.5

1. a. $E = 5000$ when $n = 0$, $E = 5100$ when $n = 1$, and $E = 7000$ when $n = 20$

  b. $(0, 5000), (1, 5100)$ and $(20, 7000)$

3. a. $(5000, 0)$ is not a solution to either equation.

  b. $(15, 24000)$ is a solution to the second equation but not the first.

  c. $(35, 40000)$ is a solution to the second equation but not the first.

5. a. $D = 3.40, 3.51, 3.62, 3.73$, and 3.84, respectively

  b. 0.11 is the slope; it represents the average rate of change of the average consumer debt per year; it is measured in thousands of dollars per year.

  c. 3.40 represents the average consumer debt when $n = 0$ years. It is measured in thousands of dollars.

7. dollars = dollars + $\frac{\text{dollars}}{\text{year}} \cdot$ years

9. a. $C(0) = \$11.00$; $C(5) = 11 + 10.50 \cdot 5 = \$63.50$; $C(10) = 11 + 10.50 \cdot 10 = \$116.00$

  b. $11.00

  c. $10.50 for every thousand cubic feet.

  d. $C(96) = \$1019$

11. a. $0; 4$　　**b.** $0; \pi$　　**c.** $0; 2\pi$　　**d.** $-17.78; 5/9$

13. a. matches **f**, since $42.50 seems the most likely cost for producing one text.

  b. matches **g**, since $0.30 seems the most likely cost to produce one CD.

  c. matches **e**, since $800 seems the most likely cost of producing one computer.

15. a. $C(p) = 1.06 \cdot p$

  b. $C(9.50) = 1.06 \cdot 9.50 = 10.07$;
  $C(115.25) = 1.06 \cdot 115.25 = 122.17$ (rounded up);
  $C(1899) = 1.06 \cdot 1899 = 2012.94$. All function inputs and outputs are measured in dollars.

17. a. hours　　**b.** miles/gallon　　**c.** calories/gram of fat

19. a. Slope = 0.4, vertical intercept = -20

  b. Slope = -200, vertical intercept = 4000

a. 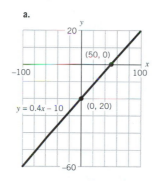　　**b.**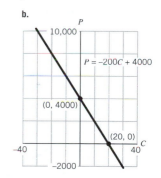

21. The equation is: $y = 3x - 2$

x	y
0	−2
1	1
2	4

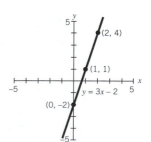

23. The equation is $y = 0 \cdot x + 1.5 = 1.5$.

x	y
−5	1.5
0	1.5
5	1.5

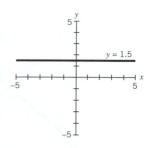

Section 2.6

Algebra Aerobics 2.6

1. $0, |-1|, |-3|, 4, |-7|, 9, |-12|$

2. **a.** The graph of $y = 6x$ goes through the origin and is two units above the graph of $y = 6x - 2$.

   b. The graph of $y = 2 + 6x$ intersects the y-axis at $(0, 2)$. The lines are parallel, so $y = 2 + 6x$ is four units above the graph of $y = 6x - 2$.

   c. The graph of $y = -2 + 3x$ has the same y-intercept $(0, -2)$ but is less steep than the line $y = 6x - 2$.

   d. The graph of $y = -2 - 2x$ has the same y-intercept $(0, -2)$ but is falling left to right, whereas $y = 6x - 2$ is rising and has a steeper slope.

3. For (a), $m = -2$, so $|m| = |-2| = 2$
   For (b), $m = -1$, so $|m| = |-1| = 1$
   For (c), $m = -3$, so $|m| = |-3| = 3$
   For (d), $m = -5$, so $|m| = |-5| = 5$

   Line d is the steepest. In order from least steep to steepest, we have b, a, c, d.

4. Answers will vary depending on the original $f(x)$. For example, if $f(x) = 2 + 2x$, then we could have:

   a. $g(x) = 5 + 2x$

   b. $h(x) = -2 + 2x$

   c. $k(x) = 2 - 2x$

   See graph below

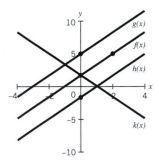

5. For $f(x) = 3x - 5$, $m = 3$.
   For $g(x) = 7 - 8x$, $m = -8$.
   $g(x)$ has a steeper slope than $f(x)$ since $|-8| > |3|$

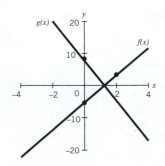

6. Answers will vary. For example,

   $$f(x) = 4 - 2x$$
   $$g(x) = 4 - 5x$$
   $$h(x) = 4 - 7x$$

   $h(x)$ has the steepest slope since $|-7| > |-5| > |-2|$

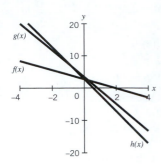

7. $f(x)$ matches Graph B, $g(x)$ matches Graph D, $h(x)$ matches Graph A, $k(x)$ matches Graph C.

Exercises for Section 2.6

1. **a.** Graph B   **c.** Graph C

   b. Graphs A, B, and D   **d.** Graphs A and D

3. Answers may vary for (b) and (c).

   a. $y = -3$ **b.** $y = x - 3$ **c.** $y = -3x + 1$

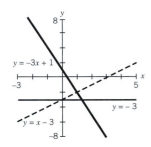

5. **a.** matches Graph B   **c.** matches Graph C

   b. matches Graph D   **d.** matches Graph A

7. Graphs may vary.

   a. Same slope   **b.** Same vertical intercept

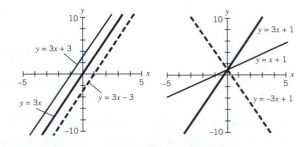

9. **a.** $R(t) = 5 + 13 - 5t = 18 - 5t$

   b. $S(t) = -3 + 13 - 5t = 10 - 5t$

   c. $T(t) = 13 + 2t$ (slope may vary)

   d. $U(t) = 12 - 5t$ (vertical intercept may vary)

   e. $V(t) = 15 + 5t$ (vertical intercept may vary)

11.

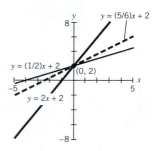

13. a. C

  b. B

  c. $m_3 < m_2 < m_1$ (Note: m_3 is negative, while m_2 and m_1 are both positive and $m_2 < m_1$.)

  d. The steepness is the absolute value of the slope. So although m_3 is negative, its absolute value is positive. Compared to the absolute values of the other slopes, we have $|m_2| < |m_1| < |m_3|$.

Section 2.7

Algebra Aerobics 2.7

1. a. $y = 1.2x - 4$

  b.

x	y
-3	-7.6
0	-4.0
3	-0.4

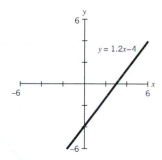

2. a. $y = 300 - 400x$

  b.

x	y
-2	1100
0	300
2	-500

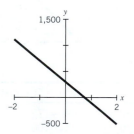

3. The vertical intercept is at $(0, 1)$, so $b = 1$. The line passes through $(0, 1)$ and $(3, -5)$, so $m = \frac{1 - (-5)}{0 - 3} = -2$. The equation is: $y = -2x + 1$.

4. a. $S = \$12,000 + \$3000x$

  b.

Year	$ Salary
0	12,000
1	15,000
3	21,000
5	27,000

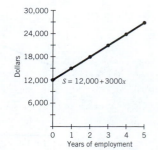

5. a.

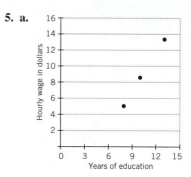

  b. Yes. $\frac{8.50 - 5.30}{10 - 8} = \frac{3.20}{2} = 1.6$, $\frac{13.30 - 8.50}{13 - 10} = \frac{4.80}{3} = 1.6$, $\frac{13.30 - 5.30}{13 - 8} = \frac{8}{5} = 1.6$. Since the rate of change is 1.6, the same over all intervals, the points lie on the same line with slope 1.6.

  c. If we choose the point $(8, 5.30) \Rightarrow x = 8, y = 5.30$. From part (b) we have $m = 1.6$.

  Solve: $5.3 = (1.6)8 + b \Rightarrow 5.3 = 12.8 + b \Rightarrow -7.5 = b$, so the equation is $y = 1.6x - 7.5$.

6. 3

7. a. $4; (1, 4)$

  b. $4; (-1, 12)$

8. a. $B = \$10,800 - 300P$

  b. $300 per month

  c. $B = \$10,800 - \$300(24) = \$3600$

  d. $0 = \$10,800 - \$300P \Rightarrow \$300P = \$10,800 \Rightarrow P = 36$ or 36 months.

9. a. $R = 7.5T$

  b. $7.50 per ticket

  c. $R = \$7.50(120) = \$900.00$

10. Graph A: $y = x$

  Graph B: $Q = 3t + 2$

  Graph C: $y = 6 - 2x$

11. a. $R = 15,000 + 12T$

  b. $R = 15,000 + 12(40,000) = \$495,000$

12. Any equation of the form $y = 6 + mx$ with three different values for m; some examples are $y = 6 + 2x$, $y = 6 - 5x$, and $y = 6 + 11x$.

13. Any equation of the form $y = -3x + b$ with three different values for b; some examples are $y = -3x + 5$, $y = -3x$, $y = -3x - 12$.

14. a. $y = 4x + 9$ **c.** $y = -10x - 7$

  b. $y = -\frac{2}{3}x + 7$ **d.** $y = 2x + 2.1$

15. a. $m = \frac{6}{2} = 3 \Rightarrow y = 3x - 1$

  b. $m = -\frac{1}{9} \Rightarrow y = -\frac{1}{9}x + \frac{5}{3}$

  c. $m = \frac{-6}{-6} = 1 \Rightarrow y = x - 5$

16. a. $y = -\frac{3}{4}x - 3$ **d.** $y = \frac{1}{2}x$

  b. $y = 7x - 5$ **e.** $y = 3x + 5$

  c. $y = -\frac{1}{4}x + \frac{1}{8}$ **f.** $y = -5x + 6$

Ch. 2

17. Note the difference in scales on all graphs.

a. Graph of $y = 2x + 5$ **c.** Graph of $y = -\left(\frac{3}{4}\right)x - 1$

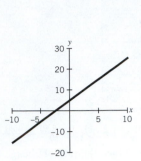

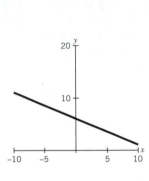

b. Graph of $y = -\left(\frac{1}{2}\right)x + 6$ **d.** Graph of $y + 3x + 9$

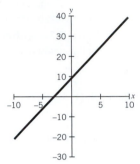

c. $y = -x + 2$ **d.** $y = 2$

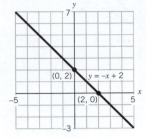

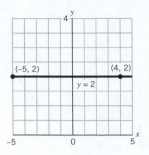

7. a. $y = \frac{2}{3}x - 2$; the slope is 2/3 and the y-intercept is -2.

b. $y = -\frac{3}{2}x + 3$; the slope is $-3/2$ and the y-intercept is 3.

c. $y = -\frac{2}{3}x + 12$; the slope is $-2/3$ and the y-intercept is 12.

d. $y = \frac{3}{2}x$; the slope is 3/2 and the y-intercept is 0.

e. $y = \frac{3}{2}x$; the slope is 3/2 and the y-intercept is 0. So the equation is equivalent to the one in part (d).

f. $y = \frac{3}{4}x + \frac{1}{4}$; the slope is 3/4 and the y-intercept is 1/4.

9. a.

x	$f(x) = 0.10x + 10$
-100	0
0	10
100	20

Graph of $f(x)$:

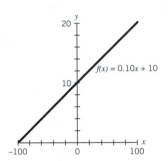

Exercises for Section 2.7

1. a. Slope $= 5$, so equation is of the form $y = 5x + b$. The line passes through $(-2, 3)$,

$\Rightarrow 3 = 5(-2) + b$

$\Rightarrow b = 13$. So $y = 5x + 13$

b. Slope $= -\frac{3}{4}$, so equation is of the form $y = -\frac{3}{4}x + b$. The line passes through $(-2, 3)$,

$\Rightarrow 3 = -\frac{3}{4}(-2) + b$

$\Rightarrow b = 1.5$. So $y = -0.75x + 1.5$

c. Slope $= 0$, so equation is of the form $y = b$. The line passes through $(-2, 3)$,

$\Rightarrow 3 = b$. So the equation is the horizontal line $y = 3$

3. a. $m = (5.1 - 7.6)/(4 - 2) = -2.5/2 = -1.25$ and $y - 7.6 = -1.25(x - 2)$ or $y = -1.25x + 10.1$

b. $m = (16 - 12)/(7 - 5) = 4/2 = 2$ and $W - 12 = 2(A - 5)$ or $W = 2A + 2$

5. a. $y = -1.5x + 6$ **b.** $y = -0.5x + 2$

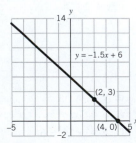

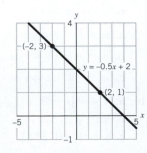

b.

x	$h(x) = 50x + 100$
-0.5	75
0	100
0.5	125

Graph of $h(x)$:

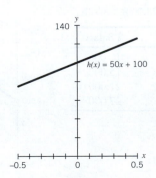

11. $C(n) = 150 + 120n$, where $C(n)$ is the cost of n credits, and 150 is measured in dollars and 120 in dollars per credit.

13. $C(n) = 2.50 + 0.10n$, where $C(n)$ is the cost of cashing n checks that month.

15. a. Annual increase $= (32{,}000 - 26{,}000)/4 = \$1500$

b. $S(n) = 26{,}000 + 1500n$, where $S(n)$ is measured in dollars and n in years from the start of employment.

c. Here $0 \leq n \leq 20$ since the contract is for 20 years.

17. a. $P = 285 - 15t$

b. It will take approximately 16.33 years to make the water safe for swimming since $40 = 285 - 15t$ implies that $15t = 245$ or $t \approx 16.33$ years.

19. a. $V(t) \approx 70{,}000 + 26{,}000t$, where $t =$ years since 1977.

b. $V(33) = \$928{,}000$.

21. The equation of the line illustrated is $y = (8/3)x - 4$. The graph in the text can be altered to appear much steeper if the graph is stretched vertically—as in the accompanying graph—where the tick marks on the x-axis are 10 units apart.

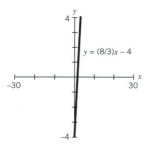

If, however, we reversed the situation and vertically compressed the graph (as in the one below, where the tick marks on the y-axis are now 10 units apart), the graph would seem less steep than the one in the text.

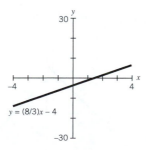

23. a. The three data points are plotted in the accompanying graph.

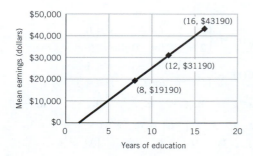

b. The relationship is linear. The slope between any two points is \$3000 per year of education.

c. The linear equation is: $M = 3000E - 4810$, where $M =$ mean earning measured in dollars and $E =$ years of education.

25. The entries in the table argue that the relationship is linear. The average rate of change in salinity per degree Celsius is a constant: -0.054. Since the freezing point, P, for 0 salinity is 0 degrees Celsius, we have that $P = -0.054S$, where S is the salinity measured in ppt and P, the freezing point, is measured in degrees Celsius.

Section 2.8

Algebra Aerobics 2.8a

1. If $(0, 0)$ is on each line, then $b = 0$ in $y = mx + b$.

a. $y = -x$.

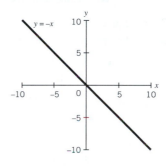

b. $y = 0.5x$.

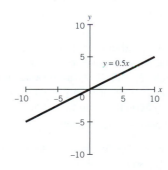

2. a. The variables x and y are directly proportional; the equation is $y = -3x$.

b. The variables x and y are not directly proportional; the equation is $y = 3x + 5$.

3. $E =$ euros; $D =$ U.S. dollars

a. $E = 0.79D$

b. $E = 0.79D + 2.50$.

c. Only (a) because it is of the form $y = mx$ for some constant m.

4. $d = 60t$. This represents direct proportionality. If the value of t doubles, the value for d also doubles. If the value for t triples, then so does the value for d.

5. Since C and N are directly proportional, $C = kN$ for some constant k. Since $50 = k \cdot 2$, then $k = \$25$, the cost per ticket. So the cost for 10 tickets is $C = 25 \cdot 10 = \$250$.

6. a. $4 = k(12) \Rightarrow k = \frac{1}{3} \Rightarrow y = \frac{1}{3}x$

   b. $300 = k(50) \Rightarrow k = 6 \Rightarrow d = 6t$

7. a. $d = kC$ **b.** $T = kI$ **c.** $t = kc$

8. a. $a = kb;$ $10 = k(15) \Rightarrow \frac{2}{3} = k;$ $a = \frac{2}{3}(6) = 4$

   b. $a = kb;$ $4 = \frac{2}{3}b \Rightarrow b = 6$

Algebra Aerobics 2.8b

1. a. $y = -5$ **b.** $y = -3$ **c.** $y = 5$

2. a. $x = 3$ **b.** $x = 5$ **c.** $x = -3$

3. a. $y = -7$ **b.** $x = -4.3$

4. Slope is -1, y-intercept is 0, so $m = -1, b = 0 \Rightarrow y = -x$.

5. $m = 360, C = 4, W = 1000$ in $W = 360C + b$. To solve for b, use the point $(4, 1000)$, so $1000 = (360)(4) + b \Rightarrow 1000 = 1440 + b \Rightarrow b = -440$. So, the equation is: $W = 360C - 440$.

6. Let $m = $ slope of given line;

   $M = $ slope of perpendicular line. So $M = -\frac{1}{m}$.

   a. $m = -3 \Rightarrow M = -\frac{1}{-3} = \frac{1}{3}$

   b. $m = 1 \Rightarrow M = -\frac{1}{1} = -1$

   c. $m = 3.1 \Rightarrow M = -\frac{1}{3.1} \approx -0.32$

   b. $m = -\frac{3}{5} \Rightarrow M = -\frac{1}{-3/5} = \frac{5}{3}$

7. a. Slope of $y = 2x - 4$ is 2, so line perpendicular to it has slope $-1/2$ or -0.5. Since it passes through $(3, -5)$, $x = 3$ when $y = -5$. So, $y = mx + b$ is: $-5 = (-0.5)(3) + b$. Solve it for b. $-5 = -1.5 + b \Rightarrow b = -3.5$. So, equation is: $y = -0.5x - 3.5$.

   b. Any line parallel to (but distinct from) $y = -0.5x - 3.5$ will be perpendicular to $y = 2x - 4$, but will not pass through $(3, -5)$. They have same slope m, (-0.5), but different values of b, in $y = mx + b$. Two examples are $y = -0.5x$ and $y = -0.5x - 7.5$.

   c. The three lines are parallel.

8. $Ax + By = C \Rightarrow By = C - Ax \Rightarrow y = (C - Ax)/B \Rightarrow$ $y = \frac{C}{B} - \frac{A}{B}x$, so $m = -\frac{A}{B}$

9. a. $2x + 3y = 5 \Rightarrow A = 2, B = 3 \Rightarrow m = -\frac{2}{3}$

   b. $3x - 4y = 12 \Rightarrow A = 3, B = -4 \Rightarrow m = -\frac{3}{-4} = \frac{3}{4}$

   c. $2x - y = 4 \Rightarrow A = 2, B = -1 \Rightarrow m = -\frac{2}{-1} = 2$

   d. $x = -5 \Rightarrow A = 1, B = 0 \Rightarrow m = -\frac{1}{0}$, which is undefined; this line is vertical

   e. $x - 3y = 5 \Rightarrow A = 1, B = -3 \Rightarrow m = -\frac{1}{-3} = \frac{1}{3}$

   f. $y = 4 \Rightarrow A = 0, B = 1 \Rightarrow m = -\frac{0}{1} = 0$ (this line is horizontal)

10. $2x + 3y = 5 \Rightarrow 3y = -2x + 5 \Rightarrow y = -\frac{2}{3}x + \frac{5}{3} \Rightarrow m = -\frac{2}{3}$. Parallel lines have slopes that are equal, so $m = -\frac{2}{3}$. If the line passes through the point $(0, 4)$ then $x = 0, y = 4$, the vertical intercept $b = 4 \Rightarrow y = -\frac{2}{3}x + 4$.

11. $3x + 4y = -7 \Rightarrow 4y = -7 - 3x \Rightarrow y = -\frac{7}{4} - \left(\frac{3}{4}\right)x \Rightarrow$ $m = -\frac{3}{4}$. Because perpendicular lines have slopes that are negative reciprocals of each other, the slope of any line

perpendicular to the given line is $m = -\frac{1}{-3/4} = \frac{4}{3}$. If the line passes through the point $(0, 3)$ then $x = 0, y = 3$, the vertical intercept $b = 3$. So the equation is $y = \frac{4}{3}x + 3$.

12. $4x - y = 6 \Rightarrow y = 4x - 6$. The slope of the given line is 4; therefore the slope of any line perpendicular to it is $-1/4$. If the line passes through $(2, -3) \Rightarrow x = 2, y = -3$. So $y = -\left(\frac{1}{4}\right)x + b$, and $(-3) = \frac{-1}{4}(2) + b \Rightarrow$ $-3 = -\frac{1}{2} + b \Rightarrow b = \frac{-5}{2} \Rightarrow y = -\left(\frac{1}{4}\right)x - \frac{5}{2}$

13. a. vertical line

   b. neither $\left(m = -\frac{2}{3}\right)$

   c. horizontal line

14. Slope of given line $= -\frac{2}{3}$, so slope of a perpendicular line $= \frac{3}{2}$. So:

   a. $y = \frac{3}{2}x + 5$

   b. Subsituting $(-6, 1)$, we have $1 = \frac{3}{2}(-6) + b \Rightarrow b = 10 \Rightarrow y = \frac{3}{2}x + 10$

15. Slope $= 2$ for the given line and for lines parallel to it. So:

   a. $y = 2x + 9$

   b. Subsituting $(4, 3)$, we have $3 = 2(4) + b \Rightarrow b = -5 \Rightarrow y = 2x - 5$

Algebra Aerobics 2.8c

1. a. Graph of $f(x)$

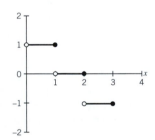

   b. Graph of $g(x)$

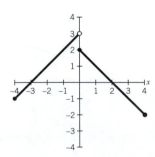

2. $Q(t) = \begin{cases} t - 2 & \text{for } -5 < t \le 0 \\ 2 - t & \text{for } 0 < t \le 5 \end{cases}$

   $C(r) = \begin{cases} r + 1 & \text{for } -3 < r \le 1 \\ 5 & \text{for } 1 < r \le 4 \end{cases}$

3. a. 2   **b.** 6   **c.** 2   **d.** -2 **d.** -15

4. a. $g(-3) = |-6| = 6;$ $g(0) = |-3| = 3;$

   $g(3) = |0| = 0;$ $g(6) = |3| = 3$

b.

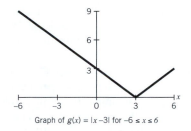

Graph of $g(x) = |x - 3|$ for $-6 \leq x \leq 6$

c. The graph of g is the graph of f shifted three units to the right.

d. $g(x) = \begin{cases} x - 3 & \text{if } x \geq 3 \\ 3 - x & \text{if } x < 3 \end{cases}$

5. a. $3 \leq t \leq 7$, which means that the values of t lie between (and include) 3 and 7.

b. $69 < Q < 81$, which means that the values of Q lie between (but exclude) 69 and 81.

6. $|T - 55°| \leq 20°$, or equivalently $35° \leq T \leq 75°$.

7. a.   **b.**

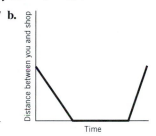

8. a. 4.25% in week 4;   5.25% in week 52;   the longest period in which the rate remained the same was week 26 through week 52.

b. Typical rate increase $= 0.25\%$. The increases are not at regular intervals.

c. Curbing inflation.

d. If $R(w) =$ the federal funds rate in week w of 2006, then we can write R(w) as:

$$R(w) = \begin{cases} 4.25\% & \text{for } 1 \leq w < 5 \\ 4.50\% & \text{for } 5 \leq w < 13 \\ 4.75\% & \text{for } 13 \leq w < 19 \\ 5.00\% & \text{for } 19 \leq w < 26 \\ 5.25\% & \text{for } 26 \leq w \leq 52 \end{cases}$$

e.

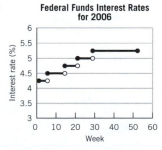

Federal Funds Interest Rates for 2006

Exercises for Section 2.8

1. a. $2 = m \cdot 10$ means $m = 0.2$

b. $0.1 = m \cdot 0.2$ means $m = 0.5$

c. $1 = m \cdot 1/4$ means $m = 4$

3. a. The slope $m = 56.92 - 42.69 = 14.23$ and since y is directly proportional to x, then $y = 14.23x$ is the equation. Thus, if $x = 5$, then $5 \cdot 14.23 = 71.15$ is the missing value.

b. The coefficient of x is the cost of a single CD.

5. a. The independent variable is the price P; the dependent variable is the sales tax T. The equation is $T = 0.065P$.

b. Independent variable is amount of sunlight S received; dependent variable is the height of the tree H. The equation is $H = kS$, where k is a constant.

c. Time t in years since 1985 is the independent variable, and salary S in dollars is the dependent variable. The equation is $S = 25,000 + 1300t$.

7. $d = 5t$; yes, d is directly proportional to t; it is more likely to be the person jogging since the rate is only 5 mph.

9. a. $m = 0$; $y = 3$

b. $m = 0$; $y = -7$

c. slope is undefined; $x = -3$

d. slope is undefined, $x = 2$

11. The horizontal line $N = 2300$, where N is measured in millions of books. (Student answers may vary.)

13. a. horizontal: $y = -4$; vertical: $x = 1$; line with slope 2: $-4 = 2 \cdot 1 + b \Rightarrow b = -6 \Rightarrow y = 2x - 6$.

b. horizontal: $y = 0$; vertical: $x = 2$; line with slope 2: $0 = 2 \cdot 2 + b \Rightarrow b = -4 \Rightarrow y = 2x - 4$.

15. a. The average rate of change is 10 lb per month.

b. $w(t) = 175$, where $t =$ number of months after end of spring training and 175 is measured in pounds. The graph of this function is a horizontal line.

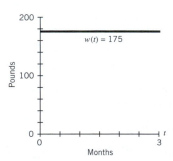

17. a. $7 = b - 3 \Rightarrow b = 10$ and therefore $y = 10 - x$.

b. $7 = b + 3 \Rightarrow b = 4$ and therefore $y = 4 + x$.

19. The lines described by: **a.** $x = 0$ **b.** $y = 0$ **c.** $y = x$.

21. a. Intercepts of one line are $(0, 4)$ and $(1, 0)$, and the intercepts of the other line are $(0, 4)$ and $(-1, 0)$. Thus the slope of the first line is -4 and the slope of the second line is 4, and thus the lines are not perpendicular.

b. Intercepts of the first line are $(0, 2)$ and $(1, 0)$ and the intercepts of the second line are $(0, 2)$ and $(-4, 0)$. So the slopes are now -2 and $1/2$, and thus the lines are perpendicular.

23. For Graph A: both slopes are positive; same y-intercept.

For Graph B: one slope is positive, one negative; same y-intercept.

For Graph C: the lines are parallel; different y-intercepts.

For Graph D: one slope is positive, one negative; different y-intercepts.

25. **a.** $y = (-A/B)x + (C/B)$, $B \neq 0$

b. The slope is $-A/B$, $B \neq 0$

c. The slope is $-A/B$, $B \neq 0$

d. The slope is B/A, $A \neq 0$

27. **a.** Graph of $f(x)$

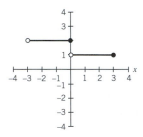

b. Graph of $g(x)$

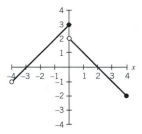

29. **a.** $g(x) = \begin{cases} -2 - x & \text{for } x < -2 \\ 2 + x & \text{for } x \geq -2 \end{cases}$

b. $g(x) = |2 + x|$

c. The graph of g is the same as the graph of f shifted horizontally two units to the left.

31. **a.** $d_A(t) = 60t$, $d_B(t) = 40t$

b. $D_{AB}(t) = 400 - d_A(t) - d_B(t) = 400 - 60t - 40t$
$= 400 - 100t$

Distance between A and B

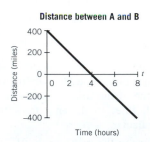

c. They will meet when the distance between them

$D_{AB}(t) = 0$ miles $\Rightarrow 400 - 100t = 0 \Rightarrow 400 = 100t$

$\Rightarrow t = 4$ hours. A will have traveled 4 hr $\cdot$ 60 miles/hr $= 240$ miles. B will have traveled 4 hr $\cdot$ 40 miles/hr $= 160$ miles. (*Note:* Together they will have traveled $240 + 160 = 400$ miles.)

d. One hour before they meet (3 hours into the trip), $D_{AB}(3) = 400 - 100 \cdot 3 = +100$ miles, which means that they are 100 miles apart and traveling *toward* each other. One hour after they meet (at 5 hours), $D_{AB}(5) = 400 - 100 \cdot 5 = -100$ miles, which means that they are 100 miles apart and traveling *away* from each other.

e. $D(t) = |400 - 100t| = \begin{cases} 400 - 100t & \text{for } 0 \leq t \leq 4 \\ 100t - 400 & \text{for } t > 4 \end{cases}$

Absolute Value of Distance between A and B

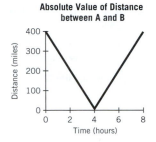

33. **a.** $S(x) = \begin{cases} 34 & \text{for } 2001 \leq x < 2002 \\ 37 & \text{for } 2002 \leq x < 2006 \\ 39 & \text{for } 2006 \leq x < 2007 \end{cases}$

b. Note that the top dotted line segment is from part (c).

Price of U.S. First-Class Stamp

c. The domain would be $2001 \leq x < 2008$, and the top (dotted) line segment would be added to the graph.

Section 2.9

Algebra Aerobics 2.9

1. **a.** In 1960 there were about 7.5 million college graduates, in 2005 about 27.5 million.

b. **Number of College Graduates**

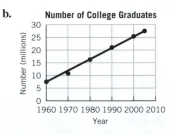

c. Two estimated points on the line are (1960, 7.5) and (2005, 27.5).

Slope $= \dfrac{(27.5 - 7.5)}{(2005 - 1960)} = \dfrac{20}{45} \approx 0.44$ million/yr.

d. $(0, 7.5)$ and $(45, 27.5)$

e. $y = 7.5 + 0.44x$

f. The number of graduates was about 7.5 million in 1960 and has been steadily increasing since—by about 0.44 million, or 440,000, persons each year.

2. a.

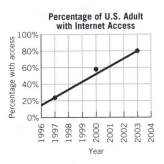

Percentage of U.S. Adult with Internet Access

b. Letting x = years from 1996, two estimated points on the line are $(0, 17)$ and $(7, 80)$. The slope $= (80-17)/(7-0) = 63/7 = 9$ percentage points per year. This tells you that the adults with access to the Internet are increasing by 9 percentage points per year.

(*Note:* You *add* 9% each year to the previous year's % value, you are not calculating a percentage.)

c. $(0, 17)$

d. If we let $I(x)$ = percent of adult population with access to the Internet x years from 1996, then $I(x) = 17 + 9x$.

e. In 1998, we have $x = 2$. So $I(2) = 17 + (9 \cdot 2) = 35\%$. In 2002, we have $x = 6$. So $I(6) = 17 + (9 \cdot 6) = 71\%$.

f. As the percentage of adults with Internet access comes close to the maximum of 100%, we would expect the rate of change to slow down. In 2003 the percent is already close to 80%, so we should not expect the linear model to continue to be appropriate.

Exercises for Section 2.9

1. Equation **a** matches with table **C.**

Equation **b** matches with table **A.**

Equation **c** matches with table **B.**

3. a. Exactly linear; $y = 1.5x - 3.5$

b. An estimated best-fit line has intercepts at $(0, 6)$ and $(2.5, 0)$, so the slope $= (6 - 0)/(0 - 2.5) = -2.4$. The equation is then $y = -2.4x + 6$. (Note that the data point $(0, 6.5)$ is not on the best-fit line.) Your answer may be somewhat different.

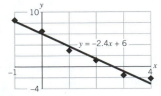

$y = -2.4x + 6$

c. Exactly linear; $y = 3x + 35$

5. a. Exactly linear; $y = 0.07$ quadrillion Btus of solar units

b. Approximately linear; farm output = 220 billion dollars

7. a. The accompanying graph shows a hand-drawn best-fit line. We can estimate two points on the line as $(1970, 10)$ and $(2000, 44)$. So the slope of the line is $(44 - 10)/(2000 - 1970) = 34/30 \approx 1.13$. So the average rate of change in the percentage of female M.D. degrees between 1970 and 2002 was about 1.13 percentage points per year.

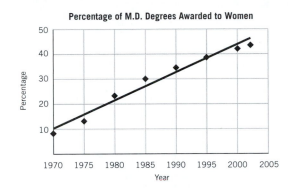

Percentage of M.D. Degrees Awarded to Women

b. To extrapolate to 2010, it helps to generate the equation of the line. If we let x = years since 1970, the slope remains the same at 1.13 percentage points per year. If $x = 0$ (at year 1970), the vertical intercept is at 10. So the equation is $y = 10 + 1.13x$ where x = years since 1970, and y = percentage of M.D. degrees awarded to women. If $y = 100\%$, we would have $100 = 10 + 1.13x \Rightarrow x = 90/1.13 \approx 80$ years since 1970 or in 2050. So if we extrapolated the model to 2050, it would predict that in that year 100% of all medical degrees would be awarded to women.

c. It is highly unlikely that this would happen. The most likely scenario is that the graph (and correspondingly the percentage of female M.D. degrees awarded) will taper off to some maximum percentage value, say at 50%. The accompanying graph fits that description. (Student answers may differ somewhat.)

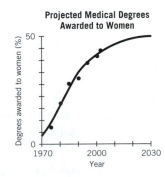

Projected Medical Degrees Awarded to Women

9. The graph on the next page gives an eyeballed best-approximation line that goes through two estimated coordinates of $(0, 150)$ and $(80, 40)$. The y-intercept is 150. The slope is $\frac{40 - 150}{80 - 0} = \frac{-110}{80} = -1.375$. Thus the equation of the graph can be estimated as $y = 150 - 1.375x$, where y stands for infant mortality rate (deaths per 1000 live births) and x stands for literacy rate (%).This means that on average for each 1% increase in literacy rate, the number of infant deaths drops by 1.375 per thousand.

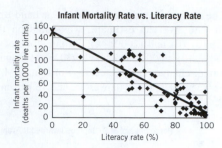

Infant Mortality Rate vs. Literacy Rate

11. a. The hand-drawn best-fit model for the years 1990 to 2003 is in the graph below.

Percentage of All U.S. Households with Cell Phones from 1990 to 2003

Slope ≈ 5.15

An eyeball estimate of two points on the best-fit line that is drawn are (1990, 5) and (2003, 72). The slope of the line between those two points is approximately $\frac{72 - 5}{2003 - 1990} = \frac{67}{13} \approx 5.15$ percentage points per year. This value indicates that on average the percent of U.S. households with cell phones grew approximately 5.15 percentage points each year between 1990 and 2003. Answers may vary slightly.

b. The percent of households having cell phones appears to be leveling off. Perhaps the saturation point has been reached.

c. Letting x = years since 1990, an estimated piecewise best-fit linear function for the combined time periods would be:

$$P(x) = \begin{cases} 5 + 5.15x & \text{for } 0 \le x < 13 \\ 70 & \text{for } 13 \le x \le 15 \end{cases}$$

where $P(x)$ = % of all households with cell phones.

13. a. Below is the graph of the data along with an estimated best-fit line. The two points (0, 5) and (40, 14) lie on the line ⇒ slope = $\frac{14 - 5}{40 - 0} = \frac{9}{40} = 0.225$

The equation of the line is $H(t) = 5 + 0.225t$

where t = years since 1960 and $H(t)$ = health care costs as % of GDP.

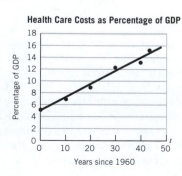

Health Care Costs as Percentage of GDP

b. 2010 corresponds to $t = 50$ years and $H(50) = 5 + 0.225 \cdot 50 = 16.25\%$ of GDP.

c. One possible reason is that the GDP has grown very large and thus the percentage for health care has not grown as much as the health care costs themselves.

15. a. Below are the graph of the data and a hand-drawn best-fit line.

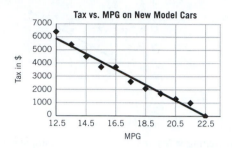

Tax vs. MPG on New Model Cars

b. The approximate coordinates of two points on this line are (17.5, 3000) and (22.5, 0) ⇒ the slope of this line is $\frac{0 - 3000}{22.5 - 17.5} = \frac{-3000}{5} = -600$ dollars per mpg. The equation of this line is $T(x) = -600x + 13,500$, where x is in mpg and $T(x)$ = tax (in dollars). The vertical intercept is very large because its value is what one would get if mpg takes the value of 0. (This value, of course, represents an impossible situation.)

c. The average rate of change for the given line is −600 dollars per mpg. As the mpg rating increases, the tax paid goes down by $600 per mpg.

Ch. 2: Check Your Understanding

1. False	**9.** True	**17.** True	**25.** False
2. True	**10.** True	**18.** False	**26.** False
3. False	**11.** False	**19.** False	**27.** False
4. False	**12.** True	**20.** True	**28.** False
5. True	**13.** True	**21.** True	**29.** True
6. True	**14.** False	**22.** False	**30.** True
7. True	**15.** True	**23.** False	
8. False	**16.** True	**24.** False	

31. Possible answer: $y = -2x + 5$, y dependent variable, x independent variable.

32. Possible answer: $D = 0$, D dependent variable, p independent variable.

33. Possible answer: $2x - 3y = 6$, y dependent variable, x independent variable.

34. Possible answer: $3x + 5y = -15$, y dependent variable, x independent variable.

35. Possible answer: $T = 37l$, T dependent variable in minutes, l independent variable in laps.

36. Possible answer: $V = 19.25 + 0.25q$, V dependent variable in dollars, q independent variable in number of quarters from now.

37. Possible answer: $C = 2T + 3$ and $C = 2(T − 2) + 3$ (or $2T − 1$), C dependent variable, T independent variable.

38. Possible answer: $y = 2x + 4$, $y = 5x + 4$, $y = 4 − 2x$, $y = 4$, y dependent variable, x independent variable.

39. Possible answer: For $m = −1/4$ (or $−0.25$), $−1/m = 4$, $d = −0.25t + 3$ and $d = 4t − 1$, d dependent variable, t independent variable.

40. Possible answer: (1990, $100), (1993, $133), (1995, $150), (2000, $175), (2003, $185).

41. True

42. False

43. False

44. False

45. True

46. True

47. True

48. False

49. False

50. False

51. False (it can be 0)

52. True

53. False (not a function)

Ch. 2 Review: Putting It All Together

1. **a.** $217,300 − $147,300 = $70,000

b. $70,000/$147,300 ≈ 0.475, or about a 47.5% increase

c. ($217,300 − $147,300)/(2006 − 2000) = $70,000/6 ≈ $11,670/year

3. **a.** The function is positive between points B and E excluding the endpoints; negative between A and B (excluding point B) and between F and G (excluding point F); zero at point B and between points E and F.

b. The slope is positive between A and C; negative between D and E and between F and G; zero between C and D and between E and F.

5. **a.** They are both linear functions of U.S. sizes since the average rates of change are constant for both.

b. The British size equals the U.S. size plus 6.

c. If U.S. sizes are denoted by U, and French sizes by F, then $F = U + 34$.

7. **a.** 34% of 296.4 = 0.34 · 296.4 ≈ 100.78 million Americans (about one-third) were without a cell phone in 2005.

b. Average rate of change =

(207.9 − 109.4) million/(2005 – 2000) =

(98.5 million)/(5 years) = 19.7 million cell phone subscriptions per year.

c. $C(t) = 109.4 + 19.7 t$, where t = years since 2000.

d. In 2010 we would have $t = 10$, and

$C(10) = 109.4 + (19.7 · 10) = 306.4$ million cell phone subscriptions. The U.S. population did exceed 300 million in fall 2006, but this extrapolation seems somewhat excessive.

9. **a.** i and ii

b. i and ii

c. iii

d. both ii (which is decreasing) and iv (which is increasing).

11. **a.** Average rate of change between weeks 1 and 31 = (3.04 − 2.22)/(31 − 1) = 0.82/30 ≈ 0.03, which means that the price of gasoline was increasing by roughly 3 cents per week.

Average rate of change between weeks 31 and 43 = (2.22 − 3.04)/(43 − 31) = −0.82/12 ≈ −0.07, which means that the price of gas was decreasing by about 7 cents per week.

Average rate of change between weeks 1 and 43 = (2.22 − 2.22)/(43 − 1) = 0/42 = 0, which means, of course, that if we look only at weeks 1 and 43, we would have the impression that gas prices had not changed at all.

b. During the summer of 2006 gasoline prices soared to an all-time high. In January the price per gallon was relatively low. By August the price per gallon had increased on average by 3 cents each week to reach a cost of $3.04 per gallon. In total the cost had increased by $0.82 per gallon or by about 37%. By October gas prices fell back to the January rate, decreasing at a rate of 7 cents per week from the August high.

13. Possible answers are:

a. $Q = 3t − 2$ or $Q = 3t$

b. $Q = −2t + 3$ or $Q = −2t$

c. $Q = 3$ or $Q = −2$

d. $Q = −2t$ or $Q = 3t$

e. $Q = 3t − 2$ or $Q = 3t$

f. $Q = −2t$ or $Q = −2t + 3$

15. **a.** (3600 people/day) · (365 days/year) = 1,314,000 people/year or, equivalently, an increase of 1.314 million people per year living in the coastal regions of the United States.

b. The coastal population $P(x)$ (in millions) can be modeled by $P(x) = 153 + 1.314x$, where x is the number of years since 2003.

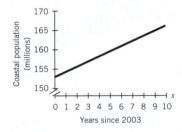

d. In 2008 $x = 5$, so $P(5) = 153 + 1.314(5) ≈ 160$ million people living in coastal regions.

17. **a.**

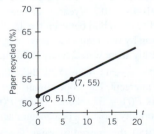

The slope = (55 − 51.5)/7 = 0.5 percentage points per year.

b. $R(t) = 51.5 + 0.5t$, where $R(t)$ is the percentage of paper recycled in t years from 2005.

c. $R(0) = 51.5\%$; $R(5) = 54\%$; $R(20) = 61.5\%$ would mean that in the year 2025, 61.5% of all paper is recycled.

19. Graph A: $y = 5$; Graph B: $x = -2$: Graph C: $y = 2x + 1$; Graph D; $y = 2x + 4$; Graph E: $y = -(1/2)x + 6$

21. a.

x	-3	-2	-1	0	1	2	3
$g(x)$	5	4	3	2	3	4	5

b.

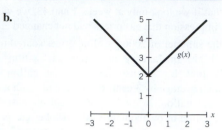

c. The graph of $g(x) = |x| + 2$ is the graph of the absolute value function $f(x) = |x|$ raised up two units.

23. Student answers will vary.

25. a. Over the last 250 years in Sweden the probability of a young child dying has steadily decreased. The child mortality rate has declined from about 40% in 1750 to less than 1% in 2000. The death rate is very similar for female and male children, though the male rates are consistently somewhat higher.

b.

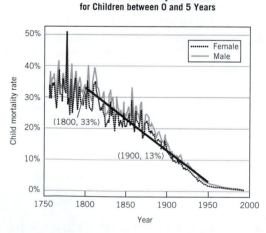

250 Years of Mortality in Sweden for Children between 0 and 5 Years

Two estimated points on the line are (1800, 33%) and (1900, 13%). So the slope of the line is $\frac{13 - 33}{1900 - 1800}$ $= -20/100 = -0.2\%$/year. This means that on average the probability of a child between the ages of 0 to 5 dying was decreasing by two-tenths of a percentage point each year, or equivalently, there were 2 fewer children dying per thousand.

c. If we let t = years since 1800, then (0, 33%) becomes the vertical intercept. The slope remains the same. So the linear model is $P(t) = 33 - 0.2t$, where $P(t)$ gives the female child mortality rate (as a percentage) at t years after 1800.

d. After 1950 the mortality rates are very low and still declining, almost approaching 0% per year. (In fact, Sweden currently has one of the lowest child mortality rates in the world.)

27. a.

Point	x-Coordinate of Point	y-Coordinate of Point	Average Rate of Change between Two Adjacent Points
A	0	4	n.a.
B	1	1	-3
C	2	0	-1
D	3	1	1
E	4	4	3

b. The function is increasing over the interval (2, 4). The average rates of change (or slopes) are increasing on that interval, which also suggests that the function is increasing.

c. The function is decreasing over the interval (0, 2). The absolute value of the average rate of change (or slope) is decreasing over that interval, which also suggests that the function is decreasing.

d. The function is concave up throughout, independent of whether the function is increasing or decreasing. The steepness of the curve (as approximated by the absolute values of the average rates of change) first decreases and then increases, also suggesting that the curve is concave up.

Exercises for EE on Education and Earnings

1. a. 0.65, 0.68, 0.07, 0.70

   b. $|-0.07|$, $|0.65|$, $|-0.68|$, and $|0.70|$

3. a. The slope = 6139 dollars per year of education past grade 8; vertical intercept = -2105 dollars; cc = 0.72

   b. On average, for each increase of a year of education past grade 8 the median personal earnings increase by $6139.

   c. $6139; $61,390

5. a. The rate of change is 10,733 dollars per year of education past grade 8.

   b. Three points are (4, 34242), (8, 77334), and (12, 10426). A sample computation: $\frac{77{,}334 - 34{,}242}{8 - 4} = \frac{43{,}092}{4} = 10{,}773$ dollars per year of education above grade 8.

   c. For each extra year of education past grade 8 the median personal total income of white males on average rises $10,733.

   d. From the FAM 1000 data:

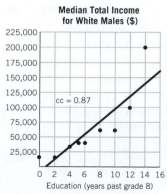

Median Total Income for White Males ($)

Median total income for white males = $-8850 + 10{,}773 \cdot$ yrs of educ. past grade 8

7. a. 0.516 is the slope of the regression line, and it indicates that, on average, each increase of an inch in the mean height of the fathers means there is an increase of 0.516 inch in the mean height of the sons.

   b. If $F = 64$, then $S = 33.73 + 0.516 \cdot 64 = 66.75$ in.

   If $F = 73$, then $S = 33.73 + 0.516 \cdot 73 = 71.40$ in.

   c. If $F = S$, then $F = 33.73 + 0.516F$ or $0.484F = 33.73$ or $S = F = 69.69$ in.

   d. For each of the data points the S value represents the mean height of all the sons whose father has the given mean height of F. There are 17 mean heights listed (from 57 to 75 inches) for the 1000 fathers.

9. a, b. Here is a graph that gives the Excel-generated regression lines, their equations, and the cc values for public and private 4-year colleges over the time period given in the table. Student estimates of the regression lines for each will vary.

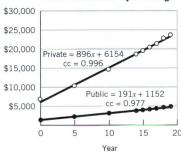

Mean Tuition and Fee Costs at Public and Private 4-year College

Private = 896x + 6154
cc = 0.996

Public = 191x + 1152
cc = 0.977

   c. In 2010 the cost at a public college will be $191 \cdot 25 + 1152 = \$5927$ and in private colleges the cost will be $896 \cdot 25 + 6154 = \$28,554$.

11. a.

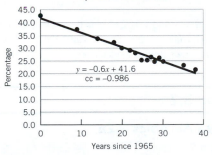

Percentage of Smokers among Total Population 18 and Older

y = −0.6x + 41.6
cc = −0.986

   i. $\frac{20.9 - 42.4}{2005 - 1965} = \frac{-22}{40} \approx -0.55$ percentage point per year for all smokers from 1965 to 2005.

   ii. $\frac{20.9 - 25.5}{2005 - 1990} = \frac{-4.6}{15} \approx -0.31$ percentage point per year for all smokers from 1990 to 2005.

   b. Student estimates of the regression line will vary. The Excel-generated regression line is shown in the diagram in part (a), and the equation is Percentage $= -0.6x + 41.6$, where $x =$ years since 1965. The average rate of change is -0.6 percentage point per year.

   c. Student estimates of the regression line will vary. The line drawn in the diagram is the Excel-generated regression line. The cc is quite good: -0.986.

   d. The regression line equation for males is: Percentage $= -0.7x + 48.7$, and the regression line equation for

females is: Percentage $= -0.4x + 35.3$, where $x =$ years since 1965 for both. The cc for males is -0.972 and for females it is -0.973, both quite good.

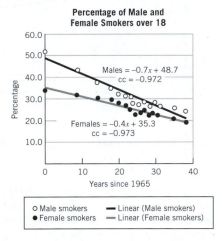

Percentage of Male and Female Smokers over 18

Males = −0.7x + 48.7
cc = −0.972

Females = −0.4x + 35.3
cc = −0.973

○ Male smokers   — Linear (Male smokers)
● Female smokers   ⋯ Linear (Female smokers)

   e. Student answers will vary. The more notable factors are the downward trend overall for both males and females and the fact that the downward trend is more rapid among males and that the percentage of smokers is always higher among males than among females. Students, in citing these factors, should also be citing actual figures.

13. a. The regression line and equation and the cc are given in the accompanying diagram.

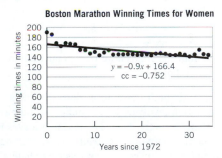

Boston Marathon Winning Times for Women

y = −0.9x + 166.4
cc = −0.752

   b. In 2010 $x = 38$ and time predicted $= -0.9 \cdot 38 + 166.4 \approx 132.2$ minutes. This does not seem to be a reasonable winning time. It is 10 minutes less than the recent winning times.

   c. Using an eyeball estimate, one gets a horizontal line that goes through approximately $y = 144$. Yes, it is more realistic than the estimate in part (b).

   d. Student answers will differ, but all should mention that the data seem to fall along a horizontal line in the constricted time span, indicating a leveling at times just above 140 minutes.

15. a. The plot of the data and the regression line are given in the accompanying diagram. The formula relating boiling temperatures in °F to altitude is $F = 211.80 - 0.0018H$, or when suitably rounded off, $F = 212 - 0.002H$, where H is feet above sea level. The correlation coefficient is -0.9999. The answer to the second part depends on where the student lives. Other factors could be something put into the water, such as salt, or variations in the air pressure.

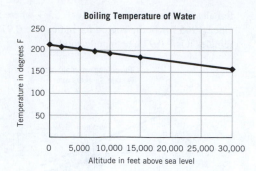

Boiling Temperature of Water

b. On Mt. McKinley water boils at $212 - 0.002 \cdot 20320 = 171.36\,°F$; in Death Valley water boils at $212 - 0.002 \cdot (-285) = 212.57\,°F$.

c. $32 = 212 - 0.002H$ or $-180 = -0.002H$ or $H = 90,000$ feet or approximately 17 miles. But this seems unreasonable since then the water would be outside earth's atmosphere layer (which goes to 9 miles above the earth).

17. a. By computer or calculator one gets that the regression line equation as:

$y = 3.5x + 26.7$, where $x =$ years since 1945 and y is measured in millions of registrations. [The cc. is 0.998].

b. On average the number of registrations goes up approximately 3.5 million per year since 1945.

c. 2004 is 59 years after 1945, and thus $y = 3.5 \cdot 59 + 26.7 = 233.2$ million. It is about 4.9 million too high.

d. 2010 is 65 years from 1945. Thus $y = 3.5 \cdot 65 + 26.7 = 254.2$ million. This seems to be too high since the entire U.S. population in 2007 was just over 300 million.

19. a. Of itself high correlation does not mean that there is causation involved. More studies would have to be done and in fact were done. The research leaves no doubt that cigarette smoking is indeed a cause of lung cancer. The high correlation coefficient was nevertheless an important factor.

CHAPTER 3

Section 3.1

Algebra Aerobics 3.1

1. Gas is the cheapest system from approximately 17.5 years of operation to approximately 32.5 years of operation. Solar becomes the cheapest system after approximately 32.5 years of operation.

2. a. $(3, -1)$ is a solution since: $4(3) + 3(-1) = 9$ and $5(3) + 2(-1) = 13$. It is a solution of both equations.

b. $(1, 4)$ is not a solution since: $5(1) + 2(4) = 13$ but $4(1) + 3(4) = 16$, not 9. So it is not a solution of $4x + 3 = 9$.

3. a. $12x - 9y = 18$ is equivalent to $4x = 6 + 3y$ because

$$\frac{1}{3}(12x - 9y) = \frac{1}{3}(18) \Rightarrow 4x - 3y = 6$$
$$\Rightarrow 4x = 6 + 3y.$$

So these equations represent the same line.

b. There is an infinite number of solutions to the system of equations in part (a) since every solution to the equation (every point on the line) is a solution to the system.

4. Graph A: $(0, 2.5)$

Graph B: $(4, -4)$

Exercises for Section 3.1

1. a. $4(5) - 3(-10) = 50$ and $2(5) + 2(-10) = -10$, and thus $(5, -10)$ does not solve the given system.

b. The coordinates must satisfy both equations.

3. A solution to a system of equations is a number (or set of numbers) that satisfies all of the equations in the system.

5. a. The coordinates are approximately (1995, $380,000)

b. To the left of the intersection point the population of Pittsburgh is larger than that of Las Vegas, and to the right the population of Las Vegas is greater than that of Pittsburgh.

7. a.   b.

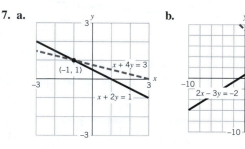

9. $y = -x - 2$ and $y = 2x - 8$ is the system, and the solution is $(2, -4)$. *Check:* $-2 - 2 = -4$ and $2 \cdot 2 - 8 = -4$, and thus the claimed solution works.

Section 3.2

Algebra Aerobics 3.2a

1. a. $y = 7 - 2x$

b. $y = \frac{6 - 3x}{5} = \frac{6}{5} - \frac{3}{5}x$

c. $x = 2y - 1$

2. a. no solution, because the lines have the same slope but different y-intercepts, so they are parallel.

b. one solution, because the lines have different slopes.

3. a. Set $y = y \Rightarrow x + 4 = -2x + 7 \Rightarrow 3x = 3 \Rightarrow x = 1$; $y = -2(1) + 7 = 5$. *Check:* $y = (1) + 4 \Rightarrow y = 5 \Rightarrow$ solution (x, y) is $(1, 5)$.

b. Set $y = y \Rightarrow -1700 + 2100x = 4700 + 1300x \Rightarrow 800x = 6400 \Rightarrow x = 8$; $y = 4700 + 1300(8) = 15,100 \Rightarrow$ solution (x, y) is $(8, 15100)$.

c. Set $F = F \Rightarrow C = 32 + \frac{9}{5}C \Rightarrow 5C = 32(5) + 9C \Rightarrow$
$-4C = 160 \Rightarrow C = -40$
$F = C \Rightarrow F = -40 \Rightarrow$ solution (C, F) is $(-40, -40)$.

4. a. Substitute $y = x + 3$ into $5y - 2x = 21 \Rightarrow 5(x + 3) - 2x = 21 \Rightarrow 5x + 15 - 2x = 21 \Rightarrow 3x = 6; x = 2;$ so, $y = (2) + 3 = 5 \Rightarrow y = 5.$ Solution (x, y) is $(2, 5)$.

b. Substitute $z = 3w + 1$ into $9w + 4z = 11 \Rightarrow 9w + 4(3w + 1)$
$= 11 \Rightarrow 9w + 12w + 4 = 11 \Rightarrow 21w = 7 \Rightarrow w = 1/3;$ so
$z = 3(\frac{1}{3}) + 1 = 2 \Rightarrow z = 2.$ Solution (w, z) is $(\frac{1}{3}, 2)$

c. Substitute $x = 2y - 5$ into $4y - 3x = 9 \Rightarrow 4y - 3(2y - 5) = 9$
$\Rightarrow 4y - 6y + 15 = 9 \Rightarrow -2y = -6 \Rightarrow y = 3;$ so
$x = 2(3) - 5 = 1 \Rightarrow x = 1.$ Solution (x, y) is $(1, 3)$.

d. Solve: $r - 2s = 5$ for r, and substitute the resulting expression for r into $3r - 10s = 13.$ $r = 2s + 5 \Rightarrow 3(2s + 5) - 10s = 13 \Rightarrow 6s + 15 - 10s = 13 \Rightarrow -4s = -2 \Rightarrow s = \frac{1}{2};$ so $r = 2(\frac{1}{2}) + 5 = 6 \Rightarrow r = 6.$
Solution (r, s) is $(6, \frac{1}{2})$.

Algebra Aerobics 3.2b

1. a. By the elimination method, add equations: $2y - 5x = -1$
and $3y + 5x = 11 \Rightarrow 5y = 10 \Rightarrow y = 2;$ so $3(2) + 5x = 11 \Rightarrow 5x = 5 \Rightarrow x = 1.$ Solution (x, y) is $(1, 2)$.

b. Multiply the equation $(3x + 2y = 16)$ by 3 and the equation $(2x - 3y = -11)$ by $2 \Rightarrow 9x + 6y = 48$ and $4x - 6y = -22.$ By the elimination method, add these equations $\Rightarrow 13x = 26 \Rightarrow x = 2,$ so $3(2) + 2y = 16 \Rightarrow 2y = 10 \Rightarrow y = 5.$ Solution (x, y) is $(2, 5)$.

c. By substitution of $t = 3r - 4$ into $4t + 6 = 7r \Rightarrow 4(3r - 4) + 6 = 7r \Rightarrow 12r - 16 + 6 = 7r \Rightarrow -10 + 5r = 0 \Rightarrow 5r = 10 \Rightarrow r = 2,$ so $t = 3(2) - 4 = 2 \Rightarrow t = 2.$ Solution (r, t) is $(2, 2)$.

d. Substitute $z = 2000 + 0.4(x - 10,000)$ into $z = 800 + 0.2x$
$\Rightarrow 2000 + 0.4(x - 10,000) = 800 + 0.2x \Rightarrow 2000 + 0.4x - 4000 = 800 + 0.2x \Rightarrow 0.2x = 2800 \Rightarrow x = 14,000;$ so, $z = 800 + 0.2(14,000) = 3600 \Rightarrow z = 3600.$ Solution (x, z) is $(14000, 3600)$.

2. a. By substitution: $2x + 4 = -x + 4 \Rightarrow 3x = 0 \Rightarrow x = 0,$ so $y = -(0) + 4 = 4 \Rightarrow y = 4.$ So solution (x, y) is $(0, 4)$.

b. By substitution of $(y = -6x + 4)$ into $(5y + 30x = 20) \Rightarrow 5(-6x + 4) + 30x = 20 \Rightarrow -30x + 20 + 30x = 20 \Rightarrow 20 = 20.$ So both equations must be equivalent. There are infinitely many solutions since both equations describe the same line.

c. $2y = 700x + 3500 \Rightarrow y = 350x + 1750.$ The slopes of the lines of both the equations are 350, but the y-intercepts are different (1500 and 1750), so the lines are parallel. There is no solution.

3. In order for a system of equations to have no solutions, they must produce parallel lines with the same slope, but different y-intercepts. One example is: $y = 5x + 10; y = 5x + 3$.

4. a. $2x + 5y = 7 \Rightarrow y = \frac{-2x + 7}{5}; 3x - 8y = -1 \Rightarrow$

$y = \frac{3x + 1}{8} \Rightarrow$ one solution since the lines have unequal slopes of $-\frac{2}{5}$ and $\frac{3}{8}$.

b. $3x + y = 6 \Rightarrow y = 6 - 3x; 6x + 2y = 5 \Rightarrow y = \frac{5 - 6x}{2}$ or
$y = \frac{5}{2} - 3x \Rightarrow$ no solution since the lines have the same slope of -3 and different y-intercepts of 6 and $\frac{5}{2}$.

c. $2x + 3y = 1 \Rightarrow y = \frac{1}{3} - \frac{2}{3}x; 4x + 6y = 2 \Rightarrow y = \frac{1}{3} - \frac{2}{3}x$
$\Rightarrow$ equivalent equations and an infinite number of solutions, since the lines have same slopes and same y-intercepts.

d. $3x + y = 8 \Rightarrow y = 8 - 3x; 3x + 2y = 8 \Rightarrow y = 4 - \frac{3}{2}x \Rightarrow$ one solution since the lines have unequal slopes of -3 and $-\frac{3}{2}$.

5. a. $6(\frac{x}{2} + \frac{y}{3}) = 6(3) \Rightarrow 3x + 2y = 18.$ Substitute $y = x + 4$
$\Rightarrow 3x + 2(x + 4) = 18 \Rightarrow 5x + 8 = 18 \Rightarrow 5x = 10 \Rightarrow x = 2;$ so, $y = (2) + 4 = 6 \Rightarrow y = 6.$ Solution (x, y) is $(2, 6)$.

b. $-60(0.5x + 0.7y) = -60(10) \Rightarrow -30x - 42y = -600.$
Add to $30x + 50y = 1000 \Rightarrow 8y = 400 \Rightarrow y = 50,$ so
$30x + 50(50) = 1000 \Rightarrow 30x + 2500 = 1000 \Rightarrow 30x = -1500 \Rightarrow x = -50,$ so the solution (x, y) is $(-50, 50)$.

6. a. $4(39) + 3q = 240 \Rightarrow 3q = 84 \Rightarrow q = 28$ gals

b. $4p + 3(20) = 240 \Rightarrow 4p = 180 \Rightarrow p = \$45$ per gal

c, d. Solve for p: $4p = 240 - 3q \Rightarrow p = 60 - 3/4q \Rightarrow p = 60 - 0.75q$.

q	p
0	60
20	45
40	30

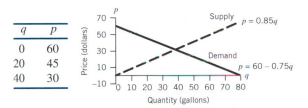

e. Solve by substitution of $p = 0.85q$ into $4p + 3q = 240 \Rightarrow 4(0.85q) + 3q = 240 \Rightarrow q = 37.5; p = 0.85(37.5) = 31.9.$ So the equilibrium point is $\sim(38, \$32),$ which means that when the price is around \$32, the demand will be around 38 gallons.

f. There is a surplus of supply because where the line $p = 39$ crosses the supply curve, it is above the demand curve, so the supply is greater than the demand.

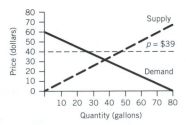

7. For this system to have an infinite number of solutions, the two linear equations should represent the same line. The slope of the line of the first equation is 2, and the y-intercept is 4. Solving the second equation for y, we get $y = -??/2x + 4.$ Thus, ??, the coefficient for which we are solving, must be $-4,$ so the slope of that line is also 2.

Exercises for Section 3.2

1. a. After approximately 9 months.

b. Approximately $8.30 per hour.

c. $W_A(m) = 7.00 + 0.15m$ and $W_B(m) = 7.45 + 0.10m$

d. The exact place where they meet is $m = 9$, and $W_A(9) = W_B(9) = \$8.35$ per hour.

e. The exact common hourly wage is $8.35 per hour.

f. Before 9 months, the monthly wage rates at company B are higher; after 9 months the monthly wage rates at company A are higher. If one had more information about the number of hours worked at each place, one could judge the companies on accumulated wages instead of hourly rates. But no such information is given.

3. a. The graph of the linear system is given with the intersection point marked.

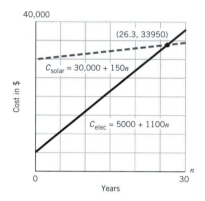

b. 1100 and 150 are the slopes of the heating cost lines; 1100 represents the rate of change of the total cost in dollars for electric heating per year since installation; 150 is the rate of change in the total cost for solar heat in the same units.

c. 5000 is the initial cost of installing the electric heat in dollars; 30,000 is the initial cost in dollars of installing solar heating. It cost a lot more initially to install solar heating than to install electrical heating.

d. The point of intersection is approximately where $n = 26$ and $C = 34,000$.

e. $n \approx 26.32$, $C \approx 33,947.37$ is a more precise answer; the values have been rounded off to two decimal places; they were obtained by setting the equations equal to each other.

f. Assuming simultaneous installation of both heating systems, the total cost of solar heat was higher than the total cost of electric heat up to year 26 (plus nearly 4 months); after that the total cost of electric heat will be greater than that of solar heat.

5. a. Setting the two equations equal gives $20,000 + 2500n = 25,000 + 2000n$ or $500n = 5000$ or $n = 10$. Plugging in that n value gives $S = 20,000 + 2500 \cdot 10 = 45,000$

b. The graphs of the two linear equations are given in the diagram below. From inspecting the graphs it seems that the intersection occurs when $n = 10$ and $S = 45,000$.

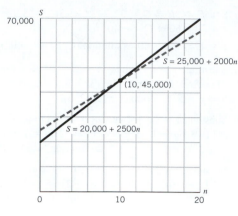

7. a. The method that is easiest is often a judgment by the person solving the problem.

    i. Either is easy   **iv.** Substitution

    ii. Elimination   **v.** Either is easy

    iii. Substitution   **vi.** Substitution

b. **i.** Setting the y values equal to each other gives $6 = -4$. Thus there is no solution.

    ii. Letting $y = 2x - 5$ in the second equation gives $5x + 2(2x - 5) = 8$ or $9x - 10 = 8$ or $9x = 18$ or $x = 2$. Thus $y = 2 \cdot 2 - 5 = -1$ and therefore the solution is $(2, -1)$. *Check:* $2(2) - (-1) = 5$; and $5 \cdot 2 + 2 \cdot (-1) = 8$.

    iii. Substituting $x = 7y - 30$ into the first equation gives $3 \cdot (7y - 30) + 2y = 2$ or $21y - 90 + 2y = 2$ or $23y = 92$ or $y = 4$ and then $x = 28 - 30 = -2$. Thus the solution is $(-2, 4)$. *Check:* $3 \cdot (-2) + 2 \cdot 4 = -6 + 8 = 2$ and $7 \cdot 4 - 30 = -2$.

    iv. Substituting $y = 2x - 3$ into second equation gives: $4(2x - 3) - 8x = -12$ or $8x - 12 - 8x = -12$ or $0 = 0$; thus the two equations have the same line as their graph. Thus all points on the line $y = 2x - 3$ are solutions.

    v. Elimination yields $0 = -6$ and thus there is no solution.

    vi. Substituting $y = 3$ into the second equation gives $x + 2 \cdot 3 = 11$ or $x = 5$. Thus the solution is $(5, 3)$. *Check:* $3 \cdot 3 = 9$ and $5 + 2 \cdot 3 = 11$.

9. a. Subtracting the first equation from the second yields $4x = -12$ or $x = -3$. Putting this value into the first equation gives $-3 + 3y = 6$ and thus $3y = 9$ and $y = 3$. Putting this value of x into the second equation gives $5(-3) + 3(3) = -6$ and thus the solution is $(-3, 3)$.

b. The graphs of the two equations and the coordinates of the intersection point are shown in the accompanying figure.

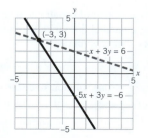

11. Let x = amount to be invested at 4% and let y = amount to be invested at 8%. Then the system of equations to be solved is $x + y = 2000$ and $0.04x + 0.08y = 100$. Substituting $y = 2000 - x$ into the second equation gives, after simplification, $x = \$1500$ and thus $y = \$500$. [*Check*: $1500 + 500 = 2000$ and $0.04 \cdot 1500 + 0.08 \cdot 500 = 100$.]

13. a. Letting $y = x - 4/3$ from the second equation and substituting this value in the first equation, we get $\frac{x}{3} + \frac{x - 4/3}{2} = 1$. Multiplying both sides by 6 gives $2x + 3(x - 4/3) = 6$ or $5x - 4 = 6$ or $x = 2$ and thus $y = 2 - 4/3 = 2/3$. [*Check*: $2 - 2/3 = 4/3$ and $2/3 + 1/3 = 1$.]

   b. Substituting $y = x/2$ from the second equation into the first equation we get $x/4 + x/2 = 9$ or $(3/4)x = 9$ or $x = 12$. Then $y = 12/2 = 6$. [*Check*: $12/4 + 6 = 9$.]

15. a. The two equations in m and b are: $-2 = 2m + b$ and $13 = -3m + b$ and the solution is $m = -3$ and $b = 4$. [*Check*: $2(-3) + 4 = -2$ and $-3(-3) + 4 = 13$.]

   b. The two equations in m and b are: $38 = 10m + b$ and $-4.5 = 1.5m + b$. The solution is $m = 5$ and $b = -12$. [*Check*: $38 = 5 \cdot 10 - 12$ and $-4.5 = 5 \cdot 1.5 - 12$.]

17. a. 12.5 is the production cost per shirt in dollars.

   b. 15.5 is the selling price of a shirt in dollars.

   c. $15.5x = 12.5x + 360$ or $3x = 360$ or $x = 120$ and $y = 15.5 \cdot 120 = \$1860$.

   d. When $x = 120$, then $C = 12.5 \cdot 120 + 360 = \$1860$ and $R = 15.5 \cdot 120 = \$1860$.

19. a. The graphs of the supply and demand equations are in the accompanying diagram.

   b. The equilibrium point is shown in the diagram. It is the spot where supply meets the demand; i.e., if the company charges \$410 for a bike it will sell exactly 4000 of them and have none left over.

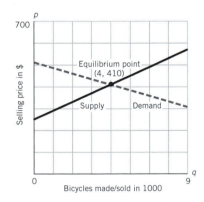

21. Two equations are equivalent if their graphs are the same, i.e., they have the same sets of solutions. An example is the system $2x + y = 1$ and $4x + 2y = 2$.

23. If we make the origin the spot on the diagram where the height (in feet) or H-axis meets the ground and let d be the distance (in feet) from the origin, then the equation of the ramp is $H = 3 - \frac{1}{12} d$. The equation to describe the rising ground is $H = \frac{1}{20} d$. Setting them equal to each other gives

$3 - \frac{1}{12} d = \frac{1}{20} d$ or $\frac{36 - d}{12} = \frac{d}{20}$ or $720 - 20d = 12d$ or $720 = 32d$ or $d = 22.5$ ft and $H = 22.5/20 = 1.125$ ft. Thus the point of meeting is where $d = 22.5$ ft from the platform and $H = 1.125$ ft above ground level.

25. a. Eliminate z; $11x + 7y = 68$ (4)

   b. Eliminate z; $9x + 7y = 62$ (5)

   c. $x = 3$ and $y = 5$ satisfy (4) and (5)

   d. Thus $z = 2 \cdot 3 + 3 \cdot 5 - 11 = 10$

   e. Thus the solution is $x = 3$, $y = 5$, $z = 10$ and the check is below:

   (1) $2 \cdot 3 + 3 \cdot 5 - 10 = 11$

   (2) $5 \cdot 3 - 2 \cdot 5 + 3 \cdot 10 = 35$

   (3) $1 \cdot 3 - 5 \cdot 5 + 4 \cdot 10 = 18$

27. Answers will vary.

   a. The system $y = x + 5$ and $y = x + 6$ has no solution.

   b. The system $y = x + 5$ and $y = -x + 5$ has exactly one solution.

   c. Algebraically: setting $x + 5 = -x + 5$ gives $2x = 0$ or $x = 0$ and thus $y = 5$; alternatively, adding the two equations together gives $2y = 10$ or $y = 5$ and thus $x = 0$. The graphs of the two lines intersecting at the point claimed is in the accompanying diagram. The answers agree.

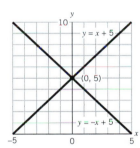

29. The system of equations has no solution if the graphs of the two equations are parallel and distinct lines. This occurs when $m_1 = m_2$ (parallel means same slope) and $b_1 \neq b_2$ (different vertical intercepts).

31. a. The equations of this pair of lines are: $y - 5 = -4(x - 2)$ and $y - 5 = 3.5(x - 2)$, or in simplified form: $y = -4x + 13$ and $y = 3.5x - 2$.

   b. The graphs of the two equations are given in the accompanying diagram. [*Check*: $-4 \cdot 2 + 13 = 5$ and $3.5 \cdot 2 - 2 = 5$.]

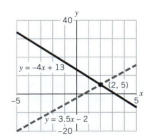

33. a. $y_B = 30, y_A = 0.625x$.

b. The common point in space that both planes will eventually occupy is where $x = 48$ and $y_B = 30$. (It is the intersection point of the graphs of $y = 30$ and $y_A = 0.625x$. These are equations for constant altitude of the flight paths of the two planes.)

c. B, in going from $(-30, 30)$ to $(48, 30)$, travels a distance of 78 miles, and this takes B 13 minutes to do (since it is traveling at 6 miles/minute). A, in traveling from $(80, 50)$ to $(48, 30)$, covers $\sqrt{(30 - 50)^2 + (48 - 80)^2} = \sqrt{400 + 1024} = \sqrt{1424} \approx 37.7$ miles, and this will take approximately 18.9 minutes (since plane A is traveling at 2 miles per minute). Thus plane A will arrive at this point nearly 6 minutes after plane B. It is a safe situation.

35. a. For a given price the new supply curve shows more items being made.

b. The requested graph is given in the diagram. In going from the old equilibrium point to the new one, the price goes down and the quantity made/sold goes up at the equilibrium point.

37. a. Higher birth rate:

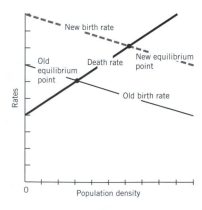

If the birth rate increases (and the death rate stays the same) the equilibrium point moves to the right and up. This means that the equilibrium point will occur at a greater population density and a higher birth rate.

b. Lower birth rate:

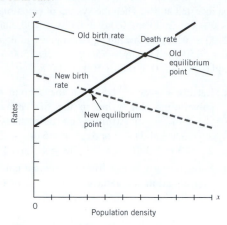

If the birth rate decreases (and the death rate stays the same) the equilibrium moves to the left and down. This means the equilibrium point will occur at a lesser population density and lower birth rate.

Section 3.3

Algebra Aerobics 3.3

1. a.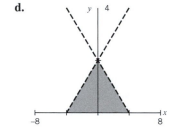

2. a. $(2, 3)$ is a solution because: $3 > 2(2) - 3 \Rightarrow 3 > 1$ and $3 \leq 3(2) + 8 \Rightarrow 3 \leq 14$ are true.

b. $(-4, 7)$ is not a solution because: $7 \leq 3(-4) + 8 \Rightarrow 7 \leq -4$ is not true.

c. $(0, 8)$ is a solution because: $8 > 2(0) - 3 \Rightarrow 8 > -3$ and $(8) \leq 3(0) + 8 \Rightarrow 8 \leq 8$ are true.

d. $(-4, -6)$ is a solution because: $-6 > 2(-4) - 3 \Rightarrow -6 > -11$ and $-6 \leq 3(-4) + 8 \Rightarrow -6 \leq -4$ are true.

e. $(20, -8)$ is not a solution because: $-8 > 2(20) - 3 \Rightarrow$ $-8 > 37$ is not true.

f. $(1, -1)$ is not a solution because: $-1 > 2(1) - 3 \Rightarrow$ $-1 > -1$ is not true.

3. A. $y \le 2 - x$ **B.** $y > 1 + 2x$ **C.** $y \ge -3$ **D.** $x > 4$

4. a. Approximately (100, $700). For sales of 100 books, the cost is equal to the revenue, which is $700.

b. The region between the two graphs to the left of the breakeven point.

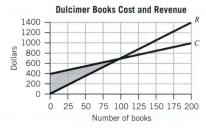

Dulcimer Books Cost and Revenue

c. $400 because that is the cost for selling 0 books (vertical intercept of the cost equation).

d. Assuming fixed costs at $400, $C_1 \ge 3x + 400$, $R_1 \le 7x$.

5. a. $C(n) = \$50{,}000 + 235n$; $R(n) = 270n$.

b. $C(n) = R(n)$ at breakeven point $\Rightarrow 50{,}000 + 235n = 270n \Rightarrow n \approx 1429$ tons. Selling about 1429 tons will yield a profit of $0 since cost = revenue at the breakeven point.

c.

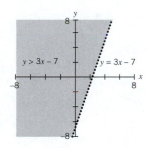

Exercises for Section 3.3

1. a.

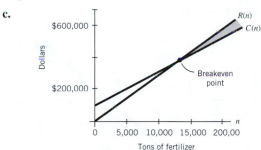

b.

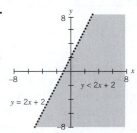

c.

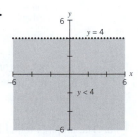

d.

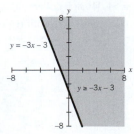

3. a. $y < \frac{2}{3}x + 2$ **b.** $y \ge -\frac{3}{2}x + 3$

5. a. Yes, (0, 0) satisfies the inequality.

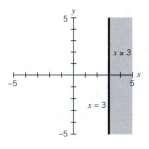

b. No, (0, 0) does not satisfy the inequality.

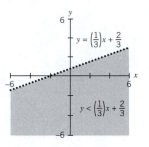

c. Yes, (0, 0) satisfies the inequality.

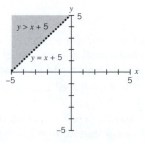

d. No, (0, 0) does not satisfy the inequality.

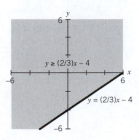

7. One looks to see if (0, 0) satisfies the inequality; it does here and thus the region is that half of the plane that contains (0, 0). In this case the shaded region is above the line.

9. a. goes with **g.**   **c.** goes with **j.**   **e.** goes with **h.**

    b. goes with **i.**   **d.** goes with **f.**

11. a. l_1 has the equation $y = 1 + 0.25x$ and l_2 has the equation $y = 3 - 1.5x$.

b. $3 - 1.5x \leq y \leq 1 + 0.2x$

13.

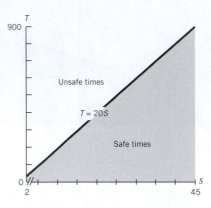

a. $T = 20S$

b. The graph is in the diagram. A suitable domain is $2 \leq S \leq 45$.

c. $T > 20S$ denotes unsafe times.

d. The shading and labels are found in the diagram.

e. The equation would be $T = 40S$ and its slope would be steeper.

15.

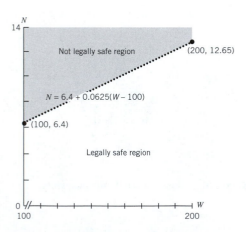

a. $N > 6.4 + 0.0625(W - 100)$, where W is measured in pounds and N measures the number of ounces of beer that gets one to the legal limit for safe driving.

b. The sketch of the shaded areas is found in the diagram for $100 \leq W \leq 200$.

c. If $W = 100$ lb, then $N = 6.4$ oz. Thus one may legally drink 6.4 oz or less; for $W = 150$ lb we have $N = 9.525$ oz and for $W = 200$ lb we have $N = 12.65$ oz.

d. $N = 0.0625W + 0.15$

e. The given rule of thumb translates into the formula $N = 6 + 0.05(W - 100) = 0.050W + 1$. Thus it starts out higher and grows more slowly than the legal one.

But its graph is lower from $W = 100$ to $W = 200$. It is a safe rule.

17.

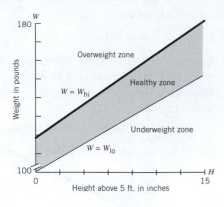

a. The two formulae and the three zones are graphed in the diagram.

b. $100 + 3.5H \leq W \leq 118.2 + 4.2H$ lb for $0 \leq H \leq 15$ in above 5 ft.

c. $W_{lo}(2) = 107$ lb and $W_{hi}(2) = 126.6$ lb. Thus the shorter woman is overweight. For the taller woman: $W_{lo}(5) = 117.5$ and $W_{hi}(5) = 139.2$. Thus the taller woman is in the healthy range.

d. $W_{hi}(4) = 135$ and $(165 - 135)/1.5 = 30/1.5 = 20$. Thus it would take 20 weeks for this woman to reach the top of the healthy range.

19.

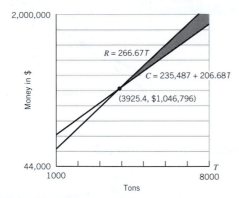

a. $C = 235{,}487 + 206.68T$ gives the cost in dollars when T is measured in tons of fertilizer produced. $R = 266.67T$ gives the revenue in dollars from selling T tons of fertilizer.

b. The graph is found in the accompanying diagram and the breakeven point is marked on the graph. It is where $T \approx 3925.4$ tons and $M \approx \$1{,}046{,}800$ dollars

c. The inequality $R - C > 0$ describes the profit region, and this occurs when $T > 3925.4$. It is shaded in the accompanying graph.

21. For Graph A: $x \geq 0$, $y \geq 0$, and $y < -1.5x + 3$
For Graph B: $x + 1 \leq y < 2x + 2$.

23.

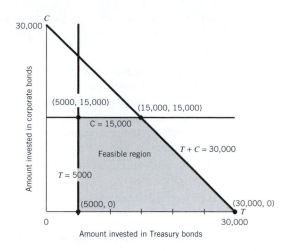

(5000, 15,000)   (15,000, 15,000)

C = 15,000

Feasible region   T + C = 30,000

T = 5000

(5000, 0)   (30,000, 0)

0   30,000

Amount invested in Treasury bonds

Amount invested in corporate bonds

a. $0 \leq T + C \leq 30{,}000$, $0 \leq C \leq 15{,}000$ and $5000 \leq T \leq 30{,}000$

b. The feasible region is the shaded area of the graph.

c. Intersection points and interpretations: (5000, 0) is where $5000 is invested in T bonds; (30000, 0) is where all $30,000 is in T bonds; (5000, 15000) is where $5000 is in T bonds and $15,000 is in C bonds; and (15000, 15000) is where $15,000 is in each kind.

Section 3.4

Algebra Aerobics 3.4

1. a.

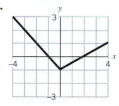

c.

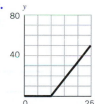

b.

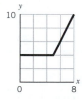

2. Graph A: $f(x) = \begin{cases} x + 3 & \text{for } x \leq 3 \\ -2x + 12 & \text{for } x > 3 \end{cases}$

Graph B: $f(x) = \begin{cases} -2 & \text{for } x \leq 3 \\ 2x - 8 & \text{for } x > 3 \end{cases}$

3. a. $P(-5) = 3$, $P(0) = 3$, $P(2) = -3$, $P(10) = -19$
   b. $W(-5) = -9$, $W(0) = -4$, $W(2) = 6$, $W(10) = 14$.

4. a. $g(i) = \begin{cases} 0.05i & \text{for } 0 \leq i \leq \$50{,}000 \\ 2500 + 0.08(i - 50{,}000) & \text{for } i > \$50{,}000 \end{cases}$

b. $g(i) = \begin{cases} 0.06i & \text{for } 0 \leq i \leq \$30{,}000 \\ 1800 + 0.09(i - 30{,}000) & \text{for } i > \$30{,}000 \end{cases}$

5. a. $f(30{,}000) = 0.0595(30{,}000)$; $g(30{,}000) = 0.055(30{,}000)$
   $= \$1785$ $= \$1650$

   Flat tax is $135 higher for $30,000 income.

b. $f(60{,}000) = 0.0595(60{,}000)$
   $= \$3570$

   $g(60{,}000) = 2761 + 0.088(60{,}000 - 50{,}200)$
   $= 2761 + 0.088(9800)$
   $= 2761 + 862.40$
   $= \$3623.40$

   Graduated tax is $53.40 higher for $60,000 income.

c. $f(120{,}000) = 0.0595(120{,}000)$
   $= \$7{,}140$

   $g(120.000) = 6263 + 0.098(120{,}000 - 90{,}000)$
   $= 6263 + 2940$
   $= \$9{,}203$

   Graduated tax is $2063 higher for $120,000 income.

6. $y = \begin{cases} 100 + 1.5x & \text{for } 0 \leq x < 200 \\ 400 & \text{for } 200 \leq x \leq 500 \end{cases}$

Exercises for Section 3.4

1. $f(-10) = 2 \cdot (-10) + 1 = -19$
   $f(-2) = 2(-2) + 1 = -3$
   $f(0) = 2 \cdot 0 + 1 = 1$
   $f(2) = 3 \cdot 2 = 6$ and
   $f(4) = 3 \cdot 4 = 12$

3. a. goes with Graph *B*.   **b.** goes with Graph *A*.

5. a. $y = \begin{cases} 1 & \text{for } 0 \leq x \leq 1 \\ x & \text{for } x > 1 \end{cases}$

b. $y = \begin{cases} 1 - x & \text{for } 0 \leq x \leq 1 \\ 1.5x - 1.5 & \text{for } x > 1 \end{cases}$

7. a. Graph of $y = h(x)$

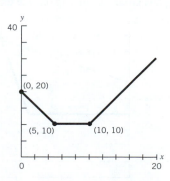

(0, 20)

(5, 10)   (10, 10)

b. Graph of $y = k(x)$

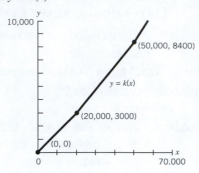

9. Answers will vary from state to state. Check student answers against the local tax form itself.

11. a. 45 mph is the speed limit.

b.

Speed (mph)	Fine ($)
40	0
45	0
50	75
55	100
60	150
65	200
70	300
75	400
80	500

c. Check the range category in which the clocked speed is found and then apply the formula for that range category.

d. In each category this number represents how much more a person is fined for each increase of 1 mph in speed.

e. $F(30) = 0$; $F(57) = 100 + 10 \cdot 2 = \$120$; and $F(67) = 200 + 20 \cdot 2 = \$240$.

f.

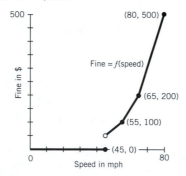

13. The graph is given here to help one see the answers:

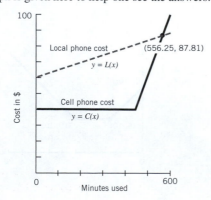

a. $C(x) = \begin{cases} 40 & \text{if } 0 \le x \le 450 \\ 40 + 0.45(x - 450) & \text{if } x > 450 \end{cases}$

$L(x) = 60 + 0.05x$ if $x \ge 0$

where x measures minutes used for long distance and $C(x)$ and $L(x)$ are measured in dollars.

The two cost functions are graphed in the diagram. They meet at $x = 556.25$ minutes and $C(x) = L(x) \approx \$87.81$. Thus the two plans cost the same at the point where one uses 556.25 minutes for long distance.

c. It would be more advantageous to use the cell phone for $0 \le x < 556.25$ minutes

d. It would be more advantageous to use the local company plan if $x > 556.25$ minutes.

15. The graphs for parts (a) and (b) are shown in the accompanying diagram.

a. For $0 \le T \le 20$: $D_{beginner} = (3.5/60)T$ or $0.0583T$ since there is 1/60 of an hour in a minute; note that T is measured in minutes and $D_{beginner}$ is measured in miles.

b. For $0 \le T \le 10$: $D_{advanced} = (3.75/60)T$ or $0.0625T$ and for $10 < T \le 20$ we have $D_{advanced} = 0.625 + (5.25/60)(T - 10) = 0.0875T - 0.25$.

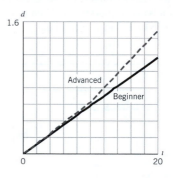

T	$D_{advanced}$	$D_{beginner}$
0	0.0000	0.0000
5	0.3125	0.2915
10	0.6250	0.5830
15	1.0630	0.8745
20	1.5000	1.1660

c. The graphs intersect only at $T = 0$.

Ch. 3: Check Your Understanding

1. False	**8.** True	**15.** True	**22.** True
2. True	**9.** True	**16.** True	**23.** True
3. True	**10.** True	**17.** True	**24.** True
4. False	**11.** False	**18.** False	**25.** False
5. False	**12.** True	**19.** False	**26.** True
6. False	**13.** False	**20.** True	**27.** False
7. False	**14.** False	**21.** False	**28.** True

29. True

30. Possible answer: $\begin{cases} 2x + 3y = 6 \\ 4x + 6y = 10 \end{cases}$

31. Possible answer: $\begin{cases} 2x + y = 7 \\ -6x - 3y = -21 \end{cases}$

32. Possible answer: $\begin{cases} y > 2x + 1 \\ y < 2x - 5 \end{cases}$

33. Possible answer: $\begin{cases} c = r + 1 \\ c = -r - 1 \end{cases}$

34. Possible answer: $\begin{cases} C = 25q + 2500 \\ R = 50q \end{cases}$

35. Possible answer: $p = 100 - 3q$

36. $\begin{cases} x > 0 \\ y < 0 \end{cases}$

37. False   **38.** False   **39.** True   **40.** True

Ch. 3 Review: Putting It All Together

1. a. Maximum: approx. 1650 MMT in 2004; minimum: approx. 1420 MMT in 2002.

   b. The three intersection points show when production equaled consumption.

   c. In 2002 there was a deficit of about 100 MMT.

   d. Original title for the graph: Grain Consumption Outstrips Production Again. (Answers will vary.)

3. a. Recall that if two lines are perpendicular to each other (and neither is horizontal) then their slopes are negative reciprocals of each other.

     $y = -4x + 26$ and $y = x/4 + 9$

   b.

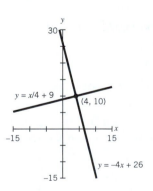

5.

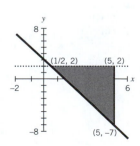

7. a. $C = 10{,}000 + 7x$; $R = 12x$, where x is the number of CDs

   b. $C = R \Rightarrow 10{,}000 + 7x = 12x \Rightarrow x = 2000$ CDs. Breakeven point is (2000, \$24000).

   c. If p is the new price per CD, then $P(1600) = 10{,}000 + 7(1600) = \$21{,}200 \Rightarrow p = \$13.25$ per CD. She would need to raise the price to \$13.25 for each CD.

   d. If $c =$ new fixed cost, then $c + 7(1600) = 12(1600) \Rightarrow c = \$8000$. She would need to reduce fixed costs by \$2000.

9. a. Estimates: production $= 2900$ thousand barrels/day and consumption $= 2300$ thousand barrels/day. The net difference is 600 thousand barrels per day. In 1990 China was producing more oil than it was consuming. It may have exported or stored this difference.

   b. 1993. The amount of oil consumed is the same as the amount produced.

   c. Estimates for 2006: production $= 3800$ thousand barrels/day and consumption $= 7400$ thousand barrels/day. The net difference is approximately 3600 thousand barrels per day. In the year 2006 China consumed almost double the amount of oil it produced. China needed to use its oil reserves or import this difference.

11. a. i. $x = -5$ and $y = 5/7$;   **ii.** $a = 3$ and $b = 0.5$

   b. Answers will vary. A system of two equations whose graphs are two distinct parallel lines will not have a solution.

13. a. $s + r \le 60$ minutes: $8s + 10r \ge 560$ calories; $s \ge 0$ and $r \ge 0$.

   b.

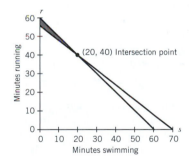

   c. There are many answers, for example: $s = 10$, $r = 50$ minutes is in the solution set and $s = 10$, $r = 40$ minutes is not in the solution set.

   d. $r + s \le 70$ minutes; $10r + 8s \ge 560$ calories; $s \ge 0$ and $r \ge 0$. The intersection point of the boundary lines changes and the shaded area representing the solution set increases.

15.

   a.

$$A(m) = \begin{cases} 39.99 & \text{for} \quad 0 \le m \le 450 \\ 39.99 + 0.45(m - 450) & \text{for } 450 < m \le 2500 \end{cases}$$

   b.

$$B(m) = \begin{cases} 59.99 & \text{for} \quad 0 \le m \le 900 \\ 59.99 + 0.40(m - 900) & \text{for } 900 < m \le 2500 \end{cases}$$

   c.

$$C(m) = \begin{cases} 79.99 & \text{for} \quad 0 \le m \le 1350 \\ 79.99 + 0.35(m - 1350) & \text{for } 1350 < m \le 2500 \end{cases}$$

d.

Number of Minutes Used/Month	Cost		
	Plan A	Plan B	Plan C
500	$39.99 + 0.45(50)$ $= \$62.49$	59.99	79.99
800	$39.99 + 0.45(350)$ $= \$197.49$	59.99	79.99
1000	$39.99 + 0.45(550)$ $= \$287.49$	$59.99 + 0.40(100)$ $= \$99.99$	79.99

17. There have been wide swings in the real (adjusted for inflation) price of crude oil since the early 1860s. Dramatic increases occurred in the early 1860s and in the 1970s, with the maximum occurring in 1864 at over $100 a barrel. In the early 1860s the price of crude oil increased over tenfold, and in 1974 the price was over three times the price in 1972. Even with the sharp rise in oil prices from 2004 to 2006, using real dollars adjusted for inflation, the price of oil in 2006 was only about two-thirds the price of crude oil in 1860s.

The nominal price or price actually paid for crude oil showed less variation than the price adjusted for inflation. It showed a huge surge during the Civil War, then remained fairly constant until the 1970s, after which the pattern more closely followed that of the price adjusted for inflation.

19. a. $C_p = 4.39N$

$C_c = 3.85N + 4.00$

$$C_i = \begin{cases} 4.99N & \text{for } 0 < N < 10 \\ 4.79N + 2.50 & \text{for } N \geq 10 \end{cases}$$

b. From the graphs of the three formulae given below, it can be seen that if one orders less than 7 bottles, then the C_p formula gives the best buy but if one orders 7 bottles or more, then the C_c formula gives the best buy.

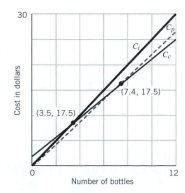

21. a. $H_b = 132 - 0.60A$

b. $H_i = 154 - 0.70A$

c. $H_a = 187 - 0.85A$

d.

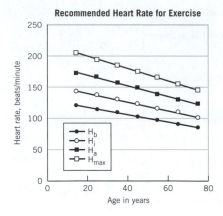

Athletes are recommended to work in the zone on and between the top two lines, H_a and H_{max}.

e. $H_b = 120$, $H_i = 140$, $H_a = 170$, $H_{max} = 200$

f. 65-year-old: $I \approx 86\%$. She is just below her $H_{max} = 200 - 65 = 135$ beats per minute.

45-year-old: $I \approx 77\%$

25-year-old: $I \approx 69\%$

CHAPTER 4

Section 4.1

Algebra Aerobics 4.1

1. a. 10^{10}: to express 10 billion as a power of 10, start with 1.0, then count the ten place values the decimal must be moved to the right, in order to produce 10 billion.

b. 10^{-14}: the decimal point in 1.0 must be moved 14 place values to the left to produce 0.000 000 000 000 01.

c. 10^5

d. 10^{-5}

2. a. 0.000 000 01

b. 10,000,000,000,000

c. 0.000 1

d. 10,000,000

3. a. 10^{-9} or 0.000 000 001 sec

b. 10^3 or 1000 m

c. 10^9 or 1,000,000,000 bytes (a byte is a term used to describe a unit of computer memory).

4. a. $7 \text{ cm} \cdot \dfrac{1 \text{ m}}{100 \text{ cm}} = 7 \cdot \dfrac{1}{10^2} \text{ m} = 7 \cdot 10^{-2}$ or 0.07 m

b. $9 \text{ mm} \cdot \dfrac{1 \text{ m}}{1000 \text{ mm}} = 9 \cdot \dfrac{1}{10^3} \text{ m} = 9 \cdot 10^{-3}$ or 0.009 m

c. $5 \text{ km} \cdot \dfrac{1000 \text{ m}}{1 \text{ km}} = 5 \cdot 10^3$ or 5000 m

5. 602,000,000,000,000,000,000,000,000

6. $3.84 \cdot 10^8$ m

7. $1 \cdot 10^{-8}$ cm

8. 0.000 000 002 m

9. a. −705,000,000   **c.** 5,320,000
  b. −0.000 040 3   **d.** 0.000 000 102 1

10. a. $-4.3 \cdot 10^7$ **c.** $5.83 \cdot 10^3$
  b. $-8.3 \cdot 10^{-6}$ **d.** $2.41 \cdot 10^{-8}$

11. a. $\dfrac{1}{100,000} = \dfrac{1}{10^5} = 10^{-5}$

  b. $\dfrac{1}{1,000,000,000} = \dfrac{1}{10^9} = 10^{-9}$

Exercises for Section 4.1

1. a. 10^6 **d.** 10^{-3}
  b. 10^{-5} **e.** 10^{13}
  c. 10^9 **f.** 10^{-8}

3. a. $1 \cdot 10^{-1}$ m   **c.** $3 \cdot 10^{12}$ m
  b. $4 \cdot 10^3$ m   **d.** $6 \cdot 10^{-9}$ m

5. gigabyte $= 10^9$ bytes; terabyte $= 10^{12}$ bytes.

7. a. $2.9 \cdot 10^{-4}$ **d.** 10^{-11} **g.** $-4.9 \cdot 10^{-3}$
  b. $6.54456 \cdot 10^2$ **e.** $2.45 \cdot 10^{-6}$
  c. $7.2 \cdot 10^5$ **f.** $-1.98 \cdot 10^6$

9. a. 723,000   **c.** 0.001   **e.** 0.000188
  b. 0.000526   **d.** 1,500,000   **f.** 67,800,000

11. a. False; $7.56 \cdot 10^{-3}$ **d.** False; $1.596 \cdot 10^9$
  b. True   **e.** True
  c. False; $4.9 \cdot 10^7$ watts   **f.** False; $6 \cdot 10^{-12}$ second

13. a. 9   **b.** 9   **c.** 1000   **d.** −1000

15. a. True   **b.** False   **c.** False   **d.** True

17. a. $|x - 1| < 5$ if $x = 5$; $|x - 1| > 5$ if $x = -5$
  b. $2|3 - x| < 10$ if $x = 5$; $2|3 - x| > 10$ if $x = -5$.
  c. $|x - 1| > 0$ whether $x = 5$ or -5
  d. $|-x| > 4$ whether $x = 5$ or -5
  e. $|2x - 1| < 11$ if $x = 5$; $|2x - 1| = 11$ if $x = -5$
  f. $|-x| < 6$ whether $x = 5$ or -5

19. Note: Since the coordinates given below come from eyeball estimates, student answers may vary from those given here.
  a. Wolcott (0.1, 1900); Sollas (0.2, 1908); and Clarke (0.3, 1921).
  b. Barrell did it in 1918; the coordinates are approximately (1.3, 1918).
  c. Estimating two points on a hand-drawn line could result in coordinates such as (0, 1910) and (4, 1956). Then the slope, m, of the line is

$$m = \frac{1956 - 1910}{4 - 0} = \frac{46}{4} = 11.5$$

  A plot of the various estimates and the graph with this slope and y-intercept are given at the top of the next column.

d. One meaning for the slope is that for each increase of a billion years in the estimated age of Earth, time advances, on average, 11.5 years. However, it would make more sense to say that on average for every 11.5 years, estimates of Earth's age increased by a billion years.

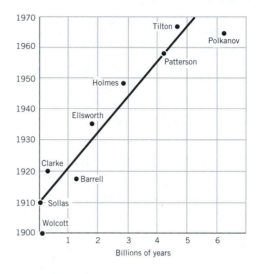

Section 4.2

Algebra Aerobics 4.2a

1. a. $10^5 \cdot 10^7 = 10^{5+7} = 10^{12}$
  b. $8^6 \cdot 8^{14} = 8^{6+14} = 8^{20}$
  c. $z^5 \cdot z^4 = z^{5+4} = z^9$
  d. Cannot be simplified because bases, 5 and 6, are different.
  e. $7^3 + 7^3 = 7^3(1 + 1) = 2 \cdot 7^3$
  f. $5 \cdot 5^6 = 5^1 \cdot 5^6 = 5^7$
  g. $3^4 + 7 \cdot 3^4 = 3^4(1 + 7) = 8 \cdot 3^4$ or $2^3 \cdot 3^4$
  h. $2^3 + 2^4 = 2^3 + 2^3 \cdot 2^1 = 2^3(1 + 2^1) = 3 \cdot 2^3$
  i. Cannot be simplified because bases, 2 and 5, are different.

2. a. $\dfrac{10^{15}}{10^7} = 10^{15-7} = 10^8$

  b. $\dfrac{8^6}{8^4} = 8^{6-4} = 8^2$

  c. $\dfrac{3^5}{3^4} = 3^{5-4} = 3^1$ or 3

  d. Cannot be simplified because bases, 5 and 6, are different.

  e. $\dfrac{5^1}{5^6} = 5^{-5}$ **f.** $\dfrac{3^4}{3^1} = 3^3$

  g. $\dfrac{2^3 \cdot 3^4}{2^1 \cdot 3^2} = \dfrac{2^3}{2^1} \cdot \dfrac{3^4}{3^2} = 2^{3-1} \cdot 3^{4-2} = 2^2 \cdot 3^2$

  h. $\dfrac{6}{2^4} = \dfrac{2 \cdot 3}{2^4 \cdot 1} = \dfrac{2^1}{2^4} \cdot \dfrac{3}{1} = 2^{1-4} \cdot 3 = 3 \cdot 2^{-3}$

3. a. $10^5 \cdot 10^6 = 10^{5+6} = 10^{11}$
  b. $10^3 \cdot 10^{-6} = 10^{3+(-6)} = 10^{-3}$
  c. $10^{-11} \cdot 10^{-5} = 10^{-11+(-5)} = 10^{-16}$
  d. $10^9 \cdot 10^{-4} = 10^{9+(-4)} = 10^5$

e. $10^6 \cdot 10^{-(-3)} = 10^{6+3} = 10^9$

f. $10^{-5} \cdot 10^{-(-4)} = 10^{-5+4} = 10^{-1}$

g. $10^{-6} \cdot 10^{-4} = 10^{-6+(-4)} = 10^{-10}$

4. a. $10^{4(5)} = 10^{20}$

b. $7^{2(3)} = 7^6$

c. $x^{4(5)} = x^{20}$

d. $(2x)^4 = 2^4 x^4$ or $16x^4$

e. $(2a^4)^3 = 2^3(a^4)^3 = 2^3 a^{12} = 8a^{12}$

f. $(-2a)^3 = (-2)^3 a^3 = -8a^3$

g. $(-3x^2)^3 = (-3)^3(x^2)^3 = -27x^6$

h. $((x^3)^2)^4 = (x^{3 \cdot 2})^4 = (x^6)^4 = x^{6 \cdot 4} = x^{24}$

i. $(-5y^2)^3 = (-5)^3(y^2)^3 = -125y^6$

5. a. $\dfrac{(-2x)^3}{(4y)^3} = \dfrac{(-2)^3 \cdot x^3}{4^3 \cdot y^3} = \dfrac{-8x^3}{64y^3} = \dfrac{-x^3}{8y^3}$

b. $(-5)^2 = (-5)(-5) = 25$

c. $-5^2 = -(5)(5) = -25$

d. $-3(yz^2)^4 = -3(y)^4(z^2)^4 = -3y^4 z^8$

e. $(-3yz^2)^4 = (-3)^4(y)^4(z^2)^4 = 81y^4 z^8$

f. $(-3yz^2)^3 = (-3)^3(y)^3(z^2)^3 = -27y^3 z^6$

6. $\dfrac{\text{(capacity of hard drive)}}{\text{(capacity of disk)}} = \text{number of disks}$

$\dfrac{4.0 \cdot 10^{10}}{7.37 \cdot 10^8} \approx 0.54 \cdot 10^2 \approx 54 \text{ disks}$

7. a. $(3 + 5)^3 = 8^3 = 512$

b. $3^3 + 5^3 = 27 + 125 = 152$

c. $3 \cdot 5^2 = 3 \cdot 25 = 75$

d. $-3 \cdot 5^2 = -3 \cdot 25 = -75$

Algebra Aerobics 4.2b

1. a. $(0.000\,297\,6)(43{,}990{,}000) \approx (0.000\,3)(40{,}000{,}000)$

$= 3 \cdot 10^{-4} \cdot 4 \cdot 10^7 = 12 \cdot 10^3 = 12{,}000$

b. $\dfrac{453{,}897 \cdot 2{,}390{,}702}{0.004\,38} \approx \dfrac{500{,}000 \cdot 2{,}000{,}000}{0.004}$

$= \dfrac{(5 \cdot 10^5)(2 \cdot 10^6)}{4 \cdot 10^{-3}} = \dfrac{10}{4} \cdot \dfrac{10^{11}}{10^{-3}}$

$= 2.5 \cdot 10^{14} \approx 3 \cdot 10^{14}$

c. $\dfrac{0.000\,000\,319}{162{,}000} \approx \dfrac{0.000\,000\,3}{200{,}000} = \dfrac{3 \cdot 10^{-7}}{2 \cdot 10^5}$

$= 1.5 \cdot 10^{-12} \approx 2 \cdot 10^{-12}$

d. $28{,}000{,}000 \cdot 7629 \approx 30{,}000{,}000 \cdot 8000$

$= 3 \cdot 10^7 \cdot 8 \cdot 10^3 = 24 \cdot 10^{10}$

$= 2.4 \cdot 10^{11} \approx 2 \cdot 10^{11}$

e. $0.000\,021 \cdot 391{,}000{,}000 \approx 0.000\,02 \cdot 400{,}000{,}000$

$= 2 \cdot 10^{-5} \cdot 4 \cdot 10^8 = 8 \cdot 10^3 = 8{,}000$

2. a. $(3.0 \cdot 10^3)(4.0 \cdot 10^2) = 12 \cdot 10^5$

$= 1.2 \cdot 10^6 = 1{,}200{,}000$

b. $\dfrac{(5.0 \cdot 10^2)^2}{2.5 \cdot 10^3} = \dfrac{25 \cdot 10^4}{25 \cdot 10^2} = 1.0 \cdot 10^2 = 100$

c. $\dfrac{2.0 \cdot 10^5}{5.0 \cdot 10^3} = \dfrac{20 \cdot 10^4}{5.0 \cdot 10^3} = 4 \cdot 10^1 = 40$

d. $(4.0 \cdot 10^2)^3(2.0 \cdot 10^3)^2 = (4^3 \cdot 10^6)(4 \cdot 10^6)$

$= 4^4 \cdot 10^{12} = 256 \cdot 10^{12} = 2.56 \cdot 10^{14}$

$= 256{,}000{,}000{,}000{,}000$

3. If we use 3.14 to approximate π:

a. Surface area of Jupiter $= 4\pi r^2 \approx 4\pi(7.14 \cdot 10^4 \text{ km})^2$

$= 4\pi(7.14)^2(10^4)^2 \text{ km}^2$

$\approx 4(3.14)(50.98)10^8 \text{ km}^2$

$\approx 640 \cdot 10^8 \text{ km}^2$

$\approx 6.4 \cdot 10^2 \cdot 10^8 \text{ km}^2$

$\approx 6.4 \cdot 10^{10} \text{ km}^2$

b. Volume of Jupiter $= \frac{4}{3}\pi r^3$

$\approx (1.3) \cdot (3.14)(7.14 \cdot 10^4 \text{ km})^3$

$\approx (4.08)(7.14)^3(10^4)^3 \text{ km}^3$

$\approx (4.08)(364)10^{12} \text{ km}^3$

$\approx 1486 \cdot 10^{12} \text{ km}^3$

$\approx 1.486 \cdot 10^3 \cdot 10^{12} \text{ km}^3$

$\approx 1.486 \cdot 10^{15} \text{ km}^3$

4. If only 3/7 of the farmable land is used, the people/sq. mi. of used farmland is:

$\dfrac{6.6 \cdot 10^9 \text{ people}}{(3/7)12 \cdot 10^6 \text{ sq.mi.}} = \left(\dfrac{7}{3}\right)\dfrac{6.6 \cdot 10^9 \text{ people}}{12 \cdot 10^6 \text{ sq.mi.}}$

$\approx 1.283 \cdot 10^3$ or 1283 people/sq.mi.

For fractions > 0, if the denominator is decreased, the value of that fraction is increased. So, one expects this ratio to be larger than the ratio of people to farmable land.

Exercises for Section 4.2

1. a. 10^7

b. $1.1 \cdot 10^4$

c. $2 \cdot 10^3$

d. x^{15}

e. x^{50}

f. $4^7 + 5^2$—this expression cannot be simplified without multiplying out the values and adding them together.

g. z^5　　　　**i.** 3^{-1} or $1/3$

h. 1 or as is　　　**j.** 4^{11}

3. a. $16a^4$　　　　**e.** $32x^{20}$

b. $-2a^4$　　　　**f.** $18x^6$

c. $-x^{15}$　　　　**g.** $2500a^{20}$

d. $-8a^3 b^6$　　　**h.** as is—nothing is simpler

5. a. $-\left(\dfrac{5}{8}\right)^2 = -\dfrac{25}{64}$

b. $\left(\dfrac{3x^3}{5y^2}\right)^3 = \dfrac{3^3 x^9}{5^3 y^6} = \dfrac{27x^9}{125y^6}$

c. $\left(\dfrac{-10x^5}{2b^2}\right)^4 = \dfrac{10^4 x^{20}}{4^2 b^8} = \dfrac{10{,}000x^{20}}{16b^8} = 625\dfrac{x^{20}}{b^8}$

d. $\left(\dfrac{-x^5}{x^2}\right)^3 = -x^9$

7. a. $(2 \cdot 10^6) \cdot (4 \cdot 10^3) = 8 \cdot 10^9$

b. $(1.4 \cdot 10^6) \div (7 \cdot 10^3) = 0.2 \cdot 10^3 = 2 \cdot 10^2$

c. $(5 \cdot 10^{10}) \cdot (6 \cdot 10^{13}) = 30 \cdot 10^{23} = 3 \cdot 10^{24}$

d. $(2.5 \cdot 10^{12}) \div (5 \cdot 10^5) = 0.5 \cdot 10^7 = 5 \cdot 10^6$

Ch. 4

9. a. $x^{13}y^4$ **c.** $-8x^9y^9$ **e.** $81x^8y^{20}$

  b. $5x^4y$ **d.** $16x^{10}y^8$ **f.** $\frac{9}{25}x^4$

11. a. $10^9/10^6 = 10^3 = 1000$

  b. $1000/10 = 10^2 = 100$

  c. $1000/0.001 = 10^6 = 1,000,000$

  d. $10^{-6}/10^{-9} = 10^3 = 1000$

13. a. Japan's population density $= (1.275 \cdot 10^8 \text{ people})/$ $(1.525 \cdot 10^5 \text{ miles}^2) \approx 836 \text{ people/mile}^2$.

  b. The U.S. population density $= (3.0 \cdot 10^8 \text{ people})/$ $(3.62 \cdot 10^6 \text{ miles}^2) \approx 83 \text{ people/mile}^2$.

  c. Japan's population density is $(836)/(83) = 10$ times larger, or one order of magnitude larger than that of the United States.

15. $(1.5 \cdot 10^4 \text{ beverages/sec})(8.64 \cdot 10^4 \text{ sec/day}) = 1.296 \cdot 10^9$ beverages/day, or over a billion Coca-Cola beverages were consumed each day worldwide in 2005.

17. a. If a is positive, then $-a$ is negative. If n is even, then $(-a)^n$ is positive; but if n is odd, then $(-a)^n$ is negative. If a is negative, then $-a$ is positive, then $(-a)^n$ is positive whether n is even or odd.

  b. This is answered in part (a).

19. a. $\left(\frac{m^2n^3}{mn}\right)^2 = (mn^2)^2 = m^2n^4$ and $\left(\frac{m^2n^3}{mn}\right)^2 = \frac{m^4n^6}{m^2n^2} = m^2n^4$

  b. $\left(\frac{2a^2b^3}{ab^2}\right)^4 = (2ab)^4 = 16a^4b^4$ and

  $\left(\frac{2a^2b^3}{ab^2}\right)^4 = \frac{16a^8b^{12}}{a^4b^8} = 16a^4b^4$

21. $\left(\frac{2a^3}{5b^2}\right)^4 = \frac{2^4a^{12}}{5^4b^8} = \frac{16a^{12}}{625b^8}$

23. Two cases are distinguished:

  If $n = 0$, then $(ab)^0 = 1$ and $a^0 \cdot b^0 = 1 \cdot 1 = 1$

  If $n > 0$, then $a^n = a \cdots \cdot a$ (n factors) and $b^n = b \cdots \cdot b$ (n factors) and thus $a^n \cdot b^n = (a \cdots \cdot a) \cdot (b \cdots \cdot b) = (ab) \cdots \cdot (ab)$ (n factors), after rearrangement, and this product is what is meant by $(ab)^n$ when $n > 0$.

25. a. Generated in the United Kingdom:

  (81 terawatt-hours)/(60.6 $\cdot$ 10^6 persons) $\approx 1.34 \cdot 10^{-6}$ terawatt-hours/person.

  (81 terawatt-hours)/(94,525 miles^2) $\approx 8.57 \cdot 10^{-4}$ terawatt-hours/mile^2.

  b. Generated in the United States:

  (780 terawatt-hours)/(3.0 $\cdot$ 10^8 persons) $\approx 2.60 \cdot 10^{-6}$ terawatt-hours/person.

  (780 terawatt-hours)/(3,675,031 miles^2) $\approx 2.12 \cdot 10^{-4}$ terawatt-hours/mile^2.

  c. The United Kingdom produces about four times more terawatt-hours per square mile than the United States, because UK/US $= (8.57 \cdot 10^{-4} \text{ terawatt-hours/mile}^2)/$ $(2.12 \cdot 10^{-4} \text{ terawatt-hours/mile}^2) \approx 4.04$.

  d. Answers will vary. The United States produces (uses) $(2.60 \cdot 10^{-6} \text{ terawatt-hours/person})/(1.34 \cdot 10^{-6}$ terawatt-hours/person) ≈ 1.9 times more nuclear energy per person than the United Kingdom, but the United Kingdom produces 4 times more nuclear energy than the United States per square mile.

Section 4.3

Algebra Aerobics 4.3

1. a. $10^{5-7} = 10^{-2} = \frac{1}{10^2}$

  b. $11^{6-(-4)} = 11^{6+4} = 11^{10}$

  c. $3^{-5-(-4)} = 3^{-5+4} = 3^{-1} = 1/3$

  d. Cannot be simplified: different bases, 5 and 6.

  e. $7^{3-3} = 7^0 = 1$

  f. $a^{-2+(-3)} = a^{-5} = \frac{1}{a^5}$

  g. $3^4 \cdot 3^3 = 3^7$

  h. $(2^2 \cdot 3) \cdot (2^6) \cdot (2^4 \cdot 3) = 2^2 \cdot 2^6 \cdot 2^4 \cdot 3 \cdot 3 = 2^{12} \cdot 3^2$

2. Time for a TV signal to travel across the

  United States $=$ (time to travel 1 kilometer) $\cdot$ (number of kilometers)

  $= (3.3 \cdot 10^{-6}) \text{ sec/km} \cdot (4.3 \cdot 10^3) \text{ km}$

  $= (3.3 \cdot 4.3) \cdot (10^{-6} \cdot 10^3) \text{ sec}$

  $= 14 \cdot 10^{-3} \text{ sec}$

  $= 1.4 \cdot 10 \cdot 10^{-3} \text{ sec}$

  $= 1.4 \cdot 10^{-2} \text{ sec or } 0.014 \text{ sec}$

So it would take less than two-hundredths of a second for the signal to cross the United States.

3. a. $x^{-2}(x^5 + x^{-6}) = x^{-2}(x^5) + x^{-2}(x^{-6})$

  $= x^3 + x^{-8}$

  $= x^3 + \frac{1}{x^8}$

  b. $-a^2(b^2 - 3ab + 5a^2)$

  $= b^2(-a^2) - 3ab(-a^2) + 5a^2(-a^2)$

  $= -a^2b^2 + 3a^{1+2}b - 5a^{2+2}$

  $= -a^2b^2 + 3a^3b - 5a^4$

4. a. $10^{(4)(-5)} = 10^{-20} = \frac{1}{10^{20}}$

  b. $7^{(-2)(-3)} = 7^6$

  c. $\frac{1}{(2a^3)^2} = \frac{1}{4a^6}$

  d. $\left(\frac{8}{x}\right)^{-2} = \left(\frac{x}{8}\right)^2 = \frac{x^2}{64}$

  e. $2^{-1}x^2 = \frac{x^2}{2}$

  f. $2x^2$

  g. $\left(\frac{3}{2y^2}\right)^{-4} = \left(\frac{2y^2}{3}\right)^4 = \frac{2^4y^8}{3^4} = \frac{16y^8}{81}$

  h. $3 \cdot (2y^2)^4 = 3 \cdot 2^4y^8 = 48y^8$

5. a. $\frac{t^{-3}(1)}{t^{-12}} = t^{-3-(-12)} = t^9$

  b. $\frac{v^{-3}w^7}{v^{-6}w^{-10}} = v^{-3-(-6)}w^{7-(-10)} = v^3w^{17}$

  c. $\frac{7^{-8}x^{-1}y^2}{7^{-5}x^1y^3} = 7^{(-8)-(-5)}x^{(-1)-1}y^{2-3}$

  $= 7^{-3}x^{-2}y^{-1} = \frac{1}{7^3x^2y}$

  d. $\frac{a(5b^{-1}c^3)^2}{5ab^2c^{-6}} = \frac{a^1 \cdot 5^2b^{-2}c^6}{5^1a^1b^2c^{-6}} = 5^{2-1}a^{1-1}b^{(-2)-2}c^{6-(-6)}$

  $= 5b^{-4}c^{12} = \frac{5c^{12}}{b^4}$

Exercises for Section 4.3

1. a. 10^1
   c. $10^{-6} = \frac{1}{10^6}$

  b. $10^{-5} = \frac{1}{10^5}$
   d. 10^5

3. a. $\frac{1}{2^2 x^4} = \frac{1}{4x^4}$
   d. $(x + y)^{11}$

  b. xy^{12}
   e. $\frac{ab^7}{c^6}$

  c. $\frac{1}{x^3 y^2}$

5. a. $4.6 \cdot 10^{10}$
   d. $5.1669 \cdot 10^{-8}$

  b. $4.07 \cdot 10^3$
   e. $2.3833 \cdot 10^{158}$

  c. $1.525 \cdot 10^{11}$
   f. $2.601 \cdot 10^{-21}$

7. a. $1 \cdot 10^{-1} \cdot 10^{-5} = 1 \cdot 10^{-6}$

  b. $5 \cdot 10^{-5}/(5 \cdot 10^4) = 1 \cdot 10^{-9}$

  c. $3/(6 \cdot 10^{-3}) = 0.5 \cdot 10^3 = 5 \cdot 10^2$

  d. $8 \cdot 10^3/(8 \cdot 10^{-4}) = 1 \cdot 10^7$

  e. $6.4 \cdot 10^{-3}/(8 \cdot 10^3) = 0.8 \cdot 10^{-6} = 8 \cdot 10^{-7}$

  f. $5 \cdot 10^6 \cdot 4 \cdot 10^4 = 20 \cdot 10^{10} = 2 \cdot 10^{11}$

9. a. -6
   c. 4

  b. -6
   d. -5

11. a. $-\frac{y^3}{2^3 x^9}$
   c. $\frac{5xy^9}{3}$

  b. $\frac{1}{2^2 x^4}$
   d. $\frac{3^6 z^{24}}{x^6}$

13. 200 times longer or $1.6 \cdot 10^{-2}$ second

15. a. $\left(\frac{1}{x^2} - \frac{1}{y}\right) \cdot (xy^2) = \left(\frac{y - x^2}{x^2 y}\right) \cdot (xy^2) = (y - x^2) \cdot \left(\frac{y}{x}\right)$

                                               $= \frac{y^2 - x^2 y}{x}$

  (This one is difficult because you need to find a common denominator.)

  b. $\frac{x^4 y^6}{5^2} = \frac{x^4 y^6}{25}$

17. $5.23 \cdot 10^{-3}$, 0.00523 and $5.23/1000$

19. $\frac{1}{x^p} = \left(\frac{1}{x}\right)^p$ for any p. Here we have $p = -n$; thus

  $\frac{1}{x^{-n}} = \left(\frac{1}{x}\right)^{-n} = (x^{-1})^{-n} = x^{(-1) \cdot (-n)} = x^n$

21. $(9 \cdot 10^9)/(35 \cdot 10^6) \approx 257$ kernels per pound.

Section 4.4

Algebra Aerobics 4.4

1. $2 \, l \cdot \frac{1 \, qt}{0.946 \, l} \cdot \frac{32 \, oz}{1 \, qt} \approx 67.65$ oz

2. $120 \, cm \cdot \frac{1 \, in}{2.54 \, cm} \approx 47.24$ in.

3. a. $12 \, in. \cdot \frac{2.54 \, cm}{1 \, in.} = 30.48$ cm

  b. $100 \, yd \cdot \frac{0.914 \, m}{1 \, yd} = 91.4$ m

  c. $20 \, kg \cdot \frac{1 \, lb}{0.4536 \, kg} \approx 44.09$ lb

  d. $\frac{\$40,000}{1 \, year} \cdot \frac{1 \, year}{52 \, weeks} \cdot \frac{1 \, workweek}{40 \, hr} \approx \$19.23$ per hour

  e. $\frac{24 \, hours}{1 \, day} \cdot \frac{60 \, min}{1 \, hour} \cdot \frac{60 \, sec}{1 \, min} = 86,400$ sec/day

f. $1 \, gal \cdot \frac{4 \, qt}{1 \, gal} \cdot \frac{946 \, ml}{1 \, qt} = 3784$ ml

g. $\frac{1 \, mile}{1 \, hour} \cdot \frac{1 \, hour}{3600 \, sec} \cdot \frac{5280 \, ft}{1 \, mile} \approx 1.47$ ft/sec

4. 1 km = 1000 m, so the conversion factor from meters to kilometers is $\frac{1 \, km}{10^3 \, m}$. Hence,

$$3.84 \cdot 10^8 \, m \cdot \frac{1 \, km}{10^3 \, m} = 3.84 \cdot 10^5 \text{ km}$$

5. 1 km = 1000 m, so the conversion factor from kilometers to meters is $\frac{10^3 \, m}{1 \, km}$. Hence

$$7.8 \cdot 10^8 \, km \cdot \frac{10^3 \, m}{1 \, km} = 7.8 \cdot 10^8 10^3 = 7.8 \cdot 10^{11} \text{ m}$$

6. 1 km = 0.62 mi, so the conversion factor to convert from km to mi is $\frac{0.62 \, mi}{1 \, km}$. Hence

$$9.46 \cdot 10^{12} \, km \cdot \frac{0.62 \, mi}{1 \, km} = 5.87 \cdot 10^{12} \text{ mi}$$

which is close to $5.88 \cdot 10^{12}$ mi.

7. 1 m = 100 cm, so the conversion factor for converting from cm to m is $\frac{1 \, m}{100 \, cm}$.

Hence $1 \, \text{Å} = 10^{-8} \, cm \cdot \frac{1 \, m}{10^2 \, cm} = 10^{-10}$ m.

8. 1 km = 0.62 mi, so the conversion factor for converting from km to miles is $\frac{0.62 \, mi}{1 \, km}$.

Hence, $218 \text{ km} = 218 \, km \cdot \frac{0.62 \, mi}{1 \, km} = 135.16$ mi or ~ 135 mi.

9. If a dollar bill is 6 in long, then two dollar bills/12 in = 2 dollars/ft. The number of dollars needed to reach from Earth to the sun is:

$$93,000,000 \, mi \cdot 5,280 \frac{ft}{mi} \cdot \frac{2 \, dollars}{ft}$$

$$\approx 9.3 \cdot 5.3 \cdot 2 \cdot 10^{7+3} \text{ dollars}$$

$$\approx 98.6 \cdot 10^{10} \approx 9.9 \cdot 10^{11}$$

or 990,000,000,000 dollar bills, almost a trillion dollars.

10. $\frac{2,560 \, mi}{4.2 \, hrs} \cdot \frac{1.6 \, km}{1 \, mi} = \frac{4096 \, km}{4.2 \, hrs} = \frac{4096 \, km}{4.2 \, hrs} \cdot \frac{1 \, hr}{60 \, min}$

$= \frac{4096 \, km}{252 \, min}$ (or approximately 16.25 km/min)

11. $5 \, \mu m \cdot \frac{1 \, m}{10^6 \, \mu m} \cdot \frac{1000 \, mm}{1 \, m} = 5 \cdot 10^{-3} \text{ mm} \Rightarrow 1$ spore is

$5 \cdot 10^{-3}$ mm in diameter. To compare the 1 mm diameter of a pencil to the diameter of a spore, we divide: $\frac{1 \, mm}{5 \cdot 10^{-3} \, mm} = 200$. The diameter of a pencil is approximately 200 times larger than the diameter of a spore.

12. a. $4.3 \text{ light-year} \cdot \frac{9.46 \cdot 10^{12} \, km}{1 \, light\text{-}year} \approx 4.07 \cdot 10^{13}$ km;

     $4.3 \text{ light-year} \cdot \frac{5.88 \cdot 10^{12} \, miles}{1 \, light\text{-}year} \approx 2.53 \cdot 10^{13}$ miles

  b. $10^8 \text{ light-years} \cdot \frac{9.46 \cdot 10^{12} \, km}{1 \, light\text{-}year} \cdot \frac{1000 \, m}{1 \, km}$

     $= 9.46 \cdot 10^{23}$ m

  c. $1.6 \cdot 10^3 \text{ light-year} \cdot \frac{5.88 \cdot 10^{12} \, miles}{1 \, light\text{-}year} \cdot \frac{5.28 \cdot 10^3 \, ft}{1 \, mile}$

     $\approx 49.67 \cdot 10^{18} = 4.97 \cdot 10^{19}$ ft

13. a. $0.5 \, \text{Å} \cdot \frac{10^{-10} \, m}{1 \, \text{Å}} = 5 \cdot 10^{-1} \cdot 10^{-10} = 5.0 \cdot 10^{-11}$ m

b. $10^5\,\mathring{A}\cdot\frac{10^{-10}\,m}{1\,\mathring{A}}=10^{-5}\,m$

c. $10^{-5}\,\mathring{A}\cdot\frac{10^{-10}\,m}{1\,\mathring{A}}=10^{-15}\,m$

14. $364\ \text{Smoot}\cdot\frac{5.6\,ft}{1\ \text{Smoot}}=2038.4\ ft$

Exercises for Section 4.4

1. a. $(50\ \text{miles})\cdot\frac{1.609\,km}{1\,mile}=80.45\ km$

b. $(3\,ft)\cdot\frac{0.305\,m}{1\,ft}\approx0.92\ m$

c. $(5\,lb)\cdot\frac{0.4536\,kg}{1\,lb}\approx2.27\ kg$

d. $(12\,in.)\cdot\frac{2.54\,cm}{1\,in.}=30.48\ cm$

e. $(60\,ft)\cdot\frac{0.305\,m}{1\,ft}=18.3\ m$

f. $(4\,qt)\cdot\frac{0.946\,\text{liters}}{1\,qt}\approx3.78\ \text{liters}$

3. a. Student estimates will vary. There are approximately 30.5 cm per foot.

b. From conversation table, 1 ft = 0.305 m or approximately 30% of a meter.

5. Converting the units to decimeters we get: $(100\,km)\cdot(250\,m)\cdot(25\,m)=10^6\cdot(2.5\cdot10^3)\cdot(2.5\cdot10^2)$ cubic decimeters $=6.25\cdot10^{11}$ cubic decimeters $=(6.25\cdot10^{11}\ \text{liters})\cdot(10^3\ \text{droplets/liter})=6.25\cdot10^{14}$ droplets

7. $1\,m\approx3.28\,ft$ and thus $9.8\,m/sec\approx9.8\cdot3.28=32.144\ ft/sec$

9. a. 186,000 miles per second or one hundred and eighty-six thousand miles per second.

b. $\frac{1.86\cdot10^5\,mi}{sec}\cdot\frac{1609\,m}{mi}\cdot\frac{60\,sec}{min}\cdot\frac{60\,min}{hr}\cdot\frac{24\,hr}{day}\cdot\frac{365\,days}{yr}$ $\approx9.438\cdot10^{15}\,m/yr$

11. 500 seconds; 500 seconds or 8 minutes and 20 seconds from now.

13. U.S. barrel of oil $=42$ gal $=42\cdot4$ qt $=168$ qt $\approx168\cdot0.946$ liter $=158.928$ liters. Thus the British barrel of oil is larger than the U.S. barrel of oil by approximately 4.727 liters.

15. a. 1 acre $=43,560\ ft^2$ and thus is $\sqrt{43,560}\approx208.71$ ft on each side.

b. $150x=1.5\cdot43,560$ or $x=435.6$ ft.; perimeter $=2\cdot(150+435.6)=1171.2$ feet or 390.4 yards.

c. 1 hectometer $=100$ meters; 1 square hectometer $=10,000$ square meters. 1 meter ≈3.28 ft. Thus 1 sq. meter ≈10.7584 square ft. Thus 1 acre $=43,560$ sq. ft $=43,560/10.7584\approx4048.929$ square meters ≈0.4049 square hectometer.

d. 1 hectare $=100$ acres $=4,356,000\ ft^2/[(5280)^2\ ft^2/mi^2]=0.15625$ sq. mi. Thus 1 acre $=0.0015625$ sq. mile. Thus in a square mile there are $1/0.0015625=640$ acres. This is 6.4 hectares.

17. Light travels $186,000\cdot5280$ ft/sec and thus it travels $186000\cdot5280/10^9\approx0.98208$ ft per nanosecond.

19. $\frac{15,000\ \text{bev}}{\text{second}}\cdot\frac{60\,sec}{1\,min}\cdot\frac{60\,min}{1\,hr}\cdot\frac{24\,hr}{1\,day}\cdot\frac{365\,day}{1\,year}\cdot\frac{1}{6.45\cdot10^9\ \text{persons}}$

≈73.34 beverages/year/person

21. a. Volume $=\pi\cdot9^2\cdot4$ cu. ft $=324\pi$ cu. ft ≈1018 cu. ft, and 1018 cu. ft $\cdot1728$ (cu. in.)/(cu. ft) $=1,759,104$ cu. in., and since there are 231 gal per cubic inch, there are $1,759,104/231\approx7615$ gal.

b. $7615\cdot24$ gal/(2500 gal/hr) ≈3.05 hr

c. $7615\cdot24$ gal/(10,000 gal/lb) ≈0.761 lb

23. a. Bush's BMI $\approx\frac{(194\,lb)/(2.2\,lb/kg)}{72\,in./39.37\,in./m^2}\approx\frac{88.18\,kg}{1.83^2m^2}\approx26.33\ kg/m^2$.

$\approx26.33\ kg/m^2$. He is slightly overweight.

b. BMI in pounds and inches is BMI $=$ (pounds/2.2)/(inches/39.37)^2 $=[(39.7)^2/2.2](lb/in^2)$, or approximately $704.5\ lb/in^2$. Thus for Bush we get $704.5\cdot194/72^2\approx26.36$, which is about the same.

c. Answers will vary, but note that $0.45\approx1/2.2$ and $0.254\approx1/39.37$.

d. A kilogram, more precisely, is approximately 2.2046 lb, and $39.37^2/2.2046\approx703.07$. Thus Kigner is correct.

25. a. $10^{-35}\,m=10^{-35}\cdot10^{-3}=10^{-38}\ km$

b. $10^{-35}\,m\approx10^{-35}\cdot0.00062=6.2\cdot10^{-39}$ mile

c. $x=10^{-35}/3\cdot10^8\approx3.33\cdot10^{-44}$ sec

Section 4.5

Algebra Aerobics 4.5a

1. a. $\sqrt{81}=9$

b. $\sqrt{144}=12$

c. $\sqrt{36}=6$

d. $-\sqrt{49}=-7$

e. not a real number

2. a. $\sqrt{9x}=3x^{1/2}$

b. $\sqrt{\frac{x^2}{25}}=\frac{\sqrt{x^2}}{\sqrt{25}}=\frac{x}{5}$

c. $\sqrt{36x^2}=\sqrt{36}\sqrt{x^2}=6x$

d. $\sqrt{\frac{9y^2}{25x^4}}=\frac{\sqrt{9y^2}}{\sqrt{25x^4}}=\frac{3y}{5x^2}$

e. $\sqrt{\frac{49}{x^2}}=\frac{\sqrt{49}}{\sqrt{x^2}}=\frac{7}{x}$

f. $\sqrt{\frac{4a}{169}}=\frac{\sqrt{4a}}{\sqrt{169}}=\frac{2\sqrt{a}}{13}=\frac{2a^{1/2}}{13}$

3. a. $S=\sqrt{30\cdot60}=\sqrt{1800}\approx42$ mph

b. $S=\sqrt{30\cdot200}=\sqrt{6000}\approx77$ mph

4. a. $\sqrt{25}<\sqrt{29}<\sqrt{36}\Rightarrow5<\sqrt{29}<6$; 5 and 6

b. $\sqrt{81}<\sqrt{92}<\sqrt{100}\Rightarrow9<\sqrt{92}<10$; 9 and 10

c. $\sqrt{100}<\sqrt{117}<\sqrt{121}\Rightarrow10<\sqrt{117}<11$; 10 and 11.

d. $\sqrt{64}<\sqrt{79}<\sqrt{81}\Rightarrow8<\sqrt{79}<9$; 8 and 9.

e. $\sqrt{36}<\sqrt{39}<\sqrt{49}\Rightarrow6<\sqrt{39}<7$; 6 and 7.

5. a. $\sqrt[3]{27} = 3$

b. $\sqrt[4]{16} = 2$

c. $\frac{1}{\sqrt[3]{8}} = \frac{1}{2}$

d. $\sqrt[5]{32} = 2$

e. $\frac{1}{27^{1/3}} = \frac{1}{\sqrt[3]{27}} = \frac{1}{3}$

f. $\frac{1}{25^{1/2}} = \frac{1}{\sqrt{25}} = \frac{1}{5}$

g. $\left(\frac{8}{27}\right)^{-1/3} = \left(\frac{27}{8}\right)^{1/3}$

$\qquad = \sqrt[3]{\frac{27}{8}} = \frac{\sqrt[3]{27}}{\sqrt[3]{8}} = \frac{3}{2}$

h. $\sqrt{\frac{1}{16}} = \frac{1}{4}$

6. a. $(-27)^{1/3} = -3$ since $(-3)^3 = -27$

b. There is no real number solution to the fourth root of a negative number, since a negative or positive number raised to the fourth power is always positive.

c. $(-1000)^{1/3} = -10$ since $(-10)^3 = -1000$.

d. $-\sqrt[4]{16} = -2$, since $2^4 = 16 \Rightarrow -\sqrt[4]{2^4} = -2$.

e. $\sqrt[3]{-8} = -2$, since $(-2)^3 = -8$.

f. $\sqrt{2500} = 50$, since $50^2 = 2500$.

7. a. $V = \frac{4}{3}\pi r^3 \Rightarrow \frac{3}{4}V = \pi r^3 \Rightarrow \frac{3V}{4\pi} = r^3 \Rightarrow$

$r = \sqrt[3]{\frac{3 \cdot 2 \text{ feet}^3}{4\pi}} = \sqrt[3]{\frac{3 \text{ feet}^3}{2\pi}} \approx \sqrt[3]{0.478 \text{ feet}^3}$

≈ 0.78 feet

We can express that with a more meaningful figure if we convert it into inches. Since 1 ft = 12 in., we can use a conversion factor of 1 = (12 in.)/(1 foot). The radius of the balloon is:

$$(0.78 \text{ ft})\left(\frac{12 \text{ in.}}{1 \text{ ft}}\right) = 9.36 \text{ in. or } \approx 9.4 \text{ in.}$$

8. a. $\sqrt{25} = 5$

b. $-\sqrt{49} = -7$

c. -5

d. $\sqrt{45} - 3\sqrt{125} = \sqrt{9 \cdot 5} - 3\sqrt{25 \cdot 5}$

$\qquad = 3\sqrt{5} - 15\sqrt{5} = -12\sqrt{5}$

9. a. $\sqrt{36} = (6^2)^{1/2} = 6^{2/2} = 6$

b. $\sqrt[3]{27x^6} = (3^3 x^6)^{1/3} = 3^{3/3} x^{6/3} = 3x^2$

c. $\sqrt[4]{81a^4 b^{12}} = (3^4 a^4 b^{12})^{1/4} = 3^{4/4} a^{4/4} b^{12/4} = 3ab^3$

10. a. $r^2 = \frac{V}{\pi h} \Rightarrow r = \sqrt{\frac{V}{\pi h}}$

b. $r^2 = \frac{3V}{\pi h} \Rightarrow r = \sqrt{\frac{3V}{\pi h}}$

c. $S = \sqrt[3]{V}$

d. $a = \sqrt{c^2 - b^2}$

e. $x = \sqrt{\frac{S}{6}}$

Algebra Aerobics 4.5b

1. a. $2^{1/2} \cdot 2^{1/3} = 2^{1/2+1/3} = 2^{5/6}$

b. $5^{1/2} \cdot 5^{1/4} = 5^{1/2+1/4} = 5^{3/4}$

c. $3^{1/2} \cdot 9^{1/3} = 3^{1/2} \cdot (3^2)^{1/3} = 3^{1/2} \cdot 3^{2/3} = 3^{7/6}$

d. $x^{1/4} \cdot x^{1/3} = x^{7/12}$

e. $x^{3/4} \cdot x^{1/2} = x^{5/4}$

f. $x^{1/3} \cdot y^{2/3} \cdot x^{1/2} \cdot y^{1/2} = x^{5/6} \cdot y^{7/6}$

2. a. $\frac{2^{1/2}}{2^{1/3}} = 2^{1/2-1/3} = 2^{1/6}$

b. $\frac{2^1}{2^{1/4}} = 2^{1-1/4} = 2^{3/4}$

c. $\frac{5^{1/4}}{5^{1/3}} = 5^{1/4-1/3} = 5^{-1/12}$ or $\frac{1}{5^{1/12}}$

d. $\frac{x^{1/2}}{x^{3/4}} = x^{-\frac{1}{4}}$ or $\frac{1}{x^{1/4}}$

e. $\frac{x^{1/3} \cdot y^{2/3}}{x^{1/2} \cdot y^{1/2}} = x^{-1/6} \cdot y^{1/6}$ or $\frac{y^{1/6}}{x^{1/6}}$

3. a. $c = 17.1(0.25)^{3/8}$ cm

$c \approx 17.1(0.59)$ cm $\Rightarrow c \approx 10.2$ cm

b. $c = 17.1(25)^{3/8}$ cm

$c \approx 17.1(3.34)$ cm $\Rightarrow c \approx 57.2$ cm

4. a. $\sqrt{20x^2} = \sqrt{2^2 \cdot 5 \cdot x^2} = 2x\sqrt{5}$

b. $\sqrt{75a^3} = \sqrt{5^2 \cdot 3 \cdot a^2 \cdot a} = 5a\sqrt{3a}$

c. $\sqrt[3]{16x^3 y^4} = \sqrt[3]{2^3 \cdot 2 \cdot x^3 \cdot y^3 \cdot y} = 2xy\sqrt[3]{2y}$

d. $\frac{\sqrt[4]{32x^4 y^6}}{\sqrt[4]{81x^8 y^5}} = \frac{\sqrt[4]{2^4 \cdot 2 \cdot x^4 \cdot y^4 \cdot y^2}}{\sqrt[4]{3^4 \cdot (x^2)^4 \cdot y^4 \cdot y}}$

$\qquad = \frac{2xy \cdot \sqrt[4]{2y^2}}{3x^2 y \cdot \sqrt[4]{y}} = \frac{2}{3x}\sqrt[4]{\frac{2y^2}{y}} = \frac{2}{3x}\sqrt[4]{2y}$

5. a. $\sqrt{4a^2 b^6} = (2^2 a^2 b^6)^{1/2} = 2^{2/2} a^{2/2} b^{6/2} = 2ab^3$

b. $\sqrt[4]{16x^4 y^6} = (2^4 x^4 y^6)^{1/4} = 2^{4/4} x^{4/4} y^{6/4}$

$\qquad = 2xy^{3/2}$

c. $\sqrt[3]{8.0 \cdot 10^{-9}} = (2^3 \cdot 10^{-9})^{1/3} = 2^{3/3} \cdot 10^{-9/3}$

$\qquad = 2 \cdot 10^{-3} = 0.002$

d. $\sqrt{8a^{-4}} = (2^3 a^{-4})^{1/2} = 2^{3/2} a^{-2}$

Exercises for Section 4.5

1. a. 10　　　**c.** $1/10 = 0.1$　　　**e.** -10

　　b. -10　　　**d.** $-1/10 = -0.1$　　**f.** -10

3. a. $\sqrt{\frac{a^2 b^4}{c^6}} = \frac{ab^2}{c^3}$　　　**c.** $\sqrt{\frac{49x}{y^6}} = \frac{7\sqrt{x}}{y^3}$

　　b. $\sqrt{36x^4 y} = 6x^2\sqrt{y}$　　　**d.** $\sqrt{\frac{x^4 y^2}{100z^6}} = \frac{x^2 y}{10z^3}$

5. a. $5\sqrt{5a}$　　**b.** $\frac{1}{2xy^3}$　　**c.** $2xy\sqrt{2x}$　　**d.** $8x^2 y^2\sqrt{y}$

7. a. $6 \cdot 10^3$　　**b.** $2 \cdot 10^3$　　**c.** $5 \cdot 10^5$　　**d.** $1.0 \cdot 10^{-2}$

9. a. 0.1　　**b.** 1/5　　**c.** 4/3　　**d.** 0.1

11. a. 4　　　**c.** $\sqrt{2} \approx 1.414$　　　**e.** $2\sqrt{2} \approx 2.8284$

　　b. -4　　**d.** not defined　　　**f.** 1

13. a. $\sqrt{3} + \sqrt{7} > \sqrt{3 + 7}$　　　**c.** $\sqrt{5^2 - 4^2} > 2$

　　b. $\sqrt{3^2 + 2^2} < 5$

15. a. 9   **b.** $\frac{1}{8}$ **c.** $\frac{1}{125}$ **d.** $\frac{1}{27}$

17. $V = (4/3)\pi r^3$ and thus if $V = 4$, then $r^3 = 3/\pi$. If one uses 3 as a crude estimate of π, then r is approximately 1 foot. (A more precise estimate, from using a calculator, is 0.985 ft.)

19. a. height $= 4\sqrt{3}$ cm ≈ 6.9 cm

   b. area $= 16\sqrt{3}$ cm^2 ≈ 27.7 cm^2

21. a. If $L \approx 2.67$ ft, then

$$T = 2\pi\sqrt{\frac{2.67}{32}} \approx 0.578\pi \text{ sec} \approx 1.816 \text{ sec.}$$

   b. Solving for L, we get: $L, = 8T^2/(\pi^2)$. So if T is 2 seconds, then $L = 8(2)^2/(\pi^2) = 32/\pi^2 \approx 3.24$ feet.

23. a. 3673 lb   **b.** 6345 lb

Section 4.6

Algebra Aerobics 4.6

1. Since $\dfrac{\text{magnitude 1988 Armenia earthquake}}{\text{magnitude 1987 LA earthquake}}$

$\Rightarrow \dfrac{10^{6.9}}{10^{5.9}} = 10^1$, the Armenian earthquake had tremors about one order of magnitude larger than those in Los Angeles.

2. Since $\dfrac{\text{magnitude Chile earthquake}}{\text{magnitude 2003 Little Rock earthquake}}$

$\Rightarrow \dfrac{10^{9.5}}{10^{6.5}} = 10^3$, the maximum tremor size of the Little Rock earthquake of 2003 was 1000 times smaller (or three orders of magnitude smaller) than the maximum tremor size of the Chile earthquake.

3. Since $\dfrac{\text{your salary}}{\text{my salary of \$100,000}} = 10^1 \Rightarrow$

your salary is \$1,000,000;

$\dfrac{\text{Henry's salary}}{\text{my salary \$100,000}} = 10^{-2} \Rightarrow$

$\dfrac{\text{Henry's salary}}{\text{my salary \$100,000}} = \dfrac{1}{10^2} \Rightarrow$ Henry's salary is \$1000.

4. a. Since $\dfrac{\text{radius of the Milky Way}}{\text{radius of the sun}} = \dfrac{10^{21}}{10^9} = 10^{21-9} = 10^{12}$,

the radius of the Milky Way is 12 orders of magnitude larger than the radius of the sun, or equivalently, the radius of the sun is 12 orders of magnitude smaller than the radius of the Milky Way.

   b. Since $\dfrac{\text{radius of a proton}}{\text{radius of the hydrogen atom}} = \dfrac{10^{-15}}{10^{-11}} = 10^{-15-(-11)} = 10^{-4}$,

the radius of the proton is four orders of magnitude smaller than the radius of the hydrogen atom, or equivalently, the radius of the hydrogen atom is four orders of magnitude larger than the radius of a proton.

5. a. \$4 million since \$400,000 $\cdot 10^1$ = \$4 million
   b. \$65,000 since \$650,000 $\cdot 10^{-1}$ = \$65,000

6. 1 km = 1000 m = 10^3 meters, so three orders of magnitude;

1 km $\cdot \dfrac{1000 \text{ m}}{1 \text{ km}} \cdot \dfrac{1000 \text{ mm}}{1 \text{ m}} = 10^6$ mm, so six orders of magnitude

7. a. $1,000,000,000$ m $= 10^9$ m $\Rightarrow$ plot at 10^9 m

   b. $0.000\ 000\ 7$ m $= 7 \cdot 10^{-7}$ m $\approx 10 \cdot 10^{-7}$ m $= 10^{-6}$ m $\Rightarrow$ plot at 10^{-6} m

   c. Plot at $10^7 \cdot 10^{-2} = 10^5$ m

Exercises for Section 4.6

1. a. 12   **b.** 5   **c.** 7   **d.** 6

3. a. 52.61 is one order of magnitude larger than 5.261.

   b. 5.261 is three orders of magnitude smaller than 5261.

   c. 526.1 is four orders of magnitude smaller than $5.261 \cdot 10^6$.

5. a. $(50 \cdot 10^3) \cdot (3 \cdot 10^8) \cdot (365 \cdot 24 \cdot 60^2) \approx 4.7304 \cdot 10^{20} =$ number of meters in 50,000 light-years. Hence, the Milky Way is $(4.7302 \cdot 10^{20})/(0.5 \cdot 10^{-4}) = 9.4608 \cdot 10^{24}$ times larger than the first life form. Thus the Milky Way is nearly 25 orders of magnitude larger than the first living organism on Earth.

   b. $(100 \cdot 10^6)/(100 \cdot 10^3) = 10^3$; thus Pleiades is three orders of magnitude older than *Homo sapiens*.

7. The raindrop is 10^{24} times heavier and this is an order of magnitude of 24.

9. a. $2.0 \cdot 10^{-5}$ in.

   b. $2.0 \cdot 10^{-5}/39 \approx 5.128 \cdot 10^{-7}$ m

   c. The name is a bit off — by two orders of magnitude (looking at meters).

   d. The tweezers would have to be made able to grasp things two orders of magnitude smaller than they can grasp now.

11. a. **i.** Radius of the moon = 1,758,288.293 meters, or about $1.76 \cdot 10^6$ meters.

   ii. Radius of Earth = 6,400,000 meters, or about $6.4 \cdot 10^6$ meters.

   iii. Radius of the sun = 695,414,634.100 meters, or about $6.95 \cdot 10^8$ meters.

   b. **i.** The surface area of a sphere is $4\pi r^2$, and thus the ratio of the surface area of Earth to that of the moon is the same as the ratio of the squares of their radii, or $[6.4 \cdot 10^6/(1.76 \cdot 10^6)]^2 \approx 13.22$. Thus the surface area of Earth is one order of magnitude bigger.

   ii. The volume of a sphere is $(4/3)\pi r^3$, and thus the ratio of the volume of the sun to that of moon is the same as the ratio of the cubes of their radii, or $[6.95 \cdot 10^8/(1.76 \cdot 10^6)]^3 \approx 6.16 \cdot 10^7$. Thus the volume of the sun is seven orders of magnitude bigger than the volume of the moon.

13. Scale (a) is additive because the distances are equally spaced, and scale (b) is multiplicative or logarithmic because the distances are spaced like the logarithms of numbers.

15. a. Being 5 orders of magnitude larger than the first atoms means that it is 10^{-5} m and thus would appear at -5 on the log scale.

   b. Being 20 orders of magnitude smaller than the radius of the sun means that it is 10^{-11} m and thus would appear at -11 on the log scale.

17. a. 1 watt = 10^0 watts

   b. 10 kilowatts = 10^4 watts

   c. 100 billion kilowatts = $10^2 \cdot 10^9 \cdot 10^3 = 10^{14}$ watts

   d. 1000 terawatts = 10^{15} watts

Section 4.7

Algebra Aerobics 4.7a

1. a. Since $10,000,000 = 10^7$, log $10,000,000 = 7$.

   b. Since $0.000\,000\,1 = 10^{-7}$, log $0.000\,000\,1 = -7$.

   c. Since $10,000 = 10^4$, log $10,000 = 4$.

   d. Since $0.0001 = 10^{-4}$, log $0.0001 = -4$.

   e. Since $1000 = 10^3$, log $1000 = 3$.

   f. Since $0.001 = 10^{-3}$, log $0.001 = -3$.

   g. Since $1 = 10^0$, log $1 = 0$.

2. a. $100,000 = 10^5$ **c.** $10 = 10^1$

   b. $0.000\,000\,01 = 10^{-8}$ **d.** $0.01 = 10^{-2}$

3. a. $\log N = 3 \Rightarrow N = 10^3 = 1000$

   b. $\log N = -1 \Rightarrow N = 10^{-1} = 0.1$

   c. $\log N = 6 \Rightarrow N = 10^6 = 1,000,000$

   d. $\log N = 0 \Rightarrow N = 10^0 = 1$

   e. $\log N = -2 \Rightarrow N = 10^{-2} = 0.01$

4. a. $\log 1000 = c, c = 3$

   b. $\log 0.001 = c, c = -3$

   c. $\log 100,000 = c, c = 5$

   d. $\log 0.000\,01 = c, c = -5$

   e. $\log 1,000,000 = c, c = 6$

   f. $\log 0.000\,001 = c, c = -6$

5. a. $x - 3 = 2$, so $x = 5$

   b. $2x - 1 = 4$, so $x = 5/2$

   c. $10^1 = x - 2$, so $x = 12$

   d. $10^{-1} = 5x$, so $x = \frac{0.1}{5} = 0.02$

Algebra Aerobics 4.7b

1. a. log 3 is the number c such that $10^c = 3$. With a calculator you can find that: $10^{0.4} \approx 2.512$ and $10^{0.5} \approx 3.162 \Rightarrow 0.4 < \log 3 < 0.5$. A calculator gives log $3 \approx 0.477$.

   b. log 6 is the number c such that $10^c = 6$. With a calculator you can find that: $10^{0.7} \approx 5.012$ and $10^{0.8} \approx 6.310 \Rightarrow 0.7 < \log 6 < 0.8$. A calculator gives log $6 \approx 0.778$.

   c. log 6.37 is the number c such that $10^c = 6.37$. Since $10^{0.8} \approx 6.310$ and $10^{0.9} \approx 7.943$, then $0.8 < \log 6.37 < 0.9$. A calculator gives log $6.37 \approx 0.804$.

2. a. Write 3,000,000 in
   scientific notation $3,000,000 = 3 \cdot 10^6$
   and substitute $10^{0.48}$ for 3 $3,000,000 \approx 10^{0.48} \cdot 10^6$

then combine powers $3,000,000 \approx 10^{6.48}$
and rewrite as a logarithm   log $3,000,000 \approx 6.48$
A calculator gives log $3,000,000 \approx 6.477121255$.

   b. Write 0.006 in scientific
   notation $0.006 = 6 \cdot 10^{-3}$
   and substitute $10^{0.78}$ for 6 $0.006 \approx 10^{0.78} \cdot 10^{-3}$
   then combine powers $0.006 \approx 10^{0.78-3} \approx 10^{-2.22}$
   and rewrite as a logarithm   log $0.006 = -2.22$
   A calculator gives log $0.006 \approx -2.22184875$.

3. a. $0.000\,000\,7 = 10^{\log 0.0000007} \approx 10^{-6.1549}$

   b. $780,000,000 = 10^{\log 780,000,000} \approx 10^{8.892}$

   c. $0.0042 = 10^{\log 0.0042} \approx 10^{-2.3768}$

   d. $5,400,000,000 = 10^{\log(5,400,000,000)} = 10^{9.732}$

4. a. You want to estimate a number x such that $10^{4.125} = x$. Since $10^4 = 10,000$, an estimate for x is 12,000. With a calculator, $x \approx 13,335$.

   b. You want to estimate a number x such that $10^{5.125} = x$. Since $10^5 = 100,000$, an estimate for x is 120,000. With a calculator, $x \approx 133,352$.

   c. You want to estimate a number x such that $10^{2.125} = x$. Since $10^2 = 100$, an estimate for x is 120. With a calculator, $x \approx 133$.

5. a. You want to estimate a value x such that log $250 = x$. $10^2 < 250 < 10^3 \Rightarrow 2 < x < 3$. With a calculator $x \approx 2.398$.

   b. You want to estimate a value x such that log $250,000 = x$. $10^5 < 250,000 < 10^6 \Rightarrow 5 < x < 6$. With a calculator, $x \approx 5.398$.

   c. You want to estimate a value x such that log $0.075 = x$. $10^{-2} < 0.075 < 10^{-1} \Rightarrow -2 < x < -1$. With a calculator, $x \approx -1.125$.

   d. You want to estimate a value x such that log $0.000\,075 = x$. $10^{-5} < 0.000\,075 < 10^{-4} \Rightarrow -5 < x < -4$. With a calculator, $x \approx -4.125$.

6. a. log $57 \approx 1.756 \Rightarrow 57 \approx 10^{1.756}$

   b. log $182 \approx 2.26 \Rightarrow 182 \approx 10^{2.26}$

   c. log $25,000 \approx 4.398 \Rightarrow 25,000 \approx 10^{4.398}$

   d. log $7.2 \cdot 10^9 \approx 9.857 \Rightarrow 7.2 \cdot 10^9 \approx 10^{9.857}$

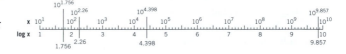

Exercises for Section 4.7

1. a. $\log(10,000) = 4$ **c.** $\log(1) = 0$

   b. $\log(0.01) = -2$ **d.** $\log(0.00001) = -5$

3. Since $\log(375) \approx 2.574$, we have that $375 \approx 10^{2.574}$.

5. a. $\log_{10}(100) = 2$

   b. $\log_{10}(10,000,000) = 7$

    c. $\log_{10}(0.001) = -3$

    d. $10^1 = 10$

    e. $10^4 = 10,000$

    f. $10^{-4} = 0.0001$

7. a. $x = 10^{1.255} \approx 17.9887 \approx 18$

    b. $x = 10^{3.51} \approx 3235.94 \approx 3236$

    c. $x = 10^{4.23} \approx 16,982.44 \approx 16,982$

    d. $10^{7.65} \approx 44,668,359.22 \approx 44,668,359$

9. a. $x = \log(12,500) \approx 4.097$

    b. $x = \log(3,526,000) \approx 6.547$

    c. $x = \log(597) \approx 2.776$

    d. $x = \log(756,821) \approx 5.879$

11. a. $x = 10^{0.82} \approx 6.607$

    b. $x = \log(0.012) \approx -1.921$

    c. $x = 10^{0.33} \approx 2.138$

    d. $x = \log(0.25) \approx -0.602$

13. a. $x - 5 = 3$ or $x = 8$

    b. $2x + 10 = 100$ or $2x = 90$ or $x = 45$

    c. $3x - 1 = -4$ or $3x = -3$ or $x = -1$

    d. $500 - 25x = 1000$ or $-25x = 500$ or $x = -20$

15. a. $1 < \log(11) < 2$ since $1 < 11 < 100$ and $\log(11) \approx 1.041$.

    b. $4 < \log(12,000) < 5$ since $10,000 < 12,000 < 100,000$ and $\log(12,000) \approx 4.079$.

    c. $-1 < \log(0.125) < 0$ since $10^{-1} = 0.1 < 0.125 < 1 = 10^0$ and $\log(0.125) \approx -0.903$.

17. a. Multiplying by 10^{-3}

    b. Multiplying by $\sqrt{10}$

    c. Multiplying by 10^2

    d. Multiplying by 10^{10}

19. a. $\text{pH} = -\log(10^{-7}) = 7$

    b. $\text{pH} = -\log(1.4 \cdot 10^{-3}) = 3 - \log(1.4) \approx 2.85$

    c. $11.5 = -\log([\text{H}^+])$; thus $[\text{H}^+] = 10^{-11.5} \approx 3.16 \cdot 10^{-12}$

    d. A higher pH means a lower hydrogen ion concentration, and one can see this because in plotting pH values one uses the numbers on the top of the given scale.

    e. Pure water is neutral, orange juice is acidic, and ammonia is basic. In plotting, one uses the top numbers to find the right spots. Thus water would be placed at the 7 mark, orange juice 85% of the way between the 2 and 3 marks, and ammonia halfway between the 11 and 12 marks.

21.

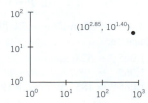

23. We measure all in seconds, the smallest time unit.

    a. 1 heartbeat ≈ 100 sec

    b. 10 minutes $= 600$ sec $\approx 10^{2.8}$ sec

    c. 7 days $= 7 \cdot 24 \cdot 3600$ sec $\approx 10^{5.8}$ sec

    d. 1 year $= 365 \cdot 24 \cdot 3600$ sec $\approx 10^{7.5}$ sec

    e. 38,000 years $= 38,000 \cdot 365 \cdot 24 \cdot 3600 \approx 10^{12.1}$ sec

    f. $2.2 \cdot 10^6 \cdot 365 \cdot 24 \cdot 3600$ sec $\approx 10^{13.8}$ sec

Ch. 4: Check Your Understanding

1. True	**8.** True	**15.** True	**22.** True
2. False	**9.** False	**16.** True	**23.** True
3. False	**10.** False	**17.** False	**24.** False
4. False	**11.** False	**18.** False	**25.** False
5. True	**12.** False	**19.** False	**26.** True
6. True	**13.** False	**20.** False	
7. False	**14.** True	**21.** False	

27. Possible answer: population of city A $= 583,240$ and population of city B $= 3615$.

28. Possible answer: $x = 150,000,000$.

29. Possible answer: $x = 0.45$.

30. Possible answer: b such that $b = \frac{1}{4}$.

31. $b = 1$.

32. Possible answer: $b = -3$.

33. False	**36.** True	**39.** False	**42.** True
34. False	**37.** False	**40.** False	
35. True	**38.** False	**41.** False	

Ch. 4 Review: Putting It All Together

1. a. 4200   **b.** -40 **c.** -16 **d.** $\frac{1}{10}$ **e.** 27

3. When x is an odd integer the statement is true.

5. $\dfrac{20 \text{ ft}}{3 \text{ sec}} \cdot \dfrac{60 \text{ sec}}{1 \text{ min}} \cdot \dfrac{60 \text{ min}}{1 \text{ hr}} \cdot \dfrac{1 \text{ mile}}{5280 \text{ ft}} \approx 4.55$ miles/hour

7. Tiger Woods makes $\$111.9 \cdot 10^6/\text{yr.} = \$111.9 \cdot 10^8/\text{yr.}$ The worker makes

$$\frac{\$6.40}{\text{hr}} \cdot \frac{40\text{hr}}{\text{wk}} \cdot \frac{52\text{wk}}{\text{yr}} = \frac{\$13,312}{\text{yr}} = \$1.3312 \cdot 10^4/\text{yr.}$$

Tiger Wood makes 4 orders of magnitude or about 10,000 times more than the minimum-wage worker.

The worker must work $\dfrac{\$1.119 \cdot 10^8}{\$1.3312 \cdot 10^4} \approx 0.84 \cdot 10^4$ years or about 8,400 years to make Tiger Wood's annual pay.

Ch. 4

Item	Value		Value in Scientific Notation
Mass-energy of electron	0.000 000 000 000 051 J		$5.1 \cdot 10^{-14}$ J
The kinetic energy of a flying mosquito	0.000 000 160 2	J	$1.602 \cdot 10^{-7}$
An average person swinging a baseball bat	80	J	$8 \cdot 10^1$ J
Energy received from the sun at the Earth's orbit by one square meter in one second	1360	J	$1.360 \cdot 10^3$ J
Energy released by one gram of TNT	4184	J	$4.184 \cdot 10^3$ J
Energy released by metabolism of one gram of fat	38,000	J	$3.8 \cdot 10^4$ J
Approximate annual power usage of a standard clothes dryer	320,000,000	J	$3.2 \cdot 10^8$ J

9. Convert to same unit of measure—for example, km².

Country	Area	Scientific Notation
Russia	17,075,200 km²	$1.70752 \cdot 10^7$ km²
Chile	$(290{,}125 \text{ mi}^2)\left(\frac{1.609 \text{ km}}{1 \text{ mi}}\right)^2$ $\approx 751{,}099$ km²	$7.51099 \cdot 10^5$ km²
Canada	$(3{,}830{,}840 \text{ mi}^2)\left(\frac{1.609 \text{ km}}{1 \text{ mi}}\right)^2$ $\approx 9{,}917{,}589$ km²	$9.917589 \cdot 10^6$ km²
South Africa	1,184,825 km²	$1.184825 \cdot 10^6$ km²
Norway	323,895 km²	$3.23895 \cdot 10^5$ km²
Monaco	$0.5 \text{ mi}^2\left(\frac{1.609 \text{ km}}{1 \text{ mi}}\right)^2$ ≈ 1.29 km²	$1.29 \cdot 10^0$ km²

a. Russia is the largest in area. Monaco is the smallest in area.

b. Russia, Canada, South Africa, Chile, Norway, Monaco

c. Russia is seven orders of magnitude larger than Monaco.

11. False. An increase in one order of magnitude is the same as multiplying by 10. A 100% increase would only double the original amount.

13. a. $\frac{1 \text{ baby}}{7 \text{ sec}} \cdot \frac{60 \text{ sec}}{1 \text{ min}} \cdot \frac{60 \text{ min}}{1 \text{ hr}} \cdot \frac{24 \text{ hr}}{1 \text{ day}} = 12{,}342.9 \frac{\text{babies}}{\text{day}} \approx$ 12,343 babies/day

b. $\frac{1 \text{ immigrant}}{31 \text{ sec}} \cdot \frac{60 \text{ sec}}{1 \text{ min}} \cdot \frac{60 \text{ min}}{1 \text{ hr}} \cdot \frac{24 \text{ hr}}{1 \text{ day}} = 2787.1 \frac{\text{immigrants}}{\text{day}} \approx$ 2787 immigrants/day

c. $\frac{1 \text{ death}}{13 \text{ sec}} \cdot \frac{60 \text{ sec}}{1 \text{ min}} \cdot \frac{60 \text{ min}}{1 \text{ hr}} \cdot \frac{24 \text{ hr}}{1 \text{ day}} = 6646.15 \frac{\text{deaths}}{\text{day}} \approx$ 6646 deaths/day

15. Volume of a sphere $= \frac{4}{3}\pi r^3$, so the volume of Earth $=$

$$\frac{4}{3}\pi(6.3 \cdot 10^6 \text{ m})^3 = \left(\frac{4}{3}\pi(6.3)^3\right) \cdot 10^{18}\text{m}^3$$
$$\approx 1047.4 \cdot 10^{18}\text{ m}^3$$
$$\approx 1.0474 \cdot 10^{21}\text{ m}^3$$

$$\text{Density} = \frac{\text{mass}}{\text{volume}} = \frac{5.97 \cdot 10^{24} \text{ kg}}{1.0474 \cdot 10^{21} \text{ m}^3} \approx 5.7 \cdot 10^3 \text{ kg/m}^3$$

17. The patient's weight in kilograms is:

$$130 \text{ lb} \cdot \frac{0.4536 \text{ kg}}{1 \text{ lb}} \approx 59 \text{ kg}.$$

daily dosage $= \frac{5 \text{ mg}}{\text{kg}} \cdot 59 \text{ kg} \approx 295$ mg

Since each tablet is 100 mg, the patient should take

$\frac{295 \text{ mg}}{100 \text{ mg/tablet}} = 2.95$, or 3 tablets each day.

19. a. $\log 1 = 0$

b. $\log 1{,}000{,}000{,}000 = \log 10^9 = 9$

c. $\log 0.000\ 001 = \log 10^{-6} = -6$

21. a. See chart at top of page for solution.

b. $\frac{3.2 \cdot 10^8 \text{ J}}{8 \cdot 10^1 \text{ J}} = 0.4 \cdot 10^7 = 4 \cdot 10^6$, or 4 million times more energy.

c. $\frac{3.8 \cdot 10^4 \text{ J}}{1.602 \cdot 10^{-7} \text{ J}} \approx 2.372 \cdot 10^{11}$, or approximately 200 billion times more energy.

23. (Requires scientific calculator.) Converting height and weight to kilograms and centimeters, respectively:

$$W = 180 \text{ lb} \cdot \frac{0.4536 \text{ kg}}{1 \text{ lb}} \approx 81.6 \text{ kg} \quad \text{and}$$

$$H = 6 \text{ ft} \cdot \frac{30.5 \text{ cm}}{1 \text{ ft}} = 183 \text{ cm}$$

$$\text{BSA} = 71.84 \cdot 81.6^{0.425} \cdot 183^{0.725} \approx 20{,}376 \text{ cm}^2$$

Converting cm² to m²:

$$20{,}376 \text{ cm}^2 \cdot \frac{1 \text{ m}^2}{(100 \text{ cm})^2} = 2.04 \text{ m}^2$$

daily dosage $= 15\frac{\text{mg}}{\text{m}^2} \cdot 2.04 \text{ m}^2 = 30.6$ mg

CHAPTER 5

Section 5.1

Algebra Aerobics 5.1

1. a. Initial value = 350
Growth factor = 5

b. Initial value = 25,000
Growth factor = 1.5

c. Initial value = 7000
Growth factor = 4

d. Initial value = 5000
Growth factor = 1.025

2. a. $P = 3000 \cdot 3^t$

b. $P = (4 \cdot 10^7)(1.3)^t$

c. $P = 75 \cdot 4^t$

d. $P = \$30{,}000(1.12)^t$

3. a. $P = 28 \cdot 1.065^0 = 28$ million or 28,000,000 people; the initial population

b. $P = 28 \cdot 1.065^{10} \approx 52.6$ million or 52,600,000 people; $P = 28 \cdot 1.065^{20} \approx 98.7$ million or 98,700,000; $P = 28 \cdot 1.065^{30} \approx 185.2$ million or 185,200,000

c. $t \approx 11$ years for P to double or reach 56,000,000.

4.

t	0	1	10	15	20
P	80	84.80	143.27	191.72	256.57

   a. 80

   b. 84.8

   c. The amount doubles (reaches 160) between $t = 10$ and $t = 15$

   d. $t \approx 12$ time periods.

Exercises for Section 5.1

1. a. 275; 3   **c.** $6 \cdot 10^8$; 5   **e.** 8000; 2.718

   b. 15,000; 1.04   **d.** 25; 1.18   **f.** $4 \cdot 10^5$; 2.5

3.

Initial Value C	Growth Factor a	Exponential Function $y = Ca^x$
1600	1.05	$y = 1600 \cdot 1.05^x$
$6.2 \cdot 10^5$	2.07	$y = 6.2 \cdot 10^5 \cdot (2.07)^x$
1400	3.25	$y = 1400 \cdot (3.25)^x$

5. a. $f(x) = 6 \cdot 1.2^x$ **b.** $f(x) = 10 \cdot 2.5^x$

7. a. 1.17   **b.** 63 cells per ml.

   c.

d	C
0	63.0
1	73.7
2	86.2
3	100.9
4	118.1
5	138.1
6	161.6
7	189.1
8	221.2
9	258.8
10	302.8

   d. C doubles somewhere between $d = 4$ and $d = 5$ days.

9. Expenditures $= 2016 \cdot (1.076)^5 \approx 2907.7$ billion dollars.

11. a. $G(t) = 5 \cdot 10^3 \cdot 1.185^t$ cells/ml, where t is measured in hours.

   b. $G(8) = 5 \cdot 10^3 \cdot 1.185^8 = 19,440.92$ cells/ml

13. a. 10 fish   **c.** 80 fish

   b. It doubled in 5 months.   **d.** approx. 35 months

Section 5.2

Algebra Aerobics 5.2

1.

t	$N = 10 + 3t$	$N = 10 \cdot 3^t$
0	10	10
1	13	30
2	16	90
3	19	270
4	22	810

$N = 10 + 3t$

For every unit increase in t, N increases by 3, so the graph is linear with a slope of 3. The vertical intercept is 10.

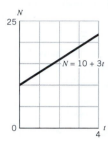

$N = 10 \cdot 3^t$

The rate of change here is not constant as in the previous problem. For example, the average rate of change between 0 and 1 is $(30 - 10)/1 = 20$ and between 1 and 2 is $(90 - 30)/1 = 60$. But the vertical intercept for the graphs of both functions is the same, 10.

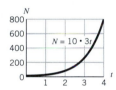

2. a.

t	$f(t) = 200 + 20t$		t	$g(t) = 200(1.20)^t$
0	200		0	200
1	220		1	240
2	240		2	288
3	260		3	345.6
4	280		4	414.72
5	300		5	497.66

   b.

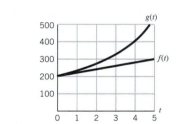

   c. The function f is linear and the function g is exponential. While both functions have the same initial value, g is growing faster than f.

3. f is linear because the rate of change is constant; that is, the slope is 6.

g is linear because the rate of change is constant; that is, the slope is 0.

h is exponential because the ratio of any two consecutive terms is constant; that is, there is a constant growth factor of 1.5 or 150%.

p is linear because the rate of change is constant; that is, the slope is 90.

r is exponential because there is a constant growth factor of 1.04 or 104%

Ch. 5

4. a. Linear: $m = \frac{620 - 500}{1 - 0} = 120$; $(0, 500) \Rightarrow b = 500 \Rightarrow$ $y = 500 + 120t$

Exponential: growth factor $= \frac{620}{500} = 1.24$; $(0, 500) \Rightarrow$ initial amount $= 500 \Rightarrow y = 500(1.24)^t$

b. Linear: $m = \frac{3.2 - 3}{1 - 0} = 0.2$; $(0, 3) \Rightarrow b = 3 \Rightarrow y = 3 + 0.2t$

Exponential: growth factor $= \frac{3.2}{3} \approx 1.067$; $(0, 3) \Rightarrow$ initial amount $= 3 \Rightarrow y = 3(1.067)^t$

5. a. growth factor $= \frac{3750}{1500} = 2.50 \Rightarrow P = 1500 \cdot 2.50^t$

b. growth factor $= \frac{82,300}{80,000} = 1.029 \Rightarrow A = 80,000 \cdot 1.029^t$

c. growth factor $= \frac{32.7}{30} = 1.09 \Rightarrow Q = 30 \cdot 1.09^t$

6. a. 2

b.

t (years)	Q
0	10
2	20
4	40
6	80
8	160

c. Let $a =$ growth factor. The 2-yr. growth factor $= 2 \Rightarrow$ $a^2 = 2 \Rightarrow a = \sqrt{2} \approx 1.41$. The annual growth factor ≈ 1.41.

d. $Q = 10 \cdot 1.41^t$

Exercises for Section 5.2

1. a. linear; 5   **c.** neither; 3   **e.** exponential; 7

b. exponential; 3   **d.** exponential; 6   **f.** linear; 0

3. a. $P(t) = 50 + 5t$

b. $Q(t) = 50 \cdot 1.05^t$

c. The graph of the two functions is in the accompanying diagram.

d. Graphing software gives that the two are equal at $(0, 50)$ and at $(26.6, 183)$. Thus the populations were both 50 people at the start and were both approximately 183 persons after 26.6 years. (Student eyeball estimates may differ.)

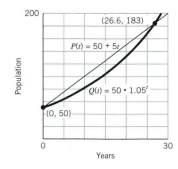

5. linear: $y = 3x + 6$; exponential: $y = 6 \cdot 1.5^x$

7. a. linear: $y = 10x + 10$; exponential: $y = 10 \cdot \left(\sqrt{3}\right)^x$

b. linear: $y = 100x + 100$; exponential $y = 100 \cdot \left(\sqrt[3]{4}\right)^x$

9. a. exponential; $f(x) = 7 \cdot 2.5^x$

b. linear: $g(x) = 0.2x + 0.5$

11.

x	$6 + 1.5x$	$6 \cdot 1.5^x$	$1.5 \cdot 6^x$
0	6.0	6	1.5
1	7.5	9	9
2	9.0	13.5	54
3	10.5	20.25	324
4	12.0	30.375	1,944
5	13.5	45.5625	11,664

The graphs are found in the accompanying diagrams.

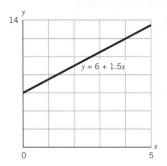

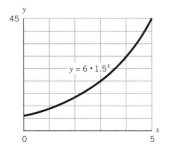

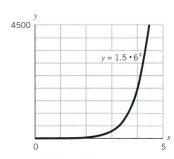

13. a. For both populations we are looking for an exponential growth factor, i.e., for the value of a, where $A(t) = C \cdot a^t$ and where t measures years from 1986.

For the entire Atlantic coast: we have $C = 5800$ and $A(16) = 14,313$ and thus $14,313 = 5800 \cdot a^{16}$ or $a = \left(\frac{14,313}{5800}\right)^{1/16} \approx 1.058$. Thus the growth factor is about 1.058.

For Massachusetts: we have $C = 585$ and $A(16) = 2939$ and thus $2939 = 585 \cdot a^{16}$ or $a = \left(\frac{2939}{585}\right)^{1/16} \approx 1.106$.

Thus 1.106 is the approximate growth factor. The growth factor in Massachusetts is somewhat larger than that of the entire Atlantic coast.

b. For the Atlantic coast, the average rate of change is the slope of the line between $(0, 5800)$ and $(16, 14313)$. This is: $\frac{14,313 - 5800}{16} = \frac{8513}{16} \approx 532$ swans/year.

For Massachusetts, the average rate of change is the slope of the line between $(0, 585)$ and $(16, 2939)$. This is $\frac{2939 - 585}{16} = \frac{2354}{16} \approx 147$ swans/year.

The average rate of change for the Atlantic coast is more than three times as large as the Massachusetts rate.

c. The linear model for Massachusetts since 1986 is $A(t) = 147t + 585$; the exponential model since 1986 for Massachusetts is $A(t) = 585 \cdot 1.106^t$.

d. Now 2010 is 24 years from 1986; thus $S(24) = 147 \cdot 24 + 585 = 4113$ swans and $A(24) = 585 \cdot 1.106^{24} \approx 6566$ swans. So the predictions are quite different for the two models.

15. a.

Years after 2004	Plan A	Plan B
0	$7.00	$7.00
1	$7.35	$7.50
2	$7.72	$8.00
3	$8.10	$8.50
4	$8.51	$9.00

b. Plan A: $F_A(t) = 7 \cdot 1.05^t$ and plan B: $F_B(t) = 7 + 0.50t$

c. $F_A(25) = 7 \cdot (1.05)^{21} \approx \$19.50$ and $F_B(25) = 7 + 0.50 \cdot 21 = \$17.50$

d. Student answers will vary, but one should note that plan B is more expensive until year 15; from then on plan A would be more expensive. Going for plan A for the short term seems like a better option.

Section 5.3

Algebra Aerobics 5.3

1. a. growth   **c.** decay   **e.** decay

    b. decay   **d.** growth   **f.** growth

2. a. $y = 2300(1/3)^t$ **c.** $y = (375)(0.1)^t$

    b. $y = (3 \cdot 10^9)(0.35)^t$

3. $y = 12(5)^{-x}$
$= 12(5^{-1})^x$
$= 12(1/5)^x$ so it represents decay.

4. a. $y = 23 \cdot 2.4^{-x} \Rightarrow y = 23(2.4^{-1})^x \Rightarrow y = 23\left(\frac{1}{2.4}\right)^x \Rightarrow$
$y = 23(0.42)^x$; decay

    b. $f(x) = 8000 \cdot (0.5^{-1})^x \Rightarrow f(x) = 8000\left(\frac{1}{0.5}\right)^x \Rightarrow$
$f(x) = 8000 \cdot (2)^x$; growth

    c. $P = 52{,}000 \cdot 1.075^{-t} \Rightarrow P = 52{,}000(1.075^{-1})^x \Rightarrow$
$P = 52{,}000\left(\frac{1}{1.075}\right)^x \Rightarrow P = 52{,}000(0.93)^x$; decay

5. a. $g(x)$ decreases more rapidly.

    b. For $g(x)$, 70% of the previous amount remains each time; that is, there is 30% less each time period. For $f(x)$, 90% of the previous amount remains each time; that is, there is 10% less each time period.

Exercises for Section 5.3

1. a. 0.43   **b.** 0.95   **c.** 1/3

3. a. $f(t) = 10{,}000 \cdot 0.4^t$ **c.** $h(x) = 219 \cdot 0.1^x$

    b. $g(T) = 2.7 \cdot 10^{13} \cdot 0.27^T$

5. a. This is not exponential since the base is the variable.

    b. This is exponential; the decay factor is 0.5; the vertical intercept is 100.

    c. This is exponential; the decay factor is 0.999; the vertical intercept is 1000.

7.

Initial Value, C	Decay Factor, a	Exponential Function $y = Ca^x$
500	0.95	$y = 500 \cdot 0.95^x$
$1.72 \cdot 10^6$	0.75	$y = 1.72 \cdot 10^6 \cdot (0.75)^x$
1600	0.25	$y = 1600 \cdot (0.25)^x$

9. a. matches graph B.   **b.** matches graph A.

11. a.

Food	Formula	Amount in 2000
Beef	$B(t) = 72.1 \cdot 0.994^t$	63.9 lb
Chicken	$C(t) = 32.7 \cdot 1.024^t$	52.5 lb
Pork	$P(t) = 52.1 \cdot 0.996^t$	48.1 lb
Fish	$F(t) = 12.4 \cdot 1.01^t$	15.1 lb

    b. Growth: chicken and fish; decay: beef and pork.

    c. Answers will vary with students. It is expected that the growth rates and per-capita consumption figures will be cited.

13. a. $f(t) = -4t + 20$

    b. $g(t) = 20 \cdot 0.25^t$

    c.

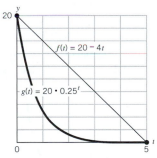

15. The accompanying diagram contains the graphs of $f(x) = 25 \cdot 5^{-x} = 25(0.2)^x$ and $g(x) = 25 \cdot 0.5^x$. It is clear that the graph of f has the faster rate of decline. Note that the decay factor for f is 0.20 and the decay factor for g is 0.50.

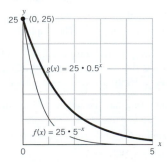

17. a. $P(t) = 100 \cdot b^t$; $99.2 = 100 \cdot b$ or $b = 0.992$ and therefore $P(t) = 100 \cdot 0.992^t$

b. $P(50) = 100 \cdot 0.992^{50} \approx 66.9$ grams; $P(500) = 100 \cdot 0.992^{500} \approx 1.8$ grams

Section 5.4

Algebra Aerobics 5.4

1. a.

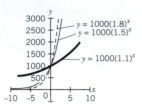

b. Yes, they intersect at (0, 1000).

c. In the first quadrant, $y = 1000(1.8)^x$ is on top, $y = 1000(1.5)^x$ is in the middle, and $y = 1000(1.1)^x$ is on the bottom.

d. In the second quadrant, $y = 1000(1.1)^x$ is on top, $y = 1000(1.5)^x$ is in the middle, and $y = 1000(1.8)^x$ is on the bottom.

2.

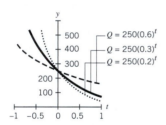

b. Yes, they intersect at (0, 250).

c. In the first quadrant: $Q = 250(0.6)^t$ is on top, $Q = 250(0.3)^t$ is in the middle, and $Q = 250(0.2)^t$ is on the bottom.

d. In the second quadrant: $Q = 250(0.2)^t$ is on top, $Q = 250(0.3)^t$ is in the middle, and $Q = 250(0.6)^t$ is on the bottom.

3. a.

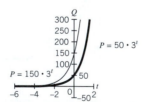

b. No, the curves do not intersect.

c. $P = 150 \cdot 3^t$ is always above $P = 50 \cdot 3^t$.

4. a. The horizontal axis is the horizontal asymptote.

b. The horizontal axis is the horizontal asymptote.

c. No horizontal asymptotes. The graph of this function is a line.

5. Comparing the growth factors: $1.23 > 1.092 > 1.06$, so the function h has the most rapid growth. Comparing the decay factors: $0.89 < 0.956$, so the function g has the most rapid decay.

6. a. The function g has the larger initial value since it crosses the vertical axis above the function f.

b. The function f has the larger growth factor since it is steeper, growing at a faster rate.

c. The function f approaches zero more rapidly as $x \to -\infty$.

d. The graphs intersect at approximately (3, 35); f is greater than g after the point of intersection.

Exercises for Section 5.4

1. Let $y_1 = 2^x$, $y_2 = 5^x$, and $y_3 = 10^x$.

a. $C = 1$ for each case; $a = 2$ for y_1, $a = 5$ for y_2, and $a = 10$ for y_3.

b. Each represents growth; y_1's value doubles, y_2's value is multiplied by 5, and y_3's value is multiplied by 10.

c. All three graphs intersect at (0, 1).

d. In the first quadrant the graph of y_3 is on top, the graph of y_2 is in the middle, and the graph of y_1 is on the bottom.

e. All have the graph of $y = 0$ (or the x-axis) as their horizontal asymptote.

f. Small table for each:

x	y_1	y_2	y_3
0	1	1	1
1	2	5	10
2	4	25	100

g. The graphs of the three functions are given in the accompanying diagram. They indeed confirm the answers given to the questions asked in parts (a) through (f).

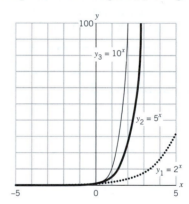

3. Let $y_1 = 3^x$, $y_2 = (1/3)^x$, and $y_3 = 3 \cdot (1/3)^x$.

a. $C = 1$ for y_1 and y_2 and $C = 3$ for y_3.

b. y_1 represents growth; y_2 and y_3 represent decay.

c. y_1 and y_3 intersect at approx. (0.5, 1.7); y_1 and y_2 intersect at (0, 1).

d. For $x > 0.5$, the graph of y_1 is on top, the graph of y_3 is in the middle, and the graph of y_2 is on the bottom. For $x < 0$, the graph of y_3 is on top, that of y_2 is in the middle, and that of y_1 is on the bottom.

e. All the graphs have the x-axis as their horizontal asymptote.

f. A small table for each:

x	y_1	y_2	y_3
0	1	1	3
1	3	1/3	1
2	9	1/9	1/3

g. The graphs of these three functions are given in the accompanying diagram, and they confirm the answers given to the questions in parts (a) through (f).

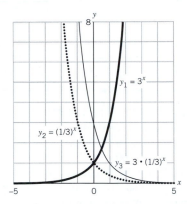

5. The function P goes with Graph C, the function Q goes with Graph A, the function R goes with Graph B, and the function S goes with Graph D.

7. a. The smaller the decay factor, the faster the descent; thus h has the smallest, then g, and finally f.

b. The point $(0, 5)$ and no other point.

c. It will approach 0 more slowly than the other two functions.

d. h is on top and will stay on top for $x < 0$.

9. The accompanying diagram contains the graphs of $f(x) = 30 + 5x$ and $g(x) = 3 \cdot 1.6^x$ with the points of intersection marked.

a. $f(0) = 30; g(0) = 3; f(0) > g(0)$

b. $f(6) = 60; g(6) \approx 50.33; f(6) > g(6)$

c. $f(7) = 65; g(7) \approx 80.53; f(7) < g(7)$

d. $f(-5) = 5; g(-5) \approx 0.286; f(-5) > g(-5)$

e. $f(-6) = 0; g(-6) \approx 0.179; f(-6) < g(-6)$

f. f and g go to infinity; $f(x) < g(x)$ for all $x > 6.5$, approximately.

g. f goes to $-\infty$ and $g(x)$ goes to 0; thus $f(x) < g(x)$ for all $x < -6$, approximately.

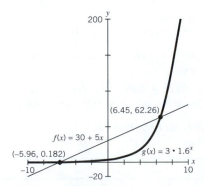

11. a. As $x \to +\infty$, $g(x)$ will dominate over $f(x)$, i.e., $g(x) > f(x)$

b. There is no one coordinate window that will display the graphs of f and g to help one see the answer to his question. Thus two displays are given: one over

$2680 < x < 2700$ and another over $-0.5 < x < 0$. From these one can see that $f(x) > g(x)$ if $-0.196 < x < 2690.51$; otherwise $g(x) > f(x)$.

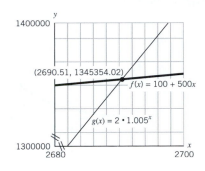

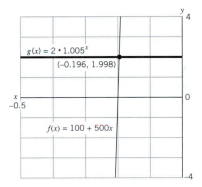

13. The graphs of the three functions are given in the accompanying diagram. The graphs of F and H are mirror images of each other and are equally steep when the absolute values of average rates of change are considered. $H(x) = 100(1/1.2)^x \approx 100(0.83)^x$ and $G(x) = 100 \cdot 0.8^x$, so $G(x)$ has a smaller decay factor and thus decays faster than $H(x)$ and is steeper than $H(x)$. Thus, the graph of G is steeper than the other two, again when considering the absolute values of the average rates of change.

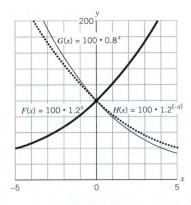

15. The graphs are given in the accompanying diagram.

a. It is clear that the graphs of f and h are mirror images of each other with respect to the y-axis.

b. The graphs of f and g are mirror images of each other with respect to the x-axis.

c. The graphs of g and h are mirror images of each other with respect to the origin.

d. The graphs of these functions are mirror images of each other with respect to the x-axis.

e. These functions are mirror images of each other with respect to the y-axis.

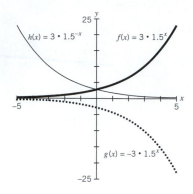

Section 5.5

Algebra Aerobics 5.5

1. a. 22%

b. 106.7% or 1.067

c. $1 - 0.972 = 0.028$ or 2.8%

d. $1 - 0.123 = 0.877$

2.

Exponential Function	Initial Value	Growth or Decay?	Growth or Decay Factor	Growth or Decay Rate
$A = 4(1.03)^t$	4	growth	1.03	0.03 or 3%
$A = 10(0.98)^t$	10	decay	0.98	0.02 or 2%
$A = 1000(1.005)^t$	1000	growth	1.005	0.005 or 0.5%
$A = 30(0.96)^t$	30	decay	0.96	0.04 or 4%
$A = 50,000(1.0705)^t$	$50,000	growth	1.0705	0.0705 or 7.05%
$A = 200(0.51)^t$	200 g	decay	0.51	0.49 or 49%

3. a. growth factor $= 1.06 = 106\%$ growth rate $= 6\%$

b. decay factor $= 0.89 = 89\%$ decay rate $= 11\%$

c. growth factor $= 1.23 = 123\%$ growth rate $= 23\%$

d. decay factor $= 0.956 = 95.6\%$ decay rate $= 4.4\%$

e. growth factor $= 1.092 = 109.2\%$ growth rate $= 9.2\%$

4. a. $\frac{\$6.00}{1.65} \approx 3.64$, so 3.64 (the growth factor) $= 1 +$ growth rate. The growth rate $= 2.64$ or a 264% increase in 30 years.

b. $\frac{0.167}{0.299} \approx 0.56$, so 0.56 (the decay factor) $= 1 -$ decay rate. The decay rate $= 0.44$ or a 44% decrease in one season.

c. $\frac{23}{17} \approx 1.35$, so $1.35 = 1 +$ growth rate. The growth rate $= 0.35$ or a 35% increase in one month.

Exercises for Section 5.5

1. a. $Q = 1000 \cdot 3^T$ **c.** $Q = 1000 \cdot 1.03^T$

b. $Q = 1000 \cdot 1.3^T$ **d.** $Q = 1000 \cdot 1.003^T$

3. a. 5%   **c.** 55%   **e.** 0.4%

b. 18%   **d.** 34.5%   **f.** 27.5%

5.

	Initial Value	Growth or Decay?	Growth or Decay Factor	Growth or Decay Rate	Exponential Function
a.	600	growth	2.06	106%	$N(t) = 600 \cdot 2.06^t$
b.	1200	growth	3.00	200%	$N(t) = 1200 \cdot 3^t$
c.	6000	decay	0.25	75%	$N(t) = 6000 \cdot 0.25^t$
d.	$1.5 \cdot 10^6$	decay	0.75	25%	$N(t) = 1.5 \cdot 10^6 \cdot 0.75^t$
e.	$1.5 \cdot 10^6$	growth	1.25	25%	$N(t) = 1.5 \cdot 10^6 \cdot 1.25^t$
f.	7	growth	4.35	335%	$N(t) = 7 \cdot 4.35^t$
g.	60	decay	0.35	65%	$N(t) = 60 \cdot 0.35^t$

7. a. $A(T) = 12,000 \cdot 1.12^T$ and $B(T) = 12,000 \cdot 0.88^T$, where $T =$ decades since 1990.

b. If $t =$ years since 1990, then $t = 10T$ or $T = t/10$, and thus $a(t) = 12,000 \cdot [1.12]^{t/10} = 12,000 \cdot [1.12^{0.1}]^t = 12,000 \cdot 1.0114^t$ people and $b(t) = 12,000 \cdot [0.88]^{t/10} = 12,000 \cdot [0.88^{0.1}]^t = 12000 \cdot 0.9873^t$ people where t is the number of years from 1990.

c. $A(2) \approx 15,053$; $a(20) \approx 15,053$; $B(2) \approx 9293$ and $b(20) \approx 9293$. $A(2)$ and $B(2)$ represent the population after two decades, whereas $a(20)$ and $b(2)$ represent the population after 20 years. Note that $A(2)$ should equal $a(20)$ and $B(2)$ should equal $b(20)$. Any small differences are because of round-off error.

9. $A(x) = 500 \cdot 0.85^x$; decay rate of 15%

11. $A(x) = 225 \cdot 1.015^x$; growth rate of 1.5%

13. a. $P(t) = 150 \cdot 3^t$ **c.** $P(t) = 150 \cdot 0.93^t$

b. $P(t) = 150 - 12t$ **d.** $P(t) = 150 + t$

15. In 2030 the population size will be approximately $200 \cdot 1.4^2 = 392$ million, and in 2060 it will be approximately $200 \cdot 1.4^3 = 548.8$ million. Using a graphing calculator to estimate the intersection point of the graphs of $y_1 = 200 \cdot 1.4^t$ and $y_2 = 1000$ (where y_1 and y_2 are measured in millions and t is measured in 30-year periods) gives $t \approx 4.78$. Thus, according to this model, sometime in the first part of 2113, the U.S. population will reach approximately 1 billion.

17. a. $A(n) = A_0 \cdot 0.75^n$, where n measures the number of years from the original dumping of the pollutant and A_0 represents the original amount of pollutant.

b. We are solving $0.1 \cdot A_0 = A_0 \cdot 0.75^n$ for n and we first divide out by A_0. Then we graph $y = 0.01$ and $y = 0.75^n$ and estimate their intersection. This gives $n \approx 16$ years.

19. a. $1.007^{12} \approx 1.087$ and thus the inflation rate is about 8.7% per year.

b. $(1 + r)^{12} = 1.05$ means that $r = (1.05)^{1/12} - 1 \approx 0.0041$. Thus the rate is about 0.4% per month.

21. a. They are equivalent, since $1.2^{0.1} = 1.0184$ to four decimal places.

b.

x	0	5	10	15	20	25
$f(x)$	15,000.0	16431.7	18,000	19,718.0	21,600.0	23,661.6
$g(x)$	15,000.0	16431.7	18,000.1	19,718.2	21,600.3	23,662.0

c. If x is the number of years, then $f(x) = 15,000(1.2)^{x/10}$ represents a 20% growth factor over a decade, and $g(x) = 15,000(1.0184^x)$ represents a 1.84% annual growth factor.

23. a. A graph with the data plotted and the best-fit exponential is given in the diagram.

Alcohol Content over Time of an Initially Legally Drunk Person

$y = 0.1 \cdot 0.67^t$

BAC g/dl — Time in hours

b. If one looks at the exponential graph it is easy to see that it fits the data quite well. Moreover, the ratios of successive g/dl values are nearly a constant 0.67. Another equation, easier to handle, is $y = 0.1 \cdot 0.67^t$, where t is measured in hours and y in g/dl.

c. In this model, the g/dl decreases each hour by about 33%.

d. A reasonable domain could be 0 to 10 hours and the corresponding range would be 0.002 to 0.100 g/dl.

e. $0.005 = 0.1 \cdot 0.67^t$ implies that t is between 7 and 8 hours. This can be obtained by computing g/dl values for successive t values using the formula given in part (b).

25. a. If the model is linear, then the slope is $\frac{(6.4 - 7.2)}{(2006 - 1996)} = -\frac{0.8}{10} = -0.08$ and the equation is $\mathrm{MRL}(y) = -0.08y + 7.2$, where y is the number of years since 1996 and MRL is the number of deaths per 1000 live births.

b. If the model is exponential, then the decay factor is $\left(\frac{6.4}{7.2}\right)^{1/10} \approx 0.988$ and thus the equation is $\mathrm{MRE}(y) = 7.2 \cdot 0.988^y$, and the units are the same as in part (a).

c. The graphs requested are in the accompanying diagram.

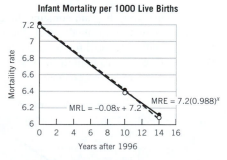

Infant Mortality per 1000 Live Births

Mortality rate — Years after 1996

$\mathrm{MRE} = 7.2(0.988)^x$
$\mathrm{MRL} = -0.08x + 7.2$

d. In 2010, $\mathrm{MRL}(14) = 6.08$ and $\mathrm{MRE}(14) \approx 6.11$; both are measured in deaths per 1000 live births.

Section 5.6

Algebra Aerobics 5.6a

1. a. Exponential growth; doubling period is 30 days since
$f(30) = 300 \cdot 2^{30/30} = 600$; Initial $= 300$;
one day later: $f(1) = 300 \cdot 2^{1/30} \approx 307.01$;
one month later: $f(30) = 300 \cdot 2^{30/30} = 600$;
one year later: $f(360) = 300 \cdot 2^{360/30} = 1{,}228{,}800$

b. Exponential decay; half-life period is 2 days since
$g(2) = 32(0.5)^{2/2} = 16$; Initial $= 32$;
one day later: $g(1) = 32 \cdot 0.5^{1/2} \approx 22.6$;
one month later: $g(30) = 32 \cdot 0.5^{30/2} = 0.00098$;
one year later: $g(360) = 32 \cdot 0.5^{360/2} =$ trace amount (almost zero)

c. Exponential decay; half-life period is one day since
$P = 32{,}000 \cdot 0.5^1 = 16{,}000$; Initial $= 32{,}000$;
one day later: $P = 32{,}000 \cdot 0.5^1 = 16{,}000$;
one month later: $P = 32{,}000 \cdot 0.5^{30} \approx 0.000\ 03$;
one year later: $P = 32{,}000 \cdot 0.5^{360} =$ trace amount (almost zero)

d. Exponential growth; doubling period is 360 days since
$h(360) = 40{,}000 \cdot 2^{360/360} = 80{,}000$; Initial $= 40{,}000$;
one day later, $h(1) = 40{,}000 \cdot 2^{1/360} \approx 40{,}077$;
one month later, $h(30) = 40{,}000 \cdot 2^{30/360} \approx 42{,}379$;
one year later, $h(360) = 40{,}000 \cdot 2^{360/360} = 80{,}000$

2. a. $70/2 = 35$ yr

b. $70/0.5 = 140$ months

c. $70/8.1 = 8.64$ yr ≈ 9 yr

d. growth factor $= 1.065 \Rightarrow$ growth rate $= 6.5\%$
$70/6.5 = 10.77 \approx 11$ yr

3. a. $\frac{70}{10} = 7$; $R \approx 7\%$ per year

b. $70/5 = 14$; $R \approx 14\%$ per minute

c. $70/25 = 2.8$; $R \approx 2.8\%$ per second

4. a. Since $a = 0.95$, $r = 0.05$, so this is decay of $R = 5\%$, and half-life is $70/5 = 14$ months.

b. Since $a = 0.75$, $r = 0.25$, so this is decay of $R = 25\%$, and half-life is $70/25 = 2.8$ sec.

c. Half-life is $70/35 = 2$ yr.

Algebra Aerobics 5.6b

1. a. $70/3 \approx 23.3$ yr   **b.** $70/5 = 14$ yr   **c.** $70/7 = 10$ yr

2. a. $70/5 = 14$; 14%   **b.** $70/10 = 7$; 7%   **c.** $70/7 = 10$; 10%

3. a. $P = \$1000(1.04)^n$ **b.** $P = \$1000(1.11)^n$
c. $P = \$1000(2.10)^n$

4. If inflation is 10%/month, then what cost 1 cruzeiro this month would cost 1.10 cruzeiros next month. We have 1 cruzeiro $\approx 91\%$ of 1.10 cruzeiros (since $1/1.1 \approx 0.91$), so a month later 1 cruzeiro would only be worth 0.91 cruzeiros or 91% of its original value. Thus, the decay factor is 0.91. So the exponential decay function $Q = 100(0.91)^n$ gives the purchasing power of 100 of today's cruzeiros at n months in the future.

When $n = 3$ months, then Q, the value of 100 of today's cruzeiros, will be $100(0.91)^3 \approx 100(0.75) = 75$ cruzeiros. When $n = 6$ months, then Q, the value of 100 of today's cruzeiros, will be $100(0.91)^6 \approx 100(0.57) = 57$ cruzeiros. When $n = 12$ months or 1 yr, then Q, the value of 100 of today's cruzeiros, will be $100(0.91)^{12} \approx 100(0.32) = 32$ cruzeiros.

With a 10% monthly inflation rate, the value of a cruzeiro will shrink by more than two-thirds by the end of a year.

5. a. Initial investment is \$10,000; growth factor $= 1.065$; growth rate $= 6.5\%$; doubling time: using the rule of 70, $70/6.5 \approx 10.8$ years

b. Initial investment is $25; growth factor = 1.08; growth rate = 8%; doubling time: using the rule of 70, 70/8 ≈ 8.8 years

c. Initial investment is $300,000; growth factor = 1.11; growth rate = 11%; doubling time: using the rule of 70, 70/11 ≈ 6.4 years

d. Initial investment is $200; growth factor = 1.092; growth rate = 9.2%; doubling time: using the rule of 70, 70/9.2 ≈ 7.6 years

6.

Function	Initial Investment	Growth Factor	Growth Rate	Amount 1 Year Later	Doubling Time (approx.)
$A = 50000 \cdot 1.072^t$	$50,000	1.072	7.2%	$53,600	9.7 years
$A = 100000 \cdot 1.067^t$	$100,000	1.067	6.7%	$106,700	10.4 years
$A = 49622 \cdot 1.058^t$	$49,622	1.058	5.8%	$52,500	12 years
$A = 3000 \cdot 1.13^t$	$3000	1.13	13%	$3390	5.4 years

Algebra Aerobics 5.6c

1. a. $x = 5$ since $2^5 = 32$

b. $x = 8$ since $2^8 = 256$

c. $x = 10$ since $2^{10} = 1024$

d. $x = 1$ since $2^1 = 2$

e. $x = 0$ since $2^0 = 1$

f. $x = -1$ since $2^{-1} = \frac{1}{2}$

g. $x = -3$ since $2^{-3} = \frac{1}{2^3} = \frac{1}{8}$

h. $x = 1/2$ since $2^{1/2} = \sqrt{2}$

2. a.    **b.** $N = 3^L$

3.

Level	# of People Called	Total Called
0	3^0	1
1	3^1	4
2	$3^2 = 9$	13
3	$3^3 = 27$	40
4	$3^4 = 81$	121
5	$3^5 = 243$	364
6	$3^6 = 729$	1093
7	$3^7 = 2187$	3280
8	$3^8 = 6561$	9841

So it will take eight levels to reach 8000 people.

Exercises for Section 5.6

1. a. Has a fixed doubling time

b. Has neither

c. Has a fixed half-life

d. Has a fixed doubling time

e. Has neither

f. Has a fixed half-life

3. b. $f(x) = 1000 \cdot 2^{t/7}$; $2^{1/7} = 1.1041$ per year; 10.41% per year

c. $f(x) = 4 \cdot 2^{t/25}$; $2^{1/25} = 1.0281$ per minute; 2.81% per minute

d. $f(x) = 5000 \cdot 2^{t/18}$; $2^{1/18} = 1.0393$ per month; 3.93% per month

5. a. $3 \cdot 2^{x/5} < 3(1.225)^x$, if $x > 0$; the inequality is reversed if $x < 0$.

b. $50 \cdot (1/2)^{x/20} \approx 50 \cdot 0.9659^x$.

c. $200 \cdot 2^{x/8} \approx 200 \cdot 1.0905^x$.

d. $750 \cdot (1/2)^{x/165} > 750 \cdot 0.911^x$ if $x > 0$, and the inequality is reversed if $x < 0$.

7.

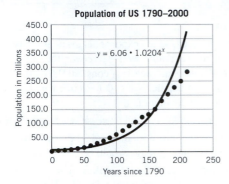

a, b. The accompanying diagram contains the graph of both the data and the best-fit exponential function, via Excel. There y = population in millions and x = years since 1790. The annual growth factor is 1.0204; the annual growth rate is 2.04%; the estimated initial population is 6.06 million. One notes that the best-fit model is lower than the plotted data from 1850 to 1930. Its values surpass those of the data from 1940 on. It is a good model for the first 160 years but is not that good thereafter.

c. The model predicts the population to be 516.2 million in 2010 and 698.8 million in 2025. These seem very high indeed.

9. a. The accompanying diagram contains the graphs of the best-fit linear and exponential models for the given data. The linear and exponential formulae are also given there. The linear model seems to fit the data better because of a slightly higher correlation coefficient.

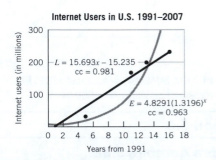

b. 2010 is 19 years after 1991. The linear model predicts $L(19) = 15.693 \cdot 19 - 15.235 \approx 283$ million users, and the exponential model predicts $E(19) = 4.8291(1.3196)^{19} \approx 938$ million users.

c. Student answers will vary.

11. The graph in the accompanying diagram goes through (0, 100) and (12.3, 50), where the first coordinate is measured in years and the second is measured in grams. Using the graphing utility gives: $A(t) = 100 \cdot 0.945^t$. Using algebra we get $A(t) = 100 \cdot (0.5)^{t/12.3}$, which gives the same formula as the best-fitting exponential formula. The graph of this function is given in the accompanying diagram.

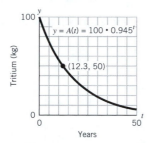

13. **a.** $0.5^5 = 1/32 = 0.03125$ and thus 3.125% of the original dosage is left.

   b. **i.** Approximately 2 hours.

   ii. $A(t) = 100 \cdot 0.5^{t/2}$ where t = hours after taking the drug and $A(t)$ is in mg.

   iii. $5 \cdot 2 = 10$ hours and $100 \cdot (1/2)^5 = 100/32 = 3.125$ milligrams. Also $A(10) = 100 \cdot 0.5^{10/2} = 3.125$.

   iv. Student answers will vary. They should mention the half-life and present a graph to make the drug's behavior clear to any prospective buyer.

15. **a.** $R = 70/5730 \approx 0.012\%$ per year

   b. $R = 70/11{,}460 \approx 0.0061\%$ per year

   c. $R = 70/5 = 14\%$ per second.

   d. $R = 70/10 = 7\%$ per second.

17. **a.**

Years from 1800	World Population (in millions)
0	980
50	1260
100	1650
150	2520
170	3700
180	4440
190	5270
200	6080
205	6480

   b. $P(t) = 799.96 \cdot 1.00955^t$ is the best-fit exponential function formula, where t is measured in years since 1800 and $P(t)$ is measured in millions. Its graph is given in the accompanying diagram.

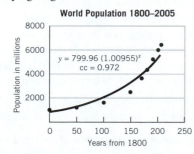

World Population 1800–2005

   c. The initial value is 799.96 or approximately 800 for the world's population in millions in 1800 (when $t = 0$). (Note that it is approximately 180 million smaller than the actual size.) The growth factor is 1.00955, which gives a growth rate of 0.955% per year. The domain of the function P technically is any real number, but concretely, its values should not go much farther back than 1800 or much past 2005. The range in that domain is about 800 million to about 5600 million.

   d. 1.00955 represents a growth rate of 0.955% per year.

   e. **i.** Using a calculator gives the $P(t)$ values in the accompanying table. Eyeball estimates from students should be close to these.

Year	t	$P(t)$
1750	−50	498.6
1920	120	2487.9
2025	225	6714.1
2050	250	8504.5

   ii. Using a calculator or the graph of the model one can see that 1 billion was reached during 1823 (or about 23.6 years after 1800); 4 billion was reached in the first half of 1970 (or about 170.2 years from 1800); and 8 billion will be reached during 2043 (about 243.5 years from 1800).

   f. As can be seen from the answers in (ii) it takes approximately 73 years for the population to double.

19. **a.** 10 half-lives gives a $0.5^{10} = 0.001$ or 0.1% of the original or a 99.9% reduction.

   b. $A(2) = 0.25A_0$, $A(3) = 0.125A_0$, $A(4) = 0.0625A_0$. After n half-lives, the amount left is $A(n) = 0.5^n \cdot A_0$.

21. **a.**

t, Time (months)	M, mass (g)
0	10
3	20
6	40
9	80
12	160

   In general $M = 10(2^{1/3})^t = 10 \cdot 1.2599^t$ after t months.

   b. Using a calculator or a graph of the model, when $M = 2000$, then $t \approx 23$ months or nearly 2 years.

   c. $2000/10 = 200$ and thus at 2000 grams, it is 20,000% of its original size.

23. **a.** **i.** $A(n) = 5000 \cdot (1.035)^n$

   ii. $B(n) = 5000 \cdot (1.0675)^n$

   iii. $C(n) = 5000 \cdot (1.125)^n$

   b. $A(40) = 19{,}796.30$ $B(40) = 68{,}184.45$
   $C(40) = 555{,}995.02$.

25. **a.** $V(n) = 100 \cdot 1.06^n$. Solving $V(n) = 200$ for n graphically gives $n \approx 12$ years, which is the approximate doubling time.

   b. $W(n) = 200 \cdot 1.03^n$. Solving $W(n) = 400$ for n graphically gives $n \approx 23$ years, which is the approximate doubling time.

c. The graphs of $V(n)$ and $W(n)$ are in the accompanying diagram. The two graphs intersect at approximately $n = 24$ years, when $V(n) = W(n) \approx \$410$. The values for $V(n)$ are larger than those for $W(n)$ after that.

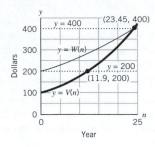

27. Since $1.00/1.06 = 0.943$, then the value of a dollar after t years of such inflation is given by $V(t) = 0.943^t$. Using a calculator or the "rule of 70," when $V(t) = 0.50$, gives $t \approx 12$ years.

29. a. and **b.** Below is a scatter plot of the data and the Excel generated exponential curve that is a best fit.

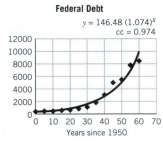

The model would predict the debt of 2008 to be 146.48 $(1.074)^{58} \approx \$9205$ billion.

c. Answers will vary depending on when the student invokes the debt clock.

31. a. In theory, the number recruited is $M_{new}(n) = 10^n$, where n measures the number of rounds and $M_{new}(n)$ measures the number of people participating in the nth round of recruiting. Note that this formula assumes that all who are recruited stay and that all recruits are distinct.

b. $M_{Total}(n) = 1 + 10 + \cdots + 10^n$.

c. $M_{Total}(5) = 111,111$, but only $11,110$ of those stem from the originator. After 10 rounds the number recruited (not including the originator) would be $11,111,111,110$, which is larger than the 2005 world population.

d. Comments will vary, but all will probably note how fast the number of recruits needed grows and how the amounts expected are not quite what one would have thought from the advertisements. If the chain is initially successful, then you would get a large number of new recruits in a short period. However, as can be seen in part (c), it quickly becomes unrealistic for each new person on the chain to recruit 10 new people.

Section 5.7

Algebra Aerobics 5.7

1. a. Judging from the graph, the number of *E. coli* bacteria grows by a factor of 10 (for example, from 100 to 1000, or 100,000 to 1,000,000) in a little over three time periods.

b. From the equation $N = 100 \cdot 2^t$, we know that every three time periods, the quantity is multiplied by 2^3 or 8.

c. Every four time periods, the quantity is multiplied by 2^4 or 16.

d. The answers are consistent, since somewhere between three and four time periods the quantity should be multiplied by 10 (which is between 8 and 16).

2. The graph of $y = 25 (10)^x$ will be a straight line on a semi-log plot.

x	$y = 25(10)^x$
0	25
1	250
2	2,500
3	25,000
4	250,000
5	2,500,000

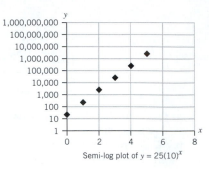

Semi-log plot of $y = 25(10)^x$

3. a. If $x = 3.5$, y is about $80,000$. If $x = 7$, $y = 250,000,000$.

Exercises for Section 5.7

1. a. Graph of given table of values for $y = 500 \cdot 3^x$

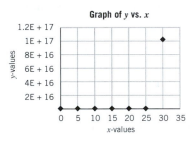

b. Graph of table for $\log(500 \cdot 3^x)$ vs. x

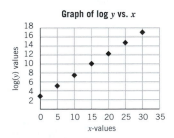

c. $y = 10^{\log(y)}$ and thus each y can be written as 10 to the power listed in the third column.

3. a. A uses the linear scale. B uses a power of 10 scale on the vertical axis and C uses a logarithm scale on the vertical axis.

b. The two graphs look the same because $\log(10^n) = n$ and the powers of 10 are spaced out like n.

c. On Graphs A and B, when $y = 1000$ is multiplied by a factor of 10 to get $y = 10,000$, x has increased by about 5.5 units.

d. On graphs B, y labels go up by factors of 10.

e. On graph A, since the scales on both axes are linear.

5. a. The accompanying graph is of the white blood cell counts on a semi-log plot. The data from October 17 to

October 30 seem to be exponential, since the data in that range seem to fall along a straight line in the plot. There does not seem to be a discernible exponential decay pattern in this graph of the data.

Semi-log plot for 5(a).

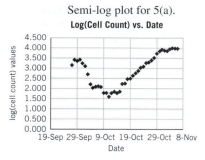

b. Below are two graphs of the *E. coli* counts; one has regular horizontal and vertical axes and the other is a semi-log plot. These data look very exponential from the third to the thirteenth time periods in the regular plot and fairly exponential in the semi-log plot (since that plot looks rather linear). For contrast, the regular linear plot is given as well.

Semi-log plot for 5(b).

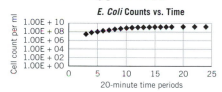

Regular plot for 5(b).

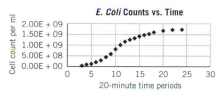

7. a. This is a time-series, semi-log plot with time scale linear.

b. It says that the growth was roughly exponential between 1993 and 2000.

c. It says that the decline was roughly exponential between 2000 and 2002.

d. Student answers will vary. Only Fidelity can give the real answer. But one might surmise that Fidelity wanted to show overall that it has done as well as the 500 Index Funds and to visually deemphasize its decline after 2000.

Ch. 5: Check Your Understanding

1. False	**7.** False	**13.** False	**19.** False
2. True	**8.** False	**14.** False	**20.** True
3. False	**9.** True	**15.** True	**21.** True
4. False	**10.** True	**16.** True	**22.** True
5. False	**11.** False	**17.** False	**23.** False
6. True	**12.** False	**18.** True	**24.** True

25. Possible answer: $P = 2.2(1.005)^t$ million people, t = years.

26. Possible answer: $P = 2.2(1.005)^{4t}$ million people, t = years.

27. Possible answer: $M = 1.4(0.977)^{t/10}$ billion dollars, t = years.

28. Possible answer: $y = 3125(0.2)^t$.

29. Possible answer: $y_1 = 300(0.88)^t$ and $y_2 = 300(0.94)^t$.

30. Possible answer: $y = -1.458x + 5$ and $y = 5(0.5)^x$.

31. Possible answer: $V = 1(0.97)^t$, t = years.

32. Possible answer: $y = 5 \cdot (1.051)^t$, t = years.

33. Possible answer: $R = 200(0.966)^t$ mg, t = years.

34. True	**37.** True	**40.** False	**43.** False
35. False	**38.** True	**41.** False	
36. True	**39.** False	**42.** True	

Ch. 5 Review: Putting It All Together

1. a. $P = 150 \cdot 2^t$ **c.** $P = 150(1.05)^t$

   b. $P = 150 - 12t$ **d.** $P = 150 + 12t$

3.

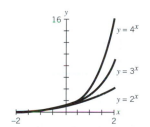

5. a. Males who just stopped smoking are about 22 times more likely (the relative risk) to get lung cancer than a lifelong nonsmoker. Females who stopped smoking 12 years ago are about three times more likely to get lung cancer than a lifelong nonsmoker.

b. One reason could be that the longer it has been since someone quit smoking, the more time the lungs have had to heal. Another reason could be that the death rate of smokers is higher than that of nonsmokers.

c. A relative risk of 1 means each group is equally likely to get lung cancer. It is highly unlikely for the relative risk of smokers vs. nonsmokers to go below 1 since that would mean smokers are less likely to get lung cancer than nonsmokers, which does not make sense.

7. Initial value = 500 and growth rate = $1.5 \Rightarrow G(t) = 500(1.5)^t$, where $G(t)$ = number of bacteria and t = number of days.

9. a. If the world population in 1999 was 6 billion people, and it grew at a rate of 1.3% per year, then it is only in the first year that there is a net addition of 78 million people. The next year the increase would be 1.3% of 6,078,000,000, which is 79,014,000. The population would continue to have an increase that becomes larger and larger than 78 million each year. The increase is a fixed amount only in a linear model.

b. Let P = world population (in millions) and t = number of years since 1999.

Linear model: $P = 6000 + 78t$

Exponential model: $6000(1.013)^t$

c. Linear model prediction for 2006:
$P = 6000 + 78(2006 - 1999) \Rightarrow P = 6546$ million

Exponential model prediction for 2006: $6000(1.013)^7 \Rightarrow$ $P \approx 6568$ million

Answers may vary for best predictor depending on current population.

The U.S Census Bureau website gives the 2006 world population as 6.567 billion, so our exponential model is a more accurate predictor for 2006.

11. If R = the annual growth rate, then using the "rule of 70" we have $70/25 = 2.8 \Rightarrow R = 2.8\%$. So the number of motor vehicles increases by almost 3% per year.

13.

			Projected	
Year	2003	2004	2005	2006
Health care expenditures (billions)	$1740.6	$1877.6	$2016.0	$2169.5
Annual growth factor	n.a.	$\frac{1877.6}{1740.6}$ ≈ 1.079	$\frac{2016.0}{1877.6}$ ≈ 1.074	$\frac{2169.5}{2016}$ ≈ 1.076
Annual percent growth rate	n.a	7.9%	7.4%	7.6%

15. The ratio of consecutive values over 5-year intervals is approximately constant at 1.09, so an exponential model would be appropriate for the data in the table. If we let $E(t)$ = energy consumption (in quadrillion Btu), where t = number of 5-year intervals since 2015, then an exponential function to model the data would be: $E(t) = 563 (1.09)^t$.

17. a. 2002 to 2015:

13 year growth factor = (2 billion/605 million) = 2,000 million/605 million $\approx$ 3.306

annual growth factor = $3.306^{1/13} \approx 1.096$

annual growth rate ≈ 0.096 or 9.6%

b. 2015 to 2040:

25 year growth factor = (3 billion/2 billion) = 1.5

annual growth factor = $1.5^{1/25} \approx 1.016$

annual growth rate ≈ 0.016 or 1.6%

19. Initial value = 20 grams gives $A(t) = 20(1/2)^{t/8}$ grams, where t is in days.

21. The graph shows a semi-log plot. When an exponential function is plotted on a semi-log plot, its graph is a straight line. The data for "Internet hosts" is plotted as a straight line and thus could be modeled with an exponential function. The data for "Pages" and "Websites" is approximately linear on this semi-log plot, and thus an exponential model may be a close approximation.

23. a. Each linear dimension of the model is one-tenth that of the actual village, so the area of the model (which is two-dimensional) would be $(1/10) \cdot (1/10) = 1/100$ or one-hundredth of A_o, the area of the actual village.

b. Here each linear dimension is one-hundredth that of the actual church. The weight depends on volume (which is three-dimensional) and would be $(1/100) \cdot (1/100) \cdot (1/100) = 1/1,000,000$, or one-millionth of Wo, the weight of the actual church.

c. $A_n = A_o \cdot \left(\frac{1}{10^2}\right)^n$

$W_n = W_o \cdot \left(\frac{1}{10^3}\right)^n$

25. a. $S(t) = 0.5 \cdot 2^t$, the tumor size in cubic centimeters, where t = number of 2-month time periods

$A(t) = 0.5 \cdot 2^{t/4}$, the tumor size in cubic centimeters, where t = number of 2-month time periods

b.

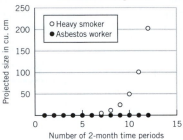

Cancer Growth for Different Doubling Times

The faster-growing cancer of the heavy smoker if untreated gets dangerously large very quickly after about 6 time periods (1 year). The slower-growing cancer of the asbestos worker after 12 time periods (2 years) is still relatively small compared to that of the smoker (the graph does not yet show an exponential curve).

c. 1 year = 6 time periods

The tumor size for the smoker after 1 year (or 6 time periods) is

$S(6) = 0.5 \cdot 2^6 = 32$ cubic centimeters $\Rightarrow$
$\text{Volume}_{\text{smoker}} = 32 = \left(\frac{4}{3}\right)\pi r^3 \Rightarrow \frac{32 \cdot 3}{4\pi} = r^3$
$$7.64 \approx r^3$$

Taking the cube root of both sides, $r \approx 1.97$ cm, or a tumor diameter of about 3.94 cm.

The tumor size for the asbestos worker after 1 year (or 6 time periods) is

$A(6) = 0.5 \cdot 2^{6/4} \approx 1.41$ cubic centimeters $\Rightarrow$
$\text{Volume}_{\text{asbestos}} = 1.41 = \left(\frac{4}{3}\right)\pi r^3 \Rightarrow \frac{1.41 \cdot 3}{4\pi} = r^3$
$0.34 \approx r^3$

Taking the cube root of both sides, $r \approx 0.70$ cm, or a tumor diameter of about 1.4 cm.

27. a. Let $M(t)$ = amount of methane gas in the atmosphere (in parts per billion), where t = number of years since 1850. The initial value in 1850 is 750 parts/billion. In 2005 the methane levels were at 1750 parts/billion, so

$$M(t) = 750 \cdot \left(\frac{1750}{750}\right)^{(1/155)t} \approx 750(1.0055)^t$$

b. Estimates from graph: Around 1980 methane gas reaches 1500 parts per billion, which is double the 1850 level of about 750 parts per billion. So the doubling time using this method is about 130 years.

Using the "rule of 70" and a growth rate of 0.55% from part (a), we have $70/0.55 \approx 127$ years. So the estimates are pretty close.

c. Since 1850 the amount of methane gas in the atmosphere has been growing exponentially at a rate of about 0.55% per year. Between 1850 and 1980, the amount of methane gas in the atmosphere doubled.

CHAPTER 6

Section 6.1

Algebra Aerobics 6.1a

1. a. When $t = 0$, $M = 250(3)^0 = 250(1) = 250$.

b. When $t = 1$, $M = 250(3)^1 = 250(3) = 750$.

c. When $t = 2$, $M = 250(3)^2 = 250(9) = 2250$.

d. When $t = 3$, $M = 250(3)^3 = 250(27) = 6750$.

2. a. $4 < t < 5$ **b.** $8 < t < 9$

3. To read each value of x from this graph, draw a horizontal line from the y-axis at a given value until it hits the curve. Then draw a vertical line to the x-axis to identify the appropriate value of x.

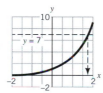

a. horizontal line $y = 7$, from y-axis to curve, then down to x-axis $\Rightarrow x \approx 1.8$

b. horizontal line $y = 0.5$, from y-axis to curve, then down to x-axis $\Rightarrow x \approx -0.5$.

4. a. $2 < x < 3$ **b.** $-1 < x < 0$

5. a. $2^3 < 13 < 2^4$ **c.** $5^{-1} < 0.24 < 5^0$

b. $3^4 < 99 < 3^5$ **d.** $10^3 < 1500 < 10^4$

Algebra Aerobics 6.1b

1. a. $\log(10^5/10^7) = \log 10^{-2} = -2 \log 10 = -2(1) = -2$ and $\log 10^5 - \log 10^7 = 5 \log 10 - 7 \log 10 = 5 - 7 = -2$. So $\log(10^5/10^7) = \log 10^5 - \log 10^7$.

b. $\log(10^5 \cdot (10^7)^3) = \log(10^5 \cdot 10^{21}) = \log 10^{26} = 26 \log 10 = 26$ and $\log 10^5 + 3 \log 10^7 = 5 \log 10 + 3(7) \log 10 = 5 + (3 \cdot 7) = 26$. So $\log[10^5 \cdot (10^7)^3] = \log(10^5) + 3 \log(10^7)$.

2. a. Rule 2: $\log 3 = \log \frac{15}{5} = \log 15 - \log 5$

b. Rule 3: $\log 1024 = \log(2^{10}) = 10 \log 2$

c. Rule 3: $\log \sqrt{31} = \log(31^{1/2}) = \frac{1}{2} \log 31$

d. Rule 1: $\log 30 = \log(2 \cdot 3 \cdot 5) = \log 2 + \log 3 + \log 5$

e. Rule 2: $\log 81 - \log 27 = \log(\frac{81}{27}) = \log 3$ or Rule 3: $4 \log 3 - 3 \log 3 = \log 3$

3. a. False   **d.** True; Rule 3 and that $\log 10 = 1$

b. False   **e.** False

c. True; Rule 3   **f.** False

4. a. $\log \sqrt{\frac{2x-1}{x+1}} = \log(\frac{2x-1}{x+1})^{1/2} = \frac{1}{2} \log(\frac{2x-1}{x+1})$
$$= \frac{1}{2}[\log(2x-1) - \log(x+1)]$$

b. $\log \frac{xy}{z} = \log(xy) - \log z = \log x + \log y - \log z$

c. $\log \frac{x\sqrt{x+1}}{(x-1)^2} = \log \frac{x(x+1)^{1/2}}{(x-1)^2}$
$$= \log x(x+1)^{1/2} - \log(x-1)^2$$
$$= \log x + \log(x+1)^{1/2} - 2\log(x-1)$$
$$= \log x + \frac{1}{2}\log(x+1) - 2\log(x-1)$$

d. $\log \frac{x^2(y-1)}{y^3 z} = \log x^2(y-1) - \log y^3 z$
$$= \log x^2 + \log(y-1) - [\log y^3 + \log z]$$
$$= 2\log x + \log(y-1) - 3\log y - \log z$$

5. $\frac{1}{3}[\log x - \log(x+1)] = \frac{1}{3}\log\frac{x}{x+1}$
$$= \log\left(\frac{x}{x+1}\right)^{1/3}$$
$$= \log\sqrt[3]{\frac{x}{x+1}}$$

6. a. $\log x = 3 \Rightarrow 10^3 = x, x = 1000$

b. $\log x + \log 5 = 2 \Rightarrow \log 5x = 2 \Rightarrow 10^2 = 5x \Rightarrow x = 20$

c. $\log x + \log 5 = \log 2 \Rightarrow \log 5x = \log 2 \Rightarrow 5x = 2 \Rightarrow x = 2/5$

d. $\log x - \log 2 = 1 \Rightarrow \log \frac{x}{2} = 1 \Rightarrow 10^1 = \frac{x}{2} \Rightarrow x = 20$

e. $\log x - \log(x-1) = \log 2 \Rightarrow \log \frac{x}{x-1} = \log 2 \Rightarrow$
$\frac{x}{x-1} = 2 \Rightarrow x = 2(x-1) \Rightarrow x = 2$

f. $\log(2x+1) - \log(x+5) = 0 \Rightarrow$
$\log \frac{2x+1}{x+5} = \log 1 \Rightarrow$
$\frac{2x+1}{x+5} = 1 \Rightarrow 2x+1 = x+5 \Rightarrow x = 4$

7. $\log 10^3 - \log 10^2 = 3\log 10 - 2\log 10 = 3 - 2 = 1$

$\frac{\log 10^3}{\log 10^2} = \frac{3\log 10}{2\log 10} = \frac{3}{2} = 1.5$

Since $1 \neq 1.5 \Rightarrow \log 10^3 - \log 10^2 \neq \frac{\log 10^3}{\log 10^2}$

Algebra Aerobics 6.1c

1. a. $60 = 10 \cdot 2^t \Rightarrow 6 = 2^t$ (dividing both sides by 10)
$\log 6 = \log 2^t$ (taking the log of both sides)
$\log 6 = t \log 2$ (using Rule 3 of logs)
$\log 6/\log 2 = t$
$0.7782/0.3010 \approx t$ or $t \approx 2.59$

b. $500(1.06)^t = 2000 \Rightarrow (1.06)^t = \frac{2000}{500} \Rightarrow$
$(1.06)^t = 4 \Rightarrow \log(1.06)^t = \log 4 \Rightarrow$
$t(\log 1.06) = \log 4 \Rightarrow t = \frac{\log 4}{\log 1.06} = \frac{0.6021}{0.0253} \approx 23.8$

c. $80(0.95)^t = 10 \Rightarrow (0.95)^t = 1/8 = 0.125 \Rightarrow$
$\log(0.95)^t = \log 0.125 \Rightarrow t \log 0.95 = \log 0.125 \Rightarrow$
$t = \frac{\log 0.125}{\log 0.95} \approx \frac{-0.9031}{-0.0223} \approx 40.5$

2. a. $7000 = 100 \cdot 2^t \Rightarrow 70 = 2^t \Rightarrow \log 70 = \log 2^t \Rightarrow$

$\log 70 = t \log 2 \Rightarrow t = \log 70/\log 2 \approx \frac{1.845}{0.301} \approx 6.13$

It will take 6.13 time periods or approximately $(6.13)(20)$ min $= 122.6$ min (or a little more than 2 hours) for the bacteria count to reach 7000.

b. $12{,}000 = 100 \cdot 2^t \Rightarrow 120 = 2^t \Rightarrow \log 120 = \log 2^t \Rightarrow$

$\log 120 = t \log 2 \Rightarrow t = \log 120/\log 2 \approx 6.907$ time periods or approximately $(6.907)(20)$ min $= 138.14$ min (or a little more than $2\frac{1}{4}$ hrs.) for the bacteria count to reach 12,000.

3. Using the rule of 70, since $R = 6\%$ per yr, then $70/R = 70/6 \approx 11.7$ yr. More precisely, we have:

$2000 = 1000(1.06)^t \Rightarrow 2 = 1.06^t \Rightarrow$

$\log 2 = \log 1.06^t \Rightarrow \log 2 = t \log 1.06 \Rightarrow$

$t = \frac{\log 2}{\log 1.06} \approx \frac{0.3010}{0.0253} \approx 11.9$ yr

4. $1 = 100(0.976)^t \Rightarrow 0.01 = (0.976)^t \Rightarrow$

$\log 0.01 = \log (0.976)^t \Rightarrow \log 0.01 = t \log 0.976 \Rightarrow$

$t = \frac{\log 0.01}{\log 0.976} = \frac{-2}{-0.0106} \approx 189$ years, or almost 2 centuries!

5. a. $30 = 60(0.95)^t \Rightarrow 0.5 = (0.95)^t \Rightarrow$

$\log(0.5) = \log(0.95^t) \Rightarrow \log 0.5 = t \log 0.95 \Rightarrow$

$t = \frac{\log 0.5}{\log 0.95} \approx \frac{-0.3010}{-0.0223} \approx 13.5$ years for the initial amount to drop in half.

b. $16 = 8(1.85)^t \Rightarrow 2 = (1.85)^t \Rightarrow$

$\log 2 = \log(1.85)^t \Rightarrow \log 2 = t \log 1.85 \Rightarrow t = \frac{\log 2}{\log 1.85} \Rightarrow$

$t \approx \frac{0.3010}{0.2672} = 1.13$ years for the initial amount to double.

c. $500 = 200(1.045)^t \Rightarrow 2.5 = (1.045)^t \Rightarrow$

$\log 2.5 = \log(1.045)^t \Rightarrow \log 2.5 = t \log 1.045 \Rightarrow$

$t = \frac{\log 2.5}{\log 1.045} \Rightarrow t \approx \frac{0.3979}{0.0191} \Rightarrow$

$t \approx 20.8$ years for the initial amount to increase from 200 to 500.

6. a. $60 = 120(0.983)^t \Rightarrow 0.5 = (0.983)^t \Rightarrow$

$\log 0.5 = t \log 0.983 \Rightarrow$

$t = \frac{\log 0.5}{\log 0.983} \Rightarrow t \approx \frac{-0.3010}{-0.0074} \Rightarrow$

$t \approx 40.6$ days

b. $0.25 = 0.5(0.92)^t \Rightarrow 0.5 = (0.92)^t \Rightarrow$

$\log 0.5 = t \log 0.92 \Rightarrow$

$t = \frac{\log 0.5}{\log 0.92} \Rightarrow t \approx \frac{-0.3010}{-0.0362} \Rightarrow$

$t \approx 8.3$ hours

c. $0.5A_0 = A_0(0.89)^t \Rightarrow 0.5 = (0.89)^t \Rightarrow$

$\log 0.5 = t \log 0.89 \Rightarrow t = \frac{\log 0.5}{\log 0.89} \Rightarrow t \approx \frac{-0.3010}{-0.0506} \Rightarrow$

$t \approx 5.9$ years

Exercises for Section 6.1

1. Student estimates will vary for each interest rate.

  a. About 38 years.   **b.** About 16 years.

3. Eyeball estimates will vary. One set of guesses is:

  a. 1.3 hrs.   **b.** 2.4 hrs.   **c.** 5 hrs.

5. a. 0.001 **b.** 10^6 **c.** 1 **d.** 10 **e.** 0.1

7. a. $2 \cdot \log 3$ **b.** $2 \cdot \log 3 + \log 2$ **c.** $3 \cdot \log 3 + \log 2$

9. a. 12 **b.** 12 **c.** 12 **d.** 2

11. If $w = \log(A)$ and $z = \log(B)$, then $10^w = A$ and $10^z = B$. Thus $A/B = 10^w/10^z = 10^{w-z}$ and therefore $\log(A/B) = \log(10^{w-z}) = w - z = \log(A) - \log(B)$, as desired.

13. a. $\log\left(\frac{K^3}{(K+3)^2}\right)$ **c.** $\log\left(T^4 \cdot \sqrt{T}\right) = \log\left(\sqrt{T^9}\right)$

  b. $\log\left(\frac{(3+n)^5}{m}\right)$ **d.** $\log\left(\frac{\sqrt[3]{xy^2}}{(xy^2)^3}\right) = \log\left([xy^2]^{-8/3}\right)$

15. Let $w = \log(A)$. Then $10^w = A$ and thus $A^p = (10^w)^p = 10^{wp}$, and then $\log(A^p) = w \cdot p = p \cdot w = p \cdot \log(A)$, as desired.

17. a. 1000

  b. 999

  c. $10^{5/3} \approx 46.42$

  d. 1/9

  e. No solution, since x can not be negative.

19. a. Solve $300 = 100 \cdot 1.03^t$ to get $t = \log(3)/\log(1.03) \approx 37.17$ years.

  b. Similarly, $\log(3)/\log(1.07) \approx 16.24$ years.

21. a. The graphs of $f(x) = 500 \cdot (1.03)^x$ and $g(x) = 4500$ are in the accompanying diagram.

  b. $x = 75$ is a good eyeball estimate.

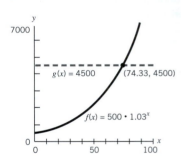

  c. $\log(4500) = \log(500) + x \cdot \log(1.03)$ or $x = [\log(4500) - \log(500)]/\log(1.03) = \log(9)/\log(1.03) \approx 74.33$

  d. The eyeball estimate and the logarithm-computed answer are very close.

23. a. Doubling time: $x = \log(2)/\log(4) = 0.5$ years

  b. Half-life $= \log(0.5)/\log(0.25) = 0.5$ years

  c. Doubling time $= \log(2)/\log(4) = 0.5$ years

  d. Half life $= \log(0.5)/\log(0.25) = 0.5$ years

25. a. $N(t) = 90(1.002)^t$, where $N(t)$ is the number of articles (in thousands) for t days after October 23.

  b. The number of articles has doubled when $N(t) = 180$ (thousand articles). So setting $180 = 90(1.002)^t$ and solving for t, we get $2 = (1.002)^t \Rightarrow \log 2 = t \log(1.002) \Rightarrow t = (\log 2)/\log(1.002) \approx 347$ days, or $347/30 \approx 11.6$ months or about one year. So if the rate of growth continues, every 12 months the number of articles will more than double.

27. a. $t = \log(2)/\log(1.5) \approx 1.71$ 20-minute time periods or about 34 min.

  b. $t = \log(10)/\log(1.5) \approx 5.68$ 20-minute time periods or about 114 min.

29. a. $B(t) = B_0(0.5)^{0.05t}$, where t is measured in minutes and $B(t)$ and B_0 are measured in some weight unit. None is specified in the problem.

b. In one hour, $t = 60$ and thus $B(60) = B_0(0.5)^{0.05 \cdot 60} = 0.125 \cdot B_0$, or 12.5% is left.

c. If its half-life is 20 minutes, then its quarter-life is 40 minutes.

d. Solving $0.10 = 0.5^{0.05t}$ for t, we get $0.05t \cdot \log(0.5) = \log(0.10)$ or $t = \log(0.1)/[\log(0.5) \cdot 0.05] \approx 66$ minutes.

31. a. $S = 300 \cdot 0.9^W$ if $0 \le W \le 10$, where W is measured in weeks and S is measured in dollars.

b. Solving $S = 150$ for W means solving $0.5 = 0.9^W$ for $W \Rightarrow W = \log(0.5)/\log(0.9) \approx 6.6$ weeks. The selling price when it would be given to charity is at $W = 10$, when $S = 300 \cdot 0.9^{10} \approx \$104.60$

33. a. If we set $4000 = 1355(1.036)^t$ then

$4000/1355 = (1.036)^t \Rightarrow 2.952 \approx 1.036^t \Rightarrow$

$\log(2.952) \approx t \log(1.036) \Rightarrow$

$t \approx \frac{\log(2.952)}{\log(1.036)} \approx 30.6$ years after 1970,

or sometime in late 2000. During that year U.S. fish production actually dropped to a little over 3000 million pounds, almost a million pounds less than the model predicted.

b. Americans are actually increasing their fish and shellfish consumption, probably spurred on by doctors' and nutritionists' advice that "eating fish is good for you." However, the amount of imported fish has grown substantially, probably cutting into U.S. fish production.

Section 6.2

Algebra Aerobics 6.2

1. a. $\$1000(1.085) = \$1085$

b. $\$1000\left(1 + \frac{0.085}{4}\right)^4 = \$1087.75$

c. $\$1000e^{0.085} = \$1088.72$

2. a. $e^{0.04} = 1.0408 \Rightarrow 4.08\%$ is effective rate

b. $e^{0.125} = 1.133 \Rightarrow 13.3\%$ is effective rate

c. $e^{0.18} = 1.197 \Rightarrow 19.7\%$ is effective rate

3. a. Principal $= 6000$; nominal rate $= 5\%$; effective rate $= 5\%$ since $1.05^1 = 1.05$; number of interest periods $= 1$

b. Principal $= 10,000$; nominal rate $= 8\%$; effective rate $\approx 8.24\%$ since $1.02^{4(1)} \approx 1.0824$; number of interest periods $= 4$

c. Principal $= 500$; nominal rate $= 12\%$; effective rate $\approx 12.68\%$ since $1.01^{12(1)} \approx 1.1268$; number of interest periods $= 12$

d. Principal $= 50,000$; nominal rate $= 5\%$; effective rate $\approx 5.06\%$ since $1.025^{2(1)} \approx 1.0506$; number of interest periods $= 2$

e. Principal $= 125$; nominal rate $= 7.6\%$; effective rate $\approx 7.90\%$ since $e^{0.076 \,(1)} \approx 1.0790$; interest is continuously compounded

4. a. $5e^{0.03t} = 5(e^{0.03})^t \approx 5(1.030)^t$

b. $3500e^{0.25t} = 3500(e^{0.25})^t \approx 3500(1.284)^t$

c. $660e^{1.75t} = 660(e^{1.75})^t \approx 660(5.755)^t$

d. $55,000e^{-0.07t} = 55,000(e^{-0.07})^t \approx 55,000(0.932)^t$

e. $125,000e^{-0.28t} = 125,000(e^{-0.28})^t \approx 125,000(0.756)^t$

5.

n	$1/n$	$1 + 1/n$	$(1 + 1/n)^n$
1	1	2	2
100	0.01	1.01	2.704 813 829
1000	0.001	1.001	2.716 923 932
1,000,000	0.000 001	1.000 001	2.718 280 469
1,000,000,000	0.000 000 001	1.000 000 001	2.718 281 827

The values for $(1 + 1/n)^n$ come closer and closer to the irrational number we define as e and are consistent with the value for e in the text of 2.71828.

6. a. about $12,000

b. about $18,000

c. about $33,000

d. about 6 years

e. 1 year: $A = \$11,255$; 5 years: $\$18,061$; 10 years: $\$32,620$. doubling time:

$1.03^{4t} = 2 \Rightarrow 4t \log 1.03 = \log 2 \Rightarrow$

$t = \frac{\log 2}{4 \log 1.03} \Rightarrow t = \frac{0.3010}{4(0.0128)} \Rightarrow t \approx 5.9$ years

7. a. $8000 \cdot \left(1 + \frac{0.08}{4}\right)^{4 \cdot (18)} = \$33,289$

b. $8000 \cdot e^{0.08 \cdot (18)} = \$33,766$

c. $8000 \cdot (1.084)^{18} = \$34,168$

8. a. $8000 \cdot \left(1 + \frac{0.08}{4}\right)^{4 \cdot (30)} = \$86,121$

b. $8000 \cdot e^{0.08 \cdot (30)} = \$88,185$

c. $8000 \cdot (1.084)^{30} = \$89,943$

9. a. continuous growth rate: 0.6 or 60%; annual effective growth rate $\approx 82.2\%$ since $e^{0.6(1)} \approx 1.822$, which is the growth factor. So $1.822 - 1 = 0.822$ is the growth rate.

b. continuous growth rate: 2.3 or 230%; annual effective growth rate $\approx 897\%$ since $e^{2.3(1)} \approx 9.97$, which is the growth factor. So $9.97 - 1 = 8.97$ is the growth rate.

10. a. continuous decay rate: 0.055 or 5.5%; annual effective decay rate $\approx 5.35\%$ since $e^{-0.055} \approx 0.946$, which is the decay factor. So $1 - 0.946 = 0.054$ is the decay rate.

b. continuous decay rate: 0.15 or 15%; annual effective decay rate $\approx 13.9\%$ since $e^{-0.15} \approx 0.861$, which is the decay factor. So $1 - 0.861 = 0.139$ is the decay rate.

Exercises for Section 6.2

1. a. 1.0353　**b.** 1.0355　**c.** 1.0356　**d.** 1.0356

e. These values represent the growth factors for compounding semi-annually, quarterly, monthly, and continuously. As the number of compoundings increase, the results come closer and closer to continuous compounding.

3. In general, if the nominal rate is 8.5%, then $A_k(n) = 10,000 \cdot (1 + 0.085/k)^{kn}$ gives the value of $\$10,000$ after n years if the interest is compounded k times per year and

Ch. 6

$A_c(n) = 10,000 \cdot e^{0.085n}$ gives that value if the interest is compounded continuously.

a. annually: $(1 + 0.085/1)^1 = 1.0850$, and thus the effective rate is 8.50%.

b. semi-annually: $(1 + 0.085/2)^2 \approx 1.0868$, and thus the effective rate is 8.68%.

c. quarterly: $(1 + 0.085/4)^4 \approx 1.0877$, and thus the effective rate is 8.77%.

d. continuously: $e^{0.085n} \approx 1.0887$, and thus the effective rate is 8.87%.

5. a. $A(t) = 25,000 \cdot (1 + 0.0575/4)^{4t}$, where $t =$ number of years

b. $B(t) = 25,000 \cdot e^{0.0575t}$, where $t =$ number of years

c. $A(5) = \$33,259.12$ and $B(5) = \$33,327.26$

d. The effective rate for compounding quarterly is $\left(1 + \frac{0.0575}{4}\right)^4 - 1 \approx 0.05875$, and the effective rate for compounding continuously is $e^{0.0575} - 1 \approx 0.05919$. Thus continuous compounding has a slightly greater effective rate.

7. a. $e^{0.045} \approx 1.046$

b. $1.0680 < 1.0704 \approx e^{0.068}$

c. $1.2690 > e^{0.238} \approx 1.2687$

d. $e^{-0.10} \approx 0.9048 > 0.9000$

e. $0.8607 \approx e^{-0.15}$

9. a. $U(x) = 10 \cdot (1/2)^{x/5}$, where x is measured in billions of years.

b. $5 = 10 \cdot (1/2)^{x/5}$, and thus $1/2 = (1/2)^{x/5} \Rightarrow 1 = x/5 \Rightarrow x = 5$ billion years.

11. a. $t = \log(2)/\log(1.12) \approx 6.12$ years

b. $t = \log(2)/\log[(1 + 0.12/4)^4] \approx 5.86$ years.

13. a. The 10-year decay factor is 0.85. The yearly decay factor is $0.85^{1/10} \approx 0.9839$. The yearly decay rate is $1 - 0.9839 \approx 0.0161$ or 1.61%.

b. $g(t) = 1.5 \cdot 0.9839^t$, where $g(t)$ is measured in millions and t in years.

c. $h(t) = 1.5 \cdot e^{-0.01625t}$, where $h(t)$ is measured in millions and t in years.

d. $g(20) \approx 1.0842$ million and $h(20) \approx 1.0838$ million. The h function decays a bit faster than the g function, but they are quite close.

15. a. $P(t) = 500 \cdot (1.0202)^t$; the continuous growth rate is 2% and the effective annual growth rate is $= 2.02\%$.

b. $N(t) = 3000 \cdot (4.4817)^t$; the continuous growth rate is 150% and the effective growth rate is 348.17%.

c. $Q(t) = 45 \cdot (1.0618)^t$; the continuous growth rate is 6% and the effective growth rate is 6.18%.

d. $G(t) = 750 \cdot 1.0356^t$; the continuous growth rate is 3.5% and the effective growth rate is 3.56%.

17. a. $0.0343 = e^r - 1$, and thus $r = \log(1.0343)/\log(e) \approx 0.0337$, so the continuous nominal interest rate is 3.37%.

b. $0.046 = e^r - 1$, and thus $r = \log(1.046)/\log(e) \approx 0.04497$, so the continuous nominal interest rate is 4.497% or 4.5% rounded off.

Section 6.3

Algebra Aerobics 6.3

1. a. $\ln e^2 = 2 \ln e = 2(1) = 2$

b. $\ln 1 = 0$

c. $\ln \frac{1}{e} = \ln 1 - \ln e = 0 - 1 = -1$

d. $\ln \frac{1}{e^2} = \ln 1 - \ln e^2 = \ln 1 - 2 \ln e = 0 - 2(1)$
$= 0 - 2 = -2$

e. $\ln \sqrt{e} = \ln e^{1/2} = \frac{1}{2} \ln e = \frac{1}{2}(1) = 1/2$

2. a. $\ln \sqrt{xy} = \ln(xy)^{1/2} = \frac{1}{2} \ln(xy) = \frac{1}{2} [\ln x + \ln y]$

b. $\ln\left(\frac{3x^2}{y^3}\right) = \ln(3x^2) - \ln(y^3)$
$= \ln 3 + 2 \ln x - 3 \ln y$

c. $\ln\left((x + y)^2(x - y)\right) = \ln(x + y)^2 + \ln(x - y)$
$= 2 \ln(x + y) + \ln(x - y)$

d. $\ln \frac{\sqrt{x + 2}}{x(x - 1)} = \ln \frac{(x + 2)^{1/2}}{x(x - 1)} = \ln(x + 2)^{1/2} - \ln x(x - 1)$
$= \frac{1}{2} \ln(x + 2) - \ln x(x - 1)$
$= \frac{1}{2} \ln(x + 2) - [\ln x + \ln(x - 1)]$
$= \frac{1}{2} \ln(x + 2) - \ln x - \ln(x - 1)$

3. a. $\ln x(x - 1)$

b. $\ln \frac{(x + 1)}{x}$

c. $\ln x^2 - \ln y^3 = \ln \frac{x^2}{y^3}$

d. $\ln(x + y)^{1/2} = \ln \sqrt{x + y}$

e. $\ln x - 2 \ln(2x - 1) = \ln x - \ln(2x - 1)^2 = \ln \frac{x}{(2x - 1)^2}$

4. growth factor $= e^r = 1 + 0.064 = 1.064 \Rightarrow$
$\ln e^r = \ln 1.064 \Rightarrow r = \ln 1.064 = 0.062$ or 6.2%

5. $50,000 = 10,000 e^{0.078t} \Rightarrow 5 = e^{0.078t} \Rightarrow \ln 5 = 0.078t \Rightarrow$
$t = \ln 5/0.078 \approx 20.6$ yr

6. a. $\ln e^{x+1} = \ln 10 \Rightarrow (x + 1)\ln e = \ln 10 \Rightarrow$
$(x + 1)(1) \approx 2.30 \Rightarrow x \approx -1 + 2.30 \Rightarrow x \approx 1.30$

b. $\ln e^{x-2} = \ln 0.5 \Rightarrow (x - 2) \ln e = \ln 0.5 \Rightarrow$
$x - 2 \approx -0.69 \Rightarrow x \approx 2 - 0.69 \Rightarrow x \approx 1.31$

7. a. true, since $\ln 81 = \ln 3^4 = 4 \ln 3$

b. false; $\ln 7 = \ln \frac{14}{2} = \ln 14 - \ln 2$

c. true, since $\ln 35 = \ln(5 \cdot 7) = \ln 5 + \ln 7$

d. false; $2 \ln 10 = \ln 10^2 = \ln 100$

e. true, since $\ln e^{1/2} = \frac{1}{2} \ln e = \frac{1}{2}$

f. false: $5 \ln 2 = \ln 2^5 = \ln 32$

8. a. $\ln 2 + \ln 6 = x \Rightarrow \ln(2 \cdot 6) = x \Rightarrow \ln 12 = x \Rightarrow x \approx 2.48$

b. $\ln 2 + \ln x = 2.48 \Rightarrow \ln 2x = 2.48 \Rightarrow$
$e^{2.48} = 2x \Rightarrow x = \frac{e^{2.48}}{2} \approx 5.97$

c. $\ln(x + 1) = 0.9 \Rightarrow e^{0.9} = x + 1 \Rightarrow x = e^{0.9} - 1 \approx 1.46$

d. $\ln 5 - \ln x = -0.06 \Rightarrow \ln \frac{5}{x} = -0.06 \Rightarrow$
$e^{-0.06} = \frac{5}{x} \Rightarrow x = \frac{5}{e^{-0.06}} \approx 5.3$

Exercises for Section 6.3

1. a. $\ln(A \cdot B) = \ln(A) + \ln(B)$ or Rule 1

  b. $\ln(A/B) = \ln(A) - \ln(B)$ or Rule 2

  c. $\ln(A^p) = p \cdot \ln(A)$ or Rule 3

  d. Rules 1 and 3

  e. Rules 1 and 3

  f. Rules 2 and 1

3. a. $10^n = 35$ **b.** $e^x = 75$ **c.** $e^{3/4} = x$ **d.** $N = N_0 \cdot e^{-kt}$

5. a. $x = 5 \cdot 2 = 10$ **d.** $x = 8 \cdot 36 = 288$

  b. $x = 24/2 = 12$ **e.** $x = 64/9 \approx 7.1$

  c. $x = 11$ **f.** $x = 16/8 = 2$

7. a. $\frac{1}{2}(\ln 4 + \ln x + \ln y)$ **c.** $\ln(3) + \frac{3}{4}\ln(x)$

  b. $\frac{1}{3}(\ln 2 + \ln x) - \ln(4)$

9. a. $\ln\left(\sqrt[4]{(x+1)(x-3)}\right)$ **b.** $\ln\left(\frac{R^3}{\sqrt{P}}\right)$ **c.** $\ln\left(\frac{N}{N_0^2}\right)$

11. a. $r = \ln(1.0253) \approx 0.025$ **c.** $x = \ln(0.5)/3 \approx -0.231$

  b. $t = \ln(3)/0.5 \approx 2.197$

13. a. $x = \ln(10) \approx 2.303$

  b. $x = \log(3) \approx 0.477$

  c. $x = \log(5)/\log(4) \approx 1.161$

  d. $x = e^5 \approx 148.413$

  e. $x = e^3 - 1 \approx 19.086$

  f. no solution [$\ln(-4/3)$ not defined]

15. Let $w = \ln(A)$ and $z = \ln(B)$. Thus $e^w = A$ and $e^z = B$. Therefore $A \cdot B = e^{w+z}$ and therefore $\ln(A \cdot B) = w + z = \ln(A) + \ln(B)$, as desired.

17. $t = [\ln(100,000) - \ln(15,000)]/0.085 \approx 22.3$ years

19. $1.0338 = e^r$, and thus the nominal continuous interest rate $r = \ln(1.0338) \approx 0.0332 = 3.32\%$.

21. a. $r = \ln(1.025) \approx 0.0247$ **c.** $r = \ln(1.08) \approx 0.0770$

  b. $r = \ln(0.5) \approx -0.6931$

23. a. growth; 37%   **c.** decay; 19%   **e.** growth; 56%

  b. growth; 115%   **d.** decay; 120%   **f.** decay; 29%

Section 6.4

Algebra Aerobics 6.4

1. a. Since $\log x^2 = 2 \log x$ (by Rule 3 of logarithms), the graphs of $y = \log x^2$ and $y = 2 \log x$ will be identical (assuming $x > 0$).

2. The graph of $y = -\ln x$ will be the mirror image of $y = \ln x$ across the x-axis.

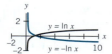

3. a. The graphs are very similar (see Figure 6.5). They intersect at $(1, 0)$, which is the x-intercept of each graph.

b. The graph of $\log x$ reaches a y value of 1 when $x = 10$ while the graph of $\ln x$ reaches a value of 1 when $x = e$ (or approximately 2.7). The graph of $\log x$ reaches 2 at $x = 100$ while the graph of $\ln x$ reaches 2 at $x = e^2 \approx 7.4$.

c. For $0 < x < 1$, both graphs lie below the x-axis and approach the y-axis asymptotically as x gets closer to 0.

d. To the right of $x = 1$, both graphs lie above the x-axis, with the $\ln x$ graph rising slightly faster than the $\log x$ graph, and thus staying above the $\log x$ graph.

4. f and its inverse f^{-1} along with the dotted line for $y = x$:

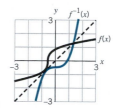

5. a. $\log 10^3 = 3$

  b. $\log 10^{-5} = -5$

  c. $3 \log 10^{0.09} = 3(0.09) = 0.27$

  d. $10^{\log 3.4} = 3.4$

  e. $\ln e^5 = 5$

  f. $\ln e^{0.07} = 0.07$

  g. $\ln e^{3.02} + \ln e^{-0.27} = 3.02 - 0.27 = 2.75$

  h. $e^{\ln 0.9} = 0.9$

6. Acidic. If $4 = -\log[H^+]$, then $-4 = \log[H^+]$. So $[H^+] = 10^{-4}$. Since the pH is 3 less than pure water's, it will have a hydrogen ion concentration 10^3 or 1000 times higher than pure water's.

7. $N = 10 \log\left(\frac{1.5 \cdot 10^{-12}}{10^{-16}}\right) = 10 \log(1.5 \cdot 10^4)$

    $= 10(\log 1.5 + \log 10^4) \approx 10(0.176 + 4)$

    $= 10(4.176) \approx 42$ dB

8. Multiplying the intensity by $100 = 10^2$ corresponds to adding 20 to the decibel level. Multiplying the intensity by $10,000,000 = 10^7$ corresponds to adding 70 to the decibel level.

Exercises for Section 6.4

1. Since $\log(5x) = \log(5) + \log(x)$ and $\log(5) > 0$, the graph of $\log(5x)$ (graph A) is above the graph of $\log(x)$ (Graph B).

3. a. $A = \log(x)$, $B = \log(x - 1)$, and $C = \log(x - 2)$; so $f(x)$ matches A, $g(x)$ matches B, and $h(x)$ matches C.

  b. $\log(1) = 0$; thus $f(1) = g(2) = h(3) = 0$.

  c. f has 1 as its x-intercept; g has 2 and h has 3.

  d. The graph moves from crossing the x-axis at $x = 1$ to crossing it at $x = k$. Assuming that $k > 0$, then $f(x - k)$ is the graph of f moved k units to the right.

5. a. The graphs of f and h are mirror images of each other across the y-axis, as are the graphs of g and k.

  b. The graphs of f and g are mirror images of each other across the x-axis, as are the graphs of h and k.

7. The table for $\log_3(x)$ is: The table for $\log_4(x)$ is:

x	1/9	1/3	1	3	9
y	−2	−1	0	1	2

x	1/16	1/4	1	4	16
y	−2	−1	0	1	2

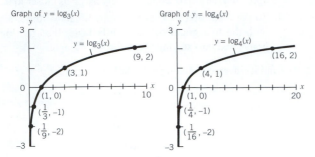

Graph of $y = \log_3(x)$

Graph of $y = \log_4(x)$

9. $dB = 10 \cdot \log(I/I_0) = 28$ implies that $I/I_0 = 10^{2.8}$ and thus $I = 10^{-13.2}$ watts/cm², and $dB = 92$ has $I/I_0 = 10^{9.2}$ and thus $I = 10^{-6.8}$ watts/cm². (Note that these answers assume, of course, that $I_0 = 10^{-16}$ watts/cm².)

11. If I is the intensity of one crying baby, then $5I$ is the intensity of five crying babies. Thus the perceived noise in decibels is $10 \cdot \log[5(I/I_0)] = 10 \cdot \log(5) + 10 \log(I/I_0) \approx 6.99 + 10 \cdot \log(I/I_0) \approx 7 +$ noise of one baby crying. Thus quintuplets crying are about 7 decibels louder than one baby crying.

13. Total volume $= 2\frac{1}{4} = \frac{9}{4}$ cups or equivalently 9 quarter cups. So lemon juice is $\frac{1}{9}$th of the mixture volume and water is $\frac{8}{9}$ths of the volume. Since the lemon juice has a pH of 2.1, its hydrogen ion concentration is $10^{-2.1}$ (moles per liter). Similarly, since the tap water has a pH of 5.8, its ion concentration is $10^{-5.8}$. So the hydrogen ion concentration of the mixture $= [H^+] = \left(\frac{1}{9}\right) \cdot 10^{-2.1} + \left(\frac{8}{9}\right) \cdot 10^{-5.8} \Rightarrow$ $pH = -\log[\left(\frac{1}{9}\right) \cdot 10^{-2.1} + \left(\frac{8}{9}\right) \cdot 10^{-5.8}] \approx 3.05$. Thus the mixture is slightly less acidic than orange juice (with a pH of 3).

Section 6.5

Algebra Aerobics 6.5

1. a. decay   **c.** growth   **e.** decay
b. decay   **d.** growth   **f.** decay

2. a. $e^k = 1.062 \Rightarrow k = \ln 1.062 \approx 0.060 \Rightarrow y = 1000e^{0.06t}$
b. $e^k = 0.985 \Rightarrow k = \ln 0.985 \approx -0.015 \Rightarrow y = 50e^{-0.015t}$

3. a. continuous nominal rate $\approx 5.45\%$ since $\ln 1.056 \approx 0.545$; effective rate $= 5.6\%$ since $e^{\ln 1.056} = 1.056$
b. continuous nominal rate $\approx 3.34\%$ since $\ln 1.034 \approx 0.0334$; effective rate $= 3.4\%$ since $e^{\ln 1.034} = 1.034$
c. continuous nominal rate $\approx 7.97\%$ since $\ln 1.083 \approx 0.0797$; effective rate $= 8.3\%$ since $e^{\ln 1.083} = 1.083$
d. continuous nominal rate $\approx 25.85\%$ since $\ln 1.295 \approx 0.2585$; effective rate $= 29.5\%$ since $e^{\ln 1.295} = 1.295$

4. a. growth factor $= e^{0.08} \approx 1.083$
b. decay factor $= e^{-0.125} \approx 0.883$

Exercises for Section 6.5

1. a. $N = 10 \, e^{(\ln 1.045)t} = 10 \cdot e^{0.0440t}$
b. $Q = 5 \cdot 10^{-7} \cdot e^{(\ln 0.072)A} = 5 \cdot 10^{-7} \cdot e^{-2.631A}$
c. $P = 500 \cdot e^{(\ln 2.10)x} = 500 \cdot e^{0.742x}$

3. a. decay   **b.** decay   **c.** growth

5. a. $t = 5$ years; $P = P_0 e^{0.1386t}$
b. $t = 25$ years; $P = P_0 e^{0.0277t}$
c. $t = 1/2$ year; $P = P_0 e^{1.3863t}$

7. If $200 = 760 \cdot e^{-0.128h}$, then $h = \ln(200/760)/(-0.128) \approx 10.43$ km.

9. The half-life is $\ln(1/2)/(-r) = [\ln(1) - \ln(2)]/(-r) = [0 - \ln(2)]/(-r) = \ln(2)/r = 100 \cdot \ln(2)/R \approx 69.3137/R$, which is approximately $70/R$.

11. a. f goes with C **b.** g goes with B **c.** h goes with A

13. a. It is a 2% nominal continuous decay rate.
b. $2500 \cdot e^{-0.02t} = 2500 \cdot 0.5^{t/n} \Rightarrow -0.02t = (t/n) \cdot \ln(0.5) \Rightarrow n = \ln(0.5)/(-0.02) \approx 34.66$.
c. It represents the half-life in whatever time units t is measured in.

15. a. Since $Q(8000) = 0.5Q_0$, the half-life is 8000 years.
b. The annual decay rate is $1 - 0.5^{1/8000} \approx 0.000\,086\,639\,6$.
c. Solving $e^r = 0.5^{1/8000}$ for r gives: $r \approx -0.000\,086\,643\,4$ as the nominal continuous decay rate.

17. Answers from students will vary. Look for an exponential curve that goes roughly through the middle of each cluster. One such curve contains the points (500, 1000) and (2000, 300). Then the best-fit exponential through these two points is $y = 1494(0.9992)^d$, where d measures depth in meters and y measures species density in an unknown unit. (Note that in base e, one gets $y = 1494e^{-0.0008d}$.)

19. Given: $n = n_0 \cdot e^{\ln(2) \cdot t/T}$ and $11 = n_0 \cdot e^{\ln(2) \cdot 2/T}$ and $30 = n_0 \cdot e^{\ln(2) \cdot 22/T}$.

Therefore: $\ln(11) = \ln(n_0) + 2 \cdot \ln(2)/T$ and $\ln(30) = \ln(n_0) + 22 \cdot \ln(2)/T$

Thus $\ln(30) - \ln(11) = \ln(2)[22 - 2]/T$

$\Rightarrow$ $T = 20 \cdot \ln(2)/[\ln(30) - \ln(11)] \approx 13.863/1.003 \approx 13.8$ seconds, the reactor period

If we substitute 13.8 for T in the equation $11 = n_0 \cdot e^{(2 \cdot \ln 2)/T}$, we get $11 = n_0 \cdot e^{(2 \cdot \ln 2)/13.8} \approx n_0 \cdot e^{0.100456} \approx n_0 \cdot 1.1057$. So $n_0 \approx 9.95$ or 10 neutrons.

21. a. Newton's Law here is of the form $T = 75 + Ce^{-kt}$ where $A = 75°$, the ambient temperature, and T is the temperature of the object at time t. When $t = 0$, $T = 160°$, so $160 = 75 + Ce^0 \Rightarrow C = 85$. So the equation becomes $T = 75 + 85e^{-kt}$. When $t = 10$, $T = 100°$ so $100 = 75 + 85e^{-10k} \Rightarrow e^{-10k} = \frac{25}{85} \approx 0.2941 \Rightarrow$ $\ln(e^{-10k}) = \ln(0.2941) \Rightarrow$ $-10k \approx -1.224 \Rightarrow k \approx 0.1224$.

So the Law of Cooling in this situation is: $T = 75 + 85e^{-0.1224t}$

b.

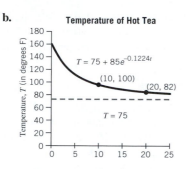

Temperature of Hot Tea

$T = 75 + 85e^{-0.1224t}$

(10, 100)

(20, 82)

$T = 75$

c. When $t = 20$ minutes, then the temperature of the tea $T = 75 + 85e^{-0.1224 \cdot 20} \approx 82°$.

Section 6.6

Algebra Aerobics 6.6

1. a. The graph of $y = 3x + 4$, a linear function, is a straight line on a standard linear plot. The graph of $y = 4 \cdot 3^x$, an exponential function, is a straight line graph on a semi-log plot. The equation $\log y = (\log 3) \cdot x + \log 4$ is equivalent to $y = 4 \cdot 3^x$, whose graph is a straight line on a semi-log plot.

b. The graph of $y = 3x + 4$ has slope $= 3$ and vertical intercept $= 4$. The equations $y = 4 \cdot 3^x$ and $\log y = (\log 3) \cdot x + \log 4$ are equivalent; their graphs have slope $= \log 3$ and vertical intercept $= \log 4$ on a semi-log plot.

2. a. $\log y = \log[5(3)^x] \Rightarrow \log y = \log 5 + x(\log 3)$

b. $\log y = \log[1000(5)^x] \Rightarrow \log y = \log 1000 + x \log 5$
$= 3 + x \log 5$

c. $\log y = \log[10,000(0.9)^x] \Rightarrow$
$\log y = \log 10,000 + x \log 0.9 = 4 + x \log 0.9$

d. $\log y = \log[5 \cdot 10^6(1.06)^x] \Rightarrow \log y =$
$\log 5 + \log 10^6 + x \log 1.06 = \log 5 + 6 + x \log 1.06$

3. a. $\log y = \log 7 + (\log 2)x \Rightarrow \log y = \log(7 \cdot 2^x) \Rightarrow$
$10^{\log y} = 10^{\log 7 \cdot 2^x} \Rightarrow y = 7 \cdot 2^x$

b. $\log y = \log 20 + (\log 0.25)x \Rightarrow$
$\log y = \log(20 \cdot 0.25^x) \Rightarrow$
$10^{\log y} = 10^{\log (20 \cdot 0.25^x)} \Rightarrow y = 20 \cdot 0.25^x$

c. $\log y = 6 + (\log 3) \cdot x \Rightarrow 10^{\log y} = 10^{6+\log (3^x)} \Rightarrow$
$y = 10^6 \cdot 10^{\log (3^x)} \Rightarrow y = 10^6 \cdot 3^x$

d. $\log y = 6 + \log 5 + (\log 3) \cdot x \Rightarrow$
$\log y = 6 + \log 5 + \log(3^x) \Rightarrow$
$10^{\log y} = 10^{6+\log 5 + \log 3^x} \Rightarrow y = 10^6 \cdot 10^{\log 5} \cdot 10^{\log 3^x} \Rightarrow$
$y = 10^6(5)3^x = 5 \cdot 10^6(3)^x$

4. a. i. slope $= \log 5$; vertical intercept $= \log 2$

ii. slope $= \log 0.75$; vertical intercept $= \log 6$

iii. slope $= \log 4$; vertical intercept $= 0.4$

iv. slope $= \log 1.05$; vertical intercept $= 3 + \log 2$

b. i. $y = 2 \cdot 5^x$

ii. $y = 6 \cdot (0.75)^x$

iii. $y = 10^{0.4}(4)^x$

iv. $y = 10^3 \cdot 2 \cdot 1.05^x = 2000(1.05)^x$

5. a. $a \approx 2.00$ since $10^{0.301} \approx 2$

b. $C \approx 524.81$ since $10^{2.72} \approx 524.81$

c. $a \approx 0.75$ since $10^{-0.125} \approx 0.75$

d. $C \approx 100,000$ since $10^5 = 100,000$

6. a. $y = 10^{0.301} \cdot 10^{0.477x} \Rightarrow y = 2 \cdot 3^x$

b. $y = 10^3 \cdot 10^{0.602x} \Rightarrow y = 1000(4)^x$

c. $y = 10^{1.398} \cdot 10^{-0.046x} \Rightarrow y = 25 \cdot (0.90^x)$

7. Exponential functions for both since the graphs on semi-log plots are approximately linear.

8. Using the point $(0, 4)$, the vertical intercept, and $(2, 5)$, the slope is $\frac{5-4}{2} = 0.5$. So $\log y = 4 + 0.5x \Rightarrow y = 10^{4+0.5x} \Rightarrow$
$y = 10^4 \cdot 10^{0.5x} \approx 10^4(3.16)^x$.

Exercises for Section 6.6

1. a. goes with **f.**   **c.** goes with **e.**

b. goes with **h.**   **d.** goes with **g.**

3. a. $\log(y) = 4.477 + 0.301x$

b. $\log(y) = 3.653 + 0.146x$

c. $\log(y) = 6.653 - 0.155x$

d. $\log(y) = 3.778 - 0.244x$

5. a. goes with Graph **B.**   **c.** goes with Graph **A.**

b. goes with Graph **C.**   **d.** goes with Graph **D.**

7. Here are the average rates of change:

a. x	$Y = \log(x)$	Avg. Rt. of Ch.	**b.** x	$Y = \log(x)$	Avg. Rt. of Ch.
0	2.30103	n.a.	0	4.77815	n.a.
10	4.30103	0.20000	10	3.52876	−0.12494
20	4.90309	0.06021	20	2.27938	−0.12494
30	5.25527	0.03522	30	1.02999	−0.12494
40	5.50515	0.02499	40	−0.21945	−0.12494

a. This is *not* exponential; the average rate of change between consecutive points keeps decreasing.

b. The average rate of change over each decade is −0.12494. Thus the plot of Y vs. x is linear and therefore y is exponential. The exponential function is $y = 60,000(0.75)^x$.

9. a. The scatter plot of points $(x, \log(y))$, where $Y = \log(y)$, is nearly linear; thus the growth is very close to being exponential.

b. The equation of the best-fit line is given in the graph and the corresponding exponential equation is approximately $y = 38.788 \cdot 1.3137^x$. The daily growth rate is 31.37%.

11. a. Approximately 12.2 micrograms/liter.

b. Yes, because its log graph is a straight line with negative slope. The decay factor is $10^{-0.0181} \approx 0.959$, and thus the decay rate is $1 - 0.959 = 0.041$ or 4.1%.

c. The exponential model is approximately $y = 12.2 \cdot 0.959^x$.

13. a. To find out the average health care costs per person, one would divide the total costs by the population estimate for each year. (See Table 2.1 for U.S. population counts.)

b. On a semi-log plot (where years are reinitialized at 1960) the data look rather linear, so it is reasonable to model the growth with an exponential function.

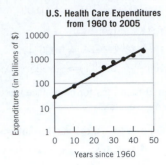

U.S. Health Care Expenditures from 1960 to 2005

Expenditures (in billions of $) vs. Years since 1960

c. The graph shows the best-fit exponential, whose equation is $H(t) = 31(1.102)^t$, where $H(t)$ is the health care costs (in billions) t years after 1960. In 2005, when $t = 45$, $H(45) = 31(1.102)^{45} \approx 2450$ billion or 2.45 trillion dollars. The estimate exceeds the actual cost in 2005 of 2016 billion by 434 billion dollars.

Ch. 6: Check Your Understanding

1. True	**8.** False	**15.** True	**22.** False
2. True	**9.** False	**16.** False	**23.** True
3. False	**10.** False	**17.** False	**24.** False
4. False	**11.** True	**18.** True	**25.** True
5. True	**12.** False	**19.** False	**26.** False
6. False	**13.** True	**20.** False	**27.** True
7. True	**14.** False	**21.** False	**28.** False

29. True

30. True

31. Possible answer: $y = \frac{1}{2}(\log x - \log 3)$

32. Possible answer: $y = 50.3(1.062)^t$

33. Possible answer: $y = 100e^{-0.026t}$

34. Possible answer: $F(t) = 3.3(0.989)^t$, t = number of years since 1966, $F(t)$ in millions

35. Possible answer: $y = 100(1.20)^x$

36. False   **37.** False   **38.** True   **39.** True

40. True

Ch. 6 Review: Putting It All Together

1. a. $100 = 50 \cdot 1.16^t \Rightarrow 2 = 1.16^t \Rightarrow \log 2 = t \log 1.16 \Rightarrow$ (log 2)/(log 1.16) $= t \Rightarrow t \approx 0.301/0.064 \approx 4.7$ months (or about 141 days) for the quantity to double.

b. $100 = 200 \cdot 0.92^t \Rightarrow 0.5 = 0.92^t \Rightarrow \log 0.5 = t \log 0.92 \Rightarrow$ (log 0.5)/(log 0.92) $= t \Rightarrow t \approx (-0.301)/(-0.036) \approx$ 8.4 months (or about 250 days) for the quantity to halve.

3. Expressions (a) and (e) are equivalent and (b), (c), and (f) are equivalent. Expression (d) does not match any other.

5. a. After one time period (20 minutes), $A(1) = 325 \cdot (0.5)^1 = 162.5$ mg (or half the original amount). After two time periods (40 minutes), $A(2) = 325 \cdot (0.5)^2 = 81.25$ mg (or one-quarter of the original amount).

b. Using the graph, it appears that after about 1.7 time periods (34 minutes) $A(t) = 100$. Using the equation, we have $100 = 325 \cdot (0.5)^t \Rightarrow 100/325 = (0.5)^t \Rightarrow 0.3077 \approx (0.5)^t \Rightarrow \log(0.3077) \approx t \log(0.5) \Rightarrow t \approx \log(0.3077)/\log(0.5) \Rightarrow t \approx 1.70$ time periods, or $1.7 \cdot 20 = 34$ minutes. So our estimate was accurate.

c. $B(t) = 81 \cdot (0.5)^t$. So $A(t)$ and $B(t)$ have different initial amounts, but the decay rate is the same.

7. a. Density for India $= (1.08 \cdot 10^9)/(1.2 \cdot 10^6) = 0.9 \cdot 10^3 = 900$ people per square mile. Density for China $= (1.30 \cdot 10^9)/(3.7 \cdot 10^6) \approx 0.35 \cdot 10^3 = 350$ people per square mile. So India's population density is about $900/350 \approx 2.6$ times larger than China's.

b. $C(x) = 1.30(1.006)^x$ and $I(x) = 1.08(1.016)^x$, where $C(x)$ and $I(x)$ are in billions and x = years since 2005.

c.

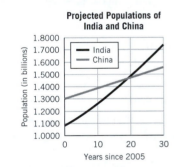

Projected Populations of India and China

Population (in billions) vs. Years since 2005

Note that India's exponential growth rate is so much higher than that for China, the graph for the Chinese population appears almost linear in comparison.

d. The projected populations are the same roughly 18 years after 2005, or in 2023. Using the models, we need the point at which $C(x) = I(x) \Rightarrow 1.30(1.006)^x = 1.08(1.016)^x \Rightarrow 1.204(1.006)^x \approx (1.016)^x \Rightarrow \log(1.204) + x \log(1.006) \approx x \log(1.016) \Rightarrow \log(1.204) \approx x [\log(1.016) - \log(1.006)] \Rightarrow 0.081 \approx x \log[(1.016)/(1.006)] \Rightarrow 0.081 \approx 0.0043x \Rightarrow x \approx 18.8$ years after 2005, that is, in late 2023 our model predicts the two populations will be the same. Evaluating $C(18.8)$, we get $1.30(1.006)^{18.8} \approx 1.455$ billion people. (To double-check you could calculate $I(18.8) \approx 1.456$ billion, with the difference due to rounding.)

9. a. $A(x) = 10,000(1.07)^x$, where $A(x)$ is the dollar amount after x years.

b. $A(18) = 10,000(1.07)^{18} \approx \$33,799$

c. $10,000(1.03)^{18} \approx \$17,024$, the difference $= \$33,799 - \$17,024 = \$16,775$

d. Investing \$10,000 at 4% for 18 years gives $\$10,000(1.04)^{18} \approx \$20,258$, which is not equivalent to the result in part (c). However it does represent the amount in the account after 18 years in current dollars (dollars that have been adjusted for inflation).

11. a. $t = \log(2.3) \approx 0.362$

  b. $t^2 \cdot 4 = 10^2 \Rightarrow t^2 = 25 \Rightarrow t = 5$ (*Note:* t = −5 is not a solution here, since log (−5) is not defined.)

  c. $2 = e^{0.03t} \Rightarrow \ln 2 = 0.03t \Rightarrow t = (\ln 2)/0.03 \approx 0.693/0.03 \approx 23.1$

  d. $\ln[(2t − 5)/(t − 1)] = 0 \Rightarrow (2t − 5)/(t − 1) = 1 \Rightarrow 2t − 5 = t − 1 \Rightarrow t = 4$

13. a. $1.5 = e^{\ln 1.5} \approx e^{0.405}$

  b. $0.7 = e^{\ln 0.7} \approx e^{-0.357}$

  c. $1 = e^0$

15. a. Matches Graph C

  b. Matches Graph D

  c, d. Both match Graphs A and B since $y = 100(10)^x \Rightarrow \log y = \log 100 + x \log 10 \Rightarrow \log y = 2 + x$.

17. a. Matches Graph B since when $x = 1$, $2 \ln x = 2 \ln 1 = 2 \cdot 0 = 0$. So the horizontal intercept is $(1, 0)$.

  b. Matches Graph A since when $x = 1$, $2 + \ln x = 2 + \ln 1 = 2 + 0 = 2$. So the graph passes through $(1, 2)$.

  c. Matches Graph D since when $x = 0$, $\ln(x + 2) = \ln 2 \approx 0.693 > 0$. So the vertical intercept is above the origin at approximately $(0, 0.693)$.

  d. Matches Graph B since the functions $y_1 = 2 \ln x$ and $y_4 = \ln(x^2)$ are the same. (*Note:* There is no match to Graph C.)

19. a. Using the table in the text, five orders of magnitude.

  b. Using the function definition, if $30 = 10 \log (I_{30}/I_0)$, where I_{30} = intensity level corresponding to 30 decibels, and $80 = 10 \log (I_{80}/I_0)$, where I_{80} = intensity level corresponding to 80 decibels. Subtracting the two equations gives:

$$80 − 30 = 10 \log(I_{80}/I_0) − 10 \log(I_{30}/I_0) \Rightarrow$$
$$50 = 10 \log[(I_{80}/I_0)/(I_{30}/I_0)] \Rightarrow$$
$$5 = \log[(I_{80}/I_0) \cdot (I_0/I_{30})] \Rightarrow 5 = \log[(I_{80}/I_{30})] \Rightarrow$$
$$10^5 = I_{80}/I_{30},$$

so I_{80} is five orders of magnitude larger than I_{30}.

21. If $R = 6.9$, we have $6.9 = \log(A/A_0) \Rightarrow 10^{6.9} = A/A_0$, so the earthquake's amplitude was 6.9 orders of magnitude (or almost 10 million times) larger than that of the base amplitude, A_0.

23. a. The data appear to be exactly linear, and two estimated points on the best-fit line are $(0, 0.3)$ and $(10, 5)$. The slope $= (5 − 0.3)/(10 − 0) = 0.47$. The equation is then $Y = 0.3 + 0.47x$.

  b. Substituting log y for Y gives $\log y = 0.3 + 0.47x \Rightarrow y = 10^{0.3+0.47x} \Rightarrow y = 10^{0.3}10^{0.47x} \approx 2 \cdot 3^x$.

  c. The function is exponential, suggesting that the graph of exponential functions is a straight line on a semi-log plot (where the log scale is on the vertical axis).

CHAPTER 7

Section 7.1

Algebra Aerobics 7.1

1. a. $S(r) = 4\pi r^2$; $S(2r) = 4\pi(2r)^2 = 4 \cdot (4\pi r^2) = 4S(r)$; $S(3r) = 4\pi(3r)^2 = 9 \cdot (4\pi r^2) = 9S(r)$

  b. When the radius is doubled, the surface area is multiplied by 4. When the radius is tripled, the surface area is multiplied by 9.

2. a. $V(r) = (4/3)\pi r^3$; $V(2r) = (4/3)\pi(2r)^3 = 8 \cdot ((4/3)\pi r^3) = 8V(r)$; $V(3r) = (4/3)\pi(3r)^3 = 27 \cdot ((4/3)\pi r^3) = 27V(r)$

  b. When the radius is doubled, the volume is multiplied by 8. When the radius is tripled, the volume is multiplied by 27.

3. The volume grows faster than the surface area, so as the radius of a sphere increases, the ratio of (surface area)/volume decreases.

4. a. $C(r) = 2\pi r$; $C(2r) = 2\pi(2r) = 2(2\pi r) = 2C(r)$; $C(4r) = 2\pi(4r) = 4(2\pi r) = 4C(r)$. So when the radius is doubled, the circumference is multiplied by 2. When the radius quadruples, the circumference is multiplied by 4.

  b. $A(r) = \pi r^2$; $A(2r) = \pi(2r)^2 = 4(\pi r^2) = 4A(r)$; $A(4r) = \pi(4r)^2 = 16(\pi r^2) = 16A(r)$. So when the radius is doubled, the area is multiplied by 4. When the radius is quadrupled, the area is multiplied by 16.

  c. The area grows faster than the circumference. So as the radius of the circle increases, the ratio of circumference/ area decreases.

5. a. The volume of the sphere is equal to the volume of the cube when $\frac{4}{3}\pi r^3 = r^3$, which is true only if $r = 0$.

  b. The volume of the cube is greater than the volume of the sphere if $r^3 > \frac{4}{3}\pi r^3 \Rightarrow 1 > \frac{4}{3}\pi$, which is never true. The volume of the cube is less than the volume of the sphere for all $r > 0$ since $r^3 < \frac{4}{3}\pi r^3 \Rightarrow 1 < \frac{4}{3}\pi$.

6. No, since if the radius is doubled from 5 to 10, the volume is increased by a factor of 4. That is, given $V = \pi r^2 h$, if $r = 5$ then $V = \pi 5^2 \cdot 25 = 625\pi$ cubic feet and if $r = 10$ then $V = \pi 10^2 \cdot 25 = 2500\pi$ cubic feet, which is a factor of 4 larger.

7. Yes; since $V = \pi r^2 h$, if the height is doubled, the volume is doubled. That is, for this example, if $r = 5$, then $V = \pi r^2 h \Rightarrow V = \pi 12^2 \cdot 5 = 720\pi$ cubic feet and if $r = 10$, then $V = \pi 12^2 \cdot 10 = 1440\pi$ cubic feet, which is twice the volume.

8. a. i. The area B of the cylinder base $= \pi r^2 \Rightarrow V = (\pi r^2)h = \pi r^2 h$

    ii. The area B of the triangular prism base $= \frac{1}{2}ab \Rightarrow V = \left(\frac{1}{2}ab\right)h = \frac{1}{2}abh$

  b. If the height is doubled, the volume is doubled since $(\pi r^2)2h = 2(\pi r^2)h = 2V$; $\left(\frac{1}{2}ab\right)2h = 2\left(\frac{1}{2}ab\right)h = 2V$

  c. If all dimensions are doubled, the volume is increased by a factor of 8 since $(\pi(2r)^2) \cdot 2h = (\pi 4r^2)2h = 8(\pi r^2)h = 8V$; $\left(\frac{1}{2}(2a)(2b)\right)(2h) = 2^3 \cdot \left(\frac{1}{2}abh\right) = 8V$

Exercises for Section 7.1

1. a. $l = \frac{V}{wh}$ **c.** $w = \frac{P - 2l}{2}$

b. $b = \frac{2A}{h}$ **d.** $h = \frac{S - 2x^2}{4x}$

3. a. $r/2$; increasing; as the radius increases, the area increases faster than the circumference.

b. $\frac{4}{3}r$; increasing; as the radius increases, the volume of the sphere increases faster than the area of its radial cross-section.

c. $\frac{2}{3}r^2$; increasing; as the radius increases, the volume of the sphere increases faster than the length of any of its great circles.

5. a. $S \approx 1.26 \cdot 10^{-19}$ m^2 **c.** $S/V \approx 3.0 \cdot 10^{10}$ m^{-1}

b. $V \approx 4.19 \cdot 10^{-30}$ m^3 **d.** ratio = $3/r$; decreases

7. a. S is multiplied by 16; V is multiplied by 64.

b. S is multiplied by n^2; V is multiplied by n^3.

c. S is divided by 9; V is divided by 27.

d. S is divided by n^2; V is divided by n^3.

9. a. $V = 2632.5$ cm^3

b. quadrupled

c. Quadruple either the length or the width or double both.

11. a. The volume doubles if the height is doubled.

b. The volume is multiplied by 4 if the radius is doubled.

13. a. $V = 3\pi r^3$; $S = 8\pi r^2$

b. Volume eventually grows faster since $V/S = 3r/8$; as $r \to +\infty$, V/S increases without bound.

15. The ratio of area to circumference is $r/2$ and its graph is labeled $h(x)$; the inverse ratio is $2/r$ and its graph is labeled $j(x)$.

Section 7.2

Algebra Aerobics 7.2

1.

Power Function	Independent Variable	Dependent Variable	Constant of Proportionality	Power
a. yes	r	A	π	2
b. yes	z	y	1	5
c. no	—	—	—	—
d. no	—	—	—	—
e. yes	x	y	3	5

2. a. y is directly proportional to x^2.

b. y is not directly proportional to x^2.

c. y is not directly proportional to x.

3. $g(x) = 5x^3$

a. $g(2) = 5(2)^3$ $g(4) = 5(4)^3$
$= 5(8)$ $= 5(64)$
$= 40$ $= 320$ So $g(4)$ is eight times larger than $g(2)$

b. $g(5) = 5(5)^3$ $g(10) = 5(10)^3$
$= 5(125)$ $= 5(1000)$
$= 625$ $= 5000$ So $g(10)$ is eight times larger than $g(5)$

c. $g(2x) = 5(2x)^3$
$= 5(8x^3)$
$= 8(5x^3)$
$= 8g(x)$ So $g(2x)$ is eight times larger than $g(x)$

d. $g\left(\frac{1}{2}x\right) = 5\left(\frac{1}{2}x\right)^3$
$= 5\left(\frac{1}{8}x^3\right)$
$= \frac{1}{8}(5x^3)$
$= \frac{1}{8}g(x)$ So $g\left(\frac{1}{2}x\right)$ is one-eighth the size of $g(x)$

4. a. $h(2) = 0.5(2)^2 = 2$ and $h(6) = 0.5(6)^2 = 18$, an increase by a factor of 9.

b. $h(5) = 0.5(5)^2 = 12.5$ and $h(15) = 0.5(15)^2 = 112.5$, an increase by a factor of 9.

c. $h(x)$ will increase by a factor of 9 since $h(3x) = 0.5(3x)^2 = 9 \cdot [0.5x^2] = 9 \cdot h(x)$.

d. $h(x)$ will decrease by a factor of 9 since $h\left(\frac{1}{3}x\right) = 0.5\left(\frac{1}{3}x\right)^2 = \frac{1}{9} \cdot [0.5x^2] = \frac{1}{9} \cdot h(x)$.

5. a. $V = k \cdot r^3$

b. $V = k \cdot l \cdot w \cdot h$

c. $f = k \cdot c$

6. a. y is equal to 3 times the fifth power of x.

b. y is equal to 2.5 times the cube of x.

c. y is equal to one-fourth of the fifth power of x.

7. a. y is directly proportional to x^5 with a proportionality constant of 3.

b. y is directly proportional to x^3 with a proportionality constant of 2.5.

c. y is directly proportional to x^5 with a proportionality constant of 1/4.

8. a. $P = aR^2 \Rightarrow \frac{P}{a} = \frac{aR^2}{a} \Rightarrow \frac{P}{a} = R^2 \Rightarrow \sqrt{\frac{P}{a}} = \sqrt{R^2} \Rightarrow R = \sqrt{\frac{P}{a}}$

b. $V = \left(\frac{1}{3}\right)\pi r^2 h \Rightarrow 3(V) = 3\left(\frac{1}{3}\right)\pi r^2 h \Rightarrow 3V = \pi r^2 h \Rightarrow \frac{3V}{\pi r^2} = \frac{\pi r^2 h}{\pi r^2} \Rightarrow h = \frac{3V}{\pi r^2}$

9. a. $f(2x) = 0.1(2x)^3 = 0.8x^3$; $2f(x) = 2(0.1x^3) = 0.2x^3$

b. $f(3x) = 0.1(3x)^3 = 2.7x^3$; $3f(x) = 3(0.1x^3) = 0.3x^3$

c. $f(2 \cdot 5) = 0.1(10)^3 = 100$; $2f(5) = 2(0.1 \cdot 5^3) = 25$

Exercises for Section 7.2

1. a. $f(2) = 20$; $f(-2) = 20$ **c.** $h(2) = -20$; $h(-2) = -20$

b. $g(2) = 40$; $g(-2) = -40$ **d.** $k(2) = -40$; $k(-2) = 40$

3. a. $k = 1/2$; $P = 288$ **c.** $k = 4\pi$; $T = 1728\pi$

b. $k = 14$; $M = 84$ **d.** $k = \sqrt[3]{18}$; $N = 6$

5. a. $y = \frac{1}{5}x$;   4

b. $p = 6\sqrt{s}$;   24

c. $A = \frac{4}{3}\pi r^3$; 64π

d. $P = \frac{1}{2}m^2$;   quartered, i.e., P is divided by 4.

7. a. $Y = kX^3$

b. $k = 1.25$

c. Increased by a factor of 125

d. Divided by 8

e. $X = \sqrt[3]{\frac{Y}{k}}$. So X is not directly proportional to Y. It is directly proportional to $\sqrt[3]{Y}$.

9. a. L is multiplied by 32.   **b.** M is multiplied by 2^p.

11. In all the formulas below, k is the constant of proportionality.

a. $x = k \cdot y \cdot z^2$ **c.** $w = k \cdot x^2 \cdot y^{1/3}$

b. $V = k \cdot l \cdot w \cdot h$ **d.** $V = k \cdot h \cdot r^2$

13. a. $C = kAt$

b. $k = 0.06$ dollars per sq. ft, per in.

c. The area of the four walls (without the door) $= 2(15 \cdot 8) + 2(20 \cdot 8) = 240 + 320 = 560$ sq. ft. The area of the door is $3 \cdot 7 = 21$ sq. ft. Thus the total wall area $= 560 - 21 = 539$ sq. ft.

d. \$129.36; \$194.04

15. a. and **b.**

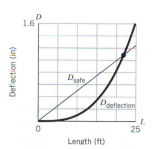

The graphs of $D_{\text{deflection}}$ and D_{safe} (with the deflections measured in inches and the plank length L measured in feet) are given in the accompanying diagram.

c. The safety deflections are well above the actual deflections for all values of L between 0 and 20. It would cease to be safe if the plank were longer than about 22 ft. (This is the L value where D_{safe} and $D_{\text{deflection}}$ meet.)

17. In all the formulas below, k is the constant of proportionality.

a. $d = k \cdot t^2$ **d.** $R = k \cdot [O_2] \cdot [NO]^2$

b. $E = k \cdot m \cdot c^2$ **e.** $v = k \cdot r^2$

c. $A = k \cdot b \cdot h$

19. a. V is quadrupled; is multiplied by 9

b. V is doubled; is tripled

c. V is multiplied by n^2

d. V is multiplied by n

21. a. $h(2) = 2$; $h(6) = 18$; the latter is nine times the former.

b. $h(5) = 12.5$; $h(15) = 112.5$; the latter is nine times the former.

c. h's value is multiplied by 9.

d. h's value is divided by 9.

Section 7.3

Algebra Aerobics 7.3

1. a. $f(2) = 4(2)^3 = 4(8) = 32$

b. $f(-2) = 4(-2)^3 = 4(-8) = -32$

c. $f(s) = 4s^3$

d. $f(3s) = 4(3s)^3 = 4(27)s^3 = 108s^3$

2. a. $g(2) = -4(2)^3 = -4(8) = -32$

b. $g(-2) = -4(-2)^3 = -4(-8) = 32$

c. $g(\frac{1}{2}t) = -4(\frac{1}{2}t)^3 = -4(\frac{1}{8})t^3 = -\frac{1}{2}t^3$

d. $g(5t) = -4(5t)^3 = -4(125)t^3 = -500t^3$

3. a. $f(4) = 3(4)^2 = 48$ **e.** $f(2s) = 3(2s)^2 = 12s^2$

b. $f(-4) = 3(-4)^2 = 48$ **f.** $f(3s) = 3(3s)^2 = 27s^2$

c. $f(s) = 3s^2$ **g.** $f(\frac{s}{2}) = 3(\frac{s}{2})^2 = \frac{3}{4}s^2$

d. $2f(s) = 2(3s^2) = 6s^2$ **h.** $f(\frac{s}{4}) = 3(\frac{s}{4})^2 = \frac{3}{16}s^2$

4. a.

x	$f(x) = 4x^2$	$g(x) = 4x^3$
-4	64	-256
-2	16	-32
0	0	0
2	16	32
4	64	256

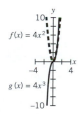

b. As $x \to +\infty$, both $f(x)$ and $g(x) \to +\infty$.

c. As $x \to -\infty$, $f(x) \to +\infty$ but $g(x) \to -\infty$.

d. The domain of both functions is all the real numbers, and the range of g is also all the real numbers; however, the range of f is only all the nonnegative real numbers.

e. They intersect at the origin and at the point $(1, 4)$.

f. All values greater than $x = 1$.

5.

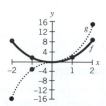

a. $f(x) = g(x)$ for $x = 0$ and $x = 1$.

b. $f(x) > g(x)$ for $0 < x < 1$ and for $x < 0$.

c. $f(x) < g(x)$ for $x > 1$

6.   **a.**   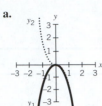   **b.**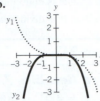

7. a. The graphs are similar; both have similar end behavior; they intersect at the origin; for positive values of x, $y_1 < y_2$; for negative values of x, $y_1 > y_2$.

b. The graphs are similar; both have similar end behavior; they intersect at the origin; for all nonzero values of x, $y_1 > y_2$.

8. a. p is even and a is positive; $a = 3$ since $(1, 3)$ is on the graph of the function.

b. p is odd and a is negative; $a = -2$ since $(1, -2)$ is on the graph of the function.

9.   **a.**   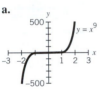   **b.**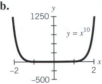

10. a. $y = x$ and $y = 4x$ and $y = -4x$:

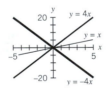

b. $y = x^4$ and $y = 0.5x^4$ and $y = -0.5x^4$:

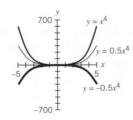

Exercises for Section 7.3

1. The graphs of the six functions are labeled in the accompanying diagram. They are all parabolas since they are graphs of functions of the form $y = ax^2$, for various values of a. The differences are due to the value of a. If $a > 0$, then the graph is concave up. If $a < 0$, then the graph is concave

down. The larger the absolute value of a, the narrower the opening of the graph.

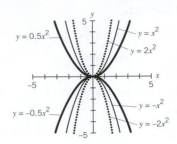

3. f goes with Graph C; g goes with Graph A; h goes with Graph D; j goes with Graph B. The graphs in B and D are mirror images of each other across the x-axis; the graph in D is steeper than that in C, the graph in A is flatter than those in C and D for $-1 < x < +1$, but the Graph in A is steeper as $x \rightarrow \pm \infty$.

5. a. $g(x) = 6 \cdot x^4$

b. $h(x) = 0.5 \cdot x^4$

c. $j(x) = -2x^4$

7. a. $f(0) = g(0) = 0$

b. If $0 < x < 1$, then $f(x) > g(x)$

c. $f(1) = g(1) = 1$

d. If $x > 1$, then $f(x) < g(x)$

e. If $x < 0$, then $f(x) > g(x)$

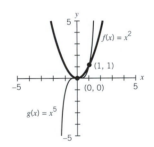

9. a. n is even   **d.** Yes, $f(-x) = f(x)$

b. $k < 0$ **e.** $-\infty$

c. Yes, $f(-2) = f(2)$ **f.** $-\infty$

11.

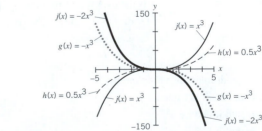

a. The constant of proportionality for f is 1; for g it is -1; for h it is $1/2$; for j it is -2.

b. g's graph is the reflection of f's graph across the x-axis.

c. j's graph is both a stretch and a reflection of f's graph across the x-axis.

d. h's graph is a compression of f's graph.

13. $g(x) = -4x^2$. Its graph is a reflection of the graph of $f(x)$ across the x-axis.

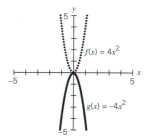

15. a. The graphs for the four functions are given in the diagram. All go through $(0, 0)$ and $(1, 1)$. The smaller the fractional power, the bigger the y value when $0 < x < 1$.

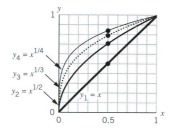

b. The graphs for the four functions are given in the diagram. All graphs go through $(0, 0)$ and $(1, 1)$. The higher the power of x, the smaller the y value when $0 < x < 1$.

Notice that over the interval $[0, 1]$ the powers ≤ 1 all have y values above or on the graph of $y = x$, but the powers ≥ 1 all would have y values below or on the graph of $y = x$.

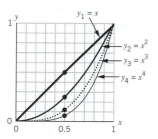

Section 7.4

Algebra Aerobics 7.4

1. a.

x	$y = 4^x$	$y = x^3$
0	1	0
1	4	1
2	16	8
3	64	27
4	256	64
5	1024	125

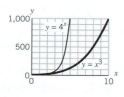

b. $y = 4^x$ dominates $y = x^3$.

In this case, $4^x > x^3$ for all values of x, but as $x \to +\infty$, the values grow farther and farther apart.

2. Graphs A and D are likely to be power functions; Graphs B and C are likely to be exponential functions.

3. a. $y = 2^x$ eventually dominates.

b. $y = (1.000\ 005)^x$ eventually dominates.

Exercises for Section 7.4

1. a. The two graphs intersect at $(2, 16)$ and $(4, 256)$. Thus, the functions are equal for $x = 2$ and $x = 4$. (See the accompanying figure.)

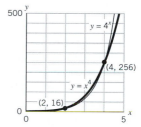

b. As x increases, both functions grow. For $0 \leq x < 2$ we have that $x^4 < 4^x$; at $x = 2$ both have a y value of 16. From $2 < x < 4$, we have $4^x < x^4$. For $x = 4$ they are again equal. For $x > 4$ we have that $4^x < x^4$. As x keeps on increasing 4^x will continue to grow faster than x^4.

c. The graph of $y = 4^x$ dominates.

3. a. $f(x) > g(x)$ as $x \to -\infty$

b. $f(x) < g(x)$ as $x \to +\infty$

c. $f(x) > g(x)$ for x in $(-\infty, -1)$

d. $f(x) < g(x)$ for x in $(-0.5, 1)$

e. $f(x) > g(x)$ for x in $(1, 6)$

f. $g(x) > f(x)$ for x in $(7, +\infty)$

5. Graphs of $y = 2^x$ and $y = x^2$

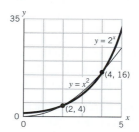

Graphs of $y = 3^x$ and $y = x^3$

In the top figure, the graphs of $y = 2^x$ and $y = x^2$ intersect at $x = 2$ and $x = 4$. The graph of $y = 2^x$ lies above the graph of $y = x^2$ for $0 \le x < 2$ and for $x > 4$.

In the bottom figure, the graphs of $y = 3^x$ and $y = x^3$ intersect at $x = 3$ and at $x \approx 2.478$. The latter intersection point can only be seen if we use a graphing device and zoom in. The graph of $y = 3^x$ is above the graph of $y = x^3$ over the intervals $(-\infty, 2.478)$ and $(3, +\infty)$. It is below that graph over the interval $(2.478, 3)$.

7. $3 \cdot 2^x > 3 \cdot x^2$ if $0 < x < 2$ and if $x > 4$. Also, $3 \cdot 2^x < 3 \cdot x^2$ if $2 < x < 4$.

9. a. $h(x)$ goes with C

   b. $i(x)$ goes with A

   c. $j(x)$ goes with B

11. The functions $f(ax)$ and $g(ax)$ are the functions $f(x)$ and $g(x)$, respectively, where the input x has been multiplied by a. Since $f(ax) = (ax)^n = a^n \cdot f(x)$, then $f(ax)$ is $f(x)$ multiplied by a^n. Since $g(ax) = b^{ax} = (b^x)^a = [g(x)]^a$, then $g(ax)$ is $g(x)$ raised to the ath power.

13. Linear function: $m = \frac{32 - 0.5}{4 - 1} = 10.5 \Rightarrow y = 10.5x - 10$.

   Exponential function: $y = \frac{1}{8} \cdot 4^x$

   Power function: $y = 0.5x^3$

15. a. True   **c.** False

    b. True   **d.** False

Section 7.5

Algebra Aerobics 7.5

1. a. x is directly proportional to y and z.

   b. y is inversely proportional to the square of x

   c. D is directly proportional to the square root of y and inversely proportional to the cube of z.

2. (a) and (d) represent direct proportionality; (b) and (c) represent inverse proportionality; (e) does not represent direct or inverse proportionality. It is an exponential function where x, the input, becomes an exponent in the output.

3. The volume will increase by a factor of 4, from 1 ft³ to 4 ft³. (See Table 7.6. in Section 7.5.)

4. a. i. $\frac{15}{x^3}$ **ii.** $\frac{-10}{x^4}$ **iii.** $\frac{3.6}{x}$ **iv.** $\frac{2}{x^2 y^3}$

   b. i. $1.5x^{-2}$ **ii.** $-6x^{-3}$ **iii.** $-\frac{2}{3}x^{-2}$ **iv.** $6x^{-3}y^{-4}z^{-1}$

5. $t = kd^{-2}$ or $t = \frac{k}{d^2}$ for some constant k

6. Intensity at 4 ft is $\frac{k}{4^2} = \frac{1}{16}k$, but at 2 ft intensity is $\frac{k}{(2)^2} = \frac{1}{4}k$. The light is four times as bright at a distance of 2 ft away than it is at 4 ft away.

7. $g(x) = \frac{3}{x^4}$

   a. $g(2x) = \frac{3}{(2x)^4} = \frac{3}{16x^4} = \frac{1}{16}\left(\frac{3}{x^4}\right)$, so $g(2x) = \frac{1}{16}g(x)$.

b. $g\left(\frac{x}{2}\right) = \frac{3}{(x/2)^4} = 3 \div \left(\frac{x}{2}\right)^4 = 3 \div \frac{x^4}{16} = 3 \cdot \frac{16}{x^4}$
$= 16\left(\frac{3}{x^4}\right)$, so $g\left(\frac{x}{2}\right) = 16(g(x))$.

8. $h(x) = -\frac{2}{x^3}$

   a. $h(3x) = -\frac{2}{(3x)^3} = -\frac{2}{27x^3} = \frac{1}{27} \cdot \left(-\frac{2}{x^3}\right) = \frac{1}{27} \cdot h(x)$, or one-twenty-seventh of the original amount.

   b. $h\left(\frac{x}{3}\right) = -\frac{2}{(x/3)^3} = -\frac{2}{x^3/27} = -2 \cdot \frac{27}{x^3} = 27 \cdot \left(-\frac{2}{x^3}\right) = 27h(x)$, or 27 times the original amount.

9. a. $x = kyz^2$ **b.** $a = \frac{kbc^3}{d}$ **c.** $a = \frac{kb^2\sqrt{c}}{de}$

10. a. $x = \frac{k}{\sqrt{y}}$; If $x = 3$ and $y = 4$, then $3 = \frac{k}{\sqrt{4}} \Rightarrow 3 = \frac{k}{2} \Rightarrow$
$k = 6$, so $x = \frac{6}{\sqrt{y}}$.

   So if $y = 400$, then $x = \frac{6}{\sqrt{400}} = \frac{6}{20} = 0.3$.

   b. $x = k \cdot y \cdot z$; If $x = 3$, $y = 4$ and $z = 0.5$, then
$3 = k \cdot 4 \cdot (0.5) \Rightarrow k = 1.5$, so $x = 1.5 \cdot y \cdot z$.
So if $x = 6$ and $z = 2$, then $6 = 1.5y(2)$ or $y = 2$.

   c. $x = \frac{ky^2}{\sqrt[3]{z}}$; If $x = 6$, $y = 6$ and $z = 8$, then

   $6 = \frac{k(6)^2}{\sqrt[3]{8}} \Rightarrow 6 = \frac{36k}{2} \Rightarrow k = \frac{1}{3}$ so $x = \frac{1}{3}\frac{y^2}{\sqrt[3]{z}}$.

   So if $y = 9$ and $z = 0.027$, then
$x = \frac{1}{3}\frac{9^2}{\sqrt[3]{0.027}} = \frac{1}{3} \cdot \frac{81}{0.3} = 90$.

Exercises for Section 7.5

1. a. y changes from $\frac{4}{1^3} = 4$ to $\frac{4}{4^3} = \frac{1}{16}$

   b. decreases

   c. y changes from $\frac{4}{2^3} = \frac{1}{2}$ to $\frac{4}{(1/2)^3} = 32$

   d. increases

   e. y is inversely proportional to x^3 and 4 is the constant of proportionality.

3. a. $h(2x) = \frac{k}{8x^3}$, $h(-3x) = \frac{k}{-27x^3}$ and $h(x/3) = \frac{27k}{x^3}$

   b. $j(2x) = \frac{-k}{16x^4}$, $j(-3x) = \frac{-k}{81x^4}$ and $j(x/3) = \frac{-81k}{x^4}$

5. a. $P = \frac{0.016}{f}$; 0.0016   **c.** $S = \frac{16}{wp}$; 2

   b. $Q = \frac{54}{r^2}$; 2/3   **d.** $W = \frac{2}{3\sqrt{u}}$; 1/6

7. a. B's value is divided by 16.

   b. Z's value is divided by 2^p.

9. Given: $I(x) = k/x^2$ and $4 = I(6) = k/36$; thus $k = 144 \Rightarrow$ $I(8) = 144/64 = 2.25$ watts per square meter and $I(100) = 144/10000 = 0.0144$ watt per square meter.

11. $I(d) = k/d^2$ and thus $I(4)/I(7) = 49/16 = 3.06$. The light intensity will be more than 3 times as great.

13. a. The volume becomes 1/3 of what it was.

   b. The volume becomes 1/n of what it was.

   c. The volume is doubled.

   d. The volume becomes n times what it was.

Ch. 7

15. a. $x = k \cdot \frac{y}{z}$

   b. Solving $4 = k \cdot \frac{16}{32}$ for k gives $k = 8$.

   c. $x = 8 \cdot 25/5 = 40$

17. a. Let L = wavelength of a wave and t = time between waves. Using the fact that the speed of the waves is directly proportional to the square root of the wave's length, we then have $t = \dfrac{L}{k \cdot \sqrt{L}} = \dfrac{\sqrt{L}}{k}$ and thus time is directly proportional to the square root of the wavelength of the wave.

   b. If the frequency of the waves on the second day is twice that of the first, then the waves will be four times as far apart as they were on the previous day.

19. a. $v = \frac{k}{l}$ **d.** decrease

   b. 24 in.   **e.** It was doubled in length.

   c. decrease

21. a. $H = k/P = k \cdot P^{-1}$, where k is a proportionality constant.

   b. The software gave $H = 87.19 \cdot P^{-0.99}$, which is very close.

   Note that the fit is quite good, as can be seen from the graph of the data and the best-fit graph as given in the accompanying diagram.

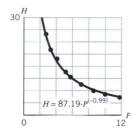

Section 7.6

Algebra Aerobics 7.6

1. a. $\frac{1}{x^2}$; $\frac{1}{(-2)^2} = \frac{1}{4}$ or 0.25

   b. $\frac{1}{x^3}$; $\frac{1}{(-2)^3} = -\frac{1}{8}$ or -0.125

   c. $\frac{4}{x^3}$; $\frac{4}{(-2)^3} = -\frac{1}{2}$ or -0.5

   d. $\frac{-4}{x^3}$; $\frac{-4}{(-2)^3} = \frac{-4}{-8} = \frac{1}{2}$ or 0.5

   e. $\frac{1}{2x^4}$; $\frac{1}{2(-2)^4} = \frac{1}{32} \approx 0.031$

   f. $\frac{-2}{x^3}$; $\frac{-2}{(-2)^3} = \frac{1}{4} = 0.25$

   g. $\frac{-2}{x^4}$; $\frac{-2}{(-2)^4} = -\frac{1}{8} = -0.125$

   h. $\frac{2}{x^4}$; $\frac{2}{(-2)^4} = \frac{1}{8} = 0.125$

2. a. Graph A: $y = \frac{a}{x^2}$ Graph B: $y = ax^3$

   b. Since the point (1, 4) lies on Graph A, $4 = \frac{a}{1^2} \Rightarrow a = 4 \Rightarrow y = \frac{4}{x^2}$.

   Since the point (2, 4) lies on Graph B, $4 = a(2)^3 \Rightarrow a = 0.5 \Rightarrow y = 0.5x^3$

3. a. $y = x^{-10}$ **b.** $y = x^{-11}$

4. a. $y = x^{-2}$ and $y = x^{-3}$

The graphs intersect at (1, 1). As $x \to +\infty$ or $\to -\infty$, both graphs approach the x-axis, but $y = x^{-3}$ is closer to the x-axis at each point after $x = 1$ and before $x = -1$

When $x > 0$, as $x \to 0$, both graphs approach $+\infty$, but $y = x^{-3}$ is steeper.

When $x < 0$, as $x \to 0$, the graph of $y = x^{-2}$ approaches $+\infty$, and the graph of $y = x^{-3}$ approaches $-\infty$.

   b. $y = 4x^{-2}$ and $y = 4x^{-3}$

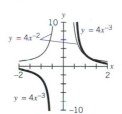

The graphs intersect at (1, 4). As $x \to +\infty$ or $x \to -\infty$, both graphs approach the x-axis, but $y = 4x^{-3}$ is closer to the x-axis after $x = 1$ and before $x = -1$ than $y = 4x^{-2}$.

When $x > 0$, as $x \to 0$, both graphs approach $+\infty$.

When $x < 0$, as $x \to 0$, the graph of $y = 4x^{-2}$ approaches $+\infty$, and the graph of $y = 4x^{-3}$ approaches $-\infty$.

5. Graph A matches $f(x)$;   Graph B matches $g(x)$;   Graph C matches $h(x)$.

Exercises for Section 7.6

1. A table for r is given below.

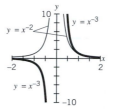

x	$R(x)$
1	6
2	3
3	2
6	1
-1	-6
-2	-3
-3	-2
-6	-1

The domain of the abstract function is all real numbers $x \neq 0$. For $x > 0$, as $x \to 0$ we have $R(x) \to +\infty$ and for $x < 0$, as $x \to 0$ we have $R(x) \to -\infty$.

3. a.

x	$g(x) = 5x$
-2	-10
-1	-5
0	0
1	5
2	10

x	$t(x) = 1/x$
-2	$-1/2$
-1	-1
0.5	2
1	1
2	$1/2$

x	$h(x) = x/5$
-2	-0.4
-1	-0.2
0	0.0
1	0.2
2	0.4

x	$f(x) = 5/x$
-2	$-5/2$
-1	-5
0.5	10
1	5
2	$5/2$

b. The graphs of g and h are both straight lines through the origin with positive slope. The y values of g's graph are 25 times those of h.

Graphs of $g(x) = 5x$ and $h(x) = x/5$

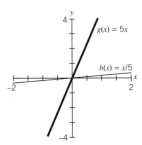

Graphs f and t both have the x- and y-axes as asymptotes, i.e., they approach but never touch these axes, and both are confined to the first and third quadrants. The y values of the graph of f are five times the y values of the graph of t.

Graphs of $t(x) = 1/x$ and $f(x) = 5/x$

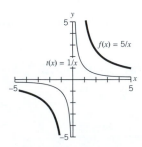

5. a. The graph of $y = x^2$ decreases for $x < 0$ and increases for $x > 0$ (as can be seen in the accompanying diagram).

The graph of $y = x^{-2}$ increases for $x < 0$ and decreases for $x > 0$ (as can also be seen from that diagram).

b. The two graphs intersect at $(-1, 1)$ and $(1, 1)$.

c. As x approaches $\pm\infty$, the graph of $y = x^2$ approaches $+\infty$. As x approaches $\pm\infty$, the graph of $y = x^{-2}$ approaches 0.

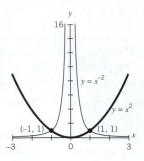

7. a. f goes with Graph A. Reason: $f(x)$ values must be positive; rapid decrease in the values of $f(x)$ when $x > 0$.

b. g goes with Graph C. Reason: There must be positive and negative values of $g(x)$.

c. h goes with Graph D. Reason: $h(x)$ values must be positive; when $x > 0$ and $x \to +\infty$ a less rapid decrease in the values in graph D than in Graph A.

d. j goes with Graph B. Reason: $j(x)$ values must all be negative.

9. a. $g(x) = 4x^{-3}$ **b.** $h(x) = \frac{1}{2}x^{-3}$ **c.** $j(x) = -3x^{-3}$

11. a. If $x > 1$, then $f(x) < g(x)$

b. If $0 < x < 1$, then $f(x) > g(x)$

c. If $x < 0$, then $f(x) < g(x)$

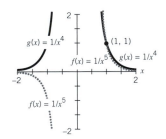

13. a. n is even **d.** yes

b. $k < 0$ **e.** 0

c. $f(-1) < 0$ **f.** 0

15. The graph of $f(x) = x^{-3}$

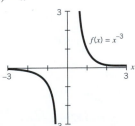

a. g has -1; h has $1/2$ and k has -2.

b. g's graph is a reflection of f's across the x-axis.

c. j's graph is a stretch and a reflection of f's across the x-axis.

d. h's graph is a compression of f's.

17. $g(x) = -4x^{-2}$

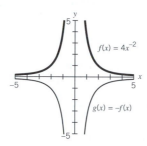

19. a. $f(-x) = \frac{1}{(-x)^4} = \frac{1}{x^4}$ and $-f(x) = -\frac{1}{x^4}$

b. $2f(x) = \frac{2}{x^4}$ and $f(2x) = \frac{1}{(2x)^4} = \frac{1}{16x^4}$

c. $g(-x) = \frac{1}{(-x)^5} = -\frac{1}{x^5}$ and $-g(x) = -\frac{1}{x^5}$

d. $2g(x) = \frac{2}{x^5}$ and $g(2x) = \frac{1}{(2x)^5} = \frac{1}{32x^5}$

e. $h(x) = -f(x) = -\frac{1}{x^4}$

f. $k(x) = -g(x) = -\frac{1}{x^5}$

21. a. $(0, 0)$ and $(1/2, 1)$ **d.** $(1, 1)$

b. $(0, 0)$ and $(1, 1)$ **e.** $(1, 4)$

c. $\left(\sqrt{\frac{1}{2}}, 2\right)$ and $\left(-\sqrt{\frac{1}{2}}, 2\right)$

Section 7.7

Algebra Aerobics 7.7

1. Let $Y = \log y$ and let $X = \log x$.

a. $\log y = \log(3 \cdot 2^x) \Rightarrow \log y = \log 3 + \log 2^x \Rightarrow$
$\log y = \log 3 + x \log 2 \Rightarrow$
$\log y = \log 3 + (\log 2)x \Rightarrow Y = 0.477 + 0.301x$

b. $\log y = \log(4x^3) \Rightarrow \log y = \log 4 + \log x^3 \Rightarrow$
$\log y = \log 4 + 3 \log x \Rightarrow Y = 0.602 + 3X$

c. $\log y = \log(12 \cdot 10^x) \Rightarrow \log y = \log 12 + \log 10^x \Rightarrow$
$\log y = \log 12 + x \log 10 \Rightarrow$
$\log y = \log 12 + (\log 10)x \Rightarrow$
$\quad\quad Y = 1.079 + x$ (since $\log 10 = 1$)

d. $\log y = \log(0.15x^{-2}) \Rightarrow \log y = \log 0.15 + \log x^{-2} \Rightarrow$
$\log y = \log 0.15 - 2 \log x \Rightarrow Y = -0.824 - 2X$

2. a. $y = 10^{0.067+1.63 \log x} \Rightarrow y = 10^{0.067}10^{1.63 \log x} \Rightarrow$
$y = 10^{0.067}10^{\log x^{1.63}} \Rightarrow y \approx 1.167x^{1.63}$

b. $y = 10^{2.135+1.954x} \Rightarrow y = 10^{2.135}10^{1.954x} \Rightarrow$
$y = 10^{2.135}(10^{1.954})^x \Rightarrow y \approx 136.458 \cdot 89.95^x$

c. $y = 10^{-1.963+0.865x} \Rightarrow y = 10^{-1.963}10^{0.865x} \Rightarrow$
$y = 10^{-1.963}(10^{0.865})^x \Rightarrow y \approx 0.011 \cdot 7.328^x$

d. $y = 10^{0.247-0.871 \log x} \Rightarrow y = 10^{0.247}10^{-0.871 \log x} \Rightarrow$
$y = 10^{0.247}10^{\log x^{-0.871}} \Rightarrow y \approx 1.766 \cdot x^{-0.871}$

3. (a), (d), and (f) represent power functions. The graphs of their equations on a log-log plot are linear.

(b), (c), and (e) represent exponential functions. The graphs of their equations on a semi-log plot are linear.

a. $y = 10^{\log 2+3 \log x} \Rightarrow y = 10^{\log 2} \cdot 10^{3 \log x} \Rightarrow$
$y = 2 \cdot 10^{\log x^3} \Rightarrow y = 2x^3$

b. $y = 10^{2+x \log 3} \Rightarrow y = 10^2 \cdot 10^{x \log 3} \Rightarrow$
$y = 100 \cdot 10^{\log 3^x} \Rightarrow y = 100 \cdot 3^x$

c. $y = 10^{0.031+1.25x} \Rightarrow y = 10^{0.031} \cdot 10^{1.25x} \Rightarrow$
$y = 1.07 \cdot (10^{1.25})^x \Rightarrow y = 1.07 \cdot (17.78)^x$

d. $y = 10^{2.457-0.732 \log x} \Rightarrow y = 10^{2.457} \cdot 10^{-0.732 \log x} \Rightarrow$
$y = 286.4 \cdot 10^{\log x^{-0.732}} \Rightarrow y = 286.4x^{-0.732}$

e. $y = 10^{-0.289-0.983x} \Rightarrow y = 10^{-0.289} \cdot 10^{-0.983x} \Rightarrow$
$y = 0.51 \cdot (10^{-0.983})^x \Rightarrow y = 0.51 \cdot (0.104)^x$

f. $y = 10^{-1.47+0.654 \log x} \Rightarrow y = 10^{-1.47} \cdot 10^{0.654 \log x} \Rightarrow$
$y = 0.034 \cdot 10^{\log x^{0.654}} \Rightarrow y = 0.034x^{0.654}$

4. a.   i. $y = 4x^3$ is a power function;
ii. $y = 3x + 4$ is a linear function;
iii. $y = 4 \cdot 3^x$ is an exponential function.

Function	**b.** Type of Plot on Which Graph of Function Would Appear as a Straight Line	**c.** Slope of Straight Line
$y = 3x + 4$	Standard linear plot	$m = 3$
$y = 4 \cdot (3^x)$	Semi-log	$m = \log 3$
$y = 4 \cdot x^3$	Log-log	$m = 3$

5. This is a log-log plot, so the slope of a straight line corresponds to the exponent of a power function.

Younger ages: slope $= 1.2 \Rightarrow$ original function is of the form $y = a \cdot x^{1.2}$, where $x = $ body height in cm and $y = $ arm length in cm.

Older ages: slope $= 1.0 \Rightarrow$ original function is of the form $y = a \cdot x^1$, where $x = $ body height in cm and $y = $ arm length in cm.

6. Graph A: Exponential function because the graph is approximately linear on a semi-log plot

Graph B: Exponential function because the graph is approximately linear on a semi-log plot

Graph C: Power function because the graph is approximately linear on a log-log plot

7. a. Estimated surface area is $10^4 = 10,000$ square cm.

b. Since $S \propto M^{2/3}$, then $S = kM^{2/3}$, where $\log k$ is the vertical intercept of the best-fit line on a log-log plot. In this case the vertical intercept $= \log 10$, so $k = 10$. So the equation is $S = 10M^{2/3}$. When $M = 70,000$g, then
$S = 10 \cdot (70,000)^{2/3} = 10 \cdot 1700 = 17,000$ cm^2.

c. Our estimated surface area was 7000 square cm lower than the calculated area.

d. Since 1 cm $= 0.394$ in, then 1 cm$^2 = (0.394$ in$)^2 = 0.155$ in^2. So 17,000 cm$^2 = (17,000)(0.155) = 2635$ in^2. Since 1 kg $= 2.2$ lb, a weight of 70 kg translated to pounds is 70 kg $= (70 \text{ kg})\left(2.2 \frac{\text{lb}}{\text{kg}}\right) = 154$ lb.

Ch. 7

8. a. power

  b. $y = ax^{-0.23}$

  c. When body mass increases by a factor of 10, heat rate is multiplied by $10^{-0.23}$ or about 0.59.

9. Estimated rate of heat production is somewhere between 10^3 and 10^4 kilocalories per day—roughly between 2000–3000 kcal/day.

Exercises for Section 7.7

1. When $x = 4$, then $100(5)^4 = 62{,}500$ and $100(4)^5 = 102{,}400$. So the exponential function $y = 100(5)^x$ in part (a) matches Graph *A*. Since an exponential function is a straight line on a semi-log plot, then the function also matches Graph *D*.

The power function $y = 100x^5$ in part (b) matches Graphs *B* and *C*, since a power function (Graph *C*) is a straight line on a log-log plot (Graph *B*).

3. An exponential function is indicated since the graph is a straight line on a semi-log plot.

5. a. $y = 4x^2$ **c.** $y = 1.25x^4$

  b. $y = 2x^4$ **d.** $y = 0.5x^3$

7. a.

x	$y = x^5$
1	1
2	32
3	243
4	1024
5	3125
6	7776

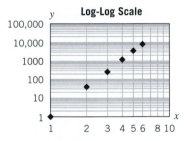

b.

x	$\log(x)$	$y = x^5$	$\log(y)$
1	0	1	0
2	0.3010	32	1.5051
3	0.4771	243	2.3856
4	0.6021	1024	3.0103
5	0.6990	3125	3.4949
6	0.7782	7776	3.8908

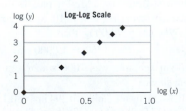

c. The graphs are exactly the same. The points lie on straight line and are plotted on equivalent scales.

9. a.

Species	Adult Mass in Grams	Egg Mass in Grams	Egg/Adult Ratio
Ostrich	113,380.0	1700.0	0.015
Goose	4,536.0	165.4	0.036
Duck	3,629.0	94.5	0.026
Pheasant	1,020.0	34.0	0.033
Pigeon	283.0	14.0	0.049
Hummingbird	3.6	0.6	0.167

Answers will vary; notable is the fact that the egg/adult mass ratio for the hummingbird is very high and it is very low for the ostrich. The other ratios are not far apart from each other.

b. The plots (where $x =$ adult mass, $y =$ egg mass (both in grams)) are, respectively,

  i. y vs. x

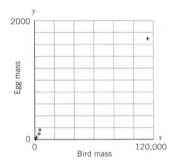

  ii. $\log y$ vs. x

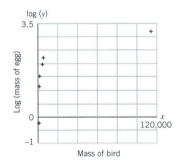

  iii. $\log y$ vs. $\log x$

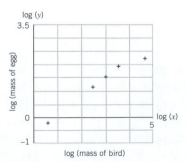

c. The log vs. log scatter diagram in part b (iii) is the most linear. Using technology the best-fit straight-line equation is: $\log(y) = \log(0.1918) + 0.7719 \cdot \log(x)$ or, in linear equation form: $Y = -0.717 + 0.7719X$ where $Y = \log(y)$ and $X = \log(x)$.

d. Regrouping, using the laws of logarithms, we get: $\log(y) = \log(0.1918 \cdot x^{0.7719})$. Thus, $y = 0.1918 \cdot x^{0.7719}$, where

x = mass of the adult and y is the corresponding mass of the egg, both measured in grams.

e. If $x = 12.7$ kg (or 12,700 grams) for the weight of an adult turkey, then $y \approx 282$ grams is the predicted weight of its egg.

f. If the egg weighs 2 grams, then the adult bird is predicted to have an adult weight $\approx$ 20.8 grams.

11. a. Since l_1 and l_2 are parallel lines they have the same slope but different vertical intercepts. Thus their equations can be written as $\log(y) = mx + \log(b_1)$ and $\log(y) = mx + \log(b_2)$, with $\log(b_1) \neq \log(b_2)$. Solving each for y in terms of x, we have $y = b_1 \cdot 10^{mx}$ and $y = b_2 \cdot 10^{mx}$ with $b_1 \neq b_2$. Thus they have the same power of 10 but differ in their y-intercepts.

b. The equation of l_3 is $\log(y) = m \cdot \log(x) + \log(b_3)$ and the equation of l_4 is $\log(y) = m \cdot \log(x) + \log(b_4)$, with $\log(b_3) \neq \log(b_4)$. Solving for y in terms of x in each gives: $y = b_3 x^m$ and $y = b_4 x^m$ with $b_3 \neq b_4$. Thus they have the same power of x but differ in their coefficients of x.

13. a.

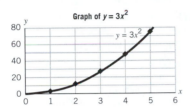

b.

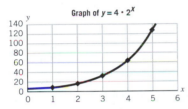

c.

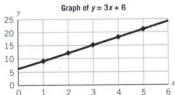

15. a. A power function seems appropriate.

b. The log-log graph is linear and we get $\log(O) = 0.75 \cdot \log(m) + c$, where O stands for oxygen consumption and m stands for body mass. Thus $O = k \cdot m^{0.75}$, a power function, where $k = 10^c$.

c. A slope of 3/4 means that if the body mass, m, is multiplied by 10^4 ($=10,000$), then the oxygen consumption, O, is multiplied by 10^3 ($=1000$). If m is multiplied by 10, then O is multiplied by $10^{3/4} \approx 5.6$.

17. a.

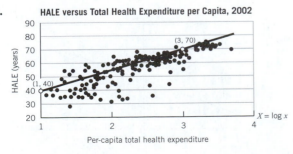

b. Two estimated points on the best-fit line are (1, 40) and (3, 70) $\Rightarrow$ slope $= \frac{(70 - 40)}{(3 - 1)} = 15$. So the equation is of the form $y = b + 15X$. Substituting in (1, 40) we get $40 = b + (15 \cdot 1) \Rightarrow 40 = b + 15 \Rightarrow b = 25$. So the equation is $y = 25 + 15X$. Substituting $\log x$ for X, we finally get $y = 25 + 15 \log x$. So HALE is a logarithmic function of per-capita health expenditure.

c. A logarithmic model makes sense, since there will always be a ceiling in life expectancy, so the results of adding funding will increase HALE but at a slower and slower rate.

d. It suggests that the functions are log functions.

Ch. 7: Check Your Understanding

1. True	**6.** False	**11.** False	**16.** False
2. False	**7.** False	**12.** True	**17.** False
3. False	**8.** True	**13.** True	**18.** True
4. False	**9.** True	**14.** True	**19.** True
5. True	**10.** False	**15.** False	**20.** False

21. True

22. Possible answer: $f(x) = 4x^8$ or $f(x) = 10x^6$

23. $g(x) = 3.2x^4$

24. Possible answer: $f(x) = -x^4$

25. Possible answer: $h(m) = m^{12}$ or $h(m) = e^m$

26. Possible answer: $y = 3^x$ or $y = 3 + x$

27. Possible answer: $y = 3x^5$

28. Possible answer: $k(x) = \frac{-1}{x^3}$

29. Possible answer: $T(m) = 3m^4$

30. False	**33.** False	**36.** True	**39.** True
31. True	**34.** False	**37.** False	**40.** False
32. False	**35.** True	**38.** True	

Ch. 7 Review: Putting It All Together

1. a. $V(x) = x(2x)(2x) = 4x^3$. So $V(X) = 4X^3$ and $V(2X) = 4(2X)^3 = 8(4X^3) = 8V(X) \Rightarrow$ the volume is multiplied by 8.

b. The surface area consists of the two top and bottom sides, which are both $(2x)(2x) = 4x^2$ in area, and the other four vertical sides, each $(x)(2x) = 2x^2$ in area. So the surface area $S(x) = 2(4x^2) + 4(2x^2) = 16x^2$. Hence $S(X) = 16X^2$ and $S(2X) = 16(2X)^2 = 4(16X^2) = 4S(X) \Rightarrow$ when the value of x is doubled, the surface area is multiplied by 4. So the volume grows faster than the surface area.

c. The ratio of (surface area)/volume $= R(x) = S(x)/V(x) = (16x^2)/(4x^3) = 4/x$. If you double the value of x, from X to $2X$, since $R(X) = 4/X$, then $R(2X) = 4/(2X) = (1/2)(4/X) = (1/2)R(X)$. So if the value of x doubles, $R(x)$, the ratio of (surface area)/volume, is multiplied by 1/2—or equivalently, cut in two. This is again confirmation that the volume grows faster than the surface area.

d. In general, as x increases, $R(x)$—the ratio of surface area to volume—will decrease. So as the rectangular solid gets larger, there will be relatively less surface area compared with the volume.

3. Think of the cake as a cylinder, with a volume of $\pi r^2 h$ where r = the radius of the cake and h = the height. The radii of the 10″ and 12″ cakes are 5″ and 6″, respectively. If the corresponding heights are h_1 and h_2, then the respective cake volumes are $25\pi h_1$ and $36\pi h_2$ cubic inches. The site claims that the volume of the 12″ cake is twice that of the 10″ cake (6 quarts vs. 3 quarts). That implies that $36\pi h_2 = 2(25\pi h_1) \Rightarrow h_2/h_1 = 50/36 \approx 1.4 \Rightarrow h_2$ is about 40% larger than h_1. So for the 12″ cake to have twice the volume of the 10″ cake, the height of the 12″ cake would have to be 40% higher than the height of the 10″ cake—which seems rather unlikely.

5. a. $V_F(t) = 3t$; domain is $0 \le t \le 5$; represents direct proportionality.

b. $V_D(t) = 15 - 0.5t$; domain is $0 \le t \le 30$; does not represent inverse proportionality.

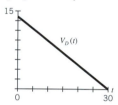

7. a. The friend is wrong.

b. If the height is increased by 50% (from 4 feet to 6 feet) or, equivalently, multiplied by 1.5 (since $1.5 \cdot 4' = 6'$), the width (2′) and the length (2′) must both also be multiplied by 1.5, giving 3′ for the new width and length. Hence the new dimensions are $6' \times 3' \times 3'$, which give a volume of 54 cubic feet. The volume of the original block of ice $4' \times 2' \times 2' = 16$ cubic feet. Since (new volume)/(old volume) = 54/16 = 3.375, the new volume = $3.375 \cdot$ (old volume) or, equivalently, a 237.5% increase in volume, almost five times what your friend predicted.

c. 10% of 54 cubic feet $= 0.10 \cdot 54 = 5.4$ cubic feet. So the volume of the melted sculpture would be $54 - 5.4 = 48.6$ cubic feet. If h is the height of the melted sculpture, then the width and length are both $0.5h$. So we have $48.6 = (0.5h)^2 h = (0.5)^2 h^3 \Rightarrow h^3 = (48.6)/(0.5)^2 = 194.4 \Rightarrow h \approx 5.8$ feet as the height of the melted sculpture.

9. a. $t = kd/r$ (the k may be needed for unit conversions)

b. $D_1 = kD_2$

c. $R = kV^2/W$

11. Graph C is symmetric across the y-axis. Graphs A, B, and D are rotationally symmetric about the origin. None of them has an asymptote.

13. a. $P(n) = 450/n$ for integer values of n between 0 and, say, 10.

b. $P(2) = 450/2 = \$225$/person; $P(5) = 450/5 = \$90$/person. The more people, the lower the cost per person.

c. If you take the function $P(n)$ out of context and treat n as any real number ($\ne 0$), then the following graph shows the result.

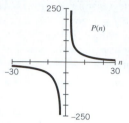

The abstract function is asymptotic to both the horizontal and vertical axes.

15. a. From the previous problem, we have $I = P/d^2$ where I is measured in foot-candles, P in candlepower, and d in feet. Setting $I = 4000$ foot-candles and $d = 3$ feet, we have $4000 = P/(3)^2 \Rightarrow P = 36,000$ candlepower.

b. From part (a) we know that the lamp has 36,000 candlepower. Setting $P = 36,000$ and $I = 2000$ foot-candles, we have $2000 = 36,000/d^2 \Rightarrow d^2 = 18 \Rightarrow d \approx 4.24$ feet or about 51 inches above the operating surface.

17. Graphs B and C are symmetric about the y-axis. Graph D is symmetric about the origin. Graphs A, B, and D all have the x-axis as a horizontal asymptote. Graphs B and D also have the y-axis as a vertical asymptote.

19. a. Y is a linear function of X.

b. Using the estimated points $(0, 0.3)$ and $(1, 3.3)$, we have $Y = 0.3 + 3X$.

c. Substituting $Y = \log y$ and $X = \log x$ in $Y = 0.3 + 3X$, we have $\log y = 0.3 + 3 \log x \Rightarrow \log y = 0.3 + \log(x^3)$. Rewriting using powers of 10, we have $10^{\log y} = 10^{(0.3 + \log(x^3))} \Rightarrow y = 10^{0.3} \cdot 10^{\log x^3} \Rightarrow y = 10^{0.3} \cdot x^3 \Rightarrow y \approx 2x^3$. Hence y is a power function and y is directly proportional to x^3.

21. To find the relationship, sketch a best-fit line to the data, where m = mass (in kg) and v = optimal flying speed (in meters per second). Since both axes use a logarithmic scale, we can replace the labels on the axes with $\log(m) = M$ and $\log(v) = V$, respectively. The horizontal labels will now read $-5, -4, -3, -2, -1, 0$ ($= \log 1$), which is now the beginning of the vertical axis, followed by 1, 2, 3, 5, 5, 6; and on the vertical axis 0, 1, 2, and $2.5 \approx \log 300$.

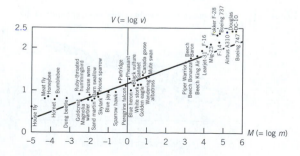

Two points on the line with coordinates of the form (M, V) are $(-1, 1)$ and $(5, 2)$. The slope is then $(2 - 1)/(5 - (-1)) = 1/6$. The vertical intercept is approximately at $(0, 1.2)$. So the equation of the line would be $V = 1.2 + (1/6)M$. Substituting in, we get $\log(v) = 1.2 + (1/6)\log(m) = 1.2 + \log(m^{1/6})$. Rewriting both sides as powers of 10 and simplifying, we get: $v = 10^{(1.2 + \log(m^{1/6}))} = 10^{1.2} \cdot 10^{\log(m^{1/6})} \approx 16m^{1/6}$. So $v \propto m^{1/6}$ $\Rightarrow$ the cruising speed v (in meters per second) is directly proportional to the 1/6 power of body mass m (in kilograms). So Professor Bejan seems to be correct.

CHAPTER 8

Section 8.1

Algebra Aerobics 8.1

1. a. The vertex at $(0, 0)$ is the minimum point. The focal point is $\frac{1}{4a} = \frac{1}{4(3)} = \frac{1}{12}$ units above the vertex at $\left(0, \frac{1}{12}\right)$.

　b. The vertex at $(0, 0)$ is the maximum point. The focal point is $\frac{1}{4a} = \frac{1}{4(-6)} = \frac{-1}{24}$ units below the vertex at $\left(0, -\frac{1}{24}\right)$.

　c. The vertex at $(0, 0)$ is the minimum point. The focal point is $\frac{1}{4a} = \frac{1}{4(1/24)} = 6$ units above the vertex at $(0, 6)$.

　d. The vertex at $(0, 0)$ is the maximum point. The focal point is $\frac{1}{4a} = \frac{1}{4(-1/12)} = -3$ units below the vertex at $(0, -3)$.

2. a. A point on the rim is $(15, 10)$, so if $y = ax^2$, then $10 = a(15)^2$, so $a = 2/45$. The focal point is $\frac{1}{4a} = \frac{1}{4(2/45)} = \frac{45}{8} = 5.625$ ft from the vertex at $(0, 0)$ on the back wall.

　b. $y = \frac{2}{45}x^2$

　c. You could not hear well if you were more than 15 ft on either side of the stage, because the sound would be traveling in straight lines from the parabolic wall. Also, the sound would not be good if the performer moved around and did not stay at the focal point.

3. a. $g(2) = 2^2 = 4$
$g(-2) = (-2)^2 = 4$
$g(0) = (0)^2 = 0$
$g(z) = (z)^2$

　b. $h(2) = -(2)^2 = -4$
$h(-2) = -(-2)^2 = -4$
$h(0) = 0^2 = 0$
$h(z) = -z^2$

　c. $Q(2) = -(2)^2 - 3(2) + 1 = -9$
$Q(-2) = -(-2)^2 - 3(-2) + 1 = 3$
$Q(0) = -(0)^2 - 3(0) + 1 = 1$
$Q(z) = -(z)^2 - 3(z) + 1$

　d. $m(2) = 5 + 2(2) - 3(2)^2 = -3$
$m(-2) = 5 + 2(-2) - 3(-2)^2 = -11$
$m(0) = 5 + 2(0) - 3(0)^2 = 5$
$m(z) = 5 + 2z - 3z^2$

　e. $D(2) = -(2 - 3)^2 + 4 = 3$
$D(-2) = -((-2) - 3)^2 + 4 = -21$
$D(0) = -((0) - 3)^2 + 4 = -5$
$D(z) = -(z - 3)^2 + 4$

　f. $k(2) = 5 - 2^2 = 1$
$k(-2) = 5 - (-2)^2 = 1$
$k(0) = 5 - 0^2 = 5$
$k(z) = 5 - z^2$

4. a. $f(x)$
　i. concave down
　ii. maximum
　iii. axis of symmetry: $x = 0$
　iv. vertex: $(0, 4)$
　v. approx. horizontal intercepts: $(-2, 0), (2, 0)$ vertical intercept: $(0, 4)$

　b. $g(x)$
　i. concave up
　ii. minimum
　iii. axis of symmetry: $x = 1$
　iv. approx. vertex: $(1, -2.25)$
　v. horizontal intercepts: $(-2, 0), (4, 0)$ vertical intercept: $(0, -2)$

　c. $h(x)$
　i. concave up
　ii. minimum
　iii. axis of symmetry: $x = 2$
　iv. vertex: $(2, -4)$
　v. horizontal intercepts: $(0, 0), (4, 0)$ vertical intercept: $(0, 0)$

　d. $k(x)$
　i. concave down
　ii. maximum
　iii. axis of symmetry: $x = -4$
　iv. vertex: $(-4, -2)$
　v. no horizontal intercepts vertical intercept: $(0, -10)$

5. a. Equation is of the form $y = ax^2$ where $a > 0 \Rightarrow$ focal length $= \frac{1}{4a}$. When $x = 12$, focal length $=$ depth $\Rightarrow$ $\frac{1}{4a} = a(12)^2 \Rightarrow a^2 = \frac{1}{4 \cdot 12^2} \Rightarrow a = \frac{1}{2 \cdot 12} = \frac{1}{24}$. So focal length $= \frac{1}{4(1/24)} = \frac{1}{(1/6)} = 6$

　b. $a = \frac{1}{24} \Rightarrow y = \frac{1}{24}x^2$.

6. $f(x) = x^2 + 2x - 15$　　　$g(x) = 2x^2 + 7x + 5$
$h(x) = 10x^2 - 80x + 150$　　$j(x) = 2x^2 - 18$

Exercises for Section 8.1

1. a. concave up; minimum; vertex at $(-1, -4)$; the line $x = -1$ is axis of symmetry; $(1, 0)$ and $(-3, 0)$ are horizontal intercepts; $(0, -3)$ is vertical intercept.

　b. concave down; maximum; vertex at $(2, 9)$; the line $x = 2$ is axis of symmetry; $(-4, 0)$ and $(8, 0)$ are horizontal intercepts; $(0, 8)$ is vertical intercept.

　c. concave down; maximum; vertex at $(-5, 3)$; the line $x = -5$ is axis of symmetry; $(-8, 0)$ and $(-2, 0)$ are horizontal intercepts; $(0, -5)$ is vertical intercept. [Students may have a different y value for vertical intercept.]

3. The graphs of the three functions are in the accompanying diagram.

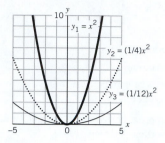

a. For y_1: (0, 1/4); for y_2: (0, 1); for y_3: (0, 3)

b. The graphs of all three functions are concave-up parabolas with vertex at (0, 0), but y_1 rises faster than y_2 and y_3.

c. Farther from the vertex and the graphs get wider.

5. a. $y = (1/24)x^2$ **c.** $y = (3/4)x^2$

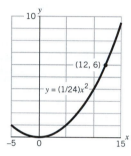

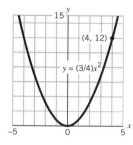

b. $y = -(1/24)x^2$ **d.** $y = 16x^2$

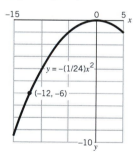

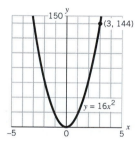

7. a. $y = (1/16)x^2$ **c.** $y = 4x^2$

b. $y = -(1/32)x^2$ **d.** $y = -6x^2$

9. a. -8 and 12   **b.** 10 and 14   **c.** 2 and -14

11. a. $y = (1/6)x^2$

b. $5 = (1/6)x^2$ implies $x = \pm\sqrt{30}$ and thus the reflector is $2\sqrt{30} \approx 10.95$ inches wide.

13. Assuming that we rotate the parabolic model to become concave up:

a. The focus is at (0, 1.25).

b. $y = 0.2x^2$

c. $2.5 = 0.2x^2$ gives $x = \pm\sqrt{12.5}$; thus the reflector should be $2\sqrt{12.5} \approx 7.07$ inches wide.

15. Student answers will vary. The general trend is that the smaller the focal length, the bigger the value of a and thus

the narrower the opening of the parabola, since focal length $= \left|\frac{1}{4a}\right|$ when the parabola has the equation $y = ax^2$.

17. a. $f(x) = x^2 + 2x - 3$ **c.** $H(z) = -z^2 - z + 2$

b. $P(t) = t^2 - 3t - 10$

19. a. $P = 2L + 2W = 200$; Thus $L = 100 - W$ and $A = LW = (100 - W) \cdot W = 100W - W^2$. So A has its maximum when $W = 50$ m, since the vertex is at (50, 50). Thus $L = 50$ m and the maximum area is 2500 m^2.

b. The same kind of argument applies. $P = 2L + 2W$; thus $L = (P/2) - W$ and $A = LW = W(P/2 - W) = (P/2) \cdot W - W^2$, and this has its vertex at $(P/4, (P/4)^2)$ and the maximum area is $(P/4)^2$ m^2.

21. a. True   **c.** False   **e.** False

b. False   **d.** True

23. a. $R(n) = (1250 + 100n) \cdot (50 - 2n)$

b. There will be no apartments rented when $n = 25$. Thus the domain is $0 \le n \le 25$.

c. The graph is given in the accompanying diagram. From inspection, the practical maximum occurs when $n = 6$. At that value $R = \$70,300$.

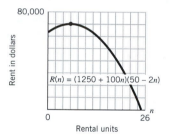

Section 8.2

Algebra Aerobics 8.2a

1. $s(x) = 2x^2 + 2$ is narrower than $r(x) = x^2 + 2$ because, looking at the coefficient of x^2, $2 > 1$. Both have a vertical intercept at (0, 2) since $c = 2$ in both equations.

2. $h(t) = t^2 + 5$ is concave up since a (=1) is positive; $k(t) = -t^2 + 5$ is concave down since a (= -1) is negative. They have the same shape because $|a| = 1$ in both. They both cross the vertical axis at 5 since $c = 5$ in both equations.

3. Both are concave down because both have $a < 0$. Both have a vertex at (0, 0) since b and $c = 0$ in both equations. $g(z)$ is flatter than $f(z)$ because $|-0.5| < |-5|$.

4. They have the same shape and are concave up since both have $a = 1$. $g(x)$ is six units higher than $f(x)$ because 8 is 6 more than 2.

5. They have the same shape and are concave down since a (= -3) is negative in both equations. $g(t) = -3t^2 + t - 2$ is three units higher than $f(t) = -3t^2 + t - 5$ since $c = -2$ is three units vertically up from $c = -5$.

6. a. $y_1 = 3x^2 + 5$ **b.** $y_2 = \frac{1}{3}x^2 - 2$ **c.** $y_3 = -2x^2 + 4$

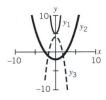

7. a. $g(x) = 3x^2$ **c.** $j(x) = 1/2x^2$
  b. $h(x) = -5x^2$ **d.** $k(x) = -x^2$

8. a. $y = 5x^2 - 2$ **c.** $y = -0.5x^2 - 4.7$
  b. $y = x^2 + 3$ **d.** $y = x^2 - 71$

Algebra Aerobics 8.2b

1. a. vertex is $(0, 0)$
  b. vertex is $(-3, 0)$
  c. vertex is $(2, 0)$

All vertices lie on the x-axis. Vertex of (b) is three units to the left of the vertex of (a). Vertex of (c) is two units to the right of vertex of (a). All have same shape and are concave up.

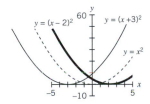

2. a. vertex is $(0, 0)$ **c.** vertex is $(-4, 0)$
  b. vertex is $(1, 0)$

All vertices lie on the x-axis. The vertex of (b) is one unit to the right of the vertex of (a). The vertex of (c) is four units to the left of vertex of (a). All have the same shape and are concave up.

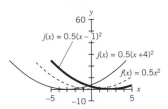

3. a. vertex is $(0, 0)$ **c.** vertex is $(0.9, 0)$
  b. vertex is $(-1.2, 0)$

All of the vertices lie on the horizontal axis. The vertex of (b) is 1.2 units to the left of the vertex of (a). The vertex of (c) is 0.9 units to the right of the vertex of (a). All have the same shape and are concave down.

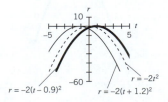

4. $y = a(x - h)^2 + k$. The value of $h = 2$ in (b), (c), and (d), so the vertices of those parabolas are two units to the right of the vertex of (a), where $h = 0$. All are concave up with the same shape since $a = 1$ in all four equations. (c) is four units above (a) and (b); (d) is three units below (a) and (b) since $k = 0$ in (a) and (b), $k = 4$ in (c), and $k = -3$ in (d).

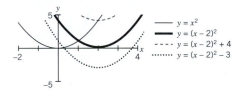

5. $y = a(x - h) + k$. All are concave down with the same shape since $a = -1$ in all four equations. (b), (c), and (d) have vertices three units to the left of vertex of (a), since $h = -3$ in those equations, and $h = 0$ in (a). $k = 0$ in (a) and (b), but $k = -1$ in (c), so (c) is one unit below (a) and (b). $k = 4$ in (d), so (d) is four units above (a) and (b).

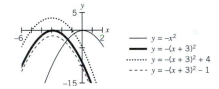

6. a. $g(x) = 3(x - 2)^2 - 1$, the vertex is $(2, -1)$
  b. $h(x) = -2(x + 3)^2 + 5$, the vertex is $(-3, 5)$
  c. $j(x) = \frac{1}{5}(x + 4)^2 - 3.5$, the vertex is $(-4, -3.5)$
  d. $k(x) = -(x - 1)^2 + 4$, the vertex is $(1, 4)$

7. a. The vertex at $(3, -4)$ is a minimum.
  b. The vertex at $(-1, 5)$ is a maximum.
  c. The vertex at $(4, 0)$ is a maximum.
  d. The vertex at $(0, -7)$ is a minimum.

Algebra Aerobics 8.2c

1. a. vertex is $(0, -4)$.
  b. vertex is $(0, 6)$.
  c. vertex is $(0, 1)$.

2. a. vertex is $(0, 3)$
  b. vertex is $(0, 3)$

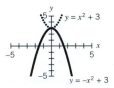

3. a. Vertex is $(0, 0)$. Parabola is concave down since $a\ (= -3)$ is negative. There is one x-intercept, $(0, 0)$, since vertex is on the x-axis.

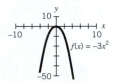

Ch. 8

b. Vertex is $(0, -5)$, which is below x-axis. Parabola is concave down since a ($= -2$) is negative. So there are no x-intercepts.

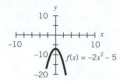

In parts (c) and (d), use the formula $h = \frac{-b}{2a}$ as the horizontal coordinate of the vertex (where $f(x) = ax^2 + bx + c$).

c. $a = 1, b = 4 \Rightarrow h = \frac{-b}{2a} = \frac{-4}{2(1)} = -2,$

$f(-2) = (-2)^2 + 4(-2) - 7 = 4 - 8 - 7 = -11$

Vertex is $(-2, -11)$. Parabola is concave up since a ($= 1$) is positive. There are two x-intercepts. They are approximately $(-5, 0)$ and $(1, 0)$.

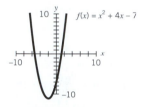

d. $f(x) = 4 - x - 2x^2 = -2x^2 - x + 4$; so $a = -2,$

$b = -1$; $h = \frac{-b}{2a} = \frac{-(-1)}{2(-2)} = \frac{1}{-4} = -\frac{1}{4}$

$f\left(-\frac{1}{4}\right) = 4 - \left(-\frac{1}{4}\right) - 2\left(-\frac{1}{4}\right)^2 = 4 + \frac{1}{4} - 2\left(\frac{1}{16}\right) =$

$4 + \frac{1}{4} - \frac{1}{8} = 4\frac{1}{8}$. So vertex is $\left(-\frac{1}{4}, 4\frac{1}{8}\right)$. Parabola is concave down since a ($= -2$) is negative. There are two x-intercepts. They are approximately $(-2, 0)$ and $(1, 0)$.

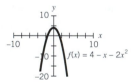

4. $y = ax^2 + bx + c$ compared with $y = x^2$ (where $a = 1$)

a. $y = 2x^2 - 5 \Rightarrow a = 2$

  i. $a > 0 \Rightarrow$ minimum at the vertex

  ii. $|a| > 1$, so parabola is narrower than $y = x^2$.

b. $y = 0.5x^2 + 2x - 10 \Rightarrow a = 0.5$

  i. $a > 0 \Rightarrow$ minimum at the vertex

  ii. $|a| < 1$, so parabola is flatter than $y = x^2$.

c. $y = 3 + x - 4x^2 \Rightarrow a = -4$

  i. $a < 0 \Rightarrow$ maximum at the vertex

  ii. $|a| > 1$, so parabola is narrower than $y = x^2$.

d. $y = -0.2x^2 + 11x + 8 \Rightarrow a = -0.2$

  i. $a < 0 \Rightarrow$ maximum at the vertex

  ii. $|a| < 1$, so parabola is flatter than $y = x^2$.

5. a. $a = 1, b = 3$; so horizontal coordinate of vertex is:

$\frac{-b}{2a} = \frac{-3}{2} = -1\frac{1}{2}$. If $x = -\frac{3}{2}, \Rightarrow$

$y = \left(-\frac{3}{2}\right)^2 + 3\left(-\frac{3}{2}\right) + 2 = \frac{9}{4} - \frac{9}{2} + 2 =$

$2\frac{1}{4} - 4\frac{2}{4} + 2 = -\frac{1}{4}$, So vertex is $\left(-1\frac{1}{2}, -\frac{1}{4}\right)$.

Vertical intercept is 2.

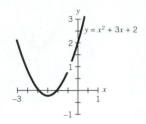

b. $a = 2$, $b = -4$, so horizontal coordinate of vertex is $\frac{-b}{2a} = \frac{-(-4)}{2(2)} = 1$. $f(1) = 2 - 4 + 5 = 3$, so vertex is $(1, 3)$. Since vertex is above x-axis, and a ($= 2$) is positive; the parabola is concave up, it does not cross x-axis, so no horizontal intercepts. Vertical intercept is 5.

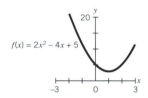

c. $a = -1, b = -4; \Rightarrow \frac{-b}{2a} = \frac{-(-4)}{2(-1)} = \frac{4}{-2} = -2$

$g(-2) = -(-2)^2 - 4(-2) - 7 = -4 + 8 - 7 = -3$, so vertex is $(-2, -3)$. Since vertex is below t-axis and a ($= -1$) is negative, the parabola is concave down; it does not cross t-axis, so no horizontal intercepts. Vertical intercept $= -7$.

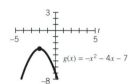

6. a. Vertex is $(-5, -11)$. y-intercept at $x = 0 \Rightarrow$ $y = 0.1(5)^2 - 11 = 0.1(25) - 11 = 2.5 - 11 = -8.5$. So y-intercept is -8.5.

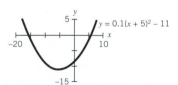

b. Vertex is $(1, 4)$. y-intercept at $x = 0 \Rightarrow y = -2(-1)^2 + 4 = -2 + 4 = 2$. So y-intercept is 2.

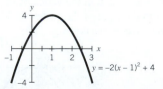

7. a. $(x + 3)^2 - 9$

b. $(x - 5)^2 - 25$

c. $(x - 15)^2 - 225$

d. $\left(x + \frac{1}{2}\right)^2 - \frac{1}{4}$

8. a. $f(x) = x^2 + 2x - 1 = x^2 + 2x + (1 - 1) - 1$
$= (x^2 + 2x + 1) + (-1 - 1) \Rightarrow$
$f(x) = (x + 1)^2 - 2$
vertex $(-1, -2)$; stretch factor 1

  b. $j(z) = 4z^2 - 8z - 6 = 4(z^2 - 2z) - 6$
$= 4(z^2 - 2z + 1) - 4(1) - 6 \Rightarrow$
$j(z) = 4(z - 1)^2 - 10$
vertex $(1, -10)$; stretch factor 4

  c. $h(x) = -3x^2 - 12x = -3(x^2 + 4x)$
$= -3(x^2 + 4x + 4) + 3(4) \Rightarrow$
$h(x) = -3(x + 2)^2 + 12$
vertex $(-2, 12)$; stretch factor -3

  d. $h(t) = -16(t^2 - 6t) \Rightarrow h(t) = -16(t^2 - 6t + 9) + 16(9)$
$\Rightarrow h(t) = -16(t - 3)^2 + 144$
vertex $(3, 144)$; stretch factor -16

  e. $h(t) = -4.9(t^2 + 20t) + 200$
$\Rightarrow h(t) = -4.9(t^2 + 20t + 10^2) + 200 + 4.9(10^2)$
$\Rightarrow h(t) = -4.9(t + 10)^2 + 690$
vertex $(-10, 690)$; stretch factor -4.9

9. a. $y = 2\left(x - \frac{1}{2}\right)^2 + 5 = 2\left(x - \frac{1}{2}\right)\left(x - \frac{1}{2}\right) + 5$
$= 2\left(x^2 - x + \frac{1}{4}\right) + 5 = 2x^2 - 2x + \frac{1}{2} + 5 \Rightarrow$
$y = 2x^2 - 2x + 5\frac{1}{2}$
vertex $(1/2, 5)$; stretch factor 2

  b. $y = -\frac{1}{3}(x + 2)^2 + 4 = -\frac{1}{3}(x + 2)(x + 2) + 4$
$= -\frac{1}{3}(x^2 + 4x + 4) + 4$
$= -\frac{1}{3}x^2 - \frac{4}{3}x - \frac{4}{3} + 4 \Rightarrow$
$y = -\frac{1}{3}x^2 - \frac{4}{3}x + 2\frac{2}{3}$
vertex $(-2, 4)$; stretch factor $-1/3$

  c. $y = 10(x^2 - 10x + 25) + 12$
$\Rightarrow y = 10x^2 - 100x + 250 + 12$
$\Rightarrow y = 10x^2 - 100x + 262$
vertex $(5, 12)$; stretch factor 10

  d. $y = 0.1(x^2 + 0.4x + 0.04) + 3.8$
$\Rightarrow y = 0.1x^2 + 0.04x + 3.804$
vertex $(-0.2, 3.8)$; stretch factor 0.1

10. a. $y = x^2 + 6x + 7 = x^2 + 6x + 9 - 9 + 7 \Rightarrow$
$y = (x + 3)^2 - 2$

  b. $y = 2x^2 + 4x - 11 = 2(x^2 + 2x) - 11$
$= 2(x^2 + 2x + 1) - 2(1) - 11 \Rightarrow$
$y = 2(x + 1)^2 - 13$

11. a. "Completing the square,"
$y = x^2 + 8x + 11 \Rightarrow y = x^2 + 8x + 16 - 16 + 11 \Rightarrow$
$y = (x + 4)^2 - 5$
So vertex is $(-4, -5)$.

  b. Using the formula,
$y = 3x^2 + 4x - 2 \Rightarrow a = 3, b = 4, c = -2 \Rightarrow$
$h = \frac{-b}{2a} = \frac{-4}{6} = -\frac{2}{3}$

When $x = \frac{-2}{3}$,
$y = 3\left(-\frac{2}{3}\right)^2 + 4\left(\frac{-2}{3}\right) - 2 \Rightarrow$
$y = 3\left(\frac{4}{9}\right) - \frac{8}{3} - 2 = \frac{4}{3} - \frac{8}{3} - 2 =$
$-\frac{4}{3} - 2 = -1\frac{1}{3} - 2 = -3\frac{1}{3}$, so vertex is at $\left(-\frac{2}{3}, -3\frac{1}{3}\right)$.
In vertex form, $y = 3\left(x + \frac{2}{3}\right)^2 - 3\frac{1}{3}$

12. a. $f(p) = -1875p^2 + 4500p - 2400 \Rightarrow$
$a = -1875, b = 4500, c = -2400 \Rightarrow$
$h = \frac{-b}{2a} = \frac{-4500}{-3750} = 1.2$
$f(1.2) = -1875(1.2)^2 + 4500(1.2) - 2400 \Rightarrow f(1.2) = 300$
So the vertex is at ($1.20, $300).

  b. The maximum daily profit is $300 at a price of $1.20 per pretzel.

13. a. The vertex is $(-1, 4) \Rightarrow h = -1, k = 4 \Rightarrow$
$y = a(x - (-1))^2 + 4 \Rightarrow y = a(x + 1)^2 + 4$
Passing through the point $(0, 2) \Rightarrow 2 = a(0 + 1)^2 + 4 \Rightarrow$
$a = -2 \Rightarrow$ the equation is $y = -2(x + 1)^2 + 4$.

  b. The vertex is $(1, -3) \Rightarrow h = 1, k = -3 \Rightarrow$
$y = a(x - 1)^2 - 3$. Passing through the point $(-2, 0) \Rightarrow$
$0 = a(-2 - 1)^2 - 3 \Rightarrow 3 = 9a \Rightarrow$ the equation is
$y = \frac{1}{3}(x - 1)^2 - 3$.

Exercises for Section 8.2

1. a. downward; the coefficient of x^2 is negative.

  b. upward; the coefficient of t^2 is positive.

  c. upward; the coefficient of x^2 is positive.

  d. downward; the coefficient of x^2 is negative.

3. The graphs of the four functions are given with their labels in the diagram below.

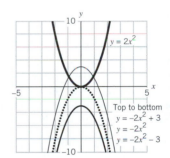

5. a. The compression factor is 0.3, the vertex is at $(1, 8)$.

  b. The expansion or stretch factor is 30 and the vertex is at $(0, -11)$.

  c. The compression factor is 0.01 and the vertex is at $(-20, 0)$.

  d. The expansion or stretch factor is -6 and the vertex is at $(1, 6)$.

7. a. $(3, 5)$; maximum   **c.** $(-4, -7)$; maximum

  b. $(-1, 8)$; minimum   **d.** $(2, -6)$; minimum

9. a. $y = 1(x - 2)^2 - 4 = x^2 - 4x$

  b. $y = -(x - 4)^2 + 3 = -x^2 + 8x - 13$

  c. $y = -2(x + 3)^2 + 1 = -2x^2 - 12x - 17$

  d. $y = 0.5(x + 4)^2 + 6 = 0.5x^2 + 4x + 14$

11. a. $h(x) = -3(x - 4)^2 + 5$ **b.** $(4, 5)$ **c.** $(0, -43)$

13. a. 9; 3   **c.** 4; 2   **e.** 1; 1

  b. 16; 4   **d.** 2.25; 1.5   **f.** 0.25; 0.5

15. a. $y = (x + 3)^2 + 4$

  b. $f(x) = (x - 2.5)^2 - 11.25$

Ch. 8

c. $g(x) = (x - 1.5)^2 + 3.75$

d. $p(r) = -3(r - 3)^2 + 18$

e. $m(z) = 2(z + 2)^2 - 13$

17. The larger in absolute value the coefficient of the x^2 term, the narrower the opening. Thus the order from narrow to broad is: d, f, a, b, c, and finally e. Technology confirms the principle.

19. a. The diagram exhibits the graph of f, its horizontal and vertical intercepts, and its vertex.

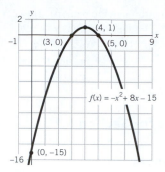

b. $f(x) = -(x - 4)^2 + 1$, and thus its vertex is at $(4, 1)$.

21. a. $y = a(x - 2)^2 + 4$ and $7 = a(1 - 2)^2 + 4$ gives $a = 3$ and thus $y = 3(x - 2)^2 + 4 = 3x^2 - 12x + 16$.

b. If $a > 0$, the graph of $y = a(x - 2)^2 - 3$ is concave up; it is concave down if $a < 0$.

c. The axis of symmetry is the line $y = (-2 + 4)/2 = 1$; since the parabola is concave downward, we have $a < 0$; also we have $y = a(x - 1)^2 + k$ with $5 = a(-2 - 1)^2 + k$ and $5 = a(4 - 1)^2 + k$. Thus we have $9a + k = 5$ or $k = 5 - 9a$. One can choose $a = -1/9$ and then $k = 6$, or one can choose $a = -2/9$ and then $k = 7$. Thus two examples are $y = (-1/9)(x - 1)^2 + 6$ or $y = (-2/9)(x - 1)^2 + 7$.

23. a. $(4, -5)$; no   **b.** 0.5   **c.** -2

25. Derivation: from the data we have:

$y = a(x - 2)^2 + 3$; and $-1 = a(4 - 2)^2 + 3$

This implies that $a = -1$.

Thus $y = -(x - 2)^2 + 3$ is its equation.

Check: $-(4 - 2)^2 + 3 = -4 + 3 = -1$

This is confirmed in the accompanying graph.

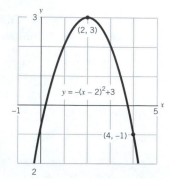

27. The maximum profit occurs when $x = -20/[2 \cdot (-0.5)] = 20$; the maximum profit is 430 thousand dollars.

29. The maximum occurs when $x = -48/(-6) = 8$ computers and the revenue from selling 8 will be 192 million dollars.

31. a. 4 ft = how high off the ground the baseball was at the instant at which it was hit.

b.

t(sec)	h(feet)
0	4
1	38
2	40
3	10
4	-52

The ball hits the ground between $t = 3$ and $t = 4$ sec, since the height at $t = 3$ is positive and the height at $t = 4$ is negative.

c. The accompanying diagram gives the graph and confirms the estimate given in part (b).

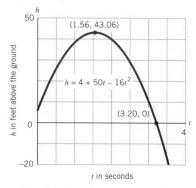

d. Negative values for h make no sense since this would mean that the ball is below ground.

e. Similarly, using technology we can determine that the ball reaches its maximum height of approximately 43.96 ft at $t \approx 1.56$ sec.

33. We are given that $2W + L = 1$ cowhide length. Thus $L = 1 - 2W$ and thus the area formula is $A = W(1 - 2W) = W - 2W^2$. Note this is at its maximum when W is at the vertex, i.e., when $W = -1/[2 \cdot (-2)] = 1/4$. At this value of W we have that $L = 1/2$. Thus, Dido was correct.

35. a. The graph is given in the accompanying diagram, with x and y marked off in miles, and with a sample point $(7.2, 9.6)$ marked on the highway along with the straight line from the origin to that point.

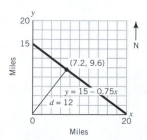

b. The highway goes through the points $(0, 15)$ and $(20, 0)$ and thus has the equation $y = -0.75x + 15$.

c, d. The distance squared:

$$d^2 = x^2 + y^2 = x^2 + (15 - 0.75x)^2$$
$$= 1.5625x^2 - 22.5x + 225$$

e. Letting $D = d^2$, we have $D = 1.5625x^2 - 22.5x + 225$; the minimum occurs at the vertex, which is at $x = 22.5/(2 \cdot 1.5625) = 7.2$. The minimum for D is 144 and thus the minimum distance, d, is $\sqrt{144}$ or 12 miles.

f. The coordinates of the point of this shortest distance from $(0, 0)$ are $(7.2, 9.6)$. [See the graph in part (a).]

37. a, b. The graph is shown below. The minimum gas consumption rate suggested by the graph occurs when M is about 32 mph, and it is approximately 0.85 gph. (Computation on a calculator gives 32.5 for M, and the corresponding gas consumption rate is 0.86 (when rounded off).)

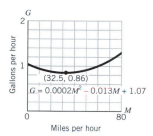

c. In 2 hours 1.72 gallons will be used and you will have traveled 65 miles.

d. If $M = 60$ mph, then $G = 1.01$ gph. It takes 1 hour and 5 minutes to travel 65 miles at 60 mph and one will have used approximately 1.094 gallons.

e. Clearly, traveling at the speed that supposedly minimizes the gas consumption rate does not conserve fuel if the trip lasts only 2 hours.

f, g.

M (mph)	G (gph)	G/M (gpm)	M/G (mpg)
0	1.07	—	0.0
10	0.96	0.09600	10.4
20	0.89	0.04450	22.5
30	0.86	0.02867	34.9
40	0.87	0.02175	46.0
50	0.92	0.01840	54.3
60	1.01	0.01683	59.4
70	1.14	0.01629	61.4
80	1.31	0.01638	61.1

For (f) we have: $G/M = (0.0002M^2 - 0.013M + 1.07)/M$.

Its graph is in the accompanying diagram. Eyeballing gives the minimum gpm at $M \approx 73$ mph.

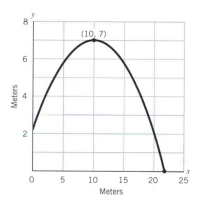

Graph for $y = G/M$

For (g) we have: $M/G = M/(0.0002M^2 - 0.013M + 1.07)$.

Its graph is given in the accompanying diagram. Eyeballing gives the maximum mpg at the same M of approximately 73 mph. This is expected since maximum = 1/minimum.

Graph for $y = M/G$

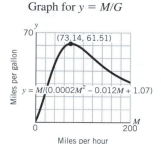

39. a. Time of release is at $x = 0$. The height then is 2 meters.

b. At $x = 4$ m, $y = 5.2$ m; at $x = 16$ m, $y = 5.2$ m.

c, d. The graph in the accompanying diagram shows the highest point, namely when $x = 10$ m and $y = 7$ m, and the point where the shot put hits the ground, namely when x is approximately 22 m.

Section 8.3

Algebra Aerobics 8.3a

1. a. $y = -2t(8t - 25) = 0 \Rightarrow -2t = 0$ or $8t - 25 = 0 \Rightarrow t = 0$ or $t = \frac{25}{8} \Rightarrow$ horizontal intercepts at $(0, 0)$ and $\left(\frac{25}{8}, 0\right)$

b. $y = (t - 5)(t + 5) = 0 \Rightarrow t = 5$ or $t = -5 \Rightarrow$ horizontal intercepts at $(5, 0)$ and $(-5, 0)$

c. $h(z) = (z - 4)(z + 1) = 0 \Rightarrow z - 4 = 0$ or $z + 1 = 0 \Rightarrow z = 4$ or $z = -1 \Rightarrow$ horizontal intercepts at $(4, 0)$ and $(-1, 0)$

d. $g(x) = (2x - 3)(2x + 3) = 0 \Rightarrow 2x - 3 = 0$ or $2x + 3 = 0 \Rightarrow x = 3/2$ or $x = -3/2 \Rightarrow$ horizontal intercepts at $(3/2, 0)$ and $(-3/2, 0)$

e. $y = (5 - x)(3 - x) = 0 \Rightarrow 5 - x = 0$ or $3 - x = 0 \Rightarrow x = 5$ or $x = 3 \Rightarrow$ horizontal intercepts at $(5, 0)$ and $(3, 0)$

f. $v(x) = (x + 1)^2 = 0 \Rightarrow x + 1 = 0 \Rightarrow x = -1$, or one horizontal intercept at $(-1, 0)$

g. $p(q) = (q - 3)(q - 3) = 0 \Rightarrow q - 3 = 0 \Rightarrow q = 3 \Rightarrow$ one horizontal intercept at $(3, 0)$

2. a. $f(x) = (5 + 4x)(1 - x) = 0 \Rightarrow 5 + 4x = 0$ or $1 - x = 0 \Rightarrow x = -5/4$ or $x = 1 \Rightarrow$ horizontal intercepts at $(-5/4, 0)$ and $(1, 0)$

Ch. 8

b. $h(t) = (8 - 3t)(8 + 3t) = 0 \Rightarrow 8 - 3t = 0$ or $8 + 3t = 0 \Rightarrow t = \frac{8}{3}$ or $t = -\frac{8}{3} \Rightarrow$ horizontal intercepts at $\left(\frac{8}{3}, 0\right)$ and $\left(-\frac{8}{3}, 0\right)$.

c. $y = (5 + t)(2 - 3t) = 0 \Rightarrow 5 + t = 0, 2 - 3t = 0 \Rightarrow t = -5$ or $t = 2/3 \Rightarrow$ horizontal intercepts at $(-5, 0)$ and $(2/3, 0)$

d. $z = (2w - 5)(2w - 5) = 0 \Rightarrow 2w = 5 \Rightarrow w = 5/2 \Rightarrow$ one horizontal intercept $(5/2, 0)$

e. $y = (2x - 5)(x + 1) = 0 \Rightarrow 2x - 5 = 0$ or $x + 1 = 0 \Rightarrow x = \frac{5}{2}$ or $x = -1 \Rightarrow$ horizontal intercepts at $\left(\frac{5}{2}, 0\right)$ and $(-1, 0)$

f. $Q(t) = (3t - 2)(2t + 5) = 0 \Rightarrow 3t - 2 = 0$ or $2t + 5 = 0 \Rightarrow t = \frac{2}{3}$ or $t = -\frac{5}{2} \Rightarrow$ horizontal intercepts at $\left(\frac{2}{3}, 0\right)$ and $\left(-\frac{5}{2}, 0\right)$

3. Product of a sum and difference:

a. $y = x^2 - 9 = (x + 3)(x - 3)$

d. $y = 9x^2 - 25 = (3x + 5)(3x - 5)$

f. $y = 16 - 25x^2 = (4 + 5x)(4 - 5x)$

Square of sum or difference:

b. $y = x^2 + 4x + 4 = (x + 2)^2$

e. $y = x^2 - 8x + 16 = (x - 4)^2$

Neither:

c. $y = x^2 + 5x + 25$

g. $y = 4 + 16x^2$

4. a. The vertical intercept is $(0, 0)$, which means at time $t = 0$, the object is on the ground. The horizontal intercepts: $h(t) = -16t(t - 4) = 0 \Rightarrow t = 0$ or $t = 4 \Rightarrow (0, 0)$ and $(4, 0)$, which means the object left the ground at time $t = 0$ seconds and returns to the ground at time $t = 4$ seconds.

b. The horizontal intercepts: $P(q) = -(q^2 - 60q + 800) \Rightarrow P(q) = -(q - 20)(q - 40) = 0 \Rightarrow q = 20$ or $q = 40$. This means that if either 20 or 40 units are sold, the profit is \$0 (or at breakeven). From the graph, $P(q) > 0$ if $20 < q < 40$. The vertical intercept is $(0, -\$800)$, which means that if no items are sold, there is a loss of \$800.

5. y_1 is Graph *C* because $(0, 0)$ and $(2, 0)$ are horizontal intercepts, and graph is concave down.

y_2 is Graph *A* because $(2, 0)$ and $(-1, 0)$ are horizontal intercepts, and graph is concave up.

y_3 is Graph *B* because $(-4, 0)$ and $(-1, 0)$ are horizontal intercepts, and graph is concave up.

6. a. $f(x) = (x - 5)(x + 6)$

b. $(5, 0)$ and $(-6, 0)$

c.

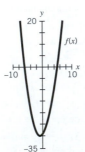

d. $f(x) = 0 \Rightarrow (x - 5)(x + 6) = 0 \Rightarrow x - 5 = 0$ or $x + 6 = 0 \Rightarrow x = 5$ or $x = -6$. So $f(x)$ has two horizontal intercepts, at $(5, 0)$ and $(-6, 0)$.

Algebra Aerobics 8.3b

1. a. $4x + 7 = 0 \Rightarrow x = -7/4$

b. $4x^2 - 7 = 0 \Rightarrow 4x^2 = 7 \Rightarrow x^2 = \frac{7}{4}$
$\Rightarrow x = \pm \sqrt{\frac{7}{4}} \Rightarrow x = \pm \frac{\sqrt{7}}{2}$

c. $4x^2 - 7x = 0 \Rightarrow x(4x - 7) = 0 \Rightarrow x = 0$ or $x = \frac{7}{4}$

d. $2x + 6 = x^2 \Rightarrow$
$0 = x^2 - 2x - 6 \Rightarrow$
$x = \frac{-(-2) \pm \sqrt{(-2)^2 - 4(-6)}}{2} \Rightarrow$
$x = \frac{2 \pm \sqrt{28}}{2} \Rightarrow x = \frac{2 \pm 2\sqrt{7}}{2} \Rightarrow$
$x = \frac{2(1 \pm \sqrt{7})}{2} \Rightarrow x = 1 \pm \sqrt{7}$

e. $(2x - 11)^2 = 0 \Rightarrow 2x - 11 = 0 \Rightarrow x = \frac{11}{2}$

f. $(x + 1)^2 = 81 \Rightarrow x + 1 = \pm\sqrt{81} \Rightarrow$
$x = -1 \pm 9 \Rightarrow x = -10$ or $x = 8$

g. $0 = x^2 - x - 5 \Rightarrow$
$x = \frac{-(-1) \pm \sqrt{(-1)^2 - 4(-5)}}{2} \Rightarrow x = \frac{1 \pm \sqrt{21}}{2}$

2. a. $a = 2, b = 3, c = -1 \Rightarrow D = (3)^2 - 4(2)(-1) = 17 \Rightarrow$ two real, unequal zeros $\Rightarrow$ two horizontal intercepts

b. $a = 1, b = 7, c = 2 \Rightarrow D = (7)^2 - 4(1)(2) = 41 \Rightarrow$ two real, unequal zeros $\Rightarrow$ two horizontal intercepts

c. $a = 4, b = 4, c = 1 \Rightarrow D = (4)^2 - 4(4)(1) = 0 \Rightarrow$ one real zero (also known as a "double zero") $\Rightarrow$ one horizontal intercept

d. $a = 2, b = 1, c = 5 \Rightarrow D = (1)^2 - 4(2)(5) = -39 \Rightarrow$ no real zeros (two imaginary zeros) $\Rightarrow$ no horizontal intercepts

3. a. $h = -4.9t^2 + 50t + 80$. The vertical intercept is the initial height (at 0 seconds), which is 80 m; coordinates are $(0, 80)$. $a = -4.9, b = 50, c = 80$. So when $h = 0$ the horizontal intercepts are:

$t = \frac{-50 \pm \sqrt{2500 - 4(-4.9)(80)}}{2(-4.9)}$

$= \frac{-50 \pm \sqrt{2500 + 1568}}{-9.8} = \frac{-50 \pm \sqrt{4068}}{-9.8} \Rightarrow$

$t = \frac{-50 + 63.8}{-9.8}$ or $t = \frac{-50 - 63.8}{-9.8}$

$= \frac{13.8}{-9.8}$ $= \frac{-113.8}{-9.8}$

$= -1.41$ $= 11.61$ seconds

Negative values of t have no meaning in a height equation, so the horizontal intercept at $(11.61, 0)$ means that the object hits the ground after 11.61 seconds.

b. $h = 150 - 80t - 490t^2$. The vertical intercept is $(0, 150)$, which means that the initial height is 150 cm. $a = -490, b = -80, c = 150 \Rightarrow$ when $h = 0$ the horizontal intercepts are at:

$t = \frac{80 \pm \sqrt{6400 - 4(-490)(150)}}{2(-490)}$

$= \frac{80 \pm \sqrt{6400 + 294,000}}{-980}$

$= \frac{80 \pm \sqrt{300,400}}{-980} = \frac{80 \pm 548}{-980} \Rightarrow$

$t = \frac{80 + 548}{-980}$ or $t = \frac{80 - 548}{-980}$

$= \frac{628}{-980}$ $= \frac{-468}{-980}$

$= -0.64$ $= 0.48$ seconds.

Discard negative solution. The object hits the ground after 0.48 seconds.

c. The vertical intercept is the height in feet at $t = 0$ seconds $\Rightarrow h = 3$ feet.

If $a = -16, b = 64, c = 3 \Rightarrow$ the horizontal intercepts are:

$$t = \frac{-64 \pm \sqrt{(64)^2 - 4(-16)(3)}}{2(-16)} \Rightarrow t = \frac{-64 \pm \sqrt{4288}}{-32}$$

$\Rightarrow t = \frac{-64 \pm 65.5}{-32} \Rightarrow t \approx 4.05$, which means the object hits the ground after about 4.05 seconds or $t \approx -0.05$ (which is meaningless in this context).

d. The vertical intercept is the height in feet at $t = 0$ seconds $\Rightarrow h = 64(0) - 16(0)^2 = 0$ feet.

$a = -16, b = 64, c = 0 \Rightarrow$ horizontal intercepts are

$$t = \frac{-64 \pm \sqrt{(64)^2 - 4(-16)(0)}}{2(-16)} \Rightarrow t = \frac{-64 \pm 64}{-32} \Rightarrow$$

horizontal intercepts are $t = 0$ and $t = 4 \Rightarrow$ the object hits the ground after exactly 4 seconds.

4. Discriminant is $b^2 - 4ac$, from $ax^2 + bx + c = y$

a. $a = -5, b = -1, c = 4 \Rightarrow$ discriminant $= 1 - 4(-5)(4)$ $= 81 > 0 \Rightarrow$ two x-intercepts. $\sqrt{81} = 9$, so roots are rational. The function has two real zeros where:

$$x = \frac{1 \pm 9}{-10} \Rightarrow$$

$$x = \frac{1 + 9}{-10} \quad \text{or} \quad x = \frac{1 - 9}{-10}$$

$$= \frac{10}{-10} \qquad\qquad = \frac{-8}{-10}$$

$$= -1 \qquad\qquad\quad = \frac{4}{5}$$

So the x-intercepts are $(-1, 0)$ and $(4/5, 0)$.

b. $a = 4, b = -28, c = 49 \Rightarrow$ discriminant $=$ $784 - 4(4)(49) = 784 - 784 = 0 \Rightarrow$ one x-intercept. $\sqrt{0} = 0$, so root is rational. The x-intercept is where $x = \frac{28 \pm 0}{8} = \frac{7}{2}$; at $\left(3\frac{1}{2}, 0\right)$.

c. $a = 2, b = 5, c = 4 \Rightarrow$ discriminant $= 25 - 4(2)(4) =$ $25 - 32 = -7$, which is negative $\Rightarrow$ no x-intercepts. The function has two imaginary zeros at

$$x = \frac{-5 \pm \sqrt{-7}}{4} = \frac{-5 \pm \sqrt{7}i}{4}$$

d. $a = 2, b = -3, c = -1 \Rightarrow$ discriminant $=$ $(-3)^2 - 4(2)(-1) = 9 + 8 = 17 \Rightarrow$ two real zeros

$$x = \frac{-(-3) \pm \sqrt{17}}{2(2)} \Rightarrow x = \frac{3 \pm \sqrt{17}}{4} \Rightarrow$$

x-intercepts are $\left(\frac{3 + \sqrt{17}}{4}, 0\right)$ and $\left(\frac{3 - \sqrt{17}}{4}, 0\right)$

e. $a = -3, b = 0, c = 2 \Rightarrow$ discriminant $=$ $(0)^2 - 4(-3)(2) = 24 \Rightarrow$ two real zeros

$$x = \frac{-0 \pm \sqrt{24}}{2(-3)} \Rightarrow x = \frac{\pm \sqrt{24}}{6} \Rightarrow x = \frac{\pm 2\sqrt{6}}{6} \Rightarrow$$

$$x = \mp \frac{\sqrt{6}}{3} \Rightarrow x\text{-intercepts are } \left(-\frac{\sqrt{6}}{3}, 0\right), \left(+\frac{\sqrt{6}}{3}, 0\right)$$

5. a. $y = a(x + 2)(x - 4)$. The point $(3, 2)$ is on the parabola $\Rightarrow$ $2 = a(3 + 2)(3 - 4) \Rightarrow a = -\frac{2}{5} \Rightarrow$ $y = -\frac{2}{5}(x + 2)(x - 4)$. The vertex is on the line of symmetry, which lies halfway between the x-intercepts or

at $x = 1 \Rightarrow y = \frac{-2}{5}(1 + 2)(1 - 4) \Rightarrow y = \frac{18}{5} \Rightarrow$ the vertex is at $\left(1, \frac{18}{5}\right)$.

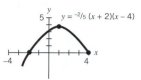

b. $y = a(x - 2)(x - 8)$. The vertical intercept is $(0, 10) \Rightarrow$ $10 = a(0 - 2)(0 - 8) \Rightarrow a = \frac{5}{8}$, so $y = \frac{5}{8}(x - 2)(x - 8)$

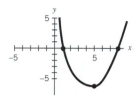

The vertex is on the line of symmetry, which lies halfway between the x-intercepts or at $x = 5 \Rightarrow y =$ $\frac{5}{8}(5 - 2)(5 - 8) \Rightarrow y = \frac{-45}{8}$ and the vertex is $\left(5, \frac{-45}{8}\right)$.

6. a. Vertex is $(1, 5)$, above x-axis; $a = 3$ is positive, so it opens up; so there are no x-intercepts.

b. Vertex is $(-4, -1)$, below x-axis; $a = -2$ is negative, so it opens down; so there are no x-intercepts.

c. Vertex is $(-3, 0)$, on x-axis; $a = -5$ is negative, so it opens down; so there is one x-intercept.

d. Vertex is $(1, -2)$, below x-axis; $a = 3$ is positive, so there are two x-intercepts.

7. Answers will vary for different values of a and will be of the form:

a. $f(x) = a(x - 2)(x + 3)$
If $a = 1$, then
$f(x) = (x - 2)(x + 3)$

b. $f(x) = ax(x + 5)$
If $a = 2$, then
$f(x) = 2x(x + 5)$

c. $f(x) = a(x - 8)^2$
If $a = -2$, then
$f(x) = -2(x - 8)^2$

8. Graph A: two real, unequal zeros $\Rightarrow$ discriminant is positive
Graph B: one real zero $\Rightarrow$ discriminant is equal to zero
Graph C: no real zeros $\Rightarrow$ discriminant is negative

Exercises for Section 8.3

1. a. $0 = (x - 3)(x + 3)$; thus $x = 3$ or -3

b. $0 = x(4 - x)$; thus $x = 0$ or 4

c. $0 = x(3x - 25)$; thus $x = 0$ or $25/3$

d. $0 = (x + 5)(x - 4)$; thus $x = -5$ or 4

e. $0 = (2x + 3)^2$; thus $x = -3/2$ twice

f. $0 = (3x + 2)(x - 5)$; thus $x = -2/3$ or 5

g. $x^2 + 4x + 3 = -1 \Rightarrow x^2 + 4x + 4 = 0 \Rightarrow (x + 2)^2 = 0$; thus $x = -2$ (a double zero)

h. $x^2 + 2x = 3x^2 - 3x - 3 \Rightarrow 2x^2 - 5x - 3 = 0 \Rightarrow$ $(2x + 1)(x - 3) = 0$; thus $x = -1/2$ or 3

3. a. $y = (x + 4)(x + 2)$

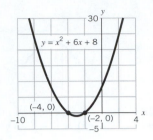

$y = x^2 + 6x + 8$, (−4, 0), (−2, 0)

b. $z = 3(x + 1)(x - 3)$

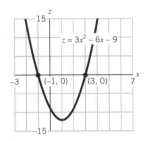

$z = 3x^2 - 6x - 9$, (−1, 0), (3, 0)

c. $f(x) = (x - 5)(x + 2)$

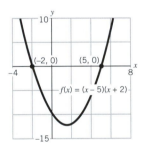

$f(x) = (x - 5)(x + 2)$, (−2, 0), (5, 0)

d. $w = (t - 5)(t + 5)$

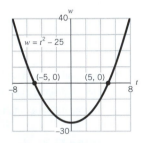

$w = t^2 - 25$, (−5, 0), (5, 0)

e. $r = 4(s - 5)(s + 5)$

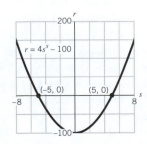

$r = 4s^2 - 100$, (−5, 0), (5, 0)

f. $g(x) = (3x - 4)(x + 1)$

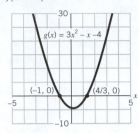

$g(x) = 3x^2 - x - 4$, (−1, 0), (4/3, 0)

5. a. $y = x^2 - 5x + 6$ has zeros at 2 and 3, as is shown in the graph below.

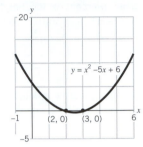

$y = x^2 - 5x + 6$, (2, 0), (3, 0)

b. $y = 3x^2 - 2x + 5$ has no real zeros, as is shown in the graph below.

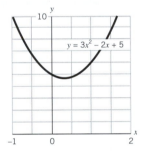

$y = 3x^2 - 2x + 5$

c. $y = 3x^2 - 12x + 12$ has a "double zero" at $x = 2$ since $y = 3(x - 2)^2$. See the graph below.

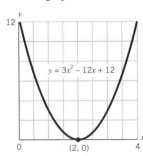

$y = 3x^2 - 12x + 12$, (2, 0)

d. $y = -3x^2 - 12x + 15$ has zeros at $x = -5$ and 1; see the graph below.

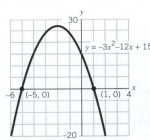

$y = -3x^2 - 12x + 15$, (−5, 0), (1, 0)

e. $y = 0.05x^2 + 1.1x$ has zeros at $x = -22$ and 0, as the graph below shows.

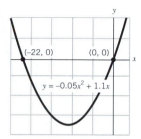

f. $y = -2x^2 - x + 3$ has roots at $x = -1.5$ and 1, as the graph below shows.

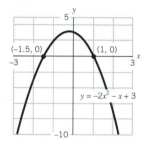

7. a. $t = \dfrac{7 \pm \sqrt{49 - 4 \cdot 6 \cdot (-5)}}{2 \cdot 6} = \dfrac{7 \pm \sqrt{169}}{12}$

$= \dfrac{7 \pm 13}{12} = \dfrac{5}{3}$ or $-\dfrac{1}{2}$

b. $x = \dfrac{12 \pm \sqrt{144 - 4 \cdot 9 \cdot 4}}{2 \cdot 9} = \dfrac{12 \pm \sqrt{0}}{18} = \dfrac{2}{3}$

c. $z = \dfrac{1 \pm \sqrt{1 - 4 \cdot 3 \cdot (-9)}}{2 \cdot 3} = \dfrac{1 \pm \sqrt{109}}{6} \approx -1.573$
or 1.907

d. $x = \dfrac{-6 \pm \sqrt{36 - 4 \cdot 1 \cdot 7}}{2 \cdot 1} = \dfrac{-6 \pm \sqrt{8}}{2} = -3 \pm \sqrt{2}$
≈ -1.586 or -4.414

e. $s = \dfrac{-17 \pm \sqrt{17^2 - 4 \cdot 6 \cdot (-10)}}{2 \cdot 6} = \dfrac{-17 \pm \sqrt{529}}{12}$

$= \dfrac{-17 \pm 23}{12} = -\dfrac{10}{3}$ or $\dfrac{1}{2}$

f. $t = \dfrac{3 \pm \sqrt{9 - 4 \cdot 2 \cdot (-9)}}{2 \cdot 2} = \dfrac{3 \pm \sqrt{81}}{4} = \dfrac{3 \pm 9}{4} = 3$
or -1.5

g. $x = \dfrac{11 \pm \sqrt{121 - 4 \cdot 4 \cdot (-8)}}{2 \cdot 4} = \dfrac{11 \pm \sqrt{249}}{8} \approx 3.347$
or -0.597

h. $x = \dfrac{12 \pm \sqrt{144 - 4 \cdot 4 \cdot 2}}{2 \cdot 4} = \dfrac{12 \pm \sqrt{112}}{8}$

$= \dfrac{12 \pm 4\sqrt{7}}{8} = \dfrac{3 \pm \sqrt{7}}{2} \approx 2.823$ or 0.177

9. a. y-intercept is -1; $y = (3x - 1)(x + 1)$ and thus x-intercepts are $1/3$ and -1.

b. y-intercept is 11; the x-intercepts are $\dfrac{6 \pm \sqrt{3}}{3} \approx 2.58$ and 1.42.

c. y-intercept is 15; x-intercepts are $5/2$ and $-3/5$.

d. The vertical intercept is $f(0) = -5$ and the x-intercepts are $\pm\sqrt{5}$.

11. Student choices for values of a, b and c will vary. Here are some choices and the accompanying graphs. The equations in the form $y = ax^2 + bx + c$ are in the graph diagrams.

a. $a = 2, b = 3, c = 1; b^2 - 4ac = 9 - 8 = 1$

b. $a = 2, b = 3, c = -2; b^2 - 4ac = 9 + 16 = 25$

c. $a = 2, b = 3, c = 4; b^2 - 4ac = 9 - 32 = -23$

d. $a = -1, b = 2, c = -1; b^2 - 4ac = 4 - 4 = 0$

e. same as in (b) above.

Graph for (a)

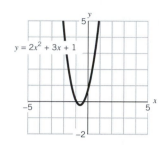

Graph for (b), (e)

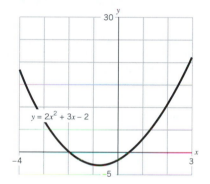

Graph for (c)

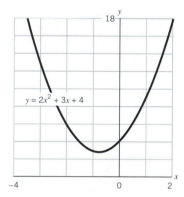

Graph for (d)

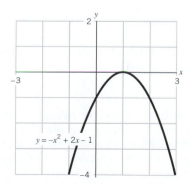

13. a. $Q(t) = -4(t + 1)^2$

b. No, since it must be of the form $Q(t) = a(t + 1)^2$ if 1 is to be a double root and a must be -4 if $Q(0) = -4$.

c. The axis of symmetry is the line whose equation is $t = -1$. The vertex is at $(-1, 0)$.

15. For Graph A: $f(x) = (x + 5)(x - 2) = x^2 + 3x - 10$

For Graph B: $g(x) = -0.5(x + 5)(x - 2) = -0.5x^2 - 1.5x + 5$

17. a. $-1 + 10i$ **b.** 1 **c.** $-1 + 3i$ **d.** $-5 + 11i$

19. $f(x) = (x -(1 + i))(x - (1 - i)) =$
$x^2 - (1 - i)x - (1 + i)x + (1 + i)(1 - i) =$
$x^2 - 2x + (1 - i^2) =$
$x^2 - 2x + 2$

21. $x = \dfrac{4 \pm \sqrt{16 - 4 \cdot 1 \cdot 13}}{2 \cdot 1} = \dfrac{4 \pm \sqrt{-36}}{2}$
$= \dfrac{4 \pm 6i}{2} = 2 \pm 3i.$

Thus $f(x) = (x - 2 - 3i)(x - 2 + 3i)$. Its roots are not real and thus there are no x-intercepts.

23. a. Factoring, $y = (x + 4)(x - 2)$ and thus the roots are -4 and 2 and their average is -1. Thus $h = -1$ and $a = 1$; therefore $y = (x + 1)^2 + k = x^2 + 2x + 1 + k = x^2 + 2x - 8$ and thus $k + 1 = -8$ or $k = -9$. Thus the equation is $y = (x + 1)^2 - 9$.

b. Factoring gives $y = -(x + 4)(x - 1)$ and thus its roots are -4 and 1 and thus $h = -1.5$ and $a = -1$; therefore $y = -(x + 1.5)^2 + k = -x^2 - 3x - 2.25 + k = -x^2 - 3x + 8$ and thus $4 = k - 2.25$ or $k = 6.25$. Thus the equation is $y = -(x + 1.5)^2 + 6.25$.

25. $y = 4(x - 2)(x - 3)$. If the quadratic is to go through $(2, 0)$ and $(3, 0)$, then it must be of the form $y = a(x - 2)(x - 3)$; and if it is to stretch the graph of $y = x^2$ by a factor of 4, then $a = 4$ must hold. But this function's graph does more. It shifts the vertex of the graph of $y = x^2$ to $(2.5, -1)$. Its graph is in the accompanying diagram.

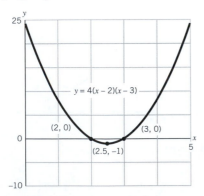

27. a. Setting the two formulas for y equal gives:
$7x^2 - 5x - 9 = -2x^2 + 4x + 9 \Rightarrow$
$0 = 9x^2 - 9x - 18 = 9(x^2 - x - 2)$
$= 9(x - 2)(x + 1)$

and thus the intersection points are where $x = -1$ and $x = 2$. If $x = -1$, then $y = 3$ and if $x = 2$, then $y = 9$ (by substitution into either original equation). The graph confirming this information is given in the accompanying diagram.

b.

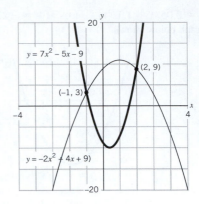

29. a. $1500 = 4W + 2L$ and thus
$L = (1500 - 4W)/2 = 750 - 2W$

b. $A(W) = W(750 - 2W) = 2W(375 - W)$

c. Domain for $A(W)$ is $0 < W < 375$

d. The area is largest when the W value is at the vertex of the parabola graph of $A(W)$, namely, when $W = 187.5$ ft. At that point $L = 375$ ft. and thus the area of the largest rectangle is $70{,}312.5$ sq. ft. See the accompanying graph for verification.

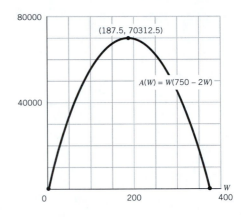

Section 8.4

Algebra Aerobics 8.4

1.

x	y	Average Rate of Change	Average Rate of Change of Average Rate of Change
-1	4	n.a.	n.a.
0	5	$\dfrac{5 - 4}{0 - (-1)} = 1$	n.a.
1	4	$\dfrac{4 - 5}{1 - 0} = -1$	$\dfrac{(-1)-1}{1 - 0} = -2$
2	1	$\dfrac{1 - 4}{2 - 1} = -3$	$\dfrac{(-3) - (-1)}{2 - 1} = -2$
3	-4	$\dfrac{(-4)-1}{3 - 2} = -5$	$\dfrac{(-5) - (-3)}{3 - 2} = -2$
4	-11	$\dfrac{(-11) - (-4)}{4 - 3} = -7$	$\dfrac{(-7) - (-5)}{4 - 3} = -2$

b. The third column tells us that the average rate of change of y with respect to x is decreasing at a constant rate, so the relationship is linear. The fourth column tells us that the average rate of change of the average rate of change is constant at -2.

2. a. The function is quadratic since the average rate of change is linear, that is, the average rate of change is increasing at a constant rate.

b. The function is linear since the average rate of change is constant.

c. The function is exponential since both the average rate of change and the average rate of change of the average rate of change are exponential, that is, are multiplied by a factor of 2 or increasing at a constant percentage.

3. a. The slope of the average rate of change $2a = 2$, giving $y = 2t + b$. The vertical intercept $b = 1$, so $y = 2t + 1$ is the equation of the average rate of change.

b. The slope of the average rate of change $2a = 6$, giving $y = 6x + b$. The vertical intercept $b = 5$, so $y = 6x + 5$ is the equation of the average rate of change.

c. The slope of the average rate of change is 10, giving $y = 10x + b$. The vertical intercept $b = 2$, so $y = 10x + 2$ is the equation of the average rate of change.

Exercises for Section 8.4

1. a. linear   **b.** positive, negative

3. a. $y = 3$; $y = -2$; $y = a$. It is horizontal line with slope 0 going through the point $(0, a)$.

b. It is a straight line with slope $= 2a$ and y-intercept b. No, the slope of linear function is constant. The slope of the quadratic is a linear function.

c. Guesses will vary. You may guess by analogy from the answers to parts (a) and (b) that its function is the quadratic $y = 3ax^2 + 2bx + c$.

5. The table and graph are given below:

t	Q	Average Rate of Change
-3	16	n.a.
-2	7	-9
-1	2	-5
0	1	-1
1	4	3
2	11	7
3	22	11

a. The function Q is quadratic. Its graph is given in the accompanying diagram.

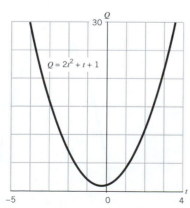

$Q = 2t^2 + t + 1$

b. The third column indicates that the average rate of change of Q is linear. For each increase of 1 in t it goes up by 4.

7. Graph A goes with Graph F   Graph B goes with Graph E
Graph C goes with Graph D

9. a. $F(t) = 6t + 1$ **b.** $G(x) = -10x + 0.4$ **c.** $H(z) = 2$

11. a. quadratic   **b.** linear   **c.** exponential

Section 8.5

Algebra Aerobics 8.5a

1. a. Degree is 5: $f(-1) = 11(-1)^5 + 4(-1)^3 - 11 =$
$11(-1) + 4(-1) - 11 = -11 - 4 - 11 = -26$

b. Degree is 4: When $x = -1$, $y = 1 + 7(-1)^4 - 5(-1)^3$
$= 1 + 7(1) - 5(-1)$
$= 1 + 7 + 5 = 13$

c. Degree is 4: $g(-1) = -2(-1)^4 - 20 = -2 - 20 = -22$

d. Degree is 2: When $x = -1$, $z = 3(-1) - 4 - 2(-1)^2$
$= -3 - 4 - 2 = -9$

2. a. degree $n = 5 + 3 + 2 = 10$

b. degree $n = 2 + 12 = 14$

3. a. degree 5; a quintic polynomial function

b. The leading term is $-2t^5$.

c. The constant term is 0.5.

d. $f(0) = 0.5 - 2(0)^5 + 4(0)^3 - 6(0)^2 - (0) = 0.5$
$f(0.5) = 0.5 - 2(0.5)^5 + 4(0.5)^3 - 6(0.5)^2 - (0.5) = -1.0625$
$f(-1) = 0.5 - 2(-1)^5 + 4(-1)^3 - 6(-1)^2 - (-1) = -6.5$

Algebra Aerobics 8.5b

1.

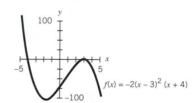

$f(x) = -2(x-3)^2(x+4)$

a. degree 3

b. two turning points

c. as $x \to +\infty$, $y \to -\infty$; and as $x \to -\infty$, $y \to +\infty$

d. $y = -2x^3$

e. The horizontal intercepts are $(3, 0)$, where the graph "touches" the x-axis and $(-4, 0)$, where the graph crosses the x-axis.

f. $y = -72$; the vertical intercept

2. Graph A: Minimum degree $= 3$ because two "bumps"; positive leading coefficient

Graph B: Minimum degree $= 4$ because three "bumps"; negative leading coefficient

Graph C: Minimum degree $= 5$ because four "bumps"; negative leading coefficient

Ch. 8

3. a. The y-intercept is at -3. The graph crosses the x-axis only once. It happens at about $x = 1.3$.

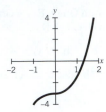

b. The y-intercept is at 3. The graph does not intersect the x-axis.

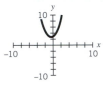

4. a. Degree is 1. x-intercept is -2.

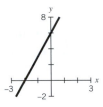

b. Degree is 2. x-intercepts are -4 and 1.

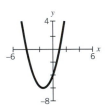

c. Degree is 3. x-intercepts are $-5, 3$, and $-2\frac{1}{2}$.

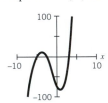

5. There are infinitely many examples of such functions. To have exactly those four x-intercepts, the functions are of the form:
$f(x) = ax^n(x + 3)^m(x - 5)^p(x - 7)^r$, where a is a real number, and n, m, p and r are positive integers.

(i) $f(x) = x(x + 3)(x - 5)(x - 7)$

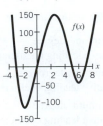

(ii) $h(x) = 2x^2(x + 3)(x - 5)(x - 7)$

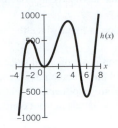

(iii) $g(x) = -20x^2(x + 3)(x - 5)(x - 7)$

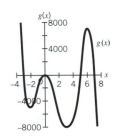

Exercises for Section 8.5

1. a. polynomial, 1
    d. not polynomial

    b. polynomial, 3
    e. polynomial, 5

    c. not polynomial
    f. polynomial, 3

3. a. $\frac{1}{8}, -\frac{1}{8}$
    c. $-\frac{1}{2}, \frac{1}{2}$

    b. $\frac{1}{2}, -\frac{1}{2}$
    d. $-32, 32$

5. a. goes with Graph A

    b. goes with Graph C

    c. goes with Graph B

7. For Graph A: **i.** 2   **ii.** 2   **iii.** minus   **iv.** 3

    For Graph B: **i.** 3   **ii.** 2   **iii.** minus   **iv.** 4

    For Graph C: **i.** 4   **ii.** 5   **iii.** plus   **iv.** 5

9. (a) and (e) are cubics; (b), (d), and (f) are quartics; (c) is by itself, since it is the only quintic.

11. a. Always negative; could have up to three turning points; has exactly one turning point (see accompanying graph).

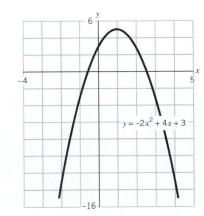

b. Always positive; in general a quartic has at most three turning points; here exactly one turning point (see accompanying graph).

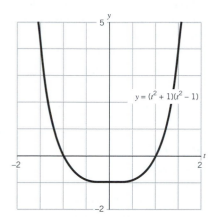

c. Negative if x is negative; positive if x is positive; at most two turning points; here none (see the accompanying graph).

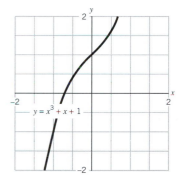

d. Positive if x is positive; negative if x is negative; in general a quintic has at most four turning points; here there are exactly four (see the accompanying graph).

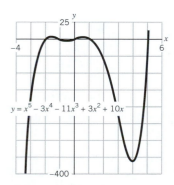

13. a. A quadratic has at most two horizontal intercepts; from the accompanying graph we see that there are two, at $x \approx -0.6$ and 2.6.

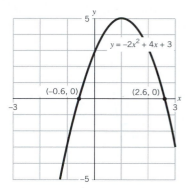

b. A quartic at most four horizontal intercepts; from the accompanying graph we see that it has two: at $t = -1$ and 1.

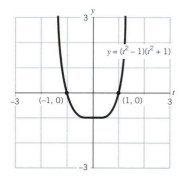

c. A cubic has at most three horizontal intercepts; from the accompanying graph we see that there is only one, at $x \approx 0.682$.

d. A quintic has at most five horizontal intercepts; from the accompanying graph we see that there are five, at $x = -2$, -1, 0, 1, and 5; in fact we have that $y = (x + 2)(x + 1)x(x - 1)(x - 5)$.

Graph for (c)   Graph for (d)

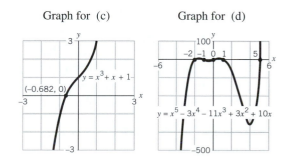

15. a. $y = 3x^3 - 2x^2 - 3$ has only one real zero at $x \approx 1.28$. (See the accompanying graph.)

b. $y = x^2 + 5x + 3$ has two real zeros:

$$x = -2.5 \pm 0.5\sqrt{13} \approx -4.302 \text{ and } -0.697.$$

(See the accompanying graph.)

Ch. 8

Graph for (a)   Graph for (b)

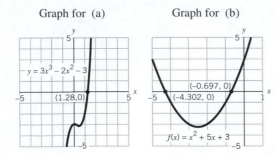

17. a. The y-values of the polynomial change sign when going from very large negative values of x to very large positive values of x and thus the graph must cross the x-axis.

b. $y = x^2 + 1$ and $y = x^4 + 1$ are such polynomials, and their graphs are both in the accompanying graph.

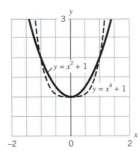

c. i. The polynomial $y = 3x^3 - 2x^2 - 3$ has exactly one real zero and its graph is given in the Solution to Exercise 15(a).

ii. The polynomial $y = (x - 1)x(x + 1)$ has three real zeros at $x = -1, 0,$ and 1, as can be seen in the following graph on the left.

d. The polynomial $y = (t^2 - 1)(t^2 + 1)$ has exactly two real zeros and its graph is shown on the right.

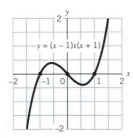
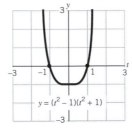

19. a. True; it is a function and 0 is in its domain.

b. True; it is a cubic.

c. True.

d. False; it is d and d can be positive, negative, or 0.

e. True, since the cubic term dominates for large positive or negative values of x.

f. False, the origin is a point on the graph if and only if $d = 0$.

21. a. $8701 = 8n^3 + 7n^2 + 0n + 1n^0$

b. $239 = 2n^2 + 3n + 9n^0$

c. The number written in base 2 as 11001 evaluates to 25 when written in base 10 notation, since $(1 \cdot 2^4) + (1 \cdot 2^3) + (0 \cdot 2^2) + (0 \cdot 2^1) + (1 \cdot 2^0) = 25.$

d. Here is one way to find the base two equivalent: find the highest power of 2 in 35. This is $2^5 = 32$. Subtracting that from 35 leaves 3, which is easily written as $2 + 1$. Thus 35, in base 10, can be written as

$(1 \cdot 2^5) + (0 \cdot 2^4) + (0 \cdot 2^3) + (0 \cdot 2^2) + (1 \cdot 2^1) + (1 \cdot 2^0)$

and this is 100011 in base 2.

Section 8.6

Algebra Aerobics 8.6

1. a. $-f(x)$ matches the graph of $g(x)$ in Graph B because it is a reflection of f across the x-axis.

b. $f(-x)$ matches the graph of $h(x)$ in Graph C because it is a reflection of f across the y-axis.

c. $-f(-x)$ matches the graph of $j(x)$ in Graph A because it is a double reflection of f across both the x- and y-axes.

2. $f(x) = 2x - 3$

a. $f(x + 2) = 2(x + 2) - 3 \Rightarrow f(x + 2) = 2x + 1$

b. $\frac{1}{2}f(x) = \frac{1}{2}(2x - 3) \Rightarrow \frac{1}{2}f(x) = x - \frac{3}{2}$

c. $-f(x) = -(2x - 3) \Rightarrow -f(x) = -2x + 3$

d. $f(-x) = 2(-x) - 3 \Rightarrow f(-x) = -2x - 3$

e. $-f(-x) = -(2(-x) - 3) \Rightarrow$
$-f(-x) = -(-2x - 3) = 2x + 3$

$f(x) = 1.5^x$

a. $f(x + 2) = 1.5^{x+2} \Rightarrow f(x + 2) = 1.5^2 \cdot 1.5^x$ or $2.25(1.5)^x$

b. $\frac{1}{2}f(x) = \frac{1}{2}(1.5^x)$

c. $-f(x) = -(1.5^x)$

d. $f(-x) = 1.5^{-x} \Rightarrow f(-x) = \frac{1}{1.5^x}$

e. $-f(-x) = -(1.5^{-x}) \Rightarrow -f(-x) = \frac{-1}{1.5^x}$

3. Graph B is symmetric across the vertical axis; Graph A is symmetric across the horizontal axis; Graph C is symmetric about the origin.

4. a. $h(t - 2) = e^{t-2}$

b. $-h(t - 2) = -e^{t-2}$

c. $-h(t - 2) - 1 = -e^{t-2} - 1$

5. a. $Q(t + 2) = 2 \cdot 1.06^{t+2}$

b. $Q(t + 2) - 1 = 2 \cdot 1.06^{t+2} - 1$

c. $-(Q(t + 2) - 1) = -(2 \cdot 1.06^{t+2} - 1)$
$\Rightarrow -(Q(t + 2) - 1) = -2 \cdot 1.06^{t+2} + 1$

6. a. $g(x)$ is a reflection of $f(x)$ across the x-axis, since $g(x) = x^2 - 5 = -(5 - x^2) \Rightarrow g(x) = -f(x).$

b. $g(x)$ is a reflection of $f(x)$ across the y-axis, since $f(x) = 3 \cdot 2^x$ and $f(-x) = 3 \cdot 2^{-x} \Rightarrow f(-x) = g(x).$

7. a. $g(x)$ is a compression of $f(x)$ by a factor of $\frac{1}{3}$ since $g(x) = \frac{1}{3x - 6} = \frac{1}{3(x - 2)} = \frac{1}{3} \cdot \frac{1}{x - 2} = \frac{1}{3}f(x).$

b. $g(x)$ is a vertical shift up of $f(x)$ by $\ln 3$ units since $g(x) = \ln 3x = \ln 3 + \ln x \Rightarrow g(x) = \ln 3 + f(x)$.

Exercises for Section 8.6

1. a. g's graph is a reflection of f's graph across the x-axis followed by a stretching of the graph by a factor of 2. Thus $g(x) = -2 \cdot f(x) = -2\sqrt{x}$.

b. g's graph is the graph of $f(x)$ shifted two units to the right. Thus $g(x) = f(x - 2) = e^{x-2}$.

c. g's graph is the graph of $f(x)$ shifted three units to the left. Thus $g(x) = f(x + 3) = \ln(x + 3)$.

3. a. i. $f(-x) = -x^3$; the original graph has been reflected across the y-axis.

ii. $-2f(x) - 1 = -2x^3 - 1$; the original graph has been first stretched by a factor of 2, then reflected across the x-axis and then lowered one unit in the y direction.

iii. $f(x + 2) = (x + 2)^3$; the graph has been shifted two units to the left along the x-axis.

iv. $-f(-x) = x^3$; the graph was reflected across both axes, but the original graph is symmetric with respect to the origin and, effectively, no visual change has occurred.

The graphs are in the diagrams below, each with the graph of the original f.

Graph for (i)	Graph for (ii)
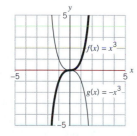	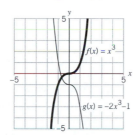
Graph for (iii)	Graph for (iv)
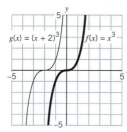	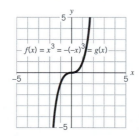

5. a. symmetric about the origin.

b. symmetric across the x-axis

c. symmetric across the y-axis

7. a. If $f(x) = a \cdot x^{2k}$, then $f(-x) = a(-x)^{2k} = ax^{2k} = f(x)$.

b. If $f(x) = a \cdot x^{2k+1}$, then $f(-x) = a(-x)^{2k+1} = -ax^{2k} = -f(x)$.

c. i. $f(-x) = (-x)^4 + (-x)^2 = x^4 + x^2 = f(x)$, so this is an even function.

ii. $u(-x) = (-x)^5 + (-x)^3 = (-x^5) + (-x^3) = -(x^5 + x^3) = -u(x)$, so this is an odd function.

iii. $h(-x) = (-x)^4 + (-x)^3 = x^4 - x^3 \neq h(x)$ and $\neq -h(x)$, and so $h(x)$ is neither even nor odd.

iv. $g(-x) = 10 \cdot 3^{-x} \neq g(x)$ and $\neq -g(x)$, and so $g(x)$ is neither even nor odd.

d. The graphs of even functions are symmetric across the y-axis and the graphs of odd functions are symmetric about the origin, as the graphs of the functions in (c)(i) and show.

9. a. $y = 20(0.5)^{x+2} - 5$

b. $y = 4(x + 2)^3 - 5$

c. $y = \log(x + 2)^{1/3} - 5$

11. a. $y = \dfrac{1}{(2 - x)^2} - 1$

b. $y = 0.5\dfrac{1}{(-t)^2} = \dfrac{0.5}{t^2}$

c. $y = \dfrac{1}{(\sqrt{s} - 3)^2} = \dfrac{1}{s - 3}$

d. $\dfrac{1}{(x + h)^2} - \dfrac{1}{x^2} = \dfrac{x^2}{x^2(x + h)^2} - \dfrac{(x + h)^2}{x^2(x + h)^2} = \dfrac{-2hx - h^2}{x^2(x + h)^2}$

13. a. $k(s - 2) = \dfrac{1}{s - 2}$

b. $\frac{1}{3}k(s - 2) = \dfrac{1}{3(s - 2)}$

c. $-\frac{1}{3}k(s - 2) = \dfrac{-1}{3(s - 2)}$

d. $j(s) = -\frac{1}{3}k(s - 2) + 4 \Rightarrow j(s) = \dfrac{-1}{3(s - 2)} + 4$

15. $f(x) = \ln x$

a. $f(x + 2) = \ln(x + 2)$

b. $\frac{1}{2}f(x) = \frac{1}{2}\ln x \Rightarrow \frac{1}{2}f(x) = \ln x^{1/2} \Rightarrow \frac{1}{2}f(x) = \ln \sqrt{x}$

c. $-f(x) = -\ln x \Rightarrow -f(x) = \ln x^{-1} \Rightarrow -f(x) = \ln\frac{1}{x}$

d. $f(-x) = \ln(-x)$

e. $-f(-x) = -\ln(-x) \Rightarrow -f(-x) = \ln(-x^{-1}) \Rightarrow -f(-x) = \ln\left(-\frac{1}{x}\right)$

$f(x) = \dfrac{1}{x^3}$

a. $f(x + 2) = \dfrac{1}{(x + 2)^3}$

b. $\frac{1}{2}f(x) = \dfrac{1}{2x^3}$

c. $-f(x) = \dfrac{-1}{x^3}$

d. $f(-x) = \dfrac{1}{(-x)^3} = -\dfrac{1}{x^3}$

e. $-f(-x) = -\left(\dfrac{1}{(-x)^3}\right) = \dfrac{1}{x^3}$

17. a. The "parent" function is $f(t) = t$. So $g(t)$ is $f(t) = t$ shifted to the right by 1 and compressed by a factor of 1/2 to get $g(t) = \frac{1}{2}f(t - 1)$.

b. The "parent" function is $f(t) = (1/2)^t$. So $g(t)$ is $f(t) = \left(\frac{1}{2}\right)^t$ shifted left by 4 and stretched by a factor of 3 to get $g(t) = 3f(t + 4)$.

c. The "parent" function is $f(t) = 1/t$. So $g(t)$ is $f(t) = \frac{1}{t}$ shifted to the right by 5, stretched by a factor of 7, reflected across the x-axis, and shifted down by 2 to get $g(t) = -7f(t - 5) - 2$.

19. a. $g(x) = 3\ln(x + 2) - 4$, so the graph of $f(x) = \ln x$ was shifted to the left by 2, with the result stretched by a factor of 3 and then shifted down by 4.

b. To find the vertical intercept, set $x = 0$ to get $g(0) = 3 \ln(0 + 2) - 4 \approx -1.92$. So the vertical intercept is at approximately $(0, -1.92)$.

To find any horizontal intercepts, set $g(x) = 0$, to get $0 = 3 \ln(x + 2) - 4 \Rightarrow 4/3 = \ln(x + 2) \Rightarrow e^{4/3} = x + 2 \Rightarrow x = e^{4/3} - 2 \approx 1.79$. So there is a single horizontal intercept at approximately $(1.79, 0)$.

21. a. If we let t = time (in hours) since the corpse was discovered and T = temperature of the corpse, then since the ambient temperature is $60°$ according to Newton's Law of Cooling, $T = 60 + Ce^{-kt}$ for some constants k and C. When $t = 0$, $T = 80$, so we have $80 = 60 + C \Rightarrow C = 20$. So the equation becomes $T = 60 + 20e^{-kt}$. When $t = 2$, $T = 75$, so we have $75 = 60 + 20e^{-2k} \Rightarrow (15/20) = e^{-2k} \Rightarrow (3/4) = e^{-2k} \Rightarrow \ln(3/4) = \ln(e^{-2k}) \Rightarrow -0.288 \approx -2k \Rightarrow k \approx 0.144$. So the full equation is $T = 60 + 20e^{-0.144t}$.

b. If we assume that the normal body temperature is $98.6°$, then to find the time of death we need to solve $98.6 = 60 + 20e^{-0.144t}$ for $t \Rightarrow (38.6/20) = e^{-0.144t} \Rightarrow \ln(38.6/20) = \ln(e^{-0.144t}) \Rightarrow 0.658 \approx -0.144t \Rightarrow t \approx -4.6$ hours. So the person died about 4.6 hours before the corpse was discovered.

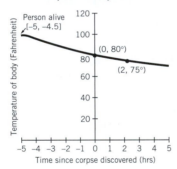

Temperature of Body Over Time

Section 8.7

Algebra Aerobics 8.7

1. a. $f(2) = 2^3 = 8$

b. $g(2) = 2(2) - 1 = 3$

c. $h(2) = \frac{1}{2}$

d. $(h \cdot g)(2) = h(2) \cdot g(2) = \frac{1}{2} \cdot 3 = \frac{3}{2}$

e. $(f + g)(2) = f(2) + g(2) = 8 + 3 = 11$

f. $\left(\frac{h}{g}\right)(2) = \frac{h(2)}{g(2)} = \frac{1/2}{3} = \frac{1}{6}$

2. a. $Q(1) + P(1) = 7 + 3 = 10$

b. $Q(2) - P(2) = 9 - 24 = -15$

c. $P(-1) \cdot Q(-1) = (-3 \cdot 3) = -9$

d. $\frac{Q(3)}{P(3)} = \frac{11}{81}$

3. a. $f(t) - h(t) = (3 - 2t) - (t^2 - 1) = -t^2 - 2t + 4$

b. $f(t) + h(t) = (3 - 2t) + (t^2 - 1) = t^2 - 2t + 2$

c. $f(t) \cdot h(t) = (3 - 2t)(t^2 - 1) = -2t^3 + 3t^2 + 2t - 3$

d. $\frac{h(t)}{f(t)} = \frac{t^2 - 1}{3 - 2t}$

4. a. $f(x + 1) = (x + 1)^2 + 2(x + 1) - 3 = x^2 + 4x$

b. $f(x) + 1 = x^2 + 2x - 3 + 1 = x^2 + 2x - 2$

c. $g(x + 1) = \frac{1}{(x + 1) - 1} = \frac{1}{x}$

d. $g(x) + 1 = \frac{1}{x - 1} + 1 = \frac{1}{x - 1} + \frac{x - 1}{x - 1} = \frac{x}{x - 1}$

5. a. horizontal intercepts at $x = 1$ and $x = -5$; vertical asymptote at $x = -3$

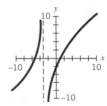

b. horizontal intercept at $x = -2/3$ and vertical asymptotes at $x = -1$ and $x = 3$

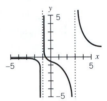

6. a. $f(r) = \frac{r^2 - 4r - 12}{r^2 - 4r + 3} = \frac{(r + 2)(r - 6)}{(r - 1)(r - 3)}$

To find any horizontal intercepts, set the numerator $= 0$ to get $(r + 2)(r - 6) = 0 \Rightarrow r = -2$ or $r = 6$. So the horizontal intercepts are at $(-2, 0)$, $(6, 0)$.

To find any vertical asymptotes, set the denominator $= 0$ to get $(r - 1)(r - 3) = 0 \Rightarrow r = 1$ or $r = 3$. So $f(r)$ is not defined at $r = 1$ and $r = 3$. So there are two vertical asymptotes at the lines $r = 1$ and $r = 3$. The graph of $f(r)$ follows.

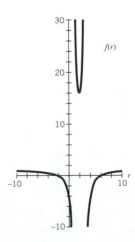

b. $g(r) = \frac{r^2 - 4r + 3}{r^2 - 4r - 12} = \frac{(r - 1)(r - 3)}{(r + 2)(r - 6)}$

To find any horizontal intercepts, set the numerator $= 0$ to get $(r - 1)(r - 3) = 0 \Rightarrow r = 1$ or $r = 3$. So the horizontal intercepts are at $(1, 0)$, $(3, 0)$.

To find any vertical asymptotes, set the denominator $= 0$ to get $(r + 2)(r - 6) = 0 \Rightarrow r = -2$ or $r = 6$. So $g(r)$ is not defined at $r = -2$ and $r = 6$. So there are two vertical asymptotes at the lines $r = -2$ and $r = 6$. The graph of $g(r)$ follows.

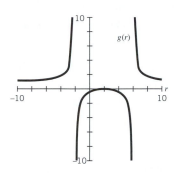

7. a. $g(x) = -\dfrac{1}{(x + 3)^2} + 1$

b. $g(x) = -\dfrac{1}{(x + 3)^2} + 1 \cdot \dfrac{(x + 3)^2}{(x + 3)^2}$

$= \dfrac{-1}{(x + 3)^2} + \dfrac{x^2 + 6x + 9}{(x + 3)^2}$

$= \dfrac{-1 + x^2 + 6x + 9}{x^2 + 6x + 9}$

$= \dfrac{x^2 + 6x + 8}{x^2 + 6x + 9} = \dfrac{p(x)}{q(x)}$

c. To find any horizontal intercepts, let the numerator $p(x) = 0$ to get $x^2 + 6x + 8 = (x + 4)(x + 2) = 0 \Rightarrow x = -4$ or $x = -2$. So the horizontal intercepts are at $(-4, 0)$ and $(-2, 0)$.

To find any vertical asymptotes, let the denominator $q(x) = 0$ to get $(x + 3)^2 = 0 \Rightarrow x = -3$. So $g(x)$ is not defined when $x = -3$. So there is one vertical asymptote at the line $x = -3$.

8. a. Horizontal intercepts   **c.** Horizontal intercept
b. Vertical intercepts   **d.** Vertical asymptote

9. $f(x) = \dfrac{2x + 6}{x - 3}$

i. To find horizontal intercept(s): set $f(x) = 0$, which is equivalent to setting the numerator $2x + 6 = 0 \Rightarrow x = -3$. So there is one horizontal intercept at $(-3, 0)$.

To find vertical intercept: evaluate $f(0) = 6/-3 = -2$. So the vertical intercept is at $(0, -2)$.

ii. To find vertical asymptote: set $x - 3 = 0 \Rightarrow x = 3$. So the vertical asymptote is the line at $x = 3$.

iii. The end behavior as $x \to \pm\infty$ is $f(x) \to \dfrac{2x}{x} = 2$. So there is a horizontal asymptote at the line $y = 2$.

iv. Graph of $f(x)$

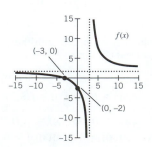

10. $g(x) = \dfrac{x^2 + 2x - 3}{3x - 1}$

i. To find horizontal intercept(s): set $g(x) = 0$, which is equivalent to setting $x^2 + 2x - 3 = 0 \Rightarrow (x - 1)(x + 3) = 0 \Rightarrow x = 1, x = -3$. So there are horizontal intercepts at $(1, 0)$ and $(-3, 0)$.

To find vertical intercept: evaluate $g(0) = 3$. So the vertical intercept is at $(0, 3)$.

ii. To find vertical asymptote: set $3x - 1 = 0 \Rightarrow x = 1/3$. So the vertical asymptote is the vertical line at $x = 1/3$.

iii. The end behavior as $x \to \pm\infty$ is $g(x) \Rightarrow \dfrac{x^2}{3x} = \dfrac{1}{3}x$. So $g(x)$ is asymptotic to the line $y = \dfrac{1}{3}x$ (called an oblique asymptote).

iv. Graph of $g(x)$

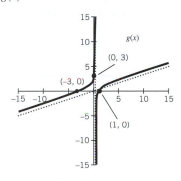

Exercises for Section 8.7

1. a. $3t^2 + 10t - 4$ **c.** $18t^3 + 27t^2 - 26t - 5$

   b. $-3t^2 + 2t + 6$ **d.** $\dfrac{3t^2 + 4t - 5}{6t + 1}$

3. a. $j(x) = 3x^5 + x^2 + x - 1$; $k(x) = 3x^5 - x^2 + x + 1$; $l(x) = 3x^7 - 3x^5 + x^3 - x$

   b. $j(2) = 101$; $k(3) = 724$; $l(-1) = 0$

5. a. $R(n) = 25n$

   b. $R(n) = 25n - 500$

   c. $R(n) = 25(n - 30) - 500 = 25n - 1250$

7. If a worker works t hours a week (where $t \geq 40$), then the worker's weekly paycheck, P (in dollars), is the sum of two terms: regular pay + overtime pay. The regular pay is $20 \cdot 40 = \$800$ a week. The overtime pay $= 30 \cdot (t - 40)$, where $t =$ total number of hours worked. So the weekly paycheck is $P = 800 + 30(t - 40)$, where $t \geq 40$.

9.

x	0	1	2	3	4	5
$h(x)$	-6	-6	-6	-6	-6	-6
$j(x)$	0	-4	-16	-36	-64	-100
$k(x)$	9	5	-55	-315	-1015	-2491

11.

x	-3	-2	-1	0	1	2	3
$f(x)$	9	4	1	0	1	4	9
$g(x)$	-4	-3	-2	-1	0	1	2
$f(x) + g(x)$	5	1	-1	-1	1	5	11
$f(x) - g(x)$	13	7	3	1	1	3	7
$f(x) \cdot g(x)$	-36	-12	-2	0	0	4	18
$g(x)/f(x)$	$-4/9$	$-3/4$	-2	undefined	0	1/4	2/9

13. **a.** $C(n) = 500 + 40n$

   b. $P(n) = \frac{500 + 40n}{n}$

   c. $P(25) = 60$; $P(100) = 45$. As the number of people attending increases, the cost per person decreases. If only 25 attend (the minimum size), the cost would be $60 per person. If 100 attend (the maximum size), the cost would be $45 per person.

15. Graph A, $f(x)$ asymptotes: horizontal, $y = -3$; vertical, $x = -2$

   Graph B, $g(x)$ asymptotes: horizontal, $y = 4$; vertical, $x = 2$

   Graph C, $h(x)$ asymptotes: horizontal, $y = -2$; vertical, $x = -2$ and $x = 3$

17. $g(x)$ has no horizontal intercepts and a vertical asymptote at the line $x = -3$. The graph of $g(x)$ verifies this.

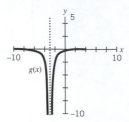

19. **a.** The domain is all real numbers.

   b. Yes, there is one horizontal intercept at $x = 0$, the origin. There are no vertical asymptotes since x is defined for all real values (so there are no singularities).

   c. As $x \to \pm\infty$, $S(x) = \frac{x}{x^2 + 1} \approx \frac{x}{x^2} = \frac{1}{x}$. So as $x \to \pm\infty$, $S(x) \to 0$. So $S(x)$ has a horizontal asymptote at the x-axis.

   d. To some the graph looks like a slithering serpent.

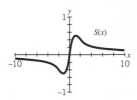

21. There are infinitely many possibilities. If a and b are nonzero real numbers, then one set of answers would be of the form $f(x) = \frac{a(x + 3)(x - 4)}{b(x - 1)(x + 5)}$. (*Note:* You could create additional rational functions by, say, squaring or cubing one or more terms.) If we let $a = b = 1$, then the graph of $f(x)$ is as follows:

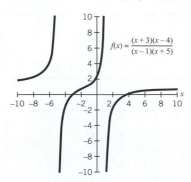

23. $f(x)$ matches Graph C; $g(x)$ matches Graph B; $h(x)$ matches Graph A.

Section 8.8

Algebra Aerobics 8.8a

1. $f(x) = 2x + 3$, $g(x) = x^2 - 4$

   a. $f(g(2)) = f(0) = 2(0) + 3 = 3$

   b. $g(f(2)) = g(7) = (7)^2 - 4 = 49 - 4 = 45$

   c. $f(g(3)) = f(5) = 2(5) + 3 = 10 + 3 = 13$

   d. $f(f(3)) = f(9) = 2(9) + 3 = 18 + 3 = 21$

   e. $(f \circ g)(x) = f(g(x)) = f(x^2 - 4) = 2(x^2 - 4) + 3$
   $$= 2x^2 - 8 + 3 = 2x^2 - 5$$

   f. $(g \circ f)(x) = g(f(x)) = g(2x + 3) = (2x + 3)^2 - 4$
   $$= 4x^2 + 12x + 9 - 4 = 4x^2 + 12x + 5$$

2. **a.** $(P \circ Q)(2) = P(Q(2)) = P(3(2) - 5) = P(1) = \frac{1}{1} = 1$

   b. $(Q \circ P)(2) = Q(P(2)) = Q(\frac{1}{2}) = 3(\frac{1}{2}) - 5 = \frac{-7}{2}$

   c. $(Q \circ Q)(3) = Q(Q(3)) = Q(3(3) - 5) = Q(4)$
   $$= 3(4) - 5 = 7$$

   d. $(P \circ Q)(t) = P(Q(t)) = P(3t - 5) = \frac{1}{3t - 5}$

   e. $(Q \circ P)(t) = Q(P(t)) = Q(\frac{1}{t}) = 3(\frac{1}{t}) - 5 = \frac{3}{t} - 5$

3. $F(x) = \frac{2}{x - 1}$, $G(x) = 3x - 5$

   a. $(F \circ G)(x) = F(G(x)) = F(3x - 5) = \frac{2}{(3x - 5) - 1}$
   $$= \frac{2}{3x - 6}$$

   b. $(G \circ F)(x) = G(F(x)) = G(\frac{2}{x - 1}) = 3(\frac{2}{x - 1}) - 5$
   $$= \frac{6}{x - 1} - 5$$

   c. $(F \circ G)(x) = \frac{2}{3x - 6} \neq (G \circ F)(x) = \frac{6}{x - 1} - 5 =$
   $$\frac{6 - 5(x - 1)}{x - 1} = \frac{11 - 5x}{x - 1}$$

4. **a.** $f(-2) = 2$ **e.** $(g \circ f)(-2) = g(2) = 2$

   b. $g(-2) = 0$ **f.** $(f \circ g)(-2) = f(0) = -2$

   c. $f(0) = -2$ **g.** $(g \circ f)(0) = g(-2) = 0$

   d. $g(0) = 1$ **h.** $(f \circ g)(0) = f(1) = -1$

5. **a.** $f(x) = x^2 - 2$

   b. $g(x) = \frac{1}{2}x + 1$

   c. $(g \circ f)(x) = g(x^2 - 2) = \frac{1}{2}(x^2 - 2) + 1$
   $$= \frac{1}{2}x^2 - 1 + 1 = \frac{1}{2}x^2$$

   d. $(f \circ g)(x) = f(\frac{1}{2}x + 1) = (\frac{1}{2}x + 1)^2 - 2$
   $$= (\frac{1}{4}x^2 + x + 1) - 2 = \frac{1}{4}x^2 + x - 1$$

6. **a.** $(g \circ f)(-2) = \frac{1}{2}(-2)^2 = 2$ and

   $(f \circ g)(-2) = \frac{1}{4}(-2)^2 + (-2) - 1 = -2$.

   So both answers agree with the answers in Problem 4, parts (e) and (f).

7. **a.** $(h \circ f \circ g)(4) = h(f(g(4))) = h(f(0)) = h(3) = 4$

   b. $(f \circ h \circ g)(1) = f(h(g(1))) = f(h(3)) = f(4) = 5$

8. **a.** $(h \circ f \circ g)(3) = h(f(g(3))) = h(f(2)) = h(-1/3) = 3$

   b. $(f \circ g \circ h)(100) = f(g(h(100))) = f(g(3)) = f(2) = -1/3$

Algebra Aerobics 8.8b

1. a.

t	$g(t)$
0	5
1	3
2	1
3	−1

t	$h(t)$
−1	3
1	2
3	1
5	0

b. $(g \circ h)(3) = g(h(3)) = g(1) = 3$; $(h \circ g)(3) = h(g(3)) = h(-1) = 3$.

c. Yes, since $(g \circ h)(t) = g(h(t)) = g\left(\frac{5-t}{2}\right) = 5 - 2\left(\frac{5-t}{2}\right) = 5 - (5 - t) = t$ and $(h \circ g)(t) = h(g(t)) = h(5 - 2t) = \frac{5 - (5 - 2t)}{2} = t$ and the domains and ranges of both g and h are all real numbers.

d. They are inverse functions.

2. $f(x) = 2x + 1$, $g(x) = \frac{x-1}{2}$

$(f \circ g)(x) = f(g(x)) = f\left(\frac{x-1}{2}\right) = 2\left(\frac{x-1}{2}\right) + 1$

$\quad = \frac{2(x-1)}{2} + 1 = x - 1 + 1 = x$

$(g \circ f)(x) = g(f(x)) = g(2x + 1) = \frac{(2x+1)-1}{2}$

$\quad = \frac{2x}{2} = x$

3. $f(x) = \sqrt[3]{x} + 1$, $g(x) = x^3 - 1$

$(f \circ g)(x) = f(g(x)) = f(x^3 - 1) = \sqrt[3]{(x^3 - 1) + 1}$

$\quad = \sqrt[3]{x^3} = (x^3)^{1/3} = x^1 = x$

$(g \circ f)(x) = g(f(x)) = g(\sqrt[3]{x} + 1)$

$\quad = \left(\sqrt[3]{x} + 1\right)^3 - 1 = [(x + 1)^{1/3}]^3 - 1$

$\quad = (x + 1) - 1 = x$

4. $f(f^{-1}(x)) = f\left(\frac{1+x}{x}\right) = \frac{1}{\frac{1+x}{x} - 1} = \frac{1}{\frac{1+x}{x} - \frac{x}{x}}$

$\quad = \frac{1}{(1/x)} = x$

$f^{-1}(f(x)) = f^{-1}\left(\frac{1}{x-1}\right) = \frac{1 + \frac{1}{x-1}}{\frac{1}{x-1}} = \frac{\frac{x-1}{x-1} + \frac{1}{x-1}}{\frac{1}{x-1}}$

$\quad = \frac{\frac{x}{x-1}}{\frac{1}{x-1}} = \frac{x}{x-1} \cdot \frac{x-1}{1} = x$

So $f(f^{-1}(x)) = x$ and $f^{-1}(f(x)) = x$.

5. Letting $f(x) = y$, we have $y = \frac{3}{x} + 5 \Rightarrow y - 5 = \frac{3}{x} \Rightarrow x = \frac{3}{y-5} \Rightarrow f^{-1}(x) = \frac{3}{x-5}$ (using the convention of designating x as the input variable).

6. Letting $g(x) = y$, we have $y = (x - 2)^{3/2} \Rightarrow x = y^{2/3} + 2 \Rightarrow g^{-1}(x) = x^{2/3} + 2$ (using the convention of designating x as the input variable).

7. Letting $h(x) = y$, we have $y = 5x^3 - 4 \Rightarrow y + 4 = 5x^3 \Rightarrow x = \left(\frac{y+4}{5}\right)^{1/3} \Rightarrow h^{-1}(x) = \left(\frac{x+4}{5}\right)^{1/3}$ (using the convention of designating x as the input variable).

8. a. Saying "no"

b. Taking the bus from home, then going to class.

c. Turning off the light, leaving the room, closing the door, and then locking the door.

d. Dividing x by 5 and then adding 3

e. Subtracting 2 from z and then dividing the result by −3

9. The functions in Graphs A and D are 1-1 since they pass the horizontal line test. The functions in Graphs B and C are not 1-1, since they fail that test.

10. a. f is not one-to-one on the domain of all real numbers (since it fails the horizontal line test), so it cannot have an inverse.

b. If we restrict the domain of f to $x \geq -2$, then f has an inverse on this new domain (since it now passes the vertical test).

c. Letting $f(x) = y$, we have

$y = (x + 2)^2 \Rightarrow x = +\sqrt{y} - 2 \Rightarrow f^{-1}(x) = \sqrt{x} - 2$ (using the convention of designating x as the input variable). The graph of $f^{-1}(x)$ is shown below.

11. a. The radius $R(t)$ as a function of time t is $R(t) = 10t$.

b. $R(2) = 10 \cdot 2 = 20$, so area $A(20) = \pi(20)^2 \approx 1257$ square feet.

c. $(A \circ R)(t) = A(R(t)) \Rightarrow A(10t) = \pi(10t)^2$

$\quad = 100\pi t^2$

Exercises for Section 8.8

1. a. $f(g(1)) = f(0) = 2$

b. $g(f(1)) = g(1) = 0$

c. $f(g(0)) = f(1) = 1$

d. $g(f(0)) = g(2) = 3$

e. $f(f(2)) = f(3) = 0$

3. a. $g(f(2)) = g(0) = 1$

b. $f(g(-1)) = f(2) = 0$

c. $g(f(0)) = g(4) = -3$

d. $g(f(1)) = g(3) = -2$

5. a. $F(G(1)) = F(0) = 1$

b. $G(F(-2)) = G(-3) = 4$

c. $F(G(2)) = F(0.25) = 1.5$

d. $F(F(0)) = F(1) = 3$

e. $(F \circ G)(x) = 2\left(\frac{x-1}{x+2}\right) + 1$

$$= \frac{(2x-2)+(x+2)}{x+2} = \frac{3x}{x+2}$$

f. $(G \circ F)(x) = \frac{(2x+1)-1}{(2x+1)+2} = \frac{2x}{2x+3}$

7. a. $A(r) = \pi r^2$, where r is measured in feet and $A(r)$ is measured in square feet.

b. $r = R(t) = 5t$, where t is measured in minutes and $R(t)$ is measured in feet.

c. $A(R(t)) = \pi 25 t^2$, where t is measured in minutes and A is measured in square feet.

d. $A(R(10)) = \pi \cdot 25 \cdot 10^2 = 2500\pi \approx 7854$ sq. ft and $A(R(60)) = \pi \cdot 25 \cdot 60^2 \approx 282{,}743$ sq. ft.

9. $r(t) = 13t$ and thus $A(r(t)) = \pi(13t)^2 = 169\pi t^2$

11. a. $T = 32 - 5s$

b. If the road is 40 feet wide, then $k = 20$ and thus $S(x) = \left[1 - \frac{1}{2} \cdot \left(\frac{x}{20}\right)^2\right] S_d = \left[1 - \frac{x^2}{800}\right] S_d$.

c. At the middle of the 40-ft road $x = 0$ and therefore $S(0) = \left[1 - \frac{0}{800}\right] S_d = S_d$. At the edge of the 40-ft road, $x = 20$, so $S(20) = \left[1 - \frac{20^2}{800}\right] S_d = \frac{1}{2} S_d$.

d. $T(S(x)) = T\left(\left[1 - \frac{x^2}{800}\right] S_d\right) = 32 - 5\left[1 - \frac{x^2}{800}\right] S_d$

$$= 32 - 5 S_d + \frac{x^2}{160} S_d$$

e. $T(S(0)) = 32 - 5 S_d$ and $T(S(20)) = 32 - 5 S_d + \frac{20^2}{160} S_d = 32 - 5 S_d + 2.5 S_d = 32 - 2.5 S_d$

13. a. $M(x) = (L \circ J \circ K)(x) = L(J(K(x))) = L(J(\log x)) = L((\log x)^3) = 1/(\log x)^3$

b. Take the log of x, cube the result, and then place it in the denominator, with 1 as the numerator.

15. If $f(x) = 4x$, $g(x) = e^x$, and $h(x) = x - 1$, then $f(g(h(x))) = f(g(x-1)) = f(e^{x-1}) = 4e^{x-1} = j(x)$

17. $f(g(x)) = f(x^2 + 1) = \sqrt{(x^2+1) - 1} = \sqrt{x^2} = x$ since $x > 0$

$$g(f(x)) = g\left(\sqrt{x-1}\right) = \left(\sqrt{x-1}\right)^2 + 1$$
$$= (x-1) + 1 = x$$

19. $f(g(x)) = f\left(\frac{x^3 - 5}{4}\right) = \sqrt[3]{4\left(\frac{x^3 - 5}{4}\right) + 5} = \sqrt[3]{x^3} = x$

$$g(f(x)) = g\left(\sqrt[3]{4x+5}\right) = \frac{\left(\sqrt[3]{(4x+5)}\right)^3 - 5}{4}$$
$$= \frac{(4x+5) - 5}{4} = x$$

21. $F(G(t)) = F(\ln(t^{1/3})) = e^{3\ln(t^{1/3})} = e^{\ln(t)} = t$ (where $t > 0$)

$G(F(t)) = G(e^{3t}) = \ln(e^{3t})^{1/3} = \ln(e^t) = t$

23.

x	$f^{-1}(x)$
5	-2
1	-1
2	0
4	1

25. a. Yes, this is a 1-1 function since each letter is associated with a unique number. The inverse function would just consist of matching each number between 1 and 26 wih its

letter equivalent. The domain of the inverse function would be the integers 1 through 26.

b. "MATH RULES"

27. a. Yes, $f(x)$ has an inverse since its graph passes the horizontal line test.

b. The domain of $f(x)$ is the interval $[-4, \infty]$. The range of $f(x)$ is the interval $[0, \infty]$.

c. $f(-4) = 0$, $f(0) = 2$, and $f(5) = 3$. This means that the points $(-4, 0)$, $(0, 2)$, and $(5, 3)$ all lie on the graph of $f(x)$.

d. Given the results in part (c), the points $(0, -4)$, $(2, 0)$, and $(3, 5)$ all lie on the graph of $f^{-1}(x)$. So $f^{-1}(0) = -4$, $f^{-1}(2) = 0$, $f^{-1}(3) = 5$.

29. $Q^{-1}(x) = \frac{3x+15}{2}$, $Q(3) = -3$, $Q^{-1}(3) = 12$

31. $Q^{-1}(x) = \frac{3}{x-1}$, $Q(3) = 2$, $Q^{-1}(3) = 3/2$

33. a.

x (cups)	4	8	16	32
$f(x)$ (quarts)	1	2	4	8

x (quarts)	2	4	8	16
$g(x)$ (gallons)	0.5	1	2	4

b. **i.** $(g \circ f)(8) = g(2) = 0.5$ gal

  ii. $g^{-1}(2) = 8$ qt

  iii. $(f^{-1} \circ g^{-1})(1) = f^{-1}(4) = 16$ cups

  iv. $(f^{-1} \circ g^{-1})(2) = f^{-1}(8) = 32$ cups

c. $(f^{-1} \circ g^{-1})(x)$ is a function that converts gallons to cups.

35. a. $W_{\text{men}}(h) = 50 + 2.3(h - 60)$, where a reasonable domain might be $60'' \le h \le 78''$; $W_{\text{women}}(h) = 45.5 + 2.3(h - 60)$, where a reasonable domain might be $60'' \le h \le 74''$.

b. $W_{\text{men}}(70) = 50 + 2.3(70 - 60) = 73$ kg, so 73 kg is the "ideal" weight of a 5'10" man. $W_{\text{women}}(66) = 45.5 + 2.3(66 - 60) = 59.3$ kg, so 59.3 kg is the "ideal" weight of a 5'6" woman.

c. $W^{-1}_{\text{men}}(77.6)$ means $77.6 = 50 + 2.3(h - 60) \Rightarrow h = 72$ inches. A man with a IBW of 77.6 kg should be 72 inches or 6 ft tall.

d. $W_{\text{newman}}(h) = \frac{50 + 2.3(h - 60)}{0.4356}$; $W_{\text{newwomen}}(h) = \frac{45.5 + 2.3(h - 60)}{0.4356}$.

e. $W^{-1}_{\text{newwomen}}(125)$ means $125 = \frac{45.5 + 2.3(h - 60)}{0.4356} \Rightarrow h \approx 63.89 \approx 64$ inches. So 125 lb is the IBW for a woman about 5'4" in height.

37. $F(G(x)) = F(\log_a(x)) = a^{\log_a(x)} = x$

$G(F(x)) = G(a^x) = \log_a(a^x) = x$

Ch. 8: Check Your Understanding

1. False	**6.** True	**11.** False	**16.** False
2. True	**7.** True	**12.** False	**17.** False
3. True	**8.** False	**13.** False	**18.** True
4. False	**9.** True	**14.** True	**19.** True
5. False	**10.** True	**15.** False	**20.** True

21. True

22. Possible answer: $y = x^4 + x^2 + 1$

23. $y = -0.25x^2$

24. Possible answer: $f(x) = (x + 1)(x - 3)(x - 4)$

25. Possible answers: $h(x) = 2(x + 1)(x - 3)(x - 4)$, $g(x) = -3(x + 1)(x - 3)(x - 4)$

26. Possible answer: $y = -(x - 1)^2 + 3$

27. Possible answer: $y = 2(x - 3)^2 - 5$

28. $y = \frac{-1}{2}(x - 2)(x + 2)$

29. $r = s^2 - s + 5$

30. $G(x) = (x + 2)^2 + 2(x + 2)$

31. $-h(t) = -(t - 2)^2$

32. Possible answer: $y = x^2$ and $y = -(x - 1)^2 + 1$

33. Possible answer: $y = (x + 4)^2$

34. $h(t) = \frac{1}{4}(x + 2)(x + 1)(x - 2)(x - 3)$

35. Possible answer: $f(x) = x^3 + 2x^2$, $g(x) = 5x - 2$

36. $H(t) = 3t + 1$, $Q(t) = \sqrt{t}$

37. Possible answer: $f(x) = \frac{x - 2}{x(x + 3)}$

38. False	**44.** False	**50.** True	**56.** True
39. True	**45.** True	**51.** False	**57.** False
40. False	**46.** False	**52.** True	**58.** False
41. True	**47.** False	**53.** False	
42. True	**48.** True	**54.** True	
43. False	**49.** True	**55.** False	

Ch. 8 Review: Putting It All Together

1. In Graph A, the parabola is concave up with an estimated minimum at (2, –4). Hence the axis of symmetry is the line $x = 2$ and there are two horizontal intercepts, at $x = 0$ and $x = 4$.

In Graph B, the parabola is concave down with an estimated maximum at (0, −3). Hence the axis of symmetry is the vertical axis (the line $t = 0$) and there are no horizontal intercepts.

3. a. Area of interior square $= x^2$ square inches; area of each of the maple strips $= (x + 1) \cdot 1 = x + 1$ square inches.

b. Cost of white oak: ($2.39/ft²) · (1 ft²/144 in²) ≈ $0.02/in²; cost of maple: ($4.49/ft²) · (1 ft²/144 in²) ≈ $0.03/in²

c. For white oak: $0.02x^2$ (in dollars); for all four maple strips: $4(0.03)(x + 1) = 0.12x + 0.12$ (in dollars)

d. $C(x) = 0.02x^2 + 0.12x + 5.12$, a quadratic function

e.

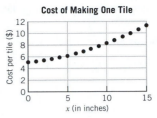

Cost of Making One Tile

f. Estimating from the graph, when $C(x) = \$7$, then $x \approx 7.5''$, so the width (and length) of the whole tile would be about 9.5''. To keep the cost/tile at $7 or below, the dimensions of a tile must be at most 9.5'' by 9.5''.

5. $g(x) = 2x^2$ and $h(x) = -0.5x^2$

7. a. Vertex for $F(x)$ is (0, 0), vertex for $G(x) = (0, 5)$, vertex for $H(x)$ is $(-2, 0)$, vertex for $J(x)$ is $(1, -5)$

b.

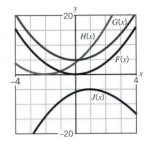

c. The graph of $G(x)$ is the graph of $F(x)$ shifted up five units. The graph of $H(x)$ is the graph of $F(x)$ shifted left two units. The graph of $J(x)$ is the graph of $F(x)$ shifted right one unit, flipped over the x-axis, and shifted down five units.

9. a. One possibility is $Q(t) = (t - 4)(t + 2) = t^2 - 2t - 8$. The vertex of $Q(t)$ is at $(1, -9)$.

b. One possibility is $M(t) = 3Q(t) = 3(t - 4)(t + 2) = 3t^2 - 6t - 24$. The vertex of $M(t)$ is at $(1, -27)$. So the vertices are not the same, though they share the same t-coordinate.

c. One possibility is $P(t) = Q(t) + 10 = t^2 - 2t + 2$. $P(t)$ has no horizontal intercepts since the discriminant $= (-2)^2 - (4 \cdot 1 \cdot 2) = 4 - 8 = -4$, which is negative.

11. Since we have set the vertex at the origin, then the equation is of the form $y = ax^2$ (where $a < 0$). We know two points on the parabola, $(d, -32)$ and $(-(100 - d), -72) = (d - 100, -72)$. Substituting each set of points into the equation $y = ax^2$, we get the two equations

$$-32 = ad^2 \quad \text{and} \quad -72 = a(d - 100)^2$$

Solving both equations for a, we get

$$a = -32/d^2 \quad \text{and} \quad a = -72/(d - 100)^2$$

Setting both expressions for a equal and solving for d gives us

$$-32/(d^2) = -72/(d - 100)^2$$

cross-multiply $\quad -32(d - 100)^2 = -72d^2$

simplify $\quad\quad\quad -32(d - 100)^2 + 72d^2 = 0$

$$-32(d^2 - 200d + 10,000) + 72d^2 = 0$$

$$40d^2 + 6400d - 320,000 = 0$$

divide by 40 $\quad\quad\quad\quad d^2 + 160d - 8000 = 0$

Ch. 8

Now we can solve for d either using the quadratic formula or factoring.

Quadratic formula (letting $a = 1, b = 160, c = -8000$) gives:

$$d = \frac{-160 \pm \sqrt{(160)^2 - 4(1)(-8000)}}{2 \cdot 1}$$

$$= \frac{-160 \pm \sqrt{25{,}600 + 32{,}000}}{2}$$

$$= \frac{-160 \pm \sqrt{57{,}600}}{2} = \frac{-160 \pm 240}{2} = -80 \pm 120$$

So $d = 40$ feet or -200 feet. Only $d = 40$ feet makes sense.

Factoring $d^2 + 160d - 8000 = 0$ gives $(d - 40)(d + 200) = 0$, which confirms that either $d = 40$ feet or -200 feet, where $d = 40$ feet is the only valid answer in this context.

Substituting $d = 40$ feet into the equation $-32 = ad^2$, we get $-32 = a(40)^2 \Rightarrow a = -32/1600 = -0.02$. So the equation for the swimming pool parabolic roof is $y = -0.02x^2$, where x and y are both in feet.

13. a. The highest point of her dive will be at the vertex of the height function (which is concave down). Letting $a = -16$, $b = 12$, and $c = 25$, the t-coordinate of the vertex is at $-12/(2 \cdot (-16)) = 0.375$ seconds. Then $H(0.375) = 25 + (12 \cdot 0.375) - 16(0.375)^2 = 27.25$ feet above water will be the highest point of her dive.

b. She will hit the water when $H(t) = 0 \Rightarrow 25 + 12t - 16t^2 = 0$. Using the quadratic formula, letting $a = -16$, $b = 12$, and $c = 25$, we have

$$t = \frac{-12 \pm \sqrt{(12)^2 - 4(-16)(25)}}{2 \cdot (-16)}$$

$$= \frac{-12 \pm \sqrt{(144 + 1600)}}{-32}$$

$$= \frac{-12 \pm \sqrt{1744}}{-32} \approx \frac{-12 \pm 41.8}{-32}$$

≈ -0.93 seconds or 1.68 seconds. Only the positive value makes sense in this context. So about 1.68 seconds (a little under 2 seconds) after she starts her dive, she will hit the water.

15. a.

x	y	Average Rate of Change	Average Rate of Change of Average Rate of Change
-1	5	n.a.	n.a.
0	0	-5	n.a.
1	-3	-3	$\frac{-3 - (-5)}{1 - 0} = 2$
2	-4	-1	2
3	-3	1	2
4	0	3	2
5	5	5	2

b. A linear function.

c. The fourth column shows that the average rate of change (of the third column with respect to x) is constant, which means that the third column is a linear function of x.

17. a. Behavior looks like a cubic polynomial. (*Note:* The graph doesn't look linear on semi-log plot, so data are not exponential.) The best-fit cubic is

$$P(t) = 1.26x^3 - 6.94x^2 + 9.22x + 50.21$$

[where $P(t)$ = price per hive in t years after 2000]. $P(t)$ is plotted on the accompanying graph along with the data.

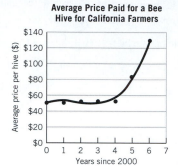

Average Price Paid for a Bee Hive for California Farmers

b. $P(10) \approx \$708$/hive

c. (2.5 hives/acre) $\cdot$ (\$708/hive) = \$1770/acre

d. Since healthy beehives would be scarce, the price per beehive would probably go up.

19.

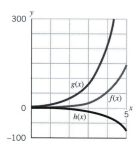

The graph of $g(x)$ is $f(x)$ vertically stretched by a factor of 4; that is, for each value of x, $g(x)$ is four times the value of $f(x)$. The graph of $h(x)$ is the graph of $f(x)$ flipped over the x-axis and then vertically compressed by a factor of 0.5.

21. a. $g(x) = (2/3)f(x - 4) - 1$. So the graph of $f(x)$ was shifted four units to the right, compressed by a factor of 2/3, then shifted down one unit to create the graph of $g(x)$.

b. $g(x) = \frac{2}{3(x - 4)} - 1 = \frac{2}{3(x - 4)} - \frac{3(x - 4)}{3(x - 4)} = \frac{-3x + 14}{3x - 12}$

c. The domain of $g(x)$ is all real numbers except 4.

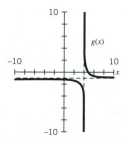

d. If $g(x) = 0$, then $0 = (-3x + 14)/(3x - 12) \Rightarrow 0 = -3x + 14 \Rightarrow x = 14/3 = 4\frac{2}{3}$. So $g(x)$ has a single horizontal intercept at $x = 4\frac{2}{3}$.

If $x = 0$, then $g(0) = 14/(-12) = -1\frac{1}{6}$. So $g(x)$ has a vertical intercept at $-1\frac{1}{6}$.

e. If the denominator $3x - 12 = 0 \Rightarrow x = 4$. So $g(x)$ is not defined when $x = 4$, but does have a vertical asymptote at the line $x = 4$.

f. As $x \to \pm\infty$, $g(x) = \frac{-3x + 14}{3x - 12} \approx \frac{-3x}{3x} = -1$. So as, $x \to \pm\infty$, $g(x) \to -1$ (but never reaches -1).

So $g(x)$ has a horizontal asymptote at the line $y = -1$.

23. a. $S(t) = 5000(1.04)^t$; $R(t) = 5000(1.10)^t$; $T(t) = S(t) + R(t) = 5000(1.04)^t + 5000(1.10)^t$ where $t = $ years since the start of the investments.

b.

Individual Return on $5000 at 4% ($S(t)$) and $5000 at 10% ($R(t)$), and Total Return $T(t) = R(t) + S(t)$

c. She would have only $S(30) = 5000(1.04)^{30} \approx \$16{,}217$.

d. Let $U(t) = 5000(1.14)^t$, where 5000 is the initial value invested at 14% per year. The following table and graph show that after 30 years the $U(t)$ account contains almost 2.5 times as much as the $T(t)$ account (or about 150% more).

Years Since Start of Investment	$T(t) =$ $S(t) + R(t)$	$U(t) =$ $5000(1.14)^t$
0	$10,000	$5,000
5	$14,136	$9,627
10	$20,370	$18,536
15	$29,891	$35,690
20	$44,593	$68,717
25	$67,503	$132,310
30	$103,464	$254,751

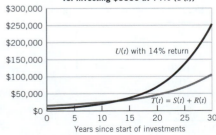

Comparing the Return on Investing $5000 at 4% plus $5000 at 10% ($T(t)$) vs. Investing $5000 at 14% ($U(t)$)

25. The function is $1-1$ since if $f(x_1) = f(x_2)$, then $(x_1 - 2)^3 + 1 = (x_2 - 2)^3 + 1 \Rightarrow (x_1 - 2)^3 = (x_2 - 2)^3 \Rightarrow x_1 - 2 = x_2 - 2 \Rightarrow x_1 = x_2$. So on an appropriate domain f^{-1} exists. Letting $y = f(x)$, we can solve for x in terms of y.

Given	$y = (x - 2)^3 + 1$
Subtract 1	$y - 1 = (x - 2)^3$
Take the cube root	$(y - 1)^{1/3} = x - 2$
Add 2, switch sides	$x = (y - 1)^{1/3} + 2$

So $f^{-1}(y) = (y - 1)^{1/3} + 2$, or since the function is abstract, we can use any name for the input variable, in particular x, to get the conventional form, $f^{-1}(x) = (x - 1)^{1/3} + 2$.

27. a. i. $\frac{761 \text{ miles}}{1 \text{ hr}} \cdot \frac{5280 \text{ feet}}{1 \text{ mile}} \cdot \frac{1 \text{ hr}}{60 \text{ min}} \cdot \frac{1 \text{ min}}{60 \text{ sec}} \approx 1116$ ft/sec.

ii. $D(t) = 1116t$, where t is in seconds and $D(t)$ is in feet from the lightning strike.

iii. Yes, the rule of thumb is reasonable since for each second after the strike, the sound thunder travels about 1116 feet.

b. i. $A(r) = \pi r^2$, where r is the radius (in feet) of the sound circle.

ii. $A(D(t)) = A(1116t) = \pi(1116t)^2 \approx 1{,}245{,}500\pi t^2 \approx 3{,}913{,}000\ t^2$, where t is in seconds and $A(D(t))$ is in square feet. $A(D(4)) = 3{,}913{,}000(4)^2 \approx 63$ million square feet or $(63 \cdot 10^6$ sq. ft$) \cdot \frac{1 \text{ sq. mile}}{5280^2 \text{ sq. ft}} \approx 2.3$ square miles.

iii. When the time doubles, the distance doubles, but the area increases by a factor of 4.

Exercises for EE on Mathematics of Motion

1.

Time (sec)	Distance (cm)	Avg. Vel. over Previous 1/30 sec. (cm/sec)
0.0000	0.00	n.a.
0.0333	3.75	113
0.0667	8.67	147
0.1000	14.71	181
0.1333	21.77	212
0.1667	29.90	243

The average velocity (over each 1/60 of a second) increases rapidly as time progresses.

3. For $d = 490t^2 + 50t$:

a. 50 is measured in cm/sec; it is the initial velocity of the object falling; 490 is measured in (cm/sec)/sec and is half the acceleration due to gravity when measured in these units.

b, c. Below is a small table of values and the graph of the equation with the table points marked on it.

t	d
0.0	0.0
0.1	9.9
0.2	29.6
0.3	59.1

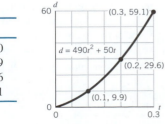

5. For $d = 4.9t^2 + 1.7t$:

   a. 1.7 is the initial velocity of the object falling; it is measured in meters per second; 4.9 is half the gravitational constant when it is measured in (meters/sec)/sec.

   b, c. Below is a small table of values, and next to it is the graph with the table points marked on it.

t	d
0.0	0.00
0.1	0.22
0.2	0.54
0.3	0.95

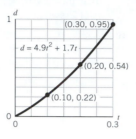

   d. The results in this question are very similar to those in earlier parts of this chapter. The shape of the graph is that of a quadratic; the coefficient of t^2 is half the gravitational constant, and the coefficient of t is the initial velocity.

7. $m = \frac{m}{sec^2} \cdot sec^2 + \frac{m}{sec} \cdot sec$

9. a. $d = 490t^2 + 50t$ and $v = 980t + 50$

   b. At $t = 1$, $d = 540$ cm and $v = 1030$ cm/sec. At $t = 2.5$, $d = 3187.5$ cm and $v = 2500$ cm/sec.

11. a. $d = 16t^2 + 12t$

   b.

t	d
0	0
1	28
2	88
3	180
4	304
5	460

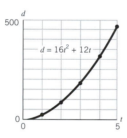

   c. The graph and table of the function are given above.

13. The distance is measured in meters if the time is measured in seconds. The use of 4.9 for half of the gravity constant is the indicator of these units.

15. a. Student answers will vary considerably.

   b. Since the velocity is changing at a constant rate, a straight line should be a good fit. The graph of this line is a representation of average velocity.

17. a. The coefficient of t^2 is one-half the gravity constant. Since the coefficient of t^2 is approximately 490, distance is measured in centimeters and time in seconds, and 490 is measured in cm/sec². The coefficient of t is an initial velocity of 7.6 cm/sec.

   b. When $t = 0.05$ sec, $d = 1.59$ cm; when $t = 0.10$ sec, $d = 5.62$ cm, and when $t = 0.30$ sec, $d = 45.99$ cm.

19. It represents an initial velocity of the object measured in meters per second.

21. a. At $t = 0$ sec, $h = 0$ m; at $t = 1$ sec, $h = 195.1$ m; at $t = 2$ sec, $h = 380.4$ m; at $t = 10$ sec, $h = 1510$ m.

   b. The graph of h over t is given in the following diagram.

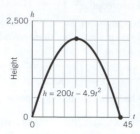

   c. The object reaches a maximum height of approximately 2000 meters after 20 seconds. It reaches the ground after approximately 40 seconds of flight.

23. For $h = 85 - 490t^2$:

   a. 85 is the height in centimeters of the falling object at the start; -490 is half the gravitational constant when measured in (cm/sec)/sec; it is negative in value since h measures height above the ground and the gravitational constant is connected with pulling objects down. This will mean subtraction from the starting height of 85 cm.

   b. The initial velocity is 0 cm/sec.

   c. Below is a table of values for this function

t	h
0.0	85.0
0.1	80.1
0.2	65.4
0.4	6.6

   d. The following diagram is the graph of the function with the table entries marked on it.

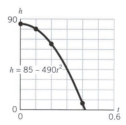

25. a. The initial velocity is positive since we are measuring height above ground and the object is going up at the start.

   b. The equation of motion is $h = 50 + 10t - 16t^2$, where height is measured in feet and t in seconds.

27. a. 980 cm/sec² since we are measuring in cm and in sec.

   b.

t	v
0	-66
1	-1046
2	-2026
3	-3006
4	-3986

   c. The graph is given below. The object is traveling faster and faster toward the ground. The increase in downward velocity is at a constant rate, as we can see from the constant slope of the graph. This constant acceleration, of course, is due to gravity.

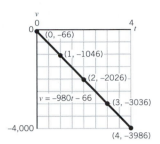

d. Ordinarily, if $t = 0$ corresponds to the actual start of the flight, then the initial condition given would indicate that the object was thrown downward at a speed of 66 meters per second. This interpretation comes from the negative sign given to the initial velocity. But this would contradict the statement in the problem to the effect that it is a "freely falling body." In this context, another interpretation is suggested by the laboratory experiment, namely that the object started being timed at a point along its downward fall.

29. a. Its velocity starts out negative and continues to be so since the object is falling; h is measured in cm above the ground; t is measured in seconds.

b. $h = 150 - 25t - 490t^2$; for $0 \le t \le 0.528$ (the second value being the approximate time in seconds it takes for the object to hit the ground).

c. The average velocity is the slope, i.e., $(15 - 150)/0.5 = -270$ cm/sec; the initial velocity is -25 cm/sec. The average velocity is 10.8 times as great in magnitude as the initial velocity.

31. Forming $\frac{d}{t} = \frac{v_0 + (v_0 + at)}{2}$ and solving for d, we get

$$d = \frac{2v_0 t + at^2}{2} = v_0 t + \tfrac{1}{2}at^2$$

This is very similar in form to the falling-body formula. The acceleration factor increases the velocity in a manner proportional to the square of the time traveled, and the initial velocity increases the distance in a manner proportional to the time.

33. a. After 5 seconds its velocity is 110 cm/sec; after 1 minute (or 60 seconds) its velocity is 660 cm/sec; after t seconds, its velocity is $v(t) = 60 + 10t$ cm/sec.

b. After 5 seconds its average velocity is $(60 + 110)/2 = 85$ cm/sec.

35. a. $v(t) = 200 + 60t$ meters/sec

b. $d(t) = 200t + 30t^2$ meters

37. a. Using units of feet and seconds, the equation governing the water spout is $h = -16t^2 + v_0 t$, where h is measured in feet and t, time, in seconds and where v_0 is the sought-after initial velocity. We are given that the maximum height reached is 120 ft. The maximum height is achieved at the vertex, i.e., when $t = -v_0/(-32) = \frac{v_0}{32}$. Substituting for t and h gives us

$$120 = -16\left(\frac{v_0}{32}\right)^2 + v_0\left(\frac{v_0}{32}\right) = -\frac{v_0^2}{64} + 2\frac{v_0^2}{64} = \frac{v_0^2}{64}$$

Thus $v_0^2 = 7680$ or $v_0 \approx 87.6$ ft per sec

b. $t = v_0/32 = 87.6/32 \approx 2.74$ sec

39. a. $d_c = v_c t + a_c t^2/2$; $d_p = a_p t^2/2$. One wants to solve for the t at which $d_c = d_p$, i.e., when $v_0 t + a_c \frac{t^2}{2} = a_p \frac{t^2}{2} \Rightarrow v_c t + a_c t^2/2 - a_p t^2/2 = 0 \Rightarrow t\,(v_c + [a_c/2 - a_p/2]t) = 0$. This occurs when $t = 0$ (when the criminal passes by the police car) and again when $t = 2v_c/(a_p - a_c)$, (when the police catch up to the criminal).

b. Now $v_c = a_c t + v_c$ and $v_p = a_p t$. One wants to solve for the t at which $v_c = v_p$, i.e., when $a_c t + v_c = a_p t$ or for $t = (a_p - a_c)/v_c$. This does not mean that the police have caught up to the criminal, but rather that the police are at that moment going as fast as the criminal is and that they are starting to go faster than the criminal.

EE

INDEX

A page number followed by *f* indicates a figure; a number followed by *t* indicates a table; a number followed by *n* indicates a footnote.